PATTY'S INDUSTRIAL HYGIENE

Fifth Edition

Volume 2

III PHYSICAL AGENTS

IV BIOHAZARDS

V ENGINEERING CONTROL AND PERSONAL PROTECTION

PATTY'S INDUSTRIAL HYGIENE

Fifth Edition

Volume 2

III PHYSICAL AGENTS

IV BIOHAZARDS

V ENGINEERING CONTROL AND PERSONAL PROTECTION

ROBERT L. HARRIS

Editor

CONTRIBUTORS

E. W. Arp
T. E. Bernard
W. A. Burgess
D. C. Byrne
H. Cember
M. A. Coffman
M. J. Dainoff
R. L. Harris
P. A. Heinsohn
K. Michael
W. R. Myers
C. A. Piantadosi
J. D. Ramsey
J. Singh
R. D. Soule
R. L. Vincent
G. M. Wilkening

A Wiley-Interscience Publication

JOHN WILEY & SONS, INC.

New York / Chichester / Weinheim / Brisbane / Singapore / Toronto

This book is printed on acid-free paper. ♾

For ordering and customer service, call 1-800-CALL-WILEY.

Library of Congress Cataloging in Publication Data:

Patty's industrial hygiene / [edited by] Robert L. Harris. — 5th ed.
[rev.]
v. ⟨ ⟩ cm.
Fourth ed. published as: Patty's industrial hygiene and toxicology.
Includes bibliographical references and index.
ISBN 0-471-29754-2 (Vol. 2) (cloth : alk. paper); 0-471-29784-4 (set)
1. Industrial hygiene. I. Harris, Robert L., 1924– .
II. Patty, F. A. (Frank Arthur), 1897– Industrial hygiene and toxicology.
RC967.P37 2000
613.6′2—dc21 99-32462

Printed in the United States of America.

10 9 8 7 6 5 4 3 2 1

Contributors

Earl W. Arp Jr., Ph.D, CIH
Lexington, Kentucky

Thomas E. Bernard, Ph.D, CIH
University of South Florida, College of Public Health, Tampa, Florida

William A. Burgess, CIH
Topsfield, Massachusetts

David C. Byrne, MS
Pittsburgh, Pennsylvania

Herman Cember, Ph.D, PE
Lafayette, Indiana

Michael A. Coffman, CIH, CSP, Hayes Wheels International, Inc.
Northville, Michigan

Marvin J. Dainoff, Ph.D., CPE
Center for Ergonomic Research, Department of Psychology, Miami University, Oxford, Ohio

Robert L. Harris, Ph.D, CIH
Raleigh, North Carolina

Patricia A. Heinsohn, Ph.D, CIH
Exponent
Menlo Park, California

Kevin Michael, Ph.D
State College, Pennsylvania

Warren R. Myers, Ph.D, CIH
West Virginia University
Morgantown, West Virginia

Claude A. Piantadosi, MD, Duke University Medial Center
Durham, North Carolina

Jerry D. Ramsey, Ph.D, PE
Department of Industrial Engineering
Texas Tech University
Lubbock, Texas

Jaswant Singh, Ph.D, CIH, Liberty Risk Services, Malaysia

Robert D. Soule, CIH, CSP
Indiana, Pennsylvania

Richard L. Vincent
New York, New York

G. M. Wilkening, CIH, Deceased

Preface

Industrial hygiene is an applied science and a profession. Like other applied sciences such as medicine and engineering, it is founded on basic sciences such as biology, chemistry, mathematics, and physics. In a sense it is a hybrid profession because within its ranks are members of other professions—chemists, engineers, biologists, physicists, physicians, nurses, and lawyers. In their professional practice all are dedicated in one way or another to the purposes of industrial hygiene, to the anticipation, recognition, evaluation, and control of work-related health hazards. All are represented among the authors of chapters in these volumes.

Although the term "industrial hygiene" used to describe our profession is probably of twentieth century origin, we must go further back in history for the origin of its words. The word "industry," which has a dictionary meaning, "systematic labor for some useful purpose or the creation of something of value," has its English origin in the fifteenth century. For "hygiene" we must look even earlier. Hygieia, a daughter of Aesklepios who is god of medicine in Greek mythology, was responsible for the preservation of health and prevention of disease. Thus, Hygieia, when she was dealing with people who were engaged in systematic labor for some useful purpose, was practicing our profession, industrial hygiene.

Industrial Hygiene and Toxicology was originated by Frank A. Patty with publication of the first single volume in 1948. In 1958 an updated and expanded Second Edition was published with his guidance. A second volume, Toxicology, was published in 1963. Frank Patty was a pioneer in industrial hygiene; he was a teacher, practitioner, and manager. He served in 1946 as eighth President of the American Industrial Hygiene Association. To cap his professional career he served as Director of the Division of Industrial Hygiene for the General Motors Corporation.

At the request of Frank Patty, George and Florence Clayton took over editorship of the ever-expanding *Industrial Hygiene and Toxicology* series for the Third Edition of Volume I, *General Principles*; published in 1978, and Volume II, *Toxicology*, published in

1981–1982. The First Edition of Volume III, *Theory and Rationale of Industrial Hygiene Practice,* edited by Lewis and Lester Cralley, was published in 1979 with its Second Edition published in 1984. The ten-book, two-volume Fourth Edition of *Patty's Industrial Hygiene and Toxicology,* edited by George and Florence Clayton, was published in 1991–1994, and the Third Edition of Volume III, *Theory and Rationale of Industrial Hygiene Practice,* edited by Robert Harris, Lewis Cralley, and Lester Cralley, was published in 1994. With the agreement and support of George and Florence Clayton, and Lewis and Lester Cralley, it is a signal honor for me to follow them and Frank A. Patty as editor of the Industrial Hygiene volumes of *Patty's Industrial Hygiene and Toxicology.*

Industrial hygiene has been dealt with very broadly in past editions of *Patty's Industrial Hygiene and Toxicology.* Chapters have been offered on sampling and analysis, exposure measurement and interpretation, absorption and elimination of toxic materials, occupational dermatoses, instrument calibration, odors, industrial noise, ionizing and nonionizing radiation, heat stress, pressure, lighting, control of exposures, safety and health law, health surveillance, occupational health nursing, ergonomics and safety, agricultural hygiene, hazardous wastes, occupational epidemiology, and other vital areas of practice. These traditional areas continue to be covered in this new edition. Consistent with the past history of *Patty's,* new areas of industrial hygiene concerns and practices have been addressed as well: aerosol science, computed tomography, multiple chemical sensitivity, potential endocrine disruptors, biological monitoring of exposures, health and safety management systems, industrial hygiene education, and other areas not covered in earlier editions.

Although industrial hygiene has been practiced in one guise or another for centuries, the most systematic approaches and the most esoteric accomplishments have been made in the past fifty or sixty years—generally in the years since Frank Patty published his first book. This accelerated progress is due primarily to increased public awareness of occupational health and safety issues and need for environmental control as is evidenced by Occupational Safety and Health, Clean Air, and Clean Water legislation at both federal and state levels.

Industrial hygienists know that variability is the key to measurement and interpretation of workers' exposures. If exposures did not vary, exposure assessment could be limited to a single measurement, the results of which could be acted upon, then the matter filed away as something of no further concern. We know, however, that exposures change. But not only do exposures change—change is characteristic of the science and practice of our profession as well. We must be alert to recognize new hazards, we must continue to evaluate new and changing stresses, we must evaluate performance of exposure controls and from time to time upgrade them. These volumes represent the theory and practice of industrial hygiene as they are understood by their chapter authors at the time of writing. But, as observed by the Greek philosopher Heracleitus about 2500 years ago, "There is nothing permanent except change." Improvements and changes in theory and practice of industrial hygiene take place continuously and are generally reported in the professional literature. Industrial hygienists, the practitioners, the teachers, and the managers, must stay abreast of the professional literature. Furthermore, when an industrial hygienist develops new knowledge, he/she has what almost amounts to an ethical obligation to share it in our journals.

One cannot ponder the rapid changes and advancements made in recent decades in science and technology, and in our own profession as well, without wondering at what the next two or three decades will bring. Developments in computer technology and information processing and exchange have greatly influenced manufacturing (robotics, computer controlled machining) and the general conduct of commerce and business in the past one or two decades. This change will only accelerate with computer speeds and capacities doubling every 18 months or so, and processing units approaching microsize. The possibility for continuously monitoring and computer storage of exposures of individual workers may become reality within the next decade. The human genome project holds promise for prevention and cure of many diseases, including some associated with conditions of work. World population continues to increase geometrically and is expected to be about eight billion in the year 2020; with improvements in preventive health care this will be an increasingly older population. Genetic engineering and highly effective pesticides are already improving yields of agricultural commodities; if all goes well in this area, and if we can avoid set-backs as might be associated with potential endocrine disruptors, feeding the expanding human population may not be a limiting factor. Globalization of manufacturing and commerce has already begun to reduce manufacturing employment in the United States and in Europe, and to expand opportunities for expanding populations in some developing nations. The United States and other developed nations are on their way to becoming world centers of information and innovation.

How will all of this affect the future practice of industrial hygiene? In the Preface to the Fourth Edition of *Patty's,* George and Florence Clayton suggested that the future of industrial hygiene is limited only by the narrowness of vision of its practitioners. More recently, Lawrence Birkner, past president of the American Academy of Industrial Hygiene, and his co-worker and spouse, Ruth McIntyre Birkner, in writing about the future of the occupational and environmental hygiene profession, say much the same thing. (See "The Future of the Occupational and Environmental Hygiene Profession" in *A.I.H.A.* Journal, pp. 370–374, 1997) Larry and Ruth report that we must be aware of the changes likely to take place in the next couple of decades, and must develop strategies now to assure the profession's full participation in protecting the health and safety of workers, and the environment, of tomorrow.

ROBERT L. HARRIS

Raleigh, North Carolina

Contents

USEFUL EQUIVALENTS AND CONVERSION FACTORS

1 kilometer = 0.6214 mile
1 meter = 3.281 feet
1 centimeter = 0.3937 inch
1 micrometer = 1/25,4000 inch = 40 microinches
  = 10,000 Angstrom units
1 foot = 30.48 centimeters
1 inch = 25.40 millimeters
1 square kilometer = 0.3861 square mile (U.S.)
1 square foot = 0.0929 square meter
1 square inch = 6.452 square centimeters
1 square mile (U.S.) = 2,589,998 square meters
  = 640 acres
1 acre = 43,560 square feet = 4047 square meters
1 cubic meter = 35.315 cubic feet
1 cubic centimeter = 0.0610 cubic inch
1 cubic foot = 28.32 liters = 0.0283 cubic meter
  = 7.481 gallons (U.S.)
1 cubic inch = 16.39 cubic centimeters
1 U.S. gallon = 3,7853 liters = 231 cubic inches
  = 0.13368 cubic foot
1 liter = 0.9081 quart (dry), 1.057 quarts
  (U.S., liquid)
1 cubic foot of water = 62.43 pounds (4°C)
1 U.S. gallon of water = 8.345 pounds (4°C)
1 kilogram = 2.205 pounds

1 gram = 15.43 grains
1 pound = 453.59 grams
1 ounce (avoir.) = 28.35 grams
1 gram mole of a perfect gas ≎ 24.45 liters
  (at 25°C and 760 mm Hg barometric pressure)
1 atmosphere = 14.7 pounds per square inch
1 foot of water pressure = 0.4335 pound per
  square inch
1 inch of mercury pressure = 0.4912 pound per
  square inch
1 dyne per square centimeter = 0.0021 pound per
  square foot
1 gram-calorie = 0.00397 Btu
1 Btu = 778 foot-pounds
1 Btu per minute = 12.96 foot-pounds per second
1 hp = 0.707 Btu per second = 550 foot-pounds
  per second
1 centimeter per second = 1.97 feet per minute
  = 0.0224 mile per hour
1 footcandle = 1 lumen incident per square foot
  = 10.764 lumens incident per square meter
1 grain per cubic foot = 2.29 grams per
  cubic meter
1 milligram per cubic meter = 0.000437 grain per
  cubic foot

To convert degrees Celsius to degrees Fahrenheit: °C (9/5) + 32 = °F
To convert degrees Fahrenheit to degrees Celsius: (5/9) (°F − 32) = °C
For solutes in water: 1 mg/liter ≎ 1 ppm (by weight)
Atmospheric contamination: 1 mg/liter ≎ 1 oz/1000 cu ft (approx)
For gases or vapors in air at 25°C and 760 mm Hg pressure:
  To convert mg/liter to ppm (by volume): mg/liter (24,450/mol. wt.) = ppm
  To convert ppm to mg/liter: ppm (mol. wt./24,450) = mg/liter

CONVERSION TABLE FOR GASES AND VAPORS[a]

(Milligrams per liter to parts per million, and vice versa; 25°C and 760 mm Hg barometric pressure)

Molecular Weight	1 mg/liter ppm	1 ppm mg/liter	Molecular Weight	1 mg/liter ppm	1 ppm mg/liter	Molecular Weight	1 mg/liter ppm	1 ppm mg/liter
1	24,450	0.0000409	39	627	0.001595	77	318	0.00315
2	12,230	0.0000818	40	611	0.001636	78	313	0.00319
3	8,150	0.0001227	41	596	0.001677	79	309	0.00323
4	6,113	0.0001636	42	582	0.001718	80	306	0.00327
5	4,890	0.0002045	43	569	0.001759	81	302	0.00331
6	4,075	0.0002454	44	556	0.001800	82	298	0.00335
7	3,493	0.0002863	45	543	0.001840	83	295	0.00339
8	3,056	0.000327	46	532	0.001881	84	291	0.00344
9	2,717	0.000368	47	520	0.001922	85	288	0.00348
10	2,445	0.000409	48	509	0.001963	86	284	0.00352
11	2,223	0.000450	49	499	0.002004	87	281	0.00356
12	2,038	0.000491	50	489	0.002045	88	278	0.00360
13	1,881	0.000532	51	479	0.002086	89	275	0.00364
14	1,746	0.000573	52	470	0.002127	90	272	0.00368
15	1,630	0.000614	53	461	0.002168	91	269	0.00372
16	1,528	0.000654	54	453	0.002209	92	266	0.00376
17	1,438	0.000695	55	445	0.002250	93	263	0.00380
18	1,358	0.000736	56	437	0.002290	94	260	0.00384
19	1,287	0.000777	57	429	0.002331	95	257	0.00389
20	1,223	0.000818	58	422	0.002372	96	255	0.00393
21	1,164	0.000859	59	414	0.002413	97	252	0.00397
22	1,111	0.000900	60	408	0.002554	98	249.5	0.00401
23	1,063	0.000941	61	401	0.002495	99	247.0	0.00405
24	1,019	0.000982	62	394	0.00254	100	244.5	0.00409
25	978	0.001022	63	388	0.00258	101	242.1	0.00413
26	940	0.001063	64	382	0.00262	102	239.7	0.00417
27	906	0.001104	65	376	0.00266	103	237.4	0.00421
28	873	0.001145	66	370	0.00270	104	235.1	0.00425
29	843	0.001186	67	365	0.00274	105	232.9	0.00429
30	815	0.001227	68	360	0.00278	106	230.7	0.00434
31	789	0.001268	69	354	0.00282	107	228.5	0.00438
32	764	0.001309	70	349	0.00286	108	226.4	0.00442
33	741	0.001350	71	344	0.00290	109	224.3	0.00446
34	719	0.001391	72	340	0.00294	110	222.3	0.00450
35	699	0.001432	73	335	0.00299	111	220.3	0.00454
36	679	0.001472	74	330	0.00303	112	218.3	0.00458
37	661	0.001513	75	326	0.00307	113	216.4	0.00462
38	643	0.001554	76	322	0.00311	114	214.5	0.00466

CONVERSION TABLE FOR GASES AND VAPORS (*Continued*)

(Milligrams per liter to parts per million, and vice versa; 25°C and 760 mm Hg barometric pressure)

Molecular Weight	1 mg/liter ppm	1 ppm mg/liter	Molecular Weight	1 mg/liter ppm	1 ppm mg/liter	Molecular Weight	1 mg/liter ppm	1 ppm mg/liter
115	212.6	0.00470	153	159.8	0.00626	191	128.0	0.00781
116	210.8	0.00474	154	158.8	0.00630	192	127.3	0.00785
117	209.0	0.00479	155	157.7	0.00634	193	126.7	0.00789
118	207.2	0.00483	156	156.7	0.00638	194	126.0	0.00793
119	205.5	0.00487	157	155.7	0.00642	195	125.4	0.00798
120	203.8	0.00491	158	154.7	0.00646	196	124.7	0.00802
121	202.1	0.00495	159	153.7	0.00650	197	124.1	0.00806
122	200.4	0.00499	160	152.8	0.00654	198	123.5	0.00810
123	198.8	0.00503	161	151.9	0.00658	199	122.9	0.00814
124	197.2	0.00507	162	150.9	0.00663	200	122.3	0.00818
125	195.6	0.00511	163	150.0	0.00667	201	121.6	0.00822
126	194.0	0.00515	164	149.1	0.00671	202	121.0	0.00826
127	192.5	0.00519	165	148.2	0.00675	203	120.4	0.00830
128	191.0	0.00524	166	147.3	0.00679	204	119.9	0.00834
129	189.5	0.00528	167	146.4	0.00683	205	119.3	0.00838
130	188.1	0.00532	168	145.5	0.00687	206	118.7	0.00843
131	186.6	0.00536	169	144.7	0.00691	207	118.1	0.00847
132	185.2	0.00540	170	143.8	0.00695	208	117.5	0.00851
133	183.8	0.00544	171	143.0	0.00699	209	117.0	0.00855
134	182.5	0.00548	172	142.2	0.00703	210	116.4	0.00859
135	181.1	0.00552	173	141.3	0.00708	211	115.9	0.00863
136	179.8	0.00556	174	140.5	0.00712	212	115.3	0.00867
137	178.5	0.00560	175	139.7	0.00716	213	114.8	0.00871
138	177.2	0.00564	176	138.9	0.00720	214	114.3	0.00875
139	175.9	0.00569	177	138.1	0.00724	215	113.7	0.00879
140	174.6	0.00573	178	137.4	0.00728	216	113.2	0.00883
141	173.4	0.00577	179	136.6	0.00732	217	112.7	0.00888
142	172.2	0.00581	180	135.8	0.00736	218	112.2	0.00892
143	171.0	0.00585	181	135.1	0.00740	219	111.6	0.00896
144	169.8	0.00589	182	134.3	0.00744	220	111.1	0.00900
145	168.6	0.00593	183	133.6	0.00748	221	110.6	0.00904
146	167.5	0.00597	184	132.9	0.00753	222	110.1	0.00908
147	166.3	0.00601	185	132.2	0.00757	223	109.6	0.00912
148	165.2	0.00605	186	131.5	0.00761	224	109.2	0.00916
149	164.1	0.00609	187	130.7	0.00765	225	108.7	0.00920
150	163.0	0.00613	188	130.1	0.00769	226	108.2	0.00924
151	161.9	0.00618	189	129.4	0.00773	227	107.7	0.00928
152	160.9	0.00622	190	128.7	0.00777	228	107.2	0.00933

CONVERSION TABLE FOR GASES AND VAPORS (*Continued*)

(*Milligrams per liter to parts per million, and vice versa; 25°C and 760 mm Hg barometric pressure*)

Molecular Weight	1 mg/liter ppm	1 ppm mg/liter	Molecular Weight	1 mg/liter ppm	1 ppm mg/liter	Molecular Weight	1 mg/liter ppm	1 ppm mg/liter
229	106.8	0.00937	253	96.6	0.01035	277	88.3	0.01133
230	106.3	0.00941	254	96.3	0.01039	278	87.9	0.01137
231	105.8	0.00945	255	95.9	0.01043	279	87.6	0.01141
232	105.4	0.00949	256	95.5	0.01047	280	87.3	0.01145
233	104.9	0.00953	257	95.1	0.01051	281	87.0	0.01149
234	104.5	0.00957	258	94.8	0.01055	282	86.7	0.01153
235	104.0	0.00961	259	94.4	0.01059	283	86.4	0.01157
236	103.6	0.00965	260	94.0	0.01063	284	86.1	0.01162
237	103.2	0.00969	261	93.7	0.01067	285	85.8	0.01166
238	102.7	0.00973	262	93.3	0.01072	286	85.5	0.01170
239	102.3	0.00978	263	93.0	0.01076	287	85.2	0.01174
240	101.9	0.00982	264	92.6	0.01080	288	84.9	0.01178
241	101.5	0.00986	265	92.3	0.01084	289	84.6	0.01182
242	101.0	0.00990	266	91.9	0.01088	290	84.3	0.01186
243	100.6	0.00994	267	91.6	0.01092	291	84.0	0.01190
244	100.2	0.00998	268	91.2	0.01096	292	83.7	0.01194
245	99.8	0.01002	269	90.9	0.01100	293	83.4	0.01198
246	99.4	0.01006	270	90.6	0.01104	294	83.2	0.01202
247	99.0	0.01010	271	90.2	0.01108	295	82.9	0.01207
248	98.6	0.01014	272	89.9	0.01112	296	82.6	0.01211
249	98.2	0.01018	273	89.6	0.01117	297	82.3	0.01215
250	97.8	0.01022	274	89.2	0.01121	298	82.0	0.01219
251	97.4	0.01027	275	88.9	0.01125	299	81.8	0.01223
252	97.0	0.01031	276	88.6	0.01129	300	81.5	0.01227

[a]A. C. Fieldner, S. H. Katz, and S. P. Kinney, "Gas Masks for Gases Met in Fighting Fires," *U.S. Bureau of Mines, Technical Paper No. 248,* 1921.

PATTY'S INDUSTRIAL HYGIENE

Fifth Edition

Volume 2

III PHYSICAL AGENTS

IV BIOHAZARDS

V ENGINEERING CONTROL AND PERSONAL PROTECTION

CHAPTER NINETEEN

Industrial Noise and Conservation of Hearing

Kevin L. Michael, Ph.D and David C. Byrne, MS, CCC-A

1 INTRODUCTION

Exposure to industrial noise and the resulting effect of occupational hearing loss is a common problem across nearly all industries. High noise levels also cause interference with verbal communication and warning signals, which can have a significant impact on safety and work performance. Finally, noise can be considered as a source of stress for workers, producing unwanted physiological and psychological effects that can lead to a degraded quality of life.

Typically, noise-induced hearing loss develops slowly, and usually goes unnoticed until a significant impairment has occurred. Fortunately, occupational hearing loss is nearly always preventable. Prevention of noise-induced hearing loss benefits the employer as well as the individual employees. An effective hearing conservation program promotes good labor-management relations, which can lead to increased morale and productivity. Employers enjoy the benefits of reduced medical expenses and worker compensation payments, and employees can expect to maintain their hearing health well into their retirement years.

1.1 Terminology

This section defines terms that are useful in discussions related to the study of noise.

Patty's Industrial Hygiene, Fifth Edition, Volume 2. Edited by Robert L. Harris.
ISBN 0-471-29754-2

Absorption Coefficient (α): The ratio of the acoustical energy absorbed by a surface to the acoustical energy incident upon that surface. The result will be a value between 0.0 and 1.0, which also can be considered as a percentage between 0% and 100%.

Acoustic Intensity: See Sound Intensity.

Acoustic Power: See Sound Power.

Acoustic Pressure: See Sound Pressure.

Ambient Noise: The overall composite of sounds in an environment.

Amplitude: An objective measure of the quantity or amount of sound generated by a source, generally described by the sound power or sound pressure measured at a specific location.

Anechoic Room: A specially-designed test room with very absorptive interior surfaces that do not reflect sound energy, and therefore prevent sound build-up or "echoes" (i.e., nearly 100% of the sound is absorbed). Any sound generated within an anechoic room is referred to as being in a free field (see definition).

Audiogram: A recording or graph of hearing levels referenced to a statistically normal sound pressure as a function of frequency.

Audiometer: An instrument for measuring hearing thresholds.

Continuous Noise: Noise with relatively small fluctuations in level within the measurement period.

Decibel (dB): A dimensionless unit that describes the logarithm of the ratio of two power-related quantities. It is normally defined as ten times the logarithm of the ratio of a measured quantity to a reference quantity. When measuring sound, different reference values have been arbitrarily established, depending on whether sound power, sound intensity, or sound pressure is to be measured.

Decibel, A-weighted (dBA): The sound level measured with the A-weighting network on a sound level meter.

Decibel, C-weighted (dBC): The sound level measured with the C-weighting network on a sound level meter.

Diffuse Sound Field: A sound field with sound pressure levels that are essentially the same throughout, with the directional incidence of energy flux randomly distributed.

Effective Sound Pressure: The sound pressure at a given location, derived by calculating the root mean square (rms) value of the instantaneous sound pressures measured over a period of time at that location.

Exchange Rate: The number of decibels that, when a sound level is increased by this amount, gives an equivalent noise exposure when the exposure duration is halved. Likewise, when a sound level is decreased by the number of decibels specified by the exchange rate, an equivalent exposure will result if the exposure time is doubled.

Free Field: A field that exists in a homogeneous isotropic medium, such as an open outdoor area without boundaries. In a free field, sound radiated from a source can be measured accurately without influence from unwanted reflections. True free-field conditions are rarely found, except in specially designed anechoic test chambers.

Frequency (f): The rate at which complete cycles of high- and low-pressure regions are produced by a sound source, measured in Hertz (Hz). In subjective terms, frequency is often referred to as "pitch," although there is not an exact correlation between the two terms.

Hertz (Hz): A unit of measurement for frequency, indicating the number of complete waves or cycles that occur in one second.

Impulse Noise: Noise that is characterized by a sharp rise and rapid decay in sound level, and is less than one second in duration.

Insertion Loss (IL): The difference in sound levels measured at a particular location before and after an acoustical treatment is installed. IL $= L_{p0} - L_{p1}$, where L_{p0} is the sound level without the attenuating device, and L_{p1} is measured at the same location with the treatment in place.

Intensity: See Sound Intensity.

Level (L_x): The value of a quantity (x) when converted into decibels (e.g., sound pressure level (L_p), sound power level (L_W), sound intensity level (L_I), etc.). The letter L indicates that the value is being described in terms of a decibel level, and the subscript denotes the type of quantity under consideration.

Loudness: An observer's impression of a sound's amplitude. Although a high-intensity sound wave is perceived as being louder than a wave of lower intensity, an exact linear relationship does not exist, and the relation between loudness and intensity is not the same for all individuals.

Noise: The terms noise and sound are often used interchangeably; however, *sound* is normally used to describe useful communication or pleasant audible signals (e.g., music), whereas *noise* is frequently considered as dissonance or unwanted sound.

Noise Reduction (NR): The difference in sound levels measured on either side of a noise-reducing wall or partition. NR $= L_{p1} - L_{p2}$, where $L_{p1}$is the sound pressure level in room 1, and L_{p2} is the sound pressure level in room 2.

Noise Reduction Coefficient (NRC): The arithmetic average of the sound absorption coefficients of a material at 250, 500, 1000, and 2000 Hz.

Noise Reduction Rating (NRR): An indication of a hearing protector's noise reduction capability (in dB).

Octave Band: A frequency bandwidth that has an upper band-edge frequency equal to twice its lower band-edge frequency.

Pascal (Pa): A unit of pressure equal to one newton per square meter.

Peak Sound Pressure Level: The maximum instantaneous level of sound that occurs over any specified period of time.

Period (T): The time (in seconds) required for one cycle of pressure change to take place (hence the period is the reciprocal of the frequency).

Pitch: The auditory sensation that depends primarily on frequency, but also on the pressure and waveform of the sound stimulus.

Power: See Sound Power.

Pure Tone: A sound wave consisting of only one single frequency.

Random Incidence Sound Field: See Diffuse Sound Field.

Random Noise: A noise made up of many frequency components whose instantaneous amplitudes occur randomly as a function of time.

Resonance: Enhancement in the response of a mechanical system to a periodic driving force when the driving frequency is equal to the natural undamped frequency of the system. A system is in resonance when any change in the frequency of forced oscillation causes a decrease in the response of the system.

Reverberation: The build-up of sound in a room due to reflections from the floor, walls, and ceiling. The reverberation time is the time required (after the source is shut off) for the sound level to decay 60 dB from its steady-state level.

Root-Mean-Square (rms) Sound Pressure: The root-mean-square value of a changing quantity, such as sound pressure, is the square root of the mean of the squares of the instantaneous values of the quantity. (See effective sound pressure).

Sound: A propagating disturbance through a physical medium. Sound is perceived by the ear as a pressure wave superimposed upon the ambient air pressure.

Sound Intensity: The average rate at which acoustic energy is transmitted through a unit area normal to the direction of propagation. The units used for sound intensity are watts per square meter (W/m^2). Sound intensity is also expressed in terms of a sound intensity level (L_I) in decibels, referenced to 10^{-12} W/m^2.

Sound Power: The total acoustic energy radiated per second by a source. Sound power is normally expressed in terms of a sound power level (L_W) in decibels. $L_W = 10 \log (W/W_{ref})$, where W is the sound power in question, and W_{ref} is the standard reference power of 10^{-12} Watts.

Sound Pressure: Oscillation of the sound wave above and below atmospheric pressure. Sound pressure level (L_p) is the rms sound pressure measured at a certain distance from a source with respect to the standard reference pressure of 20 μPa (20×10^{-6} Pa). $L_p = 20 \log (p/p_{ref})$, where p is the sound pressure in question, and p_{ref} is the standard reference pressure, which corresponds to the softest sound a normal-hearing person can detect. The symbol SPL can be used interchangeably with L_p to denote sound pressure level.

Spectrum: A distribution or range of frequencies. When performing frequency analysis, a sound may be divided into octave-band, third-octave band, or narrow-band spectrums.

Speed of Sound (c): The rate at which sound waves travel, depending on the density and elasticity of the medium. The speed of sound is usually considered to be constant under normal temperature conditions; for example, the velocity of sound is approximately 344 m/sec (1130 ft/sec) in air, 1433 m/sec (4700 ft/sec) in water, 3962 m/sec (13,000 ft/sec) in wood, and 5029 m/sec (16,500 ft/sec) in steel.

Time-Weighted Average (TWA): A single reading of noise level obtained by averaging all of the different sound levels that a worker is exposed to during the workday. The TWA represents that constant noise level in dBA that has the same severity over eight hours as the exposure to the actual workday noise.

Transmission Loss (TL): The amount of attenuation (in decibels) provided by a wall or panel, as measured in a laboratory. Transmission loss is defined as ten times the logarithm (to the base 10) of the ratio of the incident acoustic energy to the acoustic energy transmitted through a sound barrier.

Ultrasonic: Frequencies higher than those of audible sound (i.e., above 20,000 Hz).

Wavelength (λ): The length (in feet or meters) of one complete sound wave. Wavelength is calculated from known values of the frequency (f) and speed of sound (c) by: $\lambda = c/f$.

White Noise: A sound spectrum with essentially equal energy at each frequency within a specified frequency range.

2 PHYSICS OF SOUND

Sound results from oscillations in pressure in any "elastic" medium such as air, water, solids, etc., that effectively couples the sound source with the ear. When sound is transmitted through air, it is usually described in terms of variations in pressure that alternate above and below atmospheric pressure. These pressure changes are produced when vibrating objects (sound sources) cause regions of high and low pressure that propagate from the sound source. The characteristics of a particular sound depend on the rate at which the sound source vibrates, the amplitude of the vibration, and the properties of the conducting medium.

2.1 Units for Noise Measurements

2.1.1 Sound Pressure and Sound Pressure Level

The range of sound pressures encountered in many acoustical environments is usually very wide. Sound pressures as high as those produced by jet engines (about 20 Pa) are found in some work areas, whereas sound pressures down to the threshold of hearing (approximately 0.00002 Pa) are used for audiological testing. To cover this wide range of sound pressure with a reasonable number of scale divisions and to provide a scale that responds more closely to the response of the human ear, a logarithmic scale is used. This logarithmic notation was termed the Bel, in honor of Alexander Graham Bell; however, the Bel was found to be too large for practical use. Therefore, one Bel was subdivided into 10 smaller units, and the prefix "deci" was added to form *decibel* (abbreviated as dB).

Any time the word *level* is used in acoustics, decibel notation is implied. By definition, the decibel is a dimensionless unit, and is related to the logarithm of the ratio of a measured quantity to a reference quantity. The decibel can be difficult to understand and use because it is often associated with different reference quantities. Acoustic intensity, acoustic power, hearing thresholds, electric voltage, electric current, electric power, and sound pressure level may all be expressed in decibels, each having a different reference. Obviously, the decibel has no meaning unless a specific reference quantity is specified, or understood.

Most sound-measuring instruments are calibrated to provide a reading of root-mean-square (rms) sound pressures in decibels. The decibel reading taken from such an instrument is called the sound pressure level. Sound pressure level is denoted by the symbol L_p, where the capital letter L represents a decibel *level*, and the subscript p indicates sound pressure. The term level is used because the measured pressure is at a particular level above a given pressure reference. For sound measurements in air, 0.00002 Pa, or equivalently, 20 μPa, typically serves as the reference sound pressure (p_{ref}). This reference is an arbitrary pressure chosen many years ago as an approximation of the normal threshold of human hearing at 1000 Hz. Since the eardrum responds to the intensity of the sound wave, and since intensity is proportional to the pressure squared, sound pressure levels are calculated from the square of the sound pressures. Mathematically, L_p is written as follows:

$$L_p = 10 \log \left(\frac{p}{p_{\text{ref}}}\right)^2 \text{dB} \tag{1}$$

where p is the measured rms sound pressure, p_{ref} is the reference sound pressure, and the logarithm is to the base 10. Utilizing the properties of logarithms and exponents, this equation is commonly rewritten as:

$$L_p = 20 \log \left(\frac{p}{p_{\text{ref}}}\right) \text{dB} \tag{2}$$

Using this form of the equation for L_p sometimes causes confusion regarding the multiplication factor of 20 in front of the logarithm. This confusion can be eliminated by an understanding that the basic definition of the decibel is based on power-related quantities, and power is proportional to the square of the pressure (i.e., power $\propto$ pressure^2). Table 19.1 shows the relationship between sound pressure (in Pascals) and sound pressure level (in dB re 20 μPa). This table illustrates the advantage of using the decibel scale rather than the wide range of direct pressure measurements—i.e., the range of decibel values is more compact and manageable. Using decibel notation, a doubling of sound pressure is equivalent to a 6 dB change in level, and a tenfold increase in pressure raises the level by 20 dB.

2.1.2 Sound Power and Sound Power Level

Sound power was briefly mentioned in the previous section; it is defined as the amount of acoustic energy that is produced per second by a noise source. Sound power is measured in watts, and is represented by the symbol W. Power units are also frequently described in terms of decibel levels because of the wide range of powers covered in practical applications. Sound power level (L_W) is defined by:

Table 19.1. Relation between Sound Pressure and Sound Pressure Level

Sound Pressure (Pa)	Sound Pressure Level in dB re 20 μPa	Typical Environment	Average Subjective Evaluation
200	140	Near jet engine	
	130	Pneumatic chipper	Intolerable or deafening
20	120	Boiler shop; "hard rock" band	
	110	Automatic punch press; jack hammer; motorcycle	
2	100	Chainsaw; snowmobile; outboard motor	
			Very noisy
	90	Construction site; noisy urban area	
2×10^{-1}	80	Curbside of busy street; school cafeteria	
			Loud
	70	Loud radio; vacuum cleaner	
2×10^{-2}	60	Restaurant; department store	
			Moderate to quiet
	50	Conversational speech; average office	
2×10^{-3}	40	Residential area at night	
			Faint
	30	Average home	
2×10^{-4}	20	Background in TV studio	
	10	Rustle of leaves	Very faint
2×10^{-5}	0	**Normal threshold of hearing**	

$$L_W = 10 \log \frac{W}{W_{\text{ref}}} \text{ dB} \tag{3}$$

where W is the power of the source in watts (W) and W_{ref} is the standard reference power of 10^{-12} W. In air, this reference corresponds to the reference pressure of 20 μPa used for sound pressure levels.

For most hearing conservation purposes, familiarization with the concepts of sound pressure and sound pressure level is sufficient. However, it is important to understand that the measured sound pressure is actually a result of the amount of sound power generated by a source. In other words, there is a cause-and-effect relationship between sound power and sound pressure: Sound power is the "cause" of a noise, and sound pressure is the resulting "effect." Conversion from power to pressure (or vice versa) is routinely performed when developing engineering noise control treatments. Sound pressure levels can be predicted from known sound power levels, depending on the environment in which the sound source is located, the exact distance from the source, as well as other variables.

2.1.3 *Sound Intensity and Sound Intensity Level*

Sound intensity at any specified location may be defined as the average acoustic power passing through a unit area, in the direction of wave propagation. For a spherical or free-progressive sound wave, the intensity may be expressed by:

$$I = \frac{p^2}{\rho c} \frac{\text{W}}{\text{m}^2} \tag{4}$$

where p is the rms sound pressure, ρ (rho) is the density of the medium, and c is the speed of sound in the medium. Sound intensity, like sound power and sound pressure, covers a wide range of measurements, and it is often desirable to use decibel levels to compress the measuring scale. Sound intensity is converted into decibels through the use of the following equation:

$$L_I = 10 \log \frac{I}{I_{\text{ref}}} \text{ dB} \tag{5}$$

where I is the measured intensity at some given distance from the source and I_{ref} is the standard reference intensity of 10^{-12} W/m^2.

As mentioned previously, once the sound power level of a source is known, then the resulting sound pressure level can be predicted for any particular environment where that source may be placed. Unfortunately, there is no way to measure sound power directly. However, modern instruments have been developed to measure sound intensity, which is simply the amount of acoustic power radiated over a given surface area. Sound intensity instruments make it possible to determine the sound power of a source *in situ*, eliminating the need for special-purpose acoustical facilities such as anechoic rooms and reverberation chambers.

2.1.4 Combining Sound Levels

It is often necessary to combine sound levels—for example, individual frequency band levels may be combined to obtain the overall or total sound pressure level of a particular noise. Another example is the estimation of total sound pressure level resulting from adding a machine of known noise spectrum to a noise environment of known characteristics. However, because decibels are logarithmic quantities, they cannot be combined by simply summing up the individual sound pressure levels. Instead, the actual acoustic intensities represented by the logarithmic expressions must be determined by taking the antilogs of the level readings. Then the intensities can be added together, and the resulting level is determined from the logarithm of that sum. An equation commonly used for summing any number of individual sound levels is:

$$L_p(\text{total}) = 10 \log (10^{L_{p1}/10} + 10^{L_{p2}/10} + \ldots) \text{ dB} \tag{6}$$

where each level to be added is arbitrarily assigned as L_{p1}, L_{p2}, etc., for as many different levels that are to be added together. An example of decibel addition using this equation follows.

Example: Add 90 dB + 95 dB + 88 dB

$$L_p(\text{total}) = 10 \log (10^{L_{p1}/10} + 10^{L_{p2}/10} + 10^{L_{p3}/10})$$

$$L_p(\text{total}) = 10 \log (10^{90/10} + 10^{95/10} + 10^{88/10})$$

$$L_p(\text{total}) = 10 \log (4.79 \times 10^9)$$

$$L_p(\text{total}) = 96.8 \text{ dB}$$

Alternatively, a decibel addition table may be used to determine the resultant level when two or more individual levels are combined (see Table 19.2). To use a decibel addition table, determine the difference between the first two levels to be added, then find where this number falls in the left-hand column of the table. Next, move across to the right-hand column and locate the number that corresponds to this difference. Finally, determine the total sound pressure level by adding the value in the right-hand column to the higher of the two levels being added together. If more than two levels are to be added, any two may be combined first, then that result should be added to the third level, and the resultant of the three sources added to the next level, etc., until all levels have been summed. Although the order in which the levels are added usually does not matter, one typically starts with the highest levels first. This is done because once the difference between two levels is 10 dB or greater, the addition process can be discontinued since the addition of smaller values does not add significantly to the overall total.

The use of a decibel addition table may be illustrated by the following example.

Example: Add 85 dB + 87 dB + 90 dB + 71 dB

Starting with the highest levels, the difference between 90 dB and 87 dB is 3 dB. Locating this value in the left-hand column of Table 19.2 and looking across to the

Table 19.2. Table for Combining Decibel Levels of Sounds with Random Frequency Characteristics

Numerical Difference between Levels (dB)	Amount to be added to the Higher Level (dB)
0.0–0.1	3.0
0.2–0.3	2.9
0.4–0.5	2.8
0.6–0.7	2.7
0.8–0.9	2.6
1.0–1.2	2.5
1.3–1.4	2.4
1.5–1.6	2.3
1.7–1.9	2.2
2.0–2.1	2.1
2.2–2.4	2.0
2.5–2.7	1.9
2.8–3.0	1.8
3.1–3.3	1.7
3.4–3.6	1.6
3.7–4.0	1.5
4.1–4.3	1.4
4.4–4.7	1.3
4.8–5.1	1.2
5.2–5.6	1.1
5.7–6.1	1.0
6.2–6.6	0.9
6.7–7.2	0.8
7.3–7.9	0.7
8.0–8.6	0.6
8.7–9.6	0.5
9.7–10.7	0.4
10.8–12.2	0.3
12.3–14.5	0.2
14.6–19.3	0.1
19.4–∞	0.0

right side of the table, it can be seen that 1.8 dB should be added to the higher level, which gives 91.8 dB. Next, the difference between 91.8 dB and 85 dB is 6.8 dB, so according to Table 19.2, 0.8 dB should be added for a total of 92.6 dB. Finally, due to the large (i.e., greater than 10 dB) difference between 92.6 dB and 71 dB, nothing more needs to be added, and the total sound pressure level from summing all four levels is 92.6 dB.

Most industrial noises have random frequency characteristics, and can be combined as described in the preceding paragraphs. However, in the few cases when noises have a certain "pitch" or major pure tone component(s), these calculations are not accurate, and phase relationships must be considered. At zero phase difference, the resultant value of two identical pure tone sources is 6 dB greater than the value of either single level. Between 0 and 90°, the resultant is somewhere between 3 and 6 dB greater than either level. At a phase difference of 90°, the resultant is 3 dB greater than either level. Between 90 and 120°, the resultant is between 0 and 3 dB greater than either level, and with a phase difference of 120°, the resultant is equal to the individual levels. At 180°, complete cancellation of sound occurs. The fact that one sound can actually cancel out another is a very important principle in acoustics, as it forms the basis for active noise control. Noise cancellation technology, also referred to as active noise reduction, is increasingly being employed as advances in equipment and instrumentation are developed.

2.2 Frequency Analysis

The most common frequency bandwidth used for industrial noise measurements is the octave band. A frequency band is said to be an octave in width when its upper band-edge frequency f_2 is twice the lower band-edge frequency f_1:

$$f_2 = 2f_1 \tag{7}$$

Octave bands are commonly used for measurements directly related to the effects of noise on the human ear, and for general noise control purposes, because they provide the maximum amount of information with a limited number of measurements.

When more specific characteristics of a noise source are required (e.g., for pinpointing a particular noise source among a background of other sources), it is necessary to use frequency bandwidths that are narrower than octave bands. Half-octave, third-octave, and narrower bands are used for these purposes. A half-octave bandwidth is defined as a band whose upper band-edge frequency f_2 is the square root of two times the lower band-edge frequency f_1:

$$f_2 = \sqrt{2}\, f_1 \tag{8}$$

A third-octave bandwidth is a band whose upper band-edge frequency f_2 is the cube root of two times the lower band-edge frequency:

$$f_2 = \sqrt[3]{2}\, f_1 \tag{9}$$

The center frequency f_c of any of these bands is the square root of the product of the high and low band-edge frequencies (geometric mean):

$$f_c = \sqrt{f_1 f_2} \tag{10}$$

It should be noted that the upper and lower band-edge frequencies describing a frequency band do not imply abrupt cutoffs at these frequencies. These band-edge frequencies are conventionally used as the 3-dB-down points of gradually sloping curves that meet the

American National Standard Specification for Octave-Band and Fractional-Octave-Band Analog and Digital Filters (1).

3 THE EAR

Normal human ears respond to a remarkably wide frequency range that covers approximately 20 to 20,000 Hz. As might be expected, the characteristics of any individual ear are extremely complex, and large differences exist among the population. An ear's response characteristics may change as a result of physical or mental conditions, sound level, medications, environmental stresses, diseases, and other factors.

A normal healthy human ear is also capable of detecting a remarkable range of sound levels. It is sensitive to very low sound pressures that produce a displacement of the eardrum no greater than the diameter of a hydrogen molecule. At the other extreme, it can receive sounds whose sound pressures are more than one million times greater than the lower threshold value; however, exposure to high-level sounds may cause temporary or permanent damage to the hearing mechanism.

3.1 Anatomy and Physiology of the Ear

As shown in Figure 19.1, the ear can be divided into three sections: the *outer* or *external ear*, the *middle ear*, and the *inner ear* (2). Sound waves impinging on the eardrum cause it to vibrate, and these vibrations are transmitted to the fluid-filled inner ear that contains the nerve endings by which aural information is detected. The acoustic or auditory nerve carries these impulses to the hearing center in the brain.

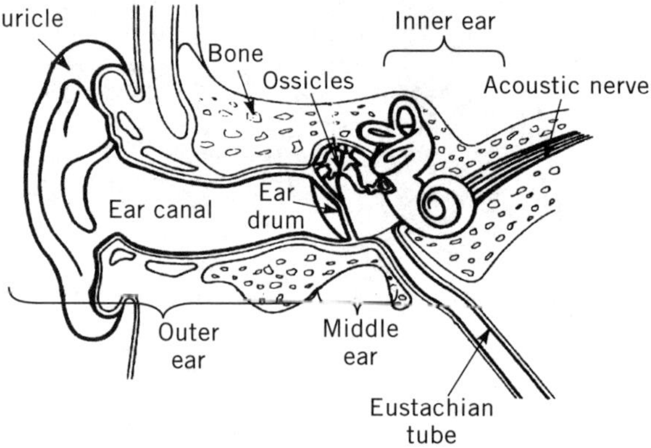

Figure 19.1. Cross-section of the ear showing the outer (external), middle, and inner ear configurations.

3.1.1 Outer Ear

Although the auricle is what most people think of when the term "ear" is used (Fig. 19.2), it plays only a limited role in the hearing process. To a certain extent, the auricle helps direct high frequency sounds into the ear canal, assisting the hearing system in determining the direction from which a sound originates. Sound waves enter the ear canal and travel towards the eardrum, which is the dividing point between the outer and middle ear regions.

Typical ear canals are about one-inch long and approximately one-quarter inch in diameter. They are seldom as straight as depicted in Figures 19.1 and 19.2, and the shape and size of ear canals differ significantly among individuals (and even between ears of the same individual). From an acoustical viewpoint, since the ear canal is closed at one end by the eardrum, it can be modeled as a cylinder with a quarter-wavelength resonance of about 3000 Hz. This resonance effectively increases the sensitivity of the ear in the higher frequencies around 3000 Hz.

Glands producing earwax or cerumen are located in the ear canal. Normally, cerumen flows outward, toward the entrance of the ear canal, carrying with it any foreign particles that may accumulate in the canal. The normal flow of wax may be interrupted by changes in body chemistry that can cause excessive wax to be produced, or the wax may become hard or "impacted." At times, the cerumen may build up to the point of totally occluding the canal, and a temporary loss of hearing will result. Any buildup of wax deep within the ear canal should be removed very carefully, to prevent damage to the eardrum and middle ear structures.

The surface of the external ear canal is extremely delicate and easily irritated. Cleaning or scratching with fingernails or other sharp objects can break the skin and cause a very painful and persistent infection. Infections can cause swelling of the canal walls and, occasionally, a loss of hearing when the canal swells shut. An infected ear should be given prompt attention by a physician.

3.1.2 Middle Ear

The air-filled space between the eardrum and the inner ear is called the middle ear (Fig. 19.3). The middle ear contains three small bones or ossicles that mechanically connect the

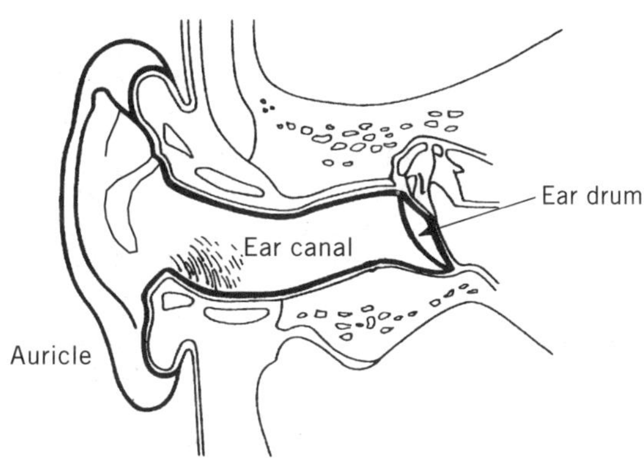

Figure 19.2. The outer or external ear.

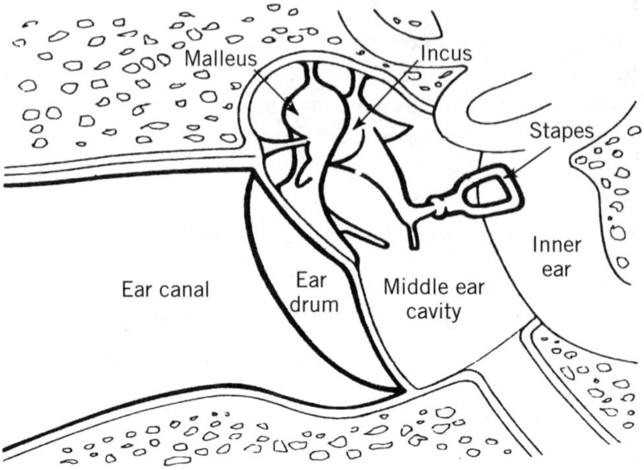

Figure 19.3. The middle ear.

eardrum to the oval window of the inner ear. Sound energy is converted into mechanical energy by the middle ear system.

The eardrum is a very thin and delicate membrane that responds to very low sound pressures at the lower threshold of hearing, yet it seldom suffers physical damage from routine exposure to continuous high-level noises. Although an eardrum may be damaged or perforated by an explosion or a rapid change in ambient air pressure, most common steady-state noise exposures do not produce a noticeable effect on the eardrum itself.

The three tiny bones or ossicles in the middle ear are commonly known as the hammer, anvil, and stirrup, although their correct anatomical names are the malleus, incus, and stapes. These three bones are connected together to form a "chain," with the handle of the malleus physically attached to the eardrum. Therefore, any movement of the eardrum is transmitted directly to the bones of the middle ear. Due to the construction of the eardrum/ossicle system, a mechanical advantage or lever action is formed, which is necessary to match the impedance of air in the ear canal with the fluid in the adjacent inner ear structure.

When an eardrum is ruptured, the attached middle-ear ossicles may be dislocated; thus the eardrum should be carefully examined immediately after the injury occurs to determine whether realignment of the ossicles is necessary. In a high percentage of cases, surgical procedures are successful in realigning dislocated ossicles or by inserting a prosthesis, so that little or no permanent loss in hearing acuity results from this type of injury.

The middle ear cavity is completely enclosed except for the small Eustachian tube that connects the middle ear to the back of the throat (see Fig. 19.1). The purpose of the Eustachian tube is to equalize the air pressure inside the middle ear space with that of the surrounding atmosphere. When flying in an airplane or when driving in mountainous territory, changes in the ambient barometric pressure are normally compensated for when the Eustachian tube briefly opens, thereby allowing air to enter or escape from the middle ear space. In this manner, the same amount of air pressure is always maintained on both sides of the eardrum.

Occasionally, the Eustachian tube will remain closed as a result of an infection or an allergy, resulting in a pressure imbalance between the middle ear and the outside air. This situation may produce a temporary loss of hearing sensitivity and extreme discomfort for the individual. Even a healthy ear may suffer a loss of hearing if the Eustachian tube becomes blocked, but this loss of hearing can often be quickly restored simply by yawning, swallowing, or chewing gum, which allows the Eustachian tube to open momentarily.

3.1.3 Inner Ear

As illustrated in Figure 19.1, the inner ear is completely surrounded by bone; a close-up of the inner ear is shown in Figure 19.4. One end of the inner ear is shaped like a snail shell, and is called the cochlea. The other end of the inner ear contains three semicircular loops or canals. The fluid-filled cochlea serves to detect incoming sound signals and to translate them into nerve impulses, which are then transmitted to the brain. Although they are located in the inner ear, the semicircular canals are balance organs, and not actually part of the hearing mechanism.

Mechanical energy from the middle ear is transferred to the inner ear by the stapes, whose footplate rests on the oval window of the cochlea (see Fig. 19.4). Movement of the stapes forces the oval window in and out with the dynamic characteristics of the incident sound, setting the fluid of the cochlea into motion, thereby transforming these mechanical vibrations into hydraulic energy. Thousands of hair cells located along the two-and-one-half turns of the cochlea detect these motions and translate them into nerve impulses. A sound is "heard" when this neural energy is sent along the acoustic or eighth cranial nerve up to the auditory center of the brain.

3.2 How Noise Damages Hearing

Noise-induced hearing loss may be temporary or permanent depending on the level and frequency characteristics of the noise, the duration of exposures, and the susceptibility of

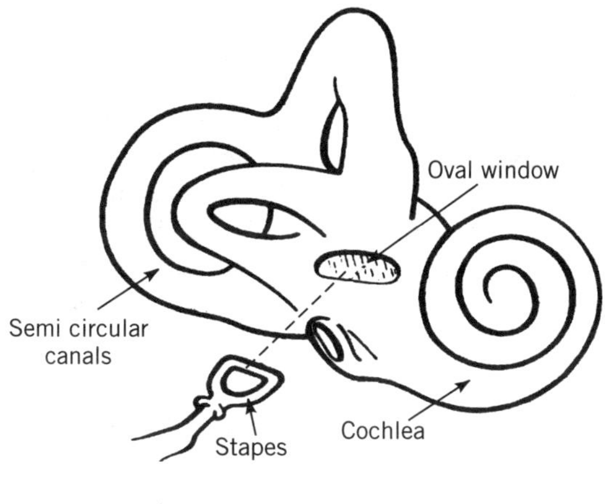

Figure 19.4. The inner ear.

the individual (3). After a temporary hearing loss occurs, the original sensitivity is usually restored within a matter of hours; however, in some cases temporary losses may last for days or weeks. Permanent losses result when these temporary losses do not recover completely. A permanent noise-induced hearing loss is not reversible and cannot be corrected by conventional medical or surgical procedures.

A noise-induced hearing impairment generally occurs with the hair cells located in the cochlea, and is called a sensorineural hearing loss. For commonly encountered noise exposures, hearing acuity is generally affected first in the frequency range from 3000 to 6000 Hz, with most affected persons showing a loss or "dip" in sensitivity at 4000 Hz. If high-level exposures are continued, the loss of hearing generally increases around 4000 Hz and spreads to adjacent frequencies. For noise exposures having significant components concentrated in narrow frequency bands below 4000 Hz, impairments usually are found about one-half to one octave above the predominant exposure frequencies.

Noise-induced hearing loss is an insidious problem, because a person does not necessarily have to experience pain or even be immediately aware that severe hearing damage has taken place. The damage may occur instantaneously or over a long period of time, depending upon the noise characteristics. Generally, impulsive or impact noises are most likely to produce significant losses with short exposure periods, and steady-state continuous noises are responsible for impairments that develop over a long period of time. Additionally, workers suffering from noise-induced hearing loss may hear a continuous "ringing" in their ears, which is called tinnitus.

Even after a significant amount of damage, a person with noise-induced hearing loss is able to hear low-frequency (vowel) sounds very well, but the high frequencies (consonants) in speech are not clearly heard. Loudness levels may be nearly normal, but intelligibility may be poor since the consonant sounds carry most of the distinguishing characteristics of speech, and degraded high frequency hearing will cause misunderstanding or confusion between similar-sounding words. A noise-induced hearing loss becomes particularly noticeable when verbal communication is attempted in noisy or reverberant areas. Speech is "masked" or obscured most effectively by background noises containing the speech frequencies, such as the sounds encountered in a crowd of people, where many are talking simultaneously.

3.3 Combined Effect of Noise and Other Agents

Research has shown that exposure to some chemicals or other industrial substances may exacerbate the effect of high-intensity noise (3). Agents such as solvents, heavy metals, and asphyxiants can increase the amount of hearing loss a worker may incur, beyond the amount expected from any one of the physical agents alone. Further research is needed to accurately define the combined effects of noise exposure and other hazardous substances.

4 HEARING MEASUREMENT

The only way to monitor the overall effectiveness of a hearing conservation program is to periodically check the hearing of all persons exposed to potentially hazardous noise. Hearing thresholds at 500, 1000, 2000, 3000, 4000, and 6000 Hz must be checked in each ear, in order to determine whether employees are losing some of their hearing ability. Hearing loss due to medical conditions unrelated to occupational noise exposure may also be identified from periodic hearing tests.

To ensure the accuracy of hearing tests, audiometric equipment must be periodically calibrated, a quiet test environment must be maintained, and a well-trained audiometric technician or hearing conservationist must perform the hearing tests. Requirements for audiometer performance and for the background noise limits in test rooms have been specified in standards published by the American National Standards Institute (ANSI S3.6-1996 and S3.1-1991) (4, 5). Guidelines for training Occupational Hearing Conservationists have been established by the Council for Accreditation in Occupational Hearing Conservation (CAOHC) (6).

4.1 Audiometers

An audiometer is the instrument used for measuring pure tone, air conduction hearing thresholds. It may be designed for manual or automatic operation; most of the newest audiometers are microprocessor-based, and can be used in conjunction with a computer for data storage.

ANSI S3.6-1996 provides the specifications that audiometers must meet to provide accurate hearing threshold information (4). Instrumentation inaccuracies are seldom obvious, and there is a strong tendency to automatically accept dial readings as being accurate. Therefore, strict attention should be given to equipment calibration. Modern audiometers are much more stable than their older counterparts; however, they may lose their specified accuracy if not cared for properly. Earphones are particularly susceptible to damage from rough handling. Unseen damage could occur during shipment, so that even a new instrument may be out of calibration when initially received. The normal aging of components and exposure to temperature extremes may also cause changes in audiometer accuracy. Dust or dirt inside the audiometer can cause dials and switches to produce electrical noise or to wear excessively. Poor electrical contacts may produce intermittent operation or poor accuracy. For this reason, dust covers should always be used to protect the instrument when not in use, and the exterior of the audiometer should be cleaned periodically to prevent dust and dirt from getting inside the case.

The Occupational Safety and Health Administration (OSHA) requires that functional or biological calibration checks (audiograms taken on persons with known stable hearing thresholds) be performed before each day's use (7). Accuracy checks should also be made any time there is a reason to suspect a problem. The person used for daily biological calibration checks does not need to have perfectly normal hearing; the main consideration is that this individual's hearing thresholds do not fluctuate. A baseline hearing test is established immediately following a comprehensive or exhaustive electro-acoustic calibration of the audiometer. Subsequent daily tests should not deviate by more than 10 dB if the audiometer remains correctly calibrated.

Hearing sensitivity may temporarily decrease as much as 20 dB because of allergies, colds, or other causes; therefore, it is strongly recommended that at least two individuals be made available for biological tests. Alternatively, an "electroacoustic ear" may be used in place of a live human subject. An electroacoustic ear (sometimes called a bio-acoustic simulator) is an instrument that is calibrated to take a hearing test and respond as a person would. The audiometer's headphones are placed on the instrument, and a normal hearing test is run. Test results are compared to the electroacoustic ear's baseline to determine whether a change in calibration has occurred, just as would be done for a human listener. The advantage to using an electroacoustic ear is that this instrument will always be available for daily biological calibration checks, and will not respond differently on successive hearing tests.

The technician should also check the audiometer periodically using the following additional procedures:

1. Check all control knobs on the audiometer to be sure that they are tight on their shafts and not misaligned, and make sure all cable connections between the audiometer and jack panel are correct and secure. Visually inspect the earphones for correct headband tension, missing parts, and serviceability of the cushions. Cushions should be replaced if they are not resilient or if cracks, bubbles, or crevices develop. A new acoustic calibration is not required when replacing earphone cushions.

2. Straighten the earphone cords so that there are no sharp bends or knots. Cords with worn or cracked insulation should be replaced. With the audiometer set to 1000 Hz and 60 dB, test the earphone cords for electrical continuity by listening while bending the cords along their length. Typical trouble spots are where the cord connects to the earphones, or at the other end where the cord attaches to the jack. Any static, intermittency, or change in test tone indicates a need for new earphone cords. A new acoustical calibration is not required when replacing only the earphone cords.

3. With the frequency set to 1000 Hz, check the linearity of the hearing level control by listening to the earphone while slowly increasing the hearing level. Each 5-dB step should produce a steady increase in loudness without changes in tone quality or any other audible extraneous noise.

4. With the audiometer set at 1000 Hz and 60 dB, test the operation of the tone interrupter by listening to the earphones and operating the interrupter several times. No audible noises (such as clicks or scratches) or changes in tone quality should be heard when the interrupter switch is used.

5. With a steady 60 dB tone coming from the right earphone, cover that earphone with the palm of your hand and listen for extraneous noises from the left earphone. No tones or noises should be heard in the non-test earphone. Repeat this procedure with the tone coming from the left earphone.

6. With the hearing level set to 60 dB, check for the presence of a tone at each test frequency. There should be a noticeable difference between each frequency, without any wavering or abrupt changes in tone quality.

According to Appendix E in the OSHA Hearing Conservation Amendment, audiometer calibration should be checked "acoustically" once per year, or sooner if the functional check indicates a deviation of 10 dB or more between the dial setting and actual output of the audiometer. The two main checks performed during an acoustic calibration are a sound pressure output check and a linearity check. When any of the measured sound levels deviate by ±3 dB at any test frequency between 500 and 3000 Hz, 4 dB at 4000 Hz, or 5 dB at 6000 Hz, an exhaustive calibration is *advised.* An exhaustive calibration is *required* if the deviations are 15 dB or greater at any test frequency.

An exhaustive calibration is typically done by an authorized repair service or the audiometer's manufacturer. This is the only stage at which changes to the audiometer are made, in order to exactly meet the requirements in the ANSI S3.6 standard. This type of comprehensive electroacoustic calibration must be performed at least every two years, regardless of whether any discrepancies have been noted in the functional or acoustical calibration checks.

4.2 Test Rooms

The sound pressure level of the background noise in rooms used for measuring hearing thresholds must be limited to prevent masking effects that cause misleading, elevated threshold values. In particular, an individual's hearing thresholds at 500 Hz may be artificially elevated (i.e., appear worse than they really are) due to interference from the background noise. Maximum allowable octave-band sound pressure levels for audiometric test rooms, as specified in Appendix D, Table D-1, of the OSHA Standard, are contained in the following chart (7):

500 Hz	1000 Hz	2000 Hz	4000 Hz	8000 Hz
40 dB	40 dB	47 dB	57 dB	62 dB

It should be noted that the maximum background levels specified by OSHA are not low enough to permit testing down to 0 dB hearing threshold level. More accurate tests can be obtained if the test room meets the criteria established in ANSI S3.1-1991, American National Standard Maximum Permissible Ambient Noise Levels for Audiometric Test Rooms (5). These background noise levels are more stringent than those specified by OSHA, but they are difficult to obtain in noisy areas, and may be below the practical limit for ambient noise in many industrial locations. Typically, audiology clinics or otologist offices are the only testing locations that meet the ANSI S3.1 background levels.

In addition to meeting OSHA's background noise level requirements, subjective tests should be made inside the test booth on-site to determine that no noises (e.g., talking and heel clicking) are audible. In particular, short impulse-type noises may be heard even though the measured sound pressure levels are below the limits in the table. Any of these extraneous noises may distract the subject and interfere with the hearing threshold measurements. If possible, these nuisance sounds should be eliminated, or the tests must be delayed until they no longer present a problem.

The noise reduction provided by a good prefabricated audiometric test booth should be adequate when the booth is placed in a reasonably quiet area of the plant or facility. However, it is recommended that the burden of on-location performance be placed on the booth supplier, to ensure that an acceptable booth is obtained. The vendor should be willing to guarantee in writing the performance of the test room or booth when installed at the specified test site, particularly if some on-site assembly is required. Attention must also be paid to proper vibration isolation for permanently installed booths.

In addition to the attenuation characteristics, several other factors should be considered when selecting a test booth or room:

1. What size booth is necessary—one person, two person, four person, etc.?
2. Will opening and closing the door result in wearing of contact material (seals), necessitating frequent replacement?
3. Are the interior surfaces durable and easily cleaned?
4. Is the door easily opened from the inside, so that subjects will not feel "locked in?"
5. Do the observation window and seating arrangement provide an easy view of the subject, but block the subject's view of the audiometer controls?
6. If the booth is equipped with a ventilation system, is the interior noise level below the limits specified for the room?
7. If the booth must be moved occasionally, is it portable?

4.3 Hearing Threshold Measurements

Normally, the purpose of audiometric testing in industrial hearing conservation programs is to monitor the effectiveness of hearing conservation procedures. Unless the test results are to be used for another purpose, there is no need for detailed diagnostic information that would require the use of sophisticated audiometric techniques and/or specialized equipment. Pure tone, air conduction hearing thresholds obtained via standard audiometric earphones are usually adequate for industrial monitoring purposes.

The conventional technique for obtaining hearing threshold levels is referred to as the modified Hughson-Westlake method (8). In this procedure, the initial tone is presented to the subject at a level well above the anticipated threshold, then decreased in 10 or 15 dB steps until it is no longer audible. Next, the level is increased in 5 dB steps until the patient responds. After a positive response, the tone is lowered by 10 dB and testing resumes, with the level again raised in 5 dB steps until another positive response occurs. This "up 5 dB, down 10 dB" approach is referred to as an ascending bracketing technique. The hearing threshold is then defined as the level at which the subject responds 50% of the time, with a minimum of two out of three positive responses at a single level. Initially, audiometers were operated manually using this technique; however, microprocessor-based audiometers programmed to conduct the hearing test in a similar manner are now more commonly used.

4.4 Records

Records must be complete and accurate if they are to have medicolegal significance. Records should be kept in ink, without erasures. If a mistake is made in recording, a line should be drawn through the erroneous entry, and the initials of the person making the correction should be placed above the line along with the date. The original entry must not be obliterated.

Whether hand-written or maintained in a computer database, the model, serial number, and calibration dates of the instrument should be recorded on each audiogram. Records should also be kept of periodic noise level measurements in the test booth. In addition to the threshold levels, an audiogram should have space provided for the recording of pertinent medical and noise exposure information. Some typical questions asked of test subjects are given on the form depicted in Figure 19.5.

5 NOISE MEASUREMENT

A wide variety of new sound-measuring equipment has been made available over the past several years. Some sound level meters provide only the basic measurements required by OSHA, while others provide a very wide range of functions including integration for dose and impulse noise measurements. Sophisticated measuring equipment can be used to obtain an enormous amount of data in a relatively short period of time. The weak link in sound

Name ______ Date ______ Recorded by ______ Employee inital ______

History	YES	NO	Comments
Have you had a previous hearing test ?			
Do you now have any trouble hearing ?			
Have you ever worked in a noisy industry ?			
Do you think you can hear better in your Right ear?			
or Left ear?			
Have you ever had noises in your ears ?			
Have you ever had dizziness ?			
Have you ever had a head injury ?			
Has anyone in your family lost his hearing before age 50 ?			
Have you ever had measles, mumps, or scarlet fever?			
Do you have any allergies ?			
Are you now taking or have you regularly taken drugs, antibiotics, or medication?			
Have you ever had an earache ?			
Have your ears ever run ? Right ear ?			
Left ear ?			
Have you been in the Military service ? Describe			
Have you been exposed to any sort of gunfire ? Describe			
Do you have a second job ? Explain			
What hobbies do you have ?			

Figure 19.5. Sample medical history form.

measuring equipment, the microphone, has also been much improved in both performance and durability in recent years. Even with this advanced technology, however, careful consideration of the objectives of the measurement must be made prior to equipment selection. In all cases, accurate sound level measurements require a well-trained operator and calibrated instruments.

5.1 The Sound Level Meter

The basic sound level meter consists of a microphone that converts the pressure variations into an electrical signal, an amplifier/filter, an exponential time-averaging circuit, a device to determine the logarithm of the signal, and an indicating meter or digital display. All instruments used for hearing conservation purposes should conform to the American National Standard Specification for Sound Level Meters (ANSI S1.4-1983 R1997) (9).

Most instruments present the sound pressure level in terms of its rms value. The rms value is useful for hearing conservation purposes because it is related to acoustic power, which forms the basis for decibel notation. Further, it must be kept in mind that an ordinary arithmetic averaging process will actually yield a value of zero for sound pressure fluctuations. The rms values cannot be used to describe prominent peak pressures of noise that extend several decibels above a relatively constant background noise; peak values are used for this purpose. On the other hand, peak readings are of relatively little value for measuring sustained noises unless the wave form is known to be sinusoidal, because the relationship between the peak reading and the acoustic power changes with the complexity of the wave. As the waveform becomes more complex, the peak value can be as much as 25 dB above the measured rms value.

The two primary classifications of sound level meters used in industrial noise monitoring are the precision (Type 1) and the general purpose (Type 2). Generally, the main difference between these instruments is simply a matter of how accurate the sound level readings are. According to ANSI S1.4, the allowable tolerances across the measurement range are approximately one decibel for the type 1 and two decibels for the type 2 instruments. Most regulations require the use of a type 2 meter; type 1 instruments are used when a greater degree of accuracy is required.

Two meter-response settings are usually available on a sound level meter. The "fast" response is used to follow rapid changes in level, while the "slow" response essentially slows the meter's display down and is intended to provide an averaging effect that makes widely fluctuating sound levels easier to read. The fast response uses a one-eighth second time constant, and the slow response has a time constant of one full second. Neither of these settings actually affect the signal itself; "fast" and "slow" refer only to the speed with which the display indicator on the meter moves. The OSHA regulation states that the slow response setting is to be used for industrial noise monitoring purposes (7). When impulse or impact noise (i.e., a transient sound less than one second in duration) is present, a measurement of the true peak sound pressure level is required. Some sound level meters have an "impulse" setting; however, this is not suitable for industrial noise measurements, because it is not designed to actually measure the true unweighted peak sound level.

General-purpose sound level meters are normally equipped with one or more filters or frequency-weighting networks, designated by the letters A, B, and C (see Table 19.3).

Table 19.3. Relative Response as a Function of Frequency for A, B, and C Weighting

Frequency (Hz)	A-Weighting Relative Response (dB)	B-Weighting Relative Response (dB)	C-Weighting Relative Response (dB)
10	−70.4	−38.2	−14.3
12.5	−63.4	−33.2	−11.2
16	−56.7	−28.5	−8.5
20	−50.5	−24.2	−6.2
25	−44.7	−20.4	−4.4
31.5	−39.4	−17.1	−3.0
40	−34.6	−14.2	−2.0
50	−30.2	−11.6	−1.3
63	−26.2	−9.3	−0.8
80	−22.5	7.4	−0.5
100	−19.1	−5.6	−0.3
125	−16.1	−4.2	−0.2
160	−13.4	−3.0	−0.1
200	−10.9	−2.0	0
250	−8.6	−1.3	0
315	−6.6	−0.8	0
400	−4.8	−0.5	0
500	−3.2	−0.3	0
630	−1.9	−0.1	0
800	−0.8	0	0
1,000	0	0	0
1,250	+0.6	0	0
1,600	+1.0	0	−0.1
2,000	+1.2	−0.1	−0.2
2,500	+1.3	−0.2	−0.3
3,150	+1.2	−0.4	−0.5
4,000	+1.0	−0.7	−0.8
5,000	+0.5	−1.2	−1.3
6,300	−0.1	−1.9	−2.0
8,000	−1.1	−2.9	−3.0
10,000	−2.5	−4.3	−4.4
12,500	−4.3	−6.1	−6.2
16,000	−6.6	−8.4	−8.5
20,000	−9.3	−11.1	−11.2

Readings obtained using any of these weighting scales are termed sound levels rather than sound pressure levels, and the particular weighting network used must always be indicated. The A, B, and C weighting curves were chosen because they approximate the response characteristics of the ear at various sound levels, and in the earliest sound level meters they could be easily produced with a few common electronic components. Empirically, the A-weighting has been found to give a good estimation of the hearing-damage risk potential from exposure to continuous noise. Therefore, OSHA specifies that the A-weighting scale be used for industrial noise measurements (7). The B-weighting is seldom used, and the C-weighting is used when assessing hearing protector effectiveness.

Figure 19.6 illustrates the frequency response characteristics of the A- and C-weighting networks. A linear or flat response, also included on some sound level meters, does not apply any correction values (i.e., it weights all frequencies equally) and would correspond to the 0 dB line in Figure 19.6.

In use, the frequency distribution of noise energy can be estimated by comparing the levels measured with each of the frequency weightings. For example, if the noise levels measured using the A and C networks are approximately equal, it can be reasoned that most of the noise energy is above 1000 Hz, because this is the only portion of the spectrum in which the weightings are similar. On the other hand, a large difference between these readings indicates that most of the energy will be found below 1000 Hz.

The rough estimate of frequency-response characteristics provided by the weighting networks of a sound level meter is not always adequate, particularly for engineering noise control work. In such cases, frequency analyzers are used. The most common types are the octave band (commonly abbreviated as 1/1), one-third octave band (abbreviated as 1/3), and narrow-band analyzers using filters or Fast Fourier Transform (FFT) digital calculations. Table 19.4 shows the range encompassed by each octave and one-third octave band, as defined by ANSI S1.11. Normally, the full octave bands provide adequate spectral information to solve most hearing conservation and noise control problems, and one-third

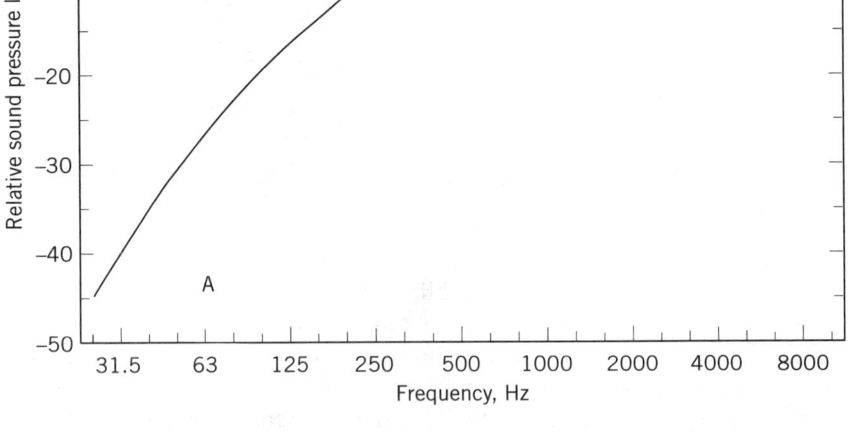

Figure 19.6. Sound level meter weighting curves.

Table 19.4. Center and Cutoff Frequencies for Full Octave and Third Octave Bands

1/1 Octave Bands			1/3 Octave Bands		
Lower	Center	Upper	Lower	Center	Upper
22.4	31.5	45	22.4	25	28
			28	31.5	35.5
			35.5	40	45
45	63	90	45	50	56
			56	63	71
			71	80	90
90	125	180	90	100	112
			112	125	140
			140	160	180
180	250	355	180	200	224
			224	250	280
			280	315	355
355	500	710	355	400	450
			450	500	560
			560	630	710
710	1000	1400	710	800	900
			900	1000	1120
			1120	1250	1400
1400	2000	2800	1400	1600	1800
			1800	2000	2240
			2240	2500	2800
2800	4000	5600	2800	3150	3550
			3550	4000	4500
			4500	5000	5600
5600	8000	11,200	5600	6300	7100
			7100	8000	9000
			9000	10,000	11,200
11,200	16,000	22,400	11,200	12,500	14,000
			14,000	16,000	18,000
			18,000	20,000	22,400

octave bands are used when more detail is required. However, significantly more measurements are necessary to cover the overall frequency range when smaller bandwidths are used; thus a compromise must be reached between the resolution required and the number of measurements that must be taken.

5.2 The Noise Dosimeter

A noise dosimeter consists of a miniature microphone connected to a small microprocessor-based sound level meter, which stores the noise data. The microphone is positioned at the top of the employee's shoulder, and the sound level meter control unit is clipped to the wearer's belt or placed in a pocket. Noise dosimeters continuously measure and process sound levels obtained near a worker's ear (i.e., hearing zone), then produce an average level of noise exposure that occurred throughout the individual's workday. Specifications for these devices are contained in ANSI S1.25-1991 (R1997), American National Standard Specification for Personal Noise Dosimeters (10).

A dosimeter is essentially identical to any other sound level meter, with the addition of an integrating function that keeps track of the noise level as well as the accumulated exposure time. Dosimeters have made it much more convenient to measure and assess employees' noise exposures, by eliminating the need for the surveyor to follow a worker throughout the workday with a sound level meter and a stopwatch to keep track of the exact amount of exposure to different noise levels. Many of the new instruments can continuously log or store the employee's noise exposure at one-minute, 10-second, or even one-second intervals. This noise exposure history information can be printed out and analyzed in any number of ways to help pinpoint periods of high noise levels or other significant occurrences during a workshift. Most regulatory inspectors use dosimeters to determine compliance with the applicable noise standards.

5.3 Instrument Calibration

It is very important to carefully read and understand the operator's manual prior to using any sound level meter or dosimeter. One of the critical points of interest is the proper calibration procedures for the particular instrument. If valid data are to be obtained, it is essential that all equipment for the measurement and analysis of sound be in calibration. When equipment is purchased from the manufacturer, it should have been calibrated to the pertinent ANSI specifications. However, it is the responsibility of the user to keep the instrument in calibration by conducting periodic checks.

Acoustic calibrators are available for checking the overall acoustical and electrical performance at one or more frequencies. Specifications for acoustical calibrators can be found in ANSI S1.40-1984 (R1997) (11). Calibration checks should be made according to the manufacturer's instructions at the beginning and end of each day's measurements, using a calibrator specifically designed for the sound level meter or dosimeter microphone. A battery check should also be done at these times. These calibration procedures cannot be considered to be of high absolute accuracy, nor will they allow the operator to detect changes in performance at frequencies other than those used for calibration. Nevertheless, they do serve as a warning for most common instrument failures, thus helping operators to avoid making invalid measurements.

Periodically, sound-measuring instruments should be sent back to the manufacturer or to a competent laboratory for a comprehensive re-calibration. This includes sound level meters, dosimeters, and their calibrators. How frequently these complete calibrations should be performed depends on the purpose of the measurements and how roughly the instruments are handled. In most cases, it is good practice to have a complete calibration performed at least once a year, and at any time an unusual reading is suspected or found.

6 ASSESSMENT OF NOISE EXPOSURE

6.1 Harmful Noise Exposures and Damage-Risk Criteria

Four major factors contribute to the development of noise-induced hearing impairment (12): (*1*) the overall noise level, (*2*) the frequency content, (*3*) the duration of exposure, and (*4*) the susceptibility of the individual. Studies leading to the development of comprehensive damage-risk criteria require complex statistical evaluations of large groups over long periods of time. Many different factors can compromise the accuracy of these studies, since there is a very large number of different combinations of noise levels, frequency component distributions, and exposure duration. In addition, there are wide differences in the susceptibility of individuals to noise-induced hearing impairment. Also, hearing losses found in a study may be due in part to noise exposures away from work that are not considered. Therefore, it is impossible to set a particular exposure level as the dividing line between safe and unsafe conditions that apply for all individuals. Damage-risk criteria are also influenced by practical limits. Typically, exposure level limits are established as a compromise between (*1*) the amount of hearing impairment that may result from a specified exposure dose and (*2*) the economic or other impact that may result from noise control expenditures.

6.2 Noise Regulations

The Safety and Health Standards for Federal Supply Contracts (Walsh-Healy Public Contracts Act), U.S. Department of Labor, was revised on May 20, 1969, to include the first national noise exposure regulation in the United States (13). This regulation stemmed from the Social Security Act of 1935, which expressed the philosophy that a worker has the right to earn a living in an environment that does not endanger his or her health. The Occupational Safety and Health Act of 1970 created the Occupational Safety and Health Administration (OSHA) and the National Institute for Occupational Safety and Health (NIOSH) (14). OSHA is responsible for the establishment and enforcement of regulations for most industries (one exception is mining, where the Mine Safety and Health Administration has responsibility). NIOSH conducts research and develops criteria in all areas of occupational safety and health for both general industry and mining.

The OSHA Occupational Noise Exposure Standard (**29** *CFR* 1910.95) was released in 1971, and was designed to protect workers exposed to hazardous noise environments from incurring permanent hearing loss (7). This regulation set the maximum levels of industrial noise to which an employee may be exposed, which is 90 dBA, measured with the slow meter response, for an 8-hour workday. Higher levels are permitted as long as the exposure time is less; a 5 dBA increase is allowed for each halving of exposure time (i.e., 95 dBA for 4 hours, 100 dBA for 2 hours, 105 dBA for 1 hour, 110 dBA for 30 minutes, and 115 dBA for 15 minutes or less). Additionally, the OSHA regulation states that exposure to impulsive or impact noise should not exceed 140 dB peak sound pressure level. A new MSHA noise regulation (that will become effective in the year 2000) is expected to closely follow the OSHA regulation, and will essentially replace the existing four different mining operation standards.

It is important to note that most noise regulations are designed to keep damage risk within "acceptable limits," which means that some people are likely to incur a hearing loss even when exposed to less than the maximum daily amount of noise specified in the regulation. NIOSH has recently revised its occupational noise exposure criteria, and, if incorporated into the OSHA standard, adherence to these criteria would reduce the "excess risk" of incurring noise-induced hearing loss from 25% down to 8% (15). Excess risk is defined as the percentage with material impairment of hearing in an occupational noise-exposed population, after subtracting the percentage who would normally incur such impairment from other causes in a population not exposed to occupational noise.

Based on the above information, many companies find it prudent to adopt the more stringent criteria and use exposure limits lower than OSHA's wherever possible. Obviously, the lowest feasible noise exposure levels are desirable for the health, safety, and well-being of workers. Moreover, these lower limits are of significant value to employers because they help maximize morale and work productivity and minimize the number of compensation claims for noise-induced hearing impairment.

When an employee's daily noise exposure is composed of two or more periods of exposure at different levels, the combined effect is determined by adding the individual contributions as follows:

$$D = \frac{C_1}{T_1} + \frac{C_2}{T_2} + \frac{C_3}{T_3} + \cdots\cdots + \frac{C_n}{T_n} \times 100\% \qquad (11)$$

where D represents the noise dose (in %), the values C_1 to C_n indicate the total time of exposure at a specific noise level, and the corresponding values of T indicate the total number of hours of exposure permitted at each of these levels. Values of T can be obtained from Table G-16 in the OSHA noise regulation, which is based on the equation

$$T_n = \frac{8}{2^{(Ln-90)/5}} \text{ hours} \qquad (12)$$

where L_n is equal to the A-weighted sound level that corresponds to each of the exposure time intervals. If the sum of the individual contributions exceeds 1.0, (i.e., the value for D exceeds 100%), then the mixed exposures are considered to exceed the overall limit value. For example, if a person were exposed to 90 dBA for 5 hours, 95 dBA for 2 hours, and 100 dBA for 1 hour during an 8-hr working day, the times of exposure are $C_1 = 5$ hours, $C_2 = 2$ hours, and $C_3 = 1$ hour. In this example, the corresponding time limits (or reference durations) for these exposures are $T_1 = 8$ hours, $T_2 = 4$ hours, and $T_3 = 2$ hours. Therefore, the total dose for this combined exposure is $5/8 + 2/4 + 1/2 = 1.625$ (about 163%), which exceeds the specified limit of 1.0 (100%).

In cases where employees are subjected to sound exceeding permissible exposure levels, feasible administrative or engineering controls must be used (7). Administrative controls usually involve an attempt to lower daily noise exposures by rotating employees through a "less noisy" work station; however, this practice is seldom feasible due to production constraints and/or employee training requirements. Engineering controls would work better in most circumstances, although they may be considerably more expensive to implement. If noise controls fail to sufficiently reduce exposure levels, personal protective equipment (i.e., hearing protection) must be provided.

A significant Hearing Conservation Amendment was added to the OSHA Occupational Noise Exposure Standard in 1983 (16). No changes were made to the original regulation—i.e., the same exposure limit still applies—and administrative or engineering controls are still required before the use of hearing protection is to be considered. The Amendment gives directions for administering a hearing conservation program, which must be implemented when employee noise exposures equal or exceed an 8-hour time-weighted average (TWA) sound level of 85 dBA. This 85 dBA TWA is referred to as the "Action Level," which is equivalent to a dose of 50%, and is computed without regard to any attenuation provided by the use of personal hearing protection. An employee's TWA may be determined from the following equation:

$$\text{TWA} = 16.61 \log \left(\frac{D}{100}\right) + 90 \text{ dBA} \tag{13}$$

where D is the percentage dose obtained from equation (11). Alternatively, the corresponding TWA for a particular dose can be obtained from Table A-1 in Appendix A of the OSHA noise regulation.

6.3 Noise Exposures Outside the Workplace

High-level noise exposures outside of the workplace can be just as harmful as those obtained in the workplace, so all exposures must be considered in order to get an accurate assessment of daily noise exposure. Potentially hazardous noises away from work may result from exposures to noises from chainsaws, snowmobiles, lawn mowers, motorboats, motorcycles, automobile or motorcycle races, farm equipment, loud music, shooting hobbies (skeet, targets, etc.), and shop tools. Even riding in a car at legal speeds with the window down may be harmful to some noise-sensitive individuals.

6.4 Ultrasound Limits

Exposure to high intensity sound in the upper sonic range (10,000 to 20,000 Hz) or ultrasonic range (above 20,000 Hz) is not common, although it does occur in certain industries. Ultrasonic equipment (e.g., cleaners, welders) will often emit subharmonics of the fundamental ultrasonic frequency, which fall into the audible range of human hearing. These subharmonics (rather than the ultrasonic sound itself) are of concern for potential noise-induced hearing loss or other health effects such as nausea and headaches. The OSHA noise exposure regulation does not specifically address ultrasonic frequencies, although technically they should be considered along with any other noise in the work environment. However, frequencies above 10,000 Hz cannot be reliably measured with a standard Type 2 sound level meter, due to its inherent design tolerances. Instead, a Type 1 or better instrument should be used for ultrasonic measurements. The American Conference of Governmental Industrial Hygienists (ACGIH) has published threshold limit values, or

TLVs, for one-third octave-bands of sound from 10,000 to 100,000 Hz (see Table 19.5) (17). These TLVs represent conditions under which it is believed that nearly all workers may be repeatedly exposed without adverse effect on their ability to hear and understand normal speech. The 8-hour TWA values are an extension of the ACGIH TLV for noise, which is an 8-hour TWA of 85 dBA for sound below 10,000 Hz.

7 NOISE CONTROL TECHNIQUES

From an engineering perspective, an ideal engineering noise control solution focuses directly on the actual source of the noise. Eliminating the noise-generating mechanism altogether obviates the need for further hearing conservation considerations. However, in cases where this is not practical nor possible, engineering controls must be oriented toward blocking the path that the sound waves travel toward employees, or a method must be provided for protecting the employees from exposure to the harmful noise. More than one

Table 19.5. Permissible Airborne Upper Sonic and Ultrasound Acoustic Radiation Exposure Levels[a]

	One-Third Octave-Band Level		
	Measured in Air in dB re: 20 μPa; Head in Air		Measured in Water in dB re: 20 μPa; Head in Water
Mid-Frequency of Third-Octave Band (kHz)	Ceiling Values	8-Hour TWA	Ceiling Values
10	105[b]	88	167
12.5	105[b]	89	167
16	105[b]	92	167
20	105[b]	94	167
25	110[c]	—	172
31.5	115[c]	—	177
40	115[c]	—	177
50	115[c]	—	177
63	115[c]	—	177
80	115[c]	—	177
100	115[c]	—	177

[a]From ACGIH, 1998.

[b]Subjective annoyance and discomfort may occur in some individuals at levels between 75 and 105 dB for the frequencies from 10 kHz to 20 kHz, especially if they are tonal in nature. Hearing protection or engineering controls may be needed to prevent subjective effects. Tonal sounds in frequencies below 10 kHz might also need to be reduced to 80 dB.

[c]These values assume that human coupling with water or other substrate exists. These thresholds may be raised by 30 dB when there is no possibility that the ultrasound can couple with the body by touching water or some other medium. [When the ultrasound source directly contacts the body, the values in the table do not apply. The vibration level at the mastoid bone must be used.] Acceleration Values 15 dB above the reference of 1g RMS should be avoided by reduction of exposure or isolation of the body from the coupling source. (g = acceleration due to the force of gravity, 9.80665 meters/second; RMS = root-mean-square.)

noise control treatment may be necessary if source treatment alone does not provide sufficient sound attenuation.

No single approach to noise control is applicable in all situations. Practical noise control solutions combine the exact science of acoustics with any production, maintenance, and economic constraints imposed upon a particular situation. Successful noise control projects are usually the result of a joint effort involving individuals from engineering, operations, and management. A compromise or balance must be struck between what is technically possible and what is truly feasible in each circumstance.

7.1 General Considerations

The first step in a noise control project is to accurately define the problem and determine its extent. In most cases, the problem will be the potential for employees to sustain a noise-induced hearing loss, although issues of verbal communication or simply complaints of annoyance often necessitate a reduction in noise levels. Area sound levels and employee noise exposures should be thoroughly measured at the outset of the project, and the data obtained will form the basis for future decisions. Questions such as "how many and/or which employees are affected?" and "exactly how much noise reduction is required?" must be answered before any other work is done.

After determining that a noise problem exists, the actual noise source(s) or "root cause" of the problem must be identified. Although this seems somewhat elementary, in reality this is often the biggest challenge encountered throughout the entire project. Noise may be produced by mechanical impacts, vibrating surfaces, exhaust gases, rotating machinery, etc. There may be hundreds of potential noise-generating mechanisms within a relatively small space in a plant or factory, and the difficulty arises when attempting to pinpoint the *exact* source of the excessive acoustical energy. Generally, it is not sufficient to simply state that the noise is coming from the "assembly line" or a particular machine—precisely what part of the equipment that is responsible for the high noise level must be known in order to implement an effective control. The services of an acoustical consultant or noise control engineer may be invaluable for this purpose; however, many noise problems can be successfully solved in-house by individuals without extensive acoustical training. The key is to systematically track down the noise source by turning different equipment (or individual components) on and off, or by using temporary controls and observing their effects.

The noise produced by one source can be distinguished from another by analyzing the differences in the octave or third-octave band spectrums. More sophisticated approaches to source identification involve using vibration measurements or performing calculations based on rotational speed, physical dimensions, or other known characteristics of the equipment in question.

Another problem encountered early in the noise control process is the prioritization of engineering control treatments. Installing a noise control device on a particular piece of equipment simply because it is "easy" or relatively inexpensive may be ineffective because the device might not affect the overall noise level, depending on what else in the immediate area is also generating noise. As a rule of thumb, when multiple noise sources are present, the loudest source(s) must be reduced first. For example, three noise sources of 91 dBA,

96 dBA, and 102 dBA combine to create a noise level of 103 dBA. If only the 91 dBA source is removed, no overall reduction would occur, since the sum of the 96 dBA and 102 dBA sources is still 103 dBA. Obviously, if the 91 dBA and 96 dBA sources are eliminated, only a one decibel reduction would occur since the overall level would be 102 dBA due to the lone remaining source. However, if the 102 dBA source is removed, the 91 dBA and 96 dBA sources added together produce 97 dBA, which results in a 6 dBA decrease from the original level of 103 dBA. When multiple noise sources are present, this type of analysis should be applied to determine how much noise reduction is necessary to meet the intended goal.

Further, it should be pointed out that successful noise control efforts do not always involve the purchase of "soundproofing" materials or products. Excessive noise levels are often produced by machinery when it is in need of adjustment, alignment, or repair (18). Therefore, restoring equipment to its optimum performance condition should be completed before investing in any noise control treatment(s). Rotating members should be dynamically balanced, with bearing and other rotating contact surfaces properly maintained. Further, no machine should be operated at an unnecessarily high speed. In many instances, a significant reduction in noise can be achieved by using a larger machine that can do the same job while operating at a lower speed. Also, it may be necessary to periodically inspect and/or replace existing noise control devices, since items such as pneumatic silencers will eventually become plugged and be rendered ineffective. Wherever possible, other maintenance-related controls should be investigated. Attempts should be made to prevent or reduce mechanical impacts between machine parts; replace metal components with quieter plastic parts; and reduce the amount of force, pressure, air velocity, etc. to the absolute minimum amount required to perform the task. Relocating noisy equipment (e.g., motors, pumps, fans) to unoccupied areas of the plant (e.g., rooftops or warehouse areas) can greatly reduce the amount of ambient noise. Even simple measures such as tightly closing access doors/panels will keep noise from "leaking out."

A successful noise control program also gives consideration to the initial procurement of noise-generating machinery and equipment. Purchase orders for new equipment should include noise limits, with the intent that noise control features will be designed and built into noisy equipment from the start. The desired sound levels may not always be met, but these specifications will provide an incentive for the development of quieter products in the future. Although the process of building quieter equipment generally increases its cost, the cost of retrofitting that piece of equipment is usually several times greater than if the same noise controls were purchased and installed in the first place.

The American National Standard Guidelines for the Specification of Noise of New Machinery (ANSI S12.16-1992, R1997) provides guidelines for obtaining noise level data from manufacturers of stationary equipment (19). This standard references existing American National Standards Institute, trade, and professional association measurement standards and techniques, to aid in formulating and submitting requests for manufacturer's noise level data. Appendices in ANSI S12.16 provide guidance for interpretation of the data received from the manufacturer.

7.2 Noise Control Materials and Products

When it is not economically feasible to replace an older, noisy machine with a new quieter unit, and it is impossible to move the noise-generating equipment away from employee workstations, many noise control problems can be effectively solved through the proper selection and installation of appropriate commercially available noise control devices, systems, or materials. Once the dominant sources are treated, the new noise environment should be evaluated to determine whether any additional noise control measures are warranted. If sufficient noise reduction is not achieved, the process should be repeated, focusing on the noise sources that have been "uncovered" by the first application of noise control treatments. Standard noise control techniques include the use of sound-absorbing materials, sound-attenuating materials/barriers, mufflers/silencers, vibration isolation devices, and damping treatments. Monthly publications such as *Sound and Vibration* periodically publish a listing of manufacturers of currently-available noise control materials (20).

7.2.1 Absorptive Materials

Absorptive materials convert sound energy into small amounts of heat. The amount of heat generated is quite small since the acoustic power is normally very small. Porous materials such as glass fiber (e.g., standard building insulation), open-cell foam, and acoustical ceiling tile are all examples of sound-absorbing materials. Acoustical performance of these materials is determined by their absorption coefficient (α), which is the fractional amount of sound energy absorbed by a surface. Absorption coefficients are determined by specific laboratory procedures, and are designated as a value between 0.0 and 1.0, which also can be considered as a percentage between 0% and 100%. Materials that are good sound absorbers have absorption coefficients approaching unity. In general, most materials perform better at higher frequencies and are less able to absorb low-frequency sound. The arithmetic average of the sound absorption coefficients of a particular material at 250, 500, 1000, and 2000 Hz provides a single-number rating of the acoustical performance of that material, called the noise reduction coefficient (NRC).

Noisy machinery is often located in large, acoustically reflective areas that reverberate and build up the sound level in the room. Sound-absorbing materials on walls and ceilings are often used to reduce this effect. However, the amount of reduction close to the actual noise source(s) may be slight, because most of the sound energy comes directly from the machines and not from the reflecting surfaces. The type, amount, configuration, and placement of absorption materials depends on the specific application. Correct selection of sound-absorbing materials can be guided by reviewing the absorption coefficients listed in manufacturer's brochures. The National Institute for Occupational Safety and Health (NIOSH) published a very useful *Compendium of Materials for Noise Control* in 1980, and this publication contains data tables as well as further explanation of acoustical absorption applications (21).

7.2.2 Noise Barriers, Enclosures, and Pipe Lagging

Sound is attenuated or "blocked" by materials that have high transmission loss (TL) values. Transmission loss is calculated in decibels, and measurements to determine the TL characteristics of a material must be done in a laboratory, according to specific test procedures. Heavy, dense materials such as concrete, steel, and thick plywood are effective as noise barriers or enclosure walls because they restrict the amount of acoustical energy that can pass through the material to the other side. Also, commercially available heavy acoustic blankets or loaded vinyl curtains are commonly used for partitions and enclosures. The main precaution that must be taken with any enclosure or barrier system is to be sure that there are no unnecessary openings. The effectiveness of an acoustical wall or barrier can be severely degraded even with only a small percentage of open area in an otherwise solid barrier.

The amount of noise reduction that can be attained with a simple noise barrier wall depends on the characteristics of the noise source, the configuration and materials used for the barrier, and the acoustic environment on each side of the barrier. Free-standing noise barrier walls are usually not very effective in indoor situations, due to the ability of the sound to reflect off of nearby surfaces (e.g., the building's walls and ceiling). Essentially, the sound bounces over the barrier with little or no attenuation. Therefore, partial or complete enclosures are used instead to contain the noise produced by both small and large noise sources inside buildings.

Noise radiated from pipelines, valves, and ducts is controlled by wrapping a barrier material around the noise source, which is essentially the same as installing a close-fitting enclosure. This technique is called acoustical lagging, which actually consists of a combination of sound absorption and sound attenuation materials. Typically, a resilient absorptive material is covered with a high transmission loss outer shell, where the absorptive layer essentially de-couples the barrier material from the pipe surface. Often, the entire length of a noisy pipeline or duct must be covered for the treatment to be effective.

7.2.3 Acoustical Silencers

Silencers, also referred to as mufflers, baffles, or sound traps, are installed in air ducts or pipelines to attenuate the noise present in a flowing medium. There are three distinct types of acoustical silencers, although two of the types may be combined into a single device to achieve a particular result. The first type, called absorptive or dissipative silencers, contains a fibrous or porous filler material that absorbs noise energy, without significantly restricting the flow through the silencer. These silencers are used to reduce broadband noise, with greater noise reduction in the higher frequencies, since the internal packing materials are typically better at absorbing high frequency sound waves.

The second type of acoustical silencer is called reactive silencers, which do not contain any kind of absorptive filler. Instead, they cause the sound waves to cancel out by reflecting them back toward the source. Reactive silencers are used in applications such as automotive exhausts, since most absorptive filler materials would eventually disintegrate or become clogged by a high-temperature, particulate-laden exhaust stream. Reactive silencers are useful for reducing low frequency noise, and are often designed or "tuned" to eliminate a certain frequency range.

The third type of commercially available silencers employs active noise cancellation technology, where loudspeakers installed in the pipe or duct produce an inverted sound wave (i.e., 180 degrees out of phase) that reduces the noise through destructive interference with the original wave pattern. These systems are best suited for very low frequency applications, and are particularly effective at reducing pure tones that may be present.

7.2.4 Vibration Isolation

Most industrial equipment and machinery vibrates during operation, and these vibrations can be felt and heard throughout a plant or factory. Vibrations can travel very long distances through solid support columns and beams, and eventually be re-radiated as sound through adjoining walls, ceilings, and floors. Therefore, it is important that the vibrating equipment be adequately isolated from the building structure. Vibration isolators are used to reduce the transmission of shaking forces to an acceptable level. A wide variety of these devices are available as off-the-shelf items or they can be specially designed for a particular application.

Three factors need to be considered when selecting the correct vibration isolator for a specific application: the size and weight of the machine, its movements and speed, and the magnitude of the vibrations it produces. Small machines might only require a cork, felt, or rubber pad for proper isolation, while a large machine may be mounted on specially designed steel springs. In any case, the product supplier or a knowledgeable engineer should be assigned to select the correct vibration mount, since the incorrect use of an isolator can actually make the vibration problem *much worse* if a resonance condition develops.

7.2.5 Vibration Damping

Metal panels or an entire machine casing may generate noise (i.e., the surface may "ring" or resonate) when excited by internal or external vibrations. A damping material may be attached to a vibrating panel or surface in order to reduce the panel motion, and thus reduce the amount of sound that can be radiated. Damping treatments may be thought of as "absorbing" solid-borne sound, by converting the vibrational energy into small amounts of heat. Both steady-state and impact-type noises can be reduced by applying the correct damping treatment.

There are two types of damping treatments: extensional (also known as free-layer or homogeneous) and constrained-layer. The first type is generally used for panels less than one-quarter inch thick, and is typically a sheet of viscoelastic material or a compound that is troweled or sprayed directly onto the vibrating surface. Vibratory energy is dissipated through extension and compression of the damping material as the panel or machine casing bends and flexes. Constrained-layer damping consists of a damping material being "sandwiched" between the vibrating panel and a stiff outer retaining layer, such as aluminum or steel. In this configuration, the damping material dissipates vibratory energy through shear deformation, rather than extension and compression. Constrained-layer systems are usually used for treating very stiff structures, or when the vibrating panel is thicker than one-quarter inch.

8 HEARING PROTECTORS

8.1 General

If personal hearing protectors are properly selected, fitted, and worn, they can provide adequate protection against noise-induced hearing impairment in a high percentage of industrial work areas. In many cases, however, the protective devices are not worn effectively, if at all. Since noise does not have to be painful to be potentially harmful, many employees do not understand the need for wearing hearing protectors. A significant effort must be made to instruct and motivate employees in order to develop and maintain an effective hearing protector management program.

8.2 Types of Hearing Protectors

Hearing protectors usually take the form of either insert-type earplugs that are inserted into the ear and seal against the ear canal walls, or muff-type devices that seal against the head around the ear. There are also concha-seated protectors that provide an acoustic seal right at the entrance to the external ear canal. There is no single "best" type for all individuals or situations. However, some types are better than others for use in specific noise environments, for some work activities, or for some environmental conditions. The 1994 publication "The NIOSH Compendium of Hearing Protection Devices" contains descriptions and laboratory attenuation data for all the HPDs currently marketed in the United States (22).

8.2.1 Insert-Type Earplugs

Several different styles of insert-type earplugs may be required to fit the wide differences in ear canal configurations. This is because ear canals differ in size, shape, and position among individuals and even within the same individual. Typically, ear canals are elliptical-shaped, although some are round, and many have only a small slit-like opening that may open into a larger diameter. Some canals are directed in a straight line toward the center of the head, but most are directed toward the front of the head, and they may bend in various ways. Ear canals can be opened and straightened by pulling the auricle upward and backward, or directly away from the head, making it possible to seat the earplug securely. All earplugs must fit snugly and have an air tight seal to be effective.

8.2.1.1 Molded Earplugs. Molded (or premolded) earplugs are constructed of soft and flexible materials that conform to the shape of the ear canal for a snug, airtight fit. Molded plugs are usually made of a silicon-type material and are available in a variety of styles. Often, these earplugs utilize one or more circular flanges to provide a seal against the canal walls (see Fig. 19.7). Different sizes of molded plugs are available from many manufac-

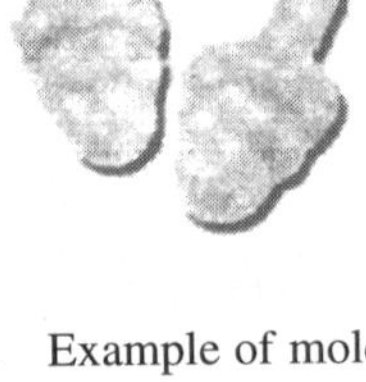

Figure 19.7. Example of molded earplugs.

turers, since a single size may not adequately fit all ear canal sizes and shapes. Molded earplugs are usually reusable, and will retain their size and flexibility over long periods of time if kept clean with soap and water.

8.2.1.2 Malleable Earplugs. Malleable (or formable) earplugs are made of materials such as silicone putty, wax-impregnated cotton, glass wool, and mixtures of these materials. Typically, a small cone or sphere of the material is hand-formed and inserted into the ear with sufficient pressure so that the material conforms to the shape of the canal and holds itself in position. The manufacturer's instructions should be followed regarding the depth of insertion of the plug into the canal. In all cases, care should always be taken to avoid deep insertions that may cause the material to touch the eardrum. Malleable plugs have the obvious advantage of fitting almost any ear canal, eliminating the need to keep a supply of various sizes, as is necessary for most other types of earplugs.

8.2.1.3 Foam-Type Earplugs. Earplugs made of cylindrical or tapered foam are very popular (see Fig. 19.8). Generally, these plugs must be rolled down and inserted so that most of the plug is in the canal, then held in place while they expand to fill the canal. This insertion process allows the plugs to provide the rated protection level. If these plugs are not inserted properly, their performance may drop significantly. It may not be possible to insert foam plugs properly into some small, sharply bending, or slit-shaped ear canals.

Foam plugs are typically composed of either vinyl or urethane-based material. The vinyl plugs are usually stamped out of sheets of foam material into cylinders, or less commonly, into hexagon-shaped plugs. The urethane plugs are usually tapered and manufactured through a molding process. Both of these materials provide excellent noise attenuation when the plugs are fitted and worn properly. Urethane plugs are typically somewhat softer than the vinyl plugs but may be more expensive due to the more complicated manufacturing process. Foam earplugs are generally comfortable to wear, and are reusable on a limited basis.

8.2.1.4 Custom-Molded Earplugs. A custom-molded earplug is manufactured from an exact impression of the wearer's ear. When made properly, these protectors can be expected to provide among the most consistent protection levels in daily use of all protector types. The initial cost of individually molded protectors is relatively high, but most should last

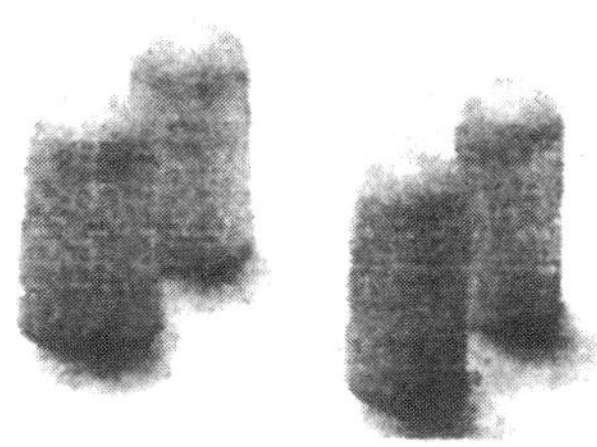

Figure 19.8. Example of foam-type earplugs.

a long period of time. A unique advantage of custom-molded hearing protectors is the capability of venting the protector to adjust the level of attenuation provided. The venting process involves manufacturing the plugs with a small hole completely through the plug, and inserting various types of dampers in the hole. These vented plugs can enhance communication in moderate-level noise environments where maximum attenuation is not required.

8.2.2 Concha-Seated Protectors

Protectors that provide an acoustic seal in the concha (i.e., the bowl part of the external ear) and/or at the entrance to the ear canal can be referred to as "canal caps," or concha-seated, semi-aural, or semi-insert, devices (Fig. 19.9). This class of hearing protectors makes use of various plug shapes attached to a lightweight headband that holds the plugs at the entrance of the external ear canal. The performance and comfort of this kind of protector varies significantly among different models.

8.2.3 Earmuffs

Many manufacturers of muff-type hearing protectors (Fig. 19.10) offer a choice of two or more models having different physical characteristics. Generally, larger and heavier earmuffs provide greater protection, but may not be comfortable to wear for long periods of time. Conversely, lightweight protectors may be accepted by the wearer more readily than

Figure 19.9. Example of concha-seated hearing protectors.

Figure 19.10. Example of earmuffs.

the larger ones, but may not provide as much attenuation. The force applied by muff suspensions is often directly related to the level of noise attenuation provided. On the other hand, the wearing comfort of a muff-type protector is generally inversely related to the suspension forces—thus a compromise must be made between performance and comfort.

Earmuff cushions are generally made of a smooth, plastic envelope filled with a foam or fluid material. Because skin oil and perspiration can have adverse effects on cushion materials, these cushions can tend to become stiff or brittle over time and require periodic replacement. Earmuffs normally provide less protection when worn over hair or glasses. If long coarse hair or other significant obstructions cannot be avoided, it may be advisable to use earplugs.

Earmuffs are sometimes chosen because their use can be monitored from greater distances than insert-type protectors. They are also easier to fit than insert types because one size fits most wearers. Comfort is also better for muff-type protectors in some work areas; however, earmuffs may be very uncomfortable to wear in hot work areas, particularly when the work involves a vigorous activity. Muff-type protectors may also be a poor choice when work is performed in areas having limited headroom.

In addition to the standard passive-attenuation earmuffs, several types of electronic earmuffs are currently available. These include muffs with electronic clipping circuits and those with noise-canceling circuits. The clipping circuits typically have a microphone on the outside shell of one or both muff cups; this microphone senses the sound outside the protector and provides input to an internal amplifier. The amplifier is connected to a speaker inside the earcup to let the wearer hear the ambient sounds at a safe level (usually limited to about 85 dBA). In some of these devices, equalization is performed on the signal to enhance the speech frequencies. Noise-canceling or active noise reduction (ANR) earmuffs use a microphone to sample the noise, then reverse the phase of the sampled signal and play back the out-of-phase signal into the earcup through a small speaker. The magnitude of the attenuation provided by noise-canceling circuits is greatest at low frequencies, and there are usually no noise-canceling effects at frequencies above about 1000 Hz. In fact, some of these devices unintentionally provide a small amount of amplification in a narrow frequency range at around 1000 Hz. Active noise cancellation is an evolving technology, and the future will likely bring better performance with a lower cost for these devices.

8.2.4 Advantages and Disadvantages of Protector Types

Earplugs, earmuffs, and canal caps all have distinct advantages and disadvantages. Insert-type protectors are convenient to carry and store, and can be worn effectively without interference from eyeglasses, headgear, earrings, or long hair. They are practical to wear in hot environments and do not restrict head movement in close quarters. The initial cost of earplugs is usually significantly less than muffs or canal caps, although in the long-term the overall cost depends on how often employees replace their earplugs. Plugs may not be appropriate for use in very dirty areas since dirt or foreign substances from one's hands may be introduced into the ear canal during insertion. For this reason, malleable and foam-type plugs (and to a somewhat lesser extent, all earplugs) may be a poor choice if the work area is excessively dirty, and intermittent noise levels make it desirable to frequently remove HPDs. Also, earplug use is more difficult for supervisors to monitor because earplugs cannot be easily seen from a distance.

Almost all insert protectors require more time and effort for proper fitting than do earmuffs, and a small percentage of employees cannot effectively wear any off-the-shelf plugs due to an improper fit. The amount of protection provided by earplugs is more variable between wearers than that of muff-type protectors. On the other hand, muffs are usually easy to fit properly and they provide relatively consistent attenuation across users. However, in addition to being uncomfortable in hot environments, muff-type protectors may be inconvenient to wear with safety glasses, welding helmets, etc. Also, earmuffs may feel cumbersome or restrict head movement in close quarters. Misuse of muffs has been documented, with users deliberately bending the headband to reduce pressure on the skull, which can lead to a significant reduction in the amount of protection. Concha-seated ear canal caps also tend to exert a substantial amount of pressure on the head, and therefore are usually better suited for brief, intermittent noise exposures. If individuals are permitted to fit themselves, they often select the type/size of hearing protector on the basis of comfort rather than on the amount of protection provided.

8.3 Hearing Protector Ratings

In most cases, hearing protector attenuation is determined by conducting tests on human subjects, where hearing tests are taken by a panel of listeners with and without the protector in place. This test method is referred to as Real-Ear Attenuation at Threshold (REAT). First, thresholds are determined with a listener's ears open, then another hearing test is administered with the subject's ears occluded by the HPD being evaluated. The decibel difference between the open-ear threshold and the occluded-ear threshold indicates the amount of sound attenuation provided at each test frequency. A detailed description of the test facility and procedures required to accurately use this technique can be found in ANSI S12.6-1997, "Methods for Measuring the Real-Ear Attenuation of Hearing Protectors" (23).

In 1975, NIOSH published a "List of Personal Hearing Protectors and Attenuation Data" which contains attenuation rating methods known as NIOSH methods #1, #2, and #3 (24). Method #1 (a.k.a "long method") is generally regarded as the most accurate, although it is the most complex method since it uses the largest amount of spectral information from the actual noise environment. The steps for calculating the amount of protection provided by a hearing protector using NIOSH method #1 are as follows:

1. Measure and record the octave band exposure levels.
2. Adjust the octave band values with the A-frequency weightings. The overall A-weighted sound exposure level (L_A) for unprotected ears is equal to the logarithmic sum of the octave band levels.
3. Record the mean hearing protector attenuation values supplied by the protector manufacturer and two times the corresponding standard deviation value.
4. Subtract the mean attenuation values from the corresponding A-weighted sound levels, then add the two standard deviations to the result to obtain the A-weighted exposure levels under the protector.

5. Combine the A-weighted exposure levels to obtain an estimate ($L_{A(\text{protector})}$) of the highest sound pressure level to which a population of wearers of this protector would be exposed 97.5% of the time if the protectors are fitted and worn properly.

A simplification of the second NIOSH method (NIOSH method #2) was adapted by the Environmental Protection Agency (EPA), and is called the Noise Reduction Rating (NRR). The NRR is a single-number rating that is required by law to be shown on the label of each hearing protector sold in the United States (25). It is intended to describe how much noise reduction a particular protector can provide. The values of sound attenuation used for calculation of the NRR are determined in accordance with ANSI S3.19-1974, "Method for the Measurement of Real-Ear Hearing Protector Attenuation and Physical Attenuation of Earmuffs" (26). It should be noted that ANSI S3.19 was revised and published in 1984 as ANSI S12.6, "Method for the Measurement of the Real-Ear Attenuation of Hearing Protectors" (27). Further, ANSI S12.6 was updated in 1997 and renamed "Methods for Measuring the Real-Ear Attenuation of Hearing Protectors," which contains the most current methodology (23). However, even the earlier version of ANSI S12.6 was adopted after the EPA hearing protector labeling laws were written, and since the EPA regulations did not make any provision for adopting newer standards, hearing protector manufacturers must still use the older S3.19 method for labeling purposes.

To determine the NRR, the experimenter (not the test subject) must fit the hearing protector for each occluded test run. The formula for calculating the NRR is

$$\text{NRR} = L_{C(\text{pink noise})} - L_{A(\text{protector})} - 3 \tag{14}$$

The first term of the equation ($L_{C(\text{pink noise})}$) is always 107.9 dBC, which is obtained by assuming a level of 100 dB SPL in each of the nine octave bands from 125 to 8000 Hz, subtracting the C-weighting corrections, and logarithmically summing to calculate the overall C-weighted level at the unprotected ear. Next, A-weighting corrections are subtracted from the pink noise octave band levels to calculate the A-weighted octave band levels at the ear. Two standard deviations are subtracted from the average attenuation at each frequency, which yields the protected A-weighted octave band level at the ear. (An adjustment of two standard deviations provides an NRR that 97.5% of the subjects will theoretically meet or exceed, provided that the wearers use the HPD in the manner that the laboratory subjects did.) The overall protected A-weighted level at the ear is determined by adding the individual octave band levels. Finally, the NRR is computed by subtracting 3 dB from the difference between the unprotected C-weighted and the protected A-weighted levels at the ear. Subtracting this additional 3 dB accounts for spectral uncertainty, since the pink noise used in the calculations may not actually match the noise environment where the hearing protector is being worn.

A worker's noise exposure when wearing the hearing protector is estimated by subtracting the NRR from the ambient C-weighted sound level (not the A-weighted sound level). For example, if a protector has an NRR of 20 dB and is used in a work area where the noise level is 95 dBC, the amount of noise entering the ear could be expected to be 75 dBA or lower in 97.5% of the cases.

Accuracy is lost when calculating hearing protection if only A-weighted sound pressure levels are known, since the NRR is based on a flat spectrum. When only A-weighted ambient noise levels have been measured, a 7 dB safety factor must be subtracted from the NRR to determine the exposure under the hearing protector. In this case, a person working in 93 dBA noise while wearing an HPD with an NRR of 20 dB can be expected to be receiving 80 dBA at the ear.

Hearing protectors are often selected with consideration being given only to the magnitude of the NRR. Although a single-number rating is useful in initiating a program when more precise octave-band noise measurement data are not available, its use requires significant safety factors because of the inherent lack of precision. A fundamental weakness of the NRR is that it can end up being based on just one or two of the nine one-third octave-band test signals used in the laboratory measurements. Generally, the controlling test signals are at or below 1000 Hz, and performance levels for other test signals may have little or no effect on the NRR. For example, two protectors may have the same NRR because of their limiting attenuation values at 250 or 500 Hz, but one of these protectors may provide 15 dB greater protection than the other at higher frequencies. The NRR values in this example would be a reasonably accurate assessment of protector performance if the primary noise source were centered at 250 or 500 Hz; however, if the noise exposures contain prominent high-frequency components, the NRR values may be very misleading.

Another limitation of the NRR is that attenuation values obtained from laboratory testing tend to over-estimate the amount of protection received by individuals during everyday use. This is not surprising, considering that well-trained and highly motivated individuals are normally used in the laboratory as subjects, and the experimenter directly fits the HPDs prior to testing. This discrepancy has led to the development of de-rating procedures that are applied to laboratory data. A prior OSHA directive called for dividing the labeled NRR by two (i.e., a 50% de-rating) before calculating exposure under the protector (28). The *Criteria for a Recommended Standard: Occupational Noise Exposure*, published by NIOSH in 1998, suggested de-rating the NRR on earmuffs by 25%, formable/foam earplugs by 50%, and molded earplugs by 70% (15). Currently, OSHA's Technical Manual permits the NRR Subject Fit (SF) value to be used for hearing protector evaluation, without any de-rating factor applied (29).

The NRR (SF) is determined from Method B procedure (which is outlined in the newest version of ANSI S12.6), and typically yields lower protection values that are expected to be closer to the "real-world" values than those obtained with ANSI S3.19. A significant difference between the older S3.19 and the S12.6 Method B procedures is that the test subjects don the protectors themselves according to the manufacturer's instructions, and testing is conducted with minimal experimenter involvement. Another feature of S12.6 Method B is that since the NRR (SF) is based on A-weighted sound levels, the 7 dB A-weighting-to-C-weighting correction is not necessary, and HPD selections can be based directly on A-weighted exposure measurements.

European hearing protector tests are performed according to the International Standard ISO 4869-1 (30). In Europe, the most common single-number rating of hearing protector performance is the Single Number Rating (SNR). The SNR has the same pitfalls as the NRR, although it is somewhat less sensitive to outlier data during the laboratory testing. The European market also utilizes a three-number rating system, called the high-middle-low, or HML, system. The three numbers in the HML rating estimate the attenuation provided in the high test frequencies, middle test frequencies, and low test frequencies, respectively. The HML values are used in conjunction with measurements of A- and C-weighted sound pressure levels to obtain the HPD attenuation that can be expected in specific noise environments.

8.4 Performance Limitations

A primary limitation to protection afforded by a hearing protector is the way it is fitted and worn. Other important limitations of a hearing protector depend upon its construction and on the physiological and anatomical characteristics of the wearer. Sound energy may reach the inner ears of persons wearing protectors by four different pathways: (*1*) by passing through bone and tissue around the protector; (*2*) by causing vibration of the protector, which in turn generates sound into the external ear canal; (*3*) by passing through leaks in the protector; and (*4*) by passing through leaks around the protector. These pathways are illustrated in Figure 19.11.

Even if there are no acoustic leaks through or around a hearing protector, some noise reaches the inner ear by bone and tissue conduction or protector vibration if noise levels are sufficiently high. The practical limits set by the bone- and tissue-conduction threshold vary significantly among individuals and among protector types, generally from about 40 to 55 dB. Limits set by protector vibration also vary widely, generally from about 25 to 40 dB, depending upon the protector type and design and on the materials used. Contact surface area and compliance of materials are major contributing factors. If hearing protec-

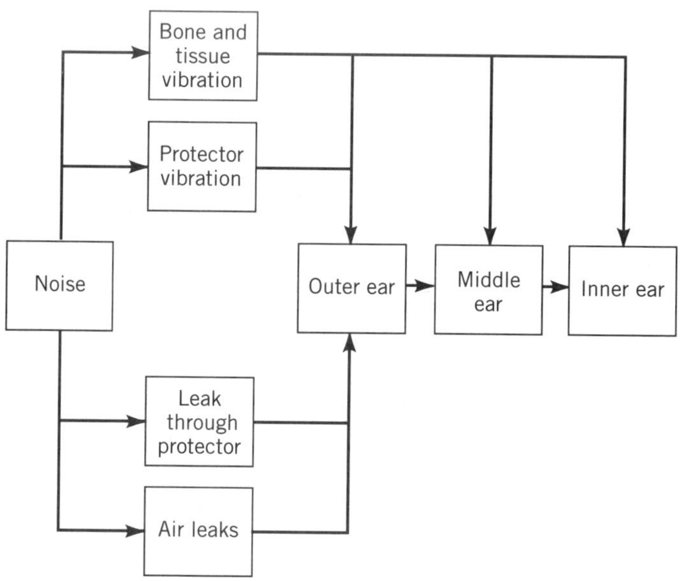

Figure 19.11. Noise pathways to the inner ear.

tors are to provide noise reduction values approaching practical limits, acoustic leaks through and around the protectors must be minimized by proper fitting and wearing.

Perhaps the major reason why hearing protective devices fail to protect employees from noise exposure is because the protector is not always worn when it should be. Workers fail to wear hearing protection for reasons of comfort or communication, or because the HPDs were misplaced or forgotten. Whatever the reason, failure to wear hearing protection for even a small fraction of a workshift will greatly reduce the effective protection. For instance, an HPD with a Noise Reduction Rating (NRR) of 25 not worn for 30 minutes out of an 8-hour workday will lose nearly 8 decibels of effective attenuation (see Fig. 19.12).

8.5 Moderate Attenuation and Flat Attenuation Hearing Protectors

Many purchasers of hearing protection have fallen into the trap of thinking that a higher NRR is always better. However, a moderate amount of attenuation rather than using a device with the highest available NRR may afford adequate protection in many industrial noise environments. In most cases, moderate attenuation protectors are more likely to be acceptable for use in noises having major components above 1000 Hz. The simplest of these devices are vented with a small hole through the protector, permitting low frequencies (below about 1000 Hz) to pass without significant attenuation.

More recently, flat (or uniform) attenuation hearing protectors have been introduced, in both muff and insert styles. A flat attenuation device "distorts" the incoming sound less than a conventional HPD since approximately equal attenuation is provided across all frequencies. This is in contrast to most hearing protectors, where more attenuation is provided in the high frequencies, and, consequently, the balance between the low and high

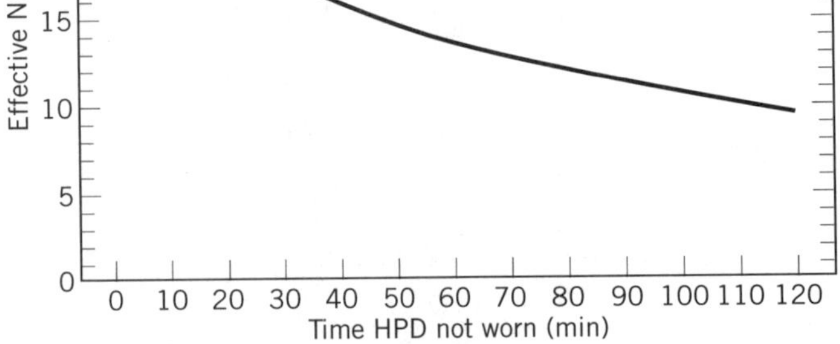

Figure 19.12. Effective NRR when an HPD with an NRR of 25 dB is not worn full-time.

frequencies is altered. Flat attenuation earplugs are sometimes referred to as musician's plugs, as they are intended to maintain the music's spectral balance while providing a moderate amount of sound attenuation. Thus, the resulting effect is that all frequency components of the music are heard, but at a uniformly reduced level. Industrial workers can also benefit from these types of HPDs, since machine/equipment sounds can essentially be heard undistorted. It is important to note that if these protectors are not fitted properly, it is likely that the low-frequency attenuation will decrease, and the overall attenuation will not be as uniform as intended.

8.6 Double Hearing Protection

The combined attenuation from wearing both a muff-type and an insert-type protector cannot be accurately predicted because of complex coupling factors. It is not correct to simply add the two NRRs to determine the overall expected noise reduction. If the attenuation values of the muff and insert protectors are about the same in a frequency band, the combined attenuation should be approximately 3 to 6 dB greater than the higher of the two individual values. If one of the two protectors provides significantly greater attenuation in a specific frequency range, the increased attenuation provided by double protection in that range will be minimal. In practice, 5 dB is typically added to the higher labeled NRR value to determine the protection afforded by wearing double hearing protection.

8.7 Hearing Protectors and Communication

Performance and safety aspects of a job often depend on a worker's ability to hear warning signals, machine sounds, and speech in the presence of high noise levels. Therefore, interference with communication is one of the most common and legitimate complaints related to hearing protector use. The effect of noise on communication depends to a large extent on the spectrum of the noise, the hearing ability of the worker, and the attenuation characteristics of the particular hearing protectors in use.

Hearing protectors attenuate both speech and background noise by equal amounts, and therefore should not adversely affect the speech reception ability of normal-hearing listeners. In fact, wearing hearing protection in high noise areas (above 85 dBA) actually improves speech recognition by lowering the overall sound level reaching a listener's ear, which reduces the potential for auditory distortion. There are situations, however, where a particular hearing protector may attenuate high frequencies substantially more than the low frequencies. In these cases, the residual low frequency sounds will mask or obscure the high frequency components, and cause the important consonant sounds to be unintelligible. Similarly, when hearing-impaired employees wear hearing protection, the higher-frequency sounds may be attenuated to a point below the level of audibility. Therefore, too much attenuation (i.e., inadequate hearing protector selection, or the use of hearing protection in areas with sound levels below 80 dBA) may be the cause of communication problems for normal-hearing as well as hearing-impaired employees. Optimal communication is usually achieved when the protector's attenuation characteristics are matched to those of the noise spectra.

To address the above issues, hearing protectors should be selected that provide sufficient protection against noise-induced hearing loss without unnecessarily inhibiting the natural perception of warning signals, machine operating sounds, speech, etc. Unfortunately, until recently, few hearing protectors were selected and purchased with any thought being given to user communication. Overprotection may be considered as acceptable, even desirable, against many health and safety hazards, but overprotection against noise exposure may cause significant communication problems and can also lead to the deliberate misuse or rejection of hearing protectors.

One European guidance document, EN 458, entitled "Hearing Protectors-Recommendations for Selection, Use, Care, and Maintenance," addresses the issue of selecting hearing protectors that maximize the ability to communicate in a noisy environment (31). This document includes several methods of HPD selection, including the octave band method, the HML method, the HML Check, and the SNR method. Recommendations for hearing protector attenuation are based on the relationship between exposure under the protector and allowable exposure criteria in dBA. Five HPD selection categories ranging from insufficient protection to overprotection are described, depending on how closely the exposure under the protector and allowable exposure criteria are matched. The underlying principle for these guidelines is that properly fitted individuals are more likely to wear their protectors consistently since they will not be unnecessarily "isolated" from other workers, and they will be less likely to intentionally disable protectors to decrease attenuation.

8.8 Field Monitoring Systems

For insert-type protectors, field monitoring systems are available that essentially replicate the laboratory tests as defined in ANSI S3.19-1974, except that the stimuli are presented via headphones instead of through loudspeakers in a diffuse sound field. Hearing thresholds of the HPD wearer are measured at selected test frequencies with and without the HPD in place, and the difference in hearing threshold at each test frequency is equal to the amount of noise attenuation provided by the hearing protector. For muff-type protectors, existing field measurement systems include a hand-held microprocessor-based unit that utilizes two microphones, one located at the entrance to the ear canal and the second located outside the cup of the muff. A digital readout displays the difference in sound pressure level between the two microphones. These units are designed to be used at the employee's workplace; therefore, the attenuation measured will be accurate while the wearer is in that particular noise exposure. Use of the protector in noises with differing spectral characteristics will affect the amount of noise attenuation provided.

Field monitoring systems for hearing protectors are designed to allow the measurement of attenuation provided by hearing protectors *on the individual wearer*. This personalized approach eliminates the need for single-number ratings (e.g., NRR) and the rough approximations associated with them. Again, the NRR is derived from laboratory test methods designed to estimate attenuation afforded to a population, not to an individual. Current laboratory performance ratings represent a best-fit situation using trained and motivated subjects under closely supervised conditions, and do not always correlate with the attenuation achieved in field situations.

Individual measurement systems are, in general, well-received by HPD wearers. The employees are typically interested in how the protectors function, and they appreciate the attention to their individual needs. Individual HPD attenuation measurements are particularly valuable as a training tool during the initial selection of insert-type HPDs. Assistance can be provided to the wearer during initial fitting, and an on-the-spot attenuation measurement can be obtained. If the measured attenuation is insufficient, the employee should re-fit the HPD and repeat the measurement procedure until a satisfactory result is achieved. These measurements document that adequate and proper protection was provided to the worker, and empower employees to take on the responsibility of effective HPD usage.

Field measurement systems perform several functions for the industrial hearing conservation program administrator, including: (*1*) training of wearers in correct fitting procedures, (*2*) random as well as routine (e.g., annual) field sampling of protector effectiveness, (*3*) documentation that training was provided and that proper protection was provided to the employee, and (*4*) identification of failing or deteriorating protectors and changes in ear physiology. Careful hearing protector selection and periodic checks with a field monitoring system should result in a successful hearing protector management program.

9 HEARING CONSERVATION PROGRAMS

9.1 Specific Requirements

The objective of a Hearing Conservation Program (sometimes referred to as a Hearing Loss Prevention Program) is to prevent noise-induced hearing loss. This obvious fact is often forgotten, or ignored, when pressures are applied for compliance with local, state, or federal noise regulations. Simply handing out earplugs or providing yearly hearing tests is not sufficient to conserve employees' hearing. As described in the OSHA Hearing Conservation Amendment, an effective hearing conservation program should provide for (7):

1. The identification of noise hazardous areas, and the determination of employee noise exposures.
2. The measurement of exposed persons' hearing thresholds through monitoring audiometry.
3. The establishment of a personal hearing protection plan, including provisions for evaluating the effectiveness of hearing protectors.
4. The training of employees about the need for hearing conservation, and the instruction of employees in the use and care of personal hearing protectors.
5. The maintenance of accurate and reliable records of hearing tests and noise exposure measurements.

9.1.1 Identification of Noise Hazardous Areas and Employee Monitoring

Noise exposure monitoring is necessary to identify employees that should be included in the hearing conservation program. Whenever employee noise exposures equal or exceed the action level (i.e., 8-hour TWA of 85 dBA, or, equivalently, a dose of 50%), they must be included in the program. Although a specific sampling strategy is not described in the regulation, the monitoring procedures must be able to identify noise-exposed employees and also enable the proper selection of hearing protectors. If employees are stationary throughout their workshift and the noise environment does not fluctuate, area monitoring is sufficient to characterize the daily noise exposure. However, a more likely scenario is that employees will be mobile and sound levels will vary somewhat, which makes personal dosimetry sampling a more practical method. All continuous, intermittent, and impulsive sound levels from 80 to 130 dBA must be integrated into the noise measurements. Monitoring must be repeated whenever a change occurs in the production process, equipment, or noise control devices, which could change the number of exposed workers or require the use of a different type of hearing protector.

9.1.2 *Audiometric Testing Program*

A baseline audiogram must be established within six months of an employee's first exposure at or above the action level. All subsequent audiograms are compared to the baseline audiogram to determine if changes have occurred in the worker's hearing ability. When mobile test vans are used to conduct hearing tests, the employer has one full year to establish the baseline audiogram. If the baseline is not obtained within the first six months of that year, employees must wear hearing protection for the remaining time period until the baseline audiogram is obtained. Employees must avoid occupational as well as non-occupational noise for at least 14 hours prior to baseline testing. If this is not possible, hearing protection should be worn to reduce any noise exposures before the baseline audiogram is conducted.

Annual audiograms must be administered to all employees who are exposed to an 8-hour TWA of 85 dBA or greater. Annual hearing tests provide the most information when conducted sometime during the middle of the workshift. The reason for this is that incorrectly fitted hearing protection may cause the employee to sustain a temporary threshold shift (TTS) that day, which can be identified when a comparison to the baseline audiogram is made. Identifying this temporary condition at the time of the annual hearing test and re-instructing the individual on the proper use of hearing protection can help prevent a permanent threshold shift (PTS) in the future.

As defined by OSHA, a Standard Threshold Shift (STS) is a persistent change in hearing threshold relative to the baseline audiogram of an average of 10 dB or more at 2000, 3000, and 4000 Hz in either ear. Allowances for age corrections are permitted when determining an STS, according to the procedure described in Appendix F of the OSHA regulation. If an STS occurs, the employer may obtain a retest within 30 days, and consider those results as the annual audiogram. In practice, a retest is justified in most cases when a STS is found, because there are many common reasons why a temporary loss of hearing sensitivity may occur. However, if the retest confirms the STS, it must be reported in writing to the affected employee within 21 days. Also, unless a physician determines that the STS is not work-related, the employer must ensure the following steps are taken:

1. Employees not currently using hearing protection must be fitted with hearing protectors, trained in their use and care, and be required to use them.

2. Employees already using hearing protection must be refitted and retrained in the use of hearing protectors, and be provided with protectors offering greater attenuation, if necessary.
3. Employees must be referred for a clinical audiological evaluation or an otological examination (as appropriate) if additional testing is necessary or if the employer suspects that a medical pathology of the ear is caused or aggravated by the wearing of hearing protectors.
4. Employees must be informed of the need for an otological examination if a medical pathology of the ear that is unrelated to the use of hearing protectors is suspected.

New baselines may be established if the audiologist or physician in charge of the hearing conservation program determines that an STS is persistent, or if the annual audiogram indicates significant improvement over the baseline audiogram.

9.1.3 Hearing Protectors

According to the OSHA standard, hearing protection must be made available to employees when their daily noise exposures equal or exceed the 85 dBA action level. The regulation specifies mandatory hearing protection for those employees with a daily noise dose equal to or greater than 100%, which is equivalent to an 8-hour TWA of 90 dBA. Hearing protection is not absolutely required for employees with TWAs less than 90 dBA; however, it is mandatory for employees with a daily exposure between 85 and 90 dBA who have experienced an STS.

Job performance, worker health and safety, morale, and the overall effectiveness of a hearing conservation program may be adversely affected if attention is not given to correct hearing protector selection. Employees must be given a choice of different types of protectors, although the regulation does not explicitly state exactly how much of a variety should be provided. Typically, an employer will offer two or more types of earplugs, and at least one style of earmuff. Whichever type is selected, the employer must ensure proper fitting and usage, and provide training and supervision to employees. Further, the employer must evaluate hearing protector attenuation for the specific noise environments in which the protector will be used. Hearing protectors must attenuate employee exposures down to a TWA of 90 dBA, or 85 dBA for workers who have previously suffered an STS. Appendix B in the OSHA regulation describes the acceptable evaluation methods that can be used. The most convenient method is to use the protector's NRR, and perform a few simple calculations either using A- or C-weighted noise measurements. The adequacy of hearing protector attenuation must be re-evaluated whenever employee exposures increase to the extent that the current devices may no longer provide sufficient attenuation.

9.1.4 Training Program

A hearing conservation program cannot be entirely effective unless each individual limits his or her noise exposure both at work and away from work. All hearing conservation programs must emphasize the need for constant awareness of noise hazards through a continuing education effort. The OSHA standard requires employers to institute a training program for all employees who are exposed to noise at or above the 85 dBA action level. This training must be repeated annually, and should be updated to be consistent with changes in available hearing protection and on-site work processes. Instruction must be provided on:

1. The effects of noise on hearing,
2. The purpose of hearing protectors, the advantages, disadvantages, and attenuation of various types, and instructions on selection, fitting, use, and care,
3. The purpose of audiometric testing, and an explanation of the test procedures.

Hearing conservation measures away from work can be encouraged or assisted by lending or giving hearing protectors to employees for use during noisy activities at home. The cost of supplying hearing protectors for non-occupational uses may be far less than potential worker's compensation costs, and often directs more positive attention to the hearing conservation program at work.

9.1.5 Recordkeeping

OSHA requires that audiograms be retained for the length of employment for the affected employee. Audiometric records must include the employee and examiner's names, employee's job classification, date of the audiogram, date of the last equipment calibration, and the employee's most recent noise exposure assessment. The employer must also keep a record of the background sound pressure levels in the audiometric test room.

According to the regulation, noise exposure measurements must be retained for two years. In many cases, companies will save records indefinitely, even after an employee is terminated. These records could be invaluable if there is ever a question of legal responsibility for an employee's hearing impairment, or if the employee is expected to return to this job in the future. Both the employer and employee benefit from having complete medical records retained after the employee is terminated, for reasons including compensation claims and general health references.

9.2 Setting up a Hearing Conservation Program

Setting up a hearing conservation program entails the development and maintenance of all of the factors mentioned in the preceeding sections. Ideally, a team made up of management, industrial hygiene, preventive medicine, safety, and production personnel will work together toward building an effective hearing conservation program. The first objective for this group would be to become aware of their specific problem areas within the facility, and to consider possible ways of limiting noise exposure levels as much as is practical.

In practice, these teams may function only as advisors on general items such as company policy. Many successful hearing conservation programs have been highly dependent upon the competence and dedication of one single person who is given the time and support required to develop and maintain the program. This individual must be motivated and knowledgeable, and be respected by both management and workers. Extensive formal training in any specific field is not usually necessary, but he or she must have a good practical knowledge of noise-induced hearing impairment, hearing testing requirements, and personal hearing protectors. If a single dedicated person is not available for this job, it may be necessary to use a hearing conservation committee to establish general policies and to appoint different individuals to be responsible for carrying out particular assignments. Generally, it is highly desirable to have well-defined tasks given to each individual.

The development of effective and practical hearing conservation programs has proven to be difficult. Many ineffective programs remain even after years of effort. Two major reasons for poor programs are

1. The responsibility for the program is given to a person within the company who is already overburdened with other tasks, and essentially no support is allotted for a hearing conservation program.

2. The responsibility for the hearing conservation program is contracted to an outside firm, and nobody within the plant acknowledges ownership of the program. One of the apparent reasons for hiring an outside firm is to transfer responsibility for the hearing conservation program to the contractor. However, problems may arise when the employer does not take the time to fully understand the requirements of a comprehensive hearing conservation program. This often leads the employer to believe that the contractor is handling the entire program, but in reality the contractor satisfies only one or two requirements (e.g., hearing tests, noise measurements), and the remainder of the program is left incomplete. Money is spent inefficiently and, without the benefit of a complete program, employees may continue to suffer hearing impairment.

Outside firms can be useful for specific jobs, such as conducting annual hearing tests, noise exposure measurements, or noise control projects, although the responsibility for developing an effective hearing conservation program must be accepted by the employer. At least one person within the company who has overall knowledge of the essential components of the company's operations must work with any outside contractors that are used. The company representative must have knowledge of internal production processes, work schedules, company politics, and so forth. An outside firm or individual cannot be expected to know these things, nor is an outside party likely to have the deep interest in the employees' or the company's well-being that an employee should have. Also, an employee is usually the only logical person to maintain the program continuously after it is first established.

Personnel and work conditions may vary widely between plants, or from one work area to another within the same facility. Significant differences are found in management/employee relationships, and in motivation, education, and communication skills of hearing conservationists from one location to another. As a result, a program that works well in one place may fail in another. Therefore, it is generally necessary to develop a unique program for each situation, if it is to be effective.

Obviously, top management must support the program, and others in middle management should be enthusiastic and knowledgeable. Perhaps even more important are the shift supervisors, safety officers, and the plant nurses. These key individuals often have the respect of all workers, and they may be the only persons at the plant with whom some employees will communicate freely. The first-line supervisor must be genuinely concerned about the health and safety of the employee and be able to answer questions about the effects of noise exposure, the use of safety equipment, and the overall importance of the program.

9.3 Hearing Conservation Program Evaluation

The OSHA noise regulation does not prescribe a method to determine absolutely whether a hearing conservation program is effective or not. At a minimum, an audit of all essential elements should be conducted, and a checklist should be used to ascertain compliance with the OSHA regulation. However, the real test of effectiveness is to look closely at the amount of hearing loss that is occurring. Ideally, the number of STSs will be zero, which generally indicates a successful program.

Evaluating the hearing conservation program from a group perspective rather than looking at each individual may help to identify an ineffective program before large threshold shifts occur for many individuals. Audiometric database analysis (ADBA) techniques have been developed for this purpose, and are presented in ANSI S12.13-1991, Draft American National Standard for Evaluating the Effectiveness of Hearing Conservation Programs (32). The scope of this standard, as well as other epidemiological methods, is to specify procedures for systematically assessing the effectiveness of hearing conservation programs in preventing noise-induced hearing loss based on the results of regular monitoring audiometry for noise-exposed employees. When ADBA results show undesirable trends, the follow-up action involves changes in overall program policies or procedures, rather than changes in the treatment of individual employees.

BIBLIOGRAPHY

1. *American National Standard Specification for Octave-Band and Fractional-Octave-Band Analog and Digital Filters*, ANSI S1.11-1986 (R 1998), American National Standards Institute, New York, 1998.
2. S. A. Gelfand, *Hearing: An Introduction to Psychological and Physiological Acoustics*, Marcel Dekker, Inc., New York, 1998, pp. 35–82.
3. T. C. Morata and D. E. Dunn, eds. *Occupational Medicine: State of the Art Reviews (Occupational Hearing Loss)*, Vol. 10, No. 3, July-September 1995, Hanley & Belfus, Inc., Philadelphia, 1995.
4. *American National Standard Specifications for Audiometers*, ANSI S3.6-1996, American National Standards Institute, New York, 1996.
5. *Ambient Noise Levels for Audiometric Test Rooms*, ANSI S3.1-1991, American National Standards Institute, American National Standard Maximum Permissible New York, 1991.
6. A. H. Suter, *Hearing Conservation Manual*, 3rd ed., Council for Accreditation in Occupational Hearing Conservation, Milwaukee, 1993.
7. Occupational Safety and Health Administration, "Occupational Noise Exposure Standard and Hearing Conservation Amendment," *Code of Federal Regulations*, Title **29**, Chapter XVII, Part 1910, Subpart G.

8. *American National Standard Method for Manual Pure-Tone Threshold Audiometry*, ANSI S3.21-1978 (R1997), New York American National Standards Institute, 1997.
9. *American National Standard Specification for Sound Level Meters*, ANSI S1.4-1983 (R1997), American National Standards Institute, New York, 1997.
10. *American National Standard Specification for Personal Noise Dosimeters*, ANSI S1.25-1991 (R1997), American National Standards Institute, New York, 1997.
11. *American National Standard Specification for Acoustical Calibrators*, ANSI S1.40-1984 (R1997), American National Standards Institute, New York, 1997.
12. *American National Standard Determination of Occupational Noise Exposure and Estimation of Noise-Induced Hearing Impairment*, ANSI S3.44-1996, American National Standards Institute, New York, 1996.
13. U.S. Department of Labor "Safety and Health Standards for Federal Supply Contracts (Walsh-Healy Public Contracts Act)," *Federal Register* **34**, 7948 (1969).
14. U.S. Department of Labor, "Occupational Safety and Health Standards (Williams-Steiger Occupational Safety and Health Act of 1970)," *Federal Register* **36**, 10518 (1971).
15. *Criteria for a Recommended Standard: Occupational Noise Exposure*, Revised Criteria 1998, DHHS (NIOSH) Publication No. 98-126, National Institute for Occupational Safety and Health (NIOSH), 1998.
16. Occupational Safety and Health Administration (OSHA), "Occupational Noise Exposure; Hearing Conservation Amendment," **29** *CFR* Part 1910, *Federal Register* **48**(42) (March 3, 1983).
17. *1998 Threshold Limit Values (TLV's) for Chemical Substances and Physical Agents and Biological Exposure Indices (BEI's)*, American Conference of Governmental Industrial Hygienists (ACGIH), Cincinnati, 1998.
18. U.S. Department of Labor, Occupational Safety and Health Administration (OSHA), *Noise Control: A Guide for Workers and Employers*, Publication No. 3048, Washington, D.C., 1980.
19. *American National Standard Guidelines for the Specification of Noise of New Machinery*, ANSI S12.16-1992 (R1997), American National Standards Institute, New York, 1997.
20. *Sound and Vibration*, **32**(7) (1998).
21. *Compendium of Materials for Noise Control*, Publication No. 80-116, (NTIS Stock No. PB298307), National Institute for Occupational Safety and Health (NIOSH), Cincinnati, 1980.
22. *The NIOSH Compendium of Hearing Protection Devices*, Publication No. 95-105, National Institute for Occupational Safety and Health (NIOSH), Cincinnati, 1994.
23. *American National Standard Methods for Measuring the Real-Ear Attenuation of Hearing Protectors*, ANSI S12.6-1997, American National Standards Institute, New York, 1997.
24. National Institute for Occupational Safety and Health (NIOSH), *List of Personal Hearing Protectors and Attenuation Data*, U.S. Dept. of HEW, Publication No. 76-120, Cincinnati, 1975.
25. Environmental Protection Agency (EPA), "Noise Labeling Requirements for Hearing Protectors," *Federal Register* Vol. 44, No. 190, **40** *CFR* Part 211, 1979, pp. 56130–56147.
26. *American National Standard Method for the Measurement of Real-Ear Protection of Hearing Protectors and Physical Attenuation of Earmuffs*, ANSI S3.19-1974, American National Standards Institute, New York, 1974.
27. *American National Standard Method for the Measurement of the Real-Ear Attenuation of Hearing Protectors*, ANSI S12.6-1984, American National Standards Institute, New York, 1984.
28. *Industrial Hygiene Technical Manual*, OSHA Instruction CPL 2-2.20A, Change 2, March 1, pp. VI-13–VI-20, Occupational Safety and Health Administration (OSHA), Washington, DC, 1987.

29. *Industrial Hygiene Technical Manual*, Section III, Chapter 5, Occupational Safety and Health Administration (OSHA), Washington, DC, 1999.
30. International Organization for Standardization (ISO), *Acoustics-Hearing Protectors-Part 1: Subjective Method for the Measurement of Sound Attenuation*, ISO 4869-1, Geneva, Switzerland, 1990.
31. *Hearing Protectors-Recommendations for Selection, Use, Care, and Maintenance*, Guidance Document, EN 458, European Committee for Standardization (CEN), Brussels, Belgium, 1993.
32. *Draft American National Standard: Evaluating the Effectiveness of Hearing Conservation Programs*, Draft ANSI S12.13-1991, American National Standards Institute, New York, 1991.

CHAPTER TWENTY

Nonionizing Radiation

George M. Wilkening, CIH (Deceased)

1 INTRODUCTION

1.1 Current Interest in Nonionizing Radiations

The development and proliferation of electronic devices that either intentionally or inadvertently emit nonionizing radiation have brought about immense interest in the subject. The promise of an increase in the number and use of such devices has concerned many persons who believe that the radiation hazards have not been sufficiently studied. Those who express concern about an inadequate understanding of the biological effects of nonionizing radiations point out that many electronic devices have already found their way into common use (e.g., microwave ovens, radar for pleasure boats, scanning lasers in supermarket checkout counters, near-ultraviolet radiation in fluorescent lighting fixtures, and a variety of high-intensity light sources). Other concerns include the many infrared, ultraviolet, microwave, and laser devices that might produce excessive occupational exposures. Because of the heightened public interest in electromagnetic radiation hazards, Congress enacted the Radiation Control for Health and Safety Act (1). The declared purpose of the act is to establish a national electronic product radiation control program, including the development and administration of performance standards to control the emission of electronic product radiation. The act covers both ionizing and nonionizing electromagnetic radiations emitted from any electronic product. This includes X-rays and gamma rays and particulate, ultraviolet, visible, infrared, millimeter wave, microwave, radio-frequency, and, interestingly enough, sonic, infrasonic, and ultrasonic radiation. Since the inception of the act, the federal government has conducted or funded research into the biological effects of radiation, with special emphasis on low-level effects. Standards have been developed and promulgated for television set receivers, medical X-rays

Patty's Industrial Hygiene, Fifth Edition, Volume 2. Edited by Robert L. Harris.
ISBN 0-471-29754-2

(amendments to existing standard), cathode ray tubes, microwave ovens, and lasers. Calibration, measurement, and product testing laboratories have been established to ensure proper evaluation of accessible radiation from electronic products, and a compliance program has been developed to obtain manufacturers' adherence to established standards. During the course of standards development efforts, the Food and Drug Administration (FDA) has been required to consult with the Technical Electronic Product Radiation Safety Standards Committee (TEPRSSC), an advisory body established under the Act. Unlike some federal advisory committees, TEPRSSC has the authority to develop and recommend its own standards directly to the Secretary of Health and Human Services.

Other federal agencies that are actively concerned with nonionizing radiation hazards include the Occupational Safety and Health Administration (OSHA), the Environmental Protection Agency (EPA), the National Institute for Occupational Safety and Health (NIOSH), and the National Institute of Environmental Health Sciences (NIEHS). The National Council on Radiation Protection and Measurements (NCRP) has enlarged its scope of interest to include nonionizing radiation; its first effort was Scientific Committee 39 on Microwaves. The product of SC39 was its NCRP Report 67 "Radiofrequency Electromagnetic Fields—Properties, Quantities and Units, Biophysical Interaction and Measurements" published in 1981. Other NCRP reports include Report 74, "Biological Effects of Ultrasound: Mechanisms and Clinical Implications (1983), and Report 86, "Biological Effects and Exposure Criteria for Radiofrequency Electromagnetic Fields" (1986). Attention is now being given to the biological effects of extremely low frequency magnetic fields, such as those associated with power transmission lines. See Section 7.2.

1.2 Nature of Electromagnetic Radiation

One important aspect of electromagnetic energy is that it may be considered to be wavelike, having characteristics such as frequency, wavelength, and velocity of propagation; or it may be considered to consist of discrete quanta or energy packets (photons) having characteristics similar to those of particles (energy, momentum, etc.). This wave–particle duality is necessary to explain such physical phenomena as the photoelectric effect or diffraction. The photoelectric effect is most readily understood by considering the incident radiation to consist of a flux of discrete photons, each with a characteristic energy E. Diffraction of electromagnetic energy, on the other hand, can be approached by assuming the incident radiation to be wavelike. However, what at first appears to be a contradiction is resolved by Planck's law, which states that the photon energy E is directly proportional to the frequency of radiation ν; that is, $E = h\nu$, where E is expressed in joules, h is Planck's constant equal to 6.625×10^{-34} J, and ν is the frequency, expressed in reciprocal seconds (s^{-1}).

In a more commonly used nomenclature, the photon energy is expressed in electron volts (eV), where 1 eV equals 1.602×10^{-19} J.

As just indicated, the photon energies of electromagnetic radiations are proportional to the frequency of radiation and hence inversely proportional to wavelength. The electromagnetic spectrum extends over a broad range of wavelengths, from less than 10^{-12} cm to greater than 10^{10} cm. The shortest wavelengths, hence the highest energies ($>10^2$ eV), are associated with X- and gamma rays, whereas the lower energies ($<10^{-5}$ eV) are associated with radio-frequency and microwave radiation. Ultraviolet, visible, and infrared radiations occupy an intermediate position. For example, radio-frequency wavelengths may range from 3×10^{10} to 3×10^{2} μm, infrared from 3×10^{2} to about 0.7 μm, visible light from approximately 0.7 to 0.4 μm, ultraviolet from approximately 0.4 to 0.1 μm, and gamma- and X-radiation below 0.1 μm. Figure 20.1 is a graphic representation of the entire electromagnetic spectrum.

The nuclear binding energy of protons may be of the order of 10^6 eV or greater; hence these energies are much greater than the binding energy of chemical bonds, which is of the order of 1 to 15 eV. The thermal energy associated with molecules at room temperature is even lower, that is, approximately 0.03 eV. Because the photon energy necessary to ionize atomic oxygen and hydrogen is of the order of 10 to 12 eV, it seems appropriate to adopt a value of approximately 10 eV as a lower limit at which ionization is produced in biological material. The electromagnetic radiations that do not produce ionization in biological systems may be presumed to have photon energies less than 10 to 12 eV and are therefore termed "nonionizing." Nonionizing radiation may be absorbed, causing changes in the vibrational and rotational energies of tissue molecules, thus leading to possible

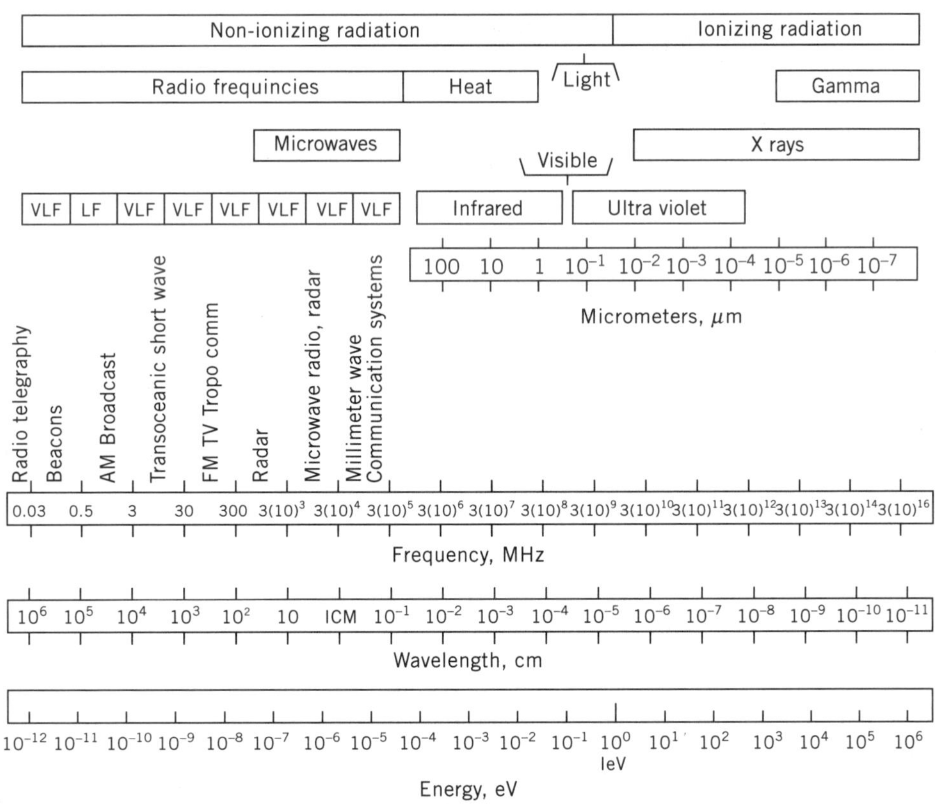

Figure 20.1. The electromagnetic spectrum. After Ref. 95.

dissociation of the molecules; or, more often, the energy is dissipated in the form of fluorescence or heat (2, 3). Exposure to very high levels of nonionizing radiation may inflict considerable thermal injury.

This chapter is arbitrarily divided into sections on ultraviolet (UV), infrared (IR), visible, laser, radio-frequency (RF), microwave, and extremely low frequency (ELF) radiation. Laser radiation is included as a separate section because of the widespread special interest in lasers, even though lasers operate at wavelengths covered in the other sections. For this reason, the reader should consider the laser section to be an extension of material contained in the ultraviolet, visible, and infrared sections. In particular, the exposure limits contained in the laser section are generally applicable to UV, visible, and IR noncoherent sources.

2 ULTRAVIOLET RADIATION

2.1 Physical Characteristics

The ultraviolet wavelength range of interest extends from the vacuum UV (0.16 μm) to the near UV (0.4 μm) as follows:

UV Region	Wavelength Range (μm)	Photon Energy (eV)
Vacuum	0.16	7.7
Far	0.16–0.28	7.7–4.4
Middle	0.28–0.32	4.4–3.9
Near	0.32–0.4	3.9–3.1

The photon energy range for wavelengths between 0.1 and 0.4 μm lies between 12.4 and 3.1 eV.

A different but commonly used UV classification scheme, oriented toward biological effects, is as follows:

UV Region	Wavelength Range (μm)	Biological Implications
UV-A	0.4–0.32	Black light region Solar UV that reaches earth's surface UV-induced fluorescence
UV-B	0.32–0.28	Erythemal (sunburn) region
UV-C	<0.28	Germicidal region

In the wavelength region below 0.16 μm, UV radiation is completely absorbed by air and materials such as quartz; this radiation can exist only in a vacuum. At somewhat longer wavelengths (0.16–0.2 μm), UV is poorly transmitted through air or quartz. At wavelengths in the range of 0.2 to 0.32 μm, UV is absorbed by ordinary window glass and by the epithelial layers of the skin and cornea; however, transmission through air and water can occur. The ozone layer above the earth's surface absorbs solar radiation at wavelengths below 0.29 μm. At wavelengths in the range of 0.3 to 0.4 μm, UV is transmitted through air but is transmitted only partially through ordinary glass, quartz, and water.

2.2 Sources of UV Radiation

The sun represents the major source of UV energy at the earth's surface, even though the atmospheric ozone layer filters out all UV wavelengths below 0.29 μm. Significant man-made sources include high- and low-pressure mercury discharge lamps, plasma torches, and welding arcs. More than 85% of the radiation emitted by low-pressure mercury vapor discharge lamps is at a wavelength of 0.2537 μm. At lamp pressures that are fractions of an atmosphere, the characteristic mercury lines predominate, whereas at high pressures (up to 100 atm), the lines broaden to produce a radiation continuum. In typical quartz lamps the amount of energy at wavelengths below 0.38 μm may be much greater than the radiated visible energy, depending on the mercury pressure. Other man-made sources include xenon discharge lamps, lasers, and certain fluorescent tubes that emit UV radiation at wavelengths above 0.315 μm, reportedly at an irradiance less than that measured outdoors on a sunny day (3).

The UV radiation in welding operations emits photons with sufficient energy (5 to 9.5 eV) to produce ozone and oxides of nitrogen from the dissociation of oxygen and nitrogen molecules (4).

2.3 Biological Interaction with UV Radiation

The penetration of UV radiation into human tissue is very limited. The deepest penetration may reach just below the epidermis if the wavelength is above 0.3 μm and the subject is essentially nonpigmented. The decrease in transmission through the skin as one proceeds from visible wavelengths to approximately 0.3 μm appears to be linear, indicating the lack of specific absorption bands in this wavelength range (5). Wavelengths between 0.28 and 0.32 μm penetrate appreciably into the corium or dermis; those between 0.32 and 0.38 μm are absorbed primarily in the epidermis, and those below 0.28 μm appear to be absorbed almost completely in the stratum corneum of the epidermis. The lens and tissues in the anterior segments of the eye may be exposed to UV if incident wavelengths are above 0.3 μm; the cornea, on the other hand, absorbs throughout the entire wavelength range of 0.1 to 0.4 μm, although absorption is much reduced between 0.32 and 0.4 μm. Because the cornea absorbs most of the energy up to 0.32 μm, radiation in this range generally produces corneal damage before lenticular effects are observed (6, 7).

The relative response of a biological system to irradiation at different wavelengths is called the "biological action spectrum" of that system. In some cases the action spectrum may be the same as the absorption spectrum; this is rarely true with complex biological systems, however. The maximum erythermal response at 0.29 μm in the biological action spectrum corresponds to absorption in tyrosine and other aromatic amino acids; the peak at 0.265 μm corresponds to absorption in pyrimidines and is thought to indicate changes in nucleic acids. The response at wavelengths longer than 0.3 μm shows no maxima or minima.

The advent of supersonic transport has raised speculation about detrimental effects to the atmospheric ozone layer caused by flight at exceptionally high altitudes. The same concern has been expressed about the interaction of halogenated chemicals (propellants) with the ozone layer. One estimate of damage is that a halving of the effective volume of the ozone layer would result in a two- to tenfold increase in the photobiological effects from UV (8).

2.4 Effects on the Skin

The immediate effects of UV radiation on the skin can be considered in terms of erythema, increased pigmentation (i.e., migration of melanin granules and the production of new melanin granules), darkening of pigment, and changes in cellular growth. Depending on the total UV dose, the latent period for the production of erythema may range from 2 to several hours. Usually an immediate erythema appears, followed by a second, more intense erythema after a period of some 2 to 10 h. The intensity of the erythema and the complicating sequelae (e.g., edema and blistering) is proportional to the UV radiation dose. Irradiation of the skin at wavelengths above 0.3 μm still produces an erythema, but the efficiency is low. Despite this relatively low efficiency, the amount of solar UV at wavelengths above 0.295 μm is sufficient to cause significant erythema in the population.

Photosensitizing agents may have biological action spectra in the UV range. The combined effect of skin contact with these agents and exposure to UV radiation may result in severe irritation and blistering. For example, it is common knowledge that workers who routinely expose themselves to coal tar products while working outdoors experience photosensitization of the skin.

The biological action spectrum for pigmentation changes is very similar to that of erythema. The first effect of UV irradiation appears to be a spreading of existing pigment granules, followed by the production of new pigment granules. Migration of melanin granules from the basal cells to the Malpighian cell layers of the epidermis may also cause a thickening of the horny layers of the skin. The process of increased pigmentation occurs immediately after irradiation. The mechanics of action are thought to be either oxidation of premelanin granules or free radical scavenger action of melanin. The period of increased pigmentation is usually followed by another period in which the skin is more vulnerable to an acute erythema induced at the same wavelength (9).

There appears to be no direct evidence that melanomas are produced by UV radiation; furthermore, melanomas are induced in experimental animals only with great difficulty. On the other hand, irradiation of persons suffering from xeroderma pigmentosum, an inherited disease caused by an insufficiency of DNA repair enzymes, will result in the early production of skin tumors (10).

Ultraviolet phototherapy has been used with infants to correct problems of icterus; however, improper use has produced cases of erythema (11). There is widespread concern about the therapeutic use of UV radiation for herpes, because this virus is already considered to be carcinogenic (12, 13).

In addition to erythema and increased pigmentation, slowing or cessation of cell mitoses in the basal and superficial layers of the epidermis often appears immediately after UV irradiation. An increase in mitoses is delayed for about 24 h postirradiation (14), after which superfluous cellular material is sloughed off.

2.5 Effects on the Eye

The most common clinical sign of overexposure to UV at wavelengths shorter than 0.3 μm is photokeratitis, appearing some 2 to 24 h after irradiation, along with acute hyperemia, photophobia, and blepharospasm if exposure is severe. The duration of such difficulties is usually of the order of one to five days. However, there are usually no residual lesions.

2.6 Cancer

Although the premise that skin cancer may be induced by UV radiation is generally accepted in scientific circles, the evidence is largely indirect (i.e., epidemiologic and statistical). The mechanism of UV-induced carcinogenesis is unknown. In spite of these uncertainties, however, certain epidemiologic studies have revealed a strong correlation between skin cancer and terrestrial solar UV-B levels found at given altitudes and ground elevations (15).

The types of skin cancer that are possibly related to UV radiation include basal cell carcinoma, spindle cell carcinoma, and melanoma, the latter generally considered to be a remote possibility. The induction of tumors is believed to occur as a result of irradiation at wavelengths below 0.3 μm (16), with the maximally effective wavelengths being in the 0.26 to 0.27 μm range, corresponding to specific interaction with nucleic acids.

The alleged high incidence of skin cancer in outdoor workers who come into contact with chemicals such as coal tar derivatives, benzopyrene, methylcholanthrene, and other anthracene compounds raises questions about the role played by UV radiation in these cases.

There is no direct evidence that the UV radiation will produce tumors in the cornea or anterior chamber of the eye; however, it is of some interest that melanoma of the eye is much more common in blue-eyed persons. Melanoma of the iris, for example, was found exclusively in blue-eyed persons in one study (17).

Abiotic effects from exposure to UV radiation occur in the spectral range of 0.24 to 0.31 μm, where most of the incident energy is absorbed by the corneal epithelium. Although the lens is capable of absorbing 99% of the energy below a wavelength of 0.35 μm, only a small portion of the radiation reaches the anterior lenticular surface. Photon energies of about 3.5 eV (0.36 μm) may excite the lens of the eye or cause the aqueous or vitreous humor to fluoresce, thus producing a diffuse haziness inside the eye that can interfere with visual acuity or produce eye fatigue. The phenomenon of fluorescence in the ocular media is not of concern from a bioeffects standpoint; the condition is strictly temporary and without detrimental effect.

An important conclusion about the effects of UV radiation on biological systems is that tissue damage appears to depend on the total energy absorbed (dose) rather than on the rate of energy absorption; hence the dose must be controlled to afford protection.

2.7 Threshold

Fairly good agreement exists on the amount of radiant exposure necessary to produce minimal photokeratitis. The threshold has been found to be 4 mJ/cm^2 for primates and humans when exposed to wavelengths between 0.220 and 0.310 μm. The minimal radiant exposure of 4 mJ/cm^2 was obtained at the most sensitive wavelength of 0.27 μm (18).

The energy necessary to elicit a barely perceptible reddening of the skin is approximately 30 mJ/cm^2. (The minimal erythema dose [MED] is the smallest radiant exposure that produces a barely perceptible reddening of the skin that disappears after 24 h. The MED depends on skin color, degree of pigmentation, age, and the irradiated body site.) Depending on skin type, the erythema thresholds may vary from 8 mJ/cm^2 for untanned skin to 50 mJ/cm^2 for tanned Caucasian skin. For white Caucasian skin with an average degree of tanning, the value for the minimal perceptible erythema is more on the order of 8 to 10 mJ/cm^2 when irradiation takes place at wavelengths between 0.24 and 0.29 μm. Any threshold number that is used should be identified with its wavelength, because some wavelengths are more effective than others in producing a given biological effect (action spectrum).

2.8 Exposure Criteria

The exposure criteria adopted by the Council on Physical Medicine of the American Medical Association (AMA) based on erythemal thresholds for 0.2537 μm radiation are as follows: 0.5×10^{-6} W/cm^2 for exposure periods up to 7 h (12.6 mJ/cm^2); 0.1×10^{-6} W/cm^2 for exposure periods up to and exceeding 24 h (8.6 mJ/cm^2). Although these criteria are thought to be stringent, they have nevertheless been successfully used for many decades. Several important qualifications must be kept in mind when the AMA guidelines are used: (*1*) since the guidelines apply only to radiation at 0.2537 μm (germicidal), corrections must be made for the biological effectiveness at wavelengths other than germicidal; and (*2*) a tenfold safety factor has been applied to the germicidal photokeratitis threshold in deriving the recommended numerical values.

The ACGIH (18) recommended threshold limit values (TLV) for UV irradiation of unprotected skin and eyes for actinic wavelengths between 0.2 and 0.315 μm (200 and 315 nm) are given in Table 20.1. It should be recognized that these TLVs have been derived on the basis of irradiating skin that has not received previous UV radiation. Note in this respect that the minimum TLV is somewhat below the photokeratitis threshold previously mentioned (i.e., 3 mJ/cm^2 rather than 4 mJ/cm^2). Conditioned or tanned individuals can tolerate skin exposure in excess of the TLV without erythemal effect. However, such conditioning may not protect persons against possible skin cancer.

Exposure to wavelengths between 0.32 and 0.4 μm should not exceed the following criteria, (the 1999 ACGIH TLV lists higher values for this region):

Wavelength, λ	TLV	Exposure Duration
0.32–0.4 μm	1 mW/cm^2	$>10^3$ sec
	1 J/cm^2	$<10^3$ sec

To determine the effective irradiance of a broad-band source weighted against the peak of the spectral effectiveness curve (270 nm), the following weighting formula should be used:

Table 20.1. Threshold Limit Values for Exposure to Ultraviolet Radiation (0.2–0.315 μm)

Wavelength (nm)	TLV (mJ/cm^2)
200	100
210	40
220	25
230	16
240	10
250	7.0
254	6.0
260	4.6
270	3.0
280	3.4
290	4.7
300	10
305	50
310	200
315	1000

$$E_{eff} = E_\lambda S_\lambda \Delta_\lambda$$

where E_{eff} = effective irradiance relative to a monochromatic source at 270 nm (0.270 μm)
E_λ = spectral irradiance $[W/(cm^2)(nm)]$
S_λ = relative spectral effectiveness (unitless)
Δ_λ = bandwidth (nm)

Table 20.2 shows relative spectral effectiveness as a function of wavelength.

Permissible exposure time in seconds for exposure to actinic ultraviolet radiation incident upon the unprotected skin or eye may be computed by dividing 0.003 J/cm^2 by E_{eff}, expressed in watts per square centimeter. The exposure time may also be determined using Table 20.3, which provides exposure times corresponding to effective irradiances in microwatts per square centimeter ($W/cm^2 \times 10^{-6}$).

The TLVs do not apply to lasers or to an exposure duration less than 0.1 s. Also, no corrections have been made in these criteria for photosensitization to chemicals, drugs, and cosmetics (18).

2.9 Measurement of UV Radiation

One device that is particularly useful for measuring UV radiation is the thermopile. The coatings on the thermopile receiver elements are usually lampblack or gold black, to approximate the properties of a black-body radiator. A special window material transparent to the UV is necessary to surround the thermopile element to minimize the detrimental

Table 20.2. Relative Spectral Effectiveness by Wavelength

Wavelength (nm)	Relative Spectral Effectiveness (S_λ)
200	0.03
210	0.075
220	0.12
230	0.19
240	0.30
250	0.43
254	0.5
260	0.65
270	1.0
280	0.88
290	0.64
300	0.30
305	0.06
310	0.015
315	0.003

Table 20.3. Permissible Ultraviolet Exposures[a]

Duration of Exposure Per Day	Effective Irradiance, E_{eff} (μW/cm^2)
8 h	0.1
4 h	0.2
2 h	0.4
1 h	0.8
30 min	1.7
15 min	3.3
10 min	5
5 min	10
1 min	50
30 s	100
10 s	300
1 s	3,000
0.5 s	6,000
0.1 s	30,000

[a]Ref. 18.

effects of air currents. Commonly used materials for windows are quartz crystals, calcium and lithium fluoride, sodium chloride, and potassium bromide.

Other UV detection devices include (*1*) photodiodes (e.g., silver–gallium arsenide, silver–zinc sulfide, and gold–zinc sulfide: peak sensitivity of these diodes is at wavelengths below 0.36 μm; the peak efficiency or responsivity is of the order of 50–70%; (*2*) thermocouples (e.g., Chromel–Alumel); (*3*) Golay cells; (*4*) superconducting bolometers; and (*5*) certain photomultiplier and vacuum photodiodes (19).

Care must be taken to use detection devices having the proper rise time characteristics (some devices respond much too slowly to permit the obtaining of meaningful measurement). Also, when measurements are being made, special attention should be given to the possibility of UV absorption by many materials in the environment (e.g., ozone or mercury vapor), thus adversely affecting the readings. The possibility of photochemical reactions between UV radiation and a variety of chemicals also exists in the industrial environment.

Table 20.4 lists the sensitivity, impedance, and response time for certain junction detectors. It is possible to obtain a low radiation intensity calibration by exposing these and other detectors to a secondary standard furnished by the National Institute for Standards and Technology.

Other devices that have been used to measure UV radiation include photovoltaic cells, photochemical detectors, photoelectric cells, and photoconductive cells. Photovoltaic and photoconductive cells are usually the same device operated in different modes (in series with an ammeter or across a load and voltmeter). It is common practice to employ selective filters, such as a monochrometer grating in front of the detecting device, to isolate the portion of the UV spectrum that is of interest to the investigator. Certain semiconductors such as copper or selenium oxide deposited on a selected metal develop a potential barrier between the layer and the metal. Light falling on the surface of the cell causes the flow of electrons from the semiconductors to the metal. A sensitive meter placed in such a circuit will record a signal proportional to the intensity of radiation falling on the cell. Some photocells used for UV measurement take advantage of the property of certain metals to exhibit quantitative photoelectric responses to specific bands in the UV spectrum. Thus, a photocell may be equipped with metal cathode surfaces that are sensitive to the UV wavelengths of interest.

One of the limitations of photocells is "solarization" or deterioration of the envelope, especially with long usage or following the measurement of high-intensity radiation. Frequent calibration of the cell is necessary under these conditions. The readings obtained with these instruments are valid only when measuring monochromatic radiation or when the relationship between the response of the instrument and the spectral distribution of the source is known. A measurement technique for which Bassi et al. (20) claim some promise involves the use of beryllium oxide and lithium fluoride thermoluminescent crystals that have been preirradiated with gamma rays. The response of the detector, after several cal-

Table 20.4. Sensitivity, Impedance, and Response Time of Junction Detectors

Type (circular)	Sensitivity [$\mu V/(\mu W)cm^2$)]	Impedance (Ω)	Response Time, 1/e (s)
1 Junction: constantan-Mn	0.005	2	0.1
4 Junction: Cu-constantan	0.025	5	0.5
4 Junction: Bi-Ag	0.05	5	0.5
8 Junction: Bi-Ag	0.10	10	1.0
16 Junction: Bi-Ag	0.20	25	2.0
(linear)	(0.05)	(10)	(0.5)
12 Junction: Bi-Ag	0.05	10	0.5

ibrated gamma predoses, appears to be a function of the UV exposure received from germicidal lamps.

For environmental measurements that are intended to correlate with a given biological effect, the ideal instrument would be designed in such a way that its spectral response closely approximates that of the biological action spectrum under consideration. Such an instrument is unavailable at this time. Because available photocells and filter combinations do not closely approximate the UV biological action spectrum, it is necessary to calibrate each photocell and meter at a number of wavelengths. Such calibrations are generally made at a great enough distance to ensure that the measuring device is in the "far field" of the source. In performing these calibrations, special care must be taken to control the temperature of the mercury lamps as secondary standards, because the spectral distribution of the radiation from the lamps depends on the pressure of the vaporized mercury.

2.10 Control of Exposure

Because UV radiations are so easily absorbed by a wide variety of materials, appropriate attenuation is accomplished in a straightforward manner. The information given in Section 2.8 ("Exposure Criteria") should be used for the specification of shielding requirements. The data of Pitts (21) may be used for UV lasers because of the narrow band UV source employed in his experiments to determine the thresholds of injury to rabbit eyes. It is important to remember that photosensitization may be induced in certain persons at levels below the suggested exposure criteria.

Personal protection against radiation at wavelengths below 0.32 μm can be accomplished through the use of eyeglasses, goggles, plastic face shields, protective clothing, or sunscreen creams or lotions. The wearing of tinted glasses or goggles is seldom recommended for protection against the UV intensities normally encountered in industry; however, such devices may be used in a supplementary manner to reduce the high visible brightness that often accompanies UV emissions.

3 INFRARED RADIATION

Infrared wavelengths extend from 0.75 to approximately 10^3 μm. This range can be arbitrarily divided into three subareas known as the near infrared (0.75 to 3 μm), the middle infrared (3 to 30 μm), and the far infrared (30 to 10^3 μm).

Infrared radiation is emitted by a large variety of sources including the sun, heated metals, home electrical appliances, incandescent bulbs, furnaces, welding arcs, lasers, and plasma torches. The energy and wavelength characteristics emitted from IR sources largely depend on the source temperature. For example, at a temperature of 1000°C, about 5% of the total energy is emitted at wavelengths below 1.5 μm, whereas at 1500 and 2000°C, the respective values are 20 and 40% (22).

Infrared radiation exchange between two heated bodies proceeds in accordance with the fourth power of the absolute temperature of each body (Stefan–Boltzmann law). When persons are placed in the vicinity of large IR sources, the transfer of the heat load to the human body can represent a severe environmental stress. The problem of heat stress on the human body resulting from convective and radiative (as exemplified by large IR sources) heat loads is covered in Chapter 22.

Infrared energy is not sufficiently energetic (<1.5 eV) to cause the removal of electrons from orbital shells; therefore, IR seldom enters directly into chemical reactions in biological systems and does not usually cause ionization. However, the possibility exists that IR energy at certain wavelengths and intensities may affect the vibrational or rotational properties of certain tissue molecules in such a way as to cause detrimental effects. Even in these cases, the consensus is that the effects produced are largely thermal. The body organs at greatest risk from IR radiation are the eye and skin.

Most biological systems are considered to be opaque to wavelengths greater than 1.5 μm; transmission of IR through the ocular media at wavelengths from 1.3 to 1.5 μm is also poor. For skin, the wavelength region of high transmission is between 0.75 and 1.3 μm, with a maximum of 1.1 μm. At a wavelength of 1.1 μm, 20% of the energy incident upon the stratum corneum will reach a depth of 5 mm.

3.1 Eye

Thermal damage to the cornea usually results from the energy absorbed in the epithelium rather than in the deep stroma. The iris is especially susceptible to radiations below 1.3 μm. Because the iris can dissipate its heat load only to the surrounding ocular media, it is often regarded as a heat sink serving to mitigate the amount of radiation reaching the lens. It is thought that a radiant exposure of 4.2 J/cm^2 within a wavelength range of 0.8 to 1.1 μm will produce a minimally perceptible lesion on the iris. To obtain a radiant exposure of 4.2 J/cm^2 at the iris, it would be necessary to irradiate the cornea at a level of 10.8 J/cm^2 (23).

The transmission characteristics of the lens apparently vary with age and nuclear sclerosis. Selective IR absorption bands in the lens exist at wavelengths from 1.4 to 1.6 μm and from 1.8 to 2.0 μm (24).

Although the principal effect of IR radiation on the eye and skin appears to be acute thermal injury, a controversy has raged throughout most of the twentieth century over the etiology of "glass blower's cataract," specifically, whether IR radiation is a cataractogenic agent. Although Dunn (25, 26) in 1950 was unable to uncover cases of cataract in employees who had been exposed to IR for more than 20 years, the evidence now seems to favor IR as the etiologic agent (27–31).

The formation of a cataract depends on initial heating of the anterior portions of the eye, especially the cornea and iris, followed by heat transfer from the iris to the lens epithelium. The elevation of temperature at the anterior portion of the lens is the primary etiologic factor in glass blower's cataract, according to Goldmann and co-workers (29, 30). In some recent studies of various glass and steel plants, a significant number of employees have shown lenticular changes after some 10 to 15 years of exposure to IR sources (22). The reported range of irradiance to which employees were exposed was 0.08 to 0.4 W/cm^2; however, no attempt was made to reconstruct exposure time–irradiance patterns for each employee. In practical cases where the eye is exposed to high temperature sources, there is often a significant near IR component as well as some visible wavelength radiation. Special care is necessary to make distinctions between visible and IR emissions when industrial hygiene or epidemiologic studies are made of infrared radiation and its possible relationship to cataract formation.

The retina is at risk from IR radiation only when near IR wavelengths are being generated by the source. The dissipation of energy within the retina is accomplished through conduction of heat to adjacent structures, notably the choroid. When the size of the retinal image and the rate of energy deposition are such that heat cannot be conducted away quickly enough to keep the tissue temperature below approximately 45°C, protein denaturation or destruction of tissue (microsteam generation and burn) may occur. More near IR energy than visible wavelength energy is required to produce a minimal lesion on the retina (threshold of damage as observed through an ophthalmoscope). For a discussion of maximum permissible exposure levels for infrared radiation, the reader is referred to Section 5 ("Lasers").

3.2 Skin

Effects on the skin include vasodilatations of the capillary beds and increased pigmentation. The skin is normally able to dissipate a heat load imposed by IR radiation because of capillary bed dilatation, increased blood circulation, the production of sweat, and ambient air movement.

The perception of warmth by the skin is related to the rate at which the skin temperature is raised. For example, for skin temperatures of 32 to 37°C, the threshold of warmth perception is reached when the rate of skin temperature increase is of the order of 0.001 to 0.002°C per second. Such a perception threshold depends on the size of the skin area irradiated by IR as well as the density of thermoreceptors in that area. The thresholds for perceiving warmth decrease with increasing area of irradiated skin. On the other hand, precooled skin can be rapidly heated without eliciting the sensation of warmth (2).

Unlike the thresholds of warmth sensation, the thresholds for pain appear to depend directly on skin temperature, despite wide individual variation in their values (16). Wertheimer and Ward (32) have recorded a range of skin temperature pain thresholds of 44.1 to 44.9°C, depending on body site. The experiments of Hardy (33) indicate that the threshold of pain is reached when the skin temperature is raised to 45°C (44.5 $\pm$ 1.3°C). The same value also appears to be critical for producing a skin burn. Pain thresholds therefore appear to be related to skin temperature, whereas the extent of tissue damage depends on skin temperature and duration of exposure.

3.3 Measurement of Infrared Radiation

An almost unlimited variety of IR instrumentation is available: spectrophotometers, radiation pyrometers, stationary or scanning radiometers, and other equipment. The various detectors usually fall into one of two categories:

1. *Thermal detectors*, where absorbed energy heats the detector, which in turn causes a change in the properties of the detector material. Response is to all IR wavelengths, but the response time is slow. Typical examples are thermopiles, thermocouples, liquid crystal, bolometers, and pyroelectric crystals.

2. *Quantum detectors*, where all incident photons are sufficiently energetic to free a bound electron. Response is to a limited range of IR wavelengths, but response time is fast. Examples of this type of detector are photodiodes, photomultiplier tubes, and semiconductor devices such as gallium arsenide diodes.

4 VISIBLE RADIATION

From the industrial hygiene standpoint, the use of visible wavelength radiations generally involves two considerations: (*1*) proper lighting practices to achieve a pleasant visual environment and the efficient performance of tasks that require optimal visual acuity, and (*2*) the control of radiation to prevent damage to the human body, usually the eyes and skin. The first subject is covered in considerable detail in Chapter 24; the second is considered in this section.

The visible portion of the electromagnetic spectrum consists of a very narrow band of wavelengths between the UV and the IR. The photon energies associated with the visible wavelength range of 0.38 to 0.75 μm lie between 3.1 and 1.65 eV. These energy levels are rather low and innocuous in terms of producing direct biological effects on the skin surfaces of the human body; however, radiation in the 0.38 to 0.75 μm range acts on a body organ that is uniquely and exquisitely designed to transduce the incident photons into intelligible vision by means of complex and incompletely understood photochemical reactions. This exquisite organ is the retina of the eye. The photoreceptors of incident photons are the rods and cones. The differences in function of these two receptor systems are worthy of note. Cones respond to higher levels of ambient illumination than do the rods; hence their level of activity is normally greater during daylight hours. The cones also come into play for the discrimination of fine visual detail. Because the rods are in greater numerical abundance in the peripheral paramacular area of the retina, they provide a wide-area peripheral view of the visual field. The rods and cones respond differently to different wavelengths: the spectral luminous efficiency for rods is at a wavelength of 0.510 μm, whereas the peak efficiency of the cones is at 0.555 μm (34). The retina is able to adapt to the wide range of light intensities, but the adaptation time varies with the time-luminous flux. Light adaptation occurs within a short time duration, usually of the order of seconds, whereas dark adaptation may require more than an hour.

Visible light is not hazardous to the eyes under ordinary circumstances, even when high-intensity light sources are encountered. For example, the adaptation of the eye to high-intensity sunlight by such mechanisms as restricted pupil size, light adaptation, partial closing of the eyelids (squinting), blinking, and shading of the eyes by the eyebrows and periorbital socket is usually sufficient to prevent excessive radiation from reaching the retina. On the other hand, certain conditions involving exposure to light may produce significant hazards for humans. For example, if light is pulsed or gated at a frequency near the alpha rhythm (brain function), certain light-sensitive individuals may experience an epileptiform seizure. The effect may be produced when such persons observe a pulsating light pattern such as flickering sunlight coming through a stand of trees as the light is observed from a moving vehicle or when watching a pulsed light pattern from a television set. In some cases, it is the pattern of light generation rather than its brightness that is critical to the production of seizures. Flash blindness is another condition that may result from exposure to high-intensity light sources. This condition is caused by the bleaching of visual pigments that produce an "afterimage" or temporary blind area (scotoma) in the field of vision. The greater the light intensity and time of exposure, the more persistent will be the afterimage. If the original stimulation by a high-intensity light source is sufficient to obliterate focused perception completely, some protracted period of time may be required before normal visual function is restored. The sun may damage the retina when viewed without proper protection; eclipse blindness is a familiar example, where retinal damage is caused by exposing the eye to a combination of excessive visible and IR radiations.

On a lesser scale, ordinary glare caused by improper placement of light sources (fixtures) may produce severe visual discomfort and may represent a safety hazard because of interference with visual performance. A more serious condition is one in which a veiling glare may be temporarily superimposed on the retinal image, obscuring one's ability to discriminate details in the object being viewed. Such a condition exists on a more permanent basis in persons with cataracts.

High-intensity light sources such as lasers, flashbulbs, spotlights, welding arcs, and carbon arcs can produce retinal burns under appropriate viewing conditions. Because of its coherent properties, a laser beam may be focused to an extremely small diameter on the retina. This results in the creation of highly intense beams of light, an obvious hazard to personnel.

Of the factors that have entered into the development of acceptable exposure levels for high-intensity light sources, three seem to be of special importance:

1. The magnification factor of the eye as visible light passes through and is focused by the cornea and lens is of the order of 10^5. This means that the acceptable irradiance (W/cm^2) and the radiant exposure (J/cm^2) at the retina must have much greater numerical values than those present in front of the cornea where the industrial hygienist makes field measurements. Permissible exposure levels are those that enter or impinge upon the eye at the cornea.

2. Acceptable exposure levels for wavelengths in the near IR and IR are somewhat less stringent than those for the visible wavelengths, whereas exposure levels for wavelengths in the blue and UV are more stringent, probably because of the photochemical mechanisms coming into play at the shorter wavelengths.

3. When exposed to a highly intense visible light source, most persons exhibit an "aversion response," that is, a tendency to blink and turn the eyes and head. Presumably, the elapsed time before the aversion response occurs gives some indication of the exposure time to be considered in establishing an exposure limit. In the case of the American National Standards Institute (ANSI) standard (35), the aversion time has been assigned a value of 0.25 s. The use of the aversion response concept in establishing safety guidelines for exposure to visible radiations has drawn criticism from some quarters because not all persons exhibit such a response. On the other hand, to establish safety standards strictly on the basis of experiments that irradiate the eyes of sedated animals whose eyelids are immobilized in an open position and receive radiation pulses in excess of 0.25-s duration means that practical conditions and normal physiological responses are ignored.

5 LASERS

Lasers are widely used for a variety of purposes such as alignment, welding, trimming, spectrophotometry, interferometry, flash photolysis, fiber optics communications systems, nuclear fusion experimentation, and surgical removal or repair procedures. A review of medical and biological applications may be found in Wolbarsht (36) and in many applied science journals.

The word “laser” is the acronym for “light amplification by stimulated emission of radiation.” When an atom has been stimulated into an excited state, it may lose its excess energy through spontaneous decay, with the random emission of electromagnetic energy at a specific wavelength. When the atoms are raised to an excited state through an applied electromagnetic field, however, the radiated wave has the same phase, direction, and polarization as the stimulating (pumping) source. This process is known as stimulated emission. Because the population of atoms at a lower or ground state is normally greater than the number existing at a higher energy (excited) state, the energy of most radiations is usually absorbed by the atoms, thus tending to reduce or attenuate the field. However, if the population of atoms can be reversed by stimulating a greater number to energy levels higher than those existing at ground state, the energy radiated by the stimulated emission will exceed the energy absorbed by the atoms; hence a phase coherent amplification of the incident radiation will occur. The net power difference (gain) will be proportional to the number of atoms that undergo a radiative decay to a lower energy level.

The energy stimulation and decay scheme just described is representative of a three-level (ground state, broad energy absorption band, and upper laser level) system, such as the ruby laser. Simply stated, atoms in a three-level system are pumped from the ground state to a broad energy absorption band, whereupon they rapidly decay by a radiation-less transition into an upper laser level. Here they remain for a relatively long residence time (long compared with the time spent within the broad energy absorption band) before they return to the ground level, with an attendant emission of radiation at a specific frequency. The frequency of the radiation is equal to the difference in energy between the metastable upper laser level and the ground state divided by Planck’s constant.

Many gas lasers and rare earth lasers (e.g., neodymium–yttrium–aluminum–garnet) operate in a four-level system, where a relatively small number of atoms (compared with a three-level system) are pumped from the ground state to achieve lasing.

Figure 20.2 depicts the energy diagram of a four-step laser system. Atoms are raised from ground state to a broad energy level (absorption band) through absorption of electromagnetic radiation, as in the ruby three-level system, or through electron impact, as in the case of most gas lasers systems. Fast nonradiative relaxation occurs before the atoms are brought to level 2, the metastable level, where relaxation time is long and an accumulation of atoms occurs. Level 1, the lower laser level, has fast relaxation time; in fact, the dumping of atoms keeps the atom population low, to ensure population inversion. In passing from level 2 to level 1, radiation is emitted at the characteristic wavelength of the laser. The relative magnitude of the relaxation time at each level and the relative populations of atoms at each level are important parameters for achieving laser action.

Stimulated emission in gaseous systems was first reported in a helium–neon mixture (37). Since the early 1960s, lasing action has been reported at hundreds of wavelengths

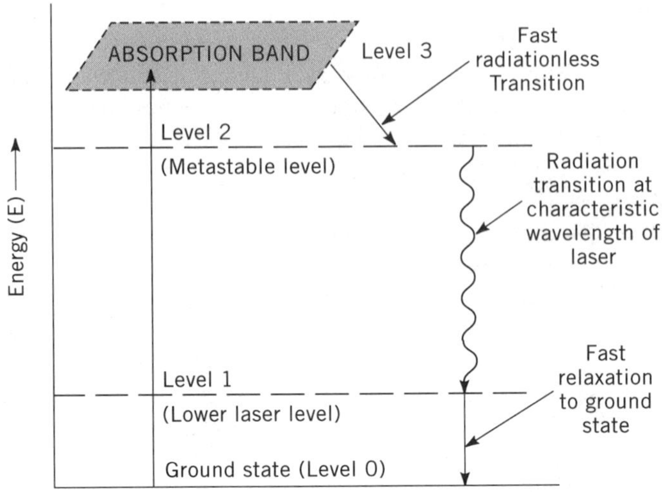

Figure 20.2. Four-level laser system.

from the ultraviolet to the far infrared. Helium–neon lasers are typical of gas systems in which stable, single-frequency operation is important. The very popular helium–neon system can be operated in a pulsed or continuous wave mode at wavelengths of 0.6328, 1.15, or 3.39 μm, depending on resonator design. Typical power output for the helium–neon systems is of the order of 1 to 500 mW.

The carbon dioxide (CO_2) laser belongs to a class known as molecular lasers. The energy for the system is derived from transitions between vibrational and rotational energy levels of CO_2 molecules. Such laser systems most frequently operate at wavelengths of 10.6 μm in either the continuous wave, pulsed, or *Q*-switched mode. A *Q*-switch is a device for enhancing the storage and quick discharge of energies to produce extremely high power pulses. The peak power output of a CO_2–N_2 system may range from several watts to greater than 10 kW. The CO_2 laser is attractive for terrestrial and extraterrestrial communication systems because of the low absorption window in the atmosphere between 8 and 14 μm. High-powered DF–CO_2 (deuterium fluoride–carbon dioxide) transfer lasers are based on the combustion of carbon monoxide and oxygen, which subsequently thermally dissociates the fluoride before supersonic injection into a laser cavity. Both the DF–CO_2 laser and high-powered CO lasers (design goal is to obtain more than 10 kJ output in subnanosecond pulses at an efficiency greater than 1%) are being used in laser fusion coupling experiments. The use of heavy ion beams, coupled with high-energy accelerators of the storage ring type, seems to give a competitive edge over laser activation in experiments on inertial confinement fusion. However, major activity continues with the development of high-power transfer lasers for use in laser fusion research.

Of major significance from the personal hazard standpoint is the capacity of far infrared lasers to radiate enormous power at wavelengths that are invisible to the human eye.

The argon ion gas systems operate in the blue wavelengths, typically 0.4578, 0.488, and 0.5145 μm, in either the continuous wave or pulsed mode. Power generation is usually less than 25 W.

Of the many ions in which laser action has been produced in solid-state crystalline materials, perhaps neodymium (Nd^{3+}) in garnet or glass and chromium (Cr^{3+}) in aluminum oxide (ruby) are most noteworthy (Table 20.5). Neodymium–glass lasers are already operational in the terawatt peak power range. Garnet or YAG (yttrium–aluminum–garnet) is an attractive host for the trivalent neodymium ion because the 1.06 μm laser transition line is sharper than in other host crystals. Frequency doubling to 0.53 μm, using lithium niobate crystals, may produce power approaching that available in the fundamental mode at 1.06 μm. Also, through the use of electrooptic materials such as potassium dihydrogen phosphate (KDP), barium–sodium niobate, or lithium tantalate, "tuning" or frequency scanning over wide ranges may be accomplished (38). Perhaps the most versatile device for obtaining a wide band of frequencies is the dye laser.

Semiconductor lasers are usually moderately low power devices (milliwatts to several watts) having relatively broad beam divergence, a factor that tends to reduce the potential ocular hazard. On the other hand, certain semiconductor lasers may be pumped with multikilovolt electron beams, thus introducing a potential ionizing radiation hazard (39). As a matter of fact, the X-rays from certain laser plasmas are already being considered for use in cancer detection.

Examples of certain ions that have exhibited lasing action appear in Table 20.5. The wavelength range of tunable infrared lasers is given in Table 20.6.

Through the use of carefully selected dyes, it is possible to tune through broad wavelength ranges. For example, if one is interested in producing lasing wavelengths in the vacuum ultraviolet (VUV), the beams from two dye lasers (one fixed at a double-quantum resonance wavelength, the other tuned over its tunable wavelength range) can be combined to obtain the VUV source. Through the use of four dye solutions, Sorokin et al. (40) were able to cover the range from 0.1578 to 0.1957 μm. Table 20.7 lists some additional lasers operating in the ultraviolet, and Table 20.8 gives the operating wavelengths and associated photon energies of certain lasers in common use.

Table 20.5. Certain Ions That Have Exhibited Lasing Action

Active Ion	Wavelength (μm)
Nd^{3+}	0.9–1.4
Ho^{3+}	2.05
Er^{3+}	1.61
Cr^{3+}	0.69
Tm^{3+}	1.92
U^{3+}	2.5
Pr^{3+}	1.05
Dy^{2+}	2.36
Sm^{2+}	0.70
Tm^{2+}	1.12

Table 20.6. Tunable Infrared Lasers

Laser Type	Wavelength (μm)
Diode	<1–34
Spin-flip Raman	5–6.5
	9–14.6
Nonlinear device, parametric oscillators	<1–11
Difference frequency generator	2–6
Two-photon mixer	9–11
Four-photon mixer	2–25
Gas lasers	
Zeeman tuned	3–9
High-pressure CO_2	9–11

Table 20.7. Recent Lasers at Wavelengths 360 nm and Below (40)

Method	Medium	Wavelength (nm)	*E*/Pulse	Efficiency[a] (%)
Nonlinear	$KB_5O_84H_2O$	217.3–234.5	0.4 μJ	2×10^{-3} (c)
	ADP (cooled)	208.0–212.4	0.5 mJ	5×10^{-3} (c)
				50–60 (p)
	Rb-Xe	364.7	30 mJ	5–10 (p)
Molecular	H_2	109.8–161.3	5 kW/cm^2	
	Xc_2	172.0	0.76 J	1 (c)
	Xe_2[b]	172.0	30 μJ	
	Ar-N_2	357.7	45 kW/cm^2	2.3 (c)

[a]Abbreviations = c, conversion; p, photon.
[b]Continuous wave.

Table 20.8. Lasers in Common Use

Laser	Wavelength (μm)	Photon Energy (eV)
Ruby (Cr^{3+})	0.69	1.79
Nd–YAG	1.06	1.17
Nd–glass	1.06	1.17
GaAs	0.90	1.47
Dye	0.36–0.65	3.5–1.9
He–Ne	0.633–3.39	1.96 (λ = 0.633)
He–Cd	0.325–0.4416	3.81 (λ = 0.325)
CO_2	10.6	0.117
Ar	0.4519–0.5145	2.54 (λ = 0.488)
CO	5.5	0.225
H_2	0.160	7.74
H_2O	7.0–220	0.0104 (λ = 118.6)
HCN	773	0.0016

5.1 Biological Effects

Despite the dramatic benefits of technology, it is axiomatic to say that the improper use or design of any apparatus can produce undesirable effects. The laser is no exception. Nevertheless, although technical achievements often move in advance of any understanding of hazards, it has been encouraging to witness from the inception the serious attention given to the evaluation of laser radiation effects on biological systems (41).

The primary hazard from laser radiation is exposure of the eye and, to a lesser extent, the skin. If the radiation levels are kept below those that damage the eye, there will be no harm to other body tissues. Therefore, the material to follow emphasizes the effects of laser radiation on ocular tissue, the radiation exposure conditions required to produce detrimental effects (usually a threshold of injury), and some practical means of preventing undue exposure. In describing the laser radiation that is incident upon tissue, it has become customary to use certain terms. For example, the output of pulsed lasers is described in terms of energy (joules), and that from continuous wave (CW) lasers in terms of power (watts). The corresponding beam density parameter of a pulsed laser is known as the "radiant exposure," expressed in joules per square centimeter (J/cm^2); that for a CW laser is known as "irradiance" and is expressed in watts per square centimeter (W/cm^2). Other parameters such as pulse duration, wavelength, spot size on the retina, and pulse repetition rate (or pulse repetition frequency) are commonly used.

Effects of laser radiation on the eye can range from an annoying glare and mild bleaching of the photoreceptors to massive damage to the foveal region of the retina. The extent of injury depends on exposure conditions and the characteristics of the laser radiation incident upon the eye. In practical situations, it is the short-term or intermittent type of exposure that is to be expected, rather than long-term, chronic exposure. Although the effects of long-term, chronic exposure have not been systematically studied, there is some indication that irreversible retinal effects may be produced when the eye is continuously bathed in a luminous flux, without interruption, for extended periods of time (41).

For the most part, work on the biological effects of laser radiation has been aimed at the elucidation of photobiological phenomena and the determination of tissue damage thresholds. "Damage" has usually been described in terms of detrimental changes that are grossly apparent or, in the case of the eye, observable with optical instruments such as microscopes and ophthalmoscopes.

The cornea, lens, and ocular media are largely transparent in the visible region (0.380 to 0.750 μm). Of the visible energy transmitted through the ocular media to reach the retina, most is absorbed in the neuroectodermal coat in the pigment epithelium. The greater part of visible radiation is absorbed in the melanin granules in the retinal pigment epithelium and choroid, which underlie the rods and cones. Figure 20.3 and 20.4 show the percentage transmission of various wavelengths of radiation through the ocular media and the percentage absorption in the retinal pigment epithelium and choroid, respectively. It is at the pigment epithelium that the greatest absorption of energy occurs; hence this thinly pigmented (10 μm) layer is the most susceptible to damage. Indeed, optical radiation damage may be produced in the pigment epithelium while the photoreceptor layer remains essentially unaffected. Figure 20.3 and 20.4 illustrate why the retina is the organ at risk with visible wavelength radiation, whereas the cornea and skin surfaces are at risk with IR and UV radiation.

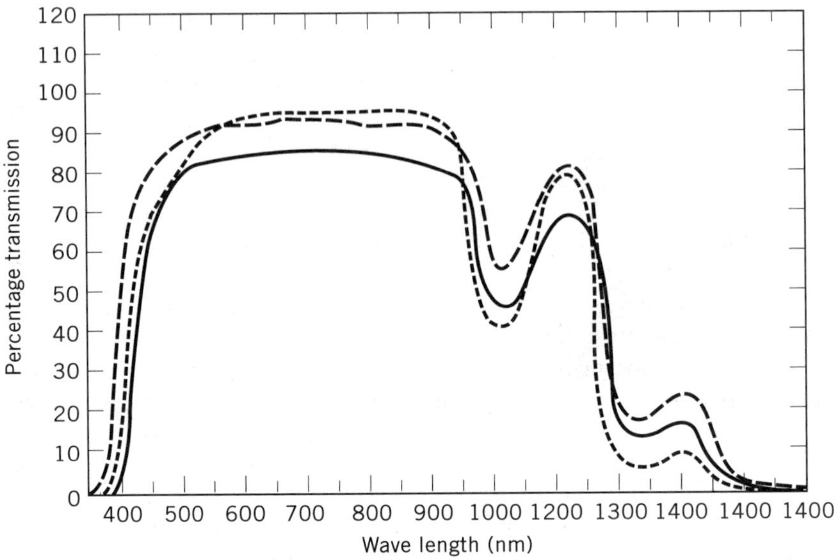

Figure 20.3. Percentage transmission for light of equal intensity through the ocular media of human (short dashes), rhesus monkey (solid curve), and rabbit (dashes). Curves are mean values.

Laser radiation damage may be attributed to thermal, thermoacoustic, or photochemical phenomena. For pulse durations of the order of 10^{-9} s and shorter, there is evidence that nonlinear mechanisms, Raman and Brillouin scattering, ultrasonic resonance, and acoustic shock waves may be brought into play (42). The wavelength shifts caused by scattering of electromagnetic radiation from acoustic modes (Brillouin) are small compared with scattering due to molecular vibrations and electron plasma oscillations (Raman). The precise mechanisms for producing damage under these short pulse conditions are not yet fully understood.

The millisecond pulses from a solid-state, visible wavelength laser may cause destruction of tissue by thermal means, whereas ocular and skin tissue may be damaged by photochemical mechanisms when exposed to UV and blue laser light. The consensus seems to be that several damage mechanisms exist. The mere stimulation of the photoreceptors caused by an irradiance that is only slightly higher than that presented to the eye under normal ambient conditions may be sufficient to increase cellular activity to the point of failure, especially if the irradiance is presented to thc eye for an extended period of time (43).

When speaking of the thermal effects caused by laser radiation, one means the denaturation of protein through the absorption of energy and the consequent conversion to heat. Thermal injury is generally considered to be a rate process; in all probability, therefore, there is no single critical temperature at which injury takes place independent of exposure time. Also, because the molecules of the melanin granules in the pigment epithelium of the retina are relatively large, a broad spectral absorption would be expected to occur.

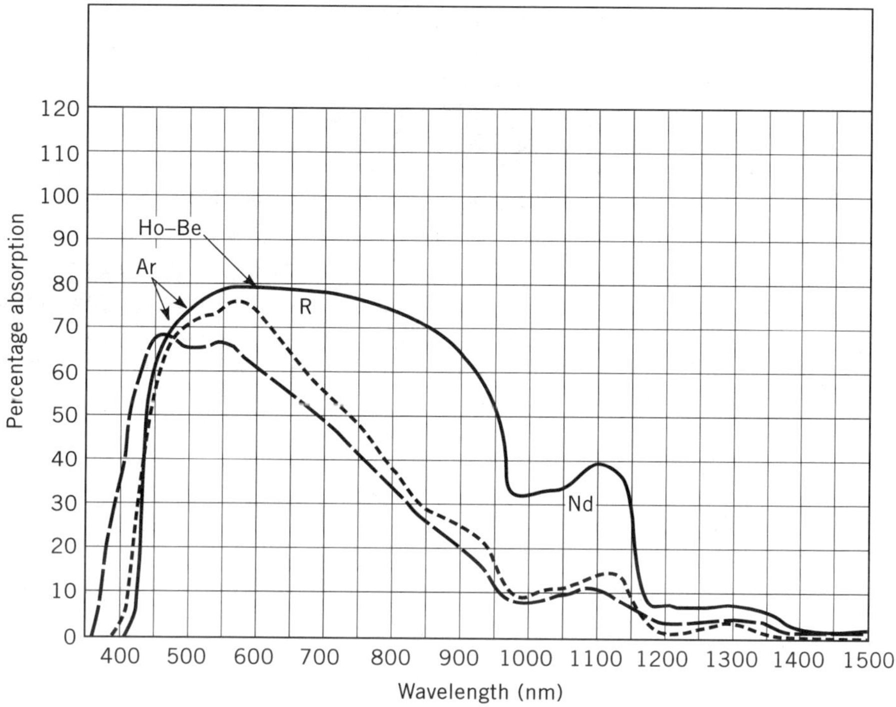

Figure 20.4. Percent absorption of light of equal intensity, incident on the cornea in the retinal pigment epithelium and choroid for Dutch chinchilla rabbits (dashes), rhesus monkey (solid curve), and humans (short dashes). Curves are mean values.

Therefore, the monochromaticity of laser radiation would not be expected to produce biological effects significantly different from those produced by radiation exposure to more conventional light sources. In fact, the coherence of the laser beam is not considered to be a significant factor in producing chorioretinal injury.

The biological response to UV laser radiation is expected to be similar to that produced by noncoherent UV sources. Photophobia, tearing, conjunctival discharge, surface exfoliation, and stromal haze accompanied by complaints of "sand in the eyes" are the expected consequences of excessive exposure. Damage to the corneal epithelium probably results from photochemical denaturation of proteins. Exposure to UV-C radiation (100 to 280 nm) and UV-B (290 to 315 nm) may produce photokeratitis. The latency period for photokeratitis varies from 30 min to as long as 24 h, depending on the severity of the exposure. Chronic high level exposure to UV-A radiation (315 to 380 nm) may produce cataracts or, for that matter, retinal changes (7).

There are three wavelength regions in the infrared that have significance from a biological standpoint: IR-A (700 to 1.4 nm), IR-B (1.4 to 3 μm), and IR-C (3 μm to 1 mm). A transition zone occurs between the far end of the visible region where retinal effects are observed and the infrared region where corneal effects are produced. The biological effects in the IR-B region are primarily lenticular and corneal damage; in the IR-C region the damage is primarily corneal, because absorption by water (a primary constituent of the cornea) is very strong at these wavelengths. For wavelengths beyond the IR-C region, in the far infrared, absorption occurs in the peripheral layers of tissue. It is in this region and in the ultraviolet A and B regions that the threshold for damage to the cornea is comparable to that of the skin.

To translate these data into the potential hazards associated with specific laser systems, we can say that lasers operating in the visible wavelengths, such as ruby (0.6943 μm) or helium–neon (0.6328 μm), produce damage primarily in the pigment epithelium of the retina. Those that operate in the ultraviolet, such as helium–cadmium (0.3250 μm) and nitrogen (0.3371 μm), produce damage primarily in the cornea, although there is some evidence that effects may be produced in the lens and retina if the radiation levels are sufficiently high (7). Lasers that emit radiation in the near infrared, such as the semiconductor type, may produce injury to the lens, although the output power is usually too low to cause damage. Far infrared lasers such as the carbon dioxide laser (10.6 μm) have the potential for producing damage to the cornea. Unlike the visible wavelengths, there is no magnification of UV or far infrared irradiation levels.

A retinal injury that occurs in the macula is more serious than one that occurs in the paramacula, because visual processes are more highly developed in the former. Injury to the paramacular region may have only a very minor effect on vision. Somewhat surprisingly, in some cases where direct injury to the macula has occurred, near-normal visual acuity has apparently been restored in a matter of months. Speculation is that optical signals reaching the retina may be refocused to various retinal sites "around the damage" site, thus permitting satisfactory acuity.

There is evidence (44) to indicate that short wavelength light (0.4416 μm) produces retinal damage in primates by means of photochemical rather than thermal processes. Also, the thresholds of retinal damage were found to be considerably lower for irradiation at 0.4416 μm than for irradiation at 1.064 μm. For example, an irradiance of 24 W/cm^2 at a retinal temperature of 23°C above the retinal ambient produced a threshold lesion in 1000 s using a wavelength of 1.06 μm (Nd–YAG laser). Using a wavelength of 0.4416 μm (He–Cd laser), however, required only 30 mW/cm^2, with a negligible temperature rise, to produce a lesion in a 1000-s exposure. The appearance of the lesion induced by the 0.4416 μm blue light was that of a light yellowish patch on the fundus, whereas the lesion induced by the 1.064 μm radiation had a typical center core "burn" characteristic. Table 20.9 supplies threshold lesion data for the wavelength range of 0.4416 to 1.064 μm (441.6 to 1064 nm).

Until recently, the thresholds for producing retinal lesions as a result of irradiation at all visible wavelengths were considered to be of the same order of magnitude (i.e., 5 to 10 W/cm^2). However, the data of Ham et al. (44) indicate a rather steep slope in the action spectrum at approximately 0.460 μm and below, particularly at exposure durations of 16 s or less. Figure 20.5 illustrates the effect of two factors (decreasing wavelength and increasing exposure duration) on an increasing retinal sensitivity to laser radiation in the visible wavelengths. It is of considerable interest to note that although retinal sensitivity seems to increase as one shifts from the visible through the blue portions of the electromagnetic spectrum, there are fewer photons reaching the retina because of absorption in

Table 20.9. Retinal Irradiance for Threshold Lesion as a Function of Wavelength and Exposure Duration[a] (44)

Laser	Wavelength (nm)	Irradiance (W/cm^2) ± SD[b] 1 s	16 s	100 s	1000 s	Transmittance (*T*)
Nd–YAG		56.1 ± 5.3	37.5 ± 4.2	32.5 ± 3.1	24.0 ± 3.0	
	1064	55°	37°	32°	23°	0.76
He–Ne		29.9 ± 1.4	15.2 ± 3.0	8.4 ± 2.8	5.4 ± 2.3	
	632.8	49°	25°	14°	9°	0.93
Ar-pumped		22.0 ± 1.9	12.3 ± 1.1	8.1 ± 1.1	5.8 ± 0.3	
dye	610	36°	20°	13°	9.5°	0.92
Laser		26.1 ± 4.5	11.5 ± 2.4	7.6 ± 0.8	4.0 ± 1.7	
	580	43°	19°	12.5°	6.6°	0.91
Ar		14.5 ± 3.3	10.3 ± 2.3	2.2 ± 0.5	0.32 ± 0.1	
	514.5	25°	18°	4°	1°	0.87
Ar		9.4 ± 1.2	6.1 ± 0.8	0.77 ± 0.2	0.15 ± 0.08	
	488	17°	11°	1°	0.25°	0.83
Ar		5.1 ± 0.8	3.2 ± 0.6	0.52 ± 0.1	0.06 ± 0.02	
	457.8	10°	6°	1°	0.1°	0.69
He–Cd		0.91 ± 0.03	0.41 ± 0.02	0.20 ± 0.02	0.03 ± 0.01	
	441.6	2°	1°	0.4°	0.05°	0.45

[a]Maximum retinal temperature (°C) above ambient estimated from mathematical model of Clarke; transmittance (*T*) through ocular media. Beam diameter on retina 500 μm to $1/e^2$ points. Entire laser beam, TEM_{00} mode, enters eye of anesthetized animal with pupil dilated to 8 mm. A beam splitter and fundus camera are used to view and direct beam on the retina. Criterion for a threshold lesion is the appearance 24 h after exposure of a funduscopically visible lesion.
[b]SD = standard deviation.

the lens and ocular media. Also, as expected, the site of absorption and damage changes from the retina to the lens.

Zuclich and Connolly (7) have reported the UV induction of cataracts at energy doses that are lower than those required to cause corneal damage. The mechanism for inducing corneal damage or cataracts is believed to be photochemical, and that for the production of lenticular cataracts is thought to be thermal. The same investigators report their finding that the cornea must be irradiated at a level tenfold its injury threshold to produce reversible clouding of the lens. On the other hand, immediate cataracts are produced from irradiation by nitrogen lasers operating at 0.3371 μm and 1 J/cm^2. This level of irradiation is reportedly an order of magnitude lower than the corneal injury threshold. Also, retinal lesions are produced using 0.3507 and 0.356 μm radiation reportedly at levels an order of magnitude below the corneal damage threshold. Therefore, it would appear that the hazard to all eye structures is increased as one proceeds into the blue and UV wavelengths. Furthermore, the damage threshold appears to depend on the presence of oxygen in that a much higher incident energy is required to produce lesions when the cornea is flushed with nitrogen.

Exposure of rhesus monkeys to ultrashort single pulses of 25 to 35 psec from an unfocused mode—locked Nd–YAG laser at a wavelength of 1.064 μm—resulted in a thresh-

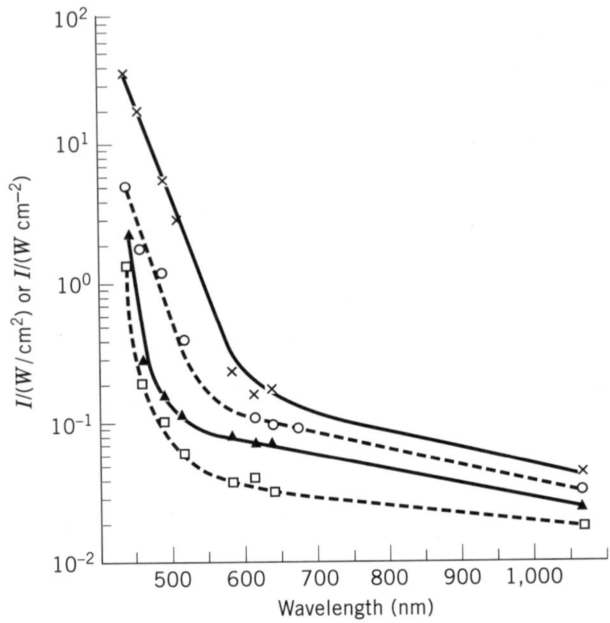

Figure 20.5. Action spectrum of retinal sensitivity to threshold damage in the rhesus monkey. Reciprocal of retinal irradiance $1/(W/cm^2)$ is plotted semilogarithmically against wavelength for four different exposure durations. Beam diameter on the retina was 500 μm to the $1/e^2$ points of the Gaussian distribution. Exposure durations (seconds): squares, 1; triangles, 16; circles, 100; crosses, 1000. From Ref. 42.

old injury to the retina at a mean energy of 13 ± 13 μJ (45). This compares with a threshold value of 68 ± 13 μJ determined by the same authors for a *Q*-switched Nd–YAG laser operating with pulse durations in the 9 to 30 ns range. Therefore, it appears that shortening the pulse durations from the nanosecond to the picosecond range—or three orders of magnitude—results in a lowering of retinal injury threshold by a factor of approximately 5. The mechanism of injury appears to be the destruction of the retinal pigment epithelium through the combined effects of acoustic and shock wave pressure and the extremely high temperature rise in the melanin granules of the pigment epithelium.

Another study (46) using ultrashort pulse trains from Nd–glass gave much lower thresholds of injury than the Ham study: 1.7×10^{-8}J versus 1.3×10^{-5}J, using similar pulse durations. Because the threshold of 1.3×10^{-5}J (13 μJ) was obtained with single pulses rather than pulse trains, the study (46) will be repeated with single pulses. These results have complicated the setting of safety standards in that the present maximum permissible exposure (MPE) levels are largely based on pulse durations no shorter than nanoseconds.

Work has also been completed on the effect of ultrashort visible radiation pulses on rhesus monkeys (47). These results indicate that the melanin granules are much more severely damaged by the visible radiation than by 1.064 μm radiation; however, the threshold of injury was higher for the visible wavelengths.

Studies of the biological effects of laser radiation have largely abated in the past 10 years. Yet, we still have a need to develop techniques that detect functional as well as histologic changes in the eye structure. The development of such techniques could have an important bearing on the viability of the ophthalmoscope as a primary means for determining ocular change. Some investigators have already observed irreversible decrements in visual performance at exposure levels that are 10% below the damage threshold determined by observations through an ophthalmoscope. McNeer et al. (48) found that at 50% of the ophthalmoscopically determined threshold, the ERG B wave amplitude was irreversibly reduced. Davis and Mautner (49) reported severe changes in the visually evoked cortical potential at 25% of the ophthalmoscopically determined threshold. In addition to the problem of determining the true acute ocular injury threshold, there is a general deficiency of knowledge concerning the effect of long-term, chronic exposure to laser radiation.

5.2 Skin

The potential damage caused by irradiating the skin with lasers is considered to be less than that caused by exposure of the eye, because skin damage is often repairable or reversible. On the other hand, exposure of the skin to high-intensity radiation can cause depigmentation, severe burns, and possible damage to underlying organs. To keep the relative eye and skin hazard potential in perspective, one must not overlook possible photosensitization of the skin caused by use of drugs or cosmetic materials. In such cases, the permissible exposure limits can be considerably below the currently recommended values.

Ultraviolet radiation in the wavelength region of 0.25 to 0.32 μm appears to be particularly injurious to the skin. Exposure to shorter (0.20 to 0.25 μm) and longer (0.32 to 0.40 μm) wavelengths is considered less harmful. Because of the structural inhomogeneities of the skin, any incident radiation will undergo multiple internal reflections; however, most radiation is absorbed in the first 3.5 mm of tissue. The shorter wavelengths are absorbed in the outer dead layer of the epidermis (stratum corneum), whereas exposure to the longer wavelengths has a pigment-darkening effect. Chronic exposure to ultraviolet radiation accelerates skin aging and probably the development of skin cancer. Please refer to Section 2, "Ultraviolet Radiation."

5.3 Exposure Limits

5.3.1 Eye

The exposure limits developed by ANSI (35) seem to have been generally accepted (with modification in certain cases) by international standards-setting organizations such as the World Health Organization (50) and the International Electrotechnical Commission (51). The ANSI limits are presented in Table 20.10, 20.11, and 20.12. Table 20.10 illustrates the exposure limits for direct ocular exposure (intrabeam viewing) to a laser beam; Table 20.11 shows the exposure limits for viewing a diffuse reflection of a laser beam or an extended source laser, and Table 20.12 gives the exposure limits for skin. The ANSI limits are used here because most laser safety criteria have been derived from those developed by the ANSI Z136 committee activities.

Table 20.10. Maximum Permissible Exposure for Direct Ocular Exposure Intrabeam Viewing, to a Laser Beam[a]

Wavelength, λ (μm)	Exposure Duration, t (s)	MPE (J/cm^2)	Notes for Calculation and Measurement
		Ultraviolet	
0.200–0.302	10^{-9} to 3×10^4	3×10^{-3}	Or $0.56t^{1/4}$ J/cm^2, whichever is lower; 1-mm limiting aperture
0.303	10^{-9} to 3×10^4	4×10^{-3}	
0.304	10^{-9} to 3×10^4	6×10^{-3}	
0.305	10^{-9} to 3×10^4	1.0×10^{-2}	
0.306	10^{-9} to 3×10^4	1.6×10^{-2}	
0.307	10^{-9} to 3×10^4	2.5×10^{-2}	
0.308	10^{-9} to 3×10^4	4.0×10^{-2}	
0.309	10^{-9} to 3×10^4	6.3×10^{-2}	
0.310	10^{-9} to 3×10^4	1.0×10^{-1}	
0.311	10^{-9} to 3×10^4	1.6×10^{-1}	
0.312	10^{-9} to 3×10^4	2.5×10^{-1}	
0.313	10^{-9} to 3×10^4	4.0×10^{-1}	
0.314	10^{-9} to 3×10^4	6.3×10^{-1}	
0.315–0.400	10^{-9} to 10	$0.56t^{1/4}$	
0.315–0.400	10^3 to 3×10^4	1	
		Visible and Near Infrared	
0.400–0.700	10^{-9} to 1.8×10^{-5}	5×10^{-7}	7-mm limiting aperture
0.400–0.700	1.8×10^{-5} to 10	$1.8t^{3/4} \times 10^{-3}$	
0.400–0.550	10 to 10^4	10×10^{-3}	
0.550–0.700	10 to T_1	$1.8t^{3/4} \times 10^{-3}$	
0.550–0.700	T_1 to 10^4	$10\ C_B \times 10^{-3}$	
0.400–0.700	10^4 to 3×10^4	$C_B \times 10^{-6}$ W/cm^2	
0.700–1.050	10^{-9} to 1.8×10^{-5}	$5\ C_A \times 10^{-7}$	
0.700–1.050	1.8×10^{-5} to 10^3	$1.8\ C_A t^{3/4} \times 10^{-3}$	
1.051–1.400	10^{-9} to 5×10^{-5}	5×10^{-6}	
1.051–1.400	5×10^{-5} to 10^3	$9\ t^{3/4} \times 10^{-3}$	
0.700–1.400	10^3 to 3×10^4	$320\ C_A \times 10^{-6}$	
1.4–10^3	10^{-9} to 10^{-7}	10^{-2}	See Table 20.17 for apertures; 1-mm limiting aperture for 1.4–100 μm; 11-mm limiting aperture for 0.1–1 mm
	10^{-7} to 10	$0.56t^{1/4}$	
	>10	0.1 W/cm^2	
1.54 only	10^{-9} to 10^{-6}	1.0	

[a] $C_A = 1$ for $\lambda - 0.400$–0.700 μm.
$C_A = 10^{2.0(\lambda 0.700)}$ for $\lambda = 0.700$–1.050 μm.
$C_A = 5$ for $\lambda = 1.050$–1.400 μm.
$C_B = 1$ for $\lambda = 0.400$–0.550 μm.
$C_B = 10^{15(\lambda 0.550)}$ for $\lambda = 0.550$ μm.
$T_1 = 10 \times 1020^{20(\lambda 0.550)}$ for $\lambda = 0.550$–0.700 μm.

Table 20.11. Maximum Permissible Exposure for Viewing a Diffuse Reflection of a Laser Beam or an Extended-Source Laser[a]

Wavelength, λ (μm)	Exposure Duration, t (s)	MPE	Notes for Calculation and Measurement
		Ultraviolet	
0.200–0.302	10^{-9} to 3×10^4	3×10^{-3} J/cm^2	Or $0.56t^{1/4}$ J/cm^2, whichever is lower; 1-mm limiting aperture
0.303	10^{-9} to 3×10^4	4×10^{-3} J/cm^2	
0.304	10^{-9} to 3×10^4	6×10^{-3} J/cm^2	
0.305	10^{-9} to 3×10^4	10×10^{-2} J/cm^2	
0.306	10^{-9} to 3×10^4	16×10^{-2} J/cm^2	
0.307	10^{-9} to 3×10^4	25×10^{-2} J/cm^2	
0.308	10^{-9} to 3×10^4	40×10^{-2} J/cm^2	
0.309	10^{-9} to 3×10^4	63×10^{-2} J/cm^2	
0.310	10^{-9} to 3×10^4	10×10^{-1} J/cm^2	
0.311	10^{-9} to 3×10^4	16×10^{-1} J/cm^2	
0.312	10^{-9} to 3×10^4	25×10^{-1} J/cm^2	
0.313	10^{-9} to 3×10^4	40×10^{-1} J/cm^2	
0.314	10^{-9} to 3×10^4	63×10^{-1} J/cm^2	
0.315–0.400	10^{-9} to 10	$0.56t^{1/4}$ J/cm^2	
0.315–0.400	10^4 to 3×10^4	1 J/cm^2	
		Visible	
0.400–0.700	10^{-9} to 10	$10t^{1/3}$ J/(cm^2)(sr)	1-mm limiting aperture or α_{min}, whichever is greater
0.400–0.550	10 to 10^4	21 J/(cm^2)(sr)	
0.550–0.700	10 to T_1	$3.83t^{3/4}$ J/(cm^2)(sr)	
0.550–0.700	T_1 to 10^4	21 C_B J/(cm^2)(sr)	
0.400–0.700	10^4 to 3×10^4	21 $C_B 10^{-3}$ W/(cm^2)(sr)	
		Near Infrared	
0.700–1.400	10^{-9} to 10	10 $C_A t^{1/3}$ J/(cm^2)(sr)	
0.700–1.400	10 to 10^3	3.83 $C_A t^{3/4}$ J/(cm^2)(sr)	
0.700–1.400	10^4 to 3×10^4	0.64 C_A W/(cm^2)(sr)	
1.4–10^3	10^{-9} to 10^{-7}	10^{-2} J/cm^2	See Table 20.17 for apertures
	10^{-7} to 10	$0.56t^{1/4}$ J/cm^2	
	>10	0.1 W/cm^2	
1.54 only	10^{-9} to 10^{-6}	10 J/cm^2	

[a] $C_A = 1$ for $\lambda = 0.400$–0.700 μm.
$C_A = 10^{2.0(-0.700)}$ for $\lambda = 0.700$–1.050 μm.
$C_A = 5$ for $\lambda = 1.050$–1.400 μm.
$C_B = 1$ for $\lambda = 0.400$–0.550 μm.
$C_B = 10^{15(0.550)}$ for $\lambda = 0.550$ μm.
$T_1 = 10 \times 1020^{20(0.550)}$ for $\lambda = 0.550$–0.700 μm.

Table 20.12. Maximum Permissible Exposure for Skin Exposure to a Laser Beam

Wavelength, λ (μm)	Exposure Duration, t (s)	MPE (J/cm^2)	Notes for Calculation and Measurement
		Ultraviolet	
0.200–0.302	10^{-9} to 3×10^4	3×10^{-3}	Or $0.56t^{1/4}$ J/cm^2, whichever is lower; 1-mm limiting aperture
0.303	10^{-9} to 3×10^4	4×10^{-3}	
0.304	10^{-9} to 3×10^4	6×10^{-3}	
0.305	10^{-9} to 3×10^4	1.0×10^{-2}	
0.306	10^{-9} to 3×10^4	1.6×10^{-2}	
0.307	10^{-9} to 3×10^4	2.5×10^{-2}	
0.308	10^{-9} to 3×10^4	4.0×10^{-2}	
0.309	10^{-9} to 3×10^4	6.3×10^{-2}	
0.310	10^{-9} to 3×10^4	1.0×10^{-1}	
0.311	10^{-9} to 3×10^4	1.6×10^{-1}	
0.312	10^{-9} to 3×10^4	2.5×10^{-1}	
0.313	10^{-9} to 3×10^4	4.0×10^{-1}	
0.314	10^{-9} to 3×10^4	6.3×10^{-1}	
0.315–0.400	10^{-9} to 10	$0.56t^{1/4}$	
0.315–0.400	10 to 10^3	1	
0.315–0.400	10^3 to 3×10^4	1×10^{-3} W/cm^2	
		Visible and Near Infrared	
0.400–1.400	10^{-9} to 10^{-7}	$2\ C_A \times 10^{-2}$	1-mm limiting aperture for C_A sec notes in Tables 20.10 and 20.11
	10^{-7} to 10	$1.1\ C_A t^{1/4}$	
	10 to 3×10^4	$0.2\ C_A$ W/cm^2	
		Far Infrared	
1.4–10^3	10^{-9} to 10^{-7}	10^{-2}	1-mm limiting aperture for 1.4–100 μm; 11-mm limiting aperture for 0.1–1 mm
	10^{-7} to 10	$0.56t^{1/4}$	
	>10	0.1 W/cm^2	
1.54 only	10^{-9} to 10^{-6}	1.0	

In the ANSI standard, exposure limits are expressed in terms of the radiant exposure (J/cm^2) or irradiance (W/cm^2) required to produce a minimal lesion, usually a retinal lesion, which is visible through an ophthalmoscope 24 h after exposure. All the ANSI criteria are based on acute exposure conditions, because very little information is available on the long-term, chronic effects of laser radiation. For this reason, caution must be exercised in applying such criteria to conditions when long-term viewing of optical sources is required. On the other hand, the use of the ANSI criteria should prove to be satisfactory for the overwhelming majority of practical situations in which accidental exposure to laser radiation occurs. Some questions have been posed regarding whether the so-called natural aversion response (the strong tendency to blink and look away from a dazzling or extremely bright light source) is truly protective under all circumstances. It is speculated that certain persons might be able to override the aversion response if sufficiently well motivated; that is, some might be able to force themselves to stare at the extremely bright laser beam. The aversion response is generally thought to occur in a period of 0.1 to 0.25 s. In the ANSI Z136 standard, a value of 0.25 s has been used in calculating the permissible exposure limits for lasers operating at visible wavelengths.

The Federal Laser Product Performance Standard establishes accessible radiation emission limits for different classes of lasers, but these are identical to the ANSI standard for all practical purposes.

The most recent revisions of the ANSI Z136.1 standard accessible emission limits for continuous wave and single pulsed lasers are given in Table 20.13 and Table 20.14.

5.4 Protection Guidelines

The objective of these guidelines is to provide reasonable and adequate guidance for the safe use of lasers and laser systems. A practical means for accomplishing this is first to classify lasers and laser systems according to their relative hazards and then to specify appropriate controls for each classification.

The basic consideration of the hazard classification scheme is the ability of the primary laser beam or reflected primary laser beam to cause biological damage to the eye or skin during intended use. A class 1 laser is one that is considered to be incapable of producing damaging radiation levels and is therefore exempt from any control measures or other forms of surveillance. Class 2 lasers (low power) are divided into two subclasses, 2 and 2a. A class 2 laser emits in the visible portion of the spectrum (0.4 to 0.7 μm), and eye protection is normally afforded by the aversion response, including the blink reflex. Class 3 lasers (medium power) are divided into two subclasses, 3a and 3b. A class 3 laser may be hazardous under direct and specular reflection viewing conditions, but the diffuse reflection is usually not a hazard. A class 3 laser is normally not a fire hazard. A class 4 laser (high power) is a hazard to the eye and skin from the direct beam and sometimes from a diffuse reflection, and it also can be a fire hazard.

Although these guidelines relate specifically to the laser product and its potential hazard, the conditions under which the laser is used, the level of safety training of individuals using the laser, and other environmental and personnel factors are important considerations in determining the full extent of safety control measures. Because such situations require informed judgments by responsible persons, major responsibility for such judgment has been assigned to a person with requisite training authority and responsibility, namely, the laser safety officer (LSO).

Lasers or laser systems certified for a specific class by a manufacturer in accordance with the Federal Laser Product Performance Standard may be considered as fulfilling all classification requirements of this standard. In cases where the laser or laser system classification is not provided or where the class level may change because of a change of engineering control measures, the laser or laser system shall be classified by the LSO.

The recommended procedure contained in the ANSI standard for evaluating the safety of lasers is as follows: (*1*) Determine the appropriate class of laser or laser system. (*2*) Comply with the measures specified for that class of laser or laser system, using the following table as a guide. This procedure will in most cases eliminate the need for mea-

Table 20.13. Accessible Emission Limits for Selected Continuous-Wave[a] Lasers and Laser Systems

Range (μm)	Emission Duration (s)	Class 1[b] (W)	Class 2 (W)	Class 3 (W)	Class 4 (W)
		Ultraviolet			
0.2–0.4	3×10^4	$\leq 0.8 \times 10^{-9}$ to $\leq 8 \times 10^{-6}$ depending on wavelength (see Tables 20.10, 20.11, and 20.12)	—	>Class 1 but ≤ 0.5 depending on wavelength (see Tables 20.10, 20.11, and 20.12)	>0.5
		Visible			
0.4–0.7	3×10^4	$\leq 0.4\ C_B \times 10^{-6}$[c]	>Class 1 but $\leq 1 \times 10^{-3}$	>Class 2 but ≤ 0.5	>0.5
		Near Infrared			
0.7–1.06	3×10^4	$\leq 0.4 \times 10^{-6}$ to $\leq 200 \times 10^{-6}$ depending on wavelength	—	>Class 1 but ≤ 0.5 depending on wavelength	>0.5
		Near Infrared			
1.06–1.4	3×10^4	$\leq 200 \times 10^{-6}$	—	>Class 1 but ≤ 0.5	>0.5
		Far Infrared			
1.4×10^2	>10	$\leq 0.8 \times 10^{-3}$	—	>Class 1 but ≤ 0.5	>0.5
		Submillimeter			
10^2 to 10^3	>10	≤ 0.1	—	>Class 1 but ≤ 0.5	>0.5

[a]Emission duration ≥ 0.25 s.
[b]When the design or intended use of the laser or laser system ensures personnel exposures of less than 10^4 in any 24 h period, the limiting exposure duration may establish a higher exempt power level.
[c]For C_B, see notes in Tables 20.10 and 20.11.

Table 20.14. Summary of Levels (Energy and Radiant Exposure Emissions) for Single-Pulsed Laser and Laser System Classification[a]

Wavelength Range (μm)	Emission Duration (s)	Class 1 (J)	Class 3 (J/cm^2)	Class 4 (J/cm^2)
		Ultraviolet		
0.2–0.4	$>10^{-2}$	$\leq 24 \times 10^{-6}$ to 7.8×10^{-3}	>Class 1 but ≤10	>10
		Visible		
0.4–0.7	10^{-9} to 0.25	$\leq 0.2 \times 10^{-6}$	>Class 1 but $\leq 31 \times 10^{-3}$	$>31 \times 10^{-3}$
		$\leq 0.25 \times 10^{-3}$	>Class 1 but ≤10	>10
		Near Infrared		
0.7–1.06	10^{-9} to 0.25	$\leq 0.2 \times 10^{-6}$ to 2×10^{-6}	>Class 1 but $\leq 31 \times 10^{-3}$	$>31 \times 10^{-3}$
		$\leq 0.25 \times 10^{-3}$ to 1.25×10^{-3}	>Class 1 but ≤10	>10
1.06–1.4	10^{-9} to 0.25	$\leq 2 \times 10^{-6}$	>Class 1 but $\leq 31 \times 10^{-3}$	>10
		$\leq 1.25 \times 10^{-3}$	>Class 1 but ≤10	>10
		Far Infrared		
$1.4–10^2$	10^{-9} to 0.25	$\leq 80 \times 10^{-6}$	>Class 1 but ≤10	>10
		$\leq 3.2 \times 10^{-3}$	>Class 1 but ≤10	>10
		Submillimeter		
$10^2–10^3$	10^{-9} to 0.25	$\leq 10 \times 10^{-3}$	>Class 1 but ≤10	>10
		≤0.4	>Class 1 but ≤10	>10

[a]There are no class 2 single-pulsed lasers known to the author at the time of writing.
[b]Wavelength dependent (see Tables 20.10, 20.11, and 20.12).

surement of laser radiation, quantitative analysis of hazard potential, or use of the MPE values.

Class	Control Measures	Medical Surveillance
1	Not applicable	Not applicable
2	Applicable	Not applicable
2a	Applicable	Not applicable
3a	Applicable	Not applicable
3b	Applicable	Applicable
4	Applicable	Applicable

Typical laser classifications for continuous wave and single pulse lasers are given in Tables 20.15 and 20.16, respectively.

5.5 Measurement of Laser Radiation

No single instrument can cover the required dynamic range, sensitivity, accuracy, precision, wavelength specificity, and stability of operation for proper measurement of the output from the wide variety of laser systems now in use. The wavelength range available in laser devices extends from the submillimeter to the VUV; power ranges from microwatts to terawatts; and operational modes encompass both the continuous wave (CW) and pulse durations from 10^{-1} to 10^{-13} s. The regulation promulgated by the Department of Health and Human Services (52) requiring the manufacturer to indicate certain operating parameters and classification of each laser device on a properly affixed label tends to reduce the need for an elaborate measurement system. If such equipment is purchased, however, it is encouraging to note that high-quality performance is readily obtainable. The output power of laser devices as specified by the manufacturer has rarely been found to differ from precision calibration data by more than a factor of 2 (3).

In choosing the appropriate detection and readout device for a specific measurement problem, one should strive to obtain (*1*) linear response over the wavelength the range of interest, (*2*) uniform response over the area of the receiving aperture, (*3*) response time that is suitable for the measured signal (e.g., time constants that are much shorter than the rise and decay times of *Q*-switched or mode-locked pulses), and (*4*) detector materials whose saturation or damage thresholds are much higher than the irradiances (W/cm^2) or radiant exposures (J/cm^2) of incident laser beams. See Table 20.17 for recommended maximum aperture diameters.

High-current vacuum photodiodes are useful for measuring the output of *Q*-switched systems and can operate with a linear response over a wide range of wavelengths (e.g., from the UV to the near IR). The response time is limited by the transit time of electrons from the photocathode to collector, but is typically 0.5 ns or less. Although the dark current is somewhat high (i.e., of the order of 5 nA at room temperature), the operation of photodiodes is very stable if they are used within the specified current limitations and if the voltage of the power supply is well maintained. Because the sensitivity of the photocathode is usually nonuniform, a large fraction of the photocathode surface area must be well illuminated; otherwise, measurement errors will be induced. The nonuniformity problem might be improved through the use of a planar photocell. Because the limiting incident power level of many vacuum photodiodes is of the order of 1 W, it may become necessary to attenuate the beam. This can be accomplished by inserting optical interference filters, neutral density filters, or diffuse reflectors into the system or by beam splitting.

Average power measurements of CW laser systems may be made with conventional thermopiles or photovoltaic cells. A typical thermopile can detect signals in the power range from 10 μW to about 100 mW. Because thermopiles are composed of many junctions, the response may be nonuniform unless the entire surface is fully illuminated by the laser beam. Bismuth–silver junctions are recommended for high sensitivity; copper–constantan junctions are better for rugged stability. Table 20.4 lists the sensitivity, impedance, and response time of typical junction detectors. Gold-black or platinum-black junctions provide the fastest response times. The use of the thermopile is not recommended for irradiance levels exceeding approximately 100 to 125 mW/cm^2. For the measurement of super high CW power (e.g., 1 to 5 kW), the use of a cone flow thermopile is recommended.

For high-energy pulsed radiation, the ballistic thermopile should be considered. In this device, the incident radiation travels down the inner surfaces of an active carbon cone, where absorption takes place. The active cone is balanced against an identical reference cone to measure the difference in temperature and to minimize drift due to ambient temperature fluctuations. These devices employ a rather long response time characteristic (e.g., 1 to 1.5 min) to integrate the pulsed signals properly.

Photovoltaic cells consist of a junction of two materials that are specially selected to produce a contact potential. The two materials can be a semiconductor and a metal or a type *n* and *p* semiconductor such as a silicon solar cell. The photovoltaic cells offer the following advantages: (*1*) the cell output is very stable with time, (*2*) external power supplies are not needed, and (*3*) the detector unit and associated circuitry is small enough to permit the construction of a portable device. Materials that are suitable as detector elements include indium arsenide and indium antimonide. In certain devices, these detector elements are installed in glass Dewar flasks to cool them with liquid nitrogen. When this is done, the detectable wavelength band is expanded to include an upper limit of approximately 6 μm. Selenium cells are good for power measurements in the near ultraviolet; silicon solar cells are preferred for near-infrared measurements, and Schottky surface barrier silicon photodiodes are preferentially used for radiation in the blue wavelengths. Many calorimeters measure total energy, but they can also be used to measure total power if the time history of the radiation is known. Other calorimeters do not measure the temperature rise in the absorber; rather, they measure small changes in polarization, which in turn produce a potential difference across a detector. This "pyroelectric effect" occurs in ferroelectrics because such materials possess a permanent polarization that is temperature dependent. The pyroelectric detectors are especially suitable for the measurement of far-infrared radiation of the carbon dioxide and hydrogen cyanide lasers; however, the input signals are usually chopped to prevent degradation of the sensing element. Table 20.18 summarizes certain characteristics of energy and power measuring devices.

For measurement throughout the infrared region, many quantum detectors are available. For example, germanium junction photodiodes are used for the near infrared up to approximately 1.8 μm; PbS and InAs photodiodes may be used in the 1.8 to 3 μm range;

Table 20.15. Typical Laser Classification—Continuous-Wave Lasers

Wavelength Range	Laser	Wavelengths	Exempt–Class 1[a] (W)	Class 2 (W)	Class 3 (W)	Class 4 (W)
		Ultraviolet				
0.1–0.28 μm	CW ND–YAG (quadrupled)	266 nm only	$\leq 0.8 \times 10^{-9}$	—	>Class 1 but ≤ 0.5	>0.5
		Ultraviolet				
0.315–0.4 μm	Helium–cadmium	325 nm only	$\leq 8 \times 10^{-6}$	—	>Class 1 but ≤ 0.5	>0.5
	Argon	351.1, 363.8 nm only				
	Krypton	350.7, 356.4 nm only				
		Visible				
	Helium–cadmium	441.6 nm only	$\leq 0.4 \times 10^{-6}$	>Class 1 but $\leq 1 \times 10^{-3}$	>Class 1 but ≤ 0.5	>0.5
	Argon (visible)	457.9, 476.5, 488, 514.5 nm, etc.				
	Helium–selenium	460.4–700 nm only, (30 lines)				
	CW Nd–YAG (doubles)	532 nm	$\leq 0.4 \times C_B \times 10^{-6}$	>Class 1 but $\leq 1 \times 10^{-3}$	>Class 1 but ≤ 0.5	>0.5
	Helium–neon	632.8 nm				
	Krypton	647.1, 530.9, 676.4 nm, etc.				

			Near Infrared			
0.7–1.4 μm	CW gallium–aluminum arsenide	0.85 μm (20°C)	$\leq 80 \times 10^{-6}$	—	>Class 1 but ≤0.5	>0.5
	CW gallium arsenide	0.905 μm (20°C)	$\leq 0.1 \times 10^{-3}$	—	>Class 1 but ≤0.5	>0.5
	CW Nd–YAG	1.064 μm	$\leq 0.2 \times 10^{-3}$	—	>Class 1 but ≤0.5	>0.5
	Helium–neon	1.08, 1.152 μm only	$\leq 0.2 \times 10^{-3}$	—	>Class 1 but ≤0.5	>0.5
			Far Infrared			
1.4–100 μm	Hydrogen fluoride Carbon monoxide Carbon dioxide Helium–neon	4–6 μm 5.0–5.5 μm 10.6 μm 3.39 μm only	$\leq 0.8 \times 10^{-3}$	—	>Class 1 but ≤0.5	>0.5
			Far Infrared			
0.1–1 mm	Water vapor Hydrogen cyanide	118 μm 337 μm	≤0.1	—	>Class 1 but ≤0.5	>0.5

[a]Assumes no mechanical or electrical design incorporated into laser system to prevent exposures from lasting to λ_{min} = 8 hr (one workday); otherwise the Class 1 AEL could be larger than tabulated.

Table 20.16. Typical Laser Classification—Single-Pulse Lasers

Wavelength Range	Laser	Wavelengths	Pulse Duration (s)	Exempt–Class 1 (J)	Class 3 (J/cm^2)	High-Power–Class 4 (J/cm^2)
Ultraviolet	Nd–YAG *Q*-sw (quadrupled)	266 nm	10 to 30 $\times 10^{-9}$ (*Q*-sw)	No criteria	≤ 10	>10
	Ruby (doubled)	347.1 nm	10 to 30 $\times 10^{-9}$ (*Q*-sw)	No criteria	≤ 10	>10
Visible 0.4–0.7 μm	Nd–YAG (doubled)	532 nm	$\sim 20 \times 10^{-9}$ (*Q*-sw)	$\leq 0.2 \times 10^{-6}$	>Class 1 but $\leq 74 \times 10^{-3}$	$>75 \times 10^{-3}$
	Ruby	694.3 nm				
	Ruby (long pulse)	694.3 nm	$\sim 1 \times 10^{-3}$	$\leq 0.2 \times 10^{-6}$	>Class 1 but ≤ 3.1	>3.1
	Rhodamine 6G (dye laser)	450–650 nm	$\sim 20 \times 10^{-9}$	$\leq 0.2 \times 10^{-6}$	>Class 1 but ≤ 0.31	>0.31
Infrared	Nd–YAG	1.064 μm	$\sim 20 \times 10^{-9}$ (*Q*a-sw)	$\leq 2 \times 10^{-6}$	>Class 1 but ≤ 0.16	>0.16
	Erbium: glass	1.54 μm	$\sim$10 to 100 $\times 10^{-9}$ (*Q*-sw)	$\leq 8 \times 10^{-3}$	>Class 1 but ≤ 10 J	>10
	Carbon dioxide	10.6 μm	1 to 100 $\times 10^{-9}$ (*Q*-sw)	$\leq 80 \times 10^{-6}$	>Class 1 but ≤ 10	>10

Table 20.17. Maximum Aperture Diameters (Limiting Aperture) for Measurement Averaging

Measurement	Exposure Duration, t (s)	Aperture Diameter (mm) vs. Wavelength: Ultraviolet (0.2–0.4 μm)	Visible and Near Infrared (0.4–1.4 μm)	Medium and Far Infrared (1.4–10^2 μm)	Submillimeter (0.1–1 mm)
Eye MPE	10^{-9} to 3×10^4	1	7	1	11
Skin MPE	10^{-9} to 3×10^4	1	1	1	11
Laser, classification[a]	10^{-9} to 3×10^4	50	50	50	50

[a]The apertures are used for the measurement of total output power or output energy for laser classification purposes, to distinguish between all classes of CW lasers or between class 1 and class 3 pulsed lasers. The use of the 50-mm apertures as shown in "Laser classification" applies only to those cases where the laser output is intended to be viewed with optical instruments (excluding ordinary eyeglass lenses) or where the LSO determines that there is some probability that the output will be accidentally viewed with optical instruments and that such radiation will be viewed for a sufficient time duration so as to constitute a hazard. Otherwise, the apertures listed for eye MPE and skin MPE are to be used.

For the specific case of optical viewing (beam collecting) instruments, the apertures listed for eye MPE and skin MPE apply to the exit beam of such devices.

Table 20.18. Devices for Energy and Power Measurements[a]

Device	Range of Operation	Typical Response Time	Surface Damage[b]
Energy			
Cone calorimeters	10^{-2} to 2×10^3 J	1–20 s	10–20 mW/cm^2 in 50 nsec
Meter disk calorimeter	10^{-2} to 10 J	10 s	50 mW/cm^2 in 50 ns
Rat's nest calorimeter	10^{-3} to 10 J	10^{-4} s	—
Wire calorimeter	10^{-2} to 0.5 J	10 s	10 mW/cm^2 in 50 ns
Liquid calorimeter	1–500 J	10–60 s	[c]
Torsion pendulum	0.5–500 J	60 s	—
Integrating photocurrent	10^{-6} to 10^{-3} J	1 s	1 mW/cm^2
Thermopile	10^{-6} to 1 J	10^{-1} s	300 mW/cm^2
Copper sphere	5×10^{-4} to 10 J	180 s	—
Power			
Phototube	10^{-8} to 10^{-3} W/cm^2	3–10 ns	1 mW/cm^2
Photodiode	10^{-4} to 6 W/cm^2	0.3–4 ns	10 W/cm^2 [d]
Nonlinear crystal	10^3 to 10^{12} W/cm^2	10^{-5} s	10^{12} W/cm^2 [e]

[a]From S. S. Charschan, ed., *Lasers in Industry*, Van Nostrand Reinhold, New York, 1972, p. 521.
[b]Surface damage may occur at the indicated power density.
[c]Local boiling of the liquid should be avoided.
[d]Above this power, response of device is nonlinear.
[e]Breakdown of quartz occurs at this power density.

PbTe, PbSe, and InSb are suitable for the 3 to 7 μm range, whereas germanium combined with a wide variety of metals (mercury, cobalt, zinc, copper, and gold) is preferred for the 7 to 40 μm range. Many of these detectors need liquid nitrogen cooling for proper performance.

Thermal detectors such as bolometers are still used where broad-band spectral response is desired. A bolometer using a platinum ribbon sensor element gives a flat response over a wavelength range of 1 to 26 μm. The change in electrical resistance due to heating by the absorbed radiation is proportional to the absorbed radiant power in these devices. Because the time constants are usually of the order to 10^{-4} s, however, they are not recommended for the measurement of *Q*-switched pulses.

The type of readout device for the various detector systems varies with individual requirements. Microammeters and microvoltmeters are often used with CW systems; microvoltmeters or electrometers are frequently employed in conjunction with oscilloscopes to measure pulsed laser system parameters. All these devices may be coupled to recorders or panel displays. For purposes of calibration, tungsten ribbon filament lamps are available from the National Institute for Standards and Technology as secondary standards of spectral radiance over the wavelength region of 0.2 to 2.6 μm. The calibration procedures for these devices permit comparisons within about 1% in the near UV and about a 0.5% in the visible. All radiometric standards are based on the Stefan–Boltzmann and Planck laws of black-body radiation. The spectral response of measurement devices should always be specified, because the ultimate use of the measurements is a correlation with the spectral response (action level) of the biological tissue receiving the radiation result.

The effective aperture (aperture stop) of any device that is used for measuring the irradiance (W/cm^2) or the radiant exposure (J/cm^2) in the visible wavelengths should closely approximate 0.7 cm (7 mm), the diameter of the fully dilated or dark-adapted pupil. Apertures other than 7 mm may be used for radiation at other wavelengths (see Table 20.17).

The ANSI reasoning behind the recommended use of a 1 mm aperture for IR and UV wavelengths is that a practical minimum aperture diameter should be employed to measure the radiant exposure or irradiance received by a relatively small area of tissue; yet the diameter should be large enough to impinge upon acceptable sensing areas of transducing elements currently available in measuring instruments. Furthermore, there is no need to specify such a relatively large aperture size (7 mm), a value that is based on the magnification of visible light levels (not UV or far infrared) on their passage from the cornea to the retina. The absorption of UV and far infrared is, after all, a surface absorption phenomena. Finally, whereas a larger area of irradiation may be necessary for proper operation of thermal detectors (e.g., calorimeters), an impingement area of 1 mm (aperture) and even smaller is well within the acceptable operating requirements of state-of-the-art quantum detectors.

5.6 Special Precautions

Probably the most significant hazard associated with lasers is electrical shock rather than exposure to optical radiation. Therefore, all pertinent provisions of the National Electrical Code should be followed, including first aid training for all persons working with lasers.

Particular attention must be paid to adequate grounding, proper discharge of energized circuits, well-designed interlocks and warning signals, and periodic educational programs.

6 MICROWAVE RADIATION

6.1 Physical Characteristics of Microwave Radiation

Microwave wavelengths vary from about 10 m to 1 mm; the respective frequencies range from approximately 3 MHz to 300 GHz. Certain reference documents, however, define the microwave frequency range as beginning as low as 10 MHz and as high as 300 GHz. The region between about 30 KHz and the infrared is generally referred to as the radio-frequency (RF) region. Reference may be made to Figure 20.1 to determine the position in the electromagnetic spectrum occupied by microwaves relative to other electromagnetic radiations. Certain bands of microwave frequencies have been arbitrarily assigned letter designations by industry (Table 20.19). However, there is no universal agreement on these letter designations. Other discrete frequencies have been assigned by the Federal Communications Commission (FCC) for industrial, scientific, and medical applications (labeled ISM bands), as shown in Table 20.20.

Table 20.19. Letter Designation of Microwave Frequency Bands

Band	Frequency (MHz)
L	1,100–1,700
LS	1,700–2,600
S	2,600–3,950
C	3,950–5,850
XN	5,850–8,200
X	8,200–12,400
Ku	12,400–18,000
K	18,000–26,500
Ka	26,500–40,000

Table 20.20. Industrial, Scientific, and Medical (ISM) Frequencies Assigned by the FCC

13.56 MHz ± 6.78 kHz
27.12 MHz ± 160 kHz
40.68 MHz ± 20 kHz
915 MHz ± 25 MHz
2450 MHz ± 50 MHz
5800 MHz ± 75 MHz
22,125 MHz ± 125 MHz

6.2 Sources of Microwave Radiation

Although microwave radiation has important applications in communications and navigational technology such as satellite communications systems, acquisition and tracking radar, air and traffic control radar, weather radar, and UHF TV transmitters, there are many commercial applications in other fields—microwave ovens, diathermy equipment, and industrial drying equipment. Some typical primary sources of microwave energy are klystrons, magnetrons, backward wave oscillators, and semiconductor transit time devices such as IMPATT diodes. Such sources may operate in a continuous mode (CW), as in the case of some communications systems; in an intermittent mode, as with microwave ovens, induction heating equipment, and diathermy equipment; or in a pulsed mode, as is the case with radar and digital communication systems. Various frequency designations and uses of microwave energy are given in Table 20.21. Of particular interest are the different designations used by the Eastern European countries, the USSR, and the Western countries.

Natural sources of RF and microwave energy can produce peak electric field strengths exceeding 100 V/m at ground level when cold weather fronts move through a given geographical area. Solar radiation intensities range from 10^{-18} to 10^{-17} W/(m^2)(Hz), but the integrated intensities at the earth's surface for the frequency range 0.2 to 10 GHz are approximately 10^{-8} mW/cm^2. This value is to be compared with an average value of 10^2 mW/cm^2 on the earth's surface attributable to the entire (UV, visible, IR, and microwave) solar spectrum.

Given the formation of our solar system, the natural background of nonionizing electromagnetic radiation in the terrestrial biosphere arises primarily from our sun and from reradiation from the earth. The maximum solar spectral irradiance (E_λ) is found at a wavelength of 0.5084 μm; the maximum for terrestrial black-body reradiation is found at 10.06 μm. For all practical purposes, the RF/microwave levels of any significance at the earth's surface arise from man-made sources. Also noteworthy is the fact that cosmic radiation and natural terrestrial radioactivity can create nonionizing electric fields of the order of 100 to 200 V/m in the earth's atmosphere, and the presence of lightning may produce fields as high as 10^3 V/m in the ELF region of 3 to 300 Hz and in the VLF region of 1 to 10 kHz (53).

6.3 Biological Effects of Microwave Radiation

When the human body is exposed to microwave radiation, the usual process of absorption, reflection, transmission, and scattering take place, depending on the wavelength and wave front characteristics. Microwave wavelengths less than 3 cm are absorbed mostly in the outer skin surface. Wavelengths between 3 and 10 cm penetrate more deeply (1 mm to 1 cm), and at wavelengths of 10 to 20 cm, penetration and absorption are sufficiently great that the potential for causing damage to internal body organs must be considered. The human body is thought to be essentially transparent to wavelengths greater than about 500 cm. At about 300 MHz (1 m wavelength), the depth of penetration changes rapidly with frequency, declining to millimeter depths at frequencies of about 3000 MHz. For purposes of comparison, the skin penetration at 3000 MHz is approximately 16 mm, whereas that for 10,000 MHz it is approximately 4 mm. One of the difficulties in establishing the

Table 20.21. Radio-Frequency and Microwave Band Designations

United States	USSR	Wavelengths	Frequencies	Typical Uses
		Radio-Frequency Bands		
Low frequency (LF)	Long (VCh)	10^4–10^3 m	30–300 kHz	Radio navigation, radio beacon
Medium frequency (MF)	Medium (HF)	10^3–10^2 m	0.3–3 MHz	Marine radiotelephone, Loran, AM broadcast
High frequency (HF)	Short (UHF)	10^2–10 m	3–30 MHz	Amateur radio, worldwide broadcasting, medical diathermy, radio astronomy, citizen bands
		Microwave Bands		
Very high frequency (VHF)	Ultra-Short (meter)	10–1 m	30–300 MHz	FM broadcast, television, air traffic control, radio navigation
Ultra high frequency (UHF)	Decimeter	1–0.1 m	0.3–3 GHz	Television, microwave point-to-point, microwave ovens, telemetry, tropo scatter and meteorological radar, mobile, telephone
Super high frequency (SHF)	Centimeter (SHF)	10–1 cm	3–30 GHz	Satellite communication, airborne weather radar, altimeters, shipborne navigational radar, microwave point-to-point radio
Extra high frequency (EHF)	Millimeter	1–0.1 cm	30–300 GHz	Radio astronomy, cloud detection radar, space research, hydrogen cyanide emission, millimeter wave communication

relationships between the characteristics of microwave fields and the bioeffects produced by those fields is our limited ability to measure and describe the incident fields as well as the special patterns of absorbed power within the tissue. Furthermore, when attempting to make environmental measurements one often encounters distortion of the field pattern when a measurement probe is inserted into that field. See Section 6.5, "Electromagnetic Fields and Absorption in Tissue," and Section 6.7, "Measurement of Microwave Radiation," for a more detailed description of measurement techniques and special problems in quantifying the deposition of microwave energy in tissue.

A review of the evidence assembled by Western investigators concerning the biological effects of microwave radiation leads to the overall conclusion that most of the reported effects are explainable on the basis of heat deposition in tissue. The photon energy of electromagnetic radiation at microwave frequencies is considered to be too low to cause the removal of electrons from molecules regardless of the number of quanta absorbed; hence ionization effects are excluded. One of the continuing sources of controversy is whether "nonthermal" effects can be induced in tissue as a result of microwave irradiation. The term "nonthermal" effect usually designates a biological change produced in the absence of any detectable rise in temperature in the test system. The phenomenon of "pearl chain" formation, an alignment of particles with the lines of force of an electromagnetic field, has been offered by some investigators as an example of a nonthermal effect. However, Sher and Schwan (54) concluded that "the implications for pearl chain formation are that on no account can biological pearl chains occur for particles smaller than 3 μm in diameter without risking overheating of tissues. Particles smaller than about 30 μm would not form pearl chains; freely moveable particles of this size are not available in the body."

The current consensus is that pearl chain formation does not occur in the human body as a result of microwave irradiation. Frey (55) reported "hearing" microwave pulses at an average power density as low as 100 $\mu W/cm^2$. The nature of the perception was described as a buzz or a ticking, depending on the pulse length and the pulse repetition rate. The subjects reported the sensation of sound within or behind the head. Selective shielding revealed that the most sensitive area was the region directly over the temporal lobe of the brain and that the greatest sensitivity was to the frequency range from 300 to 1200 MHz.

The current explanation for such microwave audition is that the microwave energy does not exert direct action on the auditory nerve or brain centers; instead it is due to induction of thermoclastic waves in the head. It is generally conceded that certain RF pulses can produce thermal gradients that in turn generate elastic shock waves at the boundaries of tissue or bone with dissimilar dielectric properties, and that the shock waves are transmitted to the cochlea and ossicular chain of the middle ear, where they are perceived as clicks. To date there is no evidence that the auditory clicks constitute a risk of injury.

Research performed in this country has generally concluded that microwave energy is capable of producing cataracts in exposed persons. This conclusion is based almost entirely on animal experimentation. One experienced ophthalmologist has openly expressed his doubts about the possibility of producing cataracts in humans as a result of microwave exposure unless a massive, extremely high level of exposure occurs. Despite this, most knowledgeable persons concede the point that microwave radiation can cause cataracts under special exposure conditions.

One of the notable early animal cataract studies was that of Carpenter and Van Ummersen (56), who investigated the effects of microwave radiation on the production of cataracts in rabbit eyes. Exposures to 2.45 GHz radiation were made at power densities ranging from 80 to 400 mW/cm^2 for different exposure durations. They found that repeated doses of 67 J/cm^2 spaced a day, a week, or 2 weeks apart produced lens opacities even though the single threshold cataractogenic dose (at a power density 280 mW/cm^2) was 84 J/cm^2. When the single dose was reduced to 50 J/cm^2, opacities were produced when the doses were administered 1 or 4 days apart, but when the interval between exposures was increased to 7 days, no opacification was noted even after five such weekly exposures. At a power density of 80 mW/cm^2 (dose of 29 J/cm^2), the lowest exposure level used in the study, no effect developed; but when this dose was administered daily for 10 to 15 days, cataracts were produced. These experiments indicate a possible cumulative effect on the lens of the eye if exposures are repeated at a rate such that the lens does not have sufficient time to recover between exposures, even at levels below the single-dose cataractogenic threshold (56).

The published allegation (57) that microwave radiation caused the death of a man has been challenged by medical authorities who claimed that the case in question was one of acute appendicitis that led to profound shock and death on the tenth postoperative day. The case was discounted as being due to microwave exposure in an Armed Forces Institute of Pathology Memorandum of July 25, 1957 (58).

A number of epidemiologic studies have been conducted on human populations that had at any time been exposed to microwave radiation. Many of these studies were designed to elicit information on the incidence of cataracts and general physiological effects. The studies of Zaret et al. (59, 60) have shown a small but statistically significant increase in the number of changes at the posterior polar surface of the lens; however, cause-and-effect relationships in each individual case have not been subsequently elucidated. Barron and Baraff (61) compared the environmental and clinical data on 335 microwave workers to those of a controlled, nonexposed population. No significant differences were found in mortality, disease, sick leave, subjective complaints, or the results of clinical analyses. Czerski et al. (62) made a similar study of some 800 Polish workers who were exposed to microwaves during the course of their work. The incidence of functional disturbances could not be correlated with exposure level (0.2 to 6 mW/cm^2) or duration of exposure. The study by Robinette and Silverman (63) of 20,000 former Navy personnel each in the exposed and control groups did not yield significant correlations between exposure to radio frequencies and long-term mortality or between exposure and various clinical manifestations that required hospitalization. The epidemiology study by Lilienfeld and co-workers of personnel who resided in the U.S. embassy in Moscow during periods when the embassy was irradiated with low levels of radio frequencies was unable to find any statistically significant differences associated with radio frequencies in total mortality or mortality from specific causes or in other illnesses (including cancer) between personnel in the Moscow embassy and control groups comprised of personnel in other Eastern European U.S. embassies (64).

The investigations in the Soviet Union and Eastern European countries have largely stressed nonthermal microwave effects at the central nervous system and cellular levels: cataract formation in rabbits has been reported after a 60-min exposure to 10 mW/cm^2

(65), a result that has not been produced in this country under the conditions described by the Soviets. It is the consensus of Soviet research that exposure to a microwave power density of the order of 10 mW/cm^2 for long periods of time constitutes a pathogenic factor (morphologic lesions in the nervous system, changes in the reproductive function, and other borderline conditions). In response to American suggestions that some of the reported results may reflect adaptive or protective reactions to microwave radiation, the Soviets claim that their results are unequivocal and cannot be regarded as harmless regulatory, adaptive, or compensatory reactions. Effects reported by the Soviets for exposure levels below 10 mW/cm^2 include lowered endurance, retarded weight gain, inhibition of conditioned reflexes, and neurosecretion and neurophysiological disturbances. Even at intensities as low as 250 to 500 μW/cm^2, biological effects such as "change in the activity of the brain" and "immunobiologic resistance" are reported to occur. An interesting insight into Soviet research philosophy is contained in the statement (66) of a Soviet investigator:

> When speaking about criteria of significance of reactions, once you keep in mind that not all reactions are of importance but some, while lacking pathologic traits under certain conditions, should nevertheless be given attention when possible harmful consequences are considered; it is therefore necessary to introduce the term "potentially harmful reaction" as a main criterion of importance of a symptom, as opposed to either a threshold biologic (regulatory) reaction in general or a threshold pathologic reaction.

Assuming this philosophy is used for the development of standards of permissible exposure, it is easy to understand why the present Soviet permissible exposure levels are so much more stringent than those in Western countries.

Soviet *in vitro* studies of 10-cm irradiation of catalase and cholinesterase have been entirely negative; however, *in vivo* investigations have shown changes in the activity of a whole range of enzymes, including cholinesterase, at exposure levels significantly lower than 10 mW/cm^2 (i.e., down to a level of 1 mW/cm^2). According to the investigators, the difference between the *in vivo* and *in vitro* experiments "results not from a direct action on molecular structures but from changes in enzyme concentrations in the tissue which are apparently related to disturbances in the neurohormonal regulation of metabolic processes" (67, 68).

Another series of Soviet studies (69) on the effects of microwave radiation on cell membranes, particularly on the selective permeability of the membranes to potassium and sodium ions when subjected to a power density of 1 mW/cm^2, has revealed statistically significant changes in the rates of transport of potassium and sodium ions across all membranes. The transport rates were apparently unrelated to a rise in the temperature of the suspension or to the radiation frequency.

The effect of 2.45 GHz continuous wave radiation on rabbit erythrocytes and rat lymphocytes has been studied in two separate cell systems. Rabbit erythrocytes were exposed at a power density of 5 mW/cm^2 for 3 h, thus duplicating the experimental conditions of a study by the Polish investigator Baranski, who reported an increase of about 1520 times in potassium efflux and a decrease in osmotic resistance. No changes were found in potassium efflux or osmotic resistance if the exposed and controlled cells were maintained at the same temperature. Although it was later learned that the cells in the Polish experiment had been washed and suspended in unbuffered saline, whereas buffered saline was used in the experiments at the NIEHS, the repeated experiments conducted at NIEHS using unbuffered saline again showed no changes in potassium efflux or osmotic resistance as long as the exposed and controlled cells were maintained at the same temperature (70). The importance of this study is highlighted by the Soviet hypothesis that the influence of microwaves on excitable cells such as erythrocytes is due to alterations in the permeability of the cell membranes and that these alterations in permeability are brought about by the transport of ions across the membranes, which in turn influences the membrane potential.

One of the experimental tools used by Soviet investigators to describe biological effects is the conditioned reflex. This fact accounts in part for the difference in threshold effects obtained by Western and Eastern European investigators. The Soviet investigator Livshits (65) comments on the dilemma:

> In conditioned reflex investigations it has been discovered that irradiation induces various disturbances of the higher nervous activity. It must be mentioned that such an effect on radiation is observed only in experiments conducted according to the schemes used by I. P. Pavlov and his followers. Among the authors that have investigated the effects of radiation upon conditioned reflexes in different animals, there is no complete unanimity in the evaluation of the phenomena that they observed and the understanding of their mechanism. The opinion of certain foreign researchers that the change in the latent period and values of the conditioned reflex expresses only or primarily a change in the excitability of the unconditioned reflex centers, in general contradicts the considerable material obtained in the laboratories of I. P. Pavlov and his followers. In radiation pathology, in particular, this question is complex and will require a special investigation (70).

In commenting on Soviet research, Dodge has stated:

> An often disappointing facet of the Soviet and East European literature on the subject of clinical manifestations of microwave exposure is the lack of pertinent data presented on the circumstances of irradiation, frequency, effective area or irradiation, orientation of the body with respect to the source, wave form (continuous or pulsed; modulation factors), exposure schedule and duration, natural shielding factors; and a whole plethora of important environmental factors (heat, humidity, light, etc.) are often omitted from clinical and hygienic reports. In addition, the physiological and psychological status of human subjects such as health, previous or concomitant medication, and mental status is also more often than not omitted. These variables, both individually and combined, affect the human response to microwave radiation (2, 71).

Whereas on the one hand Soviet researchers claim that low level ($\leqslant$10 mW/cm^2) nonthermal radiation produces behavioral effects in humans, McAfee indicates that it is thermal stimulation of the peripheral nervous system that produces the neurophysiological and behavioral changes (72, 73) and that the interaction between the peripheral nervous system and the central nervous system accounts for the heart rhythm and blood chemistry effects reported by the Soviets.

The debate over whether microwave radiation can produce cumulative effects continues unabated. As already indicated, the possibility of cumulative effects on the lens from repeated subthreshold cataractogenic exposures of rabbit eyes has been suggested by Carpenter and Van Ummersen (56). The reported cumulative effect may well be the accumulation of damage resulting from repeated exposures, each one capable of producing some degree of damage. If the repetitive exposures take place at time intervals shorter than those needed for the repair of damage, tissue damage is produced (74). Michaelson (2) claims that because this is the process that actually takes place, the suggestion of cumulative effects from microwave exposure is of questionable validity.

There is no strong support for the proposition that the blood–brain barrier (BBB) is adversely affected by brief exposures to the RF fields tested to date. The question about possible effects on the BBB with long-term exposure has not been addressed in experimental work.

With respect to possible RF interactions with the nervous system, there is evidence that some tissue interactions that produce minuscule increases in tissue temperature, that is, less than 0.1°C, can produce physiological changes that are not directly attributable to increased temperature. In such cases, RF fields may couple to tissue through amplification of incident energy at certain frequencies and power levels, that is, "frequency and power windows." Weak oscillating electric gradients in tissue have reportedly been effective in modifying cell functions by mechanisms that appear to amplify the triggering field. Such "cooperative" processes are already recognized in immunological and endocrine reactions, and they appear to play a role in neurobiological excitation, especially at the surface of the cell membrane. It is unlikely that weak extracellular fields could couple across the plasma membrane with its extremely high electric gradient; therefore some amplification of fields would almost certainly have to occur along the length and area of the membrane, and the amplification process itself may be calcium-ion dependent in a way that has yet to be elucidated (75).

6.4 Skin Sensation

Questions are often raised about whether a person can "feel" microwaves or the heat that is produced by the incident radiation. Although it is difficult to quantify spurious reports of persons who have detected increases in skin temperature, some quantitative information is available. Schwan et al. (76) found that if a person's forehead is exposed to a power density of 74 mW/cm^2 at 3000 MHz, the reaction time for experiencing a sensation of warmth varies between 15 and 73 s. The perception of warmth for a power density of 56 mW/cm^2 varies from 50 s to 3 min of exposure. Hendler's data (77, 78) show that when an area of 37 cm^2 of the forehead is exposed to microwaves, thermal sensation can be elicited within 1 s at power densities of 21 mW/cm^2 when using 10 GHz (10,000 MHz) radiation and 58.6 mW/cm^2 when using 3 GHz radiation. These data further indicate that if the entire face were exposed to 10 GHz radiation, the threshold for thermal sensation would be 4 to 6 mW/cm^2 within 5 s, or approximately 10 mW/cm^2 within 0.5 s (2). The sensation of warmth or pain from microwave heating apparently differs little from that felt from infrared heating. In the experiments of Cook (79), an average skin temperature rise of 15°C is required to achieve pain when the initial skin temperature is approximately 32.5°C.

6.5 Electromagnetic Fields and Absorption in Tissue

Before the NCRP Report 67 was published in 1981 (80), there was very little standardization of quantities and units. Most investigators were tolerating the use of terms such as "dose," "dose rate," and "absorbed power density" in the very low frequency ranges. The System Internationale (SI) was being ignored. The proposed use of the term "absorbed power density" produced sufficient revulsion in the rank and file that everyone agreed that the quantities and units had to be improved. The use of such loose terminology reflected poorly on the quality of work being performed, and it caused difficulty when trying to compare the results of different studies. Report 67 was a key forerunner in the NCRP entry into the nonionizing radiation arena in that it laid down meaningful, consistent definitions after an exhaustive review and resolution of the many disparate definitions found in electrical engineering dictionaries. Report 67 covered such fundamentals as field quantities, electrostatics, magnetostatics, electrodynamics, analogies with circuit quantities, wave propagation, polarization, modulation, reflection/refraction, standing waves, spherical waves, transmission line fields, antennas, energy transfer processes, and field and SAR measurements. The appendixes of this report contain excellent information on the natural background of nonionizing radiation in the biosphere, RF and microwave absorption in biopolymers, and molecular dynamics in the presence of electromagnetic field perturbation, to name just a few. A major contribution of the report was the introduction of the term specific absorption rate (SAR) into the RF/microwave radiation vocabulary.

SAR was defined as the rate at which electromagnetic energy is absorbed at a point in a medium, per unit mass of the medium, expressed in W/kg. Because energy absorption is a continuous and differentiable function of space and time, one may speak of its gradient and rate. Therefore, the time derivative of the incremental energy (*dW*) absorbed in an incremental mass (*dm*) contained in a volume element (*dV*) of a given density (*p*) is expressed as

$$\text{SAR} = \frac{d}{dt}\left(\frac{dW}{dm}\right) = \frac{d}{dt}\left(\frac{dW}{\rho dV}\right)$$

The term SAR should not be confused with a measure of the rate of heating of a tissue, although measuring the rate of temperature rise is one of many methods for measuring SAR. SAR is strictly a dosimetric quantity that provides no information as to interaction mechanism, that is, thermal or athermal.

It is most important to remember that the amount of microwave energy deposited and absorbed in biological tissue depends on frequency and on body size, configuration, and homogeneity. Furthermore, it is not possible to classify any of the field quantities measured exterior to an exposed subject as being hazardous, nonhazardous, thermal, or nonthermal. One must first know something about the exposure conditions, frequency, subject size, and subject geometry to permit determination of the SAR, formerly called the absorbed power density, in the subject. For subjects who are exposed to microwave fields, the equation for the time rate of change of temperature *T* (°C per second) per unit volume of subcutaneous tissue is

$$\frac{dT}{dt} = \frac{0.239 \times 10^{-3}}{c(W_a + W_m - W_c - W_b)}$$

where W_a is the specific absorption rate, W_m is the metabolic heating rate, W_c is the heat loss due to thermal conduction, W_b is the power dissipated by blood flow, all expressed in watts per kilogram, and c is the thermal conductivity, expressed in kilocalories per kilogram per degree celsius (81).

The specific absorption rate for tissue exposed to an electromagnetic field is

$$W_a = \frac{10^{-3}\,\sigma}{\rho E^2}$$

where σ is the electrical conductivity in mhos per meter, ρ is the tissue density in grams per cubic centimeter, and E is the root mean square value of the electric field in the tissue, in volts per meter. Before the tissue is exposed to electromagnetic fields, it is assumed that a steady-state condition exists where $W_a = dT/dt = 0$, requiring $W_m = W_c + W_b$. Under normal conditions, the metabolic rate W_m averages 1.3 W/kg of the whole body, 11 W/kg for brain tissue, and 33 W/kg for heart tissue (85).

According to the energy equation, one would expect to see some change in tissue temperature due to applied electromagnetic fields if the specific absorption rate W_a were of the same order of magnitude as W_m or more. The former nonwavelength-dependent safety standard of 10 mW/cm^2 of incident power was at least partially based on limiting the average value of W_a to the average resting value of W_m. Therefore, specific absorption rates that are much greater than W_m would be expected to produce marked thermal effects, whereas specific absorption rates that are much less than W_m would not be expected to produce any significant thermal effects (85).

Peak values of specific absorption rate (50 W/kg $\leq W_a \leq$ 170 W/kg) have been used to provide vigorous local therapeutic heating of deep vasculated tissues in a man treated with diathermy (82). Also, values of W_a that are equal to or much greater than 4 W/kg in the brain tissue of cats yielded measurable increases in brain temperature and decreases in the latency times of evoked potentials (83). Body temperature increases, accompanied by behavioral changes, have been observed in rats by Justesen and King (84) for values of W_a equal or greater than 3.1 W/kg and by Hunt (85) and Phillips et al. (86) for W_a values equal to or greater than 6.3 W/kg averaged over the body of the animal (81). Cataracts have been induced in rabbits by acute exposures where the specific absorption rate in the eye was greater than 138 W/kg (87).

Figure 20.6 shows the variation of specific absorption rate with frequency in a spherical muscle model with the same mass as a 70-kg man. In the frequency range from 1 to 20 MHz, the absorption varies as the square of the frequency and is due primarily to the magnetically induced electric fields. The maximum specific absorption rate induced by the incident E field is denoted by the curve marked with crosses, and that due to the incident H field is denoted by the curve with circles. In this range, the maximum specific absorption rate is only 10^{-5} to 10^{-2} W/(kg)(mW)(cm^2) of incident power. In the frequency range of 100 to 1000 MHz, internal reflections are significant for the man-sized sphere, and the

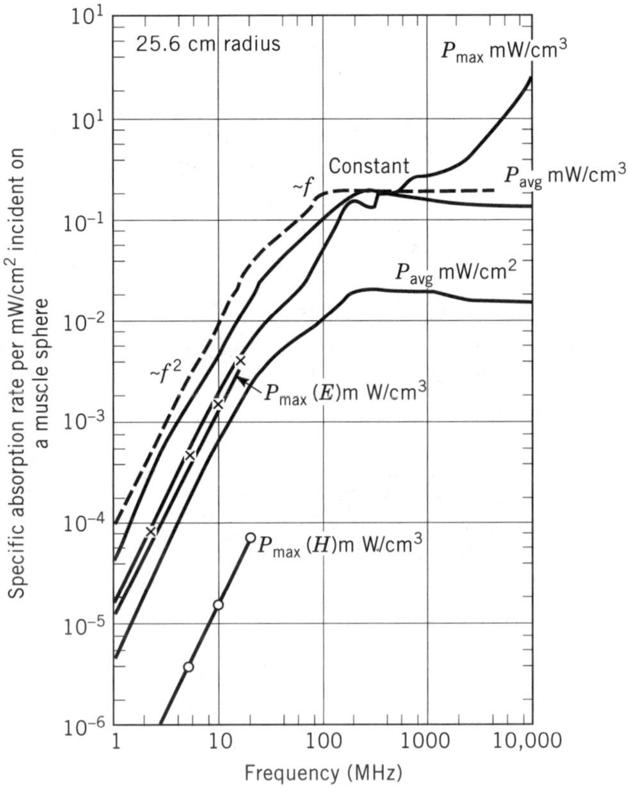

Figure 20.6. Specific absorption rate patterns versus frequency in spherical muscle model of 70-kg man exposed to plane wave 1 mW/cm^2 source. After Ref. 81.

average absorption attains a maximum of 2×10^{-2} W/(kg)(mW)(cm^2) of incident power at 200 MHz, which remains relatively constant with frequency up to 10 GHz. The maximum specific absorption rate increases with frequency above 1000 MHz, approaching that produced by nonpenetrating radiation. The dashed lines illustrate roughly the frequency dependence of the total or average absorbed power and indicate how safety standards might be relaxed as a function of frequency if the absorption characteristics in humans were the same as that for the sphere (81).

Figure 20.7 illustrates absorption patterns as a function of body size (radius of muscle sphere) for a sphere consisting of a central muscle core surrounded by concentric layers of subcutaneous fat and skin exposed to 2.45 GHz plane wave radiation at 1 mW/cm^2. Assuming that the spherical models have some relevance to the human body, Guy (85) makes the point that the peak specific absorption rate could be as high as 4.2 W/kg within the body or head of a small bird or animal but as low as 0.27 W/kg at the surface and 0.05 W/kg at a point 2.5 cm deep in the human body when all are exposed to a 1 mW/cm^2, 918 MHz source. Therefore, exposure to 10 mW/cm^2 could be of extreme signifi-

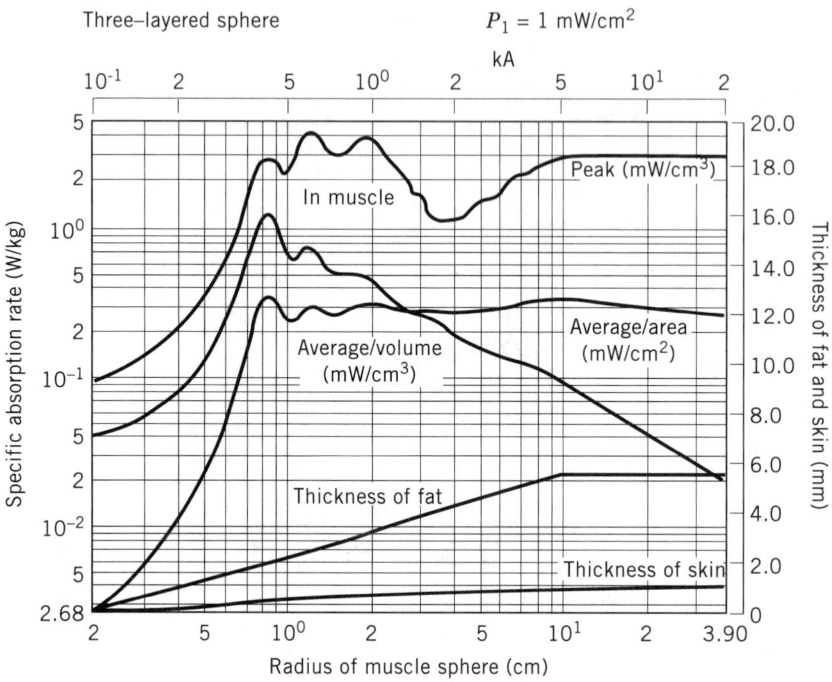

Figure 20.7. Specific absorption rate versus outside radius in spherical tissue layer model of animal exposed to 2.45 GHz 1 mW/cm^2 plane wave.

cance, and 0.5 mW/cm^2 could be of mild thermal significance to the smaller animals in comparison with their metabolic rate. For the human model, on the other hand, 10 mW/cm^2 would appear to be minimally significant in terms of thermal insult; a power density of 0.5 mW/cm^2 would have negligible thermal significance.

6.6 Exposure Criteria

Schwan and Li (88) in 1953 examined the thresholds for cataractogenesis and thermal damage to tissue. The minimum power density necessary for producing such changes was found to be approximately 100 mW/cm^2, to which they applied a safety factor of 10 to obtain a maximum permissible exposure level of 10 mW/cm^2. This number was then incorporated into many official standards. The ANSI C95.1 standard (89) subsequently required a limiting power density of 10 mW/cm^2 for exposure periods of 0.1 h or more and a limiting energy density of 1 mWhr/cm^2 during any 0.1-h period. The latter criterion permitted intermittent exposure at levels above 10 mW/cm^2 on the basis that such intermittency would not produce a temperature rise greater than 1°C in human tissue.

Later, Schwan (90) suggested that the permissible exposure levels be expressed in terms of current density rather than power density, especially when dealing with measurements in the near or reactive field, where the concept of power density loses its meaning. He proposed a permissible current density of approximately 3 mA/cm^2, because this value was considered to be comparable to a far-field value of 10 mW/cm^2. At frequencies below 100 KHz, it was believed that this value could be somewhat lower, and for frequencies above 1 GHz, somewhat higher.

The recommendations of the ANSI C95.1 committee were qualified as follows:

> Body temperature depends in part on sources of heat input such as electromagnetic radiation, physical labor, high ambient temperature and on heat dissipation capability as affected by clothing, humidity, etc. People who suffer from circulatory difficulties and certain other ailments are vulnerable. The power levels established by radiation guide numbers are related in a complicated way to power levels at which damage may occur. The guide numbers are appropriate for moderate environments. Under conditions of moderate to severe heat stress, the guide number given should be appropriately reduced. Under conditions of intense cold, higher guide numbers may also be appropriate after careful consideration is given to the individual situation. These formulated recommendations pertain to both whole body irradiation and partial body irradiation. Partial body irradiation must be included since it has been shown that some parts of the body, for example, the eyes, and testes, may be harmed if exposed to incident radiation levels significantly in excess of the recommended levels.

Mumford (91) proposed a means for adjusting the 10 mW/cm^2 exposure criterion on the basis of heat stress. According to his recommendation, the exposure limit would be 10 mW/cm^2 when the temperature–humidity index (THI) is equal to or less than 70. When the THI is between 70 and 79, the limit is (80 − THI) mW/cm^2. Thus, when the THI is equal to or greater than 79, the exposure limit is 1 mW/cm^2.

The original Soviet criterion (92) for exposure to microwave radiation stated that the intensity of the radiation within the frequency range of 300 MHz to 300 GHz should not exceed a value of 10 $\mu W/cm^2$ during the working day. For exposures not longer than 2 h per working day, the level should not exceed 100 $\mu W/cm^2$, and for exposures not longer than 15 to 20 min per working day, a power density of 1 mW/cm^2, provided the exposure level for the remainder of the working day did not exceed 10 $\mu W/cm^2$. Also, protective goggles had to be worn at the higher (1 mW/cm^2) level. From the public health standpoint, the Soviet standard stated: "In the microwave range the intensity of radiation should not exceed 1 microwatt per square centimeter (1 $\mu W/cm^2$) for places where human occupancy occurs and for persons not occupationally exposed." According to Gordon et al. (66), one of the main reasons for the relatively low permissible exposure levels was that clinical observations of microwave workers exposed for 6 to 8 years to power densities of the order of tenths of a milliwatt per square centimeter had, in their opinion, demonstrated functional disturbances, such as neural, vegetative, and asthenic syndromes.

The 1984 USSR occupational standard lists the following maximum permissible exposure limits for electric field strengths. For a frequency range of 60 kHz to 3 MHz, a value of 50 V/m is acceptable. For the range of 3 to 30 MHz, a value of 20 V/m is used. For a frequency range of 50 to 300 MHz, a value of 5 V/m is permitted. For magnetic field strengths in the frequency range of 60 kHz to 1.5 MHz, a value of 5 A/m is considered to be acceptable (Table 20.22). Values for public exposure are approximately half those

Table 20.22. USSR 1984 RF Occupational Standard

Frequency (MHz)	*E*-Field EL (V/m)	*H*-Field EL (A/m)
0.06–1.5	50	5
1.5–3	50	—
3–30	20	—
30–50	10	0.3
50–300	5	—
300–300,000	—	—

for occupational exposure. For additional reading on the development of the USSR standards, refer to References 93 and 94.

The original Bell System radiation protection guide (95) called for a three-step exposure criterion. First, power density levels in excess of 10 mW/cm^2 were considered to be potentially hazardous; hence personnel were not permitted to enter areas where major parts of the body might be exposed to such levels. Second, power density levels between 1 and 10 mW/cm^2 were permitted only for incidental, occasional, or causal exposure. And third, power density levels below 1 mW/cm^2 were considered to be acceptable for indefinite or prolonged exposure. The Bell System standard has since been changed to conform fairly closely to the present ANSI criterion.

In 1982, the Subcommittee IV of the ANSI C95 Committee adopted a frequency-dependent standard for both occupational and general-public exposure to RF radiation to replace the ANSI Radiation Protection Guide published in 1974, which was 10 mW/cm^2 or the equivalent electric and magnetic field strengths over the entire frequency range from 10 MHz to 100 GHz. The newer limits, like the older ones, are not to be exceeded for exposures averaged over any 0.1-h period. For the first time in the development of radiofrequency protection guides (RFPGs), due cognizance was taken of such factors as body size, mass, and orientation; the polarization of the incident wave; the frequency and intensity of the radiation; the presence of reflective surfaces; and whether the irradiated subject was in conductive contact with a ground plane.

Because of sparse quantitative human data on which to model an exposure criterion, Subcommittee IV of the ANSI C95 Committee conducted a review of exposure data to determine the most significant, reliable, and independently replicated biological/physiological end points produced at the lowest specific absorption rate. Thresholds of behavioral impairment (behavioral disruption) were found within a narrow range of whole-body averaged SARs, ranging from 4 to 8 W/kg (96). The corresponding range of power densities was 8 to 140 mW/cm^2. Because the thresholds of behavioral disruption in primates were found to be approximately 4 W/kg, this value was chosen as the working threshold for untoward effects in humans in the frequency range of 3 MHz to 100 GHz. (At frequencies below 3 MHz, body surface interactions and electric shock are the relevant criteria, not SAR.) The safety margin is a factor of 10 applied to a SAR value of 4 W/kg; hence compliance with the values in the tabular representation of the ANSI C95.1-1982 RFPG shown in Table 20.23 will ensure that no person will be exposed to a specific absorption rate in excess of 0.4 W/kg, averaged over any 6-min interval in the frequency range of 3

Table 20.23. ANSI (1982) Radio-Frequency Radiation Protection Guides

Frequency (f) Range (MHz)	E-Squared (V^2/m^2)	H-Squared (A^2/m^2)	Power Density (mW/cm^2)
0.3–3	400,000	2.5	100
3–30	$4000 \times (900/f^2)$	$0.025 \times (900/f^2)$	$900/f^2$
30–300	4,000	0.025	1.0
300–1500	$4000 \times (f/300)$	$0.025 \times (f/300)$	$f/300$
1500–100,000	20,000	0.125	5.0

MHz to 100 GHz. Although the frequency range of primary interest with respect to SAR is approximately 3 to 300 MHz, the so-called body resonance range, it is believed that limiting the whole body averaged SAR to 0.4 W/kg will automatically provide an additional safety factor for exposure to frequencies above this range.

The 1982 ANSI standard contains the following qualifications:

(1) At frequencies between 300 kHz and 100 GHz, the protection guides may be exceeded if the exposure conditions can be shown to produce SARs below 0.4 W/kg as averaged over the whole body, with spatial peak values not exceeding 8 W/kg as averaged over any one gram of tissue.

(2) At frequencies between 300 kHz and 1 GHz, the protection guides may be exceeded if the radiofrequency input power to the radiating device is 7 W or less. [This exclusion was provided in recognition of the fact that many low-power devices in common use by the general population may produce fields that appear to exceed the exposure guides in local regions of the body close to the devices but which would yield whole-body SARs much lower than those in the exposure guides.]

The American Conference of Governmental Industrial Hygienists published threshold limit values (ACGIH, 1984) also based on 0.4 W/kg but intended for occupational exposures only. These TLVs, to be averaged over any 6-min period, are shown in Table 20.24 (97).

One major difference between the 1982 ANSI and 1984 ACGIH standards is that the 1-mW/cm^2 value on the latter standard extends only from 30 to 100 MHz and rises with

Table 20.24. ACGIH (1984) Radio-Frequency/Microwave Threshold Limit Values[a]

Frequency (f) Range (MHz)	Power Density (mW/cm^2)	E-Squared (V^2/m^2)	H-Squared (A^2/m^2)
0.01–3	100	377,000	2.65
3–30	$900/f^2$	$3770 \times (900/f^2)$	$900/(37.7 \times f^2)$
30–100	1	3770	0.027
100–1000	$f/100$	$3770 \times f/100$	$f/(37.7 \times 100)$
1000–300,000	10	37,700	0.265

[a]See latest edition of TLVs for current values.

a slope $f/100$ at 100 MHz to 10 mW/cm^2 at 1 GHz. This difference is based on the premise that children, who have higher whole-body resonant frequencies than adults, are not likely to be occupationally exposed to radio-frequencies. Also different in the 1984 ACGIH standard is the lower frequency limit of 10 kHz instead of 300 kHz.

The standard proposed by Scientific Committee 53 of the NCRP (98) is shown in Figure 20.8. The main difference between the 1982 ANSI criteria (96) and those proposed by NCRP Scientific Committee 53 is the inclusion of a separate, more stringent general population exposure criterion in the NCRP document. ANSI proposed that its criteria apply to persons occupationally exposed in the workplace as well as persons in the general population on the basis that RF exposures produce threshold effects and that a proper exposure envelope had been included to protect all persons. NCRP countered with the rationale that occupationally exposed persons are usually well informed about potential hazards and are free to determine which risks are acceptable to them, but persons in the general population are generally unaware of RF exposure risks, have little control over exposure levels, and represent a much larger population comprised of people who are possibly more vulnerable to exposure, for example, the aged, pregnant females, infants, children, and the chronically ill. Scientific Committee 53 (98) has recommended that the exposure criterion for the general public be set at a level equal to one-fifth that for occupational groups. That is, the

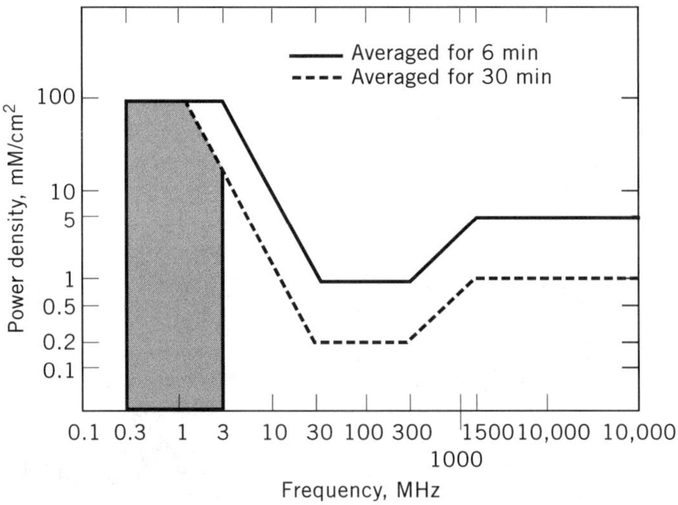

Figure 20.8. Criteria for exposure to radio-frequency electromagnetic fields. Exposure, expressed in equivalent far-field power density (mW/cm^2) for a whole-body averaged SAR of 0.4 W/kg, is shown in the solid line, taken to be the occupational exposure criterion. The dashed line, one-fifth that of the occupational criterion, is the criterion for the general population. Note the time-averaging period allowed for each criterion. The cross-hatched area represents a frequency range in which whole body SAR has limited significance. The overall frequency range for the criteria is 0.3 MHz to 100 GHz. Depending on the circumstances, use of these criteria is constrained by a number of conditions, and the criteria cannot be applied without reference to these conditions. From Ref. 98.

whole-body averaged SAR for the general population should not exceed 0.08 W/kg, and an exposure average averaging period should be 30 min. This reduction is based on the relative exposure periods of the two groups: 168 h versus 40 h/week, that is, 40/168, which equals approximately 0.2.

A few observations about these latest exposure criteria are in order:

1. The limits of exposure below 30 MHz apply to free space conditions where a person is not in contact with any object, including the ground. For other conditions, such as standing on the ground with insulated shoes and making contact with a grounded object or being grounded and touching an insulated metal object, the limits must be lowered, and a case-by-case evaluation must be made. The standard does not tell you exactly how to do this, other than to determine the applicable exposure limits through the use of three criteria: whole body SAR (0.4 W/kg); maximal local SAR (8 W/kg); and RF burns at the point of contact (200 mA). A great deal of caution is in order in applying the exposure criterion to situations where persons are exposed to high fields at frequencies below 3 MHz. Also, at higher frequencies the local or spatial peak SAR can be incredibly high while still conforming to the whole-body averaged value of 0.4 W/kg.

2. Although the whole-body average SAR may not exceed a value of 0.4 W/kg, data obtained with simulated human models indicate that the localized SAR values ("hot spots") may reach 20 times this value. Because this is also assumed to be the case for the animal species in the studies used to develop the RFPG, it was concluded that 8 W/kg in any gram of tissue is acceptable as long as the whole-body SAR does not exceed 0.4 W/kg averaged over 0.1 h (the occupational limit). More biological underpinning of the data is needed before one can be comfortable about permitting such high localized SARs. This information is needed in order to make a more quantitative assessment of the following apparently conflicting statements contained in NCRP Report No 86 (98). The first statement is: "In those cases where there are highly intense local concentrations of RF energy ('hot spots'), this knowledge should supersede whole body values and lead to a corresponding reduction in permissible exposure level." (In other words, this is a special case and permissible exposure level should be reduced.) The second statement is: "The deposition of energy in all or certain parts of the human body is a specific condition that lends itself uniquely to an evaluation through simulation modeling. Results of such procedures indicate that efficient thermal averaging over all body tissues of locally deposited RF energy will probably occur through convective heat transfer via blood flow." (In other words, this is not a special case, and average values of energy deposition may be assumed.)

3. If a given carrier frequency is modulated at a depth of 50% or greater at frequencies between 3 and 100 Hz, the NCRP recommendation is to limit the exposure criteria for occupational exposure to those that apply to exposure of the general public. Such a practice may prove to be acceptable in the long run, but one cannot know with certainty whether this action is justified or whether the general population standard will guarantee the absence of biological effects when exposures occur to modulation frequencies between 3 and 100 Hz.

4. One can still conform to a whole-body averaged SAR of 0.4 W/kg (averaged over 6 or 30 min) while exposing subjects to extremely high peak powers. There is a need to develop a limit on peak power. This is a very important research project.

5. The use of mobile and hand-held transceivers by members of the general public allows exposures that are in excess of the general population standard, but not above the occupational guidelines, as long as the user does not expose others to levels exceeding the general population standard. It is difficult to understand how the average citizen will be able to determine whether he or she is exceeding the general population standard.

6. With the curtailment of research funds by the federal government, opportunities for further investigation of mechanisms of interaction between nonionizing electromagnetic energy and biological systems will undoubtedly be limited. There is a practical need to progress from our present macroscopic understanding of the basis on which exposure limits have been set to a better understanding of the effects of fields internal to the irradiated organism. There is a need to understand these matters on a microscopic scale.

7. The ANSI-C95.1-1982 Standard and the NCRP Report No. 86 represent a significant improvement over what we had in earlier times. But given the many qualifying remarks contained in the documents, the many conditions that require professional assistance in their interpretation and application, and conditions that cannot be evaluated for the lack of data, we need improved understanding through research.

A brief chronology of events associated with the development of RF exposure limits in the United States is given in Table 20.25.

6.7 Measurement of Microwave Radiation

One of the key factors underlying the controversy over the reported biological effects of RF radiation is the lack of standardization of (or in many cases even a means of comparing) the measurement techniques used to quantify the exposure of experimental animals or humans to electromagnetic fields. For these reasons, it is heartening to see that the NCRP plans to release a very useful report entitled: "A Practical Guide to the Determination of Human Exposures to Radiofrequency Fields."

An idealized conception of microwave propagation consists of a plane wave moving in an unbounded isotropic medium where the electric $|E|$ and magnetic field $|H|$ vectors are mutually perpendicular and both are perpendicular to the direction of wave propagation. Unfortunately, the simple proportionality between the E and H fields in free space is valid only in the so-called far field (Fraunhofer region) of the radiating device. The far-field region is sufficiently removed from the source so that the angular field distribution is essentially independent of the distance to the source. The power density in the far field is inversely proportional to the square of the distance from the source and directly proportional to the product of $|E|$ and $|H|$. Therefore, the measurement of either the $|E|$ and $|H|$. Therefore, the measurement of either the $|E|$ and $|H|$ vector in the far field is all that is needed to establish the exposure conditions. Plane wave detection in the far field is well understood and easily obtained with equipment that has been calibrated for use in the frequency range of interest. Most hazard survey instruments have been calibrated in the far field to read in units of milliwatts per square centimeter. A typical device for far-field measurements is a suitable antenna coupled to a power meter.

To estimate the power density levels in the near field (Fresnel region) of large-aperture circular antennas, one has traditionally used the following simplified relationships:

Table 20.25. Chronology of Selected Events Associated with RF Exposure Limits in the United States

Prior to 1950	Fragmented Reports on Electrical Excitation of Tissue.
1952	Dr. Fred Hirsch reported eye damage in the *AMA Archives of Industrial Health* (99). Estimated "power density" to which humans were exposed: 100 mW/cm^2.
1953	Naval Medical Research Meeting; concluded that 100 mW/cm^2 was detrimental. Schwan estimated 10 mW/cm^2 as possible exposure criterion (100). Bell Laboratories estimated safe level at 0.1 mW/cm^2 (101). (Private speculation that Russians may have miscopied this value as 0.01 mW/cm^2.)
1955	General Electric established a level of 1 mW/cm^2 (102). Although many power density values were proposed, there was little scientific basis for them because of the paucity of definitive biological data.
	Mayo Clinic attendees reviewed data on RF-induced cataracts in animals (103). Schwan and Li proposed 10 mW/cm^2 at all RF frequencies (100). Basis: metabolism of food generates energy at a rate of 5 mW/cm^2; doubling this should not cause harm. No reported detrimental effects at levels of approximately 10 mW/cm^2 (104). The need for additional biological substantiation was recommended by Schwan.
1956	Microwave "hearing" reported (RADAR) (105).
1957–1958	The 10 mW/cm^2 value adopted by Bell Laboratories, GE, U.S. Army, U.S. Navy, U.S. Air Force (106). Consensus was thermal basis although nonthermal possibility discussed.
1957–1961	Tri-Service Conferences; Col. G. M. Knauf, U.S.A.F., reported international concurrence with 10 mW/cm^2 value (107)
1961	Soviet standard 0.01 mW/cm^2 maximum in 1 day (based on reported low-level CNS effects). Soviet regulations required eye protection at 1 mW/cm^2 (108). Translations of Soviet work often inaccurate. Also, few technical details available on experimental design, measurements, and statistical treatment
1962	Documentation of low-intensity microwave irradiation to the U.S. embassy in Moscow (109). Security issues.
1966	ANSI C95 Radiofrequency Standard adopted (frequency independent from 10 MHz to 100 GHz) (110). Johns Hopkins Applied Physics Laboratory attempted modeling of U.S. exposure criteria
1968	Senate Commerce Committee Meetings (111)
1968	Radiation Control for Health and Safety Act (PL 90-602) Technical Electronic Products Radiation Safety Standards Committee (TEPRSSC) established (112)
1969	Bureau of Radiological Health (BRH) standards adopted under Public Law 90–602 (113): TV, microwave ovens, electron tubes, diagnostic X-ray equipment. Symposium on Biological Effects and Health Implications of Microwave Radiation, Richmond, Virginia. Renewed interest in research. First Symposium since the Tri-Service meetings (114).
1970	Hirsch reevaluated 1952 report (115). Claimed exposure level was much greater than 100 mW/cm^2.
1971	Electromagnetic Radiation Management Advisory Council (ERMAC) established in the President's Office of Telecommunications Policy (OTC). Five-year program, multi-agency, $63 million proposed budget. One objective was the determination of the long-term effects of low-level microwave radiation.

Table 20.25. (continued)

1973	Senate Committee on Commerce. Hearings on Radiation Control for Health and Safety Act of 1968 (Public Law 90–602) (116). Consumer Union recommends against purchase of microwave ovens, claiming unknown biological effects (117). TEPRSSC recommends the 1 and 5 mW/cm^2 microwave oven emission limits to BRH. American Home Appliances Manufacturers (AHAM) make presentations in support of the manufacture of microwave ovens (117).
1974	National Academy of Science Committee on the Biosphere Effects of Extremely Low Frequency Radiation (ELF) (118). SEAFARER, formerly SANGUINE project.
1974	ANSI C95 1966 Standard reconfirmed with minor revisions (119).
1975	U.S. Air Force proposes 50 mW/cm^2 (10 kHz–10 MHz) and 10 mW/cm^2 (10 MHz–300 GHz) (120). Early U.S. evidence of a frequency-dependent exposure criterion.
1976	Irradiation of U.S. embassy continues in Moscow (121). Lilienfeld et al. epidemiologic study (122).
1981	NCRP Report No. 67—"Radiofrequency Electromagnetic Fields: Properties, Quantities and Units, Biophysical Interaction and Measurements." The report develops and proposes the quantity specific absorption rate (SAR) (80).
1981	WHO classifies East European Standards: Group I (USSR); 10 μW/cm^2 not to exceed 1 mW/cm^2; Group II (GDR, Poland, Czechoslovakia): general population standard 10–100 μW/cm^2 (123).
1982	New ANSI C95 Standard. Frequency dependent and based on a whole body averaged SAR of 0.4 W/kg (96).
1986	NCRP Report No. 86, "Biological Effects and Exposure Criteria for Radiofrequency Electromagnetic Fields" (98).

$$W = \frac{16P}{\pi D^2} = \frac{4P}{A} \text{ (near field)}$$

where P is the average power output, D is the diameter of the antenna, A is the effective area of the antenna, and W is power density. If this computation reveals a power density that is less than a specific exposure limit, no further calculation is necessary because the equation gives the maximum power density on the beam axis. If the computed value exceeds the exposure criterion, one then usually assumes that the calculated power density exists throughout the near field. The far-field power densities are then computed from the Friis free space transmission formula:

$$W = \frac{GP}{4\pi r^2} = \frac{AP}{\lambda^2 r^2} \text{ (far field)}$$

where λ is the wavelength, r is the distance from the antenna, G is the far-field antenna gain, and W, P, and A are as in the near-field equation.

The distance r from the circular antenna to the intersection of the near and far fields is approximately:

$$r_1 = \frac{\pi D^2}{8\lambda} = \frac{A}{2\lambda}$$

These simplified equations do not account for reflections from ground structures or surfaces where the power density could be as much as four times greater than the assumed free space value. If conditions are such that focusing effects can be produced, factors even greater than four can be achieved.

A number of uncertainties and errors accompany all measurements, but the uncertainties are particularly present when the measurements are made in the near field. The region closest to the radiating source is called the reactive near field where energy is stored. Between the reactive near field and the far field is the radiating near field (Fresnel region), where the phase and amplitude relationships between the E and H fields vary with distance from the source. The configuration of the source can also affect the pattern of radiation to a considerable extent. Multipath $|E|$ and $|H|$ vectors that exist in the reactive and radiative near fields may become so complex that the concept of power density is almost meaningless. For this reason there is considerable merit in measuring $|E|^2$ or $|E|$ or $|H|^2$ or $|H|$. Because $|E|^2$ and $|H|^2$ in plane waves are simply related to W, there is no problem comparable to that in the near field. Other potential measurement errors may be brought about by the characteristics of the monitoring probe.

To make measurements in the near or reactive field, the measurement device should be able to measure $|E|^2$ and $|H|^2$ fields separately. The probe sensor itself must be sufficiently small compared with wavelengths in the field, to minimize perturbation of those fields. And, the probe should be essentially nondirectional and capable of responding to all polarizations. The polarization characteristics of the device are particularly important in the near field because vertical linear, horizontal linear, elliptical, and circular polarizations may be present at any point in space regardless of source configuration. Circular polarization exists when two equal-amplitude, linearly polarized waves (e.g., one in the x-direction and the other in the y-direction) are superimposed but 90° out of phase. An elliptically polarized wave consists of two unequal-amplitude, orthogonal, linearly polarized waves that are 90° out of phase. The polarization of a wave largely determines its transmission, reflection, and scattering characteristics when impinging upon materials, including human tissue.

Many probes measure electric field strength in volts per meter by means of a thermocouple whose dc output is proportional to volts per square meter. This output may be calibrated and translated into a power density reading, usually milliwatts per square centimeter, taking into account the space impedance Z_0 that relates the $|E|$ to the $|H|$ component of the propagated wave. This proportionality constant or impedance for the far-field region has a value of 377Ω. If the space impedance is not 377Ω (the value at which the survey meter has been calibrated) at the point of measurement, a calibration error results.

In general, all devices that are designed to measure microwave radiation for the purpose of assessing hazards to personnel should possess certain minimum characteristics: the instrument impedance should be matched to the field to prevent backscatter from the probe to the source, and the probe should behave like an isotropic receiver, that is, be sensitive to all polarizations, possess a response time that is adequate for handling peak to average power, and exhibit a flat response over the frequency bands of interest.

Most microwave measuring devices are based on bolometry, calorimetry, voltage and resistance changes in detectors, rectification, and radiation pressure on a reflecting surface. The latter three methods are self-explanatory. Bolometry measurements are based on the absorption of power in a temperature-sensitive resistive element, the change in resistance being proportional to the absorbed power. For direct measurement of peak power, a bolometer is the element of choice because the typical time constant is of the order of 300 μs. Bolometers have a positive temperature coefficient (i.e., the resistance increases with temperature); thermistors have a negative temperature coefficient. Thermistors are more rugged than bolometers and have a greater power-handling capacity. They are preferred for the measurement of average power of pulsed signals because of their long time constant (0.1 to 1 s).

Crystal diode detectors are useful for the measurement of microwave power, but care must be taken to avoid burnout of the diode in the vicinity of high powered equipment.

Through the use of three orthogonal dipoles, it is possible to design a probe that is both independent of polarization and isotropic in its response. The probe consists of resistive thin-film elements that are folded back in parallel on themselves with the output terminals at one end (Figure 20.9*a*).

In the example shown in Figure 20.9, the probe elements may consist of thin films of overlapping antimony and bismuth deposited on a plastic substrate. Thermal energy is dissipated primarily at the narrow, high-resistance strips, thereby producing an increase in temperature at these points (hot junctions). The low-resistance, wider sections of the thermocouple act as cold junctions. The resultant dc voltage is directly proportional to the energy dissipated in the narrow, high-resistance strips of the thermocouple. Variation in the sensitivity of the device due to changes in ambient temperature is reportedly less than 0.05% because of close spacing of the cold junctions (124). The placement of the three mutually orthogonal sensor elements within a polystyrene sphere to obtain a 5-cm spacing between the sensor elements and the outer surface of the sphere is illustrated in Figure 20.9*b*. A representative frequency response of the probe appears in Figure 20.9*c*. Note that the midfrequency bandwidth of interest (1 dB bandwidth) is 850 MHz to 15 GHz, thereby incorporating a flat response feature at the two microwave oven frequencies of 915 and 2450 MHz (2.45 GHz). This type of broad-band radiation monitor may be used for the measurement of radiation emitted from a wide variety of sources, but the original design was fashioned to provide a useful probe for measuring the leakage radiation from microwave ovens. The 5-cm electrically transparent Styrofoam spacing device was included to ensure compliance with the microwave standard under Public Law 90-602 (Radiation Control for Health and Safety Act). Under the provision of the regulations adopted under this Act, newly manufactured microwave ovens must not emit levels in excess of 1 mW/cm^2; the permissible level throughout the useful life of the product is 5 mW/cm^2.

Certain broad-band devices in the frequency range below 300 MHz have been designed to measure the $|H|$ field. The development of such instruments has been encouraged by evidence (84) that the $|H|$ fields of low-frequency radiation produce more significant biological effects than do $|E|$ fields. The sensing probes of the $|H|$ field monitors usually contain a series of mutually perpendicular loops. Low-frequency radiation below 300 MHz is usually measured with loop or short whip antennas, tuned voltmeters, or field intensity meters.

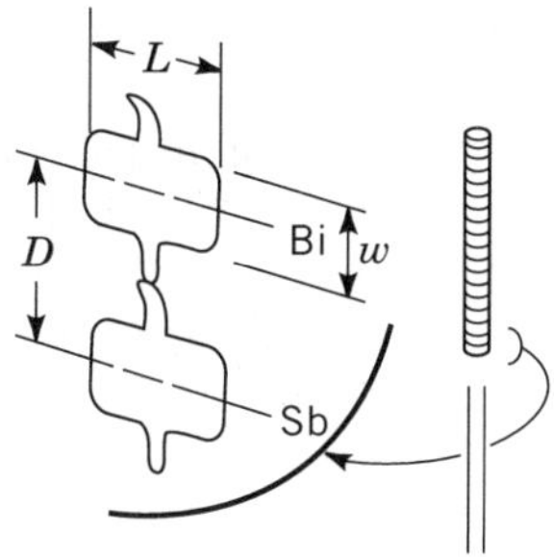

(a)

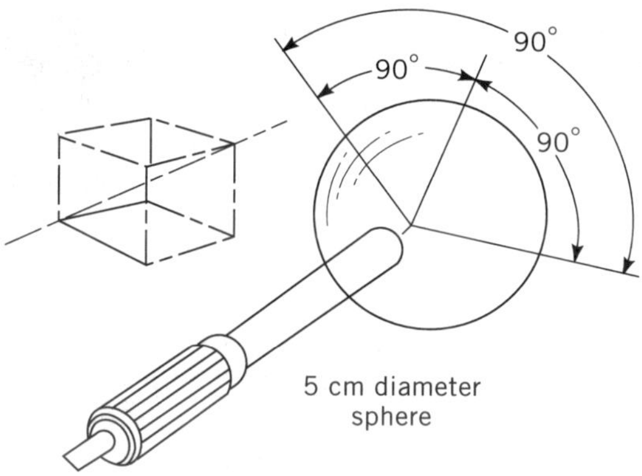

(b)

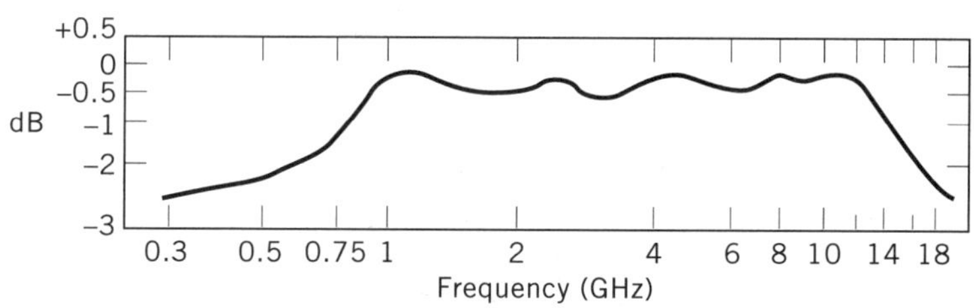

(c)

Figure 20.9. (**a**) Antenna probe element with distributed thermocouple film elements; tapered film leads reduce interaction between leads and elements. (**b**) Relationship of the three mutually orthogonal probe elements; probe elements are contained within a 5-cm sphere. (**c**) Typical frequency response characteristics. Reproduced with permission of the Narda Microwave Corporation.

When measurements must be made of electric and magnetic fields, particularly by a person who is not skilled in making such measurements, the following format is suggested: (*1*) become thoroughly familiar with the data that served as the technical basis from which the RF Protection Guideline was derived. Pay particular attention to the limitations in the data on biological effects and the field exposure measurement data, as well as the assumptions that have been made (because of the lack of data) in constructing the guideline; (*2*) become familiar with the RF equipment and exposure time patterns at the site to be assessed; and (*3*) become familiar with available measurement instrumentation, then choose those that most closely match requirements. If one understands the strengths and limitations of all three parameters, then one is in a much better position to make a realistic determination of risk.

Some examples of the usefulness of understanding the three aforementioned parameters include the following: (*1*) knowing something about how the Protective Guideline was derived will increase one's awareness of estimates, assumptions, lack of information, and degrees of precision in the data used to develop an exposure criterion; (*2*) determining whether measurements must be made in the near field, or whether a simple comparison with a Protective Guideline will suffice, or whether very precise measurements must be made in order to calculate the SAR for specific individuals, can make a big difference in approach; and (*3*) being aware of the fact that although many instruments have readout displays in units one wishes to measure and record, the device may actually respond to another parameter. For example, readouts may indicate electric and magnetic field strength but the instrument actually responds to the square of the electric and magnetic fields, or a device calibrated for the measurement of AM broadcast E fields may actually respond to the magnetic field. This is not to say that the readout is necessarily fallacious; it just means that some engineering assumption about field conditions has been made in the design of the instrument, an assumption that may not hold true for all conditions. Another fact is that many instruments exhibit a significant out-of-band response, that is, a resonance above or below the frequency range being measured. The out-of-band resonance may cause serious errors in measurements.

6.8 Control Measures

Engineering control measures should be given consideration over any use of personal protective equipment. Engineering control measures may range from the restriction of azimuth and elevation settings on radar antennas (to reduce ground level power density levels), to the complete enclosures found in microwave ovens. Because leaking radiation adversely affects the proper performance of a system, high system performance and safety are mutually reinforcing goals. Personnel who work with microwave energy should have the benefit of orientation and training sessions on the potential hazards of microwaves and the proper means for controlling undue exposures. Safe work practices should be formally prepared and instituted. Emphasis must be placed on the safe design of equipment to include shielding, enclosures, interlocks to prevent accidental energizing of circuits, dummy load terminations, warning signals, and control of electric shock potentials. Ionizing radiation (X-rays) may be produced by certain microwave generators because of the existence of accelerating voltages in excess of 16 kV.

Certain protective suits have been constructed of nylon mesh on which has been deposited a layer of metallic silver. The mesh permits a fair degree of ventilation for the body. Because the silver deposits bestow good electrical conductivity properties to the mesh, special care must be taken to avoid direct contact with exposed electrical terminals. Before using the protection garment, one should have a thorough understanding of its attenuation characteristics as a function of radiation frequency, to be certain that the external field intensity does not exceed the safe level by an amount greater than the attenuation provided by the garment. In this connection, Figure 20.10 showing the transmission loss through a wire grid may prove useful (125). Similarly, the attenuation provided by various types of material as given in Table 20.26 may be of use in designing shields or enclosures (126).

More than a decade ago, it was demonstrated that the function of certain cardiac pacemakers, particularly those of the demand type, could be seriously compromised by pulsed or intermittent microwave radiation. The most effective method of reducing the susceptibility of these devices to microwave interference was improved shielding, and the manu

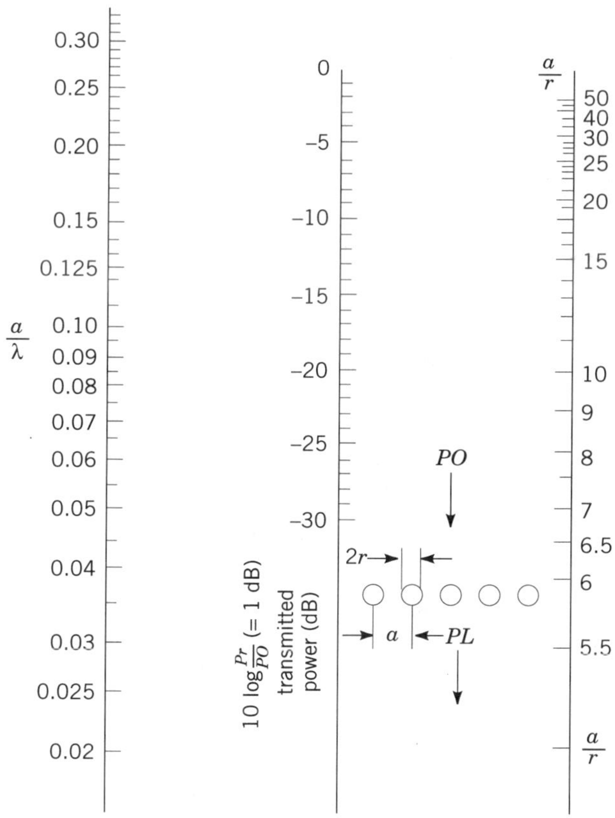

Figure 20.10. Transmission through a grid of wires of radius r and spacing a.

Table 20.26. Attenuation Factors (dB) (Shielding)

Material	Attenuation Factor (dB)			
	f = 1–3 GHz	3–5 GHz	5–7 GHz	7–10 GHz
60 × 60 mesh screening	20	25	22	20
32 × 32 mesh screening	18	22	22	18
16 × 16 window screen	18	20	20	22
0.25-in. mesh (hardware cloth)	18	15	12	10
Window glass	2	2	3	3.5
0.75-in. pine sheathing	2	2	2	3.5
8-in. concrete block	20	22	26	30

facturers of cardiac pacemakers have since instituted major programs to minimize such interference.

The judicious use of appropriate signs and labels may prove useful in alerting people to the presence of microwave sources. Figure 20.11 illustrates the RF and microwave warning signs adopted by the American National Standards Institute C95 Committee (89).

7 EXTREMELY LOW FREQUENCY (ELF) RADIATION

Until recently, the potential biological hazard associated with radio frequencies and extremely low frequencies were generally regarded as not as great as that posed by microwaves in the 800 to 3000 MHz frequency range. With the increasing concern and uncertainty about ELF magnetic fields, particularly at 60 Hz, major attention is now devoted to these portions of the electromagnetic spectrum, rather than the electric fields produced by the radio frequencies.

7.1 Biological Effects

Most of the information on the biological effects of ELF is derived from studies of the effects of radiation produced by low-power frequencies (50 to 60 Hz) and from the U.S. Navy's Project Sanguine (also called Seafarer) (127). Novitskiy et al. (128) have reported a wide range of physiological effects resulting from exposure to ELF, including changes in heartbeat and respiration, anaphylactic shock, and even death. Field strengths that produced the effects varied from approximately 50 to 5000 V/cm, with most of the effects being reported between 200 and 2000 V/cm. On the other hand, Knickerbocker et al. (129) failed to show any effects on general health, behavior, or reproductive ability of mice exposed to a 60-Hz field of 4 kV/in. (1.5 kV/cm) for 1500 h during the course of a 10.5-month study period. Soviet studies of occupations where employees have been exposed to ELF fields often report complaints such as listlessness, excitability, drowsiness, fatigue, headache, and the so-called neurasthenic syndrome. Ulrich and Ferin (130) have reviewed these studies and have concluded that the irregular shifts such as night work, dry air, and

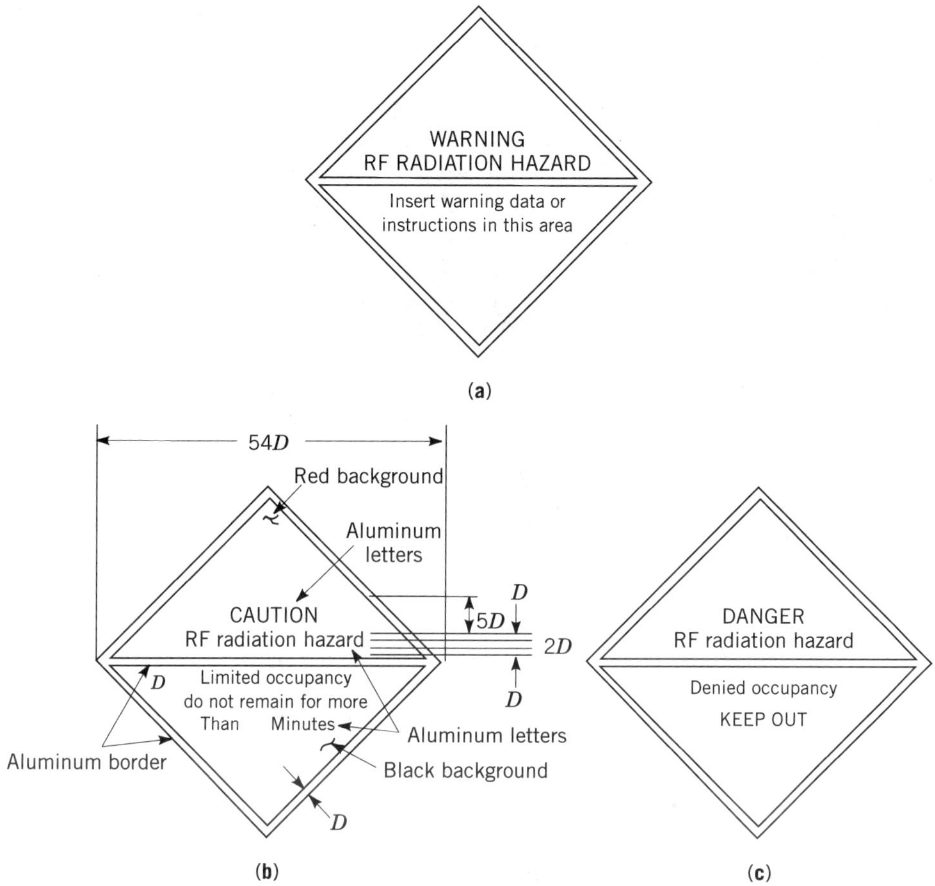

Figure 20.11. Radio-frequency signs. (**a**) General warning. (**b**) Limited occupancy. (**c**) Denied occupancy. ANSI Instructions as follows: (*1*) Place handling and routing instructions on reverse side. (*2*) Use *D*-scaling unit. (*3*) Dimensions in *b* indicate lettering ratio of letter height to thickness of letter lines. (*4*) Symbol is square, triangle, or right-angle isosceles. Derived from ANSI C95.2 Standard, American National Standards Institute, New York.

other factors were the probably causes of the signs and symptoms. Kouwenhoven et al. (131) have studied linemen in the United States who have worked on high-voltage transmission lines for many years. Observations over a 3-year period of physical, mental, and emotional health parameters failed to reveal any detrimental effects.

Project Sanguine/Seafarer (127) was begun in the 1960s to investigate the design and operation of a large antenna system that would radiate in the ELF range as a means of communicating with nuclear submarines at depths of approximately 200 m at any location in the world. Because the original proposal envisioned an antenna whose area was in the tens of thousands of square miles, the residents of the areas in which the antenna was likely to be installed became concerned about environmental and ecological effects. Later designs specified an overall antenna area of approximately 3000 square miles. In 1969, the U.S. Navy initiated what is generally regarded as the most comprehensive research program ever attempted to determine whether biological or ecological effects could be expected from exposure to ELF radiation in the immediate vicinity of the antenna and at various distances from the antenna. The electric and magnetic field levels associated with the Seafarer system operating at a frequency of 75 Hz were approximately 0.07 V/m and 0.2 G, respectively, at the surface of the earth directly above a buried antenna.

Several studies have been conducted to determine whether exposure of fruit flies to ELF electric and magnetic fields would induce mutations in the X chromosome. A comprehensive fruit fly study performed by Mittler (132) at field strengths of 10 V/m and 1.0 G at 45, 60, and 75 Hz yielded the conclusion that there was no induction of sex-linked recessive lethal genes, no translocations between II and III chromosomes, no loss of the X or Y chromosomes, and no nondisjunctions or dominant lethal genes. Exposure of *Escherichia coli* cells to field strengths of up to 20 V/m and 2.0 G at 45 and 75 Hz has failed to produce any mutagenic effect in these bacteria (133). Onion roots exposed to 20 V/m and 2 G at 45 and 75 Hz did not yield any evidence of chromosomal damage (134).

Exposure of rats through the second generation at levels up to 20 V/m and 2 G at 45 and 75 Hz failed to show any significant differences in fertility between exposed controlled populations (135). Factors measured during the study were mortality, body weight, and anatomical defects. A study (136) of blood pressure, body weight, and anatomical and physiological changes among canines was made under the same exposure conditions (i.e., 20 V/m, 2 G, at 45 and 75 Hz) with negative results except for a spurious increase in hypertension. The finding of hypertension was considered to be inconclusive.

Results were inconclusive in a study (137) of the thresholds of perception of ELF electric fields by marine animals at levels that were roughly 1000 times greater than levels proposed for Seafarer. A comprehensive study (138) of the behavior in pigeons and rats at field levels of up to 2 G and 100 V/m at 45, 60, and 75 Hz indicated no effect. Studies of reaction time in primates exposed to ELF fields showed the central nervous system to be relatively insensitive to a magnetic field of 10 G at 45 Hz (139, 140). Tests of immediate memory and general motor activity in primates exposed to 1.4 V/m and 10 G at 75 Hz had no influence on performance (141). Studies have demonstrated the exquisite sensing apparatus of sharks to ELF at a frequency of 10 Hz (142). The detection threshold for ELF radiation at 75 Hz is being investigated, but it is expected to be higher than that at 10 Hz.

Because many biological functions exhibit a 24-h periodicity, they are often referred to as circadian (*circa* = about; *dies* = day) rhythms. Although their mechanism remains obscure, circadian rhythms are endogenous to living organisms rather than the result of external influences. Also, there is some evidence that alteration of these internal timing systems may be associated with deterioration in psychological and physiological function.

When environmental time cues such as the normal day–night cycle have been eliminated, Wever (143) has demonstrated in humans a "free-running" rhythmic period of approximately 24 h for activity, rest, and body temperature. When all known and recognized environmental cues have been eliminated from the test environment, the rhythm appears to be "free running"; that is, it is not synchronized by environmental cues. Wever reports that under such conditions there is an increased tendency for the circadian rhythms to desynchronize from each other, producing what he terms "internal desynchronization." Such desynchronization apparently increases the normal circadian periods from approximately 25 h to time periods that may be twice as great. The presence of natural or artificial fields such as ELF apparently strengthens the internal coupling of biological timing mechanisms, a presumably desirable effect. Therefore, there is some implication in Wever's work that it is the absence of natural or artificial fields such as ELF, rather than their presence, that may produce undesirable physiological effects.

Halberg (144) undertook a series of investigations of ELF radiation effects on silk tree leaflets, flour beetles, and mice. Field conditions ranged from 45 to 75 Hz, 0.4 to 2 G, and 1 to 180 V/m, with the duration of exposure varying from a few days to several months. The effects of magnetic fields (60 Hz, 1 G) and electromagnetic fields (60 Hz, 0.4 G, 100 V/m; and 60 Hz, 1 G, 100 V/m) on the circadian rhythms of silk tree leaflets appeared to be negative or at best inconclusive. There was no effect of exposure to 60 to 70 Hz, 1 G fields on the circadian toxicity susceptibility of the flour beetle when exposed to an insecticide (Dichlorvos), nor did continuous exposure of mice to a 45 Hz, 136 V/m electric field for 1 week before and after ouabain injection increase susceptibility to the drug. No significant difference in mice mortality appeared to exist between control and experimental groups exposed to 75 Hz, 25 V/m electric fields. Body weight was found to be unchanged even when mice were continuously exposed to 75 Hz, 1 G fields for several months. The effects of ELF fields ranging from 1 to 10 V/m and 0.5 to 2 G, singly and in combination, on the circadian temperature rhythm, food consumption, and survival rate, were negative. Even in cases of ELF fields reportedly increasing the synchronization of biorhythms, such results must likewise be considered to be suggestive rather than conclusive.

Although most investigations into the possible biological effects of ELF radiation have yielded negative data, a few findings bear careful follow-up investigation. These include (*1*) the need to elucidate the reasons for an apparent increase in serum triglycerides in exposed Navy personnel, (*2*) the need to follow up on the exquisite sensitivity of the shark (and possibly other marine life) to ELF radiation, and (*3*) the need to clarify the effect of ELF on bird migration patterns and avian biorhythms.

7.2 ELF Magnetic Fields

7.2.1 Static and Time-Varying Magnetic Fields

Today's intense interest in the biological effects of electromagnetic fields has materialized because of a recent concern that relatively low-level fields produced by everyday electrical appliances, wiring in the home, and power transmission lines may be causally related to a number of detrimental health effects. For example, certain epidemiologic studies have claimed a statistical association between exposures to power distribution lines in the vicinity of residences and cancer in children, but an approximately equal number of studies have not shown such an association. A report by the Congressional Office of Technology Assessment stated: "The emerging evidence no longer allows one to categorically assert that there are no risks, but it does not provide a basis for asserting that there is a significant risk." In summary, the implications for public health remain unclear.

Before examining the full range of events surrounding the concern about electric and, particularly, magnetic fields, it is well to review some facts about the existence and the prevalence of magnetic fields, beginning with the field surrounding the earth. Although the earth's magnetic field varies in magnitude across the United States, its average value is approximately 0.6 G. The inclination of the earth's magnetic field ranges from 36° from horizontal in Hawaii to 81° in Alaska and from 55° in southern Texas to 76.5° in northern Minnesota. Inclination angles vary annually from 0′ in Hawaii to 7.5′ in Nantucket, Massachusetts.

The naturally occurring magnetic field has a continuous frequency spectrum, ranging from the low end of the ELF region to radio frequencies from lightning discharges. Weak 50 to 60 Hz fields exist naturally at approximately 10^{-5} mG, orders of magnitude below those created by man-made power systems. Relatively strong natural electromagnetic fields may exist for short periods during a local thunderstorm. The magnetic fields produced by lightning may be of the same order of magnitude as those produced by power systems, although the magnetic fields generated by power systems persist for much longer periods of time. On the other hand, magnetic fields around the home and on the edge of rights of way of transmission lines may be lower than the earth's natural magnetic field. The predominant sources of residential magnetic fields appear to be external to private residences, the principal contributors appearing to be the electric power distribution lines in the immediate vicinity of the home (145).

7.2.2 Applications and Sources

Electromagnetic fields are generated by a wide variety of commonly used appliances, including radio-frequency induction heaters, microwave ovens, hair dryers, and electric ranges, and mixers. The average magnetic fields in residences in various geographic areas range from about 0.5 to 1.0 mG, but a small number of homes exceed this by an order of magnitude or more. Nearby transmission lines may be an external source. Also, ground currents in proximity to the house may generate magnetic fields. Localized 60-Hz magnetic flux densities measured near (30 cm from surface) household appliances such as refrigerators, ranges, electric drills, food mixers, and shavers range from 0.3 to 270 mG, with approximately 95% below 100 mG. Measurements of highly localized fields 1.5 m from the surfaces showed that fields of 95% of appliances were below 1 mG.

7.2.3 Medical Magnetic Resonance Imaging Systems

Medical magnetic resonance imaging (MRI) systems are based on the nuclear spin resonance of the hydrogen atom. The systems generate three different fields: (*1*) a high-intensity static magnetic field (0.04 to 1.5 T) that aligns the magnetic dipoles of hydrogen nuclei, producing longitudinal magnetization that determines the intensity of the MRI signal; (*2*) transverse magnetization produced from the longitudinal magnetization by a RF signal at a frequency determined by the magnitude of the magnetic field and the magnetic moment of the nuclei (for hydrogen nuclei, a 0.35-T magnetic field requires a pulsed RF signal at 15 MHz with a pulse repetition frequency of approximately one per second); and (*3*) a ramp-like magnetic field placed across the exposed tissue (at millitesla levels) and oscillating at a few hertz to produce a magnetic gradient that shifts the spin frequency slightly, thereby providing position information as a function of spin frequency.

The RF fields used in MRI interact with tissue primarily through the deposition of thermal energy. The specific absorption rate of the fields is proportional to the electrical conductivity and E^2 in tissue, where E is the electric field intensity. When energy deposition exceeds 4 W/kg, the thermoregulatory capacity of the body is exceeded, and tissue temperature rises. However, the specific absorption rates that can produce such effects are not exceeded in the present generation of MRI equipment.

In several laboratory and epidemiologic studies of the health effects of static and time-varying fields, exposure to high-intensity fields for the brief periods typical of MRI and *in vivo* spectroscopy has not produced adverse physiological effects. However, special attention must be paid to certain specific hazards, such as the torques exerted on ferromagnetic materials such as aneurysm clips and prostheses, the heating of prosthetic devices by radio-frequency fields, pacemaker malfunctions associated with closure of the reed relay switch, and electromagnetic interference caused by time-varying fields.

Nearly all current MRI scanners employ static magnetic fields with flux densities less than 2.5 T. The time variation of the flux density in these devices is less than 3 T per second, and the whole body specific absorption rates are less than 0.4 W/kg.

7.2.4 Cancer Risk and ELF Field Exposure

A number of epidemiologic studies have suggested that chronic exposure to ELF magnetic fields may increase the risk of cancers such as leukemia and CNS tumors. Since 1979, 27 publications have appeared on residential or occupational exposure to power frequency fields. The Wertheimer and Leeper studies showed a correlation of cancer (leukemia) in children with the presence of 60-Hz magnetic fields from high-current primary and secondary wiring configurations in the vicinity of their residences (146). These investigators found similar results in adults in the same geographic area (147). In contrast, other epidemiologic studies showed no clear relation between childhood or adult cancer incidence and the presence of power lines (147–151). The Savitz et al. study of childhood cancer and exposure to 60-Hz magnetic fields (152) attempted a replication of the original Wertheimer and Leeper study using a similar case-control method. However, in addition to estimating exposure to residential 60-Hz fields (based on the high-current or low-current configuration of power distribution lines), actual measurements were made of 60-Hz electric and magnetic fields in the residences of the cases and controls. No statistically significant correlation between 60-Hz fields and the incidence of childhood cancer was found. The cancer risk was found to be correlated with power distribution line configurations, but this correlation was marginally significant. It is difficult to see how the use of codes to estimate the density of power distribution lines could be superior to the measurement of actual magnetic fields in and around the residences.

Many confounding factors, such as smoking habits and exposure to industrial pollutants of known carcinogenic potential, were ignored in nearly all the aforementioned studies.

7.2.5 Exposure Guidelines

No national standards exist for magnetic field exposures, although guidelines have been generated within certain government agencies. Most guidelines were established for limiting exposures to static magnetic fields, but the more widespread use of MRI devices has caused increasing attention to time-varying fields.

In terms of whole body 8-h exposure to static magnetic fields, it is suggested that the field be limited to 0.01 T. For exposures of less than 1 h, the field should be limited to 0.1 T, and for exposures of the order of a few minutes perhaps 0.2 T. For exposures confined to the arms or hands, the respective limits may be in the range of 0.1 to 2.0 T: the lower limit might be permitted for an 8-h exposure, whereas the upper limit could be permitted for an exposure time not to exceed 10 min. Until additional definitive data are collected on the biological effects of magnetic fields, it is suggested that peak exposures not exceed 2.0 T.

The increasing use of MRI led the National Radiological Protection Board in England and the FDA in the United States to establish guidelines for medical exposures to magnetic fields. These guidelines limit static magnetic fields to 2.5 T and time-varying magnetic fields to 20 T/s for pulses of 10 ms or longer, or a root mean square rate of change of magnetic flux density of $2/t^{1/2}$ T for shorter times, where t is the pulse duration in seconds. The average associated whole body RF specific absorption rate should not exceed 0.4 W/kg. Finally, consideration should be given to the installation of warning signs in areas near MRI devices that exceed 0.5 mT. Tentative suggested guidelines for the general public are 0.5 mT and for occupational exposures, 150 mT limited to an 8-h day.

EDITOR'S NOTE

This Chapter was prepared for the Fourth Edition of *Patty's Industrial Hygiene and Toxicology* by George M. Wilkening, CIH. Mr. Wilkening was an internationally recognized expert in the science of nonionizing radiation. Although an updating of the Chapter was undertaken, it was not possible to complete it in time for publication in this Fifth Edition. The new Chapter author and the industrial hygiene editor agreed that, although updating the Chapter is worthwhile, the content of this Fourth Edition chapter remains scientifically valid, that it still constitutes a valuable part of Patty's, and that it should not be omitted. A revision of the Chapter may be available for the first Patty's Fifth Edition CD-ROM update.

Two federal laws enacted since original publication of this Chapter, and their impacts on matters involving nonionizing radiation should be mentioned here. The 1992 Energy Policy Act (PL 102-486) directed the National Institute of Environmental Health Sciences, National Institutes of Health, and the Department of Energy to direct and manage a program of research and analysis aimed at providing scientific evidence to clarify the potential for health risks from exposure to extremely low frequency (ELF) electric and magnetic fields (EMF). A report on findings of this research is now available. The Telecommunications Act of 1996 (PL 104-104) provided federal preemption of state and local governmental regulation of personal wireless service facilities on the basis of environmental effects of radio frequency radiation. A brief comment will be made on this legislation.

Over the past two decades a large number of reports regarding the possibility that exposures to extremely low frequency electric and magnetic fields might be associated with occurrence of leukemia in humans have appeared in the scientific literature. In response to the 1992 Energy Policy Act the National Institutes of Health has published its report, *Health Effects from Exposures to Power-Line Frequency Electric and Magnetic Fields* (153). This report lists 320 references. The primary conclusion of this late 1999 report is:

> The NIEHS concludes that ELF-EMF exposure cannot be recognized as entirely safe because of weak scientific evidence that exposure may pose a leukemia hazard. In our opinion, this finding is insufficient to warrant aggressive regulatory concern. However, because virtually everyone in the United States uses electricity and therefore is routinely exposed to ELF-EMF, passive regulatory action is warranted such as continued emphasis on educating both the public and the regulated community of means aimed at reducing exposures. The NIEHS does not believe that other cancers or non-cancer health outcomes provide sufficient evidence of a risk to currently warrant concern.

Following this somewhat inconclusive conclusion the NIEHS report goes on to recommend that hypothesis-driven research in fundamental relationships of exposures to ELF-EMF and effects on health, including leukemia, should continue.

The Telecommunications Act of 1996 deals only briefly with matters related to exposures to radiofrequency (RF) radiation. Under the provisions of this Act the federal govenment through the Federal Communications Commission (FCC) has preempted state and local governments in regulating locations of telecommunications towers on basis of environmental (health) effects. With regard to radiofrequency radiation exposure limits the FCC regulations state (47 *CFR* §1.1310):

> Note . . . These limits are generally based on recommended exposure guidelines published by the National Council on Radiation Protection and Measurements (NCRP) in "Biological Effects and Exposure Criteria for Radiofrequency Electromagnetic Fields" . . . NCRP Report No. 86, 1986, and "In the frequency range from 100 MHz to 1500 MHz, exposure limits for field strength and power density are also generally based on guidelines recommended by the American National Standards Institute (ANSI) . . . (ANSI/IEEE C95.1, 1992).

The current ANSI/IEEE and NCRP safety guidelines recommend that low-power devices such as cellular hand-held telephones not cause a localized exposure in excess of a specific absorption rate (SAR) of 1.6 W/Kg. The FCC reports that studies of human head models using cellular telephones have generally found that the SAR levels are below 1.6 W/Kg as averaged over 1 gram of tissue under normal conditions of use, although some studies have reported higher peak levels under worst-case conditions (154).

BIBLIOGRAPHY

1. Public Law 90-602, Radiation Control for Health and Safety Act of 1968, Washington, DC.
2. S. M. Michaelson, "Human Exposure to Radiant Energy—Potential Hazards and Safety Standards," *Proc. IEEE* **60** (April 1972).
3. G. M. Wilkening, in *The Industrial Environment, Its Evaluation and Control*, Department of Health Education and Welfare, NIOSH, Washington, DC, 1973, chapter on "Non-Ionizing Radiations."

4. M. Kleinfeld, C. Giel, and I. R. Tabershaw, "Health Hazards Associated with Inert-Gas-Shielded Metal Arc Welding," *Am. Med. Assoc. Arch. Ind. Health* **15**, 27–31 (1957).
5. K. G. Hansen, "On the Transmission Through Skin of Visible and Ultraviolet Radiation," *Acta Radiol.*, Suppl. 71 (1948).
6. S. Lerman, "Radiation Cataractogenesis," *N.Y. State J. Med.* **62**, 3075–3085 (1962).
7. J. A. Zuclich and J. S. Connolly, "Ocular Damage Induced by Near Ultraviolet," Technology Inc., San Antonio, TX.
8. T. M. Murphy, "Nucleic Acids: Interaction With Solar UV Radiation," in *Current Topics Radiat. Res.* **10**, 199 (1975).
9. I. Willis, A. Kligman, and J. Epstein, "Effects of Long Ultraviolet Rays on Human Skin: Photoprotective or Photoaugmentative?" *J. Invest. Dermatol.* **59**, 419 (1973).
10. C. E. Keeler, *Albinism, Xeroderma Pigmentosum, and Skin Cancer*, National Cancer Institute, Monograph 10, 1963, p. 349.
11. R. L. Elder, "Phototherapy Warning," *J. Pediatr.* **84**, 145 (1974).
12. L. E. Bockstahler, C. D. Lytle, and K. B. Hellman, "A Review of Photodynamic Therapy for Herpes Simplex: Benefits and Potential Risks," Publication No. (FDA) 75-8013, U.S. Public Health Service, Bureau of Radiological Health, Rockville, MD, 1975.
13. L. F. Mills, C. D. Lytle, F. A. Andersen, K. B. Hellman, and L. E. Bochstahler, "A Review of Biological Effects and Potential Risks Associated with Ultraviolet Radiation as Used in Dentistry," Publication No. (FDA) 76-8021, U.S. Public Health Service, Bureau of Radiological Health, Rockville, MD, 1976.
14. F. Daniels, *Ultraviolet Carcinogenesis in Man*, National Cancer Institute, Monograph 10, p. 407, 1963.
15. M. Faber, personal communication, 1975.
16. G. Funding, O. M. Henriques, and E. Rekling, *Über Lichtkanzer*, Verh 3., Int. Kongress für Lichtforsch, Wiesbaden, 1936, p. 166.
17. O. A. Jensen, "Malignant Melanomas of the Uvea in Denmark, 1943–1952," Thesis, Copenhagen, 1963.
18. *Threshold Limit Values for Physical Agents*, published by American Conference of Governmental Industrial Hygienists, Cincinnati, OH, 1989. (See latest edition of TLVs for any changes of values.)
19. J. R. Richardson and R. D. Baertsch, "Zinc Sulfide Schottky Barrier Ultraviolet Detectors," in *Solid State Electronics*, Vol. 12, Pergamon Press, New York, 1969, pp. 393–397.
20. P. Bassi, G. Busuoli, L. Lembo, and O. Rimonde, *G. Fis. Sanit. Prot. Contro Radiaz.* **18**(4), 137–142 (October–December 1974).
21. D. G. Pitts, "The Ocular Ultraviolet Action Spectrum and Protective Criteria," *Health Phys. J.* **25**, 559–566 (December 1973).
22. I. Matelsky, "Non-Ionizing Radiations," in L. V. Cralley and G. D. Clayton, eds., *Industrial Hygiene Highlights*, Vol. 1, Pittsburgh, PA.
23. H. C. Weston, "Illumination and the Variation of Visual Performance with Age," *Proceedings of the International Commission on Illumination*, Vol. 2, Stockholm, 1951.
24. C. M. Edbrooke and C. Edwards, "Industrial Radiation Cataracts: The Hazards and the Protective Measures," *Arch. Occup. Hyg.* **10**, 293–304 (1976).
25. K. L. Dunn, "Cataract from Infrared Rays. Class Workers' Cataract—A Preliminary Study on Exposures," *Arch. Ind. Hyg. Occup. Med.* **1**, 166–180 (1950).

26. K. L. Dunn, "A Preliminary Study on Glass Workers' Cataract Exposures," *Trans. Am. Acad. Ophthalmol. Otolaryngol.* **54**, 597–605 (1950).
27. G. F. Keatinge, J. Pearson, J. P. Simons, and E. E. White, "Radiation Cataract in Industry," *Arch. Ind. Health* **11**, 305–315 (1955).
28. D. G. Cogan, D. D. Donaldson, and A. B. Reese, "Clinical-Pathological Characteristics of Radiation Cataract," *Arch. Ophthalmol.* **47**, 55–70 (1952).
29. H. Goldmann, H. Koenig, and F. Maeder, "The Permeability of the Eye Lens to Infrared," *Ophthalmologica* **120**, 193–205 (1950).
30. H. Goldmann, "The Genesis of the Cataract of the Glass Blower," *Ann. Ocul.* **172**, 13–41 (1935); also in *Am. J. Ophthalmol.* **18**, 590–591 (1935).
31. R. K. Langley, C. B. Mortimer, and C. McCullock, "The Experimental Production of Cataracts by Exposure to Health and Light," *Am. Med. Assoc. Arch. Ophthalmol.* **63**, 473–488 (1960).
32. M. Wertheimer and W. D. Ward, "Influence of Skin Temperature upon Pain Threshold as Evoked by Thermal Radiation—A Confirmation," *Science* **115**, 499–500 (1952).
33. J. D. Hardy, "Thermal Radiation, Pain and Injury," in S. Licht, ed., *Therapeutic Heat*, Vol. 2, New Haven, CT, 1958, pp. 157–178.
34. *IES Lighting Handbook*, 4th ed., Waverly Press, Baltimore, 1966.
35. American National Standards Institute Standard for the Safe Use of Lasers, Z136.1, New York, 1986.
36. M. L. Wolbarsht, ed., *Laser Applications in Medicine and Biology*, Vol. 1, Plenum Press, New York, 1971.
37. A. Javan, W. R. Bennett, and D. R. Herriott, "Population Inversion and Continuous Optical Laser Oscillation in a Gas Discharge Containing a He-Ne Mixture," *Phys. Rev. Lett.* **6**, 106 (1961).
38. R. C. Miller and W. A. Nordlung, "Tunable Lithium Niobate Optical Oscillator with External Mirrors," *Appl. Phys. Lett.* **10**, 53 (1967).
39. G. M. Wilkening, "The Potential Hazards of Laser Radiation," *Proceedings of Symposium on Ergonomics and Physical Environmental Factors*, Rome, September 1968, International Labor Office, Geneva, Switzerland, pp. 16–21.
40. P. P. Sorokin, J. J. Wynne, J. A. Armstrong, and R. T. Hodgson, "Resonantly Enhanced, Nonlinear Generation of Tunable Coherent, Vacuum Ultraviolet (VUV) Light in Atomic Vapors," Third Conference on the Laser, *Ann. N.Y. Acad. Sci.* **267**, 36 (1976).
41. G. M. Wilkening, "A Commentary on Laser Induced Biological Effects and Protective Measures," *Ann. N.Y. Acad. Sci.* **68**, part 3, 621–6236 (1970).
42. W. T. Ham, Jr., R. C. Williams, H. A. Mueller, D. Guerry, A. M. Clarke, and W. J. Geeraets, "Effects of Laser Radiation on the Mammalian Eye," *Trans. N.Y. Acad. Sci.* **28**(2), 517 (1965).
43. W. K. Noell, V. S. Walker, B. S. Kang, and S. Berman, "Retinal Damage by Light in Rats," *Invest. Ophthalmol.* **5**, 450 (1966).
44. W. T. Ham, Jr., H. A. Mueller, and D. H. Sliney, "Retinal Sensitivity to Damage from Short Wavelength Light," *Nature* **260**, (March 11, 1976).
45. A. I. Goldman, W. T. Ham, Jr., and H. A. Mueller, "Mechanisms of Retinal Damage Resulting from Exposure of Rhesus Monkeys to Ultrashort Laser Pulses," *Exp. Eye Res.* **21**, 457–469 (1975).
46. J. Taboada and D. W. Ebbers, "Ocular Tissue Damage due to Ultrashort 1060 μm Light Pulses from a Mode-Locked Nd:Glass Laser," *Appl. Opt.* **14**, 1759 (August 1975).

47. A. I. Goldman, report accepted for publication in *Experimental Eye Research*, 1976.
48. K. W. McNeer, M. Ghosh, W. J. Geeraets, and D. Guerry, "ERG After Light Coagulation," *Acta Ophthalmol.* Suppl. 76, 94 (1963).
49. T. P. Davis and W. J. Mautner, "Helium-Neon Laser Effects on the Eye," Annual Report, Contract No. DADA 17-69-C 9013, U.S. Army Medical Research and Development Command, Washington, DC, 1969.
50. L. Goldmann, W. Kapuscinska-Czerska, S. Michaelson, R. J. Rockwell, D. H. Sliney, B. Tengroth, and M. Wolbarsht, "Health Aspects of Optical Radiation with Particular Reference to Lasers," World Health Organization, 1978.
51. Technical Committee on Laser Products, International Electrotechnical Commission, Geneva, Switzerland, 1984.
52. U. S. Department of Health, Education and Welfare, Bureau of Radiological Health, Laser Products Performance Standard, *Fed. Reg.* **40**, 148, 32252–32265 (July 31, 1975).
53. *Radiofrequency Electromagnetic Fields: Properties, Quantities, and Units*. Biophysical Interacting and Measurements, Report No. 67, National Council on Radiation Protection and Measurements, Washington, DC, 1981.
54. L. D. Sher and H. P. Schwan, "Mechanical Effects of AC Fields on Particles Dispersed in a Liquid, Biological Implications," Ph.D. dissertation (L. D. Sher), University of Pennsylvania, under Contract AF30(602) ONR Technical Report Dissertation No. 37, 1963.
55. A. H. Frey, "Human Auditory System Response to Modulated Electromagnetic Energy," *J. Appl. Physiol.* **17**, 689–692 (1962).
56. R. L. Carpenter and C. A. Van Ummersen, "The Action of Microwave Radiation on the Eye," *J. Microwave Power* **3**, 3–19 (1968).
57. J. T. McLaughlin, "Tissue Destruction and Death From Microwave Radiation (Radar)," *Calif. Med.* **86**, 336–339 (1957).
58. T. S. Ely, "Microwave Death," letter to the editor, *J. Am. Med. Assoc.* **217**, 1394 (1971).
59. M. M. Zaret, S. Cleary, B. Pasternack, M. Eisenbud, and H. Schmidt, "A Study of Lenticular Imperfections in the Eyes of A Sample of Microwave Workers and a Control Population," New York University, Final Report No. RADC-TDR-63-125, 1963; ASTIA Doc. AD 413294.
60. M. M. Zaret and M. Eisenbud, "Preliminary Results of Studies of the Lenticular Effects of Microwaves Among Exposed Personnel," in M. F. Peytong, ed., *Proceedings of the Fourth Annual Tri-Service Conference on Biological Effects of Microwave Radiating Equipment and Biological Effects of Microwave Radiations*, Plenum Press, New York, 1961, pp. 293–308 (Technical Report No. RADC-TR-60-180).
61. C. I. Barron and A. A. Baraff, "Medical Considerations of Exposure to Microwaves (Radar)," *J. Am. Med. Assoc.* **168**, 1194–1199 (1958).
62. P. Czerski, M. Siekierzynski, and J. Gidynski, "Health Surveillance of Personnel Occupationally Exposed to Microwaves," Parts I, II, III, *Aerosp. Med.*, 1137–1148, October 1974.
63. C. D. Robinette and C. Silverman, "Causes of Death Following Occupational Exposure to Microwave Radiation (Radar) 950-1974," in D. G. Hazzard, ed., Symposium on Biological Effects and Measurement of Radiofrequency/Microwaves, Department of Health, Education and Welfare, Washington, DC, HEW Publication No. (FDA) 79-8026 (1977).
64. A. M. Lilienfeld, J. Tonascia, S. Tonascia, C. H. Libauer, G. M. Cauthen, J. A. Markowitz, and S. Weida, "Foreign Service Health Status Study: Evaluation of Status of Foreign Service and other Employees From Selected East European Posts. Final Report, July 31, 1978, Contract

No. 6025-619073 Dept. of Epidemiology, School of Hygiene and Public Health, The Johns Hopkins University, Baltimore, MD, 1978.

65. N. N. Livshits, "On the Causes of the Disagreements in Evaluating the Radiosensitivity of the Central Nervous System Among Researchers Using Conditioned Reflex and Maze Methods," *Radiobiology* **7**, 238–261 (1967).
66. Z. V. Gordon, A. V. Roscin, and M. S. Byckov, "Main Directions and Results of Research Conducted in the U.S.S.R. on the Biologic Effects of Microwaves," *Proceedings of the International Symposium on Biological Effects and Health Hazards of Microwave Radiation*, Polish Medical Publishers, Warsaw, 1973.
67. S. V. Nikogosjan, in *O biologiceskom deistvii EMP radiocastot.*, Vol. 3, Moscow, 1968, p. 97.
68. V. M. Stemler, in *Gigiena truda i biologiceskoe deistvii elektromagnitnyk voln radiocastot.*, Moscow, 1968, p. 175.
69. V. M. Stemler, in *Sbornik Bionika, 1973, Materialy 4 Vsisojuznoj konferenci po bionike*, Moscow, 1973, pp. 3, 87.
70. Fourth report on "Program for Control of Electromagnetic Pollution of the Environment: The Assessment of Biological Hazards of Nonionizing Electromagnetic Radiation," Office of Telecommunications Policy, Executive Office of the President, Washington, DC, June 1976, p. 21.
71. C. H. Dodge, "Clinical and Hygienic Aspects of Exposure to Electromagnetic Fields," in S. F. Cleary, ed., *Biological Effects and Health Implications of Microwave Radiation*, Symposium Proceedings, U.S. Department of Health, Education and Welfare, 1970, pp. 140–149.
72. R. D. McAfee, "Physiological Effects of Thermode and Microwave Stimulation of Peripheral Nerves," *Am. J. Physiol.* **203**, 374–378 (1962).
73. R. D. McAfee, "The Neural and Hormonal Response to Microwave Stimulation of Peripheral Nerves," in S. F. Cleary, ed., *Biological Effects and Health Implications of Microwave Radiation*, Symposium Proceedings, U.S. Department of Health, Education and Welfare, 1970, pp. 150–153.
74. S. M. Michaelson, "Biomedical Aspects of Microwave Exposure," *Am. Ind. Hyg. Assoc. J.* **32**, 338–345 (1971).
75. W. R. Adey, "Electromagnetic Fields, Cell Membrane Amplification and Cancer Promotion," *Proceedings of 1986*, Annual Meeting Devoted to the Nonionizing Radiations, National Council on Radiation Protection and Measurements, Washington, DC.
76. H. P. Schwan, A. Anne, and L. Sher, "Heating of Living Tissues," U.S. Naval Air Engineering Center, Philadelphia, Report No. NAEC-ACEL-534, 1966.
77. E. Hendler, "Cutaneous Receptor Response to Microwave Irradiation," in J. D. Hardy, ed., *Thermal Problems in Aerospace Medicine*, Surrey, England, 1968, pp. 149–161.
78. E. Hendler, J. D. Hardy, and D. Murgatroyd, "Skin Heating and Temperature Sensation Produced by Infrared and Microwave Irradiation," in J. D. Hardy, ed., *Temperature Measurement and Control in Science and Industry*, Part 3, Biology and Medicine, Reinhold, New York, 1963, pp. 221–230.
79. H. F. Cook, "The Pain Threshold for Microwave and Infrared Radiations," *J. Physiol.* **118**, 1–11 (1952).
80. NCRP (1981), National Council on Radiation Protection and Measurements: Report No. 67: "Radiofrequency Magnetic Fields: Properties, Quantities and Units, Biophysical Interaction and Measurements," Washington, DC.

81. A. W. Guy, "Quantitation of Induced Electromagnetic Field Patterns in Tissue and Associated Biologic Effects," *Proceedings of the International Symposium on Biological Effects and Health Hazards of Microwave Radiation*, Polish Medical Publishers, Warsaw, 1973.
82. A. W. Guy, J. F. Lehmann, and J. S. Stonebridge, "Therapeutic Applications of Electromagnetic Power," *IEEE Proc. Ind., Sci., Med. Appl. Microwaves*, special issue, January 1974.
83. C. C. Johnson and A. W. Guy, *Proc. IEEE* **60**, 692 (1972).
84. D. R. Justesen and N. W. King, "Behavioral Effects of Low Level Microwave Irradiation in the Closed Space Situation," in S. F. Cleary, ed., *Biological Effects and Health Implications of Microwave Radiation*, U.S. Department of Health, Education and Welfare, Report No. BRH/DBE 70-2 (PB 193 8 58), Rockville, MD, 1970, p. 154.
85. E. I. Hunt, "General Activity of Rats Immediately Following Exposure to 2450 MHz Microwaves," in *Digest*, 1972 IMPI Symposium, Ottawa, Canada, May 1972.
86. R. D. Phillips, N. W. King, and E. L. Hunt, "Thermoregulatory Cardiovascular and Metabolic Response of Rats to Single or Repeated Exposures to 2450 MHz Microwaves," in *1973 Symposium on Microwave Power Institute*, Ottawa, Canada, September 1973.
87. P. Kramar, A. F. Emery, A. W. Guy, and J. C. Lim, "Theoretical and Experimental Studies of Microwave Induced Cataracts in Rabbits," in *Applications in the 70s*, 1973 IEEE G-MTT International Microwave Symposium, University of Colorado, Boulder, 1973, p. 265.
88. H. P. Schwan and K. Li, *Proc. IRE* **41**, 1735 (1953).
89. C95.1 Committee of the American National Standards Institute, "Safety Level of Electromagnetic Radiation with Respect to Personnel," ANSI, New York.
90. H. P. Schwan, in S. F. Cleary, ed., *Biological Effects and Health Implications of Microwave Radiation*, U.S. Department of Health, Education and Welfare, Report No. BRH/DBE 70-2 (PB 193 8 58), Rockville, MD, 1970.
91. W. W. Mumford, "Heat Stress Due to RF Radiation," *Proc. IEEE* **57**, 171–178 (1969).
92. "Sanitarnyje normy i pravila pri raboti s istocuikami electromagnitnyck polei vysokih, ultravipokih i sverhvysokih castot," Ministerstvo Zdravohranenija, SSSR 30, 03 NS 848-70, 1970.
93. "Jednotna metodika stanoveni intensity pole a ozareni elektromagneticRymi vinami v pasma vysokych fredvenci a velmi vysokych frekvenci k hygienickym ucelum." Vynos hlavniho hygienika CsRcj. HE 344.5-3.2.70/Priloha c. 3k Informacnim zpravam z oboruhygieny prace a nemoci z povolani, Praha, kveten, 1970.
94. "Rozpovzadzenie Rady Ministrow z dnia 25.02.1972 w. spraure bezpieczenstwa i higieny pracy przy stosowaniu urzadzen wytroarzajacych pola elektromagnetyczne w zakresie mikrofalowym," Dziennik Ustaw PRL No. 21, II, Poz. 153, 1972.
95. M. M. Weiss and W. W. Mumford, "Microwave Radiation Hazards," *J. Health Phys.* **5**, 160 (1961).
96. American National Standards Institute "Safety Levels with Respect to Human Exposure to Radio Frequency Electromagnetic Fields, 300 kHz to 100 GHz." Report No. ANSI C95.1 (The Institute of Electrical and Electronics Engineers Inc., New York, 1982.
97. American Conference of Governmental Industrial Hygienists, Threshold Limit Values ACGIH, Cincinnati, OH, 1984.
98. National Council on Radiation Protection and Measurements (NCRP) Report No. 86, "Biological Effects and Exposure Criteria for Radiofrequency Electromagnetic Fields," NCRP, Bethesda, MD, 1986.
99. F. G. Hirsch, "Microwave Cataracts" presented at NSAE Conference on Industrial Health, Cincinnati, Ohio, April 25, 1952. Also F. G. Hirsch and J. T. Parker, "Bilateral Lenticular

Opacities Occurring in a Technician Operating a Microwave Generator," *AMA Arch. Ind. Health* **6**, 512–517 (Dec. 1952).

100. H. P. Schwan and K. Li, "The Mechanism of Absorption of Ultrahigh Frequency Electromagnetic Energy in Tissues, as Related to the Problem of Tolerance Dosage,"*IRE Trans of Medical Electronics*, Vol. ME-4, pp. 45–59, February 1956.
101. Central Safety Committee, Bell Telephone Laboratories, Nov. 1953.
102. B. L. Vosburgh, "Problems Which are Challenging Investigators in Industry," *IRE Trans. Medical Electronics*, Vol. ME-4, pp. 5–7, February, 1956. Also: B. L. Vosburgh, "Recommended Tolerance Levels of MW Energy; Current Views of the General Electric Company's Health and Hygiene Service," *Proc. 2nd Annual Tri-Service Conference on Biological Effects of Microwave Energy*, RADC, Griffiss AFB NY, pp. 188–125; July 8–10, 1958.
103. "Symposium of Physiologic and Pathologic Effects of Microwaves," Mayo Foundation House, Rochester, MN, Sept. 23, 24, 1955.
104. H. P. Schwan, "The Physiological Basis of Injury," *Proc. 1st Annual Tri Service Conference on Biological Hazards of Microwave Radiation*, RDAC Griffiss AFB, NY, pp. 60–63, July 15–16, 1957.
105. A. H. Frey, "Auditory System Response to Modulated Electromagnetic Energy," *J. Appl. Physiol.* **17**, 689–692, 1962.
106. U.S. Air Force, "Microwave Radiation Hazards," Rome A.F. Depot, Griffiss AFB, NY, Urgent Action Tech. Order, 31-1-511, June 17, 1957.
107. "Biological Effects of Radio Frequency Energies," 1940–1957. *Proc. 1st Annual Tri-Service Conf. on Biological Hazards of Microwave Radiation*, RADC, Griffiss AFB, NY, July 15–16, 1957. Bibliography. Appendix B, pp. 94–103.
108. Temporary Sanitary Rules for Working with Centimeter Waves, Ministry of Health Protection of the USSR, 1958, 1961.
109. The irradiation of the U.S. embassy in Moscow was reportedly known in the 1950s, but documented much later. Refer to References No. 27 and 28.
110. American National Standards Institute (ANSI) Committee 95.1, Safety Level of Electromagnetic Radiation With Respect to Personnel, C95.1, New York 1966.
111. Hearings Before the Committee on the Commerce, United States Senate, Ninetieth Congress, Serial No. 90–49, US Government Printing Office, Washington, DC, May 1968.
112. Technical Electronic Products Radiation Safety Standards Committee (TEPRSSC) established under the Radiation Control for Health and Safety Set of 1968 (Public Law 90-602).
113. Radiation Control for Health and Safety Act of 1968. (Public Law 90-602). See Annual Reports to the Congress on the Administration of DL 90-602. Bureau of Radiological Health, Washington, DC.
114. Proceedings of the Symposium on the Biological Effects and Health Implications of Microwave Radiation, Richmond, VA, Sept. 1969, S. F. Cleary, Ed., U.S. Dept. of HEW, June 1970.
115. F. G. Hirsch, "Microwave Cataracts—A Case Report Reevaluated,"*Electronic Product Radiation and the Health Physicist*, Proceedings of the 4th Annual Symposium of the Health Physics Society, Louisville, Kentucky, Jan. 28–30, 1970 HEW Publication BRH/DEP 70-26.
116. Radiation Control for Health and Safety Hearings before the Committee on Commerce, U.S. Senate, Ninety-Third Congress, Serial No. 93-24, U.S. Government Printing Office, Washington, DC, 1973.
117. Presentations of Consumers Union before the Technical Electronic Products Radiation Safety Standards Committee (TEPRSSC), Washington, DC, 1973. Presentations of American Home

Appliances Manufacturers before the Technical Electronic Products Radiation Safety Standards Committee (TEPRSSC), Washington, DC, 1973.

118. National Academy of Sciences, "Biosphere Effects of Extremely Low Frequency Radiation," (Sanguine/Seafarer) Washington, DC, 1974–1977.
119. American National Standards Institute (ANSI), "Safety Levels of Electromagnetic Radiation with Respect to Personnel," C95.1, New York, 1974.
120. U.S. Air Force, "Radiofrequency Radiation Health Hazard Control," AFR 161-42, Washington, DC, 1975.
121. U.S. Senate, Microwave Irradiation of the U.S. Embassy in Moscow. Hearings before the Committee on Commerce, Science, and Transportation Committee, Print 43-949, U.S. Government Printing Office, Washington, DC, 1979.
122. A. M. Lilienfeld, S. Tonascia, C. H. Libauer et al., "Foreign Service and Other Employees From Selected Eastern European Posts," Final Report Contract No. 6025-619073 Department of State, Washington, DC, 1978.
123. World Health Organization (WHO), *Environmental Health Criteria for Radiowaves in the Frequency Range from 100 kHz to 300 GHZ (Radiofrequency and Microwaves)*, WHO Environmental Criteria Program, United Nations, New York, 1981.
124. Narda Microwave Corporation, Catalog No. 20, Plainview, NY.
125. W. W. Mumford, "Some Technical Aspects of Microwave Radiation Hazards," *Proc. IRE* **49**, 427 (1961).
126. W. A. Palmisano, U.S. Army Environmental Hygiene Agency, presented at the American Industrial Hygiene Conference, Akron, OH, 1967.
127. Project Seafarer, U.S. Department of the Navy, Washington, DC.
128. Y. I. Novitskiy, Z. V. Gordon, A. S. Pressman, and Y. A. Kholodov, "Radio Frequencies and Microwaves, Magnetic and Electric Fields," Nassau Technical Translation No. Nassau TTF-14.021, 1971.
129. G. G. Knickerbocker, W. B. Kouwenhoven, and H. C. Barnes, "Exposure of Mice to a Strong AC Electric Field-An Experimental Study," *IEEE Trans. Power Appar. Syst.* PAS-96, 498 (1967).
130. L. Ulrich and G. Ferin, "The Effect of Working on Power Transmitting Stations upon Certain Functions of the Organisms (in Czechoslovakian)," *Pracovni Lek. (Prague)* **11**, 500 (1959).
131. W. B. Kouwenhoven, O. R. Langworthy, M. L. Singleweld, and G. G. Knickerbocker, "Medical Evaluation of Man Working in AC Electric Fields," *IEEE Trans. Power Appar. Syst.* PSD-86, 506 (1967).
132. S. Mittler, "Low Frequency Electromagnetic Radiation and Genetic Aberrations," Northern Illinois University, DeKalb, Final report, September 15, 1972 (AD 749959).
133. R. A. Pledger and W. B. Coate, "Bacteria Mutagenis Study," in *Project Sanguine Biological Effects Test Program Pilot Studies*, Hazelton Laboratories Final Report Contract No N00039-69-C-1572, Nov. 1970 (AD 717408), Chapter F.
134. W. B. Coate and S. S. Ho, "Plant Cytogenic Study," in Reference 133, Chapter G.
135. W. B. Coate and F. E. Reno, "Rat Fertility Studies," in Reference 133, Chapter C.
136. W. R. Teeters and W. B. Coate, "Canine Physiological Study," in Reference 133, Chapter D.
137. "Effects of Low Frequency Electrical Current on Various Marine Animals," Naval Air Systems Command Air Task Report No. 0410801, Work Unit 0100, June 1972 (AD 749335).
138. M. J. Marr et al., "The Effects of Low Energy, Extremely Low Frequency (ELF) Electromagnetic Radiation on Operant Behavior in the Pigeon and the Rat," Georgia Institute of Technology, Atlanta, February 28, 1973 (AD 759415).

139. J. D. Grissett et al., "Central Nervous System Effects as Measured by Reaction Time in Squirrel Monkeys Exposed for Short Periods to Extremely Low Frequency Magnetic Fields," Naval Aerospace Medical Research Laboratory Report No. NAMRL-1137, August 1971 (AD 731994).

140. J. D. Grissett, "Exposure of Squirrel Monkeys for Long Periods to Extremely Low Frequency Magnetic Fields: Central Nervous System Effects as Measured by Reaction Time," Naval Aerospace Medical Research Laboratory Report No. NAMRL-1146, October 1971 (AD 735456).

141. J. deLorge, "Operant Behavior of Rhesus Monkeys in the Pressure of Extremely Low Frequency-Low Intensity Magnetic and Electric Fields," Experiment 1, November 1972 (AD 754058).

142. A. J. Kalmijn, Woods Hole Marine Biological Laboratory, Woods Hole, MA, personal communication; also A. J. Kalmijn, "The Electric Sense of Sharks and Rays," *J. Exp. Biol.* **55**, 371–383 (1971), and A. J. Kalmijn, "Electro-Orientation in Sharks and Rays: Theory and Experimental Evidence," Scripps Institution of Oceanography, University of California, San Diego, Report Contract No. N00014-69-A-0200-6030, Office of Naval Research, November 1973.

143. R. Wever, "Einfluss schwacher elektromagnetische Felder auf die circadiane Periodik des Menschen," *Naturwissenschaften* **55**(1), 29–32 (1968).

144. F. Halberg, "Circadian Rhythms in Plants, Insects and Mammals Exposed to ELF Magnetic and/or Electric Fields and Currents," University of Minnesota, Final Report to the Office of Naval Research, August 1975.

145. W. T. Kaune, R. G. Stevens, N. J. Callahan et al. Residential Magnetic and Electric Fields, *Bioelectromagnetics* **8**, 315–335 (1987).

146. N. Wertheimer and E. Leeper, "Electrical Wiring Configurations and Childhood Cancer," *Am J. Epidemiol.* **109**, 273–284 (1979).

147. N. Wertheimer and E. Leeper, "Adult Cancer Related to Electrical Wires Near the Home," *Int. J. Epidemiol.* **11**, 345–355 (1982).

148. J. P. Fulton, S. Cobb, L. Preble et al., "Electrical Wiring Configurations and Childhood Leukemia in Rhode Island," *Am. J. Epidemiol.* **111**, 292–296 (1980).

149. M. E. McDowell, "Mortality of Persons Resident in the Vicinity of Electricity Transmission Facilities," *Br. J. Cancer* **53**, 271–279 (1986).

150. A. Myers, B. A. Cartwright, J. A. Bonnell, et al., "Overhead Power Lines and Childhood Cancer," International Conference on Electric and Magnetic Fields in Medicine and Biology, London, December 1985.

151. R. G. Stevens, R. K. Severson, W. T. Kaune et al., "Epidemiological Study of Residential Exposure to ELF Electric and Magnetic Fields and Risk of Acute Nonlymphocytic Leukemia," DOE/EPRN/NY State Power Lines Project Contractors Review, Denver, November 1986.

152. D. A. Savitz, H. Wachtel, F. A. Barnes, et al., "Case Control Study of Childhood Cancer and Exposure to 60 Hz Magnetic Fields," *Am. J. Epidemiol.* **128**, 21–38 (1988).

153. NIEHS, "Health Effects from Exposure to Power-Line Frequency Electric and Magnetic Fields," National Institute of Environmental Health Sciences, National Institutes of Health, NIH Publication No. 99-4493, 1999.

154. "Frequently Asked Questions," Office of Engineering and Technology, Federal Communications Commission, *http://www.fcc.gov/oet/rfsafety/rf-faqs.html*, 10/12/99.

CHAPTER TWENTY-ONE

Ionizing Radiation

Herman Cember, Ph.D., PE, CHP

1 INTRODUCTION

Ionizing radiation may be defined as electromagnetic (X-rays and gamma rays) and particle radiation (alpha and beta particles, electrons, neutrons, and protons) with sufficient energy to disrupt an atom or a molecule. This is done by knocking out an electron, thereby "ionizing" the atom or molecule.

Knowledge of radiation and safety problems associated with its use goes back more than 100 years, with the discovery, in 1895, of X-rays by Wilhelm Roentgen. On the same day that Roentgen announced his discovery, Emile Grubbe, a physicist working in Chicago with an apparatus similar to Roentgen's, developed severe skin burns after handling an energized cathode ray tube similar to the one used by Roentgen. This was the first work-related injury from radiation. In 1896, Antoine Henri Becquerel discovered radioactivity while working with an ore called pitchblende. Investigation revealed three different radiations that originated in the ore. Becquerel called these radiations alpha, beta, and gamma rays. (Later investigations showed that X-rays and gamma rays were the same type of radiation.)

The discovery of these radiations opened new fields of scientific investigation and uses for these radiations. Parallel with the studies of the physics and chemistry of radiation and radioactive materials were researches into their biomedical effects. With this expanded use of radiation there were further reports of harmful effects, such as skin burns and hair loss when X-rays were used in medical diagnosis. In 1899 the first case of a cancer, a basal cell carcinoma on a woman's face was cured by X-rays. In 1906, two French physiologists, Bergonie and Tribendeau, published their classical paper on the relative radiosensitivity of different cells and tissues. They found that the less differentiated a cell was, and the

Patty's Industrial Hygiene, Fifth Edition, Volume 2. Edited by Robert L. Harris.
ISBN 0-471-29754-2

more frequently it divided, the more radiosensitive it was. From then until now, we have found nothing to contradict their observations. However, much is understood now about the molecular biology basis for their findings. Since then, an enormous amount of information on the nature of the interaction of radiation with living tissue, and on the dose–effect and dose–response relationships has been amassed. Sources of information, such as the experiences of the radium dial painters, early radiologists, and the uranium miners, showed that occupational overexposures led to harmful effects. Other major sources of information include the populations that had been medically exposed to diagnostic and therapeutic radiation, the survivors of the atomic bombings in Japan, radiation accidents, epidemiological studies of populations exposed to low level radiation from nuclear facilities and from natural background. This body of knowledge forms the scientific basis for the radiation safety standards currently in use. After radiation safety guidelines were developed, occupational exposures that were within the radiation safety guidelines did not lead to harmful biomedical effects.

2 SOURCES OF IONIZING RADIATION

One is exposed to radiation from naturally occurring and from anthropogenic sources. Natural radiation background includes cosmic radiation that originates in the sun and in the stars, and terrestrial radiation that is emitted from radioactive minerals on the earth. These naturally occurring background radiations must be accounted for when making radiation measurements and decisions, and in certain instances may pose a potential health threat to workers.

2.1 Natural Radiation

Cosmic radiation, when it enters the earth's atmosphere, consists mainly (90%) of high kinetic energy protons, with the remaining 10% being alpha particles, neutrons, and electrons. Interaction of these very high energy particles with the atmosphere leads to production of certain radioactive isotopes, notably tritium (^{3}H) and radiocarbon (^{14}C), muons (extremely high energy, heavy electrons that become ordinary electrons when they lose their energy), and gamma rays. Cosmic ray intensity varies according to altitude and latitude. The atmosphere acts as a shield whose total thickness is equivalent to about 30 feet of water. Therefore, cosmic ray intensity is lowest at sea level, and increases with increasing altitude. In the United States, the average dose from cosmic rays at sea level is about 30 millirems (mrem) per year. (The mrem is a unit of radiation dose that is used for safety and regulatory purposes.) In Denver, Colorado, at an altitude of one mile, the cosmic ray dose rate is about twice that at sea level. Thus, jet aircraft crew members receive significant radiation doses in the course of their work (although very much less than the regulatory dose limit for occupational exposure).

The latitude effect is due to the earth's magnetic field. Charged cosmic ray particles that enter the earth's atmosphere in the equatorial region travel in a direction that is nearly perpendicular to the earth's magnetic field, while those charged particles that enter in the higher latitudes cross the magnetic lines of force at a glancing angle. The equatorial particles thus experience a much greater magnetic force that turns them away from their initial path towards the earth than the particles that enter at the high latitudes. The cosmic ray intensity is therefore lowest at the equator, and increases continuously as the latitude, both north and south, increases.

Terrestrial radiation comes mainly from three different groups of naturally occurring radioisotopes. Primordial radioisotopes that were present when the earth was formed are

- ^{238}U, $T_{1/2} = 4.5 \times 10^9$ years
- ^{235}U, $T_{1/2} = 7.1 \times 10^8$ years
- ^{232}Th, $T_{1/2} = 1.4 \times 10^{10}$ years
- ^{40}K, $T_{1/2} = 1.3 \times 10^9$ years

These radioactive elements are ubiquitous; average uranium and thorium concentrations in the soil are on the order of several ppm, and in potassium, which is abundant, ^{40}K constitutes 0.012%.

2.1.1 Uranium and Thorium Progeny

The primordial U and Th isotopes are the progenitors of three chains of sequentially decaying radioisotopes that have several common characteristics. Each series contains a radioisotope of the gaseous element radon, and each series terminates with a stable isotope of lead. Radon gas diffuses out of the ground, and the radon daughters, which are solids under ordinary circumstances, attach themselves to atmospheric dust particles. This naturally occurring airborne radioactivity must be accounted for when using the measured radioactivity in a dust sample to compute the concentration of an airborne radiocontaminant. Although the mean concentration of these naturally occurring radioisotopes is relatively low, the actual radioactivity concentration from this source varies according to the concentrations of uranium and thorium in the ground. Areas that have uranium concentrations much higher than the average thus may be expected to have relatively high atmospheric radon concentrations.

2.2 Anthropogenic Sources

- X-rays. Any electronic device in which electrons fall through potentials of 10 kV or more produce X-rays, and should be considered a potential source of exposure.
  Medical diagnosis and therapy
  Industrial radiography for inspection
  X-ray fluorescence and X-ray diffraction analysis in research laboratories
- Accelerators, to produce intense radiation beams for
  medical radiation therapy
  isotope production
  neutron production for neutron activation analysis
  research

- Radioisotopes

  Sealed sources for industrial uses such as gauging, inspection, and food and drug sterilization; and medically in radiation therapy

  Sealed neutron sources, whose uses include inspection, neutron activation analysis, and oil-well logging

  Unsealed radioisotopes used in medical research, diagnosis, and therapy, and generally in scientific and industrial research
- Nuclear reactors
- Technologically enhanced natural radioactivity. The concentration of radioactivity may be technologically increased to produce the so-called "TENORM" (Technologically Enhanced Naturally Occurring Radioactive Material). Since this natural radioactivity is found in water and in oil, in addition to rocks, coal, and soil, scale that is built up in pipelines in oil and gas production and in pipes carrying water and brine in industrial facilities concentrates the radioactivity . . . especially the radium. This can lead to significantly increased radiation exposure and inhalation hazards to workers when the pipe lines are cut and the scale is dispersed as a radioactive aerosol.

Another source of TENORM is the phosphate industry. Phosphorous is usually associated with uranium because the two elements bind tenaciously together. Uranium concentrations of about 20 to 300 ppm are found in phosphate rock. Radiation from phosphate slag, which is used as an aggregate in road paving materials, from phosphate fertilizers and from potash (which also contains relatively large amounts of ^{40}K), thus must be considered by the industrial hygienist when evaluating possible occupational health risks in these industries.

Radioactivity is also found in relatively high concentrations in the fly ash and in the bottom ash of coal and lignite burning boilers. Through this mechanism, many coal and lignite fired electricity generating stations release more radioactivity to the environment than do nuclear power stations.

Workers who are explicitly involved with the use and maintenance of these radiation sources are not the only ones of concern to the industrial hygienist. The sources, once produced, must be transported from the manufacturer to the user. Additionally, radioactive waste is generally generated during the manufacture of the radioisotopes, and after they have been used. Thus, one must also apply the appropriate radiation safety standards and procedures to those workers involved in the transport of the radioisotopes and in the handling of waste materials.

3 BASIC PHYSICAL PRINCIPLES

3.1 Energy

All radiation effects, whether beneficial or detrimental, as well as all radiation measuring devices, are based on the absorption of energy from the radiation source.

In health physics and in atomic and nuclear physics, energy is measured in units of electron volts (eV). One eV represents the amount of kinetic energy possessed by an electron after it is accelerated by an electrical potential of one volt. Commonly used multiples of the electron volt are the kiloelectron volt (keV), 1000 eV; and the megaelectron volt (MeV), 10^6 eV. One eV $= 1.6 \times 10^{-12}$ erg or 1.6×10^{-19} joule; 1 MeV $= 1.6 \times 10^{-6}$ erg or 1.6×10^{-13} joule.

3.2 Radioisotopes

The Bohr atomic model describes the atom as a positively charged central nucleus surrounded by electrons that revolve around the nucleus in specific radii, and are bound to the nucleus by the attractive electrical force between the positively charged nucleus and the negatively charged electron. The nucleus contains positively charged protons and electrically neutral neutrons. The attractive nuclear force between neutrons and protons and between neutrons acts to overcome the repulsive forces among the positively charged protons.

The atomic number, and hence the chemical nature of the element, is determined by the number of protons within the nucleus. Most of the naturally occurring elements contain more than one complement of neutrons. Nuclei of the same element that have different numbers of neutrons are called isotopes. Copper, for example, whose atomic number is 29, is a mixture of two isotopes. About 69% of Cu atoms contain 34 neutrons, and hence has 63 particles (nucleons) within its nucleus, and about 31% have 36 neutrons, or 65 nucleons in the nucleus. The number of nucleons is called the atomic mass number. Symbolically, isotopes are written with the chemical symbol and the atomic mass number as a superscript to the left of the symbol: ^{63}Cu and ^{65}Cu in the case of copper 63 and copper 65, respectively. For an isotope to remain stable, the neutron/proton ratio must lie within a relatively narrow range. If the ratio is outside this range, the nucleus is unstable, and achieves stability by the spontaneous nuclear transformation that corrects the neutron/proton ratio. Such an unstable isotope is called a radioisotope. Radioisotopes emit highly energetic particles from the nucleus; this process is called radioactive transformation or radioactive decay. In most cases, the transformed nucleus is a different element, one whose neutron/proton ratio is closer to stability than the unstable parent's nucleus.

All radioisotopes are uniquely identified by a set of three different characteristics:

- Type of radiation that is emitted.
- Energy of the emitted radiation.
- Rate at which the radioisotope decays. This rate is described either by the half life, or by the fraction that decays per unit time (the decay rate constant).

3.3 Rate of Decay

The half-life is defined as the period of time during which one half of the radioisotope decays. After a second half-life period, one half of the remaining isotope decays, leaving one fourth of the original activity, and after the next half life only one eighth of the original activity is left. More than 99% of the original activity is gone after seven half lives. Half-lives of radioisotopes range from microseconds to billions of years.

The decay rate constant is defined as the instantaneous fractional decrease of the radioisotope. In the case of ^{226}Ra, whose half-life is 1620 years, the decay rate constant is 0.000428 per year. This means that it is decreasing at a rate of 0.0428% per year. Short half-lived isotopes decay at a much faster rate. For example, ^{131}I, with a half-life of 8 days, decays at a rate of 8.7% per day. Generally, the relationship between an initial amount of activity, A_o, and the amount of activity left after a time t, and whose decay rate constant is λ per unit time, is given by

$$A = A_0 e^{-\lambda t} \tag{1}$$

The half-life, $T_{1/2}$, is related to the decay rate constant by

$$T_{1/2} = 0.693/\lambda$$

The half-time $T_{1/2}$ and the decay constant λ must be in the same time units when calculating with these decay equations.

3.4 Quantity of Radioactivity

Since radioactive isotopes are used because of their radioactivity, the unit for expressing the quantity of radioactive material is based on the disintegration rate. Thus, for example, 3.7×10^{10} atoms in one gram of ^{226}Ra decay in one second, while 1.7×10^{14} atoms in 1 gram ^{210}Po decay in one second. One gram of ^{210}Po is much more radioactive than 1 gram of ^{226}Ra. Clearly, weight is not an most appropriate unit for specifying the quantity of a radioisotope. The traditional unit for the quantity of a radioisotope is the curie, symbolized by Ci. The Ci is defined as that quantity of a radioactive material in which 3.7×10^{10} atoms are transformed in 1 second. The commonly used sub-units are the millicurie, mCi (10^{-3} Ci), microcurie, μCi (10^{-6} Ci), nanocurie, nCi (10^{-9} Ci), and picocurie, pCi (10^{-12} Ci).

In the SI system, the unit for the quantity of radioactivity is the becquerel, Bq. One Bq is defined as that quantity of radioactive material in which 1 atom disintegrates in 1 second. Since the Bq is such a small quantity of radioactive material, the kilobecquerel, kBq (10^3 Bq), and the megabequerel, MBq (10^6 Bq) are frequently used. One MBq = 27 μCi.

3.5 Specific Activity

The concentration of radioactivity is called the specific activity. Specific activity may be designated in traditional units of curies per unit mass or unit volume. In the SI system, the analogous quantities are given in MBq per kilogram or in MBq per cubic meter.

The specific activity, in Ci/gram of a carrier free radioisotope is given by:

$$SA_i = \frac{T_{Ra} \times 226}{T_i \times A_i} \text{ Ci/g} \tag{2}$$

where T_{Ra} and T_i are the half-lives, in the same time units, of ^{226}Ra (T = 1620 years) and of the radioisotope, and A_i is the atomic mass number of the radioisotope. In the case of ^{131}I, for example, whose half life is 8 days, the specific activity is

$$SA(^{131}\text{I}) = \frac{1620y \times 365 \frac{d}{y} \times 226}{8d \times 131} = 1.3 \times 10^5 \frac{\text{Ci}}{\text{g}}$$

4 PARTICLE RADIATION

Radiation can be considered to be energy carried by one of several different kinds of subatomic particles. When these particles interact with living tissue they transfer their energy to the tissue. The energy that is absorbed by the tissue is dissipated in the disruption of atoms and molecules, thus leading to the biological effects associated with ionizing radiation. The particles of radiation are discussed below.

4.1 Alpha Particles

An alpha particle is a highly energetic nucleus of an ordinary helium atom (which consists of an assembly of two positively charged protons and two electrically neutral neutrons, and is designated by the symbol ^4_2He) that is emitted from the nucleus of certain radioactive isotopes. When an alpha particle is emitted, the alpha emitting parent is transformed into a daughter element whose atomic number is 2 less, and whose atomic mass number is 4 less than that of the parent. In the case of ^{238}U, for example, which is transformed to ^{234}Th:

$$^{238}_{92}\text{U} \Rightarrow {}^4_2\text{He} + {}^{234}_{90}\text{Th}$$

Two important properties of alpha radiation are the high rate of energy transfer to the medium through which the alphas travel and the consequent very small penetration into the absorbing medium. Although most alpha particles from radioisotopes have kinetic energies in the range of 3 to 7 MeV, they can penetrate only a few centimeters of air before exhausting all their kinetic energy. Tissue and other solid materials are almost opaque to alpha rays. A 5.3 MeV alpha particle from ^{210}Po, for example, whose range in air is about 4 centimeters, can penetrate only about 0.005 centimeters of soft tissue. This thickness is less than that of the dead outer layer of skin. For this reason, alpha radiation originating from alpha emitting radioisotopes outside the body is not considered hazardous. However, alpha radiation from an internally deposited radioisotope transfers its kinetic energy directly to viable tissue. The high concentration of absorbed energy, together with the microscopic distribution of the absorbed energy makes internally absorbed alpha radiation more toxic, per unit amount of absorbed energy, than energy absorbed from the other radiations.

4.2 Beta Particles

Beta particles are ordinary electrons that are ejected with a high kinetic energy from certain radioisotopes. Emission of a beta particle results in the transformation of the beta emitter into another element whose atomic number is one greater than that of the parent, and whose atomic mass number remains unchanged. In the case of strontium-90 (^{90}Sr), for example, a pure beta emitter, the daughter element is yttrium-90 (^{90}Y), the next higher element after Sr in the periodic table. In this case, the reaction does not stop with the production of ^{90}Y. The yttrium daughter is also radioactive, and decays by beta emission to zirconium-90 (^{90}Zr), which is stable. These transformations are written symbolically as:

$$^{90}_{38}\mathrm{Sr} \Rightarrow {}^{0}_{-1}\beta + {}^{90}_{39}\mathrm{Y}$$

$$^{90}_{39}\mathrm{Y} \Rightarrow {}^{0}_{-1}\beta + {}^{90}_{40}\mathrm{Zr}$$

In the radioactive decay of many radioisotopes, gamma rays accompany the beta particles; these radioisotopes are called beta-gamma emitters. Radionuclides that emit only beta rays without any accompanying gamma radiation are called pure beta emitters.

Beta particles emitted by any particular radioisotope have a spread of energies, or a spectral distribution, that extends from near zero to a maximum that is characteristic of that radioisotope, as illustrated in Figure 21.1.

These maximum energies vary from one radioisotope to another, and span a range of energies extending from several keV to several MeV. The average energy beta from any particular radioisotope is, in most cases, approximately one third the maximum energy.

The depth of penetration, or the range of the beta radiation in matter, increases as the energy of the radiation increases. In air, very low energy betas have a range of several centimeters, while high energy betas travel about 3 meters in air per MeV of energy. Figure 21.2 shows the relationship between range and energy for beta particles.

The range of the beta radiation in Figure 21.2 is expressed in units of *density thickness*. Density thickness is related to linear thickness by

$$\textit{density thickness}, \frac{\text{grams}}{\text{cm}^2} = \textit{density}, \frac{\text{grams}}{\text{cm}^3} \times \textit{linear thickness}, \text{cm} \qquad (3)$$

For example, a sheet of aluminum 1 mm (0.1 cm) thick, density = 2.7 gm/cm^3, has a density thickness of

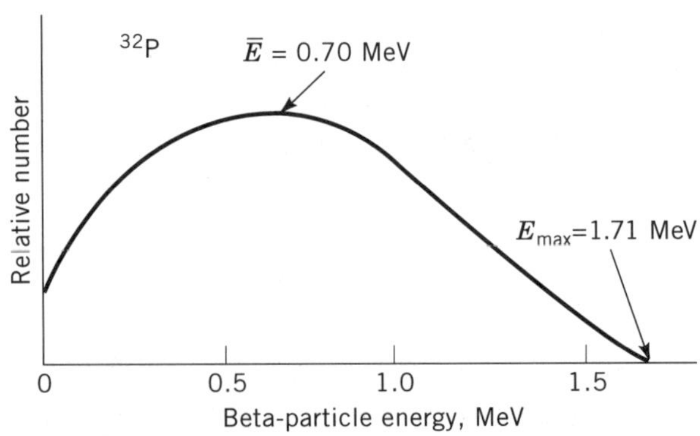

Figure 21.1. Beta ray energy spectrum for ^{32}P.

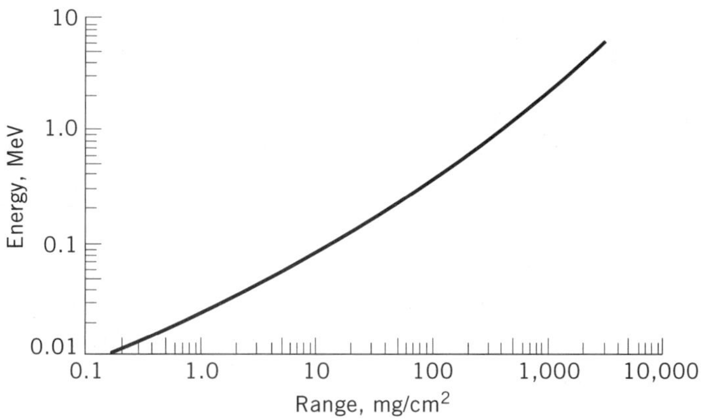

Figure 21.2. Range-energy relationship for beta particles. From Ref. 1.

$$\textit{density thickness} = 2.7 \frac{\text{g}}{\text{cm}^3} \times 0.1 \text{ cm} = 0.27 \frac{\text{g}}{\text{cm}^2} \tag{4}$$

The concept of density thickness is useful because different materials are about equivalent in their ability to stop beta radiation if their density thicknesses are equal. In this context, a sheet of graphite, whose density is 2.2 gm/cm^3, is equivalent to 0.1 cm Al if its linear thickness t is

$$t = \frac{\textit{density thickness}}{\textit{density}} = \frac{0.27 \dfrac{\text{g}}{\text{cm}^2}}{2.2 \dfrac{\text{g}}{\text{cm}^3}} = 0.12 \text{ cm} \tag{5}$$

4.2.1 Interaction with Matter

The physical mechanisms by which charged particles, such as alpha and beta particles, transfer their kinetic energy to matter are relatively well understood. The most likely occurrence is a collision between the charged particle and one of the extra-nuclear orbital electrons in the energy absorbing matter with which the radiation interacts. When the orbital electron is struck with sufficient force, it is knocked out of the atom, and the atom is said to be ionized. Ionization is the process of separating an electron from an atom, thereby upsetting the electrical neutrality of the atom and producing a pair of electrically charged particles—the ejected electron, which is the negative ion, and the residual atom, which is a positive ion. Work must be done by the ionizing particle, the alpha or beta particle, to separate the electron from its atom. An alpha particle loses an average of about 35.5 eV of energy per ion pair produced in air or in soft tissue, while a beta particle loses an average of about 34 eV per ionizing collision. These "collisions" are in reality interactions between the electric fields associated with the ionizing particles and the electrons in the absorbing media. Because an alpha particle has a large mass, moves slowly, and is doubly charged, it has a high rate of ionizing collisions, and consequently a high rate of energy loss. This high rate of energy loss explains the very low penetration of alpha radiation into matter. The linear rate of ion production is called the specific ionization. In air, the specific ionization of alpha particles is on the order of 10^4 to 10^5 ion pairs per centimeter.

Beta particles have only a single charge and travel at speeds near that of light. As a consequence, the specific ionization of beta particles is relatively low. Whereas an alpha particle produces on the order of 50,000 ion pairs per centimeter in air, a beta particle in air produces only about 100 ion pairs per centimeter. This difference in the specific ionization between the two types of radiation is important in health physics for several reasons. First, it accounts for the higher penetrating power of beta rays than that of alpha radiation. For example, a 1.71 MeV beta penetrates tissue to a depth of about 8 mm, while a 5.3 MeV alpha, which has three times the energy, penetrates only about 0.005 mm of tissue. Second, it is utilized in the design of radiation measuring instruments sensitive to this difference in order to be able to distinguish between the two types of radiation.

4.3 Bremsstrahlung

Bremsstrahlung (from the German meaning "braking radiation") is the production of X-rays when a charged particle undergoes a sudden change in velocity. When a high speed electron collides with an atomic nucleus, there is an abrupt change in the particle's velocity, and a fraction of the particle's kinetic energy is converted into X-rays. This fraction is extremely small for low energy betas and for low atomic numbered absorbers, but it increases with increasing energy and with increasing atomic number. For this reason too, beta ray shields are made of materials of low atomic number. In practice, beta ray shielding material of atomic number higher than 13 (Al) are seldom used.

Bremsstrahlung production is of importance in two cases. First when it is deliberately used to generate useful X-rays. In this application, electrons are emitted from the cathode in a specially designed high vacuum diode, Figure 21.3, and are accelerated across a high voltage ($\sim$100 kV in medical diagnostic X-ray units). When they strike the high atomic numbered tungsten anode, some of their kinetic energy is converted into electromagnetic X-ray energy. The intensity of the X-rays increases as the beam of accelerated electrons increase and as the accelerating voltage increases; the penetrating power of the X-rays depends only on the high voltage, and increases with increasing high voltage. To protect against unwanted bremsstrahlung, X-ray tubes are enclosed in lead shields that have shuttered apertures through which the useful beam escapes.

Secondly, bremsstrahlung are an unwanted side effect of shielding betas or in an instrument or other device in which electrons are accelerated across high voltages, such as an electron microscope, a klystron microwave generator, or an electron beam metallurgical furnace. Since these devices are not intended to be used as an X-ray source, these unwanted X-rays can pose a serious hazard if the user or if the industrial hygienist is unaware of their existence.

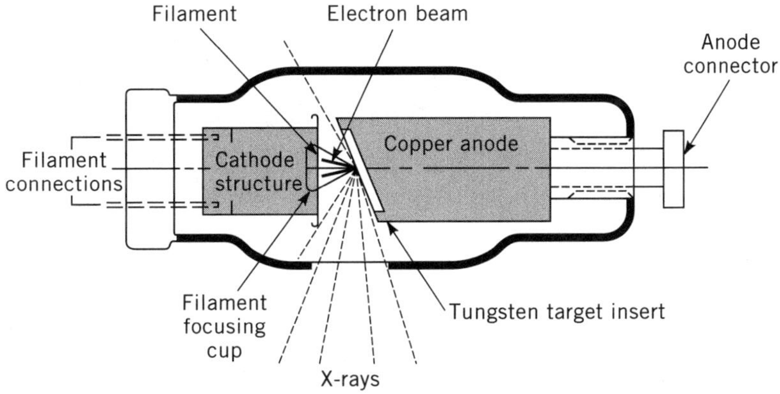

Figure 21.3. X-ray tube in which X-rays are generated by the bremsstrahlung process.

4.4 Electromagnetic Radiation

X-rays and gamma rays are electromagnetic radiations that occupy the high energy end of the continuous electromagnetic spectrum that includes radio waves, microwaves, infrared rays, visible light, and ultra-violet radiation. X-rays and gamma rays are qualitatively the same; they differ only in their manner of origin. Accordingly, the two terms may be used interchangeably in the context of radiation safety. Gamma rays are very penetrating; they pass through matter fairly easily, and can travel long distances in air. All electromagnetic radiation travels through space at the same speed, 3×10^8 meters per second. Gamma and X-rays are sufficiently energetic to generate ions by knocking electrons out of atoms, while the other portions of the electromagnetic spectrum are not energetic enough to generate ions. Accordingly, X-rays and gamma rays are called ionizing radiation, while the other parts of the electromagnetic spectrum are called nonionizing radiation.

According to Maxwell's theory of electromagnetism, a changing electric field is always associated with a changing magnetic field, and a changing magnetic field is always associated with a changing electric field. The intensity of these associated changing electric and magnetic fields as a function of time can be represented by a sine wave, thus leading to the wave model for electromagnetic radiation, Figure 21.4. The wave length, λ, and the frequency, f, of these waves are related to their speed, c, by

$$c = f \times \lambda \tag{6}$$

Wavelength is frequently expressed in angstrom units (1 angstrom unit $= 10^{-10}$ m). According to the wave model, the energy carried by the waves is proportional to the square of the amplitude of the electric and magnetic field strengths.

The wave model of electromagnetic radiation is useful for explaining many, but not all physical phenomena, and is the basis for understanding the effects of nonionizing electromagnetic energy. Phenomena that are not amenable to explanation by the wave theory are explainable by the *quantum theory*. According to the quantum theory, electromagnetic

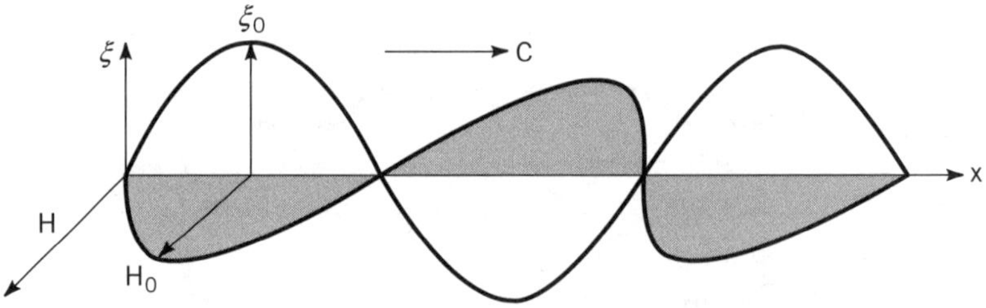

Figure 21.4. Schematic representation of an electromagnetic wave.

radiation behaves as if it consists of particles of energy, called photons, which travel through space at the speed of light (3×10^8 meters per second). Each particle contains a discrete quantity, or "quantum" of electromagnetic energy. Ionizing radiation includes photons whose energy exceeds 12 eV. The photon's energy content is proportional to the frequency of the radiation when the radiation is considered as a wave, and is given by

$$E = hf = h\frac{c}{\lambda} \tag{7}$$

where h, is Planck's constant and has a value of 6.626×10^{-34} joule-second and the frequency f is in hertz.

4.4.1 Interaction with Matter

Gamma rays interact with absorbing media by several different competing mechanisms, depending on the quantum energy and on the atomic number of the absorber. Two of these mechanisms, photoelectric absorption and Compton scattering involve an interaction between an orbital electron and the photon which results in the ejection of the electron. For photons whose energy exceeds 1.02 MeV, interaction with the absorber nucleus by the pair production mechanism, which is the direct conversion of energy into mass according to Einstein's theory, results in the transformation of the photon's energy into two particles: a negatively charged electron and a positively charged electron. In each of these types of interactions high speed electrons which carry a part of the photon's energy are produced. These energetic electrons, which are called primary ionizing particles, then transfer their energy by ionizing collisions with the orbital electrons of the absorbing media. The primary ionizing particles are the means by which energy is transferred from X-rays to the absorbing media.

5 DOSIMETRY

The magnitude of radiation effect is directly related to the concentration of absorbed energy in tissue (or in any other medium that is being irradiated). Other factors which must be considered in assessing the biological effect of radiation include the type of radiation (alpha, beta, or gamma) and the radiosensitivity of the irradiated tissue.

The first dosimetric quantity used in radiation safety, the roentgen (R) as originally defined, is actually a measure of exposure to X-rays. The amount of energy absorbed from a given roentgen exposure depends on the nature of the absorber. Soft tissue absorbs about 97 ergs per gram per R of X-ray exposure. Because of the dependence of absorbed dose on the nature of the absorber, the traditional dosimetric unit, the *rad*, was introduced. The rad is defined as that dose in which 100 ergs of energy are absorbed per gram of absorber.

$$1 \text{ rad} = 100 \text{ ergs/g}$$

For purposes of radiation safety control, it is customary to use the sub-unit millirad (mrad), which represents 0.001 rad. The rad is applied to all types of radiation dosimetry—to external radiation as well as to internally deposited radionuclides and to doses from alpha, beta, and gamma radiation. The traditional radiation unit is being replaced by the SI system unit, called the *gray*, symbolized by Gy. One gray is defined as that dose in which one joule of energy is absorbed in one kilogram of the absorbing medium:

$$1 \text{ Gy} = 1 \text{ J/kg}$$

Since $1 \text{ J} = 10^7$ ergs and since $1 \text{ kg} = 1{,}000$ g,

$$1 \text{ Gy} = 100 \text{ rads}$$

For radiation safety purposes and for regulatory purposes, a dose unit is used that is normalized for type of radiation and for the relative radiosensitivity of the irradiated tissue. The relative radiosensitivity is important when dealing with nonuniformly irradiated tissues and organs. The normalizing factor that accounts for the type of radiation is called by the U.S. Nuclear Regulatory Commission (USNRC) the *quality factor*, Q, and by the International Commission on Radiological Protection (ICRP) the *radiation weighting factor*, W_R. Relative radiosensitivity of irradiated tissue is considered by the *tissue weighting factor*, W_T, which ranges from 1 for uniform whole body radiation to 0.03 for the thyroid. The traditional dose equivalent unit is called the *rem*, H. For whole body irradiation, the rem is defined as the product of the quality factor and the dose:

$$H, \text{ rems} = \text{rads} \times Q$$

In the SI system, the unit for the dose equivalent is called the *sievert, Sv*, and is defined by:

$$H, \text{ Sv} = \text{grays} \times W_R$$

Values used for the normalizing factor for several radiations are listed in the Table 21.1.

In the belief that the probability of a stochastic effect should be the same whether the whole body is uniformly irradiated or whether the radiation is non-uniformly distributed,

Table 21.1. Values for the Quality Factor, Q, and the Radiation Weighting Factor, W_R

Radiation	Q or W_R
X-rays	1
Gamma rays	1
Beta rays	1
Alpha rays	20

the ICRP introduced the concept of the *effective dose equivalent*. The safety standards assume that the probability of harm to any tissue or organ is proportional to the dose to that tissue. However, because of the differences in radiosensitivity among the various tissues, the value for the proportionality factors differs among the tissues. Generally, the higher the mitotic rate of a cell line, and the less differentiated a cell line is, the more radiosensitive is that cell line. (This observation was first enunciated by two French physiologists, Bergonie and Tribendeau in 1906.) The differences in radiosensitivity are accounted for by *tissue weighting factors, W_T*, Table 21.2. The effective dose equivalent is given by

$$H_E = \sum_T W_T H_T \tag{8}$$

Example: A laboratory technician had accidentally inhaled ^{131}I. Using data from whole body counts and bioassay measurements, the health physicist calculated a thyroid dose

Table 21.2. Values for Tissue Weighting Factors, W_T[a]

Tissue or Organ	W_T, ICRP 26	W_T, ICRP 60
Bladder	Not given	0.05
Breast	0.15	0.05
Bone marrow (red)	0.12	0.12
Bone surface	0.03	0.01
Colon	Not given	0.12
Esophagus	Not given	0.05
Gonads	0.25	0.20
Liver	Not given	0.05
Lung	0.12	0.12
Skin	Not given	0.01
Stomach	Not given	0.12
Thyroid	0.03	0.05
Remainder	0.30	0.05

[a]The values are based on a reference population of equal numbers of both sexes and a wide range of ages. In the definition of effective dose, they apply to workers, to the entire population, and to both sexes.

equivalent of 6000 millirems (60 mSv) and a whole body dose equivalent of 15 millirems. What was the technician's effective dose according to the USNRC criterion?

Substituting into the equation for effective dose equivalent,

$$H_E = 0.03 \times 6000 + 0.97 \times 13 = 193 \text{ millrems}$$

6 RADIATION BIOEFFECTS

Quantitatively the harmful effects of radiation overexposure are the same whether the radiation was delivered from a source outside the body (external exposure) or from radionuclides within the body (internal exposure). The effects of a given dose are the same whether the dose was from an external radiation exposure or from an internally deposited radionuclide.

There are two qualitatively different categories of radiation bioeffects. *Deterministic (or non-stochastic)* effects are those for which there is a threshold dose, for which the magnitude of the effect depends on the size of the dose, and for which there is an unequivocal relationship between the radiation exposure and the effect (such as skin burns and hair loss). For deterministic effects, the threshold dose for any harmful effect must be exceeded before that harmful effect is observed. These threshold doses are very large relative to radiation safety standards. For example, the threshold dose for radiation illness exceeds 100 rads of acute whole body irradiation. Threshold doses for some other effects are shown below in Table 21.3.

The second category of radiation bioeffects is called *stochastic* effects, and it includes cancer and genetic effects. (It should be emphasized that, although genetic effects have been observed in experimental animals, we have observed no increase in genetic effects in any human population exposed anywhere at any time at any dose level.) Stochastic effects occur by chance; and they occur to unexposed persons as well as to exposed individuals. Stochastic effects are therefore not unequivocally related to the exposure. Exposure to a carcinogen or to a mutagen increases the likelihood of expression of the

Table 21.3. Threshold Doses for Some Clinically Harmful Deterministic Effects[a]

Organ	Injury at 5 Years	1–5% Dose	25–50% Dose
Ovaries	Sterilization	200–300 rads	600–1200 rads
Testes	Sterilization	500–1500	2000
Lens	Cataract	500	1200
Kidney	Nephrosclerosis	2300	2800
Liver	Failure	3500	4500
Thyroid	Hypothyroidism	4500	15000
Breast	Atrophy	<5000	<10000
Bladder	Ulcer	6000	8000
Bone	Necrosis	6000	15000

[a]From Ref. 3.

effect in proportion to the size of the dose. At no time, however, regardless of the size of the dose, can we be certain that a cancer will develop. If cancer does develop, it cannot be certain that the cancer was due to the exposure. No pathologist can say with certainty that the cancer would not have occurred if the person had not been exposed to the carcinogen. The best that one can do is to calculate the probability, based on the person's exposure history and dose, that the cancer was due to the exposure. It has been observed that there are no increases in leukemia rates at acute whole body doses less than 10 rads, and no increase in solid tumor frequency at acute whole body doses of 40 rads. Nevertheless, for purposes of setting radiation safety standards one postulates that there is a zero threshold for stochastic effects, and that the probability of a stochastic effect increases linearly with the size of the dose. This model for setting radiation safety standards is called the *Linear No Threshold (LNT)* model.

7 SAFETY STANDARDS

Radiation safety standards are designed to prevent deterministic effects and to limit the probability of a stochastic effect to a societally acceptable level in return for the benefits to society that result from the use of the radiation. Radiation safety standards and practices are recommended by the International Commission on Radiological Protection (ICRP), a nongovernmental scientific body. These recommendations form the basis for the radiation safety standards promulgated by the various international and national regulatory agencies. On an international level, the United Nation's International Atomic Energy Agency (IAEA) is the leading organization. Its safety standards are published in the *Safety Series* under the title *International Basic Safety Standards for Protection against Ionizing Radiation and for the Safety of Radiation Sources* (4). In the United States, the Environmental Protection Agency (EPA) sets the basic standards, and the several regulatory agencies, including the Nuclear Regulatory Commission (NRC), the Occupational Safety and Health Administration (OSHA), the Mine Safety and Health Administration (MSHA), the Department of Energy (DOE), and the food and Drug Administration (FDA) issue specific regulations for those activities over which they have legal jurisdiction. The American safety standards are based on ICRP Publications 26 and 30 (5, 6).

7.1 Nuclear Regulatory Commission (NRC)

The NRC's task is to oversee the safe use of nuclear energy and its radioactive byproducts. To this end, it regulates the production, possession, and use of:

- Source materials: Uranium and thorium, and their ores containing at least 0.05% U or Th.
- Special nuclear material (SNM): Plutonium, ^{233}U, and uranium enriched in ^{233}U or ^{235}U.
- By-product material: Any material, except SNM, produced or made radioactive incident to making or using SNM.

7.1.1 Primary Radiation Safety Standards (Dose Limits)

For occupational exposure, the NRC's radiation safety standards are

- To limit stochastic effects: 5,000 mrems (50 mSv) total effective dose in 1 year.
- To prevent deterministic effects:
  50,000 mrems (500 mSv) in 1 year to any tissue except the lens of the eye.
  15,000 mrems (150 mSv) to the lens.
- To protect a conceptus: 500 mrems during the entire pregnancy.
- To protect individual members of the public: 100 mrems in 1 year.

7.1.2 ALARA

It must be emphasized that these limits are maximum doses that may not be exceeded by design (except in special cases, such as a rescue operation following an accident). In practice, the ICRP recommends, and the regulatory agencies require the application of the ALARA principle, which states that operations be designed so that the work is done in such a manner that the resulting dose is **A**s **L**ow **A**s **R**easonably **A**chievable. NRC regulations require a licensee to demonstrate that the ALARA principle is being applied.

7.1.3 Derived Radiation Safety Standards

The primary safety annual dose limits include the sum of external doses and doses due to radionuclides within or on the surface of the body. To achieve the primary standard, secondary standards are prescribed which establish an annual limit of intake, ALI, for all radionuclides in air and in potable water. The ALI is the maximum intake of radionuclide, either by inhalation or by ingestion, that results in the primary annual dose limit to the worker if radiation exposure were limited to that radionuclide taken in through that portal of entry. Thus, the ALI must be proportionally decreased if a worker is exposed to more than one radionuclide and/or external radiation.

Tertiary limits are expressed as radionuclide concentrations in air and in water. In the case of occupational exposure to airborne radioactivity, the Derived Air Concentration, DAC, is the inhalation ALI divided by the volume of air, 2,400 m^3, that the worker is assumed (according to the Reference Man model) to inhale during the course of a working year. Effluent limits to the air and to the water environments are derived in a similar manner using the dose limit for the public rather than the occupational dose limit.

These standards are enforced through a system of licensing, inspection, and fines for violations. The NRC is authorized to delegate its licensing and inspection of by-product material to Agreement States. An Agreement State is a state that has signed an agreement with the NRC to regulate the use of radioactive materials within that state to at least the same degree as the NRC. The state must show the NRC that it has the resources to meet this obligation. The agreement may be terminated if the state does not meet the NRC's criteria. At this time (1999), there are 30 Agreement States.

It should be pointed out that the NRC does not regulate naturally occurring radioactive materials (NORM) that are not classified as "Source Material", nor does it regulate radionuclides produced in accelerators that are not contractors or research grantees from the Federal government. Also, the NRC does not regulate the manufacture or use of radiation producing machines, such as X-ray machines and particle accelerators.

The NRC's regulations are published in Title 10, *Code of Federal Regulations*. To assist applicants for NRC licenses, or to assist licensees in complying with the regulations, the NRC publishes Regulatory Guides. These Guides describe procedures and methods that are acceptable to the NRC for the purpose of demonstrating compliance with the regulations.

7.2 Occupational Safety and Health Administration (OSHA)

OSHA is a Federal regulatory agency whose function is to assure safe workplaces. It does this by establishing occupational safety and health standards and by inspecting workplaces to determine whether they are in compliance with the regulations. Radiation exposure from all nonmining radiation sources that are not regulated by the NRC, such as X-ray machines, naturally occurring and accelerator-produced radioactive materials (NARM), are subject to OSHA regulation. Because of OSHA's limited resources, however, safety regulation of these radiation sources in Agreement States is left to the state. It should be noted that although OSHA is authorized to regulate and inspect to determine compliance with the regulations, it is not authorized to license any facility. OSHA's radiation regulations are found in **29** *CFR* 1910.96.

7.3 Mine Safety and Health Administration (MSHA)

Radiation exposure in mines is regulated by MSHA, a Federal regulatory agency located within the Dept. of Labor. MSHA's regulatory authority extends to underground and to surface mines, as well as to the mining infrastructure, such as ore processing facilities and tailings. In the case if uranium, MSHA shares mining health and safety responsibilities with the NRC. Contributions of uranium tailings to the off-site environment are regulated by the EPA. MSHA's health and safety regulations are published in **30** *CFR* 57.

7.4 Department of Energy (DOE)

Rather than being regulated by the NRC, Federal law requires that the DOE establish its own radiation safety standards for its workers. These standards, which are consistent with EPA guidance to Federal agencies, are very similar to those of the NRC. DOE radiation safety standards are published in **10** *CFR* 835.

8 PRINCIPLES OF RADIATION SAFETY

The objective of health physics practice is to keep the radiation dose to the worker as low as reasonably achievable below the statutory limits. The practice of radiation protection is a special application of the control of the working environment by engineering means. In principle, if the potentially hazardous source cannot be eliminated, one must first try to isolate the source. If that is not sufficient, an attempt is made to control the environment to minimize the worker's exposure, and if that still is not good enough, the worker is isolated. The exact manner of applying these general principles to radiation safety depends on the individual situation. In health physics practice, the problem is broken down into protection from external radiation and protection from personal contamination resulting from inhaled, ingested, or tactilely transmitted radioactivity.

8.1 External Radiation

External radiation originates in machines designed to produce radiation, such as X-ray machines and accelerators, sealed sources such as those used as gauges, in industrial radiography, in medical radiotherapy, and in other devices specifically designed to produce radiation; and unsealed radionuclides, such as those used in research laboratories and in medicine. Of special importance to the industrial hygienist are those devices in which X-ray production (bremsstrahlung) is a side effect in an instrument or device in which electrons are accelerated across a high voltage. If it is not feasible to eliminate the radiation source, then exposure must be controlled by application of one or more of the following three techniques:

- Minimizing exposure time.
- Maximizing distance from the source.
- Shielding the radiation source.

8.1.1 Time

For the purpose of control of occupational radiation dose, the reciprocity relationship

$$\text{Dose rate} \times \text{Exposure time} = \text{Total dose}$$

is valid. Thus, if work must be performed in a relatively high radiation field, restriction of exposure time, so that the product of the dose rate and exposure time does not exceed the maximum allowable total dose, allows the work to be done in compliance with the radiation safety criteria. If the job cannot be finished within the restricted time slot, then the operation must be redesigned (by shielding the source, for example) in order to decrease the intensity of the radiation field, or the worker must be better trained so that he can do the job more efficiently, or another worker must be substituted when the first worker reaches the dose limit.

8.1.2 Distance

It is intuitively evident that radiation dose decreases with increasing distance from the source. In the case of a "point source" (For most practical purposes, a source appears like a point at a distance equal to about 10 times the longest physical dimension of the source.), the dose rate decreases rapidly according to the inverse square law:

$$\frac{I_1}{I_2} = \frac{d_2^2}{d_1^2} \tag{9}$$

where I_1 and I_2 are the dose rates at distances d_1 and d_2 respectively. Thus, a dose rate of 100 mrads/hr at a distance of 0.5 m from a point source decreases to 25 mrads/hr at 1 m, and to 6.25 mrads/hr at 2m.

For the case of a line source, such as a brine carrying pipe that has a buildup of radium containing scale, the dose rate decreases more slowly with increasing distance than from a point source. The same is true for an area source, such as a spill of a radioactive liquid. However, at distances approaching 10 times the maximum linear dimension of the spilled area, the dose rate begins to fall off as though the spilled area were a point.

8.1.3 Shielding, Gamma

Shielding a radiation source decreases the gamma ray dose rate to a level given by

$$I = I_0 e^{-\mu t} \tag{10}$$

where I_0 is the dose rate incident on the shield, I is the emergent dose rate, μ is the attenuation coefficient for the shielding material and for the gamma ray energy (μ is a function of the shielding material and of the gamma ray energy), and t is the shield thickness. (Values for μ maybe found in appropriate handbooks and textbooks). For example, if one wishes to use a lead shield to reduce the dose rate from 100 mrems/hour to 2.5 mrems/hour at a certain distance from a ^{137}Cs source (0.661 MeV gamma), calculate the shield thickness after finding that μ (Pb, 0.662 MeV) = 0.058 per cm:

$$\begin{aligned} I &= I_0 e^{-\mu t} \\ 2.5 &= 100 e^{-0.058 \times t} \\ t &= 6.4 \text{ cm} \end{aligned}$$

8.1.4 Shielding, Beta

Two factors must be considered when designing a shield for beta rays: the beta energy and the generation of bremsstrahlung. To minimize the bremsstrahlung, choose a beta ray shield of low atomic number. To stop the betas, choose a shield thickness, using the range–energy curve, Figure 21.2, that is equal to the range of the beta particle. For example, calculate the thickness of a Lucite shield, whose density is 1.12 g/cm^3, to stop all the betas from ^{32}P.

Phosphorus-32 emits a 1.71 MeV beta. From Figure 21.2 one finds the range of a 1.71 MeV beta to be 800 mgm/cm^2 (0.8 gm/cm^2). The required thickness is

$$t = \text{density thickness/density}$$

$$t = 0.8 \text{ gm/cm}^2 \div 1.12 \text{ gm/cm}^3 = 0.71 \text{ cm} = 7.1 \text{ mm}$$

8.2 Internal Radiation

In the context of radiation safety, the radiation dose from an internally deposited radionuclide is no different from the same dose absorbed from external radiation. Therefore it must be emphasized that a *millirem of dose is a millirem, regardless of whether it was delivered from an internally deposited radionuclide or from an external source.*

Radioactive substances, like other noxious agents with which the industrial hygienist deals, may gain entry into the body through three pathways:

1. Inhalation—by breathing radioactive aerosols or gases.
2. Ingestion—by drinking contaminated water, eating contaminated food, or tactilely transferring radioactivity into the mouth.
3. Absorption—by transcutaneous absorption or through wounds.

Measures to prevent internal exposure therefore are designed to either block a portal of entry or to interrupt the transmission pathway. These preventive measures can be effected either at the source by enclosing and confining it, by environmental control by ventilation, or through the use of protective clothing and respiratory protective devices. It should be noted that these measures are the same as those employed in the practice of industrial hygiene in a nonradiation environment. However, because of the nature of radiation and radioactive materials, the degree of control required for radiation safety greatly exceeds the requirements for chemical safety. For example, the PEL for inorganic mercury is listed as 0.1 mg/m^3. For a commonly used radioisotope of mercury, ^{203}Hg, the NRC's limit of 4×10^{-11} μCi/cc corresponds to a mass concentration of 3×10^{-8} mg/m^3. A consideration when dealing with radioactive materials that is absent in the practice of industrial hygiene in a nonradioactive environment is whether or not to use a respirator. This question arises when a worker is exposed simultaneously to external radiation and to radiaoctive aerosols. Using a respirator offers protection against the airborne radioactivity and therefore limits the worker's internal dose. However, wearing a respirator decreases the worker's efficiency and thus increases the time necessary to complete the job, and therefore increases his external dose. The decision regarding the use of a respirator depends on the effect of the respirator on the worker's total dose . . . the sum of the external dose and the internal dose from the inhaled radioactivity.

9 SAFETY ASSESSMENT

The effectiveness of radiation hazard controls is assessed in a surveillance program similar to that for the assessment industrial hygiene controls for nonradiation hazards. Such a surveillance program includes the observation of both workers and the environment, and may employ a variety of techniques, depending on the nature of the hazard and the consequences of a breakdown in the system of controls. These techniques may include personal monitoring for external radiation exposure, environmental radiation surveys, including measurement of external radiation levels, air sampling and continuous air monitoring, and contamination surveys; and estimation of internally deposited radioactivity by bioassay and whole body counting (*in vivo* bioassay) methods. Since humans have no sensory response to radiation, one must rely completely on instruments for our radiation safety assessment. Instruments are used in the practice of health physics for a variety of different purposes. Nevertheless, all the instruments fall into one of two categories: particle counting instruments and dose measuring instruments. Within each of these categories, instruments have been designed for specific applications, such as measurement of low energy X-rays, alpha radiation, beta rays, measurement in high backgrounds, and neutrons of various energies.

10 PARTICLE COUNTING INSTRUMENTS

Particle counting instruments are widely used as portable instruments for surveying and in the laboratory for measuring the activity in an air or water sample. Particle counting instruments are usually very sensitive detectors of radiation; they literally respond to a single ionizing particle. They therefore are widely used in searching for unknown radiation sources, leaks in shielding, and contaminated areas. Generally, they are not designed to measure dose. The radiation detector in a particle counting survey meter may be either a gas, as in the case of a Geiger counter, or a solid crystal in the case of a scintillation counter. In both detector types passage of an ionizing particle through the detector results in a burst of ionization that is converted into an electrical output signal that actuates a readout device to register a count.

10.1 Geiger-Muller (G-M) Counters

Geiger counters are among the most used and the oldest of all the portable survey instruments. This popularity is due to their relative simplicity, which makes them comparatively inexpensive, and to their wide range of usefulness. G-M counters are available in a variety of sizes and shapes, and they can be designed to respond to alpha, beta, and gamma radiation through the choice of window thickness or wall thickness. For detecting alpha and low energy beta particles, the window through which the radiation must pass in order to get into the counter's sensitive volume must be very thin. Geiger counters are very efficient in counting those alphas or betas that get into the sensitive volume. However, the inherent efficiency of G-M counters for gamma rays is very low. Gamma rays are very highly penetrating and thus can easily pass through the sensitive volume in a G-M tube. Use of a relatively thick walled G-M tube greatly increases its sensitivity for gamma rays. The gamma ray interacts with the thick wall by knocking an electron out of one of its atoms, and the electron enters into the tube's sensitive volume and triggers the counting mechanism. Of course, if the wall is too thick, the electron will not reach the sensitive volume, and the photon will not be counted. It is thus seen that the radiation response of G-M counters is energy dependent. Two of the factors in the choice of a G-M instrument, therefore are the type and energy of radiation to be detected.

There is a finite time after a particle is counted, called the resolving time, during which the G-M counter will not respond to a second particle. For G-M counters, the resolving time is ~100 μsec. If particles enter the counter at a faster rate than can be resolved, then the counter becomes paralyzed, and the meter reading falls to zero. Therefore, the counter should be turned on before entering a radiation area. If there is a very high radiation field, then the counter will respond, and the count rate will continue to increase (thereby signaling the surveyor of the possible very high radiation level) until it becomes paralyzed. Thus, the radiation level at which the counter becomes paralyzed is an important parameter in choosing an instrument.

Another important parameter in choosing or in using a portable surveying instrument is the instrument's response time. A finite time is required for the meter to give the true value of the radiation field. This response time is determined by the time constant of instrument's electronic circuitry. The time constant is defined as the time required for the meter reading to reach 63% off its final reading. Thus, the instrument must be exposed to the radiation field for several time constants in order to obtain a true measurement of the radiation field. This is of importance when scanning an area too quickly, or in measuring radiation fields of short duration. The time constant for most instruments is on the order of several seconds. Furthermore, if a surveying instrument has several ranges, the time constant increases as the sensitivity of the instrument increases. This consideration is significant, for example, in measuring the radiation level from a dental or diagnostic medical X-ray machine at the technician's position because the X-ray exposure is only a small fraction of a second. In this case, the meter would read much less than the true radiation level. Because of the inertia in the metering system of a G-M counter, most Geiger counters have an audio output that is not limited by the meter's time constant. It is therefore imperative that the health physics surveyor use the audio output, either through earphones or through a loudspeaker that is incorporated in many monitoring and surveying instruments.

A G-M counter is not designed for measuring radiation dose. However, it may be used for dosimetry only for the energy or for the isotope for which it has been calibrated. Typical full scale readings of portable G-M survey meters are about 500–50,000 counts per minute, or (when calibrated with ^{60}Co or ^{137}Cs) 0.1–10 mrads/hour.

10.2 Scintillation Counters

A scintillation detector is a transducer that converts the energy of the radiation into a flash of light, which is then converted by a photomultiplier into a voltage pulse. Whereas the inherent gamma ray detection efficiency of G-M counters is very low, usually less than 1%, the gamma ray detection efficiency of solid scintillation detectors is very high. Thus, a scintillation detector is very much more sensitive for detecting gamma radiation than a gas filled G-M detector. Furthermore, since the intensity of the flash of light in the scintillation detector is proportional to the energy of the incident radiation, a scintillation counter can, with the aid of the proper electronics, be used to distinguish among various gamma ray energies. Using this principle, scintillation survey meters are available that can be used to identify the gamma emitting isotope. The resolving time of a scintillation counter is very much less than that of a G-M counter. Scintillation counters therefore can be used in much higher radiation fields than G-M counters. However, in return for these advantages, scintillation counters are relatively expensive and relatively fragile as well as being temperature dependent. Because of its high gamma ray sensitivity, the typical scintillation survey meter has a full scale response of 0.02 mrads/hour (from ^{137}Cs or ^{60}Co). Like G-M counters, scintillation counters generally are not designed for measuring dose. Table 21.4 shows the dose rate response of G-M counter and scintillation counters that had been calibrated with radium to read 1 mrem/hr in a radiation field of that magnitude.

11 DOSE MEASURING INSTRUMENTS

The response of dose measuring instruments is proportional to the amount of energy absorbed from the radiation, rather than to the radiation flux. Dose measuring instruments used in the practice of health physics fall into two categories: Portable survey instruments and personal monitoring devices.

11.1 Portable Survey Meters

The most commonly used survey meter is of the ion chamber in which the ionization current is measured. Since a specific amount of energy is expended in creating a single ion (34 eV per ion in air), the ion current generated by the radiation within the ion chamber is proportional to the rate of energy transfer to the gas within the chamber. The same caveats regarding energy dependence and response time that applied to G-M counters also apply to ionization chambers. Generally, survey meters are relatively energy independent over a wide range of energies, from about 50 keV to about 3 MeV. The energy range can be extended downward by incorporating several windows of varying thickness into the instrument. For example, a window thickness of 10 mg/cm^2 enables an ionization chamber to measure dose from 10 keV X-rays. Ordinarily, ionization chambers are sensitive down to about 0.5 mrad/hour. However, the sensitivity can be greatly increased by increasing the gas pressure in the ion chamber. This presents the radiation with more atoms that can be ionized, thereby increasing the ion current from a given radiation field.

11.2 Personal Dosimeters

The International Commission on Radiological Protection (ICRP) recommends monitoring individual workers whose annual external doses are likely to exceed 1 mSv (100 mrems),

Table 21.4. Response of G-M and Scintillation Counters Calibrated with ^{226}Ra[a]

Isotope	Gamma Energy, MeV	G-M Counter, mrem/h	Scintillation Ctr., mrem/h
^{226}Ra	0.84	1.0	1.0
^{60}Co	1.25	1.2	0.6
^{137}Cs	0.661	0.9	1.4
^{198}Au	0.411	0.8	2.7
^{203}Hg	0.279	1.3	7.5
^{141}Ce	0.145	2.4	14.1
^{241}Am	0.06	6.0	9.8

[a]From Ref. 7.

and that workers whose annual doses are likely to exceed 5 mSv (500 mrems) "should certainly be monitored". The U.S. Nuclear Regulatory Commission (NRC) requires personal monitoring if a worker's annual external dose is likely to exceed 10% of the dose limit. Despite the legal requirement for monitoring only those workers whose dose is likely to exceed 10% of the limit, it is reasonable to monitor all radiation workers in case a question should arise in the future about the possible relationship between the worker's ill health and his occupational radiation dose. In addition to monitoring for legal reasons, knowledge of personal doses and dose distribution can contribute to the control of operating procedures and to the design of facilities. Personal monitoring for external radiation is accomplished by personal dosimeters that are worn by individual workers. These dosimeters include:

11.2.1 Film Badges

Exposure to penetrating ionizing radiation of a film wrapped in light-tight paper darkens the film, and the degree of darkening is proportional to the radiation dose. After developing the film, the degree of darkening is measured and the dose is read from a calibration curve that relates the degree of darkening to the dose. The lower limit of quantitative dose measurement with film is considered to be 10 mrems. Film badges are worn for periods ranging from 1 week to 1 month, depending on the anticipated degree of exposure.

11.2.2 Thermoluminescent Dosimeters (TLD)

Several different crystals emit light when they are heated after having been exposed to ionizing radiation. The intensity of the emitted light is proportional to the amount of energy, and hence to the radiation dose, absorbed by the crystal. This effect is called *thermoluminescence*, and dosimeters based on this principle are called *thermoluminescent dosimeters*. Absorption of energy from the radiation excites the atoms in the crystal, and the excitation energy remains trapped in the crystal until the crystal is heated and the trapped excitation energy is released as light. For personal monitoring a small piece of TL material is placed into a small holder that is worn by the worker being monitored. After being worn for the prescribed period of time, the TLD material is electrically heated and the resulting luminescence is measured with a photomultiplier tube. The instrument is calibrated by measuring the intensity of light from thermoluminescent phosphors that had been exposed to known doses of radiation. Phosphors used to monitor radiation workers can measure doses down to about 1 mrem. Other phosphors are available for use in environmental monitoring that are sensitive to doses as low as 0.01 mrem.

11.2.3 Pocket Dosimeters

Pocket dosimeters are the size and shape of a fountain pen. Two different types of pocket dosimeters are in common use. One of these, a condenser type, is simply a capacitor that is fully charged before exposure to radiation. Ionization produced by the radiation neutralizes the charge on the capacitor, and thus reduces the voltage across the capacitor. This reduction in voltage is directly proportional to the radiation dose. The instrument is charged before exposure, and the dose read after exposure, in a charger-reader. Commercially available condenser type dosimeters measure integrated X- or gamma ray doses up to 200 mrems, with an accuracy of about ±15% for quantum energies of about 50 keV to 2 MeV.

The second type of pocket dosimeter is direct reading (and therefore is often called an "SRD", a Self Reading Dosimeter). It operates on the principle of a gold-leaf electroscope. A gold-plated quartz fiber is displaced electrically by charging it to a potential of about 200 V at which time the instrument reads zero. An image of the fiber is focused on a scale by an optical system within the SRD and is viewed through a lens at the other end of the instrument. Exposure to radiation discharges the fiber, thus allowing it to move back towards its uncharged position. The amount of discharge, and therefore the position of the fiber, is directly proportional to the radiation dose. These dosimeters too read up to 200 mrems of X- or gamma radiation with an accuracy of ±15% across an energy span of about 50 keV to 2 MeV.

11.2.4 Electronic Dosimeters

Electronic dosimeters are small, lightweight (~100 g), accurate (~ ±10%), dosimeters that measure and display instantaneous dose rate and integrate over time to obtain, display, and store the cumulative dose. Personal electronic dosimeters are approximately energy independent from about 60 keV to about 1.5 MeV, can measure dose rates of 0.1 mrem (1 μSv) per hour to 1000 mrems (10 mSv) per hour, and store doses from 0.1 mrem (1 μSv) to 1000 rems (10 Sv). They also alarm at any pre-set dose rate or integrated dose, and can be downloaded into a computer for dose tracking and record keeping purposes.

12 RADIATION SAFETY SURVEYS

A survey is a systematic set of environmental measurements made in order to determine one or more of the following:

- An unknown radiation source.
- Dose rate from known sources.
- Surface contamination.
- Atmospheric contamination.

The surveyor must choose the proper instruments and use them correctly in order to make these determinations.

12.1 Choosing a Health Physics Instrument

The choice of a health physics surveying instrument for a specific application depends on a number of factors. Some general requirements are the same as those for other industrial hygiene surveying instruments, such as portability, mechanical ruggedness, and reliability. They also should:

- Be calibrated for the radiation that they are designed to measure.
- Be sufficiently sensitive and appropriate for the radiation levels to be detected and measured.
- Be responsive to the radiation being measured. This important point can best be clarified with a practical example. A commonly used beta-gamma probe has a 30 mg/cm^2 wall. This probe would be worse than useless if one wished to survey for low beta energy contaminant, such as ^{14}C or ^{35}S, or for an alpha emitting contaminant such as ^{210}Po. The range of the radiation from each of these radioisotopes is less than the probe's wall thickness, and therefore would not be detected. If, in fact, there would be severe contamination, use of this probe would falsely indicate safe conditions.
- Be used to make measurements appropriate to the instrument's response time. A short time constant signifies that the instrument has a short response time. This means that the instrument responds to rapid changes in radiation level, such as would be experienced when passing the probe over a small area of contamination or over a crack in a radiation shield. A fast response time, however, means a decrease in sensitivity. Furthermore, a fast response time may result in rapid fluctuations in the meter reading, thus making it difficult to obtain an average reading. Many instruments offer a range of response times, the appropriate one being selected by the surveyor.

11.3 Surface Contamination

Surface contamination can be located with a sensitive detector, such as a thin end-window G-M counter. After finding a contaminated area, a dose measuring instrument may be employed to measure the dose rate from the contamination. The main concerns about contamination are transmission of the contaminant into the body via inhalation or ingestion following tactile transfer to the skin, or transmission of the contaminant to other clean areas. To estimate the transmission hazard, a *smear test* is performed to determine whether the surface contamination is fixed or whether it is loose, and therefore transmissible. A smear test consists of wiping the suspected area with a piece of filter paper and then measuring the activity on the paper. The area to be smeared is determined by the extent of the contamination and the physical conditions under which the survey is made; an area of 100 cm^2 is usually smeared. It must be emphasized that a smear test is a qualitative test whose chief purpose is to allow an estimate to be made regarding the degree of transmissibility of the contaminant. If significant transmissible contamination is found, then prompt decontamination procedures are recommended.

11.4 Air Sampling

Air sampling in a radioactive environment is essentially the same as sampling in the ordinary industrial environment. The air sampling systems used in the two environments are exactly the same. First an appropriate collecting device, such as a filter, a liquid absorber, or a charcoal adsorbing medium, depending on the nature of the contaminant is used. Then a metering device for measuring the volume of air sampled, and finally a pump for sucking the air through the collector is used. Although the overall sampling strategy is the same in non-radioactive and in radioactive environments, there are some differences in the details. The size of the sample is determined by the analytical capabilities of the laboratory. In the case of radioactivity, the half life of the contaminant may influence the sample size. If the half life is very short, the collected activity will decay while the sample is still being collected, and may reach a steady state where the quantity of activity that decays is equal to the quantity collected. Under these conditions, increasing the sampling time does not increase the sample size. If this steady state activity is less than the lower limit of detection by the analytical procedure, then the volumetric sampling rate (not the sampling time) must be increased.

Another air sampling detail deals with the actual quantity of the contaminant. For example, OSHA's permissible exposure level (PEL) for airborne lead is 0.05 mg/m^3. For 1 micron mass median aerodynamic diameter (MMAD) particles, the PEL corresponds to 3.13×10^8 particles/m^3. NRC's derived air concentration (DAC) for lead-210 is 1×10^{-4} μCi/m^3, which corresponds to a 1 micron sized particle concentration of 8 particles/m^3. If the radiolead were uniformly dispersed throughout the air, and if one wished to determine 50% of the DAC, that is, 5×10^{-5} μCi/m^3, and if the lower limit of detection of the counting system is 5×10^{-6} μCi, the required sample volume would be calculated as:

$$V = \frac{5 \times 10^{-6}\ \mu\text{Ci}}{5 \times 10^{-5}\ \dfrac{\mu\text{Ci}}{\text{m}^3}} = 0.1\ \text{m}^3$$

However, because the radiolead is not continuously dispersed throughout the air, but is randomly distributed as a very small number of discrete particles, we must consider the probability of catching a particle in our sample. If one wishes to detect 50% of a DAC, or 4 particles/m^3 within ±25% at the 95% CI (confidence interval), the required sample size is 16 m^3, calculated as follows:

$$25\%\ \text{of 4 particles} = 1\ \text{particle}$$
$$\text{The 95\% Cl} = \pm 1.96\sigma = 1\ \text{particle/m}^3$$
$$\sigma = 1/1.96 = 0.5\ \text{particle/m}^3$$

According to Poisson statistics:

$$C \pm \sigma = \frac{N}{V} \pm \frac{\sqrt{N}}{V}$$

$$\sigma = \frac{\sqrt{N}}{V} = \sqrt{\frac{N}{V} \times \frac{1}{V}} = \sqrt{\frac{C}{V}}$$

$$0.5 = \sqrt{\frac{4}{V}}$$

$$V = 16\ \text{m}^3$$

12 ASSESSMENT OF INTERNAL RADIOACTIVITY

Internally deposited radionuclides are monitored by a bioassay program. Two different, complementary bioassay techniques are employed to determine internally deposited radioactivity. *In vivo* bioassay means direct determination of internal radionuclides by scanning the whole body, or a selected part of the body, with a sensitive detector. This method is useful only for gamma-emitting radionuclides or high energy beta emitters that generate bremsstrahlung within the body. *In vitro* bioassay involves the analysis of body fluids, excreta, and exhaled air for the purpose of estimating the intake. Although a single measurement may suffice for determining whether or not radioactivity is present in the body, generally, repeated measurements are necessary if one wishes to achieve the best estimate of the intake and of the dose.

The rationale underlying the practice of *in vitro* bioassay is that a quantitative relationship exists among inhalation or ingestion of a radionuclide, the resulting body burden, and the rate at which the radionuclide is eliminated. From measurements of radioactivity in urine or feces, therefore, one should be able to infer the body burden, and with this information one should be able to estimate the radiation dose. Unfortunately, the kinetics of metabolism in any particular person is influenced by many factors. Thus there is a great deal of uncertainty about the exact quantitative relationships among elimination rates, body burden, and radiation dose. In most instances, therefore, bioassay data give only a reasonable estimate of the intake and dose.

13 CONCLUSION

Ionizing radiation and radioisotopes are used safely today in industry, science, and medicine because of the vast amount of knowledge about radiation physics and biology that has been gained since discovery about a century ago. Occupationally and medically exposed populations have been observed: 270,000 survivors of the nuclear bombings in Japan, those who had been exposed to fallout from nuclear weapons tests and from nuclear reactor accidents in England in 1956 and Chernobyl in 1986 (there was no exposure to the public during the Three Mile Island accident), and populations living in environments with high levels of naturally occurring radiation. More is known about radiation bioeffects than is known for most other environmental stressing agents. Coincident with the expansion of knowledge about radiation bioeffects was the development of sensitive radiation measuring instruments that enabled us to measure accurately radiation levels and quantities of radioactive materials that are far below those at which harmful radiation effects are seen. This vast amount of dose–response data enables health physicists and industrial hygienists to control the radiation environment in the workplace so that no harmful radiation effects are seen. Thus, medical, scientific, and industrial applications of radiation technology continue at levels of risk no greater than, and often less than, those associated with other applications of science and technology that are generally considered by society to be safe.

BIBLIOGRAPHY

1. *Radiological Health Handbook*, Rev. Ed., Public Health Service, U.S. Govt. Printing Office, Washington, DC, 1970.
2. H. Cember, *Introduction to Health Physics*, 3rd. ed., McGraw Hill, New York, 1996.
3. A. C. Upton, *Nonstochastic Effects of Ionizing Radiation, NCRP Proceedings No. 6*, National Council on Radiation Protection and Measurements, Bethesda, MD, 1985.
4. International Atomic Energy Agency (IAEA), *International Basic Safety Standards for Protection* against Ionizing *Radiation and for the Safety of Radiation Sources*, Safety Series No. 115-1, Vienna, 1994.
5. International Commission on Radiological Protection (ICRP), *Recommendations of the International Commission on Radiological Protection*, ICRP Publication 26, *Annals of the ICRP* (**1**) (3) (1977).
6. International Commission on Radiological Protection (ICRP), *Limits for Intakes of Radionuclides by Workers, ICRP Publication 30, Annals of the ICRP*, (**2**) (3–4) (1979).
7. D. H. Pierson, *Physics in Medicine and Biology* **7**, 450 (1963).

GENERAL REFERENCES

A. Brodsky, *Radiation Risks & Uranium Toxicity*, RSA Publications, Hebron, CT, 1996.

H. Cember and T. Johnson, *Health Physics Solutions Manual*, PS&E Publications, Silver Spring, MD, 1999.

K.F. Eckerman, A.B. Wolbarst, and A.C.B. Richardson, *Limiting Values of Radionuclide Intake And Air Concentration and Dose Conversion Factors for Inhalation, Submersion, and Ingestion*. Federal Guidance Report No. 11, EPA-520/1-88-020, U.S. Environmental Protection Agency, Washington, DC, 1988.

Environmental Protection Agency (EPA), *Radiation Protection Guidance to Federal Agencies for Occupational Exposure*, U.S. Environmental Protection Agency, Office of Radiation Programs, Washington, 1987.

N. H. Harley, "Toxic Effects of Radiation and Radioactive Materials" in C. D. Klaassen, ed., *Casarett & Doull's Toxicology*, 5th ed., McGraw Hill, New York, 1996.

International Commission on Radiological Protection (ICRP), *Report of the Task Force on Reference Man*, ICRP Publication 23, Pergamon Press, Oxford, 1975.

International Commission on Radiological Protection (ICRP), *Radionuclide Transformation Energy and Intensity of Emission*, ICRP Publication 38, *Annals of the ICRP* **11–13** (1983).

International Commission on Radiological Protection (ICRP), *1990 Recommendations of the International Commission on Radiological Protection*, ICRP Publication 60, *Annals of the ICRP* **21** (1–3) (1990).

International Commission on Radiological Protection (ICRP) *Annual Limits on Intake of Radionuclides by Workers Based on the 1991 Recommendations*, ICRP Publication 61, *Annals of the ICRP* **21** (4) (1990).

E.T. Lessard, et al., *Interpretations of Bioassay Measurements*, NUREG/CR-4884, Brookhaven National Laboratory, Upton, NY 1987.

F.A. Mettler, Jr., C.A. Kelsey, and R.C. Ricks eds., *Medical Management of Radiation Accidents*, CRC Press, Boca Raton, FL, 1990.

National Council on Radiation Protection and Measurements (NCRP), *Management of Persons Accidentally Contaminated with Radionuclides*, NCRP Report No. 65, NCRP Publications, Bethesda, MD, 1979.

National Council on Radiation Protection and Measurements (NCRP), *General Concepts for the Dosimetry of Internally Deposited Radionuclides*, NCRP Report No. 84, NCRP Publications, Bethesda, MD, 1985.

National Council on Radiation Protection and Measurements (NCRP), *Use of Bioassay Procedures for Assessment of Internal Radionuclide Deposition*, NCRP report No. 87, NCRP Publications, Bethesda, MD, 1987.

National Council on Radiation Protection and Measurements (NCRP), *Exposure of the Population in the United States and Canada from Natural Background Radiation*, NCRP Report No. 94, NCRP Publications, Bethesda, MD, 1987.

National Council on Radiation Protection and Measurements (NCRP), *Calibration of Survey Instruments Used in Radiation Protection for the Assessment of Ionizing Radiation Fields and Radioactive Surface Contamination*, NCRP Report No. 112, NCRP Publications, Bethesda, MD, 1991.

National Council on Radiation Protection and Measurements (NCRP), *Maintaining Radiation Protection Records*, NCRP Report No. 114, NCRP Publications, Bethesda, MD, 1992.

National Council on Radiation Protection and Measurements (NCRP), *Limitation of Exposure to Ionizing Radiation*, NCRP Report No. 116, NCRP Publications, Bethesda, MD, 1993.

National Council on Radiation Protection and Measurements (NCRP), *Use of Personal Monitors to Estimate Effective Dose Equivalent and Effective Dose to Workers for External Exposure to Low LET Radiation*, NCRP Report No. 122, NCRP Publications, Bethesda, MD, 1995.

National Council on Radiation Protection and Measurements (NCRP), *Deposition, Retention and Dosimetry of Inhaled Radioactive Substances*, NCRP Report No. 125, NCRP Publications, Bethesda, MD, 1997.

National Council on Radiation Protection and Measurements (NCRP), *Operational Radiation Safety Program*, NCRP Report No. 127, NCRP Publications, Bethesda, MD, 1998.

Nuclear Regulatory Commission, *Acceptable Concepts, Models, Equations, and Assumptions for a Bioassay Program*, Regulatory Guide 8.9, U.S. Nuclear Regulatory Commission, Washington, DC, 1993.

Nuclear Regulatory Commission, *Standards for Protection Against Radiation*, **10** *CFR* Part 20, U.S. Nuclear Regulatory Commission, Washington, DC, 1996.

O.G. Raabe, ed., *Internal Radiation Dosimetry*, Medical Physics Publishing, Madison, WI, 1994.

C.E. Roessler, ed., *Management and Administration of Radiation Safety Programs*, Medical Physics Publishing, Madison, WI, 1998.

United Nations Scientific Committee on the Effects of Ionizing Radiation (UNSCEAR), *Sources, Effects, and Risks of Ionizing Radiation*, Publication E.88.IX.7, United Nations, New York, 1988.

United Nations Scientific Committee on the Effects of Ionizing Radiation (UNSCEAR), *Sources and Effects of Ionizing Radiation*, Publication E.94.IX.2, United Nations, New York, 1993.

CHAPTER TWENTY-TWO

Heat Stress

Jerry D. Ramsey, Ph.D., PE and Thomas E. Bernard, Ph.D., CIH

1 INTRODUCTION

This chapter provides an explanation of heat stress and heat strain. Heat stress is considered the aggregate of external heat from the thermal environment, and internal heat from physical work activities, which collectively impose thermal load on a person. Heat strain is represented as the human responses to heat stress. The goals of this chapter are to create an understanding of the development and current status of heat stress standards and guidelines; the physiological basis for worker-environment thermal exchange; the measurement and interpretation of the thermal environment; the assessment of heat stress and heat strain; and the application of controls.

A list of abbreviations used in this chapter is provided in Table 22.1.

1.1 Significance of Heat Stress in Industry

Heat represents one of the classical physical stresses in the occupational environment. Overexposure to heat has been a problem for centuries. As industry developed, progressed through the Industrial Revolution and evolved into our present highly technological society, on-the-job potential for injury and illness from acute exposure to heat has increased far beyond that known earlier to home-centered craftsmen. Among the more dangerous original industrial vocations were those using molten materials such as glass and metal. In these first hot industries, the ever-present danger of burns, explosions and spills of molten material were well-known and accepted, as were potential illness and death from very hard physical work in excessively hot environments.

Historically, heat stress likely manifested itself first as a serious occupational hazard among armies operating in warm climates and slave laborers. Interestingly, present-day

Patty's Industrial Hygiene, Fifth Edition, Volume 2. Edited by Robert L. Harris.
ISBN 0-471-29754-2

Table 22.1. List of Abbreviations Used in this Chapter

C	Rate of heat exchange by convection between an individual and the environment.
CET	Corrected effective temperature index for estimating the effect of temperature, humidity, and air movement on the subjective sensation of warmth using globe temperature rather than air temperature. Also known as ETCR (effective temperature corrected for radiation).
CL	Ceiling limit (WBGT) recommended by NIOSH for all workers in hot jobs.
E	Rate of heat loss by evaporation of water from the skin.
$E_{\max}$	Maximum rate of evaporative heat loss at prevailing climatic conditions.
E_{req}	Rate of evaporative heat loss required to maintain body heat balance.
ET	Effective temperature, an index used to estimate the effect of temperature, humidity, and air movement on the subjective sensation of warmth.
F_{cl}	Adjustment to account for the effects of clothing on heat transfer due to convection and radiation (dry heat transfer).
F_{pcl}	Adjustment to account for the effects of clothing on heat transfer due to evaporation.
HSI	Index of heat stress derived from the ratio of E_{req} to $E_{\max}$.
h_c	Heat transfer coefficient for convection.
h_e	Heat transfer coefficient for evaporation.
h_r	Heat transfer coefficient for radiation.
M	Rate of metabolic energy production.
P_a	Partial pressure exerted by water vapor in the air.
P_{sk}	Vapor pressure exerted by water on the skin.
R	Rate of heat exchange by radiation between a person and the surrounding surfaces.
RAL	Recommended alert limits (WBGT) specified by NIOSH for protection of unacclimatized, healthy workers.
REL	Recommended exposure limits (WBGT) specified by NIOSH for protection of acclimatized, healthy workers.
RH	Relative humidity—the ratio of the water vapor pressure in the ambient air (P_a) to the water vapor pressure in saturated air at the same temperature.
S	Rate of change in heat content of the body.
T_a	Temperature of the air surrounding a body (also known as dry bulb temperature).
T_{db}	Temperature of the air (called dry bulb temperature).
T_g	Temperature inside a blackened, hollow, thin copper globe measured by a thermometer whose sensing element is at the center of the sphere (called globe temperature).
TLV®	Threshold limit values specified by the American Conference of Governmental Industrial Hygienists.
T_{nwb}	Natural wet bulb temperature measured under conditions of the prevailing (natural) air movement.
T_r	Mean radiant (surface) temperature of the material and objects totally surrounding an individual.
T_{sk}	Mean of skin temperature, usually taken at several locations and weighted for surface area.
TWA	Time-weighted average.
T_{wb}	Psychometric wet bulb temperature.
V_{air}	Air speed.
V_{eff}	Effective air speed accounting for body movement.
ULPZ	Upper limit of the prescriptive zone, basis for setting WBGT limits
WBGT	Wet bulb globe temperature, an empirical index of heat stress obtained by weighting T_{nwb} and T_g, (and T_a when outdoors with solar load).
WGT	Wet globe temperature, an empirical index of heat stress as measured within a 3-in. copper sphere covered by a black cloth kept at 100% wettedness (called also Botsball temperature).

standards of good practice for control of heat stress in the workplace stem largely from experience in the armed forces, including several studies at military training centers in the United States (1–4).

The mining industry has also provided much information on heat disorders and tolerance limits, not only because of traditionally substantial mortality and morbidity from overexposure, but also because of concern for reduced productivity that accompanies excessive thermal exposure in the workplace (5–9).

Workers in most hot jobs have typically been a highly select (survivor) population. Those who cannot cope with the prevailing hot conditions seek less demanding employment. As a result of this self-selection process, the majority of workers in hot jobs perform at a high level and are highly adaptable to work in heat (10).

In addition to ever-increasing levels of mechanization, there are widespread, often undocumented work practices in industry that help relieve workers of heat strain on excessively hot days. Such practices include the following:

1. Performing only unavoidable operations and postponing other tasks.
2. Reassigning workers in auxiliary jobs to help those who work in the hottest areas.
3. Substituting younger and more fit workers to relieve the older and less fit.
4. Reducing work output expectations to allow workers to reduce their work pace when needed.

In the so-called hot-dry industries, such as steel mills, forge shops and glass manufacturing plants, the thermal load on a worker increases from the sensible heat that escapes from process equipment and operations into the surrounding work space and from the radiant heat from the surfaces of tanks, hot metals and the like. While each contributes, peak stress conditions typically occur when both sensible and radiant heat sources are present. In this type of hot industry, little or no moisture is added to the air, except by changing weather conditions. Therefore, little or no decrement occurs in the evaporative cooling capacity of the work environment. Nevertheless, the requirements for sweating and evaporation are considerable because of the radiant and convective thermal loads.

In the warm-moist industries, such as laundries, papermaking and mines, in which large amounts of water are used, workplace humidity rises from wet processes or escaping steam. Although relatively little sensible heat transfers from the process to the work environment, high humidities greatly reduce the evaporative cooling capacity. Consequently, workers may be unable to dispose adequately of metabolic and environmental heat.

Today one would expect deaths and heat-related illnesses from involuntary overwork to be rare in the United States. One of the characteristic features of industrial exposure to heat today is learning over time to match the body's capability to maintain a thermal balance with the environment, including voluntary reduction in the metabolic generation of heat when necessary, to the demands. The Bureau of Labor Statistics in their analysis of 1997 occupational injuries reported 40 fatalities (0.6% of total fatalities) due to "contact with temperature extremes." This category primarily includes burns from contact with hot metals and fluids, with no cases definitely attributed to heat strain (11). Of these, about 15 per year are suspected to be fatal heat strokes (12). In contrast, nonoccupational heat-related casualties are commonly encountered during periods of extended heat waves in the United States, but these appear to be dominated by premature rather than extra mortality among the elderly. The National Center for Health Statistics reported an annual average of 229 deaths from excessive heat during the six years between 1989–1995 (13), where an excessive heat casualty was a death concurrent with a body temperature in excess of 40°C.

1.2 Development of Heat Stress Standards

The search for means of determining heat stress dates back to 1905, when Haldane reported wet bulb temperature preferable to dry bulb temperature as an indicator of physiological effects (14). The first American standard for work in hot environments dates back to 1947, when the Committee on Atmospheric Comfort published its report, *Thermal Standards in Industry* (15). The criteria and permissible exposure limits in this standard were intended only as a guide. The committee recommended that each industry develop its own standard because of the complexity of industrial work and the variation in tolerance to heat among individual workers. As general criteria, the committee cited the comfort and health of individuals, their work output, and their physiological and psychological reaction to work. It used an index known as effective temperature (ET) as the basis of its guidelines, but the guidelines included exposure limits for only two levels of work. The higher level was 500W, and the lower was given only in the qualitative terminology, i.e., "light sedentary activities."

1.2.1 Current Status—OSHA

The only federal standard in the United States for occupational exposure to heat stress is that implicit in the general duty clause of the Occupational Safety and Health Act of 1970 (16). This legislation stipulates that each employer has a general duty to provide each employee "employment free from recognized hazards causing or likely to cause physical harm." The Occupational Safety and Health Administration (OSHA) has received recommendations from both the National Institute for Occupational Safety and Health (NIOSH) and a statutory Standards Advisory Committee on Heat Stress, but a specific OSHA standard covering exposure to heat has not been issued. Compliance officers, however, have guidance in the evaluation of heat stress in their technical manual (17).

Any heat stress standard proposed by OSHA would likely be a work practices standard. This means that when the value of an environmental index, such as wet bulb globe temperature (WBGT), reaches a threshold level based on various levels of work load, certain actions and procedures would be required to minimize the effect of exposure.

1.2.2 NIOSH Recommendations

The National Institute for Occupational Safety and Health (NIOSH) first published its *Criteria for a Recommended Standard. . . . Occupational Exposure to Hot Environments* in 1972 (10). This initial document was based on the generally accepted premise of the relationship between combinations of environmental heat and metabolic heat that would allow physiological control of body core temperatures below 38°C (100.4°F) for extended periods of time. This region of physiological response is called fully compensable heat stress.

Because the technology for physiological monitoring was limited at the time, NIOSH experts selected the common strategy of industrial hygienists to assess exposure rather than body burden. The scheme must rely on the measurement of workplace factors that predict the acceptable limits of physiological responses to heat. Of the several environmental indices of heat stress that were available, the WBGT was the one most commonly used for assessing heat stress for exposed workers.

NIOSH proposed use of the WBGT index as a reasonable way to measure environmental heat, but it assumed that the metabolic heat was too difficult to include in a regulatory standard. It established a threshold limit at 26°C (79°F) based on the metabolic heat associated with the most stressful condition, i.e., a heavy workload. NIOSH recommended that when exposure exceeds this level, any one or a combination of work practices should be initiated to ensure that body core temperature does not exceed 38.0°C (100.4°F). In addition, the Criteria Document outlined other requirements for management when exposures exceed 26°C (79°F) WBGT. These included preplacement and periodic medical examinations, WBGT profiles for each workplace, and detailed record keeping.

The NIOSH Criteria Document received widespread negative response from the hot industries, especially in the southern states, where outdoor WBGT values exceed 79°F nearly every day during the warm seasons (18). Thus it proved easy to determine, but it was an extremely conservative limit. In 1973, the U.S. Secretary of Labor appointed a Standards Advisory Committee on Heat Stress to help resolve some of the discrepancies in and objections to the NIOSH Criteria Document. The Standards Advisory Committee recommendations modified those of the NIOSH Criteria Document in several different aspects (19). Committee recommendations for use of a two-hour time-weighted average and higher limits for increased air velocity were directed at effectively raising the limits as proposed by NIOSH. They also felt it important to clarify that the proposed limits were not upper limits, but rather threshold or beginning levels where the risk of heat strain starts increasing and protective work practices would be indicated. The Committee voted 10 to 5 to support these draft changes, but public hearings generated considerable data and evidence that occupational heat stress was not a major problem in U.S. industry (19).

In 1979, NIOSH convened a workshop to which experts from industry, labor, OSHA and academia were invited to review advances in instrumentation and knowledge concerning the affects of heat stress on workers safety and health (20). This workshop established a cadre of review consultants who were used extensively by NIOSH in a 1986 revision, *Criteria for a Recommended Standard . . . Occupational Exposure to Hot Environments* (21). In its revised recommendations, NIOSH relied on the use of recommended alert limits (RALs) and recommended exposure limits (RELs). It asserted that nearly all healthy workers who are not acclimatized to working in hot environments and who are exposed to combinations of environmental and metabolic heat less than the NIOSH RALs should be able to work without increased risk of acute adverse health effects. Healthy workers who are acclimatized to working in hot environments and who are exposed to combinations of environmental and metabolic heat less than the NIOSH RELs should be capable of working without adverse effects. The estimates of both environmental and metabolic heat are expressed as 1-hr time-weighted averages (21).

NIOSH also recommended that no worker be exposed at combinations of environmental and metabolic heat exceeding its ceiling limits without adequate and appropriate personal protection.

In the revised Criteria Document, healthy workers referred to those not excluded from placement in hot-environment jobs by any of several criteria. These exclusionary criteria are qualitative in that the epidemiological parameters of sensitivity, specificity and predictive power of the evaluation methods were not documented fully. The recommended exclusionary criteria represented the best judgment of NIOSH based on the available data and comments of peer reviewers. These include both absolute and relative indicators related to age, gender, percent body fat, medical and occupational history, specific chronic diseases or therapeutic regimens, and the results of clinical tests such as maximum aerobic capacity, electrocardiography, pulmonary function, and chest X-rays.

A brief summary of key recommendations in the revised NIOSH Criteria Document follows (21):

Acclimatized Workers: Total heat exposure to workers should be controlled so that unprotected healthy workers who are acclimatized to working in hot environments are exposed to combinations of metabolic and environmental heat no greater than the applicable REL shown in Figure 22.1.

Unacclimatized Workers: The applicable RAL to be used for workers not yet acclimatized is depicted as a curve below the REL shown in Figure 22.1. The RAL is located 2°C below at 175W, 3°C below at 400W and 4°C below at 580W.

Ceiling Limits: No worker should be exposed to combinations of metabolic and environmental heat exceeding the applicable ceiling limits (CL) in Figure 22.1 unless provided with and properly using appropriate and adequate heat-protective clothing and equipment.

Environmental Heat Exposures: These are assessed by the WBGT method or an equivalent technique with conversion to WBGT values. When air- or vapor-impermeable protective clothing is worn, the dry bulb temperature or the adjusted dry bulb temperature is a more appropriate measurement than WBGT. Environmental heat measurements should be taken at or as close as possible to the work area where exposure takes place. When a worker in a hot job moves between two or more areas with variable levels of environmental heat, or when the environmental heat varies substantially in the single hot area, the environmental heat exposures must be measured at each area and during each period of constant heat levels where employees are exposed. Hourly time-weighted average (TWA) WBGTs must be calculated for the combination of jobs (tasks), including all scheduled and unscheduled rest periods.

Modifications of Work Conditions: Environmental heat measurements must be taken at least hourly during the hottest portion of each work shift, during the hottest months of the year, and when a heat wave occurs or is predicted. If two such sequential measurements exceed the applicable RAL or REL, work conditions must then be modified by use of appropriate engineering controls, work practices or other measures until two sequential measures conform with the exposure limits of the NIOSH-recommended standard.

1.2.3 Threshold Limit Values

In 1971, the American Conference of Governmental Industrial Hygienists (ACGIH) proposed and later adopted threshold limits values (TLVs®) for heat stress (22). These TLVs

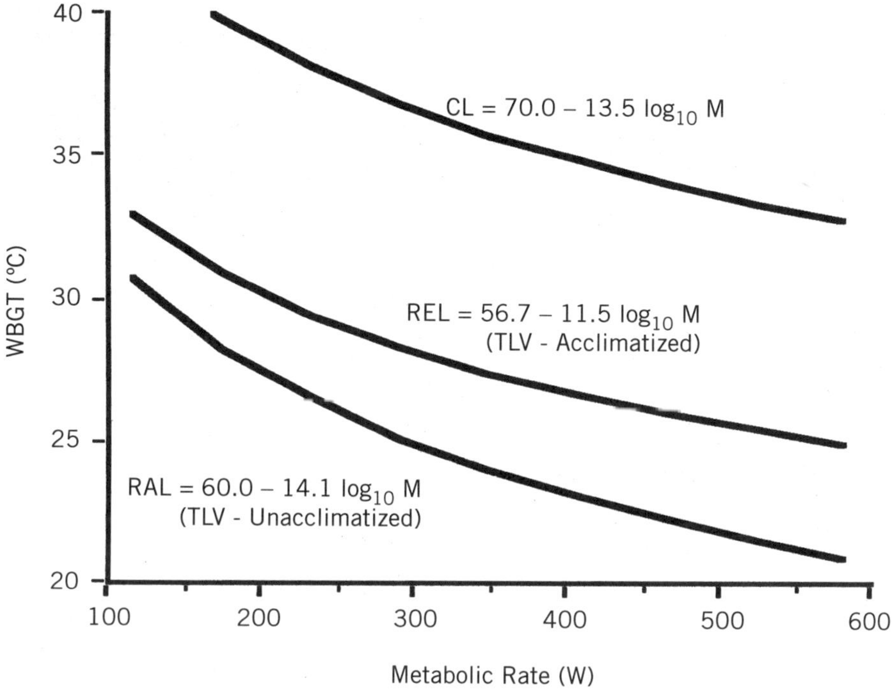

Figure 22.1. NIOSH recommended heat stress alert limits (RAL) and recommended exposure limits (REL) as well as the ceiling limit (CL). (Note: The TLV for unacclimatized workers is equivalent to the RAL, and the TLV for acclimatized workers is the same as the REL.). (This figure is provided courtesy of the National Safety Council, Itasca, Illinois.) (40).

have remained essentially unchanged in subsequent annual updates through 1999 (23). These TLVs also rest on the assumption that an acclimatized, fully clothed worker whose deep body temperature is less than 38°C (100.4°F) is not subjected to excessive heat stress. Although both the NIOSH and ACGIH utilized the same basic relationships between environmental heat (WBGT) and metabolic heat (M) to determine the threshold exposure limits, the TLVs included the information in tabular format as well as the graphic form depicted in Figure 22.1 (21). The table extracted from this graph, however, was represented in its most conservative interpretation in that WBGT TLVs were associated with the maximum metabolic rate in the range. This meant for example the TLV of 25°C for all heavy work (between 400 W and 580 W) was selected at an M of 580 W, a level which probably never occurs as an hourly time-weighted average in a workplace in the United States.

The 1999 Notice of Intended Change to the ACGIH TLVs proposed a change of title from "Heat Stress" to "Heat Stress and Heat Strain," and also reflects important changes in content and format (23).

Table 22.2 shows the current TLVs for heat exposure and the 1999 Notice of Intended Change. The proposed TLV values have been increased for each combination of metabolic heat; for example the heavy workload TLV of 25°C has been raised to 26.5°C as shown in Table 22.2. The increases are consistent with previous ACGIH and NIOSH graphs showing environmental and metabolic heat (see Figure 22.1). The difference comes from extracting the TVL table value at a metabolic rate somewhat less than the maximum of the work rate range rather than at the previously used maximum value. This provides a rational and appropriate, but less conservative TLV.

The TLV also provides information such as: the measurement and calculation of WBGT; correction factors for different clothing ensembles; methods for assessment of work load; water and salt supplementation; acclimatization; and adverse health effects. The proposed changes for 1999 present a simplified format for this information and add more detailed WBGT evaluations and rational heat stress assessments as well as physiological criteria for heat-stressed workers, which better addresses the protection of those wearing impermeable clothing.

1.2.4 International Standards and Recommendations

There has been considerable activity in the development of international standards concerning thermal environments. The European Community through its CEN (Comité Européen de Normalisation) is authorized to establish standards which will eventually replace the numerous national standards from France, Germany, United Kingdom, Belgium, Denmark, Greece, Italy, Luxembourg, Portugal, and Spain. Other countries such as Austria, Finland, Iceland, Norway, Sweden, and Switzerland are also actively involved in CEN standards (24).

The International Organization for Standardization (ISO) during the past 10 years or so has developed and has continued to work on environmental ergonomic standards through its technical committee ISO/TC 159, Ergonomics. Most of these ISO standards will be

Table 22.2. Current and Proposed[a] (in brackets) TLVs for Heat Stress Exposure[b] where WBGT Criterion Values for Work Rate Categories are Based on Acclimatization State and Percent of Work in a One to Two Hour Cycle

	Acclimatized			Unacclimatized		
Activity	Light	Moderate	Heavy	Light	Moderate	Heavy
100% Work	30.0 [30.0][a]	26.7 [28.0]	25.0 [26.5]	26.5 [27.0]	24.5 [25.0]	22 [22.5]
75% Work: 25% Rest	30.6 [31.0]	28.0 [29.0]	26.9 [27.5]	— [28.5]	— [26.5]	— [24.5]
50% Work; 50% Rest	31.4 [32.0]	29.4 [30.0]	27.9 [28.5]	— [30.0]	— [28.0]	— [26.0]
25% Work; 75% Rest	32.2 [32.5]	31.1 [31.5]	30.0 [30.5]	— [31.0]	— [29.5]	— [27.5]

[a]Values in Notice of Intended Change 1999.
WBGT values are expressed in °C and are rounded to the nearest half degree.
Note: If work and rest environments are not the same, an hourly time-weighted average WBGT must be calculated and compared to table value.
[b]Ref. 23.

proposed directly as CEN standards. Table 22.3 includes a list of ISO environmental ergonomic standards, many of which are available in the USA from American National Standards Institute (ANSI) as ANSI/ISO Standards. More detailed descriptions of the ISO standards shown in Table 22.3 are also available (25, 26).

ISO 7243 (27) which uses WBGT to estimate heat stress on working man, closely resembles the ACGIH TLV and the 1986 NIOSH Criteria Document. The basic premise for each is that no worker be exposed to any combination of environmental heat and physical activity that would cause the body core temperatures to exceed 38°C (100.4°F). The threshold values recommended are all very similar except that the ISO standard provides for differential WBGT limits based on air movement for metabolic rates greater than 360 W. These WBGT reference values are shown in Table 22.4.

Table 22.3. ISO Environmental Ergonomic Standards

Standards	References
Hot, Moderate and Cold Environments	
ISO 7243—Estimation based on WBGT Index	27
ISO 7933—Required sweat rate for hot environments	28
ISO 7730—Moderate thermal environments	29
ISO/TR 11079—Cold environments-required clothing insulation	30
ISO/DIS 11399—Principles and applications of standards	31
General Environmental Ergonomics	
ISO 7726—Measuring physical environments	32
ISO 8996—Metabolic heat	32
ISO 9886—Physiological measurements	34
ISO 9920—Clothing	35
ISO 10551—Subjective assessment	36
ISO/CD 12894—Medical supervision	37

Table 22.4. ISO Reference Values of the WBGT Heat Stress Index[a]

Metabolic Rate Class	Total Metabolic Rate M (W)	Reference Value of WBGT (°C)			
		Acclimatized		Unacclimatized	
0 (resting)	$M < 117$	33		32	
1	$117 < M < 234$	30		29	
2	$234 < M < 360$	28		26	
3	$360 < M < 468$	25[b]	26[c]	22[b]	23[c]
4	$M > 468$	23[b]	25[c]	18[b]	20[c]

[a]Ref. 27.

Note: Total metabolic rate for a standard man with a surface area of 1.8 m^2

[b]No sensible air movement

[c]Sensible air movement

A comparison of threshold (reference) values for acclimatized workers as proposed by ACGIH, AIHA, OSHA, NIOSH and ISO can be found in the chapter on thermal standards and measurement techniques in the *Occupational Environment—Its Evaluation and Control* (38).

2 HEAT EXCHANGE BETWEEN THE WORKER AND THE ENVIRONMENT

Normal body core temperature is 37°C (98.6°F) with a daily variation of ±1°C (1.8°F). Core temperature is maintained within this narrow band through physiological adjustments that modulate the rate of heat exchange between the body and the ambient environment. The heat exchange mechanisms are governed by the fundamental laws of thermodynamics. In essence, heat stress is a problem of removing the heat generated by the body during work. This section describes the basic principles of heat exchange.

2.1 Heat Balance Equation

The basic heat balance within the body and between the body and the environment can be described by the following equation (24, 28, 39, 40).

$$S = M + C + R + E \tag{1}$$

where S = rate of change in body heat content
M = metabolic rate or rate of heat produced within the body
C = rate of convective heat exchange
R = rate of radiant heat exchange
E = rate of evaporative heat loss (E is always negative for evaporative loss.)

The equation forms the basis for understanding heat stress both qualitatively and quantitatively. For a quantitative analysis, measurements or reliable estimates of metabolic heat production, air temperature, air water vapor pressure (humidity), air velocity, and mean radiant temperature are necessary.

2.2 Metabolic Heat Production

A logical first step is to look at the amount of energy released inside the body. The energy required to sustain life and function comes from the exothermic oxidation, under enzymatic control, of food-derived fuel: carbohydrates, fats and proteins (41). Although such oxidation occurs at rates and temperatures much different from those during equivalent combustion in the external environment, the thermodynamics of the oxidation is the same in both instances. Thus the same quantities of food substrates yield essentially the same thermal energy when oxidized at low temperature inside the body or at high temperature outside the body. This equivalency provides a basis for indirect calorimetry, allowing metabolic heat production to be measured indirectly by the rate of oxygen uptake of the body. One liter of oxygen supports the release of 21,000 J of energy in the form of high energy phosphate bonds (i.e., ATP) and heat. Breaking the high energy phosphate bond to support cellular function including muscular work releases more energy as heat. An average man at rest consumes about 0.3 liter of oxygen per minute, equivalent to a metabolic heat production of 100 W. During work, the rate of heat generation can exceed 500 W.

The energy cost of an activity (metabolic heat) is the major contributor to body heat gain, which must be dissipated to the environment to prevent heat storage. The rate of metabolic heat generation (M) can be either measured indirectly through determination of oxygen consumption or estimated by observation. Most often, it is estimated. There is a tendency for industrial hygienists to place the energy expenditure into one of three or four broad categories (e.g., light, moderate, heavy, very heavy). This approach leads to considerable imprecision in the process and, in practice, there is a tendency toward overestimating the demands. Estimation with tables of reference activities is a frequently used alternative, but still requires professional judgment. The most appropriate method to estimate metabolic rate for any analysis beyond a cursory assessment is a task analysis. The estimate of metabolic rate for a job is the time-weighted average of the tasks within each job.

NIOSH (21) recommended, and the ACGIH (23) refers to, an estimation method, which was borrowed from discussions on what was eventually described in an ISO standard (33). The method uses a component analysis to examine each task within a job. Following this method, the total metabolic rate of a specific task is the sum of five components following Equation 2 with assigned values as described in Table 22.5.

$$M_{\text{task}} = B + P + W + H + V \tag{2}$$

where B = basal metabolic rate (fixed value)
P = metabolic rate associated with maintaining a sitting or standing posture (one value for each posture)

Table 22.5. Components for Estimating Metabolic Rate[a]

Component	Subcomponent	Metabolic Rate in Watts by Component			
Basal (B)					80
Posture (P)	Sit				20
	Stand				45
Work[b] (P)		Light	Moderate	Heavy	Very heavy
	None	0	0	0	0
	Hand(s) only	25	55	70	80
	One arm	65	100	135	170
	Both arms	115	155	190	230
	Whole body	225	340	505	700
Horizontal (H)	For average horizontal distance (m) traveled in 1 minute (h): H = 3.3 h				
Vertical (V)	For average vertical distance (m) traveled in 1 minute (v): V = 56 v				

[a]Refs. 33.
[b]Effort level:
Light, Can be performed indefinitely with ease.
Moderate, Can be performed indefinitely with some effort.
Heavy, Can be performed for 30–60 minutes before a break.
Very heavy, Can be performed for about 15 minutes before a break.

W = metabolic rate associated with the general activity (value depends on degree of body involvement and degree of effort)
H = metabolic rate for average horizontal movement in one minute (value is multiple of the average rate of horizontal travel) (e.g., $H = 3.3$ (m/min) watts)
V = metabolic rate for average vertical movement in one minute (value is multiple of the average rate of vertical travel) (e.g., $V = 56$ (m/min) watts)

Alternative methods for estimating the metabolic rate from task analysis have been proposed by others (42–44). Inter-individual variation in metabolic heat production is relatively small for sedentary or light activities, but wide differences exist for more demanding work, depending largely on body weight and amount of external work accomplished (41).

2.3 Environmental Heat Exchange

In terms of heat load on a person, metabolic heat is purely an internal source. Because workers exist within an ambient environment, they will also exchange energy with the environment. The major avenues of heat exchange between a worker and the environment are convection, radiation and evaporation. Other than for brief periods of body contact with hot surfaces (e.g., tools, equipment, floors) conduction plays a minor role in occupational heat stress.

2.3.1 Convection

The rate of convective heat exchange between a person and the ambient air is a function of (*1*) the temperature difference between the ambient air (dry bulb temperature) (T_{db}) and the mean skin temperature (T_{sk}), (*2*) the coefficient of convection (h_c) and (*3*) the degree of isolation due to clothing (F_{cl}) (24).

$$C = F_{cl} h_c (T_{db} - T_{sk}) \tag{3}$$

The temperature difference establishes the direction of heat flow (toward the lower temperature) as well as the maximum flow rate. Mean skin temperature is usually assumed to be 36°C. When $T_{db} > 36$°C (97°F), there is a gain in body heat from the ambient air by convection (positive values of C); when $T_{db} < 36$°C, heat is lost from the body to the ambient air by convection (negative values of C). The factor h_c relates the energy flow rate to the temperature difference as a function of effective air speed around the person (V_{eff}). Values for h_c can be calculated from information provided in Table 22.6. F_{cl} is a multiplier with values from 0 to 1.0 that describes what fraction of the maximum convection associated with a nude person is permitted by the clothing. If the person is totally isolated convectively from the ambient environment the value of F_{cl} would be zero. Values for F_{cl} for some clothing ensembles is provided in Table 22.7 (45).

2.3.2 Radiation

Radiant exchange (R) is a function of the difference between the mean radiant (black body) temperature of the solid surroundings (T_r) and skin temperature. Radiant heat exchange is

Table 22.6. Computation of Coefficients of Convection (h_c) and Radiation (h_r)[a] for a Standard Man with Body Surface Area of 1.8 m^2

Adjust air velocity (V_{air}) in m/s for work activities if M (W) >100

$V_{eff} = V_{air} + 0.0029(M - 100)$

Limit the increase due to metabolic rate to 0.7 m/s

Compute Coefficient for Convection (h_c) (W/°C)

$h_{cn} = 4.28(|T_{db} - 36|)^{0.25}$

If $V_{eff} \leq 1$ m/s: $h_{cf} = 6.3 + 9.4V_{eff}$

If $V_{eff} > 1$ m/s: $h_{cf} = 16V_{eff}^{0.6}$

h_c = Greater value of $[h_{cn}, h_{cf}]$

Compute Mean Radiant Temperature (T_r)

ζ = Greater value of $[0.4(|T_g - T_{db}|)^{0.25}, 2.5V_{air}^{0.6}]$

$T_r = 100(((T_g + 273)/100)^4 + \zeta(T_g - T_{db}))^{0.25} - 273$

Compute Coefficient of Radiation (h_r) for a standing person (W/°C)

$h_r = 7.4 \times 10^{-8}[(T_{sk} + 273)^4 - (T_r + 273)^4]/(T_{sk} - T_r)$

[a]Refs. 40.

Table 22.7. Experimentally Derived Clothing Factors for some Common Clothing Ensembles[a]

Clothing	F_{cl}	F_{pcl}	WBGT Adj (°C)
Cotton work clothes w/ undershirt	0.85	0.53	0
Cotton coveralls w/ hood	0.64	0.40	3.5
Coveralls w/ hood of DuPont Tyvek® 1422A	0.82	0.44	2
Coveralls w/ hood of SMS polypropylene	0.82	0.62	−1
Coveralls w/ hood of microporous film laminate	0.83	0.39	2.5
Flame-retardant cotton shirt and pants w/ undershirt	0.81	0.45	1

[a]Ref. 45.

a function of the fourth power of the absolute temperatures of the surfaces involved, but a first-order approximation is sufficiently accurate for estimating R in the applications outlined here (24).

$$R = F_{cl}h_r(T_r - T_{sk}) \quad (4)$$

Similar to convection, radiant heat exchange depends on the temperature difference to establish the direction of heat flow (toward the lower temperature) as well as the maximum flow rate. Generally, when $T_r > 36$°C (97°F), there is a heat gain (positive values) and when $T_r < 36$°C, heat is lost. The factor h_r relates the energy flow rate to the temperature difference. Table 23.6 describes the computation of h_r. Again, F_{cl} is the same multiplier as for convection with values from 0 to 1.0.

2.3.3 Evaporative Cooling

The evaporation of water (sweat) from the skin surface results in significant heat loss from the body. To avoid heat storage, the net gain of heat from the metabolic demands plus the dry heat exchange from the environment must be balanced by evaporative cooling (24). The required evaporative cooling (E_{req}) is

$$-E_{req} = M + C + R \tag{5}$$

During heat stress conditions, the body will adjust the sweat rate to achieve the required amount of evaporative cooling. There are two limits to evaporative cooling. The maximum evaporative capacity ($-E_{\max}$) is a physical limit and is a function of (*1*) the water vapor pressure difference between the ambient air (P_a) and the wetted skin (P_{sk}), which is the saturation pressure at skin temperature, (*2*) the coefficient of evaporation (h_e) which is a function of air motion (V_{eff}), and (*3*) the reduction due to the water vapor permeability of the clothing (F_{pcl}). The equation defining this relationship is

$$-E_{\max} = h_e F_{pcl}(P_{sk} - P_a) \tag{6}$$

The nominal water vapor pressure on skin at 36°C is 5.94 kPa (45 mmHg). The value of $E_{\max}$ is always negative, meaning that evaporative cooling represents a loss of heat from the body. As the humidity (or water vapor pressure) in the ambient air increases, the gradient for evaporative cooling decreases. The thermodynamic coefficient for evaporative cooling is equal to the Lewis Number (L) times the convective heat transfer coefficient (i.e., $h_e = Lh_c$) (24). For all practical purposes, the Lewis Number is a constant (and, in SI units, equal to 16.7 °C/kPa). The restriction on water vapor movement through the clothing is represented by F_{pcl}. It has a value of 1.0 for the unrestricted (nude) condition and zero for a complete loss of evaporative capability. Estimates of F_{pcl} for some ensembles are provided in Table 22.7 (45).

The other limit to evaporative cooling is physiological. There is a maximum rate of sweat production that can be achieved. The maximum rate of sweat production for an acclimatized person is about 1 L/hr, equivalent to an evaporative heat loss of about 400 W/m^2 or 720 W for the standard man of 1.8 m^2.

2.4 Effects of Clothing

The principal purpose of clothing in a work environment is to provide a barrier against hazardous chemical, physical, and biological agents. But clothing also alters the rate of heat exchange between the skin and the ambient air by convection, radiation and evaporation; and this effect can be substantial. When estimating worker-environment heat exchange, correction factors must be used to reflect the type, amount and characteristics of the clothing being worn. The clothing corrections for dry heat exchange (F_{cl}) and evaporative cooling (F_{pcl}) are nondimensional and can be estimated from bench tests and wear tests (24, 45). Table 22.7 provides some estimates based on wear tests in a controlled environment (45).

The recent TLVs include some basic guidelines for handling this effect (25). Adjustments to the measured WBGT for clothing ensembles other than ordinary work clothes can be made. Some of these are provided in Table 22.7, but some of these suggested corrections have not been substantiated by long-term industrial experience.

3 PHYSIOLOGY OF HEAT STRESS

Our current understanding of the physiological effects of working under hot conditions has emerged from classic laboratory and field studies (see Ref. 46 for a comprehensive overview of thermal physiology and Ref. 47 for the basics). Regardless of the thermal environment, humans attempt to maintain steady heat content and body temperature through a set of involuntary compensating mechanisms. In that context, several physiological systems influence body temperature regulation, and suggest the basis for heat strain, risk of heat disorders, and individual tolerance factors.

3.1 Body Temperature Regulation

As homeotherms, humans must maintain an internal body temperature within a narrow range near 37.0°C. Hyperthermia results when body core temperature exceeds 40°C; and death will likely occur when body core temperature exceeds 42°C.

Humans inherent thermal homeostasis provides temperature regulation within suitable limits by the involuntary control of blood flow from sites of metabolic heat production (muscles and deep tissues) to the cooler body surface (skin). Here, heat loss takes place through the mechanisms of radiation, convection and evaporation (see Section 2.3). Thus, a worker reacts dynamically to changes in the thermal environment, always striving to maintain a core body temperature within the critical range.

3.1.1 Hypothalamic Regulation of Temperature

The temperature regulating center for humans lies in a region at the base of the brain known as the hypothalamus. The hypothalamus responds to increases in its own temperature as well as to incoming nerve impulses from temperature receptors in the skin. It activates heat loss through increased blood flow to the skin. Such physiological reactivity to elevated body temperature leads to interaction with the thermal environment so as to offset the increase, thus maintaining body temperature within an acceptable range.

3.1.2 Muscular Activity

Muscles represent the largest single group of tissues in the body, typically accounting for about 45% of body weight. At rest, however, the muscles account for about 20 to 25% of the total metabolic heat production.

The heat produced by the muscles during exercise or work can be many times higher than at rest. That is, most of the energy released in the body during work is released in the muscles. This heat must be moved to the skin convectively by blood flow and then dissipated to the environment to maintain a heat balance.

3.1.3 Circulatory Regulation

The circulatory system provides the transport mechanism needed to deliver oxygen and foodstuffs to all tissues and to convey unwanted metabolites and heat away from the tissues. The autonomic nervous system and endocrine system control the allocation of blood flow among competing organ systems. Nevertheless, the heart may not provide enough cardiac output to meet the peak needs of all of the body's organ systems as well as the need for dissipation of body heat.

During exercise or physical work, there is widespread sympathetic vasoconstriction throughout the body, even in the cutaneous bed. The increase in blood supply to the active muscles is assured by the action of locally produced vasodilator substances and by neural input. In inactive vascular beds, there is progressive vasoconstriction with increasing intensity of the muscle activity. As body core temperature increases, there is an increase in skin blood flow. This raises skin temperature to facilitate losses (or reduced gains) by convection and radiation. With the onset of sweating, the heat is also dissipated by evaporative cooling. The circulation to the skin is matched to deliver to the skin surface the amount of excess heat necessary to maintain thermal equilibrium, if it can be maintained.

3.1.4 Sweating and Fluid Replacement

Sweating occurs over a wide range of thermal exposure in amounts sufficient to achieve enough evaporative cooling ($-E_{req}$) to offset the total heat load represented by $M + R + C$. The rate of secretion of individual sweat glands as well as the number of active glands determines the sweat rate. They are stimulated by cholinergic sympathetic nerves and secrete a hypotonic watery solution onto the surface of the skin. The central drive for sweating is determined by metabolic rate; however, actual sweat output is modulated by skin temperature to meet the evaporative requirements up to the limits for sweating capacity. Under maximum thermal load, some 2.5 million sweat glands secrete at peak rates of more than 3 kg/hr for up to an hour; highly acclimatized persons can maintain rates of 1 to 1.5 kg/hr for several hours. Sweat production at rates of about 1 L/hr have been recorded frequently in industrial work.

Evaporative cooling is then regulated to balance the heat load. This represents a large potential source of cooling if the sweat can be evaporated, where each liter of sweat evaporated from the skin surface represents a loss of about 2400 kJ of heat to the environment. Because the difference in water vapor pressure on the skin surface and the air controls the rate of sweat evaporation, environments with high humidity limit the amount of sweat that can evaporate. Sweat not evaporated drips from the skin and does not result in any heat loss from the body. Sweating efficiency is the ratio of sweat evaporated to the total produced. When $-E_{req}$ is less than 50% of $-E_{max}$, sweating efficiency is virtually 100%. Above 50% of $-E_{max}$, the efficiency drops, reaching 50% when E_{req} equals E_{max}

(24, 28). The drops in efficiency can be costly, because it represents a loss of water from the body that has provided no benefit to the person.

Replacing the water lost by sweating is imperative. It is not uncommon for workers in hot environments to lose 6 to 8 quarts of sweat (about 1 L/hr) during a work shift. If this fluid is not replaced, there is a progressive decrease of body water with a shrinkage not only of the extracellular space and interstitial and plasma volumes but also of water in the cells. There is clear evidence that the rate of sweat production depends on the state of hydration. Progressive dehydration results in lower sweat production and a corresponding increase in body temperature, a potentially dangerous situation.

Sweat lost in large quantities is often difficult to replace completely as the work shift proceeds, and it is not uncommon for individuals to register a water deficit of 2 to 3% of body weight. Because the normal thirst mechanism is not sensitive enough to ensure a sufficient water intake, every effort should be made to encourage individuals to drink water or a low-sodium noncarbonated beverage. One pint (two cups) should be consumed for every pound of weight lost by sweating. The fluid should be as palatable as possible at 10 to 15°C (50 to 60°F). Taking small quantities at frequent intervals, about 150 to 200 mL every 15 to 20 min, is a more effective regimen for practical fluid replacement than taking 750 mL or more once an hour.

Sodium and potassium are constituents of sweat. In most circumstances, a sodium deficit does not occur because normal dietary intake provides about 10 to 15 grams of NaCl per day. Nevertheless, the salt content of sweat in unacclimatized individuals may be as high as 4 g/L. Acclimatized individuals will show lower concentrations, 1 g/L or less. Thus it is possible for a heat unacclimatized individual with a low salt intake to develop a negative salt balance. In view of the high incidence of elevated blood pressure in the U.S. population and the relatively high salt content of the typical U.S. diet, recommending increased salt intake is probably not warranted (21).

3.1.5 Acclimatization to Heat

Acclimatization refers to a set of adaptive physiological and psychological adjustments that occur when an individual accustomed to working in a temperate environment undertakes work in a hot environment. These progressive adjustments occur over periods of increasing duration and reduce the strain experienced on initial exposure to heat. This enhanced tolerance allows a person to work effectively under conditions that might have been unendurable before acclimatization.

When workers are initially exposed to hot work environments, they can show signs of distress and discomfort; develop increased core temperatures and heart rates; complain of headache, giddiness, or nausea; and present other symptoms of incipient heat exhaustion. Yet on repeated exposure, there is a marked adaptation in which the principal physiological benefit appears to result from an increased sweating effectiveness (earlier onset, greater sweat production, and lower electrolyte concentration) and a concomitant stabilization of the circulation. After heat exposure on several successive days, the individuals perform the same work with a much lower core temperature and heart rate and higher sweat rate (reduced thermoregulatory strain), and with none of the distressing symptoms that may be experienced initially. Acclimatization to heat is a remarkable example of physiological adaptation that is well demonstrated in laboratory experiments and field experience. Full heat acclimization occurs with relatively brief daily exposure to working in the heat. The minimum exposure time for achieving heat acclimatization is a continuous exposure of about 100 min daily. Guidance for acclimatization to heat stress exposures is provided in Table 22.8 (40). The schedule for a new employee is more gradual than for an experienced employee based on the assumption that a new employee must also learn about the job and its related hazards (21). In this way, new employees have a somewhat reduced risk for accidents and making unsafe decisions.

Heat acclimatization represents a dynamic state of conditioning rather than a long-term change in innate physiology. The level of acclimatization is relative to the initial level of physical fitness and the total heat stress experienced by an individual. Thus a worker who does only light work indoors in a hot climate will not achieve the level of acclimatization needed to work outdoors (with the additional heat load from the sun) or to do harder physical work in the same hot environment indoors. Also, failure to replace the water lost in sweat will retard or even prevent the development of the physiological adaptations characteristic of acclimatization.

Although acclimatization will be reasonably well maintained for a few days of nonheat exposure, absence from work in the heat for a week or more results in a significant loss in the beneficial adaptations. Nevertheless, heat acclimatization usually can be regained in

Table 22.8. Guidelines for Acclimatization and Reacclimatization as a Percent of Effort[a]

Acclimatization Guidelines

Day	Activity (percent of full work assignment): Experienced	New
1st	50%	20%
2nd	60%	40%
3rd	80%	60%
4th	100%	80%
5th		100%

Re-Acclimatization Guidelines

Days Away from Heat-related Jobs Routine		Exposure Sequence (percent of full work assignment)			
Absence	Illness	1st	2nd	3rd	4th
<4	—	100%			
4–5	1–3	R/E[b]	100%		
6–12	4–5	80%	100%		
12–20	6–8	60%	80%	100%	
>20	>8	50%	60%	80%	100%

[a]Ref. 40. Table is courtesy of the National Safety Council, Itasca, Ill.
[b]Reduced Expectations by the worker and supervision.

two to three days upon return to a hot job. Table 22.8 provides some guidance on re-acclimatization for different periods away from heat exposures whether due to routine reasons such as reassignment, vacation or injury or for illness (40).

3.2 Indices of Heat Strain

Heat strain is a reactive physiological manifestation of environmental heat stress. The burden of balancing bodily heat gain and heat loss—a balance needed to prevent elevation in body temperature—falls primarily on the sweating mechanism and the circulatory system. Several indices of heat strain have been identified and studied in the evaluation of the thermal environment. The most common indices of heat strain include sweat rate and loss, heart rate and core temperature. This section provides a rationale for the indices and some guidance on how to employ them.

3.2.1 Sweat Rate and Loss

The effectiveness of cooling by sweat evaporation depends on the clothing and the climatic conditions (see Section 2.3.3). When the evaporation of sweat is restricted, skin temperature (T_{sk}) rises above that observed under less humid conditions at the same air temperature. When this occurs, more sweat glands are activated, thereby increasing the fraction of wetted body surface. If such a response achieves the degree of cooling needed to balance the heat load, the body core temperature remains essentially unchanged.

At higher levels of humidity or lower air velocities, the fraction of the body surface that is wetted increases until the body is completely wetted. Any further increase in sweat production does not contribute to cooling because the liquid perspiration drips off the body and is wasted as a coolant. With the restrictions on evaporation, body heat is not dissipated, thus raising the temperature of the skin as well as the deep body temperature. The response is a greater central drive for sweating, with evaporative rate increasing with increases in water vapor pressure on the skin. As a result, a new thermal balance may be established, but at a higher body temperature and at the cost of increased thermoregulatory strain. A modest rise is physiologically acceptable, but any substantial increase in body temperature is accompanied by serious heat strain (41).

Sweat rate thus represents an index of heat stress and sweat loss over a given time period serves as a time-weighted average of the heat stress. Sweat loss is measured by the loss in body weight, plus the weight gained by water and food intake and minus the weight lost in urine and feces (41). Over relatively short periods of time, sweat rate can be estimated by loss in body weight plus amount of fluids consumed, where meal gains and excretion losses are noted or controlled. A loss of more than 1 L/h for acclimatized personnel is potentially excessive. If the work is likely to demand more than 5 liters of sweat over the work day, dehydration becomes a significant risk.

An alternative method to assess dehydration is to monitor body weight at the beginning and end of a shift, making sure that the clothes are the same and dry. Losses of more than 1.5% of body weight is an indicator of dehydration.

If there is any concern that an individual may not have adequate salt replacement, a 24-hour urine collection should be made. If the sodium concentration is less than 50 millimoles, then there is an inadequate salt replacement. (23).

3.2.2 *Heart Rate and Circulatory Strain*

Maintaining a thermal balance without rise in body core temperature requires an effective physiological response to increase the transport of metabolic heat to the skin. This occurs by augmented blood flow through the dermal vessels, but under extreme conditions, this can tax the capacity for cardiac output.

Heart rate represents the major mechanism for adjustments in cardiac output required by working muscles and the added circulatory strain imposed by heat exposure; therefore it is a useful index of total heat load. The heart rate of a subject reflects the combined demands of environmental heat, work level, elevation of body temperature, and individual cardiovascular fitness. When the heat stress is no longer compensable (i.e., thermal equilibrium cannot be achieved), heart rate increases disproportionately with heat load. This pattern reflects the body's increasingly futile effort to avoid a rise in core temperature as heat load rises beyond the compensable zone. As the external heat load increases, skin temperature also rises, decreasing the thermal gradient between body core and skin temperature and thereby demanding an increased rate of skin blood flow.

Because heart rate reflects the combined influence of environmental heat, work level, elevation of body temperature, and individual cardiovascular fitness, a given heart rate has a relatively consistent physiological meaning among workers. For example, a rate of 180 to 200 beats per minute (bpm) represents the maximum capacity for most adults and is sustainable for only a few minutes (47). Further, a given heart rate signifies approximately the same level of strain regardless of the degree of an individual's fitness, because a person with high cardiovascular fitness will reach a given heart rate at a higher level of thermal stress than an unfit person. To avoid excessive cardiovascular strain in the short term, the ACGIH is recommending limiting sustained heart rates of several minutes to a rate equivalent to 180 minus the individual's age in years (23). Minard et al. (48) noticed a cardiovascular deficit at the end of a day if the average heart rate exceeded about 115 bpm. This is a rate supported by others for an average heart rate for an 8-hour day. Goldman (20) and Bernard and Kenney (49) also argue for intermediate thresholds for shorter periods of time. Any of these approaches can be easily accomplished through the use of available heart rate monitors.

Brouha proposed a simple guide based on heart rates during recovery to ensure that persons performing intermittent work in heat will remain in thermal balance for a full work shift without cumulative effects of strain (50). The method was further examined by Fuller and Smith (51). Based on recommendations from the four agency document for hazardous waste sites (52) and observations by Bernard and Kenney (53), a useful threshold for recovery heart rate is 110 bpm at one minute of recovery, which can be assessed by palpation or electronically between 55 and 65 seconds after sitting down.

3.2.3 *Body Core Temperature*

The steady-state core temperature in a temperate climate is determined by an individual's metabolic rate. As the total heat load increases, sweat rate and blood flow rate increase proportionally, with body core temperature maintained at a uniform level determined only by metabolic rate. Lind (54) termed this level of thermal equilibrium as the prescriptive zone, to indicate the range of thermal environments without excessive strain on homeostatic control of body core temperature.

For prolonged exposures to heat stress, the safe upper limit of core temperature is 38°C (100.4°F). In the prescriptive zone (below the WBGT limits proposed by NIOSH and the ACGIH), the mechanisms of circulatory control and sweating are so effective that body core temperature is maintained below 38°C. For intermittent exposures above the REL or TLV, there may be uncompensable heat stress with a resulting increase of core temperature above 38°C. While the World Health Organization panel suggested that 39°C was an acceptable limit (55), professional practice is tending toward 38.5°C to allow an operational margin (25, 49). These limits are for body core temperature.

While socially acceptable measures of core temperature may become routinely available for evaluation and control of heat stress exposures, surrogate measures are the most frequently used. Of the surrogate measures, oral temperature has the longest history. It is best measured at least 15 minutes after the last drink and with the mouth closed. The rule of thumb is to add 0.5°C to the measured value.

Ear canal temperature (as opposed to tympanic temperature, which is not recommended for safety and comfort reasons) provides some insight to core temperature. It, however, is easily influenced by environmental conditions, but in a protective fashion. That is, hot environments tend to raise ear canal temperature more than expected from an increase in core temperature.

Passive infrared devices are available for measuring tympanic temperature in a less intrusive manner are available. They have been designed for application in a climate-controlled environment. On some devices, the offset for rectal temperature may have been selected based on data for infants. Care must be taken to use them in a workplace, although they may produce protective results due to the influence of hotter ambient conditions.

Bernard and Kenney (49) have proposed an insulated surface sensor as a surrogate measure with a threshold set to include 75% of readings when core temperature reaches 38.5°C and all at 39°C. While Reneau and Bishop (56) have pointed out the potential variability in this approach alone, Bernard and Kenney emphasize the value of using heart rate and symptoms as a three prong approach to assessing exposure so that a miss in one parameter may be caught by another co-varying parameter.

3.3 Heat Disorders and Tolerance Factors

Environmental heat and the inability to remove metabolic heat lead to well-known reactions in humans, including increased cardiovascular activity, sweating, and increased body core temperature. Viewed from a wider perspective, three overlapping responses to heat can occur: psychological, psychophysiological, and physiological.

In addition to the physiological effects of heat, subclinical effects may modify performance or behavior, or potentiate response to other stresses imposed simultaneously. There is no demonstrable effect on health at low levels of thermal stress. As heat load increases, however, there appears to be a higher order of psychophysiological disturbance, an increase in the frequency of errors, a higher frequency of accidents, and a reduction in efficiency in the performance of skilled physical tasks. At even higher levels of exposure increasingly well-defined disturbance of physiological well-being occurs, with strain on the heart and circulatory system, and overloading of the mechanisms of electrolyte and water balance in the body (10). Nevertheless, one of the most striking features of heat stress and tolerance to such stress is the high degree of individual susceptibility to heat. This section examines the various heat disorders in more detail and focuses on several personal factors that can influence heat tolerance.

3.3.1 Psychophysiological Effects

Under conditions of heat stress, increased demands are made for blood flow to the periphery of the body, diverting some of the cardiac output and rendering it unavailable to active muscles. Accordingly, as upper limits of tolerance are reached, work output must be reduced (57). Wyndham et al. demonstrated that real limits of endurance exist (58).

Effects of heat stress at levels lower than limiting levels include lack of efficiency in performing heavy tasks (59) and interference with skilled manipulations or psychological tasks (57). Pepler has reviewed the effects of heat on skilled tasks and mental tasks; qualitatively, there is no doubt that heat interferes with these types of activity (60, 61). Furthermore, it is common experience that heat exposure accelerates the onset of fatigue. One study showed that the production of delicately assembled items (based on incentive pay) included higher scrap rates when the subjects were working in heat, and although production of good pieces was maintained, lower efficiency resulted (62). In another study, Pepler (63) showed that relatively slight increases in environmental temperature affect classroom learning adversely. Increased bodily temperature and discomfort also increase irritation, anger, and other untoward emotional reactions (64, 65).

3.3.2 Interactive Effects

Much of our insight into the combined effects of heat and other stresses, such as noise and vibration, is expected to accumulate from the space program and related aerospace medical research. Crew members in aircraft and space vehicles are exposed to combinations of environmental stresses that traditionally have been studied one-by-one in laboratory research.

The combination of heat and carbon monoxide (CO) has been shown to have a deleterious effect greater than that due to either stress alone (66). The subjects reported persistent headaches, anorexia, irritability, depression and general malaise; the effects were more pronounced in women than in men. These symptoms were markedly more severe after exposure to heat and CO than after exposure to either alone. It is interesting to note that these physiological disturbances were more severe in the hours following the exposures.

Renshaw (67) investigated the effects of noise and heat on performance on a five-choice serial reaction task. The effect of heat was found to be statistically significant. Performance was poorer at 32°C Effective Temperature than at 22°C Effective Temperature when the noise level was held constant.

3.3.3 Acute Heat Disorders

A variety of heat disorders can occur when individuals are exposed to excessive heat. These illnesses range from simple postural heat syncope (fainting) to the life-threatening complexities of heatstroke. The heat disorders are interrelated and seldom occur as discrete entities. A common feature of most heat-related disorders is some degree of elevated body temperature and/or dehydration. The prognosis depends on the absolute level of the elevated body temperature, the promptness of treatment to lower the body temperature, and the extent of deficiency or imbalance of fluids and electrolytes. A summary of the classification, clinical features, prevention and treatment of common acute heat disorders is shown in Table 22.9 (21, 41, 65). Some important highlights of the disorders are mentioned in the following paragraphs.

3.3.3.1 ***Heatstroke.*** Heatstroke is a serious illness that is linked to failure of the thermoregulatory center. The classical description of heatstroke includes (*1*) a major disruption of central nervous function (unconsciousness or convulsions), (*2*) a lack of sweating, and (*3*) a core (rectal) temperature in excess of 41°C (105.8°F) (46, 65). Dry skin is frequently mentioned as a distinguishing characteristic of heatstroke, but the skin may be wet due to residual sweat or early stages when sweating has not stopped. The local circumstances of metabolic and environmental heat loads that give rise to the disorder are highly variable (21).

Heatstroke calls for emergency medical action. Any procedure from the onset that will cool the patient improves the prognosis. For instance, placing the patient in a shady area, removing outer clothing and wetting the skin, and increasing air movement to enhance evaporative cooling as well as placing them in a shower or cool water bath are all needed and appropriate until transport to a hospital is available. By the time a patient is admitted to a hospital, the heatstroke may progress to a multisystem lesion affecting virtually all tissues and organs (68). In the typical clinical presentation, the central nervous system is disorganized, and there is commonly evidence of fragility of small blood vessels, possibly coupled with the loss of integrity of cellular membranes in many tissues. The blood-clotting mechanism is often severely disturbed, as are liver and kidney functions. Postmortem evaluation indicates that few tissues escape pathological involvement. Early recognition of the disorder, together with appropriate and aggressive treatment, reduces the risk of death considerably as well as the extent of organ and tissue involvement (21).

3.3.3.2 ***Heat Exhaustion.*** Heat exhaustion is a milder form of heat disorder linked to depletion of body fluids and electrolytes. The critical event is low arterial blood pressure caused partly from inadequate cardiac output and partly from widespread vasodilation. The chief factors leading to heat exhaustion include: (*1*) increased vascular dilation with decreased capacity of circulation to meet the demands for heat loss to the environment, exercise and digestive activities, and (*2*) decreased blood volume due to dehydration, gravitational edema, vasodilatation, or lack of salt (electrolyte imbalance).

Heat exhaustion usually yields readily to prompt treatment. This disorder has been encountered frequently in experimental assessment of heat tolerance. Characteristically, it is accompanied by a small increase in body temperature (38–39°C or 100.4–102.2°F). The symptoms of headache, nausea, vertigo, weakness, thirst and giddiness are common both to heat exhaustion and the early stage of heatstroke. There is wide individual variation in the ability to tolerate increased body temperature; some individuals cannot tolerate core temperatures of 38 to 39°C, and others continue to perform well at even higher rectal

Table 22.9. Classification, Medical Aspects, and Prevention of Heat Illness[a]

	Heatstroke
Clinical features	(1) Hot dry skin; red, mottled, or cyanotic; (2) high and rising core temperature, 40.5°C and above; (3) mental confusion, loss of consciousness, convulsions, or coma, as core temperature continues to rise. Fatal if treatment delayed
Underlying physiological disturbance	Failure of the central drive for sweating leading to loss of evaporative cooling and an uncontrolled, accelerating rise in core temperature. Failure of sweating may be partial rather than complete
Predisposing factors	(1) Sustained exertion in heat by unacclimatized workers, (2) obesity and lack of physical fitness, (3) recent alcohol or drug intake, (4) dehydration, (5) individual susceptibility, (6) chronic cardiovascular disease
Treatment	Immediate and rapid cooling by immersion in chilled water with massage, or by wrapping in wet sheet with vigorous fanning with cool, dry air. Avoid overcooling. Treat shock if present
Prevention	Medical screening of workers. Placement based on health and physical fitness. Acclimatization for 5–7 days by graded work and heat exposure. Monitoring workers during sustained work in severe heat
	Heat Syncope
Clinical features	Fainting while standing erect and immobile in heat
Underlying physiological disturbance	Pooling of blood in dilated vessels of skin and lower parts of the body
Predisposing factors	Lack of acclimatization
Treatment	Remove to cooler area. Recovery typically prompt and complete
Prevention	Acclimatization. Intermittent activity to assist venous return to heart
	Heat Exhaustion
Clinical features	(1) Fatigue, nausea, headache, giddiness; (2) skin clammy and moist, complexion pale, muddy, or with hectic flush; (3) may faint on standing with rapid, thready pulse and low blood pressure; (4) oral temperature normal or low but rectal temperature usually elevated (37.5 to 38.5°C). Water-restriction type: urine volume small, highly concentrated. Salt-restriction type: urine less concentrated: chlorides less than 3 g/l.
Underlying physiological disturbance	(1) Dehydration from deficiency of water and/or salt intake, (2) depletion of circulating blood volume, (3) circulatory strain from competing demands for blood flow to skin and to active muscles
Predisposing factors	(1) Sustained exertion in heat, (2) lack of acclimatization, (3) failure to replace water and/or salt lost in sweat
Treatment	Remove to cooler environment. Administer fluids by mouth or give intravenous infusions of normal saline (0.9%) if patient is unconscious or vomiting. Keep at rest until urine volume and content indicate water and electrolyte balances have been restored

Table 22.9. continued

Prevention	Acclimatize workers using a breaking-in schedule for 5 to 7 days. Supplement dietary salt only during acclimatization. Ample drinking water to be available at all times and to be taken frequently during work shift
	Heat Cramps
Clinical features	Painful spasms of muscles used during work (arms, legs, or abdominal). Onset during or after work hours
Underlying physiological disturbance	Loss of body salt in sweat. Water intake dilutes electrolytes. Water enters muscles, causing spasm
Predisposing factors	(1) Heavy sweating during hot work, (2) drinking large volumes of water without replacing electrolytes
Treatment	Salted liquids by mouth, or more prompt relief by intravenous infusion
Prevention	Adequate salt intake with meals. In unacclimatized workers, provide salted (0.1%) drinking water or equivalent electrolyte replacement beverage
	Heat Rush
Clinical features	Profuse tiny raised red vesicles (blisterlike) on affected areas. Pricking sensations during heat exposure
Underlying physiological disturbance	Plugging of sweat gland ducts, with retention of sweat and inflammatory reaction
Predisposing factors	Unrelieved exposure to humid heat and skin continuously wet with unevaporated sweat
Treatment	Mild drying lotions. Skin cleanliness to prevent infection
Prevention	Cool sleeping quarters or convection to allow skin to dry between heat exposures
	Heat Fatigue—Transient
Clinical features	Impaired performance of skilled sensorimotor, mental, or vigilance tasks, in heat
Underlying physiological disturbance	Discomfort and physiological strain
Predisposing factors	Performance decrement greater in unacclimatized and unskilled workers
Treatment	Not indicated unless accompanied by other heat illness
Prevention	Acclimatization and training for work in the heat

[a]Adapted from References 41 and 65.

temperatures. Cases of heat exhaustion can be expected to occur about 10 times more frequently than cases of heatstroke (69).

3.3.3.3 Heat Cramps. Heat cramps are fairly common among individuals who work vigorously in the heat. Heat cramps are painful spasms in the voluntary muscles used while working, such as arms, legs, and abdomen. They are associated with a continued loss of salt in the sweat, accompanied by copious intake of water without appropriate replacement of salt. Other electrolytes such as Mg^{2+}, Ca^{2+}, and K^{+} may also be involved. Cramps often occur in the muscles principally used during work and can be readily alleviated by rest, ingestion of water, and correction of electrolyte imbalance in body fluids (21).

On a short-term basis, sweating workers drinking large volumes of unsalted water may develop heat cramps. Treatments of the clinical symptoms of water and salt depletion are similar, namely, the replacement of water and/or salt by oral ingestion of salted liquids in mild cases or the intravenous infusion of saline in more serious cases. Any excess of salt or water over actual needs is readily controlled by urinary excretion, assuming a healthy heart and kidneys.

3.3.3.4 Heat Rashes. The most common heat rash is prickly heat (*miliaria rubra*), which appears as red papules, usually in areas where the clothing is restrictive, and gives rise to a prickling sensation, particularly as sweating increases. It occurs in skin that is persistently wetted by unevaporated sweat, apparently because the keratinous layers of the skin absorb water, swell, and mechanically obstruct the sweat ducts (46, 70). The papules may become infected unless they are treated.

Another skin disorder (*miliaria crystallina*) appears with the onset of sweating in skin previously injured at the surface, commonly in sunburned areas. The damage prevents the escape of sweat with the formation of small to large watery vesicles which rapidly subside once sweating stops, and the problem ceases to exist after the damaged skin is sloughed.

Miliaria profunda occurs when the blockage of sweat ducts occurs below the skin surface. This rash may also follow sunburn injury, but has been reported to occur without clear evidence of previous skin injury. Discrete and pale elevations of the skin, resembling gooseflesh, are present.

In most cases,the rashes disappear when the individuals are returned to cool environments. It seems likely that none of the rashes occur (or if they do, with greatly reduced frequency) when a substantial part of the day is spent in cool and/or dry areas so that the skin surface can dry.

3.3.4 Personal Risk Factors

3.3.4.1 Age. Older, healthy workers perform well in hot jobs if allowed to proceed at a self-regulated pace. Under demands for sustained work output in heat, however, an older worker is at a distinct disadvantage compared with younger colleagues. First, the older worker has less cardiocirculatory reserve. Second, under levels of heat stress above the prescriptive zone, an older worker compensates for heat loads less effectively than do younger persons, as indicated by higher core temperature and peripheral blood flow for the same work output. This occurs because of a delay in the onset of sweating and a lower sweat-rate capacity with age, thus resulting in greater heat storage during work and longer time required for heat recovery.

Aging also leads to a more sluggish response of the sweat glands, which leads to less effective control of body temperature. Aging also results in a greater level of skin blood flow associated with exposure to heat (21). When two groups of male coal miners of average age 47 and 27 years worked in several comfortable or cool environments, they showed little difference in their responses to heat near the REL with light work. In hotter environments, however, the older men showed substantially greater thermoregulatory strain than their younger counterparts. The older men also had lower aerobic work capacities (47). In analyzing the distribution of 5 years' accumulation of data on heatstroke in South African gold mines, Strydom (71) found a marked increase in heatstroke with increasing age of the workers. Thus men over 40 years of age represented less than 10% of the mining population, but they accounted for 50% of the fatal and 25% of the nonfatal cases of heatstroke. The incidence of cases per 100,000 workers was 10 or more times greater for men over 40 years than for men under 25 years of age. In all the experimental and epidemiologic studies described above, the workers had been medically examined and were considered free of disease. Total body water decreases with age, which may be a factor in the observed higher incidence of fatal and nonfatal heatstroke in the older group (21).

3.3.4.2 Gender. Because of a lower aerobic capacity, the average woman, similar to a small man, is at a disadvantage when she has to perform the same hot job as the average-sized man. Although all aspects of heat tolerance in women have not been fully examined, gender-related differences in thermoregulatory capacities have been reviewed extensively. Lower sweat rates for females are widely reported in the literature, and related differences in physiological response by males and females are well established (46). Nevertheless, when working at similar proportions of their maximum aerobic capacity, women perform similarly or only slightly less well than men (46).

3.3.4.3 Body Fat. It is well established that obesity predisposes individuals to heat disorders (65). The acquisition of fat means that additional weight must be carried, thereby calling for a greater expenditure of energy to perform a given task and use of a greater proportion of the aerobic capacity. In addition, the body surface to body weight ratio becomes less favorable for heat dissipation. Probably more important is the lower physical fitness and decreased maximum work capacity and cardiovascular capacity frequently associated with obesity.

3.3.4.4 Water and Electrolyte Balance. Sustained, effective work performance in heat requires, among other things, a replacement of body water and electrolytes lost through sweating. A fully acclimatized worker weighing 70 kg may secrete up to 6 to 8 kg of sweat per 8-hr shift. If this water is not replaced by drinking, continued sweating will draw on water reserves from both tissues and body cells, leading to dehydration (41).

Water lost from sweating of 1 kg (1.4% of body weight in a 70-kg person) can be tolerated without serious effects. Water deficits of 1.5 kg or more during work in the heat reduce the volume of circulating blood, resulting in signs and symptoms of increasing heat strain, including elevated heart rate and body temperature, thirst, and severe heat discomfort. At water deficits of 2 to 4 kg (3 to 6% of body weight), work performance is impaired; continued work under such conditions leads to heat exhaustion (21, 41). Therefore, sweating workers should drink at frequent intervals, at least two or three times per hour, to assure adequate fluid replacement.

Sodium chloride lost in sweat must be replaced. A typical American diet (10 to 15 g/day of salt) supplies the needs of an acclimatized worker producing 6 to 8 kg of sweat during a single shift, for whom 1 kg of sweat contains 1 to 2 g of salt. During acclimatization, however, workers require additional salt. Although maximal sweating rates in unacclimatized persons are lower, salt concentrations are higher than after acclimatization. Thus an unacclimatized person may lose 18 to 30 g of salt per day. Salt supplements in the form of tablets should never be used. A preferable practice is to use extra salt on foods.

If workers replace water losses without adequate salt intake, a progressive dehydration occurs because thermoregulatory controls in the human body are geared to maintain a balance between electrolyte concentration in tissue fluids and body cells. Deficient salt intake with continued intake of water dilutes the tissue fluid, which in turn suppresses the release of antidiuretic hormone (ADH) by the pituitary gland. The kidney then fails to reabsorb water and excretes dilute urine containing little salt. Under these conditions, homeostasis maintains the electrolyte concentration of body fluids but at the cost of depleting body fluids and ensuing dehydration. Under continued heat stress, the symptoms of heat exhaustion (elevated heart rate and body temperature plus severe discomfort) develop similarly to those resulting from water restriction, but signs of circulatory insufficiency are more severe and there is notably little thirst. An excellent diagnostic tool for salt deficiency is the presence of a very low level of chloride (less than 3 g/L) in the urine (41).

3.3.4.5 Alcohol and Drugs. Alcohol has been associated as a contributing factor in the occurrence of heatstroke (65). This drug interferes with central and peripheral nervous function and is associated with dehydration by suppressing ADH production. Notwithstanding the potential hazards from central nervous system (CNS) depression, the ingestion of alcohol before or during work in the heat should not be permitted, because it reduces heat tolerance and increases the risk of heat illnesses.

Many prescription and over-the-counter drugs prescribed or taken for therapeutic purposes can interfere with thermoregulation (70). Almost any drug that affects CNS activity, cardiovascular reserve, or body hydration could potentially affect heat tolerance. Any worker subject to heat stress who requires drug therapy should be under the supervision of a physician who understands the potential ramifications of medications on heat tolerance. A worker taking therapeutic medications who is exposed even intermittently or occasionally to a hot environment should seek the guidance of a physician.

It is difficult to separate the heat-disorder implications of drugs used therapeutically from those which are used socially. Nevertheless, there are many drugs other than alcohol that are used on social occasions. Some of these have been implicated in cases of heat disorder, sometimes leading to death.

3.3.4.6 Physical Fitness. Physical conditioning alone does not assure heat acclimatization, but persons with a high level of physical fitness have a distinct advantage (41, 46,

47). Physical training, without heat exposure, improves heat tolerance as indicated by somewhat lower heart rates and core temperatures when compared with persons who are exposed to heat before conditioning. Sweat rates do not increase, and skin temperature remains high. Physical conditioning enhances heat tolerance by increasing the functional capacity of the cardiocirculatory system. Increased heat tolerance occurs by two important changes. First, an increase occurs in the number of capillary blood vessels relative to muscle mass, providing a larger interface between blood and muscle for the exchange of oxygen and waste products. Second, the increased tone of small veins from nonmuscle tissue reduces their volumetric capacity during exercise and thus increases pressure on large, central veins returning blood to the heart. This increase in venous return causes cardiac output to increase during work with less need to accelerate the heart. Therefore a physically conditioned person, by virtue of having a higher maximum cardiovascular capacity, has a wider margin of safety in coping with the added circulatory strain of working under heat stress.

4 MEASUREMENT OF THE THERMAL ENVIRONMENT

Heat stress refers to the aggregate of environmental and physical factors that constitute the total heat load imposed on the body. Assessment of heat stress always includes some theoretical or empirical combination of environmental variables to describe the severity of the thermal environment and the capacity of that environment to facilitate heat exchange with a worker. As discussed in Section 2, both the degree of discomfort and the level of stress caused by environmental heat depend on air temperature, relative humidity, velocity of air movement, and temperature of the surrounding surfaces with which the body exchanges heat by radiation.

4.1 Indices of Heat Stress

Investigators in this field have proposed numerous indices for assessing heat stress, ranging from as simple a parameter as dry bulb temperature to algebraic combinations of multiple environmental variables, some with correction factors and modifications. Goldman (72) identified 32 different indices for rating heat stress. Several indices of thermal stress have survived or emerged as those most commonly encountered in the evaluation and control of industrial heat stress. Beyond dry bulb and wet bulb temperatures, these include: effective temperature (ET), defined by Houghton and Yaglou (73); corrected effective temperature (CET), developed by Bedford (74) as equivalent effective temperature corrected for radiation (ETCR); the Belding and Hatch heat stress index (HSI) (39); wet bulb globe temperature (WBGT) (1), Botsford's wet globe temperature (WGT) (75); and the ISO's required sweat rate (28).

Investigators who have tried to develop optimal heat stress indices have approached the problem in four general ways (72): (*1*) physical indices based on one or more of the physical factors of the environment (e.g., WBGT); (*2*) subjective indices based on assessment of thermal sensation (e.g., ET and CET); (*3*) rational indices based on the human heat balance equations (e.g., HSI and Required Sweat Rate); and (*4*) physiological indices based on physiological strain (e.g., sweat rate, heart rate or core temperature).

4.1.1 Dry Bulb and Wet Bulb Temperatures

The dry bulb air temperature (T_{db}) is the temperature of the ambient air as measured with a thermometer or equivalent instrument (see Section 4.2). It is the simplest to measure of climatic factors. The SI temperature units are degrees Celsius (°C) and degrees Kelvin (K = °C +273).

The natural wet bulb temperature (T_{nwb}) is the temperature measured by a thermometer or equivalent sensor that is covered by a wetted cotton wick and exposed only to the naturally prevailing air movement. Accurate measurement of T_{nwb} requires use of a clean wick and distilled water.

The psychrometric wet bulb temperature (T_{pwb}) is obtained when the wetted wick covering the sensor is exposed to a high forced air movement. The T_{pwb} is commonly measured with a psychrometer, which consists of two mercury-in-glass thermometers mounted alongside each other on the frame of the psychrometer. One thermometer is used to measure the T_{pwb} by covering its bulb with a clean cotton wick wetted with water, and the second measures the dry bulb temperature (T_{db}). The air movement is obtained manually with a sling psychrometer or mechanically with a motor-driven fan. Both sensors in a psychrometer are shielded from radiant heat.

While these simple temperature parameters are convenient to obtain, they provide less-than-complete information on the thermal exchange between a worker and the occupational environment. Sole use of dry bulb temperature when the temperature exceeds the comfort zone is somewhat justified only for situations where the worker is wearing vapor- or air-impermeable encapsulating protective clothing and radiant heat is not present.

4.1.2 Effective Temperature

Effective temperature (ET) has a long history and wide application as a useful index of thermal comfort. Effective temperature emerged in the search for design criteria for thermal comfort in occupied space and was introduced in 1923 by Houghton and Yaglou (73). Effective temperature is an empirical sensory index combining dry and wet bulb temperatures and air speed to yield a thermal sensation equivalent to that at a given temperature of still, saturated air. To develop the effective temperature scale, subjects were exposed to one combination and then another of the various parameters of air temperature, air motion and humidity. On the basis of a large number of trials, nomograms were developed that characterized equivalent environments, expressed in terms of the temperature of a still, saturated environment.

Wyndham, in studying gold miners, who were acclimatized but frequently performed strenuous work, concluded that below 26°C ET the risk of heat stroke was negligible and that the risk of fatal heat stroke began at 28°C ET (9). Up to 32°C ET there was a slow, steady rise in the risk of both fatal and non-fatal cases of heat stroke but above 33°C ET the risk increased sharply. In a past publication of the American Industrial Hygiene Association (76), the authors suggested upper permissible limits for heat acclimatized men as: 32.2°C (90°F) ET for light work; 29.4°C (85°F) ET for moderate work; and 26.7°C

(80°F) ET for heavy work. Similar though lower ET values were reported by Brief and Confer in a comparison of heat stress indices (77).

4.1.3 *Corrected Effective Temperature*

Bedford (74) developed the equivalent effective temperature corrected for radiation (ETCR), now referred to as the corrected effective temperature (CET). He showed that the ET scale could be corrected to account for radiation from hot surroundings by using the black globe temperature (see Section 4.2.4) in place of the dry bulb temperature. Present use of the CET scale incorporates such modifications into a single chart.

In spite of widespread use, both the ET and CET have serious limitations as indices of heat stress (65):

1. ET was developed on the basis of transient thermal sensations. This emphasis tended to neglect the importance of the sorption and desorption of moisture from the subject's clothing.
2. The ET scale was developed using clothed subjects in the dress of that era.
3. The subjects were sedentary. Later modifications were made to include the effective metabolic rate. Nevertheless, the scale was designed primarily for environments reasonably near the comfort zone, and extrapolation to thermally stressful environments is tenuous.

4.1.4 *Wet Bulb Globe Temperature*

The wet bulb globe temperature (WBGT) index was intended originally as a simple expression of heat stress for use in military training where men were exercising outdoors in conditions of high solar radiation (1). It proved very successful in monitoring heat stress and minimizing heat casualties in the United States and has been widely adopted. The WBGT index provides a convenient method to assess quickly, with a minimum of operator skills, conditions that pose threats of thermal strain. Because of its simplicity and close correlation with CET, it was adopted in 1971 (and has prevailed) as the principal index for the TLV for heat stress established by the ACGIH (22).

Fundamentally, the WBGT index is an algebraic approximation of the effective temperature concept. As such, it has the built-in limitations of the effective temperature, but it also has the advantage that air or wind velocity does not have to be measured directly to calculate the intensity of WBGT. For outdoor use with solar load, the index is derived from the formula:

$$WBGT = 0.7T_{nwb} + 0.2T_g + 0.1T_{db} \tag{7}$$

where T_{nwb} = natural wet bulb temperature
T_g = globe temperature
T_{db} = dry bulb (air) temperature

For indoor use, the weighted expression becomes:

$$WBGT = 0.7T_{nwb} + 0.3T_g \tag{8}$$

Although direct measurement of air movement is not required for the computation of WBGT, allowances are made for this factor by the use of the naturally convected wet bulb sensor.

NIOSH endorses WBGT as the preferred measure of severity of occupational exposures to heat stress (10, 21). The main criteria used by NIOSH for the selection of a suitable index were that (*1*) the measurements and calculations must be simple and (*2*) index values must be predictive of the physiological strain of heat exposure.

There is little dispute that the WBGT index is convenient and simple to use, but WBGT is not a perfect predictor of physiological strain. Ramanathan and Belding, among others, have shown clearly that environmental combinations yielding the same WBGT levels result in different physiological strains in individuals working at a moderate level (78). This suggests that WBGT has limitations as a heat stress index, especially at high levels of severity. In addition, when impermeable clothing is worn, the WBGT will not be a relevant index, because evaporative cooling (wet bulb temperature) will be limited. Nevertheless, WBGT has become the index most frequently used and recommended for use throughout the world.

4.1.5 Wet Globe Temperature (WGT)

The wet globe thermometer includes a hollow, 7.6 cm (3-in.) copper sphere covered by a black cloth that is kept at 100% wet from a water reservoir. The sensing element of a thermometer is located at the inside center of the copper sphere, and the temperature inside the sphere is read on a dial on the end of the stem. The wet sphere exchanges heat with the environment by the same mechanism that a person with a totally wetted skin would use in the same environment; that is, heat exchange by convection, radiation and evaporation is integrated into a single instrument reading (75). The stabilization time of the instrument ranges from about 5 to 15 min depending on the magnitude of the heat-load differential (21).

The WGT has been used in many laboratory studies and field situations where it has been compared with the WBGT (79–83). In general, the correlation between the two is high. Nevertheless, the relationship between the two is not constant for all combinations of environmental variables. Correction factors ranging between 1°C (1.8°F) and 7°C (12.6°F) have been suggested. A simple approximation of the relationship is $WBGT = WGT + 2°C$ for conditions of moderate radiant heat and humidity. These approximations are likely adequate for general monitoring in industry (21).

4.1.6 Heat Stress Index and Required Sweat Rate

The heat stress index (HSI), developed by Belding and Hatch at the University of Pittsburgh in the mid-1950s (39), combines the environmental heat exchange components of radiation R and convection C with metabolic heat M in an expression of stress in terms of the required sweat evaporation, E_{req} (see eq. 5). The resulting physiological strain is determined by the ratio (times 100) of the stress (E_{req}) to the maximum evaporative capacity of the environment (E_{max}) (see eq. 6) with a limit of 760 W for a maximum rate of cooling by sweat evaporation. That is,

$$HSI = 100\ E_{req}/E_{max} \tag{9}$$

Belding and Hatch suggested an HSI of 40 to be the onset level for severe heat strain, 70 to be the onset level for very severe heat strain, and over 100 to be above the maximum strain tolerated daily by fit, acclimatized young men. The HSI loses some applicability at very high heat stress conditions. Because of the greatly reduced efficiency of evaporative cooling between 70 and 100, the 100 limit is unrealistically high as a general limit. It also does not identify correctly the heat stress differences resulting from hot-dry and hot-humid conditions. The strain resulting from metabolic versus environmental heat may not be differentiated because E_{req}/E_{max} is a ratio that may disguise the absolute values of the two factors (21).

The required sweat rate analysis recommended by the ISO (28) is a second generation rational approach, which evolved from the concepts contained in the HSI. Equations 3 and 4 are the methods for estimating convective and radiant heat exchange. The use of required sweat rate as a means to evaluate heat stress is described in the ISO standard (28) and elsewhere (24, 40).

A principal advantage of the HSI and the required sweat rate in the practice of industrial hygiene is that the components of a rational approach offer an excellent starting point for specifying corrective measures when heat exposure is excessive. The relative values of the convective exchange C, radiant exchange R, and evaporative capacity E_{max} not only provide a rigorous way to estimate heat stress, but also offer a basis for a rational approach to corrective engineering measures. This subject is developed further in Section 6.

4.2 Sensing Instruments

The four primary environmental factors that determine the rate of heat transfer between a worker and the surroundings are air temperature, humidity, air velocity, and radiation (solar and infrared). This section discusses the instruments used and techniques involved in measuring these four variables. Technology is advancing steadily in this aspect, and coverage here is limited to instruments that are generally available and adaptable to field use.

4.2.1 Air Temperature

Air temperature can be measured by a variety of instruments, and each has advantages under specific circumstances.

Liquid-in-Glass Thermometers: A widely used instrument for measuring air temperature is a glass thermometer in which mercury or alcohol is the expanding liquid. This type of instrument is relatively inexpensive and is available in various temperature ranges and with varying degrees of accuracy. The response time of a thermometer depends primarily on the bulb size and the air velocity at the bulb. Transient temperature readings should not be recorded until the thermometer reaches steady state.

Thermocouples: A thermocouple is a junction of wires of two different metals. Such thermojunctions release an electromotive force (emf) that varies with the temperature of the junction. In a circuit containing two thermocouples, the emf in the unit depends on the temperature difference between the two junctions. If one of the junctions is held at a constant reference temperature, the temperature of the other junction can be determined by the measured emf on the circuit. The emf in a thermocouple circuit may be measured accurately with a potentiometer or a high resistance millivolt meter.

Although use of thermocouples entails the initial cost of a read-out device, thermocouples can give remote readings. In such applications, simultaneous readings may be taken at one place for several work sites. In addition, the equilibration time required with varying temperatures is much less than that of mercury-in-glass thermometers.

Thermistors: A thermistor or resistance thermometer is a semiconductor (typically a metal wire) that exhibits a substantial change in resistance with a small change in temperature. Because the resistance of a thermistor is on the order of thousands of ohms, incremental resistance added by lead wires up to about 25 m is immaterial. Therefore, like thermocouples, thermistors offer the advantage of remote monitoring.

Difficulties with Air Temperature Measurements: Thermal radiation can cause serious errors in the measurement of air temperature. When the surrounding surfaces are warmer or cooler than the air in a space, radiation effects will cause a sensor used to measure air temperature to become warmer or cooler than the air. Because of their smaller size, thermocouples and thermistors are less affected by radiation than mercury-in-glass thermometers. The temperature distortion due to radiation can be reduced by shielding, by increasing the velocity of air movement over the sensor, or by a combination of the two.

4.2.2 Humidity

Humidity refers to water vapor in the atmosphere. Relative humidity refers to the percentage saturation of water vapor in air at a given air temperature. The level of humidity can be expressed quantitatively in several different ways, each expression being applicable to certain processes or problems. The terms specific humidity or dew-point temperature are useful to an air-conditioning engineer. In heat stress calculations, vapor pressure and partial pressure are convenient terms. Only dew-point temperature and relative humidity can be measured directly by ordinary instrumentation. The other expressions of humidity must be derived from other humidity-related measurements.

Psychrometers: All thermodynamic properties of mixtures of air and water vapor can be determined readily from knowledge of the dry and wet bulb temperatures. These temperatures are measured with a psychrometer, an instrument with two sensors, one of which is enclosed in a wetted wick or sock. The sensor in the wetted wick is cooled by evaporation, depressing its temperature below that of the dry bulb sensor. The air velocity over the wetted sensor must be at least 3.0 m/s (600 ft/min) to reach the true psychrometric wet bulb temperature (84). The psychrometric wet bulb temperature (T_{pwb}) is sometimes referred to as the vented wet bulb.

A sling psychrometer uses a pair of liquid-filled thermometers, one being covered by a wetted cotton wick. The operator holds the instrument by a handle and whirls the pair of thermometers to effect the required air velocity over the wetted bulb. After repeatedly whirling the sling psychrometer until no further depression of wet bulb temperature is noted, the final readings of the two thermometers define dry bulb and psychrometric wet bulb temperatures. When these two temperatures are known, the relative humidity and vapor pressure can be determined through use of a psychometric chart.

Aspirated psychrometers include two stationary temperature-sensing elements, one dry and one covered with a wetted wick. Airflow over them is produced by a small fan or a squeeze bulb. The sensors of an aspirated psychrometer are usually shielded to prevent radiation-related errors in the indicated temperatures.

Natural wet bulb temperature (T_{nwb}) is used in the determination of the wet bulb globe temperature (WBGT) index. The T_{nwb} is obtained by exposing a thermometer with a wetted wick to natural air movement without regard to the minimum air velocity. Therefore, T_{nwb} is always more than or equal to psychrometric wet bulb temperature.

Hygrometers: Hygrometers are classified as one of two types, relative humidity and dew point. A direct-reading device of long standing for measuring relative humidity is an organic hygrometer. The sensing element of a hygrometer can be made of an organic material that undergoes dimensional changes as a function of the relative humidity. Although this type of sensor does not provide a high degree of accuracy, it is relatively inexpensive and was widely used in self-contained humidity recorders and in control instruments. New methods for direct readings of relative humidity have entered the market. These provide better stability and more precision than the organic devices.

Dew point hygrometers provide a direct measure of dew point temperature. This is accomplished by noting the temperature of a highly polished surface at the time condensation starts to form. The condensing plate can be cooled thermoelectrically or by other means, such as the evaporation of a refrigerant at the back of the plate. The onset of condensation can be determined by visual inspection or by use of a photoelectric cell and light source. The temperature at the surface is usually measured by thermocouples affixed to it. The dew point hygrometer is not a widely used field instrument for measuring humidity in the industrial setting. Nevertheless, it is useful in the laboratory and finds specific application in estimating the moisture content of air or other gases at elevated temperatures.

4.2.3 Air Velocity

Any instrument used to measure air movement in the ambient work environment must be capable of measuring very low air velocities and must be nondirectional. These two requirements can be fulfilled only by some type of thermal anemometer. Higher air velocities can be measured adequately by vane anemometers.

4.2.3.1 Thermal Anemometers. A thermal anemometer has an electrically heated temperature-sensing element that is cooled by air passing over it. Because a thermal anemometer responds to mass flow over the sensor, a correction is required when the gas density differs greatly from standard air density.

Heated Thermocouple Anemometer: A heated thermocouple anemometer consists of heated and unheated thermocouples connected in series and exposed to the airstream. The resultant emf (voltage) is a function of temperature difference between the heated sensor and the air. The response of the instrument is calibrated to indicate air velocity.

Hot-Wire Anemometer: The sensor in a hot-wire anemometer is a resistance thermometer element of fine wire that is heated. Its temperature and electrical resistance vary with the velocity of air passing over it. The air velocity is calibrated to and determined by the measurement of resistance.

Heated-Bulb Thermometer Anemometer: The heated bulb-thermometer anemometer includes two matched mercury-in-glass thermometers. The temperature of one thermometer is raised by passing a known current through a fine resistance wire wound around the bulb. The two thermometers are exposed to the air simultaneously, and their temperatures are read. The air velocity is calibrated to and determined by the temperature difference.

4.2.3.2 Vane Anemometers. Vane anemometers include rotating vane and swinging vane anemometers. A rotating vane anemometer consists of a light, rotating wind-driven wheel enclosed in a ring. Using recording dials, it indicates the number of revolutions of the wheel or the linear distance in meters or feet. In order to determine air or wind velocity, a stopwatch must be used to record the elapsed time, although newer models have a digital readout. The swinging anemometer consists of a vane enclosed in a case that has inlet and outlet air openings. The vane is placed in the pathway of the air, and the movement of the air causes the vane to deflect. This deflection can be translated to a direct reading of the wind velocity by means of a gear train. Rotating vane anemometers typically are more accurate than swinging vane anemometers (21).

Another type of rotating anemometer consists of three or four hemispherical cups mounted radially from a vertical shaft. Wind from any direction causes the cups to rotate the shaft, and wind speed is then determined from the shaft speed (85).

4.2.4 Radiation

Radiant heat sources in the workplace include infrared sources indoors and solar radiation outdoors. The radiant heat exchange between a worker and the surrounding surfaces of the work space can be computed, at least theoretically, if the absolute temperatures and emissivities of all surfaces are known. In practical applications, however, such an analytical solution would be extremely tedious. Therefore an empirical approximation of radiant heat exchange has been developed and is used widely for indoor sources of radiation.

The black globe thermometer is used widely for assessing thermal radiation in the occupational environment (86, 87). A temperature sensor is located at the center of a thin-walled copper sphere. The outer surface of the globe is painted flat black. When the globe thermometer is used in a space in which surface temperatures are higher than the air temperature, the globe temperature (T_g) rises above the air temperature because of the radiant heat adsorbed by its black surface. As this occurs, the globe begins to lose heat by convection. Its temperature stabilizes when the rate of heat gained by radiation equals the rate of heat lost by convection. Usually a minimum of 15 min is required for the globe temperature to reach steady state. The Vernon globe thermometer, sometimes called the standard 15 cm (6-in.) black globe, is recommended by NIOSH to measure black globe temperature (21).

Mean radiant temperature (T_r) represents the uniform black body temperature of an imaginary enclosure, equivalent to the actual environment, with which a worker exchanges radiant heat. With at least some air motion and following Table 22.6, the T_r can be calculated from the globe thermometer temperature, air velocity, and air temperature for a specific location using the following equation (24, 28):

$$T_r = 100(((T_g + 273)/100)^4 + 2.5V_{\text{air}}^{0.6}(T_g - T_{db}))^{0.25} - 273$$

where T_r = mean radiant temperature (°C)
T_g = globe temperature (°C)
V_{air} = air velocity (m/sec)
T_{db} = dry bulb (air) temperature (°C)

Solar or outdoor radiation can be classified as direct, diffuse or reflected. Direct radiation comes from the solid angle of the sun's disk. Diffuse radiation (sky radiation) refers to the scattered and reflected radiation coming from the whole hemisphere after shading the solid angle of the sun's disk. Reflected solar radiation is that reflected from the ground or water surface(s). The total solar radiant load is the sum of the direct, diffuse and reflected solar radiation as modified by clothing worn and position of the body relative to the solar radiation (88).

Direct solar radiation can be measured with a pyrheliometer. A pyrheliometer includes a tube directed at the sun's disk as well as a thermal sensor. Generally, a pyrheliometer with a thermopile as sensor and a view angle of 5.7° is recommended (89, 90). Two varieties of pyrheliometers are widely used: the angstrom compensation pyrheliometer and the Smithsonian silver disk pyrheliometer, each of which uses a slightly different scale factor.

Diffuse and total solar radiation can be measured with a pyranometer. Pyranometers can be used for measuring solar or other radiation between 0.35 and 2.5 μm, which includes the ultraviolet, visible and infrared range (21). A pyranometer for measuring diffuse radiation is fitted with a disk or a shading ring to prevent direct solar radiation from reaching the sensor. The receiver usually resembles a hemispherical dome to provide a 180° view angle for total sun and sky radiation. The receiver is used in an inverted position to measure reflected radiation. The thermal sensor could be a thermopile, a silicon cell or a bimetallic strip.

4.3 Integrating Instruments

Many attempts have been made to develop a single-reading heat stress instrument for assessing and integrating the environmental factors of air temperature, air speed, humidity, and radiant temperature. Among the instruments that have been proposed are the globe thermometer (91), the heated globe thermometer (92), the wet Kata thermometer (93), the eupathescope (94), the thermointegrator (95), and the wet globe thermometer (75). Several of these instruments have been largely abandoned, and the search continues for improved single-reading instruments.

4.3.1 Requirements and Limitations

In an excellent review of the design requirements and limitations of a single-reading heat stress meter, Hatch showed that despite the mathematical equivalence of human and instrument, the capacity of heat exchange by convection and evaporation is inherently greater for a physical instrument of simple geometric shape and small size than for a much larger person (96).

Single-reading, integrating instruments have the appeal of convenience that surely will continue, but their use in the context of today's state of the art in heat stress evaluation may offer convenience at the cost of knowing the levels of the key stress components of the thermal environment. Often there is a dual purpose in assessing heat stress: (*1*) to determine the level of stress and (*2*) to determine in what manner and to what extent control procedures can be employed to reduce the stress to an acceptable level. There is a parallel in industrial hygiene with respect to exposure levels estimated on the basis of integrated samples, such as full-shift personal samples, and exposure estimates constructed from discrete exposure intervals based on the temporal and spatial variability of contaminant levels in the work environment. Although a single, time-weighted exposure level obtained from an integrated sample is convenient, the convenience comes at the expense of the diagnostic information needed to assess the causes for exposure and the basis for a rational, cost-effective approach to control strategy.

4.3.2 Wet Globe Temperature

As reviewed earlier, Botsford has developed a wet globe temperature instrument consisting of a small copper sphere fitted with a black cotton wick and a water reservoir (75). This instrument is available from Howard Manufacturing, Inc., Bethlehem, PA. This device not only integrates the effects of air temperature, mean radiant temperature, and air motion, it also includes the effect of fractional wetness. The objective of the Botsball is to provide an index and model that simulates a worker's exchange of heat with the environment by convection, radiation, and evaporation. Although the relative values of the three components of heat transfer from the wet globe and the human body may differ considerably, a number of investigators have reported good correlation between wet globe temperature and other well-known heat stress indices (79–83, 97–99).

4.3.3 WBGT Index

The TLVs for heat stress as well as the recommendations to OSHA from both NIOSH (1972 and 1986) and the Standards Advisory Committee (1974) specify the use of the WBGT index. Based on the equations for calculating WBGT (Section 4.1.5), one must know the natural wet bulb temperature (T_{nwb}), the globe temperature (T_g), and, in the case of a solar load, the dry bulb air temperature (T_{db}). The instruments and techniques for measuring these parameters have been reviewed in Section 4.2. A common configuration is to mount all these instruments on a stand or tripod which can be located in the area of interest.

Several WBGT instruments have been developed to sense and indicate dry bulb, natural wet bulb, and globe temperatures, as well as to integrate the values to yield WBGT. Electronic, direct-reading instruments are now available commercially to measure, display

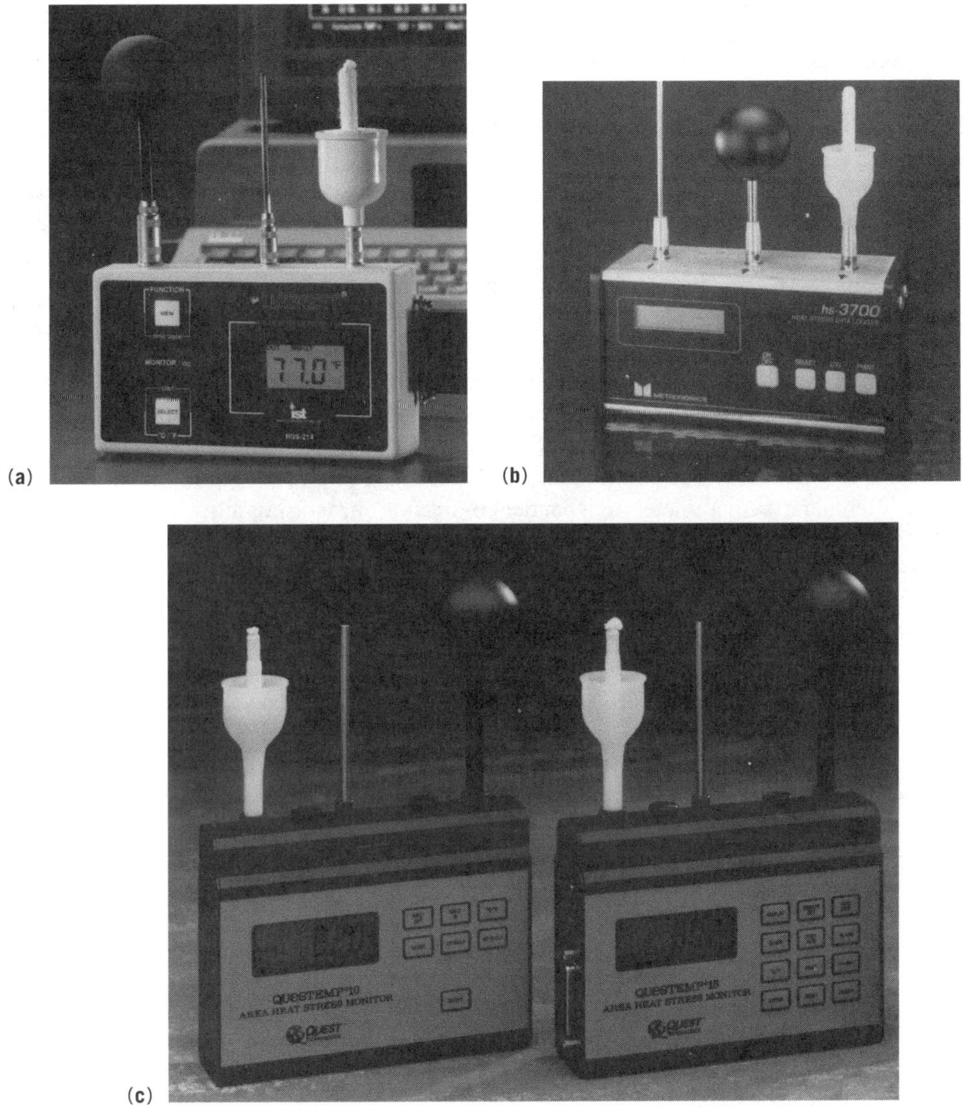

Figure 22.2. Typical electronic, direct-reading WBGT instruments. Courtesy of Metrosonics Inc., Quest Technologies TSI, Inc.

and record WBGT. Typical electronic instruments as shown in Figure 22.2 can be obtained from sources such as Metrosonics, Quest and IST. These electronic WBGT instruments are easy to use but as with most instruments require careful handling and cannot be exposed to the elements (rain, sandstorms, etc.).

4.4 The Effect of Weather

The potential severity of heat stress is determined largely by weather conditions even in the hot industries. The prevailing meteorological conditions, the onset of hot weather, and heat wave episodes can have a very significant effect on the heat stress experienced by workers.

Attempts have been made by several industries and institutions to establish qualitative or semiquantitative relationships between outdoor weather conditions and the degree of environmental heat stress imposed on workers. Studies reporting methods for predicting indoor temperatures and heat stress limits in naturally ventilated aluminum reduction plants have been described by Akbar-Khanzadeh and Ramsey (100) and by Bernard with Cross (101) and with Pourmoghani (102).

A 1967 study (103) offers a limited functional relationship between outdoor and indoor climates developed on the basis of afternoon measurements of climatic variables taken on four days during the summer. Curves of the temperature differences between indoors and outdoors were developed for individual work areas in a shop, forming a family of hyperbolas. A temperature parameter, independent of outside climatic conditions, was developed based on the indoor and outdoor dry bulb temperatures. This parameter is determined as the mean value of temperature differential of two pairs of measurements and depends on the heating value of the work area and the rate of ventilation in the shop. Limitations of the model include insensitivity to rapid changes in outside temperature and a narrow applicable time span (1 to 4 pm) during which deterministic measurements are taken.

Other investigators (104, 105) used a thermal circuit concept to estimate indoor air temperature and relative humidity, for these are affected by outdoor air temperature and solar radiation. Because no indoor heat source was considered, however, inside climate was estimated directly from outdoor meteorological conditions, using only building structure and materials as the critical variables.

4.4.1 National Weather Service Information

The National Weather Service in the USA and most countries around the world maintain details on environmental measurements that can be a useful supplement to the climatic factors measured at a work site. The National Weather Service data include daily observations at 1 and 3-hr intervals for air temperature, wet bulb temperature, dew point temperature, relative humidity, wind velocity, sky cover, ceiling, and visibility. A summary of daily environmental measurements includes maximum, minimum and average temperatures as well as wind velocity (direction and speed), extent of sunshine and sky cover. These data, where available, can be used for approximate assessment of the worksite environmental heat load for outdoor jobs or for some indoor jobs where air conditioning is not in use.

4.4.2 Predictions of WBGT

In the mid-1970s, NIOSH commissioned a major study to develop predictive models that can be used to calculate WBGT from estimates of a minimum number of conventional meteorological variables (106). The resulting models were based on data acquired during

27 week-long surveys conducted over a 2-year period in 15 representative hot industries in three climatic regions of the United States.

Using the recommended methodology and three sets of relatively simple equations, WBGT can be calculated directly from independent variables that quantify outdoor weather conditions. The first set of models is predictive in nature and is designed to estimate future levels of WBGT from weather forecasts. The second series can be used to estimate current levels of WBGT at a work site from current values of meteorological variables reported routinely by a nearby office of the National Weather Service. The third set also estimates current levels of WBGT, but uses outdoor meteorological variables measured locally in the immediate vicinity of the workplace of interest.

For each of the three sets of models, one of seven equations is selected as the most appropriate for any given work site, depending on the range of differences between inside and outside dry bulb temperature. The appropriate model is then used to estimate WBGT from either direct knowledge of, or forecasts on, prevailing weather conditions (106).

5 ASSESSMENT OF HEAT STRESS AND STRAIN

The specification of workplace standards or guidelines for limiting heat stress requires an understanding of the relationship between environmental heat stress and physiological strain. The objective of a heat stress standard is to limit the level of health risk (level of strain and the danger of incurring heat-related illnesses) associated with the total heat load (environmental and metabolic) imposed on a worker in a hot environment. The methodology of risk estimation has improved but still lacks a high level of accuracy (21). Although several studies show generally good correlation between some index of environmental heat stress and one or more indices of physiological strain, no single stress index and no single strain index have proved sufficient to characterize fully the complex stress-strain relationship over the entire range of conditions expected in industry.

5.1 Correlation of Stress and Strain

As research continues on the environmental and physiological aspects of heat stress, a primary goal is to define direct and indirect physiological limitations in the context of the industrial environment, and to translate these acceptable physiological limits to corresponding levels of environmental heat stress. Many investigators share credit for advancing our level of understanding on these issues.

The basic approaches used to correlate heat stress and physiological strain fall into three types: theoretical or semitheoretical calculations, empirical evaluations under controlled conditions, and epidemiologic studies conducted on workers exposed in occupational settings.

5.1.1 Calculated Limits of Heat Exposure

Gagge et al. (107) defined an upper limit of tolerable heat exposure at which the body heat balance is just maintained, at a given work rate, without a significant rise in skin temperature. Starting with the fundamental heat balance, Haines and Hatch (108) derived a thermal balance line for skin temperature on the basis of the ratio of E_{req}, the rate of evaporative cooling required to maintain thermal balance, to E_{max}, the maximum rate of evaporative cooling possible in a given environmental situation, using either the operative temperature (109) or the Vernon globe temperature (110).

Belding and Hatch (39) combined the two limiting requirements to define the Heat Stress Index, (HSI), the percentage ratio of E_{req} to E_{max}. They related several levels of HSI to expected physiological or psychological effects. Unfortunately, the HSI does not provide the complete solution to the problem of measuring and predicting heat stress and strain (111). Experience showed that the effects did not always occur as predicted. Therefore modifications were attempted to increase the accuracy of HSI and to use the heat balance concept to derive better correlative indices of heat stress and strain (for instance, the required sweat rate (28)).

5.1.2 Empirical Studies in Controlled Conditions

Although several new or modified heat stress indices have appeared in the literature, they usually have been tested in experimental programs designed to evaluate their relation to selected physiological responses. As Peterson (112) has indicated, many physiological parameters have been related to limited measures of heat stress but not necessarily to any of the heat stress indices.

In a comprehensive project typical of the empirical approach identified earlier, Peterson designed and executed a study to relate several heat stress indices to several measures of physiological response. His objective was to identify an optimal indicator of strain as well as the physiological responses that are sensitive to change as a function of heat stress. Although WBGT was not included among the heat stress indices considered, the results demonstrated that at least three indices of stress were necessary to describe the probable response of man to the thermal environment. Sweat rate was the biothermal strain best correlated with all stress indices except the equivalent operative ambient temperature. This suggests that sweat rate is a suitable parameter in terms of total strain. Neither the actual nor predicted sweat rate correlated well, however, with ear, rectal or skin temperatures, suggesting that the total-strain concept probably has limited utility (112).

Another study of three fit subjects under controlled environmental conditions evaluated the relationship among environmental parameters, physiological strain parameters, and heat stress indices (78); WBGT was included as a heat stress index at levels of 68, 85, and 89°F WBGT. The authors concluded that although a given level of WBGT has meaning dependent on environmental conditions, higher levels of the WBGT scale do not always signify greater strain, especially when the environment is relatively dry. Furthermore, the WBGT values failed to parallel the increased physiological strain of varying humidity, air speed, or radiant heat. A more serious inconsistency was that a higher WBGT level did not produce greater strain than a lower WBGT level in dry environments. Therefore WBGT would seem to have limited value as a predictor of physiological strain at the higher heat stress levels that may be encountered in industry (78).

The same study showed that the Heat Stress Index differentiates between environmental conditions in the correct order of physiological strain, although the resulting strain seems to be overemphasized. The wet globe temperature, however, did not differentiate between two widely different levels of physiological strain in the participants (78).

5.1.3 Workers Exposed in Occupational Settings

Since the 1920s researchers have focused on the relationships between stress and strain for workers in hot industries (5, 15, 101, 113–116). Nevertheless, few epidemiologic studies have been reported that relate the length and intensity of heat exposure to the long-term health experience of workers. Health experience statistics, although appearing in several studies, tend to relate qualitative differences in some dependent variable representing morbidity or mortality to some single categorical or independent environmental variable (6, 7, 65, 117–120).

Much information relating heat stress and strain has accumulated over the past 40 years in the South African deep mines. From data derived from laboratory studies, a series of curves has been prepared to predict the probability of a worker's body temperature reaching dangerous levels when working under various levels of heat stress (121, 122). Based on these data and on epidemiologic data on heatstroke from miners, estimates of the probabilities of reaching dangerously high rectal temperatures have been made. If a body temperature of 40°C (104°F) is accepted as the threshold temperature at which a worker is in imminent danger of irreversible heatstroke, the estimated probability of reaching this body temperature is 10^{-6} for workers exposed to an effective temperature (ET) of 34.6°C (94.3°F), 10^{-4} at 35.3°C (95.5°F), 10^{-2} at 35.8°C (96.4°F), and $10^{-0.5}$ at 36.6°C (97.9°F). These ET correlates were established for conditions with relative humidity near 100%; whether they are equally valid for these same ET values for low humidities has not been proven. Probabilities of body temperature reaching designated levels at various ET values are also available for unacclimatized men (121, 122). Although these estimates have proven to be useful in preventing heat casualties under the conditions of work and heat found in the South African mines, their direct application to general industrial environments has not been validated (21).

Present understanding of the effects of heat on health and safety suggests that standards of exposure should be based on the acute effects. It is clear, however, that the physiological consequences of exposure to heat are not directly proportional to the intensity of heat stress over the entire tolerable range. Physiological strain increases exponentially in the upper range of heat stress. Thus a small incremental increase at high levels of heat stress can result in a large increase in strain. Furthermore, many factors mitigate the relationship between exposure and effects, as discussed in Section 3.3. Any exposure limits stipulated for heat stress must reflect a recognition that the specific environment-worker-job situation defines a total stress that must remain within acceptable levels by adequate control of one or a combination of factors.

5.2 Rationale for a Heat Stress Standard

Many persons and agencies have proposed several expressions of permissible exposure limits for work in hot environments. These recommendations or standards typically include some heat stress index, and most account for the four crucial climatic factors: air temperature, humidity, radiant heat and air velocity. Several indices consider the work load as well. A World Health Organization (WHO) panel of experts has attested to the shortcomings in virtually all the indices as well as the proposed upper limits set forth for each of these parameters (55).

Many of the recommended standards for thermal exposure have been derived from laboratory experience rather than from studies on industrial workers. This means the experimental subjects are not necessarily representative in age and fitness of an employed work force. In addition, the severity of physiological strain is known to correlate poorly with various levels of each of the heat stress indices. In a study previously referenced, the use of at least three indices simultaneously was recommended to obtain an adequate evaluation of the heat strain of an exposed worker (112).

The WHO panel of experts recommended that a deep body temperature of 38°C be considered the limit of permissible exposure for continuous work in heat (55). This is consistent with the observation that body core temperature in excess of 38°C increases the likelihood of heat disorder or illness (54, 123).

5.2.1 Criteria for a Standard Heat Stress Index

The NIOSH Criteria Documents set forth several reasonable criteria for any heat stress index to be used in industry (10, 21):

1. Feasibility and accuracy must be proven through industrial use.
2. All important factors (environmental, metabolic rate, clothing, physical condition, etc.) must be considered.
3. Required measurements and calculations must be simple.
4. Index exposure limits must be supported by corresponding physiological and/or psychological responses which reflect an increased risk to safety and health.
5. The index must be applicable for setting limits under a wide range of environmental and metabolic conditions.

Several indices satisfy the first criterion, including ET, WBGT, WGT and HSI. An estimate of the work load is included in HSI, and this confers an advantage over WBGT, WGT and ET with respect to the second criterion. The measurements and calculations are much simpler for ET, WGT, and WBGT than for HSI, however; and even WBGT is somewhat simpler to use than ET. With respect to the fourth criterion, all indices have shortcomings. Considering the applicability for use in regulatory limits, HSI has the advantage of being used to calculate an allowable exposure time and minimum recovery time for a given heat stress condition. Nevertheless, WBGT and WGT are the most convenient indices with which to monitor levels of exposure because they can be read directly from an integrating instrument or, in the worst case, WBGT can be derived easily from sets of two or three environmental measurements.

Even though HSI offers several distinct supplementary advantages with respect to the engineering analysis of a given heat stress exposure, WBGT remains the preferred, prevailing environmental index for heat stress.

5.2.2 Prescriptive and Environment-Driven Zones

The prescriptive zone refers to the range of environmental conditions in which deep body temperature is determined by work intensity only. The prescriptive zone has been defined empirically in terms of effective temperature by a series of experiments (rectal vs. effective temperature) at three levels of metabolic rate: 210, 350, and 490 W (54). Such studies show clearly that steady-state core temperature is not a function of effective temperature until critical levels of effective temperature (dependent somewhat on M) are reached, at which point core temperature increases sharply. The range of environmental heat stress within which core temperature rises sharply with elevations in climatic conditions is called the environmental driven zone (EDZ). The value of an environmental heat stress index at the interface of the prescriptive zone and the EDZ is called the upper limit of prescriptive zone (ULPZ). The value of the ULPZ varies among individuals. It is higher for workers who are acclimatized to heat, but it is lower when more clothing is worn (10).

To be certain that 95% of a heat-acclimatized population wearing work clothing will not have core temperatures in excess of 38°C, the level of environmental heat stress at which a 5th percentile person will reach the ULPZ must be established (10). This value must then be corrected for the level of acclimatization and clothing. Lind and Liddell found the ULPZ to be 27°C (80.5°F) ET for a group of 128 men of average physical fitness (124).

Permissible exposure limits for heat stress cannot be based on 8-hr average values because exposures in excess of 1 hr may cause a worker to accumulate enough heat that body temperature will rise sufficiently to induce an acute heat disorder. Thus in continuous heat exposure, hourly averages are necessary. If the exposure is intermittent, the accumulation of heat will be slowed, allowing the use of a 2-hr average exposure (10).

Another study by Lind justifies the use of time-weighted average hourly work load values for intermittent work (125). The results show that for ULPZ, a certain hourly amount of work can be performed either continuously at a lower rate or at a higher rate interrupted with rest periods. In another study, the ULPZ was found to be the same for men of different ages; thus no correction for age is required (126). When older persons are exposed to a strenuous heat load, increased caution is advisable because of lower physiological capacities and higher overall susceptibility to disease.

5.2.3 Work Practices Versus Environmental Standard

Lind's work on the prescriptive zone represents a sound basis for the development of an environmental heat stress standard. It combines both the climatic and work load conditions that are imposed on the worker, and it can be monitored with a convenient index such as WBGT. Nevertheless, a number of practical shortcomings and unresolved questions remain (10, 21).

It is important that the upper limit of prescriptive zone concept be confirmed with data from a representative industrial work force. More data are required on the age and sex distribution of the work force. Also to be clarified is the effect of the self-selection process that normally occurs in an industrial situation when a worker self-determines the ability to endure high levels of heat stress. This consideration alone may have resulted in past heat stress standards that were unrealistic for an industrial work force. The lack of data regarding intermittent exposures to heat is another major unresolved question of the effect of heat stress on the workforce. In addition, differences in sweat loss under a wide variety of industrial conditions still have not been thoroughly studied. The highly variable work loads and the intermittency of strenuous work loads that are normal in industrial operations may also have a major effect on heat stress (10, 21).

The answers to these questions require additional research to validate the techniques presently proposed for the evaluation of heat stress conditions. Nevertheless, our current level of understanding justifies a work practices standard rather than an environmental standard for heat stress. Clearly the available environmental indices are sufficient to estimate threshold limits of heat stress that can be used to initiate work practices to protect the industrial worker adequately against heat.

5.3 Management of Heat Stress Exposures

The management of risk for heat stress exposures centers around two principles: general controls and job-specific controls. General controls are implemented when there is a reasonable chance of exposure above the TLV for unacclimatized workers. That is, at a point that is very protective of the workers. The general controls included employee and supervisor training, heat stress hygiene practices, and medical surveillance. When the exposures are likely to exceed the TLV for acclimatized workers, a greater depth of analysis is advised and job-specific controls should be considered (40). These are directed to specific manifestations of heat stress and follow the traditional hierarchy of engineering controls, administrative controls and personal protection.

Before selecting control measures, it is crucial to identify the components of heat stress to which workers are exposed currently or expected to be exposed in new operations. Only then, by examining the alternatives, can the most effective means of control be selected. A review of the heat balance equation described in Section 2.1 suggests that total heat stress can be reduced only by modifying one or more of the following factors: metabolic heat production, heat exchange by convection, heat exchange by radiation, or heat exchange by evaporation. Environmental heat load (C, R, and E) can be reduced by engineering controls (e.g., ventilation, air conditioning, screening, insulation, process or operational modification), as well as protective clothing and equipment. Metabolic heat production can be modified by application of labor-reducing devices and work practices.

Clearly, heat stress control depends on understanding the contributions of all the elements of the exposure. It is important, however, to be aware of the full range of options for control in a specific situation. Because engineering controls are the first consideration, they are developed more fully in the next section.

6 ENGINEERING CONTROL OF HEAT STRESS

The control of heat in industry to ensure that exposures fall within acceptable limits requires the application of feasible engineering procedures. Cost-effective engineering control of heat stress calls for an understanding of the physiological response of man in a hot environment. The underlying worker–environment thermal balance (Section 2) suggests that when a worker is exposed to elevated temperatures, the rate of heat loss decreases. The role of engineering controls is to help sustain a rate of bodily heat loss equal to the imposition of heat from the environment. In general this can be accomplished by increasing the velocity of air across the body, a technique useful within certain limits of temperature and humidity. Above such limits it may be necessary to reduce the surrounding temperature. Barriers may be necessary when radiant heat is a factor. In practice, it is often necessary to combine some or all of these approaches to achieve an acceptable thermal work environment.

The important role of engineering methods in regulating heat stress is underscored by the recommendations of the Department of Labor Standards Advisory Committee on Heat Stress (19). In its report, the committee sets forth special work practices needed to bring the 2-hr average heat exposure level within the limits of its recommended threshold WBGT values. The first such practice is that "the employer should adopt engineering controls which are appropriate for reducing and controlling the level of heat exposure" (19).

An engineering approach to the reduction of heat stress generally parallels the same control strategy applied for other workplace environmental stresses such as airborne gases, vapors, and dusts. Among the alternative control methods are substitution, control at the source, local controls, and general ventilation. Another approach to an analysis of engineering controls for heat stress focuses on the important environmental components of the thermal balance: convection, evaporative capacity, and radiation. This section combines both viewpoints and presents control methods that not only reflect a traditional approach to engineering control alternatives, but add appropriate emphasis to the aspects of heat control that characterize this specific physical stress.

6.1 Control at the Source

The first and most fundamental approach to engineering control of heat in the workplace is to examine options for eliminating heat at its point of generation. Occasionally, one feasible alternative is to change the operation or substitute a process component of lower temperature for one of higher temperature. Two well-known examples are the use of induction heating rather than direct-fired furnaces for certain forging operations, and the use of microwave ovens rather than convective ovens for certain heating operations.

As with most environmental stresses, heat is controlled most effectively if it is regulated at the source, to prevent contamination of the space occupied by workers. In general, the options for control of heat at the source are isolation, reduction in emissivity, insulation, radiation shielding, and local exhaust ventilation.

6.1.1 Isolation

The most practical method for limiting heat exposure from hot processing operations that are difficult to control or for operations that are extremely hot is to isolate the heat source. Such operations might be partitioned and separated from the rest of the facility, located in a separate building, or relocated outdoors with minimal shelter. A typical example is an industrial boiler, invariably segregated from the other operations in the same facility.

6.1.2 Reduced Emissivity of Hot Surfaces

The rate at which heat is radiated from the surface of a hot source can often be lowered if the emissivity of the source is reduced through surface treatment. When an oven, boiler, or other hot surface is covered with aluminum paint, the reduced emissivity of its surface offers two advantages. Less heat radiates to workers nearby; and heat is conserved inside the unit representing a substantial savings in energy costs (76).

The emissivity of a hot source can also be lowered by sheathing or by covering the source with sheet aluminum. Other metals, such as galvanized steel, have relatively low emissivities but are more expensive to use than aluminum. In addition, the emissivity of galvanized steel increases faster with aging than with aluminum sheet (76).

Because of reduced emissivity, structural steel members will radiate less heat if they are painted with aluminum paint. Even though aluminum-painted surfaces have a higher emissivity than aluminum sheet, they do not radiate as much heat as oil-based painted steel at a given temperature.

6.1.3 Insulation

Insulation is not mutually exclusive of isolation in the context of engineering control of heat stress. Insulation also prevents the escape of sensible and radiant heat into the work environment. A well-known example of insulation that also has implications for energy conservation is that of pipe-covering insulation on steam lines. By reducing the escape of heat into the environment, such insulation clearly helps conserve energy and fuel resources.

In addition to reducing radiative exchange, insulation reduces the convective heat transfer from hot equipment to the work environment by minimizing local convective currents that form when air that contacts very hot surfaces is heated.

6.1.4 Radiation Shielding

Shielding against radiant heat represents an extremely important control measure. The characteristics of radiant heat are quite different from those of high air temperatures, and the difference must be understood if engineering controls are to be effective. Radiant heat passes through air without heating the air; it heats only the objects in its path that are capable of absorbing it. Shielding of radiant heat sources means putting a barrier between the worker and the source to protect the worker from being a targeted receptor of the radiant energy. Radiation shielding can be classified into reflecting, absorbing, transparent, and flexible shields.

Reflective Shields. Reflective shields are constructed from sheets of aluminum, stainless steel, or other bright-surface metallic materials. Aluminum offers the advantage of about 90% reflectivity. It is also used as shielding in the form of foil with insulative backing, and in aluminized paint, with reduced effectiveness.

Successful use of aluminum as shielding requires an understanding of certain principles (76):

1. There must be an aluminum-to-air surface; the shield cannot be embedded in other materials.

2. The shield should not be painted or enameled.
3. The shield should be kept free of oil, grease, and dirt to maximize reflectivity.
4. When used to enclose a hot source, the shield should be separated from the source by several inches.
5. Corrugated sheeting should be arranged so that the corrugations run vertically rather than horizontally, to help maintain a surface free of foreign matter.

Absorptive Shielding. Absorption shielding absorbs infrared radiation readily. This type of shielding, preferably flat black, is constructed typically of two or three sheets separated by air spaces. Heat can then be removed by water flowing between two metal plates in the shield, transferring heat from the shield by conduction (76). The surface(s) of absorptive shielding exposed to work areas should be constructed of aluminum or aluminized to reduce emissivity.

Transparent Shielding. Transparent shielding consists of two general types: special glass and metallic mesh. Special glass reduces transmission of infrared radiation because it is either heat absorptive or infrared reflecting. Infrared reflecting glass is used commonly in the windows of control rooms amid excessive heat sources. Metallic mesh shielding uses chains and wire mesh to provide partial reflectance and to help reduce the radiant heat reaching an operator (76). Such use is warranted where manual operations prevent use of a solid barrier.

Flexible Shielding. Flexible shielding uses fabrics treated with aluminum. When worn as aprons or other items of clothing, they protect against radiant heat by reflecting up to 90%. Reflective garments are useful for protection against very localized and directional radiant sources.

6.1.5 Local Exhaust Ventilation

Canopy hoods with natural draft or mechanical exhaust ventilation are used commonly over furnaces, hot equipment or hot processes. Where the process is hot and wet, exhausting the moisture has the added benefit of lowering the relative humidity and enhancing evaporative heat loss of workers in the area. Although the benefits of such an application are obvious, because heated air has a natural tendency to rise, it must be remembered that local ventilation removes only convective heat. Radiant energy losses, whose magnitude often overrides convective losses, are not controlled by local exhaust hooding. Radiation shielding must be used as well to control what is likely to be the larger fraction of the total heat escaping from the hot process (127).

6.2 Localized Cooling at Workstations

When practical considerations limit the feasibility of heat control at the source or throughout the general work area, relief can be provided at localized areas. In such instances, cool air is introduced in sufficient quantity to surround the worker with an independent atmospheric environment. This local relief, often termed spot cooling, can serve two functions, depending on the relative magnitudes of the radiant and convective components in the total heat load.

If the overall thermal load is primarily convective in the form of hot air surrounding the worker, the local relief system displaces the hot air immediately around the individual with cooler air having a higher velocity. If such air is available at a suitable temperature and is introduced without mixing with the hot ambient air, no further cooling of the worker is necessary.

When there is a significant radiation load, however, the local relief system must provide some actual cooling to offset the radiant energy that penetrates the mass of air surrounding the worker. Here the temperature of the supplied air must be low enough to make the convection component C negative in the heat exchange model to offset R, the radiation component.

Air movement within the local relief zone is desirable to maintain adequate evaporative cooling. Nevertheless, the use of high air velocities (increased evaporative cooling) to offset radiation may be counterproductive if the convective load is increased (where air temperature exceeds skin temperature), causing a greater sweating demand. This defines a rationale for not using simple human-cooling fans, which are often used inappropriately throughout industry.

A major problem in the design of localized cooling systems is the introduction of supplied air in such a way that minimal mixing takes place with the surrounding hot air. High-velocity jets encourage mixing, yet are necessary to throw the air into the work zone from an offsite supply duct.

In extreme heat, workers should be stationed inside an insulated, locally cooled observation booth or relief room to which they can return after brief periods of high heat exposure. The practical utility of such a protected room will vary from industry to industry. Air-conditioned crane cabs represent one application of this concept now in common use.

6.3 General Ventilation

A common method for heat removal in the hot industries is general ventilation, making use of wall openings for the entrance of cool outside air and roof openings, commonly of the gravity type, for the discharge of heated air.

Although a higher fraction of heat loss from hot sources occurs by radiation, such radiant heat is absorbed by walls and other solid structures. These heated surfaces then become secondary sources of convective heat and also act as secondary radiators. Eventually, most of the heat must be removed from the enclosed space by ventilation; the balance is lost by conduction through walls. Therefore general ventilation is an essential part of an overall heat control strategy. It cannot offset direct radiant heat exposures, however, and even in a secondary role, often fails to function as needed because of inadequate supply of cooler make-up air. Insufficient area of openings and poor location of inlets often result in improper distribution or too little air within the building (128).

6.3.1 Fresh Air Supply

The use of uncooled outside air is sometimes effective in controlling heat, but a more practical approach for a new facility is to design a general ventilation system with the capability of cooling outside air before distributing it throughout the plant.

Evaporative Cooling. Evaporative cooling is a well established method for lowering the dry bulb (air) temperature. An evaporative cooler provides intimate contact between the incoming air and water by using sprays or wetted filters, and upon such contact,the air is cooled adiabatically. This means that no appreciable energy is transferred to or from the air. By converting sensible heat in the airstream to latent heat, dry bulb temperature is caused to drop as the relative humidity increases. Even though additional humidity is added to the incoming air, the effect is not necessarily negative because the new dry bulb temperature may be sufficiently below skin temperature to allow the worker to be cooled by convection. Evaporative cooling can be a highly effective cooling method in regions of the country where ambient relative humidities are typically low.

Chilled Coil Systems. In a chilled coil system, air passes over coils containing a medium whose temperature is sufficiently below that of the air to result in satisfactory cooling. These systems are usually one of the following types (128):

1. Water-Cooled Coil. In this system, the simplest, water is circulated directly through the coil while air from the conditioned space is passed over the coil to remove heat and moisture. As a general rule, water must enter the coil at not more than 11°C, and it is wasted after leaving the coil unless it is used for process work.

2. Cooled-Water System. When water is not available from a sufficient supply or at a satisfactory temperature, it can be cooled artificially and recirculated. This system is often used when the coolant must be piped long distances and a variety of loads must be overcome. Usually the heat is removed from the water by mechanical refrigeration, but in some intermittent operations it is removed by the melting of ice.

3. Refrigerated Coils. As the name implies, the coils are cooled by the direct vaporization and expansion of a liquid refrigerant. This method is widely used in small units as well as large central systems.

6.3.2 Distribution of Make-up Air

Convective flow around hot bodies is naturally upward. For an ideal system of general ventilation, combined with radiation shielding, the inlet air should enter near floor level, be directed toward the workers, and flow toward the hot equipment. In this way the coolest air available is received by the workers before its temperature is increased by mixing with warm building air or circulation over hot processes. This air then flows toward the hot equipment and, as its temperature increases, rises and escapes through vent openings in the roof. The combination of the motion and the cool temperature of the air yields the maximum comfort that can be obtained for the worker without artificially cooling the air. Contrary to general practice, provisions for proper distribution of the air supply should receive the same careful consideration that is given to the selection of exhaust equipment (128).

6.3.3 Removal of Heated Air

Increasingly, buildings are constructed to provide for the natural removal of heated air. For example, glass manufacturing facilities usually have large gravity roof ventilators for exhaust purposes. Unfortunately many hot industries are characterized by large, flat building configurations. Exhaust fans or gravity ventilators without a forced air supply may not provide satisfactory ventilation patterns in the building's workspace.

The basic strategy when general ventilation is used to remove heated air is to position exhaust openings, either natural draft or mechanically operated, above the sources of heat and as close to them as practical (76).

6.4 Moisture Control

The importance of moisture control is underscored in the warm-moist industries where high temperatures and high relative humidities prevail. In some industries relative humidity is deliberately maintained at a high level to maintain product quality or to prevent static electricity. Examples are textile mills, munitions plants, coal mines, and flour mills. Whenever feasible, dehumidification can help to offset heat stress because it reduces the partial pressure of water vapor in air, which increases the evaporative capacity (Section 2.3).

In a general sense, moisture control includes both prevention of increased humidity and the use of dehumidification procedures. Such unsophisticated yet effective controls as enclosing hot water tanks, local exhausting of water vapor from hot processes, covering drains carrying hot water, and repairing leaky joints and valves in steam piping offer direct measures to help alleviate heat stress in warm-moist industries.

Dehumidification, aimed at reducing workplace humidity rather than preventing its increase, can be accomplished by refrigeration, absorption, or adsorption. In the context of occupational heat exposures, refrigeration is the most widely used technique to condition the air in relief areas, operating booths, or other local or regional portions of an industrial facility. Although it is clearly impractical to air condition some operations, such as a hot rolling mill in a steel plant, the use of air conditioning is useful as a heat control method, especially in combination with a work-rest regimen for a given job function. The concept of mechanical refrigeration to reduce the temperature (and humidity) of supply air is mentioned explicitly in the list of engineering controls of the Standards Advisory Committee on Heat Stress (19).

6.5 Typical Examples

The first step in the engineering control of heat stress is to identify the magnitude of the various components of stress to which the workers are exposed in existing or expected conditions. Only by knowing the relative magnitude and the factors contributing to the heat stress components, can strategies be selected to help solve the problem with a cost-effective program. The examples that follow were adapted from other sources (76, 129) and summarized in Table 22.10.

6.5.1 Increasing Ventilation in a Warm and Humid Environment

Table 22.10 gives environmental data typical for a hypothetical laundry, where a small wall fan moves air at a relatively low velocity. Here WBGT is at 32.5°C, showing that some control measures are necessary. Note that R and C are both negligible because the air and skin temperatures are similar.

Table 22.10. Examples of Engineering Controls as Interventions to Reduce Heat Stress[a]

		Warm and Humid		Hot and Dry		High Radiant Heat	
		Before Fan	After Fan	Before Evaporative Cooler	After Evaporative Cooler	Before Shielding	After Shielding
Clothing	Units	Work Clothes	Work Clothes	Work Clothes	Work Clothes	Work Clothes	Work Clothes
T_{db}	°C	35.0	35.0	42.8	29.4	47.8	43.3
T_{awb}	°C	31.4	31.1	24.2	24.4	36.3	29.8
T_g	°C	35.0	35.0	46.1	35.0	71.7	43.3
V_{air}	m/s	0.25	1.00	2.00	2.00	3.00	3.00
T_r	°C	35.0	35.0	55.3	51.7	127.0	43.3
P_a	kPa	4.24	4.24	1.65	2.69	3.23	3.23
M	W	230	230	230	230	230	230
$WBGT$	°C	32.5	32.3	30.8	27.6	46.9	33.9
R	W	−4	0	170	140	1000	70
C	W	−5	−1	160	−110	210	220
$-E_{req}$	W	220	230	560	260	1400	520
$-E_{max}$	W	160	240	960	350	860	630
HSI	—	140	96	58	74	170	83
SR^*_{req} (28)	L/h	0.6	0.6	0.9	0.4	4.3	1.1

[a]Adapted from Ref. 29.

In this case, increasing air speed from 0.25 to 1.0 m/s (15 to 60 m/min) produces an $-E_{max}$ that better matches $-E_{req}$, a condition necessary to avoid bodily accumulation of heat and its consequent strain. In the before situation, the workers are dripping sweat but still cannot achieve thermal balance because a deficit of 60 W. Increasing the air speed effectively increases the evaporative cooling to at least match E_{req} to E_{max}.

It should be noted that even under these conditions WBGT is at 32.3°C, an unacceptable level and virtually no change occurs due to increased air speed. Nevertheless, an acclimatized person could continue to work in this environment with less strain. At higher air velocities, additional cooling would occur. There is an effective limit of about 3 m/s as well as a practical limit where air speed could interfere with operations.

A more effective, permanent control method would be to reduce the moisture content of the air. This should be done as close as possible to the source of water vapor when the source is within the workplace. If the source of humidity were outside the workroom, mechanical air conditioning would likely be required to reduce the temperature and moisture level of the work space.

6.5.2 Evaporative Cooling in a Hot and Dry Environment

Table 22.10 shows data for a hot, dry environment. Because the air speed is already at 2 m/s (120 m/min), incremental increases in air velocity would not be expected to make major improvements in evaporative capacity.

With low humidity (P_a = 1.65 kPa or 10 mm Hg), a good solution to the problem would be the installation of an evaporative cooler. Assume that in the situation described, the inside temperature is usually 5°C hotter than outdoors owing to process heat and building insulation. If the outside temperature does not exceed 40°C and P_a is about 1.65 kPa, outside air could be drawn through a water spray and the temperature reduced to that of the prevailing outdoor wet bulb temperature (23°C). P_a would then be increased to 2.7 kPa, and the temperature of the workplace could be reduced to 29°C. The after calculation of the heat components shows a negative value for the convective heat component with an $-E_{req}$ of only 260 W. Because $-E_{max}$ exceeds $-E_{req}$, the sweat rate of 0.4 L/hr could be accomplished easily without undue heat strain.

6.5.3 Radiant Heat Shielding

When the main component of heat stress comes from a radiant source, such as a furnace wall, the first control priority should be shielding of the radiant energy. Increased air velocity does little to improve such situations. Table 22.10 represents another before and after situation taken from the work of Leihnard et al. (130). Here the task requires skimming dross from molten bars of aluminum. Manipulation of a ladle at a fixed station, with moderate use of shoulder and arm muscles, suggests a metabolic rate of about 230 W. Air is directed at the worker from an overhead duct at a velocity of 3 m/s (240 m/min). The humidity is quite high, and WBGT is 46.9°C, an extremely stressful situation. Calculation of the heat components confirms this, and the worker would have to take frequent work breaks to maintain thermal equilibrium.

Despite use of clothing and a face-shield, the workers were able to perform this task only a few minutes at a time, and heat exhaustion was common. Control of the high heat exposure was achieved by placing a finished aluminum sheet between the heat source and the worker. Infrared reflecting glass at face level enabled the workers to see the work. Spaces were left in the aluminum sheet to permit workers to operate the ladle. After shielding was installed, T_g and T_a were both reduced to 43.3°C, and WBGT was lowered to 33.8°C. $-E_{max}$ was then greater than $-E_{req}$, and it was possible for a worker to evaporate about 1.1 liter of sweat per hour, thus maintaining a thermal balance. The previous sweat rate was an impossible to sustain level of 4.3 L/hr. Although there may be large errors in the estimate of R at extremely high globe temperatures, the maximum relief that could be expected from shielding was actually achieved in the case cited (130).

It should be noted that a polished metallic surface will not maintain its shielding effectiveness (reflectivity) if allowed to become dirty. This is true even for fabrics coated with very fine metallic particles. Even a thin layer of grease or oil can change the emissivity of a polished surface from 0.1 to 0.9. To provide shielding that does not interfere with performance of the work task, a curtain of metal chains can be installed; the chains reduce radiation but can be parted as required. Another approach is the use of a mechanically activated door, opened only during ejection or manipulation of a product. Also, remote-operated tools can be used. Partial barriers can also be used effectively because radiant heating from an open portal is limited to the line of sight.

ACKNOWLEDGMENTS

This Chapter is based upon and updates Chapter 21, *Heat Stress: Its Effects, Measurement, and Control*, by John E. Mutchler, C.I.H., which appears in *Patty's Industrial Hygiene and Toxicology*, Vol. 1, Part A, 4th ed., 1991.

BIBLIOGRAPHY

1. C. P. Yaglou and D. Minard, "Control of Heat Casualties at Military Training Camps," *Arch. Ind. Health*, **16**, 302–316 (1957).
2. D. Minard, Research Report No. 4, Contract No. MR 005.01-0001.01, Naval Medical Research Institute, Bethesda, MD, 1961.
3. D. Minard and R. L. O'Brien, Research Report No. 7, Contract No. MR 005.01-0001.01, Naval Medical Research Institute, Bethesda, MD, 1964.
4. D. Minard, "Prevention of Heat Casualties in Marine Corps Recruits. Period of 1955–60 With Comparative Incidence Rates and Climatic Heat Stresses in Other Training Categories," *Mil. Med.*, **126**, 261–272 (1961).
5. T. Bedford and C. G. Warner, "Observations on Working Capacity of Coal Miners in Relation to Atmospheric Conditions," *J. Ind. Hyg. Toxicol.* **13**, 252–260 (1931).
6. H. M. Vernon, T. Bedford, and C. G. Warner, Industrial Fatigue Research Board Report No. 62, Her Majesty's Stationery Office, London, 1931.
7. C. P. Yaglou, "Abnormal Air Conditions in Industry: Their Effects on Workers and Methods of Control," *J. Ind. Hyg. Toxicol.* **19**, 12–43 (1937).
8. C. H. Wyndham, "Research in the Human Sciences in the Gold Mining Industry," *Am. Ind. Hyg. Assoc. J.* **35**, 113–136 (1974).
9. C. H. Wyndham, *Ergonomics* **5**, 434–444 (1962).
10. Criteria for a Recommended Standard—Occupational Exposure to Hot Environments, U.S. Department of Health, Education and Welfare, NIOSH, HSM-72-10269, 1972.
11. "National Census of Fatal Occupational Injuries, 1997," in News, Bureau of Labor Statistics, US Dept. of Labor, USDL 98-336, (Aug. 12, 1998).
12. R. McCunney, Introductory remarks at Advances in Heat Stress Monitoring and Mitigation Conference, Boston, May 1999.
13. National Safety Council, *Accident Facts*, 1997 Edition, Itasca, IL, 1997.
14. J. S. Haldane, "The Influence of High Air Temperatures," *J. Hyg.* **5**, 494–513 (1905).
15. C. P. Yaglou, Thermal Standards in Industry, in Report of the Committee on Atmospheric Comfort, Year Book Publishers, Chicago, 1947.
16. Occupational Safety and Health Act PL91-596, 91st Congress, S.2193, U.S. Department of Labor, Washington, DC, 1970.
17. OSHA Technical Manual (TED 1-0.15A), Heat Stress (Section III, Chapter 4) (see www.osha-slc.gov/dts/osta/otm/otm_iii/otm_iii_4.html)
18. B. A. Hertig, Work in Hot Environments: Threshold Limit Values and Proposed Standards, in *Industrial Environmental Health-The Worker and the Community*, Academic Press, New York, 1975.
19. J. D. Ramsey, *National Safety News* **68**, 89–95 (June, 1975).

20. Proceedings of a NIOSH Workshop on Recommended Heat Stress Standards, US Department of Health and Human Services, DHHS (NIOSH) Pub. 81-108, 1980.
21. Criteria for a Recommended Standard—Occupational Exposure to Hot Environments, Revised Criteria 1986, U.S. Department of Health and Human Services, NIOSH, 1986.
22. American Conference of Governmental Industrial Hygienists (ACGIH), Threshold Limit Values for Chemical Substances and Physical Agents in the Workroom Environment with Intended Changes for 1971, ACGIH, Cincinnati, OH, 1971.
23. American Conference of Governmental Industrial Hygienists (ACGIH), 1999 Threshold Limit Values and Biological Exposure Indices for Chemical Substances and Physical Agents, ACGIH, Cincinnati, OH, 1999.
24. K. C. Parsons, *Human Thermal Environments*, Taylor and Frances, Ltd., London, 1993.
25. K. C. Parsons, "International Standards for the Assessment of the Risk of Thermal Strain on Clothed Workers in Hot Environments," *Annals of Occupational Hygiene*, **56**, 297–308 (1999).
26. J. D. Ramsey, Environmental Ergonomic Standards: USA and International, in B. Das and W. Karwowski, eds., *Advances in Occupational Ergonomics and Safety II*, IOS Press, 1997, pp. 553–556.
27. Hot Environments-Estimation of Heat Stress on Working Man Based on the WBGT Index (Wet Bulb Glob Temperature), ISO 7243-1989 (1989).
28. Hot Environments-Analytical Determination and Interpretation of Thermal Stress Using Calculation of Required Sweat Rate, ISO 7933-1989 (1989).
29. Moderate Thermal Environments-Determination of the PMV & POD Indices and Specification of Conditions for Thermal Comfort, ISO 7730-1994 (1994).
30. Evaluation of Cold Environments-Determination of Required Clothing Insulation (IREC), ISO/TR 11079-1993 (1993).
31. Ergonomics of the Thermal Environment-Principles and Applications of International Standards, ISO/DIS 11399 (1994).
32. Thermal Environments-Instruments for Measuring Physical Quantities, ISO 7726-1996 (1996).
33. Ergonomics-Determination of Metabolic Heat Production, ISO 8996-1990 (1990).
34. Evaluation of Thermal Strain by Physiological Measurement, ISO 9886-1992 (1992).
35. Ergonomics of the Thermal Environment-Estimation of the Thermal Insulation and Evaporative Resistance of a Clothing Ensemble, ISO 9920-1995 (1995).
36. Ergonomics of the Thermal Environment-Assessment of the Influence of Thermal Environment Using Subjective Judgment Scales, ISO/DIS 10551 (1995).
37. Ergonomics of the Thermal Environment-Medical Supervision of Individuals Exposed to Extreme Heat or Cold Environments, ISO/CD 12894 (1994).
38. J. D. Ramsey and M. Y. Beshir, "Thermal Standards and Measurement Techniques," Chapter 25 in S. R. DiNardi, ed., *Occupational Environment—Its Evaluation and Control*, AIHA Press, 1997.
39. H. S. Belding and T. F. Hatch, "Index for Evaluating Heat Stress in Terms of Resulting Physiological Strain," *Heat, Pip. Air Cond.* **27**, 129–135 (1955).
40. T. E. Bernard, "Thermal Stress," in B. Plog (ed) *Fundamentals of Industrial Hygiene*, 4th ed. Chicago: National Safety Council, Chapt. 12, 1995.
41. D. Minard, "Physiology of Heat Stress," in *The Industrial Environment—Its Evaluation and Control*, U.S. Department of Health, Education and Welfare, NIOSH, 1973.

42. A. Garg, D. B. Chaffin and G. D. Herrin, "Prediction of Metabolic Rate for Manual Materials Handling Jobs," *American Industrial Hygiene Association Journal* **39**, 661–674 (1978).
43. F. Tayyari, C. L. Burford and J. D. Ramsey, "Guidelines for the Use of Systematic Workload Estimation," *International Journal of Industrial Ergonomics* **4**: 61–65 (1989).
44. T. E. Bernard and B. S. Joseph, "Estimation of metabolic rate using qualitative job descriptors," *American Industrial Hygiene Association Journal* **55**: 1021–1029 (1994).
45. D. W. Barker, S. Kini and T. E. Bernard, "Thermal Characteristics of Clothing Ensembles for Use in Heat Stress Analysis," *American Industrial Hygiene Association Journal* **60** 32–37 (1999).
46. K. B. Pandolf, M. N. Sawka and R. R. Gonzalez, eds. *Human Performance Physiology and Environmental Medicine at Terrestrial Extremes*, Cooper Publishing Group, Carmel IN, 1986.
47. W. D. McArdle, F. I. Katch and V. L. Katch, *Exercise Physiology* 4th ed. Baltimore: Williams & Wilkins, Baltimore, 1996.
48. D. Minard, R. Goldsmith, P. H. Farrier, and B. J. Lambiotte, "Physiological evaluation of industrial heat stress," *American Industrial Hygiene Journal* **32**, 17–28 (1971).
49. T. E. Bernard and W. L. Kenney, "Rationale for a Personal Monitor for Heat Strain," *American Industrial Hygiene Association Journal* **55**, 505–514 (1994).
50. L. A. Brouha, *Physiology in Industry-Evaluation of Industrial Stresses by the Physiological Reactions of the Worker*, Pergamon Press, Elmsford, NY, 2nd ed. 1967.
51. F. H. Fuller and P. E. Smith, "Evaluation of Heat Stress in a Hot Workshop by Physiological Measurements," *American Industrial Hygiene Association Journal* **42**, 32–37 (1981).
52. NIOSH, OSHA, USCG and EPA, Occupational Safety and Health Guidance Manual for Hazardous Waste Site Activities. Washington, DC, DHHS(NIOSH) 85-115, (1985).
53. T. E. Bernard and W. L. Kenney, "Heart Rate Recovery," American Industrial Hygiene Conference, 1988.
54. A. R. Lind, "A Physiological Criterion for Setting Thermal Environmental Limits for Everyday Work," *Journal of Applied Physiology* **18**, 51–56 (1963).
55. World Health Organization (WHO), Health Factors Involved in Working under Conditions of Heat Stress, Technical Report Series 412, WHO Geneva, 1969.
56. P. D. Reneau and P. A. Bishop, "Validation of a Personal Heat Stress Monitor," *Am. Ind. Hyg. Assoc. J.* **57**, 650–657 (1996).
57. H. S. Belding, "Resistance to Heat in Man and Other Homeothermic Animals," in A. H. Rose, ed., *Thermobiology*, Academic Press, London, 1967.
58. C. H. Wyndham, N. B. Strydom, H. M. Cooke, and J. S. Martiz, "Studies on the Effects of Heat on Performance of Work," Applied Physiology Laboratory Reports, Transvaal and Orange Free State Chamber of Mines, South Africa, 1959.
59. C. W. Suggs and W. E. Splinter, "Some Physiological Responses of Man to Workload and Environment," *J. Appl. Physiol.* **16**, 413 (1961).
60. R. D. Pepler, "Performance and Well-Being in Heat," in J. D. Hardy, ed., *Temperature—Its Measurement and Control in Science and Industry*, Vol. 3, Part 3, Reinhold, New York, 1963.
61. R. D. Pepler, "Psychological Effects of Heat," in *Heat Stress and Heat Disorders*, F. A. Davis, Philadelphia, 1964.
62. B. C. Duggar, "Trial of an Assembly Job for Evaluating the Effects of Heat on Human Performance," M.S. Thesis, University of Pittsburgh, 1956.

63. R. D. Pepler, "Variations in Students' Test Performance and in Classroom Temperatures in Climate Controlled and Non-Climate Controlled Schools," *ASHRAE Trans.*, 47 Part II, 35–42 (1971).
64. J. B. Moses, "The Physiological Effects of Heat," *Ind. Med. Surg.* **22**, 20 (1960).
65. C. S. Leithead and A. R. Lind, "Heat Stress and Heat Disorders," *Heat Stress and Heat Disorders*, F. A. Davis, Philadelphia, 1964.
66. B. A. Hertig, D. W. Badger, P. J. Schmitz, and L. D. Siler, "Interaction of Carbon Monoxide and Heat," *Am. Ind. Hyg. Assoc. J.* **32**, 4 (1971).
67. F. Renshaw, "Combined Effects of Heat and Noise on Work Performance," *Am. Ind. Hyg. Assoc. J.*, Abstr. Suppl. 38 (1971).
68. F. N. Dukes-Dobos, "Hazards of Heat Exposure: A Review," *Scand. J. Work Environ. Health* **7**, 73–83 (1981).
69. M. Khagali and J. S. R. Hayes, *Heatstroke and Temperature Regulation*, Academic Press, Sydney, 1983.
70. E. Nadel and M. R. Cullen, "Thermal Stressors," in L. Rosenstock and M. R. Cullen, eds., *Textbook of Clinical Occupational and Environmental Medicine*, W.B. Saunders Co., Philadelphia, (1992).
71. N. B. Strydom, "Age as a Causal Factor in Heat Stroke," *J. S. Afr. Inst. Mining Metall.* **72**, 112–114 (1971).
72. R. F. Goldman, "Standards for Human Exposure to Heat," in I. B. Mekjavic, E. W. Banister, and J. B. Morrison, eds., *Environmental Ergonomics*, Taylor and Francis, London, 1988, pp. 99–136.
73. F. C. Houghton and C. P. Yaglou, "Determining Lines of Equal Comfort," *J. Am. Soc. Heat. Vent. Engr.* **29**, 165–176 (1923).
74. T. Bedford, Medical Research Council War Memo No. 17, London, 1946.
75. J. H. Botsford, "A Wet Globe Thermometer for Environmental Heat Measurement," *Am. Ind. Hyg. Assoc. J.* **32**, 1–10 (1971).
76. *Heating and Cooling for Man in Industry*, 2nd ed., AIHA, Akron, OH, 1975.
77. R. S. Brief and R. G. Confer, "Comparison of Heat Stress Indices," *Am. Ind. Hyg. Assoc. J.*, **32**(1), 11–16 (1971).
78. N. L. Ramanathan and H. S. Belding, "Physiologic Evaluation of the WBGT Index for Occupational Heat Stress," *Am. Ind. Hyg. Assoc. J.* **34**(9), 375–383 (1973).
79. V. M. Ciricello and S. N. Snook, "The Prediction of WBGT from the Botsball," *Am. Ind. Hyg. Assoc. J.* **38**, 264–271 (1977).
80. A. T. Johnson and G. D. Kirk, "Correlation of WBGT and Botsball Sensors," *Am. Ind. Hyg. Assoc. J.* **41**, 361–366 (1980).
81. M. Y. Beshir, J. D. Ramsey, and C. L. Burford, "Threshold Values for the Botsball: A Field Study of Ergonomics," *Ergonomics* **25**, 247–254 (1982).
82. M. Y. Beshir, "A Comprehensive Comparison Between WBGT and Botsball," *Am. Ind. Hyg. Assoc. J.* **42**, 81–87 (1981).
83. R. D. Parker and F. D. Pierce, "Comparison of Heat Stress Measuring Techniques in a Steel Mill," *Am. Ind. Hyg. Assoc. J.* **45**(6), 405–415 (1984).
84. ASTM Standard E337: Standard Test Method to Measure Humidity with a Psychrometer (1990).
85. *ASHRAE Handbook*, 1981 Fundamentals, The American Society for Heating, Refrigeration and Air Conditioning, Atlanta, 1981.

86. T. Bedford, *Basic Principles of Ventilation and Heating*, K. J. Lewis, London, 1948.
87. T. Bedford and C. G. Warner, "On Methods of Measuring Skin Temperatures," *J. Hyg.*, **34**(2), 81–98 (1934).
88. W. L. Roller and R. F. Goldman, "Estimation of Solar Radiation Environment," *Int. J. Biomeoterol.*, **11**, 329–336 (1967).
89. A. P. Garg, Treatise on Solar Energy, in *Fundamentals of Solar Energy*, Vol. 1, John Wiley & Sons, Inc., New York, 1982.
90. R. W. Allen, M. D. Ellis, and A. W. Hart, *Industrial Hygiene*, Prentice-Hall, Englewood, NJ, 1976, pp. 87–114.
91. H. M. Vernon, "The Measurement of Radiant Heat in Relation to Human Comfort," in *J. Physiol.—Proceedings of the Physiological Society*, pp. xv–xvii (1930).
92. C. P. Yaglou, "Heated Globe Thermometer for Evaluating Environmental Conditions for Comfort and For Studying Radiation-Convection Effects," *J. Ind. Hyg.* **17**(9), 185–198 (1935).
93. L. Hill, Medical Research Council Special Report No. 32, Part I, London, 1919.
94. A. F. Dufton, *Physiol. Mag.*, **9**, 858 (1930).
95. C. E. A. Winslow and L. Greenburg, *Am. Soc. Heat. Vent. Eng. Trans.* **41**, 149 (1935).
96. T. F. Hatch, "Design Requirements and Limitations of a Single-Reading Heat Stress Meter," *Am. Ind. Hyg. Assoc. J.* **34**(2), 66–72 (1973).
97. J. E. Mutchler and J. L. Vecchio, "Expirical Relationships Among Heat Stress Indices in 14 Hot Industries," *Am. Ind. Hyg. Assoc. J.* **38**(6), 253–263 (1977).
98. J. W. Jabara, "Comparison of Wet Bulb Globe Temperature and Modified Botsball Temperature in an Outdoor Environment," *Appl. Ind. Hyg.* **3**, 3–309 (1988).
99. R. C. Jensen and D. A. Heines, "Relationship Between Several Prominent Heat Stress Indices," U.S. Department of Health, Education, and Welfare, DHEW (NIOSH) Pub. 77-109, 1976.
100. F. Akbar-Khanzadeh and J. D. Ramsey, "The Prediction of Temperatures and Heat Stress Limits in the Workplace with Natural Ventilation," *Am. Ind. Hyg. Assoc. J.* **48**(4), 396–399 (1987).
101. T. E. Bernard and R. L. Cross, "Heat Stress Management: Case Study in an Aluminum Smelter;" *International Journal of Industrial Ergonomics* **23**, 609–620 (1999).
102. T. E. Bernard and M. Pourmoghani, "Prediction of workplace wet bulb global temperature," *Applied Occupational and Environmental Hygiene* **14**, 126–134 (1999).
103. V. Basus, A Temperature Parameter Useful in Evaluation of Summer Microclimatic Conditions in Spacious Manufacturing Shops and Smaller Individual Working Places, *STS Translation*, from *Zdrav. Tech. Vzduchotech.* (Prague), **10**(1), 29–36 (1967).
104. A. G. Loudon, *J. Inst. Heat Vent. Eng.* (London), **37**, 280–292 (1970).
105. G. V. Parmelee, *Bull. Am. Meteorol. Soc.* **36**, 256–264 (1955).
106. J. E. Mutchler, D. D. Malzahn, J. L. Vecchio, and R. D. Soule, "An Improved Method for Monitoring Heat Stress Levels in the Workplace," *Am. Ind. Hyg. Assoc. J.* **37**(3), 151–164 (1976).
107. A. P. Gagge, L. P. Herrington, and C. E. A. Winslow, "Thermal Interchanges Between Human Body and Its Atmospheric Environment," *Am. J. Hyg.* **26**, 84 (1937).
108. G. F. Haines, Jr. and T. F. Hatch, *Heating Venti.* **49**, 93–104 (1952).
109. C. E. A. Winslow and L. P. Herrington, *Temperature and Human Life*, Princeton University Press, Princeton, NJ, 1949.

110. H. M. Vernon, "Radiation Experienced in Factories and Houses," *J. Ind. Hyg. Toxicol.*, **19**, 498 (1937).
111. H. S. Belding, "Work in Hot Environments," in *Industrial Hygiene Highlights*, Vol. 1, Industrial Hygiene Foundation of America, Pittsburgh, 1968.
112. J. E. Peterson, "Experimental Evaluations of Heat Stress Indices," *Am. Ind. Hyg. Assoc. J.* **31**, 305–317 (1970).
113. H. C. Bazett and J. B. S. Haldane, *J. Physiol.* **4**, 252 (1921).
114. D. B. Dill, *Life, Heat, and Altitude*, Harvard University Press, Cambridge, MA, 1938.
115. C. K. Drinker, "Effects of Heat and Humidity Upon Human Body," *J. Ind. Hyg. Toxicol.* **18**, 471 (1936).
116. J. H. Talbott, "Heat Cramps," *Medicine* **14**, 323 (1935).
117. H. M. Vernon, T. Bedford, and C. G. Warner, Industrial Fatigue Research Board Report No. 51, Her Majesty's Stationery Office, London, 1928.
118. H. M. Vernon and E. A. Rusher, Industrial Fatigue Research Board Report No. 5, Her Majesty's Stationery Office, London, 1920.
119. R. H. Britten and L. R. Thompson, U.S. Public Health Service Bulletin No. 163, 1926.
120. C. R. Bell, in "Hot Environments and Performance," in the *Effects of Abnormal Physical Conditions at Work*, E. M. Davis, P. R. Davis, and F. H. Tyrer, Eds., Livingston, London, 1967.
121. J. M. Stewart, "The Use of Heat Transfer and Limiting Physiological Criteria as a Basis for Setting Heat Stress Limits," Second International Mine Ventilation Congress, Reno, Nevada, 1979.
122. C. H. Wyndham and A. J. Heyns, "The Probability of Heatstroke at Different Levels of Heat Stress," *Arch. Sci. Physiol.* **27**, A545–A562 (1973).
123. F. H. Fuller and L. Brouha, *ASHRAE J.* 39 (1966).
124. A. R. Lind and F. D. K Liddell, "The Influence of Individual Variation on the Prescriptive Zone of Climates," National Coal Board Memo, London, 1963.
125. A. R. Lind, "Physiological Effects of Continuous and Intermittant Work in Heat," *J. Appl. Physiol.* **18**, 57–60 (1963).
126. A. R. Lind, P. W. Humphreys, K. Collins, and K. Foster, "The Influence of Aging on the Prescriptive Zone of Climates," National Coal Board Memo, London, 1963.
127. B. A. Hertig, "Control of the Thermal Environment," in *Ergonomics and Physical Environmental Factors*, International Labor Office, Occupational Safety and Health Series No. 21, Geneva, 1970.
128. T. F. Hatch, "Heat Control in the Hot Industries," in F. A. Patty, Ed., *Industrial Hygiene and Toxicology*, Vol. 1, Wiley-Interscience, New York, 1958.
129. R. S. Brief and R. G. Confer, *Med. Bull.* **33**, 229–253 (1973).
130. W. F. Leihnard, R. S. McClintock, and J. P. Hughes, "Appraisal of Heat Exposures in an Aluminum Plant," Paper No. 8-3, presented at the 13th International Congress on Occupational Health, New York, 1960.

CHAPTER TWENTY-THREE

Physiological Effects of Altered Barometric Pressure

Claude A. Piantadosi, MD

1 INTRODUCTION

Scientists of the twentieth century have been fortunate to participate in the most exciting period of discovery in human environmental physiology. New biomedical knowledge coupled with a technological revolution has given human beings the opportunity to visit and explore the depths of the ocean and the frontiers of space. Such extreme environments present the physiologist with the challenge of defining reasonable parameters for physiological tolerance so that life support systems can be engineered to maximize our opportunity to interact with the environment. Among the most troublesome environmental factors are heat and cold, exposure to radiation, effects of acceleration and microgravity, hypoxia and hyperoxia, and extremes of barometric pressure. Nonetheless, humans can survive for many weeks or months in many exceptional environments ranging from high altitude near the vacuum of space to simulated undersea depths of more than 70 atmospheres of pressure absolute (ATA).

Abrupt or extreme changes in barometric pressure account for many of the injuries and deaths connected with the practice of aviation and diving. Physiological problems arise most commonly in association with rapid decompression from a higher pressure to a lower one, such as rapid ascent from depth under water, or to high altitude. These problems may arise from technical or procedural failures and from our incomplete understanding of inert gas transport and gas elimination from body tissues.

High-pressure environments have an important role in modern society for many reasons. The earliest engineering application of high pressure, to maintain dry working conditions

Patty's Industrial Hygiene, Fifth Edition, Volume 2. Edited by Robert L. Harris.
ISBN 0-471-29754-2

in caissons and tunnels, continues today, and it extracts a toll in decompression sickness and aseptic bone necrosis from workers who undergo daily decompression after long shifts in compressed air. As the demand to exploit undersea resources such as oil has increased, so has the working depth and time, and hence the medical vulnerability, of the diver. In relatively shallow diving with air and mixed gases, many commercial and military divers rely on surface supplied gear to provide respirable gas and continuous communication with surface tenders. These diving rigs provide an unlimited supply of breathing gas to the working diver in open, semiclosed, or closed (rebreathing) circuits. In addition, thousands of commercial, military, and scientific divers also have been trained to use self-contained underwater breathing apparatus (SCUBA) to work in shallow water. For some 50 years, the availability of lightweight, open-circuit SCUBA and low-resistance regulators has opened up recreational diving possibilities for thousands of sports divers. Finally, the growth of therapeutic hyperbaric oxygen as well as and the need to provide recompression therapy to the injured diver results in the routine exposure of patients, physicians, nurses, and other medical personnel to hyperbaric environments.

2 PRESSURIZED ENVIRONMENTS

2.1 General Principles

In order to understand the physiological responses to pressurized environments, the physical behavior of gases and the mechanical effects of pressure must first be made clear. The units of pressure measurement, force divided by area, can be expressed in a variety of ways. (Table 23.1). Normal atmospheric pressure is approximately 760 torr or 14.7 pounds per square inch (psi). The pressure exerted by a column of water varies linearly with its height and must be added to the normal atmospheric pressure to obtain the absolute pressure. As shown in Table 23.1, a column of seawater 33.1 feet deep (fsw) exerts the same pressure as the normal atmosphere at sea level. Thus, a diver immersed in seawater at 33.1 feet is exposed to a total pressure of 2 atmospheres absolute (ATA).

In diving, the pressure of a breathing gas must be increased in proportion to the absolute pressure in order to allow the diver to inhale against the pressure of the water column

Table 23.1. Pressure Equivalents

		Pressure		
Condition	Altitude or Depth	Torr	psig[b]	Atmospheres (ATA)[a]
Altitude	+18,000 feet	380	—	0.5
Sea level	0	760	0	1.0
Seawater	33 fsw	1,520	14.7	2.0
Seawater	66 fsw	2,280	29.4	3.0
Seawater	330 fsw	8,360	147.0	11.0
Seawater	660 fsw	15,960	294.0	21.0

[a]ATA, atmospheres absolute; note that depth in ATA = (fsw + 33)/33.
[b]psig, pounds per square inch gauge; note that a pressure gauge at sea level reads zero.

above him or her. As a result, more gas molecules must occupy the lungs and other gas-containing cavities of the body to maintain constant volume at a given temperature. The relationship between pressure (P), volume (V), temperature (T), and the number of moles of gas (n) can be described by the ideal gas law:

$$PV = nRT$$

where R is the universal gas constant (0.082 L ATA)(mol K)$^{-1}$. Several limited gas laws relevant to diving can be derived from the ideal gas law. For a constant number of moles of gas under condition 1,

$$P_1V_1/T_1 = nR$$

After changing one of the variables on the left side of the equation to create condition 2, we have

$$P_2V_2/T_2 = nR \text{ and}$$

$$P_1V_1/T_1 = P_2V_2/T_2$$

Three special cases of the ideal gas law, where one of the three variables is held constant are relevant to diving and are shown in Table 23.2.

All respirable gases contain oxygen and usually one or more other molecular species. In a mixture of gases, such as air, the total pressure is equal to the sum of the partial pressures of the individual gases (Dalton's Law). This means that each type of molecule in the mixture behaves as though it alone occupies the entire space available. The quantity of gas in the body is determined primarily by three factors: diffusion of gas into or out of blood from the alveolar spaces, its partial pressure (by Henry's Law), and the solubility of the gas in blood and tissues. The amount of gas taken up by tissues also depends upon the delivery of the gas to the tissue by the circulation and its diffusion into or out of the tissue in accordance with its concentration gradient. In general, the physiological effects of diving gases, such as inert gas narcosis and oxygen toxicity, correlate directly with the partial pressure of the gas in body tissues.

2.2 Immersion and Breath-Hold Diving

Profound adjustments in several physiological parameters are accompanied by immersion in water, including responses by the respiratory, cardiovascular, renal, and endocrine sys-

Table 23.2. Special Gas Laws

Boyle's Law	$P_1V_1 = P_2V_2$
Charles' Law	$V_1/T_1 = V_2T_2$
Gay Lussac's Law	$P_1/T_1 = P_2/T_2$

tems (1, 2). The immediate physiological consequences of immersion are pulmonary restriction, a sustained increase in cardiac output, and diuresis. These responses are related to the effects of hydrostatic pressure and to the high density of water relative to air. During head-out immersion, the extrathoracic blood vessels are supported by the water, and the body is exposed to a hydrostatic pressure gradient proportional to the vertical distance from the water surface. The hydrostatic pressure surrounding the body compresses the abdomen and thorax, causing negative-pressure breathing of approximately -20 cm H_2O. These hydrostatic effects displace the diaphram upward, thereby decreasing the thoracic gas volume and expiratory reserve volume. The pressure gradient across the diaphram created by immersion, together with hydrostatically decreased venous capacitance in the legs, increases the volume of blood in the lungs and heart. The increase in intrathoracic blood volume during immersion is also facilitated by the high density of water, which largely abolishes venous pooling. Central blood volume also may be augmented by peripheral arterial vasoconstriction if the water temperature is not thermoneutral (34°C).

Cardiovascular distension during immersion activates cardiac merchanoreceptors that normally respond to hypervolemia. Although the circulating blood volume is actually constant, the apparent hypervolemia is detected at the hypothalamus via vagal afferents and leads to an immersion response. There are two components to this response, which include diuresis and natriuresis, respectively. The diuresis and natriuresis after immersion have different time profiles, indicating that they operate by different mechanisms. Peak diuresis occurs in the first 1–2 hours, while peak natriuresis occurs after 4 to 5 hours of immersion. Fluid restriction and the administration of aqueous vasopressin before immersion prevent diuresis but do not affect natriuresis.

During immersion, the central blood volume increases by 500 to 700 ml. The amount of distension of the central circulatory organs correlates with the excretion of urinary sodium (2). The mechanism of the response to increased water loss during immersion is suppression of antidiuretic hormone (ADH) release. This is also known as the Gauer-Henry response. The mechanism of the natriuresis is related to decreased tubular reabsorption of sodium and not to an increase in filtered sodium. The factors in the natriuresis appear to be (*1*) aldosterone suppression via decreased renin-angiotensin activity, (*2*) increased release of atrial natriuretic factor, (*3*) increased release of renal prostaglandins, and (*4*) decreased sympathetic activity.

The increase in central blood volume during immersion distends the cardiac chambers and enhances ventricular diastolic filling, which increases cardiac output (3). The increase in cardiac output with immersion is attributable almost entirely to an increase in stroke volume, because resting heart rate usually remains unchanged or decreases slightly. The stroke volume and cardiac output may double. The increase in cardiac output is persistent and occurs with no measurable increase in systemic oxygen uptake. The mechanism of the increase in stroke volume appears to be primarily an increase in cardiac preload, although changes in cardiac contractility may also occur.

During breath-hold diving, the lungs act as a reservoir for exchange of oxygen (O_2) for carbon dioxide (CO_2) in pulmonary capillary blood (4). The P_{O_2} in the lungs decreases with time, and its rate of change is a function of the decline in the P_{O_2} in the blood returning to the lungs (mixed venous blood). During the first seconds of apnea, the rate of O_2 uptake by the body is relatively constant. Shortly thereafter, O_2 transfer from the lungs can no longer support the needs of the body, the O_2 uptake falls, and anaerobic metabolism increases. Carbon dioxide enters the lungs during apnea in proportion to both the blood flow through the lungs and the diffusion gradient for CO_2, i.e., the difference between mixed venous P_{CO_2} and alveolar P_{CO_2}. The rate of CO_2 transfer is high initially, but decreases rapidly because the gradient for diffusion of CO_2 decreases as alveolar P_{CO_2} approaches the mixed venous value. As CO_2 production continues, mixed venous P_{CO_2} increases and causes alveolar P_{CO_2} to rise again. The point at which high P_{CO_2} causes breathing to resume is called the break point. The time to the break point can be extended by maneuvers that lower P_{CO_2} or raise P_{O_2}, such as hyperventilation, oxygen breathing, or both. Notably, the body's O_2 store is not increased appreciably by hyperventilation because the increase in alveolar P_{O_2} made possible by lowering alveolar P_{CO_2} increases blood O_2 content only slightly. Therefore, hyperventilation, associated with a greatly extended breath-hold time, can produce hypoxia-induced unconsciousness before the CO_2 rises to the break point.

During breath-hold diving, gas exchange is altered by actual descent because lung volume decreases from thoracic compression, and the partial pressures of O_2, CO_2, and N_2 increase in the lungs. As the descent progresses, alveolar O_2 and CO_2 concentrations decrease because of a greater transfer of those gases to pulmonary capillary blood compared with N_2, which is metabolically inert and not very soluble in blood. The alveolar P_{O_2} is also higher during breath-hold diving than during simple breath-holding, and the transfer of CO_2 during early descent is opposite normal, i.e., CO_2 moves from alveoli to pulmonary capillary blood. During ascent, the lung re-expands, and the alveolar P_{O_2} and P_{CO_2} decline. Near the surface, the alveolar P_{O_2} may be almost as low as the mixed venous P_{O_2}. If the dive has been of sufficient magnitude, reverse O_2 transfer from mixed venous blood to alveoli may occur during ascent. Carbon dioxide retained in the blood during the dive also leaves the blood on ascent as the alveolar P_{CO_2} decreases. Carbon dioxide elimination usually continues postdive, as does elimination of the small amount of N_2 that entered the blood during the dive. This amount of N_2 is so small that it is not a significant decompression risk; however, repetitive breath-hold dives, if frequent enough, theoretically may allow enough N_2 retention to produce decompression sickness.

During breath-hold diving, a dangerous situation may arise if hyperventilation is used to extend the dive. During the dive, the alveolar P_{O_2} increases as a result of compression of the lung. Thus, the need to return to the surface is signaled by the rising P_{CO_2}. When the diver hyperventilates before the dive, the arterial P_{CO_2} begins at a lower level, and the time to the breath-hold break point is prolonged. The O_2 reserve is more severely depleted during the longer dive, and when the thorax re-expands as the diver approaches the surface, life-threatening hypoxia may occur. This results in shallow-water blackout and causes a significant number of drownings.

The responses to breath-hold diving in humans can be modified by a rudimentary diving response induced by apnea and facial immersion, particularly in cold water. This diving response, manifested primarily as bradycardia, is most pronounced in young children. The diving response has been interpreted as an O_2-conserving mechanism that extends breath-holding time by redistributing blood flow from organs that are relatively resistant to hypoxia to organs with continuous requirements for oxygen such as heart and brain. The validity of this interpretation in humans is uncertain, although a diving response may contribute to survival of children resuscitated after more than an hour of submersion in cold water.

2.3 Diving with Compressed Air and Mixed Gas

The first invention to attempt to circumvent limitations of breath-hold diving was the snorkel. This device was recognized almost immediately, however, to be limited to only 100 cm in length because of the inability of the respiratory muscles to expand against a greater load. This problem was not solved until the nineteenth century, when reliable diving systems were invented to provide the diver with a continuous supply of pressurized breathing gas at any depth. Diving in shallow water, e.g., 135 fsw or less, is usually carried out with compressed air as the breathing gas because air is inexpensive and can be obtained readily, even at remote dive sites. Most air diving today is conducted using SCUBA. For deeper air diving, the diver may be supplied by an umbilical to the surface. The tethered diver has a number of advantages over the SCUBA diver in that access is provided to continuous air supply, to surface communication, and to power for thermal protection. The umbilical also allows the diver to be retrived; however, it also can be an encumbrance and has occasionally caused a diver to become trapped by an underwater obstruction.

The recommended maximum safe depth for SCUBA is approximately 135 fsw, and for surface-supplied air diving, approximately 200 fsw (5). These limitations are brought about by three factors: nitrogen narcosis, decompression time, and oxygen toxicity. In order to dive beyond the practical range of compressed air, special gas mixtures must be supplied. To extend dive time at relatively shallow depths, oxygen-enriched air (nitrox) is frequently used to decrease nitrogen uptake. For deeper diving, where nitrogen narcosis is problem, helium is substituted for nitrogen as the diluent gas, although other nonnarcotic inert gases such as hydrogen have been used successfully. In early helium dives, decompression sickness was a serious obstacle. In 1937 the first safe helium decompression schedules were developed, and helium-oxygen (heliox) mixtures were proven to work effectively for deeper diving. Heliox was first used successfully in 1939 by the U.S. Navy to salvage the sunken submarine Squalus from 243 fsw (6). In current heliox operations, the oxygen pressure is usually maintained at a constant 0.4 to 1.0 ATA, and the diver often operates out of a diving bell. The helium may be recycled by gas reconditioning equipment located on the bell. Surface-supplied heliox diving is expensive, but effective and relatively safe to depths to 400 fsw.

2.4 Saturation Diving

The ability to dive deeper with helium-oxygen mixtures led inevitably to longer decompression time relative to effective working time for the diver. In an effort to solve this problem, U.S. Navy Medical Officers, led by George Bond in 1957, explored the feasibility of saturation diving (6). The physiological principle was that after prolonged exposure (12 to 24 hours) at any depth, a finite limit of gas uptake by tissues would occur (saturation). Thereafter, no further decompression penalty would be incurred. The diver residing at that depth could work for long intervals and would require only a single, slow decompression.

Using these principles, divers have lived and worked for extended periods at ocean depths of 1500 fsw and at simulated depths to 2400 fsw in dry chambers.

Saturation diving is routine practice for many military and commercial divers when extended periods of time are needed to complete a task at 400 fsw or greater. Saturation techniques are widely employed in the offshore oil drilling industry in the North Sea and Gulf of Mexico. The divers generally live in a deck decompression system mounted on board the ship or the drilling platform. The divers work in shifts and are transported to their working depth in a pressurized transfer capsule or bell that mates with the deck system. At the working depth, the hatch of the capsule is opened and the divers in their rigs are free to leave the bell to work. The divers are tethered to the bell by umbilicals containing safety cables, communications, and water lines for hot water suits. The bell or capsule systems are managed by an inside tender with communications to the mother ship or platform. After each shift, the divers return to the bell, the hatch is closed, and the bell is returned to the ship. This procedure allows shipboard personnel to tend the divers continuously and obviates the need to supply large amounts of power to great depths 24 hours a day.

2.5 Hyperbaric Chambers

Hyperbaric chambers are pressure vessels designed for controlled compression and decompression of atmosphere gas. Hyperbaric chambers for human occupancy may be of either multiplace or monoplace design. These chambers are used primarily to administer hyperbaric oxygen as a therapy for certain specific or refractory medical problems. Multiplace chambers are large and expensive and require considerable expertise to operate; however, they offer significant advantages over monoplace chambers in caring for very sick patients. The monoplace chamber, by definition, holds only one person, the patient, and is usually compressed with 100% oxygen. Therefore, procedures that require an attendant or tender cannot be accomplished in the monoplace setting. Multiplace chamber complexes are also used to simulate the diving environment and, therefore, have been invaluable to scientists conducting diving-related research.

The safe operation of hyperbaric chambers requires a thorough understanding of the equipment, procedures, and potential hazards of the hyperbaric environment. This requires specialized training and extensive knowledge of principles of pressure integrity, gas handling, built-in-breathing systems (BIBS), electrical systems, and fire safety. Hazards of chamber operation include abrupt changes in gas pressure, gas contamination, hypoxia, thermal problems, oxygen toxicity, and fire. The occurrence of fire in oxygen-enriched atmospheres is a special hazard. The ability of oxygen to support combustion increases as the oxygen concentration increases. At O_2 concentrations above 25% by volume, ultrarapid combustion of flammable materials can occur inside a chamber, with disasterous consequences. Over the past 40 years more than two dozen chamber fires have resulted in nearly 80 deaths, most of them patients receiving hyperbaric oxygen therapy.

Safe chamber operation requires careful, full-time supervision by a qualified operator. Operational and emergency procedures must be established in advance and be readily available on site at all times. Special care must be taken to eliminate sources of ignition and minimize the amount of flammable material in oxygen-rich environments. Careful operations and maintenance records should be maintained for each facility. The U.S. Navy and ASME have established standards for certification of hyperbaric chambers for human occupancy.

2.6 Tunnels and Caissons

The use of compressed air to keep water out of shafts and tunnels evolved during the middle of the nineteenth century. The engineering problems associated with providing a continuous supply of compressed air were solved first in France in the 1830s by the engineer Triger. Successful caissons containing compressed air were used for bridge construction in the 1850s and for tunnels in the 1880s. Compressed-air workers were exposed daily to pressures as high as 4.5 ATA. Similar air pressures are still used today in construction of tunnels, bridges, and mine shafts (7).

Traditional practice for the construction of underwater shafts is to force compressed air into the entire working chamber at a pressure high enough to keep water out of the face of the shaft. The workers usually work one shift per day. The length of the shift is governed by the maximum working pressure, the decompression profile, and local regulations. For example, in the United Kingdom, the maximum working pressure permitted by law is 50 psig (4.5 ATA) and the maximum time at pressure cannot exceed 10 hours in any 12 hour period. Thus, a compressed-air worker at 50 psig is allowed just over 4 hours at depth because the decompression time approaches 6 hours. Work shifts in the United States have been kept relatively short to limit the incidence of aseptic bone necrosis; however, this practice is not completely effective.

The working chamber at each site is fitted with a number of air locks that serve different purposes. Separate air locks for people are required for timed decompressions. Equipment and supplies pass through muck locks, and medical locks for the treatment of decompression sickness are required when the working pressure is 2 ATA or greater. In some places, decant locks are used for decompression; in these cases workers are rapidly decompressed in the human lock and then recompressed in the decant lock for completion of the appropriate decompression schedule.

Compressed-air workers often face difficult working conditions, which range from cool, damp environments to hot, humid atmospheres sometimes containing potentially toxic gases and fumes. Although very large quantities of clean air are supplied to tunnels and caissons by on-site compressors, welding fumes and other contaminants such as CO tend to accumulate in pockets where ventilation is inadequate. Current guidelines require that at least 10 ft^3 of air at working pressure per minute be supplied per person in the tunnel or caisson. Other immediate hazards encountered in caisson or tunnel environments are flooding, heat exhaustion, decompression sickness, and hypoxia. Hypoxia may occur with influx of oxygen-depleted air from soil beds upon reduction of the pressure inside the chamber. The risk of fire is enhanced greatly in compressed-air environments because of the elevated partial pressure of oxygen. The most troublesome long-term occupational health problem for tunnel workers has been aseptic necrosis of bone (see Section 4.4).

2.7 Submarines and Submarine Escape

The ambient environment of the nuclear-powered submarine is sealed but not pressurized. The most significant problem in submarine atmosphere control remains the metabolic production of CO_2 by the crew (8). Atmospheric CO_2 concentrations are allowed to approach 1% for the duration of a submarine patrol. Continuous exposure to increased ambient CO_2 levels is associated with small increases in respiratory minute ventilation and cyclic changes in blood bicarbonate and pH. The acid-base disturbance correlates with cycles in calcium homeostasis, probably related to the uptake and release of CO_2 from bone. Other minor atmospheric contaminants include CO, freons, and various aerosols. The only detectable health effect of these contaminants has been a slight increase in hematocrit attributed to low-level CO exposure.

Though modern submarines do not operate routinely at pressures above 1 ATA, they can be pressurized by virtue of their pressure-tight internal hulls and gas storage capacity. This capability in U.S. submarines is approximately 6 ATA, but it provides a measure of protection against accidental flooding, which in peacetime is most likely to result from collisions in shallow coastal waters. The occurrence of submarine disasters in shallow waters initiated the development by Momsen, Tibbels, and Hobson in 1929 of an apparatus to facilitate escape from U.S. submarines. The Momsen lung consists of a rebreathing bag containing a canister for CO_2 absorption. The device is charged with air or oxygen, and excess pressure in the lungs and apparatus during ascent is relieved by exhaust of gas through a ventral flutter valve. Simulated submarine escapes were practiced by U.S. submariners for many years from 100 foot training towers. Thousands of safe practice ascents have been made, but occasionally pulmonary overinflation accidents occurred resulting in air embolism. As a result, the routine practice of free ascents for submariners has been abandoned by the U.S. Navy. The Momsen lung and its variant, the Steinke hood, have been superseded by a hooded pressure suit developed by the British Royal Navy. Royal Navy divers have made escapes safely from depths of 600 feet with this type of equipment.

The current operational plan for rescue of submarine crews in the U.S. Navy is centered around the Deep Submergence Rescue Vessel (DSRV) (6). This maneuverable submersible is designed to interface with the hatches of a sunken submarine. The DSRV can operate to 5000 fsw and can be pressurized in accordance with the capacity to pressurize the inner hull of the submarine. There are a number of potential operational and physiological problems associated with use of DSRVs, and the system has never been put to actual operational use.

3 PHYSIOLOGICAL EFFECTS OF COMPRESSION AND PRESSURIZED GASES

3.1 Compression Arthralgias

Compression of the body in air- and mixed-gas diving can produce a variety of joint symptoms (9). These symptoms are most prominent during rapid compression to great depth and include both pain on movement and joint popping or cracking. The former is more common in helium-oxygen diving and the latter in nitrogen-oxygen diving, although both symptoms occur during compression with helium. Compression arthralgias most often involve the shoulders, knees, hips, and back, and the symptoms resolve spontaneously after a few hours at maximum depth.

The etiology of compression arthralgias is unknown. The two leading hypotheses about their origin relate to direct effects of hydrostatic pressure, and osmotic shifts of fluid out of the synovium during the uptake of inert gas by well-perfused tissues near the joint space. Whether the joints are damaged by compression is unknown; however, the presence of symptoms raises the concern that divers may be predisposed to degenerative joint disease. Compression arthralgia can be ameliorated by using slow rates of compression.

3.2 Barotrauma of Descent

The physical pressure-volume relationships of gases have important consequences for how gas-containing body cavities equalize during diving. Body structures that normally contain gas in equilibrium with ambient air include the paranasal sinuses, lungs, middle ear space, and hollow viscera of the gastrointestinal tract. Any gas-filled cavity that is not fully collapsible must equilibrate with ambient pressure during both descent and ascent. If the pressure within the space does not equalize, barotrauma may occur to the tissues. Common examples of barotrauma of descent include middle ear squeeze, sinus squeeze, and lung squeeze. Middle ear squeeze is the most common form of barotrauma of descent and may result in a range of injuries including tympanic membrane perforation. More severe otic injuries may also occur, including rupture of the round or oval windows (10). Lung squeeze occurs in either deep breath-hold diving or when the pressurized gas supply of the diver cannot be maintained during rapid descent. Mask squeeze may also occur in SCUBA diving if the diver fails to equalize the space in his facemask on descent. This sometimes produces dramatic periorbital hemorrhage, which is of little clinical consequence. Barotrauma also may occur on ascent, usually in compressed air or gas environments when an equilibrated space fails to vent during the ascent. Common examples of barotrauma of ascent include reversed middle ear and reversed sinus squeeze, alternobaric vertigo of ascent (rarely on descent), and pulmonary barotrauma syndromes (see Section 4.3). One important but rare consequence of middle ear overpressurization during ascent is baroparesis of the facial nerve (11). Facial baroparesis is a transient, ischemic paresis of the seventh cranial nerve in divers with defects in the wall of the facial canal. Similar palsies of the fifth cranial nerve due to its compression in the maxillary sinus also have been reported. These cranial nerve findings are important because they may be confused with neurological decompression sickness (see Section 4.2) and/or arterial gas embolization (see Section 4.3).

3.3 Inert Gas Narcosis and High-Pressure Nervous Syndrome

For more than a century, divers have known that compressed air has narcotic properties when breathed at pressures greater than 3 ATA. Compressed air narcosis is related directly to the partial pressure of N_2, therefore, its manifestations increase in direct proportion to the ambient pressure. Symptoms of nitrogen narcosis, such as euphoria and impaired cognitive function, are similar to those produced by alcohol and the early stages of anesthesia. At depths beyond 10 ATA, N_2 may result in unconsciousness, thus limiting the useful depth of compressed air as a breathing gas. Narcosis is a property of nitrogen shared with other noble gases such as neon, argon, and xenon. The narcotic properties of these inert gases correlate well with their solubility in lipids. For example, helium, which is the least-narcotic noble gas, has a lipid solubility and narcotic potency only one-fourth that of nitrogen, while xenon is about 25 times more lipid soluble and narcotic as N_2. Carbon dioxide also has narcotic properties and at one time was proposed erroneously as the cause of compressed air narcosis.

The precise biological mechanisms of inert gas narcosis have not yet been delineated (12, 13). Most investigators favor mechanisms related to the critical volume hypothesis of general anesthesia. According to this hypothesis, anesthetic gas molecules are absorbed by lipid layers in nerve cell membranes, thereby causing an increase in volume. This increased volume is thought to alter the electrical and ionic permeability characteristics of neuronal membranes, particularly at sites of synaptic transmission. Volatile gases with high lipid solubility, e.g., nitrous oxide (N_2O), achieve the critical volume effect at lower partial pressures less than narcotic gases such as N_2. The critical volume hypothesis is supported indirectly by pressure reversal of anesthesia, wherein increased hydrostatic pressure or compression with a gas of low narcotic potency, such as helium, ablates the narcotic effect.

Pressure reversal of anesthesia appears to be related to the broader phenomenon of increased excitability in the CNS produced by direct effects of high pressure (12). In practical terms, this neuronal hyperexcitability has posed a barrier to rapid compression and very deep diving. The effects of rapid compression and very high pressure have become known collectively as the high-pressure nervous syndrome (HPNS), although they may well have different biological mechanisms. HPNS is experienced by both humans and animals at pressures above 15–20 ATA. The neurophysiologic basis of the syndrome is poorly understood but may involve decreased membrane fluidity and altered presynaptic neurotransmitter release, and/or decreased postsynaptic binding or responsivity at high pressure. The hallmarks of HPNS in humans are rapid tremor (6–12 Hz), incoordination, myoclonus, microsleep, and various electroencephalographic abnormalities. Convulsions and death occur in animals at extremely high pressures, depending on species and individual susceptibility. In accordance with the critical volume hypothesis, some of the neurophysiologic manifestations of HPNS can be ameliorated by increasing the concentration of a narcotic diluent gas in the breathing gas mixture. This approach has resulted in the successful use of trimix (helium, nitrogen, and oxygen) for very deep human dives. At this time, however, HPNS, more than any other single physiological factor, limits the maximum depth at which humans can dive.

3.4 Pulmonary Function and Gas Exchange

Although the physical work capacity of humans in hyperbaric conditions is normal at shallow to moderate pressures, oxygen uptake ultimately becomes limited as pressure increases, primarily because of limitations in pulmonary mechanical function (14, 15). These mechanical limitations are related more closely to the increased density of the breathing gas than to either increased hydrostatic pressure or increased gas viscosity. The density (ρ) of the gas that a diver breathes increases in proportion to both the molecular weight of the gas and ambient pressure. Thus, N_2, which has a density of 1.1 g/L at BTP, is about seven times as dense as helium under the same conditions. Helium-oxygen breathing gases at 50 ATA, however, are approximately seven times as dense as air at 1 ATA.

The important respiratory consequences of breathing gas at high density are an increase in resistance to airflow, an increase in the functional residual capacity (FRC), and a decrease in the ventilatory response to CO_2. The first two factors translate to greater work for breathing needed to sustain the high levels of ventilation accompanying heavy exercise. The third effect, a depressed ventilatory response to CO_2, is probably secondary to the increased resistance to airflow. High resistance to airflow decreases the amount of ventilation produced by the muscular response to a particular CO_2 level (16). Ultimately, the capacity to exercise at high gas densities becomes limited by ventilation, unlike in the normal environment where maximum oxygen uptake is usually limited by cardiovascular factors. Gas density also has three other potentially important although less well studied effects on respiratory function. These are increased respiratory heat loss, increased inertance of flowing gas, and decreased gas-phase diffusivity. The latter effect may impair alveolar gas mixing and decrease effective alveolar ventilation in divers by increasing the physiological dead space and therefore, contribute to CO_2 retention. The potential ventilatory limitation to exercise is also exacerbated underwater by the external resistance added by the breathing apparatus.

As gas density increases, resistance to airflow increases linearly (at moderate flow rates) for densities up to 25 g/L. The increase in airways resistance produced by breathing dense gases is mechanically different from the resistance produced by chronic obstructive pulmonary disease or breathing through an external resistance; however, the mechanical effects of dense gas may be understood in much the same way as the other processes (17). During a forced expiration, the maximum rate of airflow is determined by lung volume and dynamic compression of airways. Dynamic airway compression during expiration is a direct result of the high pressures generated by the expiratory muscles, which adds force to the normal lung elastic recoil pressure. As the pressure inside the airways drops enroute to the larger airways, the sum of the expiratory plus elastic recoil pressure outside the airways equals (equal pressure point) and eventually exceeds the pressure inside the airways and any resistance to deformation by the tissue. At the equal pressure point, the airways collapse, creating a flow-limiting segment. Beyond this point, airflow is independent of effort, i.e., it will not increase further with additional expiratory force. In the human respiratory system, flow-limiting segments are of sufficient radius that the airflow is primarily turbulent. Therefore, airflow is inversely related to gas density. Simple calculations yield the useful relationship that airflow through flow-limiting segments is proportional to the reciprocal of the square root of gas density. This square root relationship describes how high gas density decreases maximum respiratory flow rates; it also approximates the strong inverse correlation between gas density and maximum voluntary ventilation (MVV). If the MVV is limited by dynamic airway compression, and the MVV at depth determines maximum exercise ventilation, then gas density will become the limiting factor for the amount of aerobic exercise that a diver can perform.

During exercise, the minute ventilation ($\dot{V}_E$) increases in proportion to the amount of oxygen consumed ($\dot{V}_{O_2}$). The ventilatory equivalent, $\dot{V}_E$(BTPS)/$\dot{V}_{O_2}$ of approximately 30, generally maintains the arterial P_{CO_2} near 40 torr. If the maximum exercise ventilation is decreased by dense breathing gas, then maximal O_2 uptake can occur only at a smaller ventilatory equivalent and, thus, a higher P_{CO_2}. Hypercapnia is exaggerated further by the depressed ventilatory response to CO_2. Respiratory acidosis and high sustained levels of respiratory work may produce respiratory muscle fatigue that limits exercise at high gas densities. In fact arterial P_{CO_2} values above 60 torr have been reported in working divers just before they cease exercise (18).

As noted previously, breathing dense gas can increase the FRC during exercise; however, this change in FRC may not be significant during immersed exercise. When the FRC is larger, exercise ventilation is done at higher lung volumes where airflow resistance is less. The pressure-volume characteristics of the respiratory system, however, require a large increment in the inspiratory work needed to overcome elastic recoil forces at high lung volumes. Ventilation at high lung volume is also inefficient because the angle of apposition between the diaphragm and chest wall may widen and decrease inspiratory efficiency, thereby resulting in inspiratory muscle fatigue.

3.5 Oxygen Toxicity

Oxygen concentrations sufficient to cause pulmonary and CNS toxicity are encountered frequently in hyperbaric environments. Toxic effects of high concentrations of O_2 have been known since 1878 when Paul Bert reported convulsions in animals exposed to hyperbaric oxygen (19). Pulmonary oxygen toxicity was discovered more than 20 years later by Lorrain-Smith. High concentrations of oxygen are used to accelerate inert gas elimination during decompression in mixed-gas diving. Pure oxygen is used in closed-circuit diving with a CO_2 scrubber to prevent detection of exhaust bubbles in shallow-water military operations. During hyperbaric oxygen therapy, patients routinely breathe pure oxygen at treatment depths of 2 to 3 ATA.

Oxygen is toxic to all living cells at high concentrations; however, significant oxygen poisoning is encountered only in three tissues; the lungs, the CNS, and the retina of the newborn. The toxic effects of O_2 can be avoided if the hazards of oxygen are appreciated and efforts are made to limit the exposure time to potentially toxic doses. As a general rule, the rate at which O_2 toxicity develops increases as a hyperbolic function of the inspired partial pressure of the gas (20, 21). This principle is illustrated in Figure 23.1 for both pulmonary and CNS oxygen toxicity. The most effective practice for avoiding O_2 toxicity has been to use empirical tables to limit exposure time in inverse proportion to partial pressure. There are several ways to estimate the physiologic effects of a known dose of oxygen (22); however, such calculations have not proven useful in practice because of large differences in individual susceptibility to O_2 toxicity. When high levels of inspired oxygen are needed, the O_2 tolerance time can be extended greatly by interrupting the exposure briefly with air or other gas containing a nontoxic concentration of oxygen.

The most dangerous form of oxygen toxicity for the diver is CNS O_2 toxicity, and its most serious manifestation is O_2-induced convulsions. Although oxygen convulsions are transient, they may occur without warning at P_{O_2} at or above 1.4 ATA. Such an event may have disastrous consequences, particularly for free-swimming divers. The most common signs and symptoms of CNS oxygen toxicity are given in Table 23.3 (23).

The mechanisms of O_2 toxicity have been clarified in recent years by elegant experimental work. Hyperoxia accelerates the production of reactive species (ROS), which mediate of the toxic effects of O_2 produced on tissues (24). The exact relationship between ROS generation and clinical signs and symptoms of O_2 poisoning, however, are still being defined. Many intracellular enzymes and electron transport systems catalyze single electron

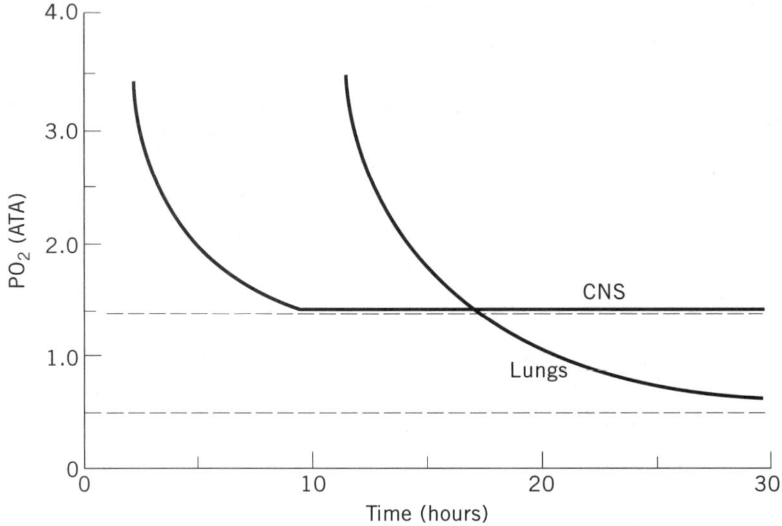

Figure 23.1. Hyperbolic relationships between O_2 concentration and exposure time for oxygen toxicity. Curves for pulmonary and CNS toxicity are illustrated.

Table 23.3. Common Signs and Symptoms of CNS Oxygen Toxicity

Visual symptoms
Loss of acuity
Scotomata
Constriction of visual fields
Acoustic symptoms
Tinnitus
Music
Unpleasant odors and tastes
Twitching of lips and face
Irritability
Hiccoughs
Nausea and vomiting
Syncope
Convulsions

transfers to dioxygen as by-products of normal metabolic processes. These reactions produce superoxide anion ($^{\cdot}O_2-$) or hydrogen peroxide (H_2O_2). The risk of cellular injury from ROS is reduced by dismutation of $^{\cdot}O_2-$ to less reactive H_2O_2. This process occurs spontaneously or via a reaction catalyzed by the superoxide dismutases (SODs). Hydrogen peroxide, which also may be toxic to cells, is converted catalytically to water plus O_2 by the enzymes catalase and glutathione peroxidase. There is also the potential for generating a hydroxyl radical ($OH^{\cdot}$) from H_2O_2 and ferrous iron (Haber-Weiss reaction). Under normal conditions, SODs, peroxidases and low molecular weight nonenzymatic free-radical scavengers, such as reduced glutathione and ascorbic acid, provide sufficient antioxidant defenses to limit free-radical-mediated damage such as lipid peroxidation, protein and nucleic acid oxidation, and thiol depletion. During hyperoxia, production of ROS is accelerated, and the rate of ROS generation may exceed the ability of antioxidant systems to contain them.

In the brain, the ROS hypothesis of O_2 toxicity is linked closely to the effects of hyperoxia upon cellular energy metabolism (25). During hyperbaric oxygenation, brain tissue P_{O_2} increases despite compensatory vasoconstriction of cerebral blood vessels. Tissue hyperoxia is accompanied by oxidation of sulfhydryl-containing enzymes and mitochondrial pyridine nucleotides (NADH). The intramitochondrial events precede O_2-induced convulsions and, together with evidence of enhanced glycolysis, have been interpreted to indicate that cerebral energy metabolism is activated by hyperbaric oxygen. Thus, the increased rate of mitochondrial electron transport may be needed to supply energy in the form of ATP; however, it may also promote ROS formation. These events are accompanied by return of cerebral blood flow to higher values, leading to critical oxidation events in specific regions of the brain. By mechanisms yet to be worked out, these biochemical events lead to greater neuronal excitation and to seizures.

Lung injury from ROS has also been associated with activation of lung metabolism. The resulting pathophysiologic response has been divided into three distinct phases: the early, or initiation, phase; the inflammatory phase; and the late, or destructive, phase (26). During lethal O_2 exposures the ROS generation is not associated with early structural changes in the lung. Later, the biochemical processes initiate inflammation, leading to damage of lung parenchyma. If the O_2 exposure is less severe, repair processes supervene and a proliferative phase follows. The proliferative response diminishes the destruction of lung parenchyma, but it may produce pulmonary fibrosis.

There are a number of ways to increase or decrease human tolerance to oxygen. Several physiological factors are important amplifiers of CNS oxygen toxicity in humans, including CO_2, immersion, exercise, body temperature, and sympathetic stimulation. The most practical approach to the extension of human oxygen tolerance has been to decrease the inspired oxygen concentration to nontoxic levels for brief (5–10 min) intervals of time during prolonged exposure.

3.6 Thermal Problems and Energy Balance in Diving

Most diving activities are conducted in water that is cold relative to the thermoneutral temperature for the immersed human (about 34°C at rest). This rather high thermoneutral point in water is a result of its high density and thermal conductivity relative to air. Surface water temperatures range from 4°C near the poles to about 30°C near the equator; at 600 fsw, water temperatures are approximately 4°C to 10°C. Thus, the physiological responses to body cooling and the principles of thermal protection are important factors in most diving operations. Human thermoregulation has been studied extensively; however, only those physiological responses to cold stress with relevance to diving are summarized here. A more comprehensive discussion of this topic has been provided by Webb (27).

Heat transfer from or to the human body obeys the second law of thermodynamics, i.e., transfer of heat occurs from a system of higher temperature to one of lower temperature in direct relation to the temperature difference between the two systems. There are four modes of heat transfer, including conduction, convection, radiation, and evaporation. Conduction, or transfer of heat from two physical systems in direct contact, and convection, or heat transfer in moving fluids, account for the bulk of the heat exchange from the skin during underwater immersion and in hyperbaric environments. Radiant heat transfer, or transmission of heat between two bodies at a distance, is less important in diving, and evaporative heat transfer, which is the heat required to change a liquid into a vapor at constant temperature, is usually limited to insensible respiratory heat losses by the diver. Respiratory heat losses can be quite significant in deep diving, and the convective component, which increases as a function of gas density and temperature, supplants evaporative heat loss as the major component of respiratory heat loss at great depth.

The exchange of heat between the body and its environment can be expressed in the form of a simple heat balance equation:

$$M \pm C \pm R - E = S$$

where M is metabolic heat production, C is convective (and conductive) heat loss or gain, R is radiant heat loss or gain, E is evaporative heat loss, and S is heat loss or gain (storage). The respiratory heat losses contribute to both the evaporative and convective terms in the heat balance equation. In general, however, convective losses from the skin (C_{sk}) account for most of the heat lost from the body during immersion in both cold water and cold dense gas, e.g., dry hyperbaric habitats. Convective heat loss (C_{sk}) from the body surface can be expressed as

$$C_{sk} = h_c(T_{\overline{sk}} - T_a)$$

where h_c is the convective heat transfer coefficient in W/(m^2 °C), $T_{\overline{sk}}$ is the mean skin temperature in °C, and T_a is the ambient temperature in °C. The convective heat transfer coefficient (h_c) is affected by the velocity, density (ρ), and the specific heat of the fluid (liquid or gas) surrounding the body. In gas-filled hyperbaric habitats, the density of the gas is proportional to the ambient pressure; thus, as the depth increases, the convective heat loss increases linearly for any temperature difference ($T_{\overline{sk}} - T_a$). This heat loss problem can be compensated for partly by both vasoconstriction, which decreases $T_{\overline{sk}}$, and by decreased evaporative heat loss. The practical solution, however, is to avoid the physiological responses to cold, including shivering. This is accomplished by increasing T_a as a function of depth to maintain the habitat within the thermal comfort zone. This principle can be appreciated by examining the list of optimum temperatures for thermal comfort in hyperbaric helium-oxygen in Table 23.4 (28). Even this practice, however, does not appear to be entirely successful in preventing energy expenditure in excess of energy intake in saturation divers. Many saturation divers do not appear to maintain proper energy balance despite high caloric intakes during a dive (29). This effect has resulted in significant

Table 23.4. Thermal Comfort Zone for Hyperbaric Helium

Depth (ATA)	Comfort Zone (°C)
1	24–26
10	26–28
20	28–30
30	30–31
40	31–32
50	32–33

decreases in body weight in some individuals in long saturation dives. Hyperbaric diuresis also contributes to weight loss, and its pathogenesis is poorly understood. It has been variably attributed to convective cooling, suppression of evaporative heat loss, osmotic gas gradients, and direct effects of hyperbaric pressure.

Convective heat loss from the body is the major thermal problem in both wet and dry hyperbaric environments, and it can be defended against to some extent by thermoregulatory responses such as vasoconstriction of the skin, decreased evaporative heat loss, and increased metabolic heat production (shivering). The physiological responses to cold stress also warns the diver about his thermal balance. In addition, passively insulated garments, e.g., wet suits and dry suits, offer reasonably good thermal protection in shallow water. Thermal protection of the hands and feet, however, remains a serious problem in passive garments at water temperatures below approximately 10°C. Active thermal protection systems, e.g., hot water suits, are used for deep diving because of compression and loss of insulation in passive garments such as wet suits. These actively heated systems generally rely on a continuous flow of hot water to the diver through an umbilical line.

There are three other notable thermal concerns in deep diving operations: slow cooling, respiratory heat loss, and the lost bell situation (30). These thermal problems are significant because each may lead to lethal consequences for the diver. The problem of slow cooling has been recognized primarily through the work of Webb, who has reported that humans may lose up to 300 kcal without significant shivering responses (27). Shivering is one of the diver's early defenses against hypothermia, and its absence may allow the diver to cool to the point of serious fatigue or decrement in performance or judgement. A similar problem may occur with high convective respiratory heat losses that are not readily detected by thermoregulatory sensory mechanisms. At 600 fsw in 4°C water, the entire metabolic heat production of the body may be lost directly from the body core by respiratory convection. Such massive respiratory heat losses cannot be made up by shivering or exercise because of the accompanying increase in ventilatory requirements. Hazardous cooling of the respiratory tract can be averted by appropriate heating of the inspired gas. The accidental loss of a diving bell results in loss of power to heat the dense gas in the diver's environment. Hence, the gas in the bell cools rapidly to ambient temperature, and trapped divers will survive only as long as their passive thermal protection will prevent lethal hypothermia. The survival time is directly related to the depth (gas density), the temperature of the water, and respiratory minute ventilation.

4 DECOMPRESSION FROM PRESSURIZED ENVIRONMENTS

4.1 Principles of Decompression

The use of compressed air in diving during the last century led inevitably to physiological problems caused by release of excess gas from tissues after diving. The process of inert gas elimination from the body after a decrease in pressure during ascent from diving or to altitude is known as decompression. Although the basic concept that rapid decompression could produce decompression sickness (DCS) was formulated by Pol and Watelle in 1854, it was not until 1878 that Paul Bert reported that bubbles consisting mainly of nitrogen were formed in tissue during rapid decompression (19). Guidelines for safe decompression did not exist, however, until 1908 when Boycott, Damant, and Haldane published empirical decompression tables based on exponential half-times for elimination of nitrogen from body tissues (31). This seminal paper has provided a scientific basis for much of modern decompression theory.

The rate of uptake or elimination of nitrogen or other inert gas from the body after a change in pressure is exponential with respect to time. Inert gas exchange is determined primarily by gas solubility in blood and tissue, blood flow, and the volume of tissue. This assumes that the diffusion resistance between blood and tissue is inconsequential for nitrogen and other inert gases. Tissues that behave this way are perfusion limited, and the characteristics of their inert gas exchange can be defined by a half-time. Because the tissues of the body are perfused differently and nitrogen is more soluble in fat than other tissues, the half-times for different tissues may vary considerably. This principle was used to calculate the original 1908 tables for safe decompression. It assumed that gas bubbles would form and decompression sickness would occur only if the tissue was supersaturated with inert gas. The degree of critical supersaturation for nitrogen was believed to be about twice as great as the absolute pressure. This approach has been modified over the years, particularly with respect to the magnitude and duration of allowable supersaturation. These adjustments have been based on observations of the occurrence of DCS at depths as shallow as 25 fsw after prolonged exposures. Many decompression tables are still based on multiple parallel exponential models, although a variety of other modeling approaches also have been used successfully (32). A variety of empirical decompression algorithms have been programmed into wrist-worn decompression computers to provide divers continuous information about their decompression obligation. Most of these devices work relatively well; however, they are designed to maximize the dive time and have not been shown to be associated with a decreased risk of DCS.

Although the primary variable affecting uptake and elimination of inert gas is perfusion, the rate of inert gas exchange may be limited by diffusion under some conditions. Diffusion limitation may occur if the rate of exchange of inert gas by perfusion is extremely rapid, if the tissue perfusion is very low, or if part of the tissue such as the capillary endothelium has a high diffusion resistance. Diffusion may also be important when two adjacent tissues have very different rates of perfusion. In such a situation, the elimination of inert gas from the faster tissue may allow additional inert gas to enter by diffusion from the slower tissue. Thus, the faster tissue may maintain supersaturation for a longer time. Diffusion is also important after gas bubbles have formed in a tissue during decompression. Gas bubbles contain large amounts of inert (N_2) gas, which can be removed by tissue perfusion only after the N_2 diffuses back into the tissue. The rate at which N_2 diffuses away from a gas bubble is determined by the bubble surface area, the intrabubble pressure, and the partial pressure difference between the bubble and the tissue. It is noteworthy that decompression also can be modeled successfully using slab diffusion, and the resulting decompression profiles work reasonably well.

During decompression, bubble formation appears to occur at specific nucleation sites in the body. The number and location of nucleation sites seems to vary according to physiological circumstances. For example, exercise may increase the number of bubbles formed by tribonucleation, a mechanism that creates bubbles in areas such as joint capsules, where large negative pressures can be generated by traction between surfaces lubricated by a liquid. Bubbles created by this so-called vacuum phenomenon may arise from either preexisting gas nuclei or by de novo formation. Tiny unstable bubbles, also called gas nuclei, may have long residence times in the body if they stabilize at hydrophobic sites. Such gas nuclei may grow into bubbles during decompression. There is little doubt that gas bubbles contribute to DCS, although clinically silent bubbles are frequently detected in the circulation by ultrasonic devices during normal decompression. There is still uncertainty, however, about the mechanisms of bubble formation in the body and how they are related to the diverse manifestations of DCS.

Bubble formation occurs most commonly after a decrease in ambient pressure, but bubbles also may form whenever the sum of the dissolved gas pressures in tissue exceeds ambient pressure (33). Supersaturation does not require a decrease in hydrostatic pressure if a diver, saturated with one inert gas, suddenly changes breathing gas at constant pressure, or if a diver breathes a soluble inert gas while surrounded by a less soluble gas. These forms of inert gas exchange are called isobaric, and they may lead to supersaturation and bubble formation by various mechanisms. For example, body tissues do not exchange all inert gases at the same rate. If a diver saturated with a slowly exchanging inert gas switches to a more rapidly exchanging gas, the new gas will enter the tissue faster than the slow gas leaves, and transient supersaturation of the tissue will occur. Other mechanisms of isobaric inert gas exchange are counterperfusion and counterdiffusion. These forms of isobaric inert gas exchange require that two adjacent layers of tissue with dissimilar gas exchange characteristics be exposed to different gases on the unopposed surfaces. This situation may arise when a diver breathes one inert gas while surrounded by another less soluble gas. Each gas diffuses across the two tissues independently at rates proportional to the diffusion gradients, which differ as a function of the diffusive resistances of the dissimilar tissues. Therefore, it is possible for isobaric supersaturation to occur across the tissues, particularly at tissue interfaces.

4.2 Decompression Sickness

Understanding the factors that govern the formation of bubbles during decompression has helped investigators develop safe tables for decompression. The computation of depth-time profiles for decompression usually assumes that DCS is related to the breach of some critical supersaturation threshold. This assumption conflicts with clinical data that indicate that DCS often occurs as a statistical rather than as a threshold event (34). Therefore, even proper use of well-tested decompression tables is associated with a small but definite risk of decompression sickness for dives of approximately 25 fsw or deeper. Most serious DCS, however, is the result of omitted decompression after deep air or mixed-gas dives.

The clinical manifestations of DCS are related to growth of bubbles in body tissues. Primary sites of bubble growth are the joint spaces, tendon sheaths, and periarticular tissues including peripheral nerve endings. Bubbles in these locations produce pain that is known as type I decompression sickness, or bends. Bubble formation also occurs in the spinal cord and other areas of the nervous system. The latter is called type II, or serious decompression sickness. DCS involving the audiovestibular system also occurs, but it is relatively rare (35). The most common clinical manifestations of DCS are summarized in Table 23.5. Reports of DCS in military and commercial divers indicate a 4:1 predominance of type I DCS. Altitude bends, or dysbarism, is generally accompanied by mild to moderate type I DCS. Tunnel workers have a reported incidence of DCS of 0.7 to 1.5% with a great predominance of type I symptoms. These symptoms involve the knee and lower leg in two-thirds of the cases (36). In contrast, recent statistics from recreational divers suggest a 3:1 predominance of type II DCS in these individuals (37). This effect is shown in Figure 23.2. The reasons for this difference in type of DCS are uncertain, although underreporting of type I DCS is likely. In some instances prolonged delays in the diagnosis and treatment of type I DCS may allow it to evolve into type II disease. Also, recreational divers are more likely to omit part of their decompression obligation than are professional divers, thus increasing the probability of encountering serious DCS.

One of the most troublesome forms of serious DCS is spinal cord injury (38). The precise mechanisms of spinal cord DCS are not yet fully understood. In dogs, spinal cord injury may be precipitated by intravascular bubbles that form in the low-pressure, epidural venous plexus of the spinal cord. This venous lake is susceptible to bubble formation because of its slow, bidirectional blood flow. Bubbles therein resolve slowly, and bubble-induced thrombi can easily obstruct venous outflow from the plexus, leading to

Table 23.5. Clinical Manifestations of Decompression Sickness

Type I (mild DCS)
Limbs: pain (bends), niggles, lymphatic obstruction, numbness and paresthesias usually involving the large joints, e.g. shoulders, elbows, knees
Skin: mottling, itching, rash, pallor, urticaria, edema
Type II (Serious DCS)
CNS
Brain: headache, seizures, loss of consciousness, visual disturbances, hemiparesis, aphasia, tremor, ataxia (staggers)
Spinal cord: low back or pelvic girdle pain, paraparesis, urinary retention, incontinence
Audiovestibular DCS: tinnitus, vertigo, nystagmus, decreased hearing, nausea and vomiting.
Cardiopulmonary: substernal pain, cough, tachypnea, asphyxia (chokes)
Systemic: extreme fatigue, hemoconcentration, shock

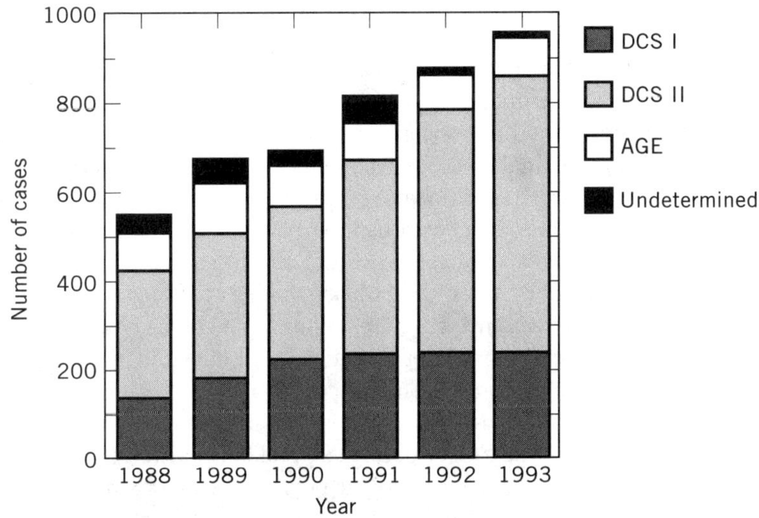

Figure 23.2. Number and type of DCS cases reported to the Divers Alert Network 1988–1993. Note the high occurrence of type II DCS in recreational divers.

ischemic injury of the spinal cord. Despite evidence for bubble formation in the spinal venous plexus, intravascular formation of bubbles appears to be otherwise uncommon. Most intravascular bubbles probably originate at the tissue-blood interface and stream into the circulation where they are absorbed by the lungs. In the circulation, the surface active effects of bubbles at the blood-to-bubble interface may cause a number of hematologic changes. These include complement activation, platelet aggregation and destruction, mediator release, and procoagulant activity. The precise role of these factors in the pathogenesis of decompression sickness is still under investigation. Although the bubble hypothesis of DCS certainly explains many of the physiological problems encountered during decompression, there are diving-related conditions, such as aseptic bone necrosis, that are not well understood in terms of the physical properties and behavior of gas bubbles.

When massive numbers of bubbles are released into the circulation, they may overwhelm the pulmonary circulation and cause chest pain, shortness of breath, and cough, a potentially fatal syndrome called the chokes. Some of these bubbles may also cross the pulmonary capillary bed as micro-gas-emboli. It has been pointed out that intravenous bubbles may pass into the arterial circulation in a significant number of divers with type II DCS by a right-to-left intracardiac shunt such as a patent foramen ovale (PFO) (39). This cardiac anomaly is present in about 20% of the normal population, and its presence during decompression stress may lead to a higher risk of serious forms of DCS involving obstruction of systemic blood vessels with gas bubbles. The relative risk of DCS in diving with a PFO is increased; however, precise figures are not available, and controversy remains about whether to screen diving candidates for the condition.

4.3 Pulmonary Barotrauma and Arterial Gas Embolism

Pulmonary barotrauma is a potentially serious consequence of the failure of expanding gases in the lung to escape from alveoli or acini during ascent. Overinflation of the lung may lead to alveolar rupture and pulmonary interstitial emphysema. These problems may cause pneumothorax, mediastinal or soft-tissue emphysema, pneumopericardium, and/or arterial gas embolization (AGE). Pulmonary overinflation during ascent is most likely to occur with breath-holding, loss of consciousness, or local obstruction of airways with trapping of gas, or after explosive decompression of aircraft at altitude. Lung rupture during a high-risk ascent also depends upon several physiologic factors including pulmonary compliance, transrespiratory pressure, and lung volume. Dahlback and Lundgren have also suggested that airway closure and air trapping induced by immersion in the upright position might increase the risk of lung rupture in free ascent (40).

When the pressure within the lung exceeds that at the body surface by about 100 cm H_2O, the lung will rupture. During breath-holding at total lung capacity (TLC), the pressure difference between the alveoli and the ambient pressure is approximately 50 cm H_2O. Thus, ambient hydrostatic pressure outside the body during ascent must fall another 50 cm H_2O for the lung to rupture. If the compliance of the lung and chest wall at TLC is 15 mL/cm H_2O, then the lung volume during ascent must increase by 15 mL $\times$ 50 cm H_2O, or 750 mL, before the lung will rupture. If the TLC is 6250 mL, then using Boyle's Law we can determine the approximate ambient pressure from which a diver must ascend for pulmonary rupture during a breath-hold. Thus, at the surface, $P_1 = 1.0$ ATA and $V_1 =$ 7000 mL. At P_2, $V_2 = 6250$ mL. Therefore,

$$(1.0)(7000 \text{ mL}) = P_2(6250 \text{ mL})$$

and $P_2 = 1.0(7000/6250) = 1.12$ ATA, or 4.0 feet. This calculation explains how pulmonary overinflation and arterial air embolization may occur in shallow water when an inexperienced diver breathing compressed gas ascends too rapidly. The example also points out that fractional volume changes are greatest at low hydrostatic pressures. For example, 1 L of gas at 1 ATA decreases in volume to 0.5 L at 2 ATA. One liter of gas at 2 ATA, however, only decreases to 0.67 L at 3 ATA. This principle operating in reverse during ascent from depth accounts for the fact that the amount of gas released into the arterial circulation after pulmonary overpressurization can be quite great. The distinction between AGE and DCS is based upon the release of expanding gas from the airspaces of the lung into the pulmonary venous circulation during AGE. In DCS, bubbles are formed from gas dissolved physically in tissues, which comes out of solution during decompression.

The clinical onset of AGE is quite sudden and generally presents with ischemic symptoms involving the brain. Neurological symptoms and signs usually appear within a few seconds to a few minutes after the ascent. Initial symptoms of AGE rarely appear more than two hours after ascent, and later onset of symptoms usually indicates DCS. Common clinical findings of acute AGE include headache, pleuritic chest pain, confusion, loss of vision, hemiparesis, seizures, and unconsciousness.

4.4 Aseptic Necrosis of Bone

Aseptic necrosis of bone, or dysbaric osteonecrosis, is seen frequently in compressed-air workers and saturation divers. This poorly understood condition is felt by most investigators to be related to decompression stress, and it may actually represent subclinical DCS caused primarily by the delayed occlusive effects of bubbles (41). The uncertainty about the pathogenesis of aseptic necrosis is because of the lack of a definite relationship between the site of the bone lesions and the location of previous episodes of type I DCS. Aseptic necrosis of bone is also associated with a variety of other medical conditions, which may confound its appearance in a diver (see Table 23.6). Most other causes of bone necrosis, however, can be excluded by a thorough clinical evaluation.

Aseptic bone necrosis in compressed-air workers and divers is usually restricted to the humerus, femur, and tibia. In compressed-air workers, the shoulders and hips are affected approximately equally, while in divers, the shoulders are affected most often. At present, the diagnosis of aseptic necrosis is established by radiographic criteria because clinical symptoms appear only in advanced disease. The sites of necrosis are divided into those located next to an articular surface (juxta-articular, or JA lesions) and those located away from articular surfaces (head, neck, and shaft, or HNS lesions). In 1966, the Decompression Sickness Panel of the Medical Research Council proposed a standard system for classifying the radiographic changes of aseptic necrosis (42). This classification is summarized in Table 23.7. Radiographic diagnosis of aseptic necrosis of bone is still the gold standard, although there are other diagnostic modalities such as bone scintigraphy and magnetic resonance imaging that are useful on establishing the diagnosis.

The early diagnosis of juxta-articular bone necrosis is of paramount importance to the compressed-air worker or diver because these lesions may collapse and give rise to clinical symptoms and various degrees of disability. Approximately 8%–10% of JA lesions will progress to the point of collapse and irregularity of the articular surface. At this time, however, there is no way to determine which lesions will progress and which will stabilize. Since further diving may increase the risk of progression, workers with JA lesions should be prohibited from further diving. In contrast, HNS lesions are usually inconsequential and rarely become symptomatic.

5 RECOMPRESSION THERAPY AND HYPERBARIC OXYGEN

Gas bubbles in tissues will gradually resolve, but the rate at which they are removed can be enhanced greatly by oxygen and recompression. Oxygen recompression is primary

Table 23.6. Common Causes of Aseptic Necrosis of Bone

Hyperbaric exposure	Rheumatoid arthritis
Chronic alcoholism and cirrhosis	Gout
Steroid therapy	Syphilis
Sickle cell anemia	Radiation
Diabetes	Trauma

Source: After Walder 1984 (41).

Table 23.7. Radiographic Classification of Aseptic Bone Necrosis

Juxta-articular Lesions	
A1	Dense areas, with intact articular cortex
A2	Spherical segmental opacities
A3	Linear opacities
A4	Structural failure
A4a	Transradiant subcortical bands
A4b	Collapse of articular cortex
A4c	Sequestion of articular cortex
A5	Osteoarthritis
Head, Neck, and Shaft Lesions	
B1	Dense areas
B2	Irregular areas of calcification
B3	Transradiant areas

therapy for both DCS and AGE (43). The clinical management of DCS and AGE is beyond the scope of this chapter, and the reader is referred to more thorough discussions of these topics (44–46). The scientific basis for treatment of gas bubble disease, however, is pertinent to this chapter.

The resolution of bubbles is related to their size and the partial pressure difference that exists between the gas cavity and respiring tissue. This partial pressure difference is due to the inherent unsaturation of venous blood, which in turn is a result of the difference in solubility of O_2 and CO_2 in tissue. CO_2 is 20 times more soluble in body tissues than is O_2. As O_2 is consumed, it is replaced by CO_2 from substrate oxidation at a ratio of about 0.8 moles of CO_2 for each mole of O_2 reduced. Thus, oxygen entering a tissue at an arterial P_{O_2} of 100 torr leaves the venous capillary at a P_{O_2} of 40 torr. In contrast, CO_2 enters at 40 torr and leaves at only 46 torr. The remainder of the gas pressure in the tissues is primarily N_2, which at equilibrium has the same partial pressure, 573 torr, on both sides of the circulation. Therefore, the sum of the partial pressures in the arterial system is 54 torr less than in the venous system. This so-called oxygen window provides a potential pressure gradient for elimination of N_2. As a gas pocket or bubble collapses from removal of O_2 by metabolism, the internal P_{N_2} must rise above tissue P_{N_2} because the total pressure in the bubble remains in equilibrium with ambient pressure. In this way, the inert gas is resorbed gradually. The oxygen window can be expanded during decompression at the surface and after recompression by administration of high partial pressures of O_2. Oxygen decreases the tissue P_{N_2} and increases the partial pressure gradient for N_2 between the bubble and tissue. Surface oxygen administration at the scene of a diving accident before recompression therapy can be initiated is clearly beneficial in the management of DCS and AGE.

There are a variety of recompression tables available to treat DCS, provided the treatment is begun promptly. Treatment tables that employ minimal recompression and hyperbaric oxygen have become the treatment of choice in most situations. U.S. Navy Treatment Table 6, which employs intermittent hyperbaric oxygen at a maximum depth of 2.8 ATA (60 fsw), has become a therapeutic standard for primary DCS as well as for recurrent symptoms. Also, this U.S. Navy Table 6 or its equivalent is probably adequate initial therapy for arterial gas embolization, although deeper recompression to 165 fsw, e.g., U.S. Navy Table 6A, may be appropriate for very recent occurrence of AGE or when the quantity of intravascular gas is large and greater reduction of intravascular bubble size is desired.

In some instances of DCS, oxygen recompression is not available, and air recompression tables must be used. Air treatment tables are longer and less effective than oxygen tables, and they are more likely to produce DCS in the attendants inside the chamber. For compressed-air workers, recompression with air is usually carried out to the minimum depth of relief. The use of hyperbaric oxygen for DCS in compressed-air workers becomes more accepted with institution of good fire safety precautions (7).

6 FITNESS TO DIVE

6.1 Individual Variation, Acclimatization, and Physical Fitness

The question of who is fit to dive represents a complex issue in industrial and occupational health. Few work places expose the worker to such an inhospitable environment and such a wide range of physiological stresses. According to current OSHA regulations, divers are grouped into five categories with different requirements (Table 23.8) (47). As a result, there is no universally accepted consensus on physical qualifications to dive. Standards for diving fitness have been tailored to different diving situations and to the required level of performance. It is generally agreed that work in hyperbaric environments is associated with a higher risk of injury, e.g., DCS, for individuals with certain physical attributes. There are also many pre-existing medical conditions that constitute contraindications to exposure to alterations in barometric pressure.

There is considerable individual variation in susceptibility to pressure-related diseases such as aseptic necrosis of bone and DCS (48). For DCS, several factors have been con-

Table 23.8. Current OSHA Diving Standards

Category	Requirements
Commercial	29 CFR Chapter XVII Part 1910 Subpart T: Commercial Diving Operation[a]
Scientific	Exempt
Instructor	Exempt
Search and Rescue	Exempt
Experimental	Governed by 45 CFR Part 46 (Protection of Human Subjects, U.S. Department of Health and Human Services)

[a]See Reference 47.

nected routinely with enhanced susceptibility including age, weight, body fat, and lack of acclimatization. Acclimatization or habituation is well described in compressed-air workers who become less susceptible to DCS after daily exposure to the hyperbaric environment. Acclimatization requires approximately 14 days for maximum effect, while ceasing to work in compressed air leads to deacclimatization. The mechanism of such acclimatization, however, has never been determined.

The roles of age, weight, and body fat in susceptibility to DCS also have been evaluated. During World War II, Gray's analysis of thousands of decompressions in altitude chambers revealed that relative susceptibility to dysbarism based on mean values increased about 11% per year between ages 18 and 28 (49). Similarly, mean susceptibility was correlated highly with increase in body weight per unit of height over the same age range. The role of body fat as a risk factor for DCS has been well described. Inert gases such as N_2 are highly soluble in lipid, while the total perfusion of a fat diver is approximately the same as for a thin diver. Hence, large amounts of inert gas may be stored in body fat during long dives. The inert gas in fatty tissues is eliminated more slowly during decompression, and the risk of critical supersaturation is increased during decompression. This rationale also has been used to suggest that women should be more susceptible to DCS than men because they average about 10% more body fat per unit of mass. A male-female difference in susceptibility to DCS has not been proven to be the case. There are a number of other well-known risk factors for DCS. These include physical deconditioning, heavy exercise before and during the dive, and diving in cold water. The effects of exercise during decompression, however, may lessen the risk of DCS by enhancing inert gas elimination. It also has been proposed that individuals whose blood is sensitive to complement activation by gas bubbles are more susceptible to DCS (50).

6.2 Physical Evaluation of Divers

The physician who examines the diver for fitness to dive requires a thorough familiarity with diving physiology and an understanding of the particular environment, the equipment, and type of work the individual will perform. There are special requirements and waivers that pertain to military and commercial divers. For instance, the Association of Diving Contractors in the United States has published a set of consensus physical examination standards for commercial divers (51). For the recreational diver, it is important to remember that the diving activity is a sport, hence any medical condition that could predispose to DCS, AGE, or other injury should be disqualifying. Any diving candidate with chronic disease, significant physical disability, or requirement for long-term medication should not be cleared physically for diving until he or she is evaluated by a physician trained in diving medicine. In addition, the Divers Alert Network (DAN) offers a free consultative service in recreational diving medicine.

The need for ancillary laboratory studies in routine screening of asymptomatic diver candidates is moot, although a baseline chest radiograph and audiogram are usually recommended by the author. Requirements for blood counts, blood chemistry, and urinalysis vary among experts. Requirements for electrocardiograms and exercise stress tests for older divers also vary but in general are prudent as part of initial certification to dive for individuals over 40 years of age. A set of recommended medical standards for unrestricted

Table 23.9. Medical Contraindications to Diving

1. Eye
   Recent ocular surgery
   Uncorrected loss of visual acuity
2. Ear, Nose and Throat
   External ear obstruction, e.g., exostoses
   Chronic middle ear disease, e.g., otitis media, otorrhea
   Tympanic membrane perforation or tympanoplasty
   Middle ear or mastoid surgery
   Inner ear disease, e.g., deafness, tinnitus, or vertigo
   Nasal or sinus obstruction, e.g., infection or nasal polyps
3. Pulmonary
   Upper airway obstruction
   Asthma
   Chronic obstructive pulmonary disease
   Pulmonary interstitial diseases
   Spontaneous pneumothorax
   Thoracic surgery involving the lung or pleura
4. Cardiovascular
   Congestive heart failure
   Post myocardial infarction or coronary artery bypass graft with persistent left ventricular dysfunction
   Angina pectoris
5. Musculoskeletal Diseases
   Metabolic or traumatic bone or joint diseases with compromised circulation
   Infectious diseases of the bone
6. Miscellaneous Medical Disorders
   Chronic renal failure
   Diabetes mellitus with insulin or oral agent dependence
   Hemoglobinopathies, e.g., HbS disease
   Thrombocytopenia
7. Nervous System
   Epilepsy
   Severe migraine headaches
   Serious head injury with a history of prolonged unconsciousness
   Stroke or other cerebrovascular disease
   Craniotomy
   Spinal cord diseases
   Spinal surgery
   Type II DCS with residual neurological abnormalities
8. Psychiatric Disorders
   Alcohol or other drug dependence
   Schizophrenia, severe depression, and personality disorders
   Need for psychotropic medication

Source: Adapted from Reference 52.

sports diving, published by the Undersea and Hyperbaric Medical Society (UHMS) in 1987, is summarized in Table 23.9 (52). This table is not a comprehensive list of medical contraindications, and some of the conditions can be considered relative contraindications. Specific questions should be referred to a specialist in diving medicine.

BIBLIOGRAPHY

1. M. Epstein, A. De Nunzio, and M. Ramachandran, "Characterization of the renal response to prolonged immersion in normal man. Implication for an understanding of the circulatory adaptation to manned space flight," *J. Appl. Physiol. Resp. Environ. Exercise Physiol.*, **49**, 184–188 (1980).
2. M. Epstein, "Water immersion and the kidney: implications for volume regulation." *Undersea Biomed. Res.*, **11**, 113–121 (1984).
3. Y. C. Lin, "Circulatory functions during immersion and breath-hold dives in humans," *Undersea Biomed. Res.*, **11**, 123–138 (1984).
4. J. C. Mithoefer, "Breath holding," in W. Fenn and H. Rahn, Eds., *Handbook of Physiology, Respiration*, Section 3, Vol. II, American Physiological Society, Washington, DC, 1965, pp. 1011–1025.
5. *U.S. Navy Diving Manual*, Vol. 1, *Air Diving*, Revision 3, (NAVSEA 0994-LP-001-9110), U.S. Department of the Navy, Washington, DC, 1993.
6. *U.S. Navy Diving Manual*, Vol. 2, *Mixed Gas Diving*, Revision 3, (NAVSEA 0994-LP-001-9120), U.S. Department of the Navy, Washington, DC, 1991.
7. E. P. Kindwall, "Caisson decompression," in R. D. Vann, Ed., *The Physiological Basis of Decompression, Proceedings of the Thirty-eighth Undersea and Hyperbaric Medical Society Workshop*, UHMS Publication #75, Bethesda, MD, 1989, pp. 375–395.
8. K. E. Schaefer, "Physiological stresses related to hypercapnia during patrols on submarines," *Undersea Biomed. Res.*, **6**(Suppl.), 515–547 (1979).
9. J. A. Kylstra, I. S. Longmuir, and M. Grace, "Dysbarism: osmosis caused by dissolved gas?" *Science*, **161**, 289–291 (1968).
10. J. C. Farmer, Jr., "Ear and sinus problems," in A. A. Bove, Ed., *Diving Medicine*, WB Saunders, Philadelphia, 1997, pp. 235–257.
11. O. I. Molvaer and S. Eidsvik, "Facial baroparesis: a review," *Undersea Biomed. Res.*, **14**, 277–295 (1987).
12. W. L. Hunter and P. B. Bennett. "The causes, mechanisms and prevention of the high pressure nervous syndrome," *Undersea Biomed. Res.*, **1**, 1–28 (1974).
13. R. A. Smith, B. A. Dodson, and K. W. Miller, "The interactions between pressure and anaesthetics," *Phil. Trans. R. Soc. Lond. B.*, **304**, 69–84 (1984).
14. E. H. Lanphier and E. M. Camporesi, "Respiration and exertion," in P. B. Bennett and D. H. Elliott, Eds., *The Physiology and Medicine of Diving*, 4th ed., WB Saunders, London, 1993, pp. 77–120.
15. H. D. Van Liew, "Mechanical and physical factors in lung function during work in dense environments," *Undersea Biomed. Res.*, **10**, 225–226 (1983).
16. J. Milic-Emili and J. M. Tyler, "Relation between work output of respiratory muscles and end tidal CO_2 tension," *J. Appl. Physiol.*, **18**, 497–504 (1963).

17. N. B. Pride, S. Permutt, R. L. Riley, and B. Bromberger-Barnea, "Determinants of maximum expiratory flow from the lungs," *J. Appl. Physiol.*, **23**, 646–662 (1967).
18. J. V. Salzano, B. W. Stolp, R. E. Moon, and E. M. Camporesi, "Exercise at 47 and 66 ATA," in A. J. Bachrach and M. M. Matzen, Eds., *Proceedings of the Seventh Symposium on Underwater Physiology*, Undersea Medical Society, Bethesda, MD, 1981, pp. 181–196.
19. P. Bert, *Barometric Pressure: Researches in Experimental Physiology*, English translation by Hitchcock and Hitchcock, Long College Book Co., Columbus OH, 1943.
20. J. M. Clark, and C. J. Lambertsen, "Pulmonary oxygen toxicity: a review," *Pharmacol. Rev.*, **23**, 37–133 (1971).
21. J. M. Clark, "Oxygen toxicity," in P. B. Bennett and D. H. Elliott, Eds., *The Physiology and Medicine of Diving*, 4th ed., WB Saunders, London, 1993, pp. 77–120.
22. A. L. Harabin, L. D. Homer, P. K. Weathersby, and E. T. Flynn, "An analysis of decrements in vital capacity as an index of pulmonary oxygen toxicity," *J. Appl. Physiol.*, **63**, 1130–1135 (1987).
23. K. W. Donald, "Oxygen poisoning in man," *Br. Med. J.*, **1**, 712–717 (1947).
24. B. A. Freeman and J. D. Crapo, "Free radicals and tissue injury," *Lab. Invest.*, **47**, 412–426 (1982).
25. A. Mayevsky, "Brain oxygen toxicity," in A. J. Bachrach and M. M. Matzen, Eds., *Proceedings of the Eighth Symposium on Underwater Physiology*, Undersea Medical Society, Bethesda, MD, 1984, pp. 69–89.
26. R. J. Folz, C. A. Piantadosi, and J. D. Crapo, "Oxygen Toxicity," in R. G. Crystal, J. B. West, E. R. Weibel, and P. J. Barnes, Eds., *The Lung: Scientific Foundations*, Raven Press, New York, 1996.
27. P. Webb, "Thermal problems," in P. B. Bennett and D. H. Elliott, Eds., *The Physiology and Medicine of Diving*, 3rd ed., Best Publishing, San Pedro, CA, 1982, pp. 297–318.
28. L. W. Raymond, E. D. Thalmann, G. Lindgren, et al., "Thermal homeostasis of resting man in helium-oxygen at 1–50 ATA," *Undersea Biomed. Res.*, **2**, 51–68 (1975).
29. S. K. Hong, R. M. Smith, P. Webb, and M. Matsuda, "Hana Hai II: a 17-day dry saturation dive at 18.6 ATA," *Undersea Biomed. Res.*, **4**, 211–220 (1975).
30. L. A. Kuehn, "Thermal effects of the hyperbaric environment," in A. J. Bachrach and M. M. Matzen, Eds., *Proceedings of the Eighth Symposium on Underwater Physiology*, Undersea Medical Society, Bethesda, MD, 1984, pp. 413–439.
31. A. E. Boycott, G. C. C. Damant, and J. S. Haldane, "The prevention of compressed air illness," *J. Hyg. Camb.* **8**, 342–443 (1908).
32. R. D. Vann, "Decompression theory and application," in P. B. Bennett and D. H. Elliott, Eds., *The Physiology and Medicine of Diving*, 3rd ed., Best Publishing, San Pedro, California, 1982, pp. 352–382.
33. B. G. D'Aoust, "Inert gas exchange and counterdiffusion in decompression sickness and diving medicine," in A. J. Bachrach, and M. M. Matzen, Eds., *Proceedings of the Eighth Symposium on Underwater Physiology*, Undersea Medical Society, Bethesda, MD, 1984, pp. 159–170.
34. P. K. Weathersby, L. D. Homer, and E. T. Flynn, "On the likelihood of decompression sickness," *J. Appl. Physiol.*, **57**, 815–825 (1984).
35. J. C. Farmer, "Inner ear decompression sickness," in C. W. Shilling, C. B. Carlston, and R. A. Mathias, Eds., *The Physician's Guide to Diving Medicine*, Pleum Press, New York, 1984, pp. 312–316.

36. T. H. Lam, and K. P. Yau, "Manifestations and treatment of 793 cases of decompression sickness in a compressed air tunneling project in Hong Kong," *Undersea Biomed. Res.*, **15**, 377–388 (1988).
37. Divers Alert Network, *Report on Recompression Illness and Diving Fatalities*, 1998, Divers Alert Network, Durham, NC, 1998.
38. J. M. Hallenbeck, A. A. Bove, and D. H. Elliott, "Mechanism underlying spinal cord damage in decompression sickness," *Neurology*, **25**, 308–316 (1975).
39. R. E. Moon, E. M. Camporesi, and J. A. Kisslo, "Patent foramen ovale and decompression sickness in divers," *Lancet*, **1**, 513–514 (1989).
40. G. O. Dahlback, and C. E. G. Lundgren, "Pulmonary air-trapping induced by immersion," *Aerospace Med.* **43**, 768–774 (1972).
41. D. N. Walder, "Osteonecrosis," in C. W. Shilling, C. B. Carlston, and R. A. Mathias, Eds., *The Physician's Guide to Diving Medicine*, Pleum Press, New York, 1984, pp. 397–405.
42. R. I. McCollum, D. N. Walder, R. Barnes, et al., "Bone lesions in compressed air workers with special reference to men who worked on the Clyde Tunnels 1958 to 1963," *J. Bone Joint Surg.* (London), **48B**, 207–235 (1966).
43. N. B. Hampson, Ed., *Hyperbaric Oxygen Therapy Committee Report, 1999*, Undersea and Hyperbaric Medical Society, Kensington, MD, 1999.
44. D. R. Leitch and J. M. Hallenback, "Neurological forms of decompression sickness," in C. W. Shilling, C. B. Carlston, and R. A. Mathias, Eds., *The Physician's Guide to Diving Medicine*, Pleum Press, New York, 1984, pp. 316–327.
45. R. R. Pearson, "Diagnosis and treatment of gas embolism," in C. W. Shilling, C. B. Carlston, and R. A. Mathias, Eds., *The Physician's Guide to Diving Medicine*, Pleum Press, New York, 1984, pp. 333–367.
46. R. E. Moon and D. F. Gorman, "Treatment of the decompression disorders," in P. B. Bennett and D. H. Elliott, Eds., *The Physiology and Medicine of Diving*, 4th ed., WB Saunders, London, 1993, pp. 506–541.
47. *Occupational Safety and Health Administration Code of Federal Regulations*, 29CFR Chapter XVII, Part 1910, Subpart T: Commercial Diving Operations, 1997.
48. M. L. Dembert, "Individual factors affecting decompression sickness," in R. D. Vann, Ed., *The Physiological Basis of Decompression, Proceedings of the Thirty-eighth Undersea and Hyperbaric Medical Society Workshop*, Publication #75, Bethesda, MD, 1989, pp. 355–367.
49. J. S. Gray, "Constitutional factors affecting susceptibility to decompression sickness," in J. Fulton, Ed., *Decompression Sickness*, W.B. Saunders, Philadelphia, 1951, pp. 182–191.
50. C. A. Ward, P. K. Weathersby, D. McCullough, and W. D. Fraser, "Identification of individuals susceptible to decompression sickness," in A. A. Bove, A. J. Bachrach, and L. J. Greenbaum, Jr., Eds., *The Ninth International Symposium on Underwater and Hyperbaric Physiology*, Undersea and Hyperbaric Medical Society, Bethesda, MD, 1987, pp. 239–247.
51. Association of Diving Contractors (ADC), "Physical Examination Standards," in *ADC Consensus Standards for Commercial Diving Operations*, 4th ed., Change 1, ADC, 1994.
52. P. G. Linaweaver and J. Vorosmarti, Jr., "Fitness to Dive," in *Proceedings of the Thirty-fourth Undersea and Hyperbaric Medical Society Workshop*, Publication #70, UHMS, Bethesda, MD, 1987.

CHAPTER TWENTY-FOUR

Lighting for Seeing and Health

Richard L. Vincent

1 INTRODUCTION

Nearly 2000 years ago it was stated, "The eye is the lamp to the body" (Matthew 6:22, Holy Bible, New International Version). Today a holistic approach to the understanding of the role of light continues. The objectives of this chapter are to relate the current status of our knowledge of how the absorption of light stimulates both seeing and health and thus influences human performance.

Most of life's activity is guided by the perception of visual cues that are processed by the thousands and millions during the waking hours. Many of these cues are unconsciously handled and result in involuntary activity. For example, the unconscious detection by our peripheral vision of sudden movement may cause us to move out of the way of a bike or a bus while crossing the street. All of this can happen in a split second. A much smaller number of visual cues involve conscious processing in carrying out productive work. These are concerned with critical details that must be handled successfully for production and economic returns. There are a multitude of details in industrial processes that are critical for the production of good products; there are a tremendous number of symbols to be recognized and individually processed in the office. The lighting practitioner calls each of these critical details a visual task. Each detail involves not only its own configuration but also the background against which it is seen. The contrast of the detail with its background determines its ability to be seen.

The total ability to see the details depends on (*1*) the characteristics of the visual task and (*2*) the characteristics of the observer. The characteristics of the task depend on the following:

Patty's Industrial Hygiene, Fifth Edition, Volume 2. Edited by Robert L. Harris.
ISBN 0-471-29754-2

1. The configuration of the detail
2. The narrowest size of the configurated detail or gap in the detail
3. The contrast of the detail with its background
4. The time of exposure of the detail to the pause of the eye or the glance of the observer
5. The luminance (or brightness) of the background to which the eyes become adapted
6. The specularity of the detail and its background

The characteristics of the observer depend on another set of factors:

1. The sensitivity of the individual visual system to
   a. Size
   b. Contrast
   c. Time
   d. Blur
   e. Wavelength
2. Transient adaptation
3. Glare
   a. Disability
   b. Discomfort
4. Age
5. Motivational factors
6. Physiological factors

The ability to maintain a proper luminous input for biological health depends on understanding the interaction and role of light for the following:

1. Tissues (i.e., of the eye and skin)
2. Circadian rhythms (i.e., the daily regulation of body functions based on a light/dark cycle)
3. Hormone production and secretion

These biological responses are based on these factors:

1. Irradiance
2. Wavelength
3. Duration
4. Time of exposure

2 THE LUMINOUS ENVIRONMENT

We have been provided with a luminous environment that ranges from the sky, to the land, to the sea. All of these natural elements send light to the eyes of the observer. Even on the darkest night, light is radiating from all elements of the field of view. Prior to the invention of electrically generated light nearly 100 years ago, most productive work ceased at sunset. Since then productive activity has steadily increased such that today we live in an around-the-clock 24-hour society. Questions being raised for the design of man-made interiors today include—how the designer should proportion the light, for example, for optimal seeing of critical details, for motivational values in the creative and energetic pursuit of tasks, for working longer periods without fatigue, for maintaining an optimum for health and safety, or for maintaining vitality for a growing, aging population.

Research is being carried out to find the answers in terms of the relationship of light to the characteristics of the task and the characteristics of the observers, to yield ultimately the optimal visual environment (1–8). Although much is known about providing an appropriate environment for quite accurate seeing, further work is needed to validate and predict the influence of lighting on overall task performance (9–13). In addition to the role of lighting for seeing, new frontiers are being forged as we try to understand our need for light to regulate body rhythms, moods, and health. This is increasingly important for extending an aging population's productivity, enjoyment, and safety as we live and work in a competitive, global, 24-hour society (14–18).

3 LIGHTING TERMS AND UNITS

There are many terms and units used in illuminating engineering and lighting design. However, for this discussion it is necessary only to have a clear understanding of certain basic ones (9–13, 19, 20).

Adaptation: The process that takes place as the visual system adjusts itself to the amount of color or light in the visual field. The term is also used, usually qualified, to denote the final state of the visual system when it has become adapted to a very low luminance.

Ambient lighting: General lighting for an interior space.

Arousal: The state of alertness of the human organism.

Candela (cd): The System Internationale (SI) unit of luminous intensity, equal to 1 lm/sr.

Candlepower: A luminous intensity expressed in candelas. An ordinary candle about 1 inch in diameter produces about 1 cd in a horizontal direction.

Circadian rhythms: The 24-hour oscillations of human physiological function that are normally reset (entrained) daily by the natural light–dark period.

Color rendering: A general expression for the appearance of surface colors when illuminated from a given source compared, consciously or unconsciously, with their appearance under light from some reference source.

Color rendering index: A measure of the degree of color shift that objects undergo when illuminated by a light source as compared with the color of those same objects when illuminated by a reference source of comparable color temperature (see CIE Publication 13.2).

Contrast: A term used *subjectively* and *objectively. Subjectively*, it describes the difference in appearance of two parts of a visual field seen simultaneously or successively. The difference may be one of brightness or color or both. *Objectively*, the term expresses the luminance difference between an object and its immediate background. This luminance difference may be expressed in different ways depending on the relationship of the object to the background.

For periodic stimuli, such as gratings, contrast is defined as

$$\text{contrast} = \frac{L_{\text{max}} - L_{\text{min}}}{L_{\text{max}} + L_{\text{min}}}$$

where L_{max} = maximum luminance of the grating
L_{min} = minimum luminance of the grating

For small objects on large uniform backgrounds contrast is defined as

$$\text{contrast} = \frac{L_{\text{t}} - L_{\text{b}}}{L_{\text{b}}}$$

Sometimes for small luminance increments on a large background,

$$\text{contrast} = \frac{L_{\text{t}}}{L_{\text{b}}}$$

where L_{t} = luminance of the object
L_{b} = luminance of the background

Contrast rendering factor (CRF): The ratio of visual task contrast in a given lighting environment to its contrast in reference-lighting conditions, that is, perfectly diffuse and unpolarized lighting by CIE standard illuminant A.

Critical flicker frequency (CFF): The frequency of intermittent stimulation of the eye at which flicker disappears.

Diffuse reflection: The process by which incident flux is redirected over a range of angles.

Direct lighting: Where all the light produced by a luminaire is focused downward.

Direct/indirect lighting: Where a proportion of the lighting produced by the luminaire is aimed upward toward the ceiling and the remaining proportion is aimed downward toward the floor.

Disability glare: Glare that reduces visual performance and visibility and may be accompanied by discomfort.

Discomfort glare: Glare that causes visual discomfort but does not necessarily interfere with visual performance or visibility.

Flicker: Visible oscillation in luminous flux.

Footcandle (fc): A non–System Internationale (SI) unit of illuminance having the same value as the lumen per square foot. One footcandle = 10.76 lm/m^2.

Footlambert (fL): A non–System Internationale (SI) unit of luminance equal to $1/\pi$ cd/ft^2, or to the uniform luminance of a perfectly diffusing surface emitting or reflecting light at a rate of 1 lm/ft^2. The brightness in footlamberts is equal to the illuminance on a surface multiplied by its reflectance. *Note*: This term, although widely used in the literature, is being deprecated in current use.

Glare: The sensation produced by luminances within the visual field that are sufficiently greater than the luminances to which the eyes are adapted, causing annoyance, discomfort, or loss in visual performance and visibility. Types of glare related to lighting include discomfort, disability, and reflected.

Illuminance: The luminous flux density at a surface, that is, the luminous flux incident per unit area.

Illumination: The act of illuminating or state of being illuminated.

Infrared radiation: For practical purposes, any radiant energy within the wavelength range 770 to 10^6 nm. This radiation is arbitrarily divided as follows: near (short-wavelength) infrared, 770 to 1400 nm; intermediate infrared, 1400 to 5000 nm; far (long-wavelength) infrared, 5000 to 1,000,000 nm.

Irradiance: The radiant flux density at a surface, that is, the radiant flux incident per unit area of the surface.

Lamp: Any electric light-producing device, also interchangeably used with light source.

Light: Radiant energy that is capable of exciting the retina and producing a visual sensation. The visible portion of the electromagnetic spectrum extends from about 380 nm to 770 nm.

Lumen (lm): The System Internationale (SI) unit of luminous flux, used in describing a quantity of light emitted by a source or received by a surface. A small source that has a uniform luminous intensity of 1 cd emits a total of 4π lm in all directions and emits 1 lm within a unit solid angle.

Luminaire: A lighting *fixture* that controls the distribution of light given by a lamp or lamps and that includes all the components necessary for fixing and protecting the lamps and for connecting them to the supply circuit.

Luminance: The physical measure of the stimulus that produces the sensation of brightness; it is measured by the luminous intensity of the light emitted or reflected in a given direction from a surface element, divided by the area of the element in the same direction. The System Internationale (SI) unit of luminance is the candela/meter^2.

Luminance contrast: (*See objective use of contrast.*)

Luminance ratios: The ratio between the luminances (photometric brightness) of any two areas in the visual field.

Luminous efficacy: The quotient of the total luminous flux emitted by a lamp and total lamp power input. It is expressed in lumens per watt, which is analogous to miles per gallon of gas for a car.

Luminous flux: The light emitted by a source or received on a surface. The quantity is derived from radiant flux by evaluating the radiation in accordance with the spectral sensitivity of the standard eye, as described by the CIE Standard Photometric Observer.

Luminous intensity (candlepower) distribution: The distribution of the luminous intensity of a lamp or luminaire in all spatial directions. Luminous intensity distributions are usually shown in the form of a polar diagram or a table for a single vertical plane, in terms of candelas per thousand lumens of lamp luminous flux.

Lux (lx): The System Internationale (SI) unit of illuminance; 1 lx equals 1 lm/m^2. One footcandle $\times$ 10.76 $=$ 1 lx.

Mesopic vision: Vision mediated by both rod and cone photoreceptors. It is generally associated with adaptation to luminances in the range 3.4 to 0.034 cd/m^2.

Metamers: Two colors that are indistinguishable under a given light source but that have different spectral compositions.

Optical radiation: That part of the electromagnetic spectrum from 100 nm to 1 mm.

Photopic vision: Vision mediated essentially or exclusively by the cone photoreceptors. It is generally associated with adaptation to luminance of at least 3.4 cd/m^2.

Photoreceptors: The sensors in the retina of the eye that convert light to electrical signals. There are two types, rods and cones, the classification being made on the basis of their shape, range of operation, and photochemical constituents.

Purity: A measure of the proportions of the amounts of the monochromatic and specified achromatic light stimuli that, when adequately mixed, match the color stimulus.

Radiance: At a point on a surface, the quotient of the radiant intensity emitted from an element of the surface in a given direction by the area of the element in the same direction.

Radiant efficiency: The ratio of the radiant flux to the power consumed.

Radiant exposure: At a point on a surface, the product of the irradiance and its duration.

Radiant flux: The power emitted, transferred, or received as radiation.

Radiant intensity: Of a source in a given direction, it is the quotient of the radiant flux emitted in a narrow cone containing the direction, by the solid angle of the cone.

Reflectance: The ratio of the luminous flux reflected from a surface to the luminous flux incident on it. Except for mat surfaces, reflectance depends on how the surface is illuminated, especially the direction of the incident light and its spectral distribution.

Reflected glare: Bright luminous elements being reflected from a glossy surface toward the eyes, causing discomfort or disability (loss of visibility) for horizontal surfaces (i.e., a glossy desk top) or vertical surfaces (i.e., vidco display terminal screens).

Retinal illumination: The luminous flux falling on the retina of the human eye. Retinal illumination is usually calculated from the luminance of the object and the pupil area of the viewer.

Scotopic vision: Vision mediated essentially or exclusively by the rod photoreceptors. It is generally associated with adaptation to a luminance below about 0.034 cd/ m^2.

Solid angle: A measure of that portion of space about a point bounded by a conic surface whose vertex is at the point. It can be measured by the ratio of intercepted surface area of the sphere centered on that point to the square of the sphere's radius. It is expressed in steradians.

Specular reflection: Reflection without diffusion, in accordance with the laws of optical reflection, as from a mirror.

Steradian (sr): A solid angle subtending an area on the surface of a sphere equal to the square of the sphere radius.

Stroboscopic effect: An illusion caused by oscillation in luminous flux that makes a moving object appear as stationary or as moving in a manner different from that in which it is truly moving.

Task/ambient lighting: A nonuniform lighting strategy that lights the general working space with a lower level of light for circulation. The work station lighting is supplemented with a localized task light (either furniture integrated or movable). This provides the light level needed to perform the visual task (i.e., data entry from a paper into a computer).

Task performance: The performance of a complete task, expressed in measures of output such as speed and/or accuracy.

Threshold: The value of a variable of a physical stimulus that permits the stimulus to be seen a specific percentage of the time or at a specific accuracy level. In many psychophysical experiments, thresholds are presented as accurate perception 50% of the time.

Ultraviolet radiation: For practical purposes, any radiant energy within the wavelength range 100 to 380 nm. For convenience in describing biological effects of certain wavelength bands this range is broken down further:

- UV-C 100–280 nm, germicidal
- UV-B 280–315 nm, actinic (tanning)
- UV-A 315–380 nm, ocular effects (black light)

Veiling reflections: Specular reflections superimposed on diffuse reflections from an object that partially or totally obscures the details to be seen by reducing the contrast. This effect is sometimes called *reflected glare*.

Visibility: The quality or state of being perceivable by the eye. In many outdoor applications, visibility is defined in terms of the distance at which an object can be just perceived by the eye. In indoor applications it usually is defined in terms of the contrast or size of a standard test object, observed under standardized viewing conditions, having the same threshold as the given object.

Visual acuity: A measure of the ability to discriminate fine details. Quantitatively it can be expressed as the reciprocal of the angular separation, in minutes of arc, between two lines or points that are just separable by the eye.

Visual angle: The angle subtended by an object or detail at the point of observation. It is usually measured in minutes of arc.

Visual field: The full extent in space of what can be seen when looking in a given direction.

Visual performance: The quantitative assessment of the performance of a task, taking into consideration speed and accuracy.

Visual task: Conventionally designates the details and objects that must be seen for the performance of a given activity, including the immediate background of the details or objects.

4 LIGHTING AND HUMAN PERFORMANCE

Increasingly in developed countries, the workforce is moving from manual labor base in industry to a hi-tech, knowledge base. With this societal shift, workers in all areas—from the finance to computer chip manufacturers to healthcare industries—are working in a global, 24-hour timeframe, with more time being spent viewing video displays, longer work hours, and accompanying stresses. Workplace ergonomics, an aging population, impacts on health, and the continued electrification of the economy are all factors that must be considered for keeping the workplace optimized as well as energy efficient (12, 14–19). Lighting as one aspect of the built environment can play a significant role in providing comfort, arousal, and alertness for creativity, while allowing the processing of visual information. To better understand the role of lighting in human productivity, a framework was developed (Table 24.1) that charts the direct and indirect affects of light on the human organism. The direct effects include how light is absorbed and processed by the visual system for ease of seeing by changing either the visual stimulus or the system's operating state. The direct effects include vision and visual performance, task performance, color vision, age and individual differences, and fatigue effects (sustained performance). The indirect effects from lighting include discomfort (glare, flicker), light as an attention stimulus, light and arousal, light and mood, lighting's influence on behavior, and lighting and (hormone) biology. In order to design and predict the impact of lighting on these various direct and indirect aspects, metrics are being developed and refined to produce standards and recommended practices.

5 VISUAL PERFORMANCE

As a designer evaluates how to light a space, there are two goals that relate to the performance of work in the space: (*1*) the overall spatial order and architectural feeling to be

Table 24.1. Lighting and Human Performance Matrix[a]

Category	Quantitative Knowledge	Qualitative Knowledge	Possible Relationships	Research Agenda 1989	Typical Practical Questions
Visual performance	Yes	Yes	Yes	Test accuracy of models. Extend models.	How does performance change if illuminance is changed?
Task performance	—	Yes	—	Extend models to tasks that are only partly visual.	What is the cost-effectiveness of better lighting?
Color vision	Yes	—	—	Examine performance of tasks requiring color judgment.	What is the best illuminant to use?
Visual search	Yes	Yes	Yes	Test accuracy of model. Examine use of visual cues.	How important is luminance uniformity to visual search?
Age and individual differences	Yes	—	—	Test accuracy of models for different age groups. Study how age affects practical visual ability.	How should places used by elderly be lit?
Fatigue effects (sustained performance)	Yes	Yes	—	Develop measures of fatigue. Use them to examine effects of lighting.	Should lighting for sustained work be different?
Discomfort (glare, flicker, etc.)	Yes	—	Yes	Identify lighting which causes discomfort. Establish if discomfort affects performance.	Does uncomfortable lighting matter?
Light as attention stimulus	—	Yes	—	Identify what aspects of lighting attract attention and how consistent that effect is.	Can lighting be used to improve performance of tasks requiring attention over many areas?
Light and arousal	—	Yes	Yes	Identify what aspects of lighting affect arousal. Examine if the effect is repeatable.	How can each place be made an exciting place to be in?
Light and mood	—	Yes	Yes	Establish a "menu" of how lighting affects impression. Examine the stability of impression and how it affects mood	Does creating an impression change people's motivation?
Lighting influence on behavior	—	Yes	Yes	Explore how lighting modifies behavior.	Can lighting be used to direct people along a desired route?
Lighting and (hormone) biology	—	Yes	Yes	Identify the route by which light affects hormone balance. How important is spectral composition?	Is lighting important to people's health in ways other than allowing them to see?

[a]From Ref. 12.

reinforced and (2) the quantity and quality of light necessary to perform the visual work. For now we will look at the extent to which the quality and quantity of lighting can change performance of the task. Visual performance is the measurement of the observer response to the characteristics of the task. It may be the response to the visibility of the details of the task or the overall response to carrying out a given sequence of acts involving the visibility of the details.

5.1 General Visual Response to Different Task Parameters

It is helpful to know the general response of the visual system to each of the elemental characteristics of a task size, contrast, time of seeing, and the luminance of the background of the detail. For more than a century, eye specialists have measured the response of human visual capability by having a person read the smallest line of letters on an "eye chart" bearing a graduated series of lines of different-sized letters. A series of test objects that are symbolic of the details found in practical workplace tasks has evolved into the configurations shown in Figure 24.1. Each series gives a different curve of response, but the curves for each are largely parallel, showing the same general nature of the response (21–23). The ability to see decreasing size with changing illumination for black Landolt rings on white background (high contrast) is represented in Figure 24.2. The ability to see

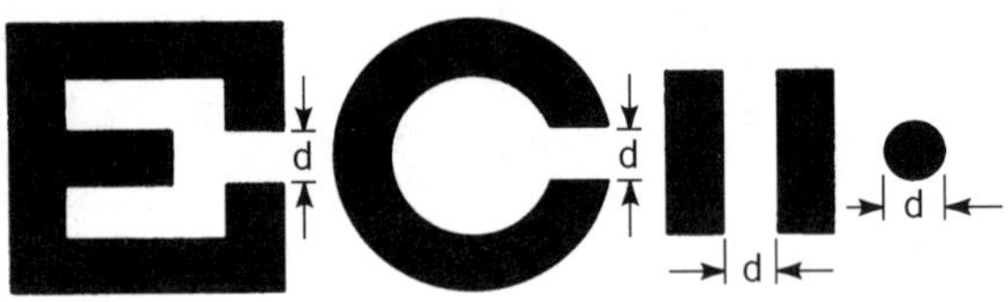

Figure 24.1. Commonly used test objects for determining size discrimination and visual acuity: Snellen E, Landolt ring, parallel bars, and disk (24).

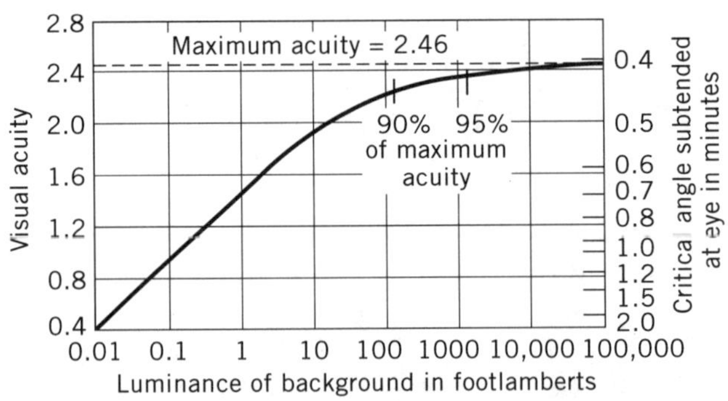

Figure 24.2. The variation in visual acuity and visual size with background luminance for a black object (Landolt ring) on a white background (25).

decreasing contrast with increasing illumination for a disk with visual angle of 4′ subtense and 0.1-s exposure is shown in Figure 24.3. The ability to see with decreasing amounts of time required as illumination increases can be understood from Figure 24.4. A further illustration of how these elemental characteristics influence the seeing ability with changing illumination is given by the eye charts in Figure 24.5, illustrating the effects of size, contrast, and reflectance of background. Of course reflectance of background makes a marked difference in seeing the detail because the sensitivity of the observer depends on the adaptation luminance of the background (27), as further illustrated in Figure 24.6. Fry has pointed out the need to recognize that the image received and processed by the retina of the eye is strongly influenced by blur. For the most part, blur of the retinal image can

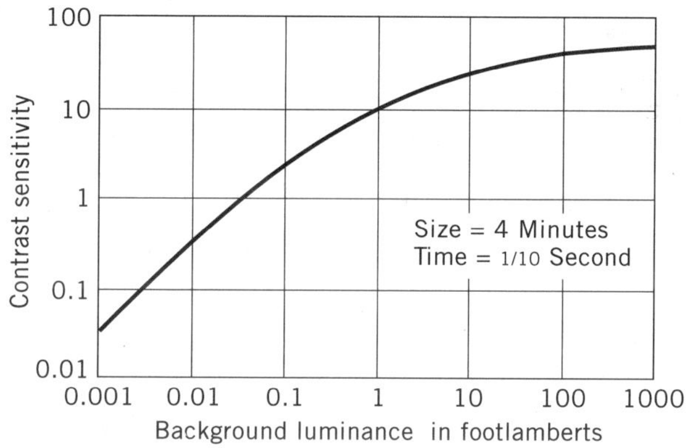

Figure 24.3. Variation of contrast sensitivity with background luminance (25).

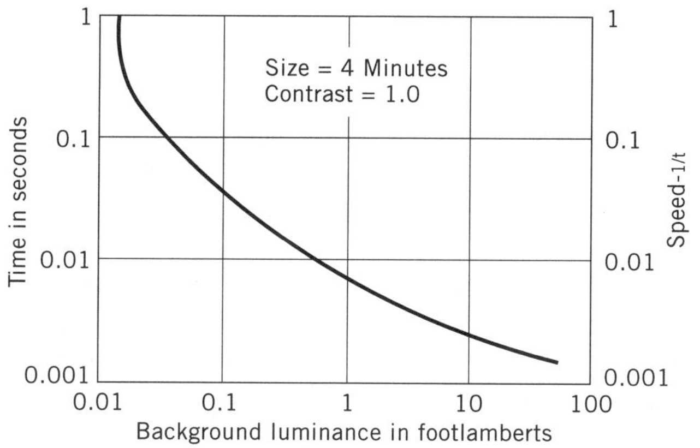

Figure 24.4. The effect of background luminance on the time required to see a high-contrast test object; right-hand scale represents speed of vision (25).

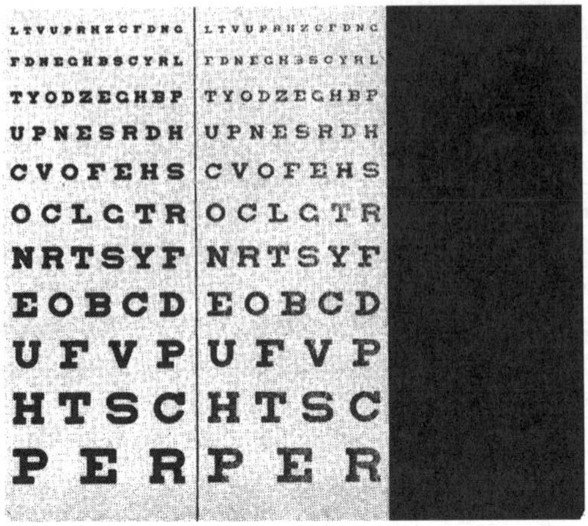

Figure 24.5. Series of eye charts showing effects of size, contrast, and reflectance. Black on white required 1 fc; gray on white, 120 fc; black or dark gray 520 fc (26).

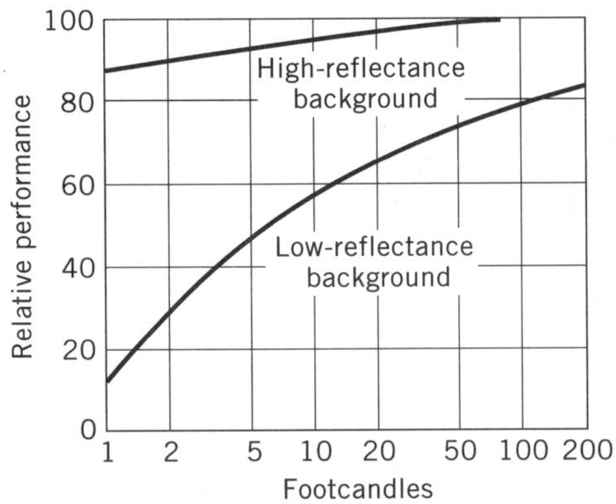

Figure 24.6. The effect of level of illumination on relative visual performance when the task involves high contrast (upper curve) and low contrast (lower curve) (27).

be minimized by using optical aids, yet even when eyesight is optimal (with or without such aids) a certain amount of residual blur can occur. We need a better understanding of blur related to chromatic aberration, the ability of a person to focus, and the limits blur imposes on visibility (28). Research is being conducted to understand the role of blur and to quantify how the spectral distribution of light sources affects blur and visual performance. At interior lighting levels, light source spectrum impact on visual performance was enhanced by blur (29). At night, road sign legibility was improved for both young and older drivers when the image was slightly blurred. Legibility for older drivers improved because the slight blurring of the image exploited their ability to process low-frequency information (30).

5.2 Effect of Surroundings

The ability to see size and contrast is influenced not only by the luminance of the background surrounding the detail but by the differing luminance and size of immediate surroundings around the background of the task, as illustrated in Figure 24.7, Figure 24.8, and Figure 24.9. These effects appear to come from static viewing of the details involved, but it has been reasoned by some that the eyes could be stealing glances to the surroundings, and the effect of sensitivity losses because of the differing surround luminance could be due to transient adaptation, discussed below (31, 32).

5.3 Simulating Practical Tasks

Generally the productivity of practical tasks in commerce and industry is related to the speed and accuracy of carrying out those tasks. Correlations between productive performance of tasks at the work site have proved difficult to analyze because there are a number of factors that cannot be controlled. Thus laboratory tasks simulating practical tasks have been designed to measure the speed of response to the processing of test objects that would approximate the size and contrast of the range of details found in the field. Such a series,

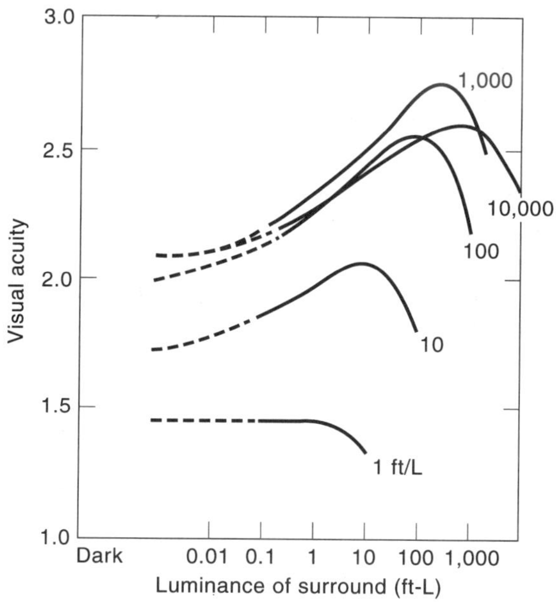

Figure 24.7. Relation between visual acuity and surround luminance for various central field luminances (120° surround to central field) (31).

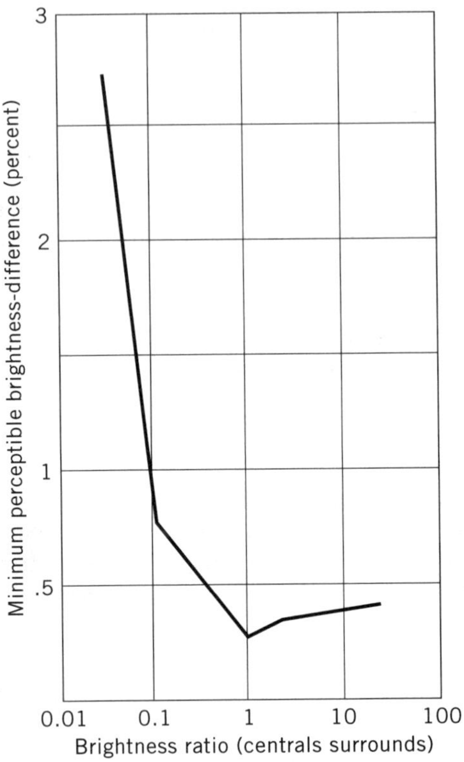

Figure 24.8. Relation between minimum perceptible brightness difference and the ratio of the brightness of the task to the brightness of the surrounding field (34).

under the auspices of the Committee on Industrial Lighting of the National Research Council (1920s), appears in Figure 24.10, Figure 24.11, and Figure 24.12. Unfortunately, the study did not cover values of illumination beyond 100 fc.

A further practical simulation of a task vitally used in industry was the reading of a steel vernier rule to the nearest 1/1000 inch (36). In carrying out the simulation, it was found that the speed of making the complete reading of the tenth-inch numbers, the quarter divisions within the tenth, and the exact position of the vernier divided into 25 parts varied with the luminance of the highlight of the background of the rule, against which the black divisions were seen in bold relief (Fig. 24.13). The speed of reading the vernier with luminance is shown in Figure 24.14. Again the study did not carry values of luminance beyond 200 fL.

In the 1970s further work was conducted on simulated industrial tasks. The threading of needles (Fig. 24.15) represented fine detail work in industry; the reading of date digits on Lincoln pennies (Fig. 24.16) represented metalworking details; and the finding of symbols on a circuit board (Fig. 24.17) illustrated intricate electronic assembly. Some of the results are given in Figure 24.18, Figure 24.19, and Figure 24.20. When one considers

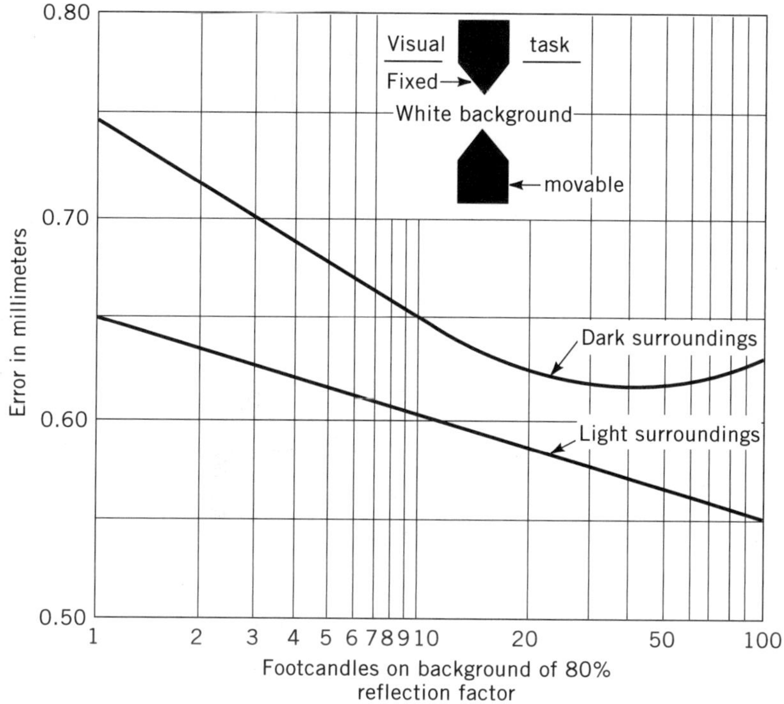

Figure 24.9. The effect of surroundings and illumination on the precision of performing a mechanical task guided by vision (33).

labor costs, space ownership or rental, lighting costs, and electricity rates, it is found in many cases that monetary savings in productivity and a decrease in errors are positive in relation to lighting levels (37, 38) (Fig. 24.21). Researchers continue to look at the fundamental relationships of light and visual tasks by looking at reading speed, size and contrast, and the effect of spectral power distribution (SPD) at interior lighting levels to see where improvements in the actual task or a designer spectrum might enhance visibility; so far experts are debating the merits of enhanced SPD (39, 40). At low light levels, where vision is rod based, because of a shift in the eye's sensitivity toward the short wavelength (blue), significant differences exist between indoor and outdoor lighting conditions (41–45).

These productivity graphs are all in positive relation to the visibility of the detail. In other words, the better the visual cues, the faster the speed of processing and the fewer the errors, but at a diminishing rate until a maximum is reached. Figure 24.22 illustrates this principle.

There are field tasks that involve much mechanical activity or manipulation along with the visual piloting of the work. This activity may occupy so much of the time of doing the overall processing that the productivity curves (time and speed) reveal little or no

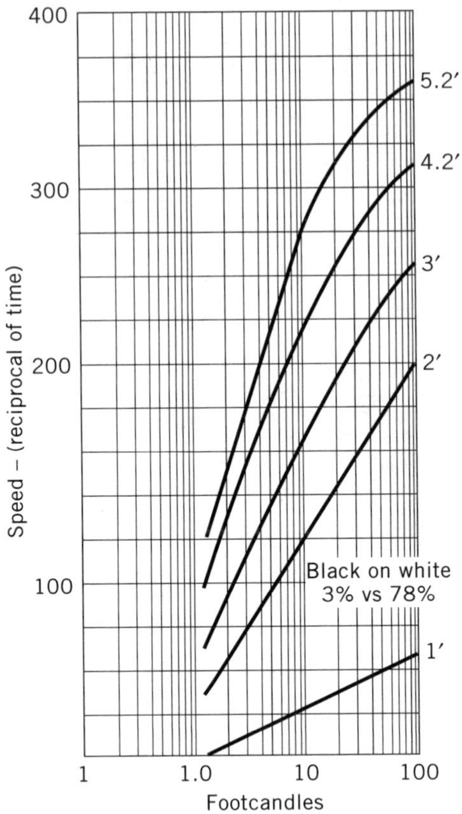

Figure 24.10. Speed of vision versus illumination for the task of identifying the position of a Landolt ring. Data from Ferree and Rand (35), under the auspices of the National Research Council.

change with illumination level or increasing visibility. The designer is then faced with a further consideration at this point. Either the emphasis is placed on providing good visual efficiency, regardless of the output curves, or visual efficiency is depreciated on the basis of the output curve, letting the economics force a lower level of illumination. This would result in a lower visibility than would be used if the visual component were the major work item. Past research described above is of short time duration. Lowered visual efficiency induced by money-saving measures might result in lowered output over a period of time. Although this might be especially true of older workers whose visual capability is much less than that of college students, who were the subjects of previous tests. Data are now becoming available on possible deterioration of performances over a long period, such as several hours or a day. Researchers are seeking the long-term effects on sustained performance. For older drivers, the ability to cope with the mental load of driving at night improves past the point where visibility alone is adequate for vision. This begins to address the "effort of seeing" (46, 47).

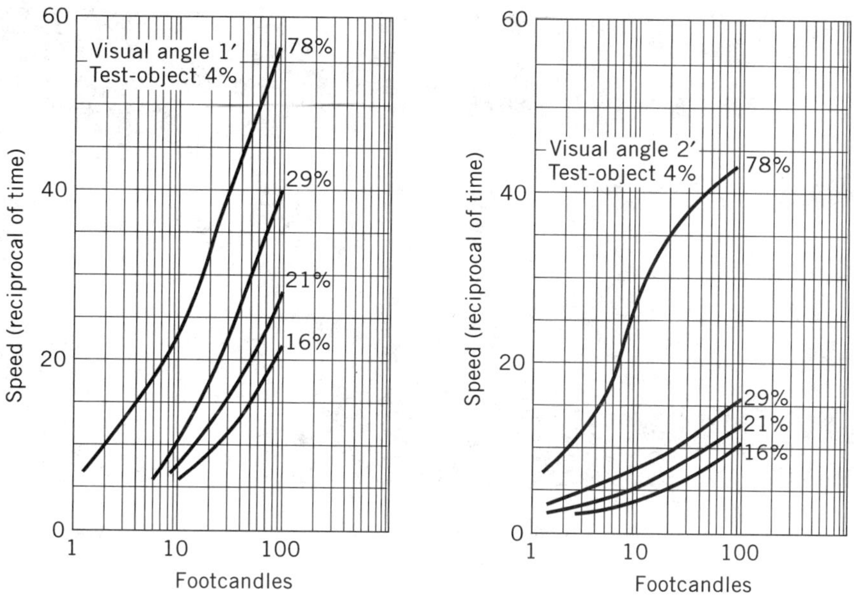

Figure 24.11. Speed of vision versus illumination for task involving varying contrasts (35).

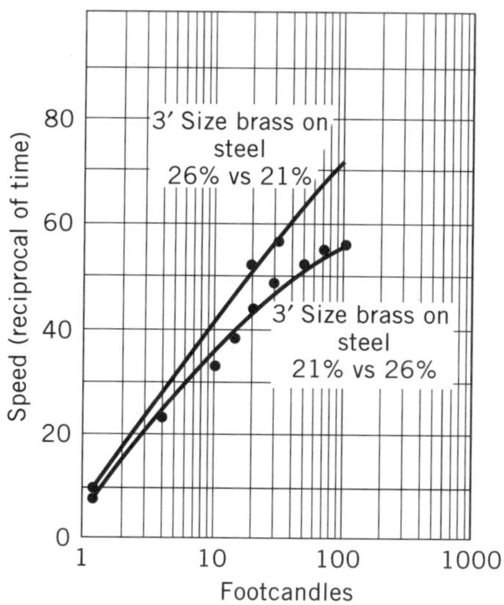

Figure 24.12. Discrimination of a brass test object on a steel background and a steel test object on a brass background (36).

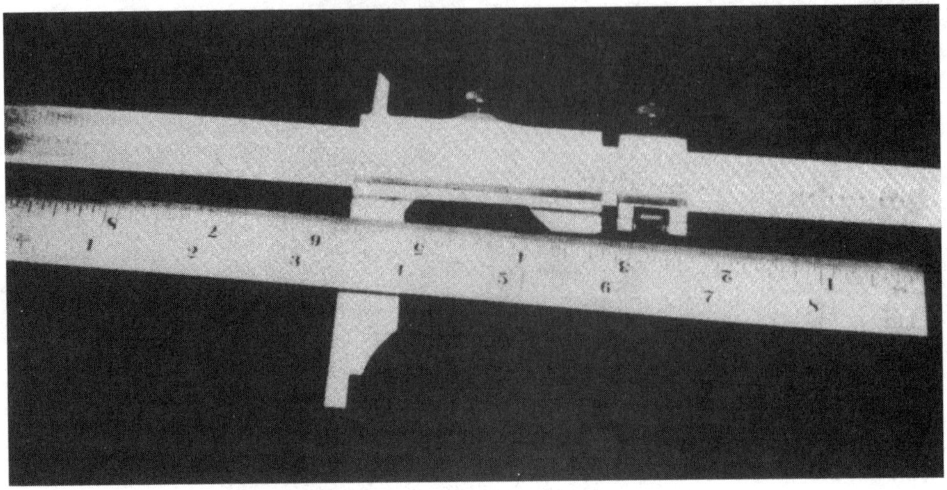

Figure 24.13. A steel vernier rule used in the practical simulation of a task down to the nearest 1/1000 in.

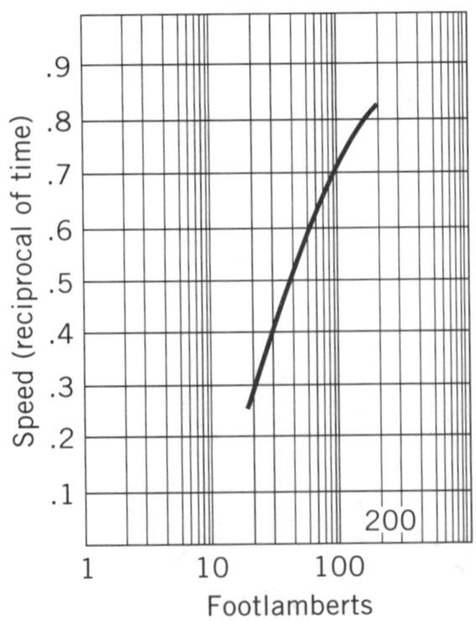

Figure 24.14. Reading a steel vernier rule: perception 1/1000 inches out of alignment.

Moreover, there is the issue of the workers' morale. The same visual details may be processed at different locations in a plant or office with differing visual versus mechanical components. Shall the designer furnish much more illumination to workers where the visual component is the larger, and much less to those who do more mechanical work on the same visual details?

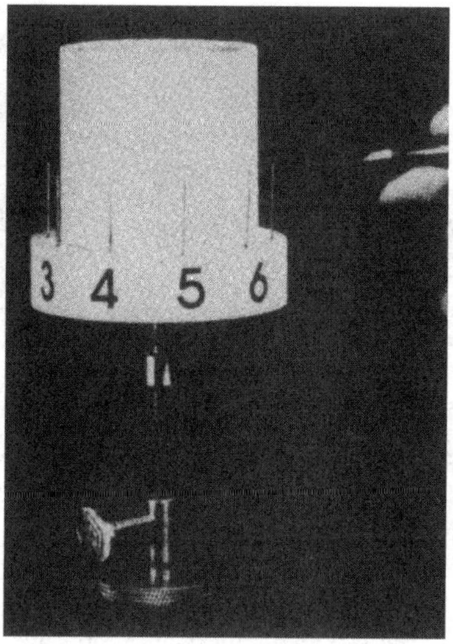

Figure 24.15. Simulated industrial task: the threading of needles represents fine detail work (37–38).

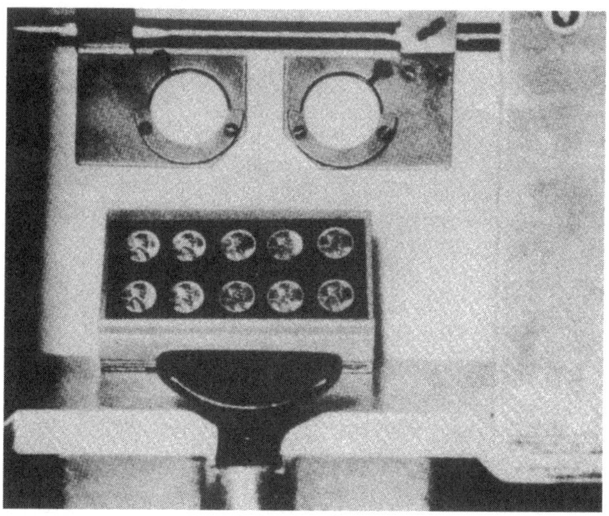

Figure 24.16. Simulated industrial task: reading date digits on Lincoln pennies represents metalworking details (37–38).

Figure 24.17. Simulated industrial task: a circuit board represents an intricate electrical assembly (37–38).

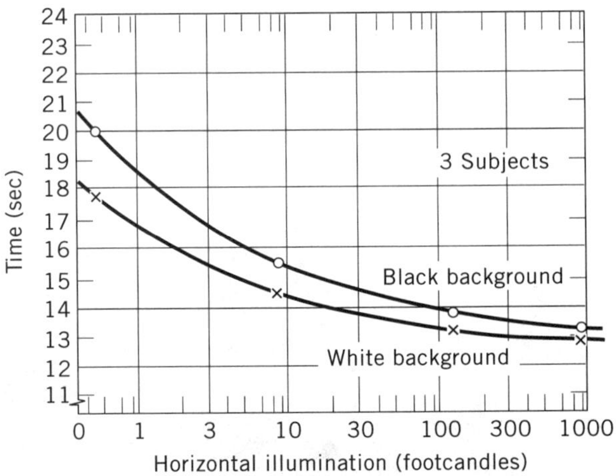

Figure 24.18. Relation between performance, time, and illumination to the needle probe task.

Finally, an accumulating body of recent research to be described below indicates there may be a certain basic amount of lighting needed to stimulate biologically the worker to maintain alertness and prevent industrial accidents (14–18).

The same concepts apply to office lighting. In the middle 1970s tests similar to those described earlier for simulated tasks were conducted. These involved proofreading (48, 49) of mimeographed material having good, fair, and poor quality copy (Fig. 24.23),

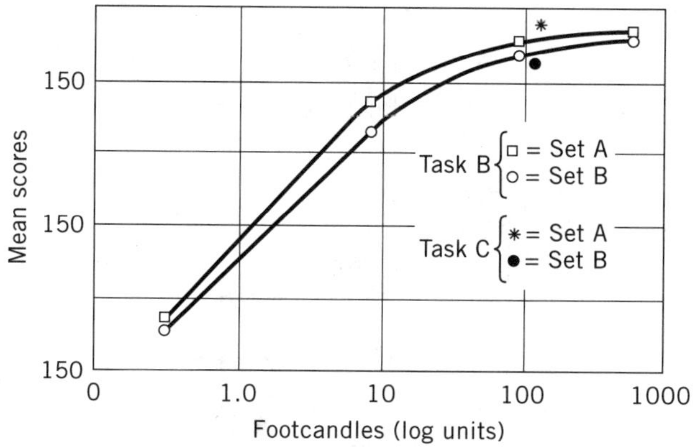

Figure 24.19. Results of response to the reading of date digits on Lincoln pennies.

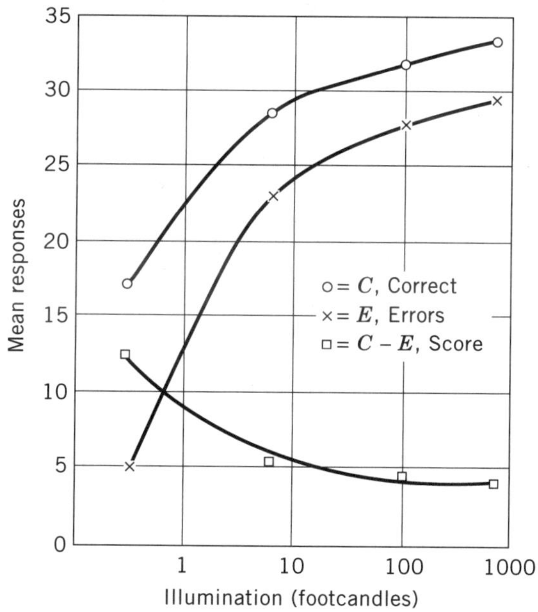

Figure 24.20. Results of response to the reading of a circuit board (finding of symbols versus illumination).

reading handwritten bank checks of varying quality (48, 49), comparing dollar amounts against an adding machine tape listing that included systematic errors (Fig. 24.24), reading and typing (48, 49) from printed copy having 12-point and 6-point characters printed in black on white and light gray on white (Fig. 24.25), and reading the Davis Reading Test

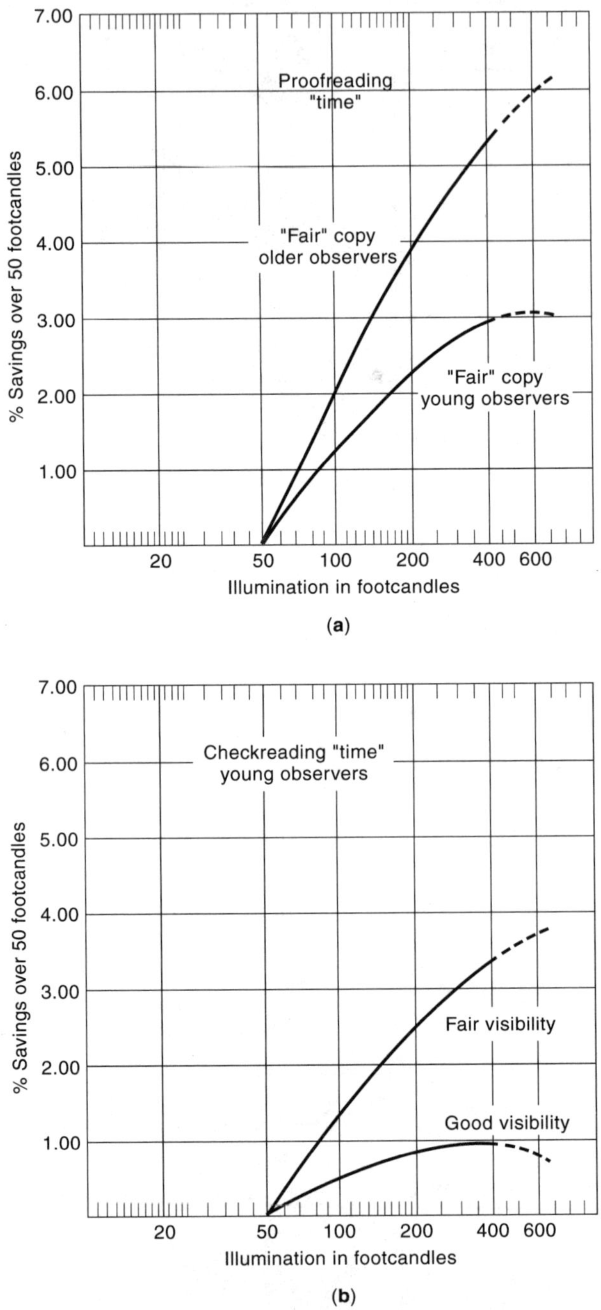

Figure 24.21. Percentage money savings from productivity response to lighting levels. (**a**) Proofreading time for older observers and young observers using "fair" copy. (**b**) Check reading time for young observers with fair visibility and good visibility.

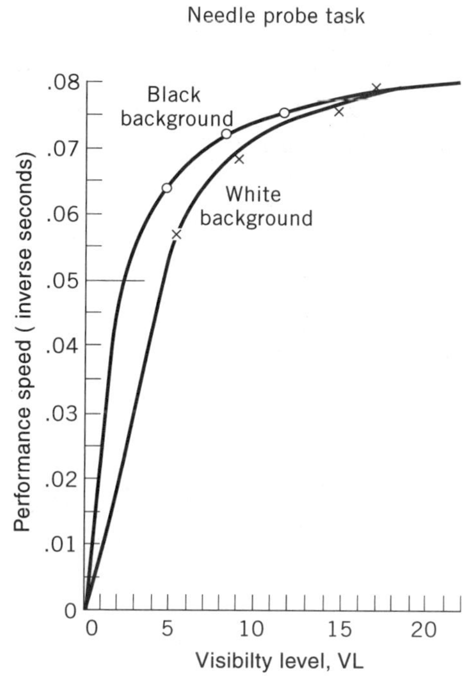

Figure 24.22. Relation between performance speed and visibility level (37–38).

in black and white and poor reproduction copy (Fig. 24.26) for comprehension and speed of comprehension. Some of the results are presented in Figure 24.27, Figure 24.28, and Figure 24.29.

The comprehension and speed of comprehension of the Davis Reading Test did not change with changes of illumination even for the poorer copy. Analysis of the results indicates that study and thinking occupied so much time that the gain in visual efficiency was masked. Here again the designer is faced with a choice. Because the overall performance remained the same under 1 fc and 454 fc, should the designer use 1 fc (used in street lighting); put in some amenity lighting, such as 10 to 30 fc, to make the task look lighted; or put in what visual efficiency would call for, 50–100 fc depending on the age of the observer? Furthermore, should the designer place 1–10 fc on the desk with the Davis Reading type of work but put, based on age, 50–100 fc on the desk with the "reading-typing" task that involves approximately the same visual details as the Davis Reading type of task? Compare Figure 24.25 and Figure 24.26. In the 1930s Weston was faced with similar questions in England. He was director of research in occupational optics for the Medical Research Council. For many years he had studied the role of illumination in various types of industry, but he came to the conclusion that one could not isolate the gain in visual performance from the many contributing factors to industrial plant productivity. Under the aegis of his advisory committee, therefore, he set up a series of laboratory tests

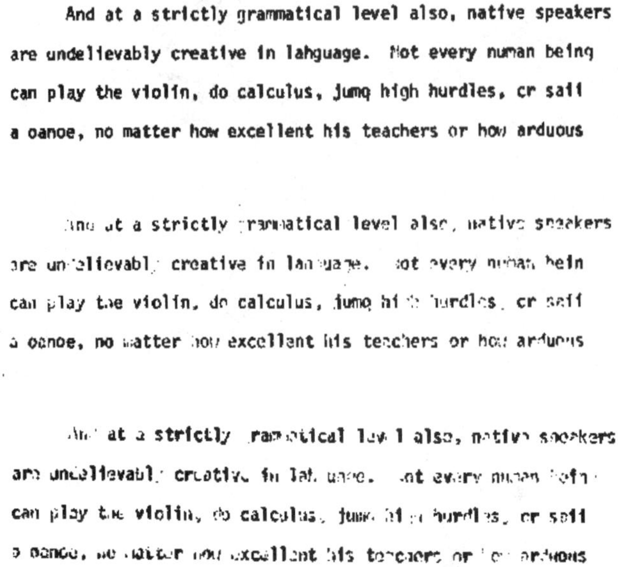

Figure 24.23. Simulated office task showing good, fair, and poor quality copy.

that would simulate industrial tasks, eliminating the many extraneous factors and determining the true function of "visual efficiency" in relation to illumination (50, 51). He recognized that mechanical, or "action" time, which varies greatly in field tasks, dilutes or masks the true visual efficiency; therefore he measured and subtracted this component to arrive at the net visual time and its relationship to illumination. This represented the true visual performance as related to illumination. Under the leadership of Crouch and Fry, since the early 1950s through the early 1980s in the United States, the Illuminating Engineering Research Institute used the Weston concept as it began its research program on required illumination for good visual performance. The concept of suprathreshold visibility for seeing was first introduced by Luckiesh and Moss as an attempt to quantify how well a stimulus can be seen. This was later used extensively by Blackwell, who, starting with the threshold of visibility for a base, extended to suprathreshold using visibility to unify his findings on size, contrast, time, and luminance (Fig. 24.30). Fry and Enoch (52), studying Weston's work (58–61), began relating it to patterns of eye movements. In conducting their visual performance experiments using the Landolt C, a frequency of seeing curve similar to that in Figure 24.31 was derived. At the same time Blackwell was using the frequency of seeing concept in his studies. Fry noted that Blackwell's studies using frequency of seeing showed a good correlation with Weston's work. Ultimately this work culminated in an international model: CIE 19 and CIE 19/2 (10, 53) relating lighting, visibility, and visual performance, as described below.

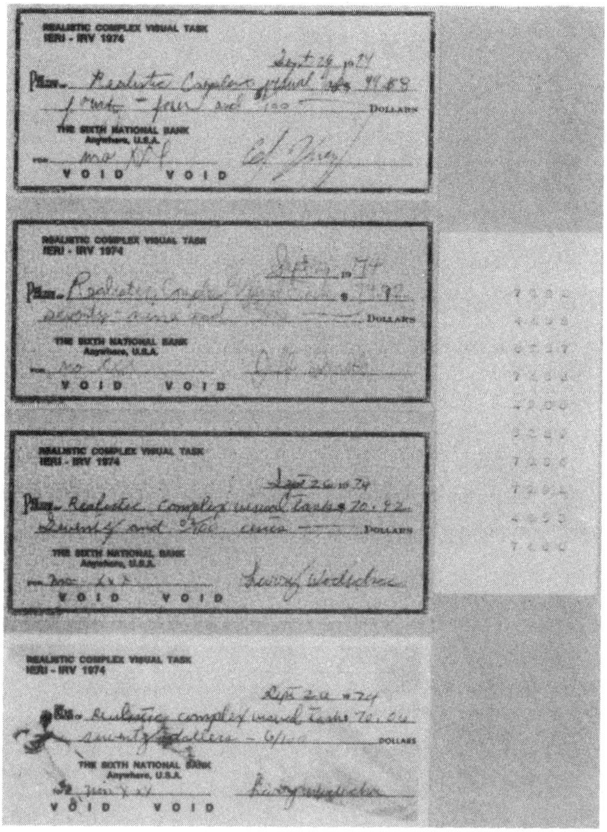

Figure 24.24. Simulated office task: variation in visibility of handwritten checks being compared with adding machine tape.

5.4 Visual Performance Models

What is a model and why is it important? A model is a mathematical framework for understanding existing relationships. In its most useful form it not only conforms to existing data but also provides a rational basis for designers and engineers to predict what might happen in a work space before it is built. Two major forms of models exist: (*1*) an empirical model that summarizes experimental data without offering a theoretical basis and (*2*) an analytical model that combines knowledge of the process being modeled as well as empirical data (12).

5.4.1 The CIE Visual Performance Model

The CIE 19/2 Visual Performance Model is an analytical model. It is based on the empirical studies of the measurement of speed and accuracy of real or simulated tasks versus illuminance or luminance. Some of these studies were already described in Section 4.3. The

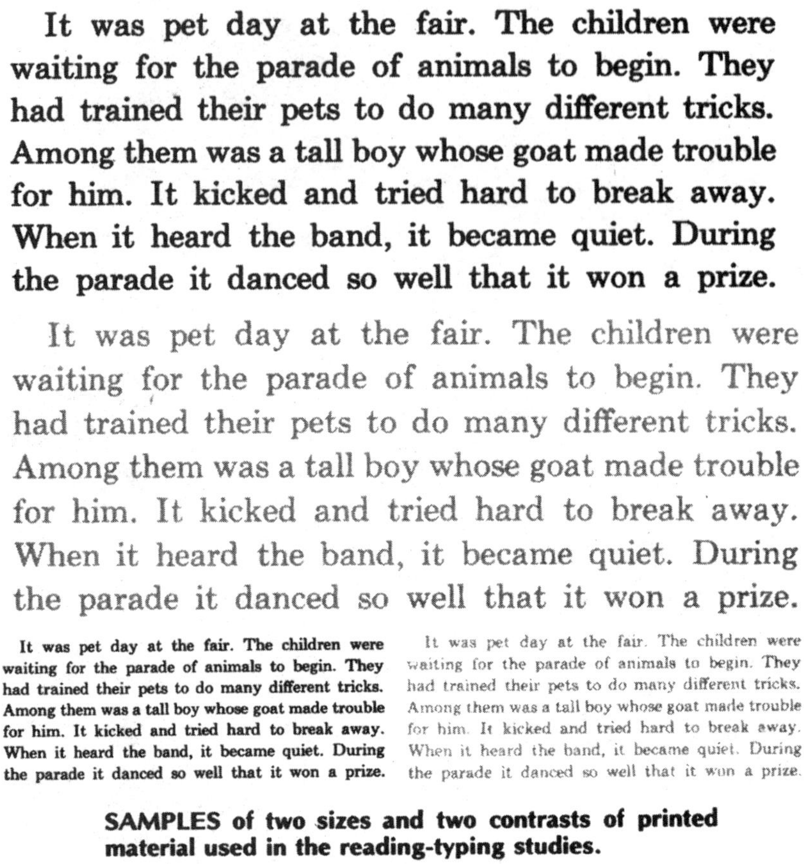
It was pet day at the fair. The children were waiting for the parade of animals to begin. They had trained their pets to do many different tricks. Among them was a tall boy whose goat made trouble for him. It kicked and tried hard to break away. When it heard the band, it became quiet. During the parade it danced so well that it won a prize.

It was pet day at the fair. The children were waiting for the parade of animals to begin. They had trained their pets to do many different tricks. Among them was a tall boy whose goat made trouble for him. It kicked and tried hard to break away. When it heard the band, it became quiet. During the parade it danced so well that it won a prize.

It was pet day at the fair. The children were waiting for the parade of animals to begin. They had trained their pets to do many different tricks. Among them was a tall boy whose goat made trouble for him. It kicked and tried hard to break away. When it heard the band, it became quiet. During the parade it danced so well that it won a prize.

It was pet day at the fair. The children were waiting for the parade of animals to begin. They had trained their pets to do many different tricks. Among them was a tall boy whose goat made trouble for him. It kicked and tried hard to break away. When it heard the band, it became quiet. During the parade it danced so well that it won a prize.

SAMPLES of two sizes and two contrasts of printed material used in the reading-typing studies.

Figure 24.25. Simulated office task showing variation in type size with contrast variability.

analytical portion involves measurements of visual processes versus illuminance or luminance. First it uses a 4′ luminous disk as a standardized target because its configuration is simple. Then to simulate more closely the visual performance of tasks, the Landolt C was chosen as the standardized test object. The CIE 19/2 model then mathematically combines the results of the practical performance of tasks with the present understanding of the visual response to light, contrast, size, and other parameters.

To provide a baseline of understanding of visual performance, laboratory studies were first conducted under idealized conditions of uniform, glare-free light and using normally sighted 20- to 30-year-old college students. Later, real-life considerations such as the effects of glare, age, and nonuniform lighting were included. In summary form, the model first describes the effect of light on the visibility of the visual task and then the effect of visibility on visual performance. Visibility then becomes the transforming link for the effect of light on visual performance. The following equations describe these relationships:

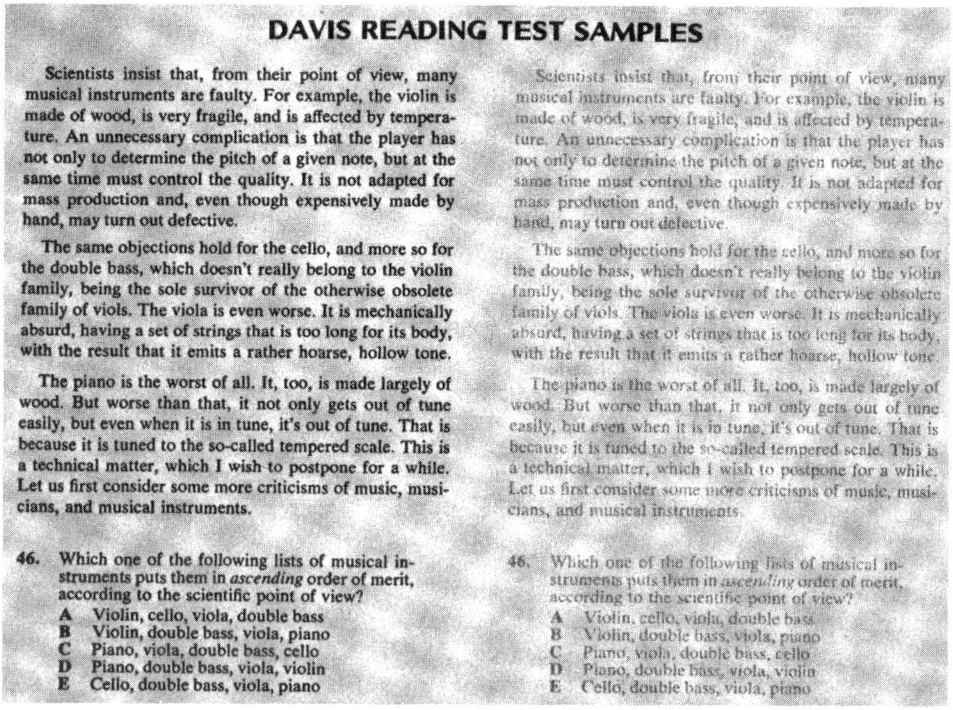

DAVIS READING TEST SAMPLES

Scientists insist that, from their point of view, many musical instruments are faulty. For example, the violin is made of wood, is very fragile, and is affected by temperature. An unnecessary complication is that the player has not only to determine the pitch of a given note, but at the same time must control the quality. It is not adapted for mass production and, even though expensively made by hand, may turn out defective.

The same objections hold for the cello, and more so for the double bass, which doesn't really belong to the violin family, being the sole survivor of the otherwise obsolete family of viols. The viola is even worse. It is mechanically absurd, having a set of strings that is too long for its body, with the result that it emits a rather hoarse, hollow tone.

The piano is the worst of all. It, too, is made largely of wood. But worse than that, it not only gets out of tune easily, but even when it is in tune, it's out of tune. That is because it is tuned to the so-called tempered scale. This is a technical matter, which I wish to postpone for a while. Let us first consider some more criticisms of music, musicians, and musical instruments.

46. Which one of the following lists of musical instruments puts them in *ascending* order of merit, according to the scientific point of view?
A Violin, cello, viola, double bass
B Violin, double bass, viola, piano
C Piano, viola, double bass, cello
D Piano, double bass, viola, violin
E Cello, double bass, viola, piano

Scientists insist that, from their point of view, many musical instruments are faulty. For example, the violin is made of wood, is very fragile, and is affected by temperature. An unnecessary complication is that the player has not only to determine the pitch of a given note, but at the same time must control the quality. It is not adapted for mass production and, even though expensively made by hand, may turn out defective.

The same objections hold for the cello, and more so for the double bass, which doesn't really belong to the violin family, being the sole survivor of the otherwise obsolete family of viols. The viola is even worse. It is mechanically absurd, having a set of strings that is too long for its body, with the result that it emits a rather hoarse, hollow tone.

The piano is the worst of all. It, too, is made largely of wood. But worse than that, it not only gets out of tune easily, but even when it is in tune, it's out of tune. That is because it is tuned to the so-called tempered scale. This is a technical matter, which I wish to postpone for a while. Let us first consider some more criticisms of music, musicians, and musical instruments.

46. Which one of the following lists of musical instruments puts them in *ascending* order of merit, according to the scientific point of view?
A Violin, cello, viola, double bass
B Violin, double bass, viola, piano
C Piano, viola, double bass, cello
D Piano, double bass, viola, violin
E Cello, double bass, viola, piano

Figure 24.26. The Davis Reading Test. Right, with poor reproduction; left: in original form.

Equation 1: Visibility level (VL)

$$VL = \frac{\tilde{C} \times RCS \times CRF \times DGF \times TAF}{0.923 \times m_1}$$

Equation 2: Relative visual performance (RVP)

$$RVP = w_1P_1 + w_2P_2 + w_3P_3$$

This equation can be replaced by an equation in which its terms are related to VL:

$$RVP = \int \left[w_1\left(\frac{\log VL - \bar{\alpha}_1}{0.145 + [0.278(\bar{\alpha}_1 - 0.050)]}\right) + w_2\left(\frac{\log VL - \bar{\alpha}_2}{0.180}\right) + w_3\left(\frac{\log VL - \bar{\alpha}_3}{0.180}\right)\right]$$

Equation 3: Relative task performance (RTP)

$$RTP = w_{123}RVP + w_4P_4$$

A discussion of the mathematical variables that must be calculated for these formulas follows.

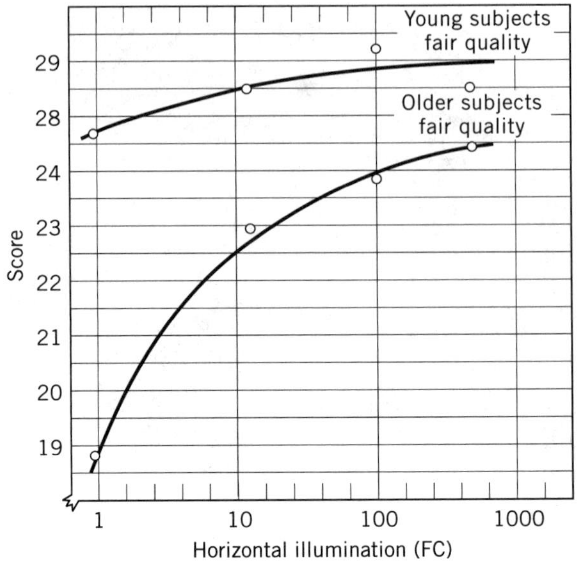

Figure 24.27. Results of mimeograph proofreading experiment of Figure 24.23. Score is a combination of speed and penalty for errors.

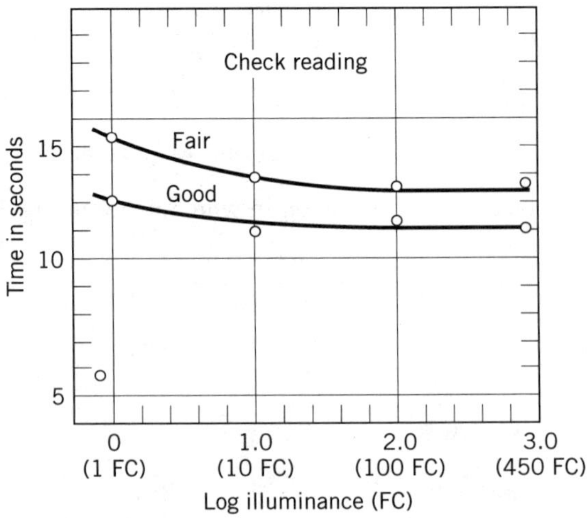

Figure 24.28. Results of check reading experiment of Figure 24.24.

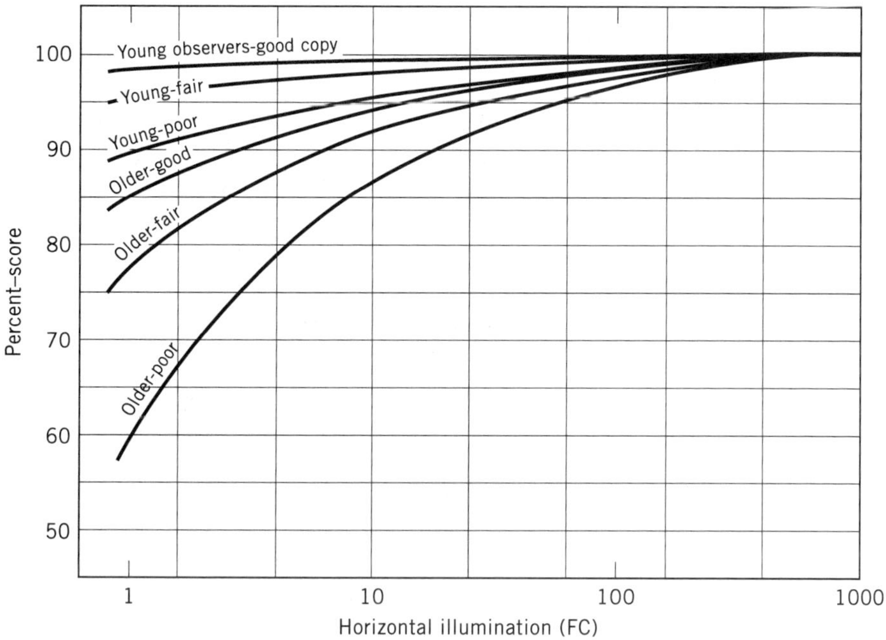

Figure 24.29. Relative results of "mimeograph" proofreading experiment from Figure 24.23 showing effect of quality of mimeograph material and age of observers.

Visibility level. The key concept of the CIE model is visibility described in terms of visibility level (VL). At its most simple form, it is the contrast C of the visual detail divided by its contrast brought down to threshold $\bar{C}$, the border between visibility and invisibility: $VL = C/\bar{C}$. This $\bar{C}$ value varies with individuals; therefore, studies using the concept of threshold were conducted to find what kind of average function an entire population of observers of the same age would produce. The first studies focused on sizes of details that are critical for seeing in commerce and industry. These were found to be in the range of $1'$ to $10'$ of arc subtended at the eye at the usual viewing distance (Fig. 24.30). It was found over a 50-year span (16–18, 24, 26, 33–36, 50) that $3'$ to $4'$ size (*Readers' Digest* type or No. 2 pencil handwriting) represents the weighted average of the critical seeing tasks in working situations. The average eye pause to assimilate information in reading is 0.2 s. Using this information a curve of threshold contrast versus luminance for a $4'$ circular disk exposed for 0.2 s was derived (Fig. 24.32). Threshold defines the probability of seeing an object 50% of the time. Because no one would want to work under threshold conditions, suprathreshold levels of performance in the range of practical use were determined through empirical studies that produced curves of similar shape and were for the most part parallel to the threshold function. These formed the basis of the function for determining the illumination required for quite accurate seeing (53) for many years (Fig. 24.33).

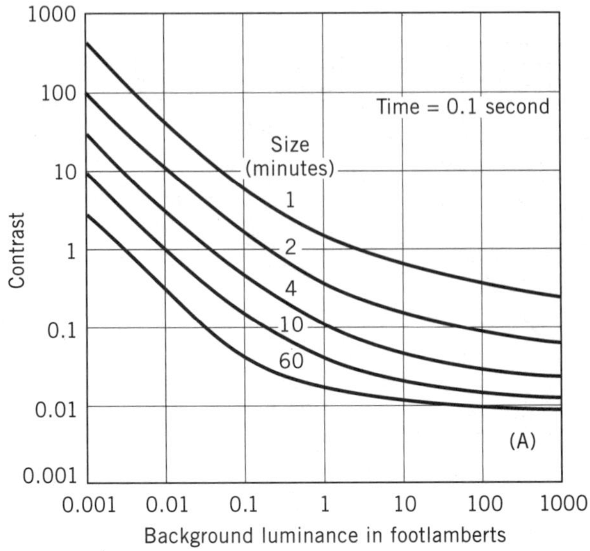

Figure 24.30. Relation between threshold contrast and background luminance for objects of various sizes (9).

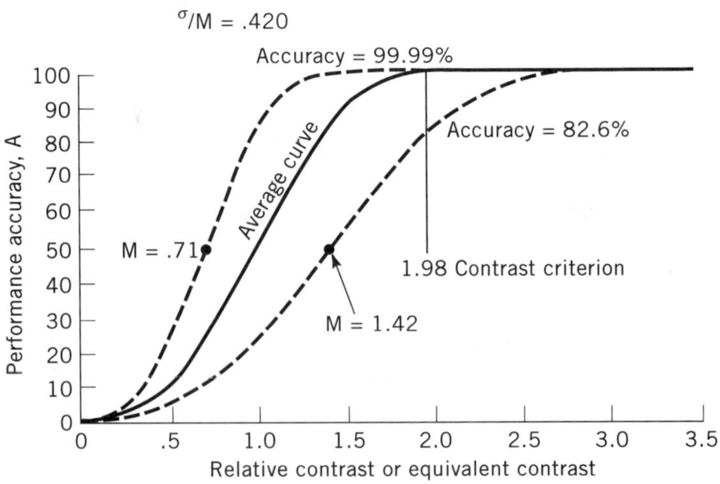

Figure 24.31. Variations in individual accuracy due to variation in *M*. Dashed curves represent limits of individual variability (i.e., limits of young population). M = midpoint (50% probability) of greater or less accurate performance.

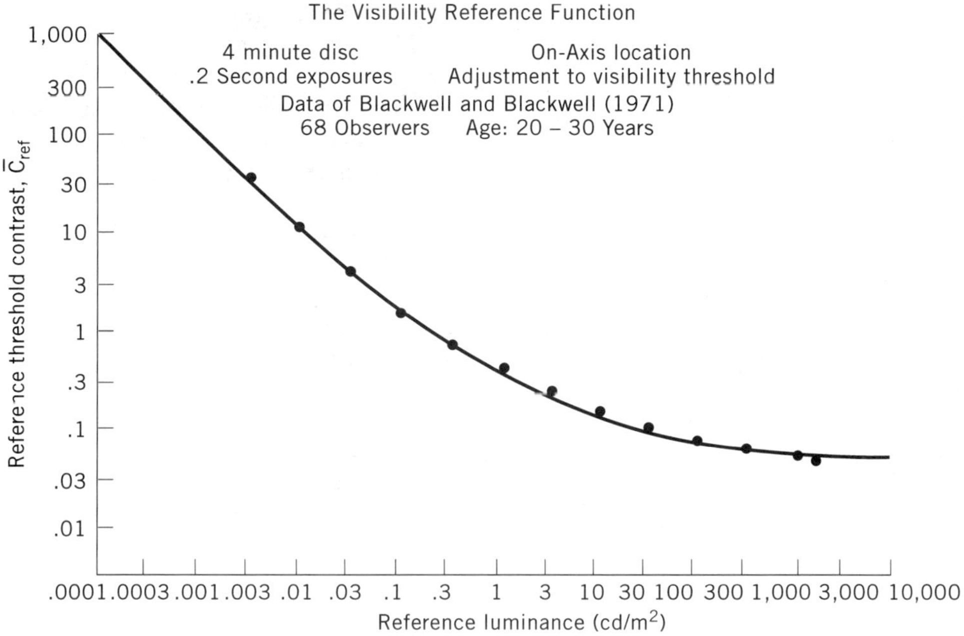

Figure 24.32. Curve shows 50% probability threshold for young normal sighted observers (52, 53).

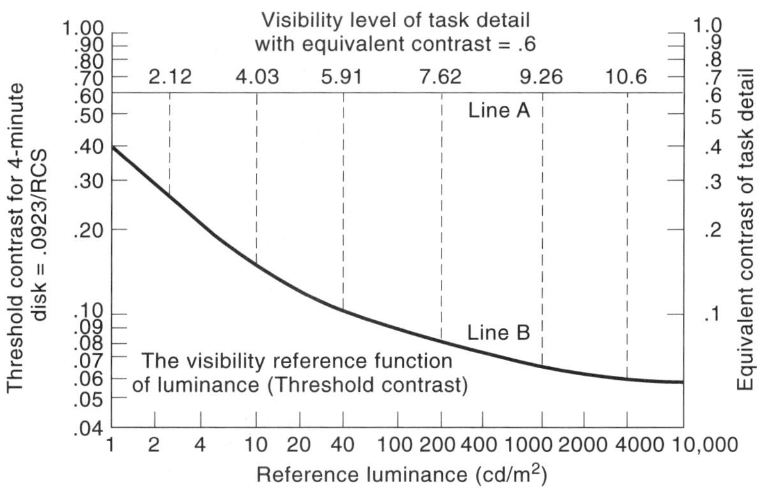

Figure 24.33. Visibility level increases for a given object as the luminance increases and threshold decreases (as in Fig. 24.32) (53).

Equivalent contrast of 4′ Disk. To link the laboratory data to the visibility of the visual detail being studied in the field, a measure of equivalency was derived: C, equivalent contrast. It is the measure of the intrinsic difficulty of the detail being studied. At threshold all objects are equal in visibility. While the eye maintains a constant state of adaptation, the contrast of the visual task in the field is reduced to threshold, usually using a visibility meter constructed with variable filters and optics described below. The degree of reduction of the task to its threshold point is the measure of how great its visibility is. The equivalent contrast then is a comparison of the visibility of a task measured in the field versus the visibility of the 4′ disk brought to its threshold under idealized reference conditions and at the same adaptation level of luminance. This is illustrated in Figure 24.33.

Relative contrast sensitivity. This variable is a measure of how sensitive the visual system is to the perception of contrast versus the illuminance or luminance provided (Fig. 24.34). It represents the inverse of the threshold curve shown in Figure 24.33, which is put on relative basis as a matter of mathematical convenience. At first the relative contrast sensitivity (RCS) function represented only the average response of normally sighted 20- to 30-year-old college students; however, further research resulted in a family of curves that can be derived based on age, whether the task is viewed on or off the line of sight, and the size of the detail being studied, or combinations of these factors, as shown in Figure 24.35, Figure 24.36, and Figure 24.37.

Visibility threshold multiplier. The visibility threshold multiplier, m_1, accounts for the change in contrast sensitivity, and hence the loss of visibility with increasing age (Fig. 24.38).

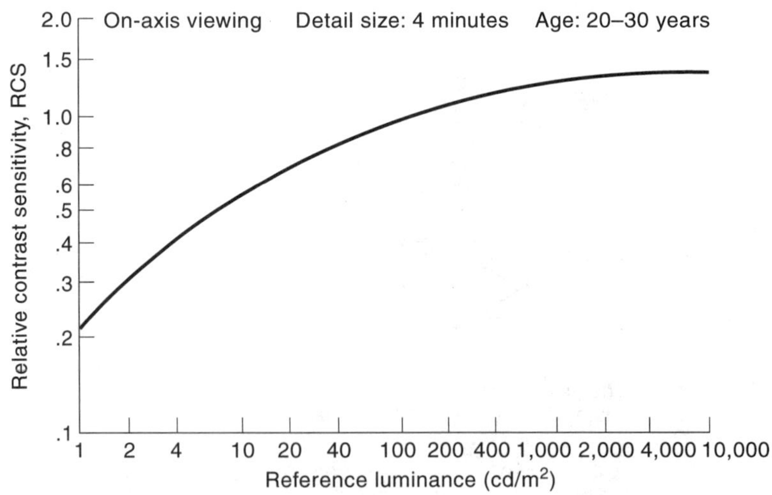

Figure 24.34. Relative contrast sensitivity (RCS); the measure of the eyes' sensitivity to contrast. This is the inverse of the threshold curve shown in Figures 24.32 and 24.33 mathematically normalized at 100 cd/m^2 (52, 53).

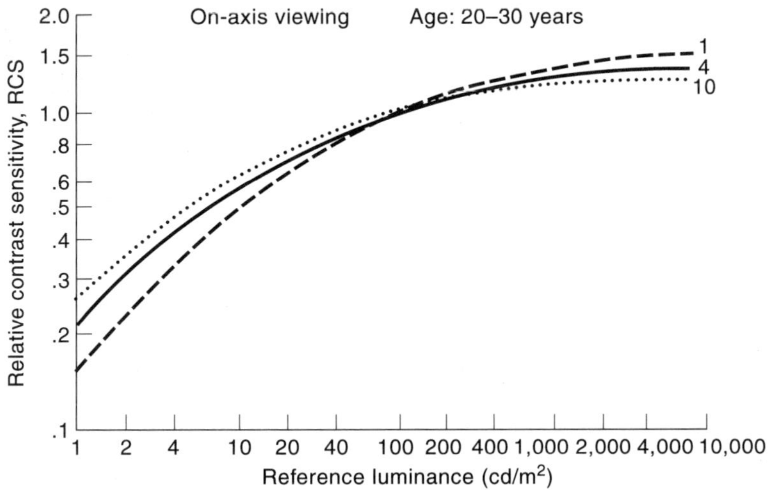

Figure 24.35. The RCS function changes based on the size of the object being viewed in terms of the angular size subtended at the eyes (here 1, 4 and 10 min) (53).

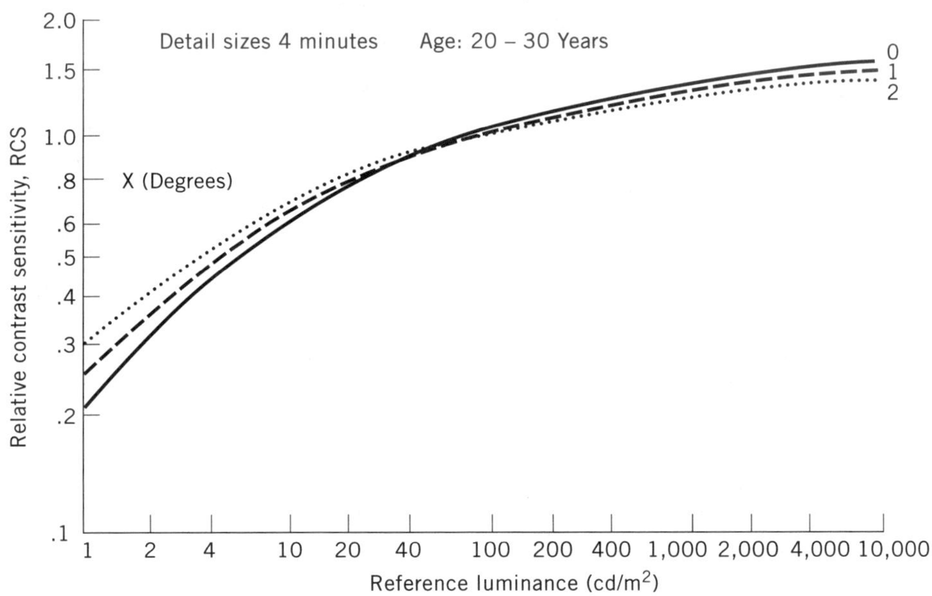

Figure 24.36. The RCS function changes based on the how far off the line of sight the object is being seen (here 0°, 1°, and 2°) (53).

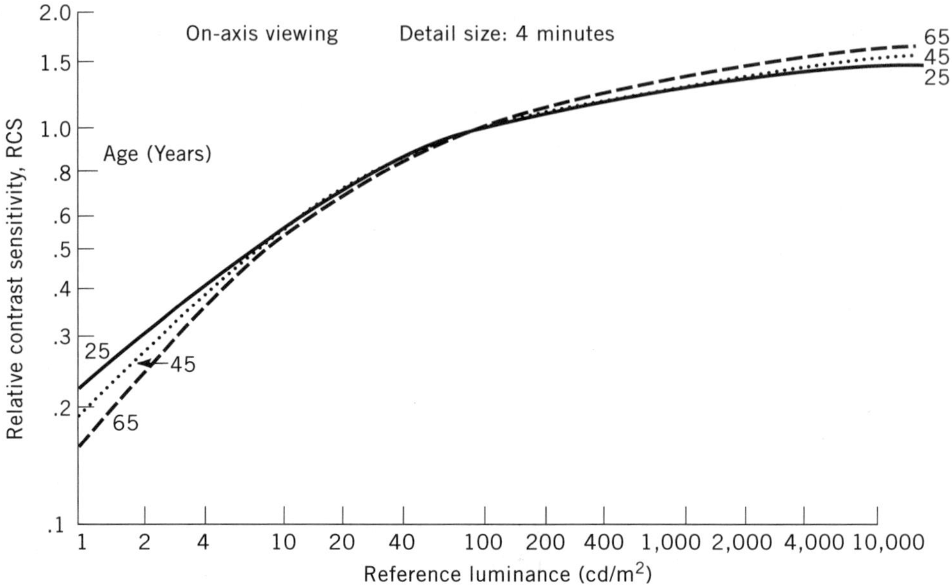

Figure 24.37. The effect of age of the observer on the RCS function (53).

Visibility threshold constant for the 4′ disk. The value 0.0923 is the visibility threshold for the 4′ standard disk at 100 cd/m^2, based on an average of the 20- to 30-year-old population data.

Contrast rendition factor. The contrast rendition factor (CRF) accounts for the degree of loss of visibility due to the reduction in contrast of the visual detail by light reflected from specular components of the task to the eyes of the observer, for example, from glossy photographs, magazines, or highly reflective metals. It is the ratio of the visibility of a task under an actual lighting system compared with the visibility using completely diffuse uniform reference lighting produced under laboratory conditions, usually an integrated sphere.

Disability glare factor The disability glare factor (DGF) accounts for the loss of visibility due to summation of the focused and scattered light in the field of view. The overall effect of disability glare is to produce an overlay of brightness on the image to be seen. This overlay of brightness is light being scattered in the interior of the eye from the glare sources in the field of view. DGF is different from the factor studied in earlier works of disability glare assessment in that it accounts for two different effects of ocular stray light produced by glare sources: (*1*) the reduction of the contrast of the image being viewed and (*2*) the change in RCS due to the summation of focused and stray light. While the contrast of the image is being degraded by the glare, the higher adaptation luminance level produced by the glare increases the eye's sensitivity to the perception of contrast. Therefore the RCS factor modifies the total loss of visibility that would be determined by just con-

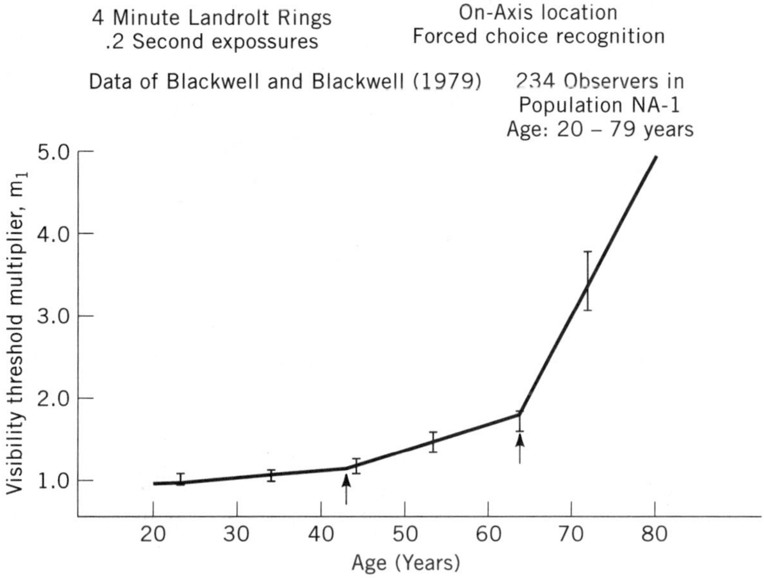

Figure 24.38. The visibility level varies inversely with the age of the observer and is expressed by the visibility threshold multiplier m_1 and the visibility level (VL) formula (53, 54).

sidering the reduction of contrast produced by scattered light. For the interested reader, an extensive treatment and procedures for measurement of DGF are covered in the CIE 19/2 report (10, 52, 53, 54–57). The significant influence of disability glare on age is discussed below.

Transitional adaptation factor. The eye for the most part operates in a dynamic mode in processing visual information. The transient adaptation factor (TAF) takes into account the dynamic eye movement and resulting changes in adaptation to the luminances encountered in these patterns of movement. As the eye moves from one point of reference to another in gathering information, it may encounter different luminances from its current adapted state. As transitional adaptation occurs, it results in differences of visibility. Work concerning this factor was pioneered by Boynton and Boss and Rinalducci and Stearn and used by Blackwell and Tsou to develop the TAF. Ideally, one would like to be able to measure the continuous change in adaptation; to date, however, such a general model does not exist. Instead, only two dynamic conditions are covered: (*1*) glance transient adaptation factor, where an observer while looking at a visual task takes brief glimpses (totaling less than 2 s) away to a brighter portion of the visual field; and (*2*) readaptive transient situation, where the observer looks away from the primary visual task for 10 s or more, thus creating two visual tasks. This is an area under investigation today (53–57). Greater understanding is especially needed to to assess luminance difference encountered between processing paper tasks as well as visual display terminals (VDT) tasks.

Relative visual performance. This second transfer function relates visibility to visual performance. In Blackwell's early studies the basic visual response was determined using the standardized 4′ luminous disk, where the criterion used was the detection of the presence of the disk (the observer responded to whether it was present or not present). Later Blackwell chose the international test object used by Weston, the Landolt C, to conduct experiments, the results of which he used to develop performance curves; an observer scanned either a single Landolt C or a display of Landolt Cs to determine the orientation of the gap in the C under different conditions of size, contrast, luminance, and so on. This resulted in a family of S-shape performance curves called ogives, shown in Figure 24.39. The next step was to attempt to relate these laboratory studies to the performance of real field tasks or simulated field tasks. Fortunately a series of performance studies had been completed both in the United States and abroad. These data were transformed into ogive curves and compared with the laboratory performance curves. The results are shown in Figure 24.40 and Figure 24.41. In most instances the practical performance curves and the laboratory performance curves were quite similar. However, some of the ogive curves were of a slightly different shape, which challenged the hypothesis that there is only one process of seeing. Subsequent studies determined that there are at least four processes; the first three are considered critical to gathering information, and the fourth is a noncritical mechanical process. Relative visual performance depends on the following three processes and their weighting coefficient, w: (*1*) a sensory process, w_1P_1, which determines the detail to be seen; (2) fixation maintenance, w_2P_2, which guides and keeps the eye at a viewing

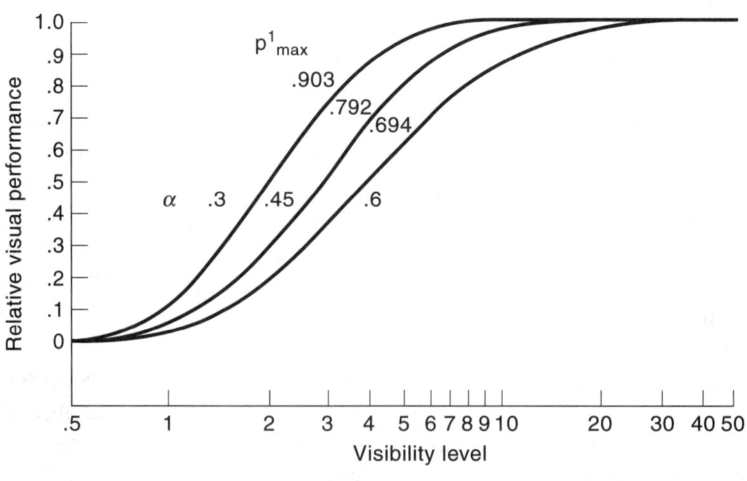

Figure 24.39. Family of ogives or S-shaped curves developed from scanning studies of the Landolt rings. The flattest curve—that farthest to the right—represents the most difficult task, the steepest curve the easiest task (53). P' upper asymptote of original performance curves before mathematical normalization.

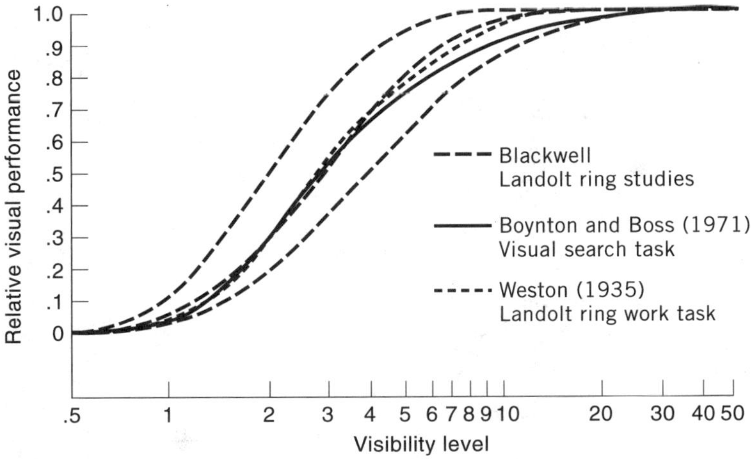

Figure 24.40. Meshing the ogives from the simulated tasks of two field researchers [Weston (50) and Boynton and Boss (53)] with Blackwell laboratory Landolt-ring ogives (53).

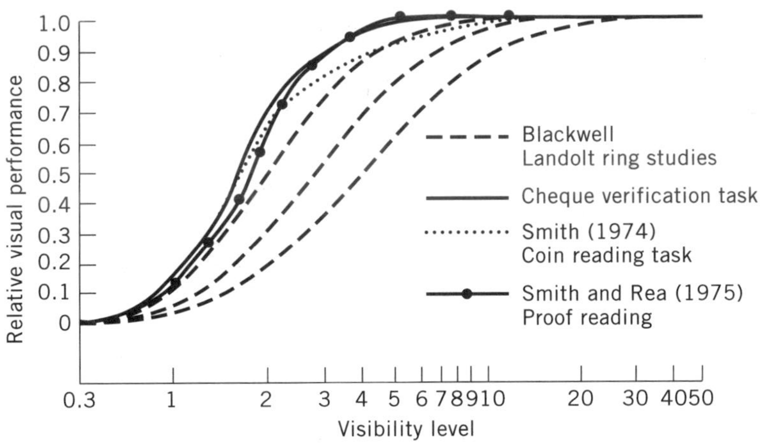

Figure 24.41. Meshing of ogives from industrial and task studies with Blackwell's laboratory Landolt-ring ogives. Coin reading simulates industrial task (53).

point while gathering information; and (*3*) saccades, w_3P_3, the pattern of eye movements caused by the eye muscles responsible to move the eyes to turn to view a task and to sweep from one detail to the next. See equation 2 and Figure 24.42.

Relative Task Performance. Equation 3 shows the fourth process, the noncritical motor response w_4P_4 with its weighting coefficient. This process is composed of two subprocesses (Fig. 24.43). As a hypothetical example of how these processes may work, one may rep-

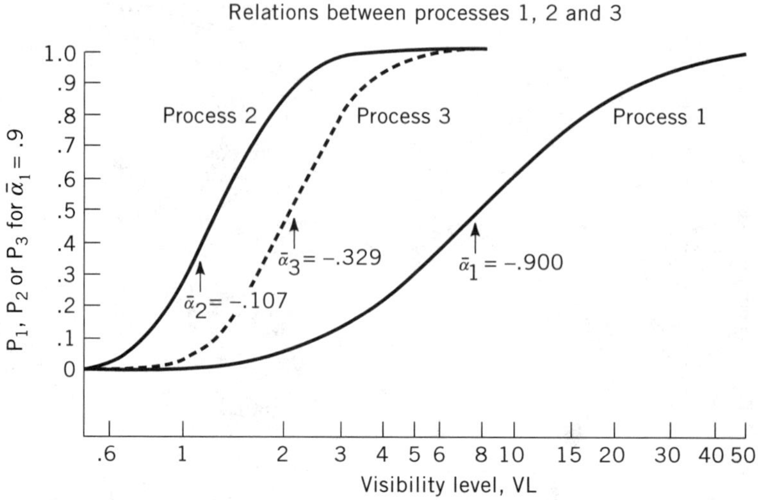

Figure 24.42. The first of three processes is called the sensory process, the seeing task. The next two are fixation maintenance and saccades; these are eye-directing adjustments. Ogives reflect a specific task (53).

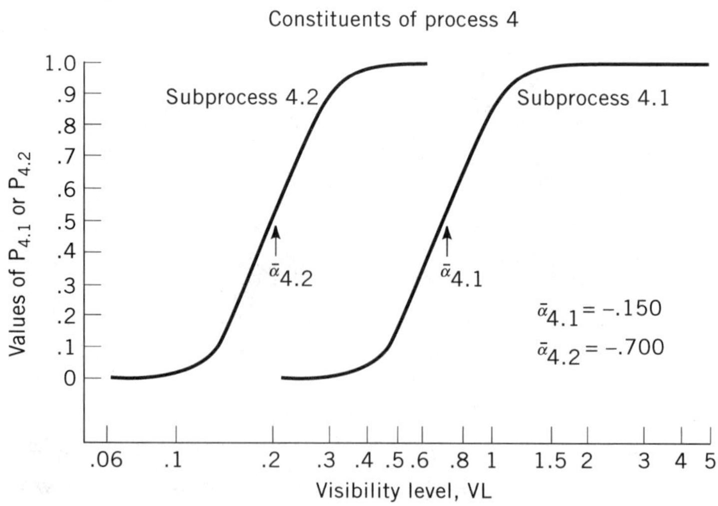

Figure 24.43. Subprocess 4.1 represents guidance to the point of work. Subprocess 4.2 represents execution of the work (53).

resent the guidance of a pencil to accomplish a task, while the other may indicate the cognitive act of accomplishing the task. This process is shown in relation to the others in Figure 24.44. The combination of all evaluates the total human performance. Although the International CIE committee that adopted this performance model placed a strong emphasis on relative visual performance, they also recognized that for any cost-benefit analysis, the relative task performance (RTP) values would have to be considered.

5.4.2 Other Visual Performance Models

Today an international committee is developing the next generation of the CIE 19/2 model, and other models of vision and visibility are being developed at lighting research laboratories (9, 58–70). In recent years several researchers (including Adrian, Indistsky, and Rea) have developed visual performance models to describe the influence of lighting parameters on visibility and hence performance (58–61). Of these, Adrian and Rea's models are the most applicable to interior lighting conditions. Adrian's first report of a model in 1969 has been extended in recent years and is quite complete and advanced. Although it is primarily being applied for roadway lighting purposes, it covers the range of luminances found in interior applications as well. Visibility as a characteristic of roadway lighting is being considered in the development of roadway lighting standards in North America. Of various models tested, Adrian's model most accurately predicts the visibility of small targets (20 cm × 20 cm) on a roadway as measured in the field using visibility (contrast reduction) meters, among other techniques (62–65). The elements considered in the Adrian model include background luminance, veiling luminance (disability glare), object size, observer

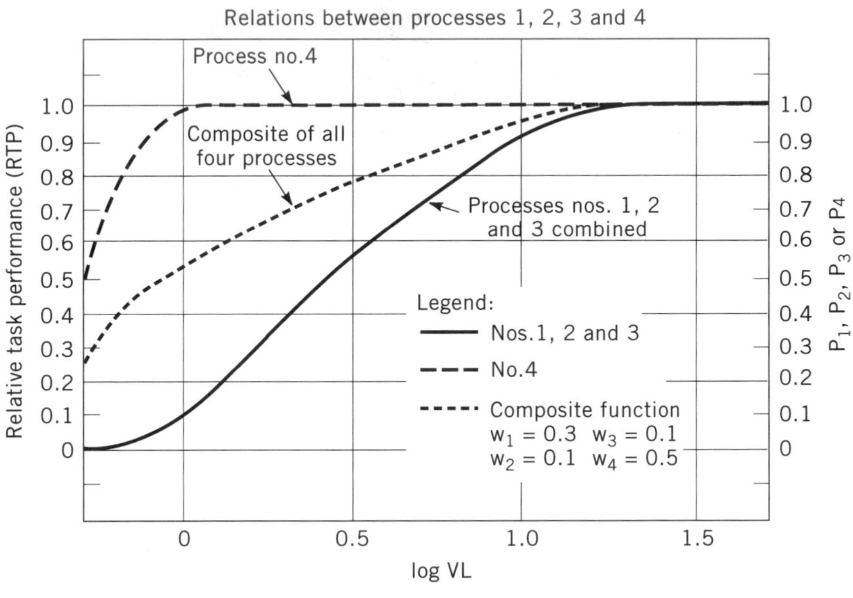

Figure 24.44. Combination of processes 1–4 based on specific task shown in Figure 24.43 (53).

age (334, 335), and viewing time. Rea (60) has developed a performance surface model based on a numerical verification task that basically tests the performance of observers in detecting errors by comparing a standard column of numbers with another column of varying contrast with errors introduced. The results of his tests have been plotted three dimensionally, as shown in Figure 24.45. This plot suggests that there is a wide range of physical conditions that can produce a high level of visual performance (as denoted by the flatter surface); however, once you move from this area, performance falls off rapidly. In comparing the model with empirical studies, Rea found that the work of McNelis could be predicted at the high levels of contrast but not at the lower levels. Further refinements of the model are being incorporated based on a series of reaction time studies in response to a square target of varying size, contrasts, polarity (dark letter on light background or vice versa), and luminances.

5.5 Limitations

Each of the models described above has particular limitations. Our discussion has represented the state of the art according to the best research knowledge available, yet the situation never remains constant because new knowledge is being obtained by continuing research. New improved techniques are being developed. The models described need refinement and validation so that they can be used subject to the knowledge of their limi-

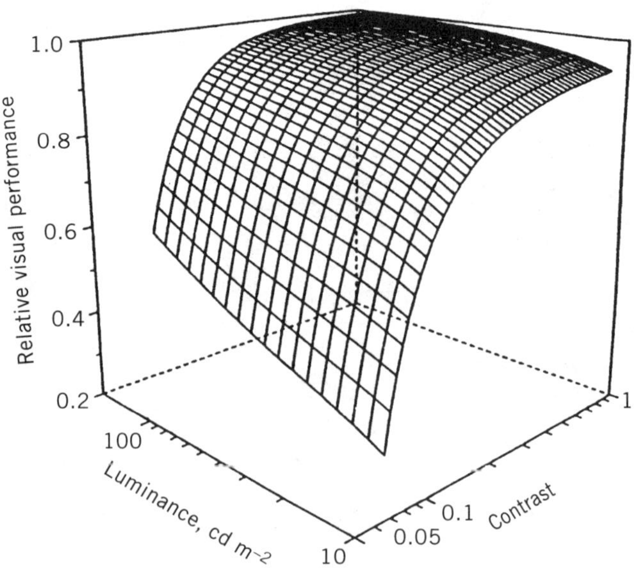

Figure 24.45. Three-dimensional surface of relative visual performance derived from performance study on the numerical verification task. Courtesy of *Lighting Design and Application*, New York, May 1987, p. 10.

tation. Probably the most universal criticism of each is its ability to predict actual performance accurately. We have basically discussed lighting in terms of enhancing the visibility of practical tasks, but it should be understood that some tasks have such poor inherent visibility that no amount of ordinary lighting will give an adequate visibility level. In this case the contrast might be improved by special direction or color of light, to enhance the contrast. If there are indentations or raised characteristics to the detail, light striking these configurations obliquely will cause highlights and shadows that will increase the contrast. Magnification is another method of increasing visibility. Where all of these techniques fail, consideration should be given to altering the task if possible. Crouch and Vincent (73, 74) have found that for some difficult lighting applications (strip mining and underground mining equipment, aircraft carriers, school stairwells, etc.) visibility of objects to be seen, for example, stairs, was improved by using highly reflective materials that produced greater contrast and better visibility.

5.6 Field Application

A number of investigators (73–83) have found that the visibility of unknown details in the field can be assessed by using a contrast threshold meter. Such a meter puts a variable veiling luminance over the detail of regard until its contrast has been reduced to threshold. The amount of veil needed to reduce the task to its threshold denotes its degree of visibility (above threshold), provided the luminance to which the eyes are adapted is maintained constant. This is done by coupling an absorbing filter for the external view with the mechanism for increasing the veil, so that viewed luminance through the instrument stays constant. Furthermore, because the eyes while scanning pause to take a picture at 200 ms, it is important to determine the threshold of visibility on the basis of exposure of the field for 200 ms. Once threshold has been reached for the field detail, then at the same luminance of the field object, the 4′ luminous disk is reduced to threshold. Because all objects at threshold are equal in visibility, there is now an equivalency between the 4′ disk with its resulting contrast and the field detail. Such instruments are now available, as illustrated in Figure 24.46. Attachments to these meters have been successfully used to determine the disability glare factor (DGF) (84, 85). More portable streamlined versions are being developed and used in field measurement. These refined visibility meters use the ambient light available to produce the veiling luminance and adaptation luminance. The goal of taking visibility meter measurements is to obtain an equivalent contrast of the standard 4′ test object designated C-bar. When this is obtained, the value of luminance required can be obtained using calculational methods described in the CIE 19/2 or by referring to the current illuminance selection procedure of the Illuminating Engineering Society of North America (IESNA) (88, 89). New video camera are being adapted to perform digital image photometry. This is another quick technique for describing and recording the luminances of the scene for detailed contrast analysis (86, 87).

5.7 Current Illumination Determination Method

Research continues to further the understanding of the various visual performance and vision models, but only by field testing and evaluation will these models become useful

Figure 24.46. The Visual Task Evaluator designed to use 0.2-s exposures that permit more realistic field measurements.

to the practitioner. Detailed assessments of the various pros and cons of each model were reviewed by the IESNA as it sought to develop a science-based method for determining illuminance necessary for a level of visual performance (9, 66–70). The IESNA roadway lighting committee has taken this approach the furtherest in development of a visibility-based metric of performance (62–65). Only when this approach is tried for interior lighting will a further step in validation be accomplished. Professional standardizing bodies throughout the world have developed experience-driven consensus bases for recommending light levels for various tasks. In the United States, IESNA has developed an illuminance selection procedure in which a range of illuminance required for a task is recommended (67, 86, 88, 89, 90). A designer selects what target level of illuminance is to be provided for the workplace based on the age of the occupants, an estimation of the difficulty of tasks involved, the importance of speed and accuracy, and so on. A combined weighting factor then shows whether the illuminance to be provided is at the low, middle, or high end of the range. Additionally, allowance is made for measurements with a visibility (contrast threshold) meter to determine the difficulty of specific tasks. A detailed description of the IES method is shown in Appendix A along with examples of target illuminance levels for various industrial tasks. It should be borne in mind that the recommended light level is for the task, not necessarily for the whole space. On the other hand, just lighting the overall space does not guarantee that the task will be lighted to the degree recommended. The designer must carefully determine the balance in the overall lighting design program. In order to develop a procedure in the lighting design process where this balance is considered, IESNA is developing a new Lighting Design Guide: IESNA Lighting Handbook, 9th Edition. A preview of a section of this guide is shown in Appendix C for warehouse lighting (89). There are a number of energy-efficient as well as pleasing ways of providing the necessary light level without uniformly lighting the entire space. This is discussed below.

5.8 Industrial Tasks

Although many industrial tasks have remained the same, a revolution is taking place in the area of factory automation. The workforce and production methods are being streamlined. As a result new tasks are being introduced, such as computer-aided design and drafting (CADD) screens for easy reference to working drawings. These computerized tasks must be treated with care from a lighting standpoint in order to eliminate or minimize distracting reflections on the screen (91–96). Figure 24.47 illustrates the problem of how luminous elements (whether the ceiling or luminaires) obscure the details to be seen by producing either a uniform luminance over the detail or an annoying pattern of reflections. Newly developed luminaires have sharp cutoffs to eliminate the majority, if not all, of the light in the offending zone Figure 24.48. Careful guidelines for balancing the lighting needed for paper-based tasks being performed in conjunction with VDT work have been

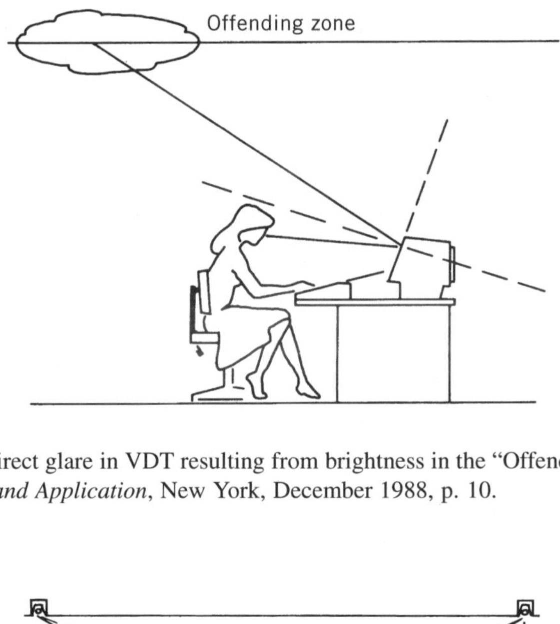

Figure 24.47. Indirect glare in VDT resulting from brightness in the "Offending Zone." Courtesy of *Lighting Design and Application*, New York, December 1988, p. 10.

Figure 24.48. Illustration of luminaire cutoff distribution to eliminate or minimized indirect glare in VDT screen. Courtesy of *Lighting Design and Application*, New York, December 1988, p. 10.

developed (94). A commonsense guideline for VDT use, such as next to windows, would suggest placing the VDT and desk perpendicular to the window. Adrian and Fleming (96) have conducted a current review of vision, VDTs, and lighting interaction. The review covers research on screen polarity and contrast of characters, polarity and adaptation, screen flicker, the effects of color, the effects of VDT use on accommodation and its impacts on fatigue, the effects of rest breaks on fatigue, and lighting systems for VDT use. From a lighting standpoint, further research is needed to better determine the luminance ratios to be used between the VDT task, surrounding tasks, other nearby surrounds, and the luminous elements related to the ceiling. Attempts to understand the impact of lighting on computer screen visibility have resulted in identification of lighting and display parameters that need to be addressed for mutual compatibility. Lloyd et al. summarized these variables and their direction of influence on lighting and displays:

- Luminance of sources of specular reflections: lower is better
- Modulation (contrast) of specular reflections sources: lower is better
- Illumination on displays: lower is better
- Modulation (contrast) of display: higher is better
- Display polarity: light background is better
- Specularity of display: lower is better
- Diffuse reflectance of display: lower is better
- Modulation (contrast) of the reflected image: lower is better

Beyond the specialized VDT task, there are almost innumerable industrial processes and products; it would be difficult to describe all the critical details from start to finished products. A number of details in various industries are described in lighting literature (98–101). The illumination required for various industries is given in the current edition of the *IES Lighting Handbook* (88, 89).

5.8.1 Type of Tasks

Surface characteristics vary from mat to highly glossy finishes and from two-dimensional to three-dimensional shapes. Mat finishes do not present a lighting problem, although if there are three-dimensional objects to be seen, some unidirectional light beamed together with diffused overall lighting will bring out highlights and shadows that identify the three-dimensional form. Some mat surfaces have slight wrinkles or ridges in them that do not show up with general overhead lighting but do appear when a light beam is directed very obliquely across the surface (Fig. 24.49). Some objects have a surface gloss that overlays the grain or textured base material. The glossy reflections of light room surfaces near the line of sight interfere with view of the texture beneath. In this case, light directed to the surface obliquely eliminates the reflections from the line of sight and penetrates the surface to illuminate the texture beneath. The metalworking industry largely involves glossy surfaces that act as mirrors. In this case, the use of large, low-brightness overhead luminaires serves to reflect low-brightness highlights in the surface, against which are seen any markings as interruptions of the highlight (Fig. 24.50).

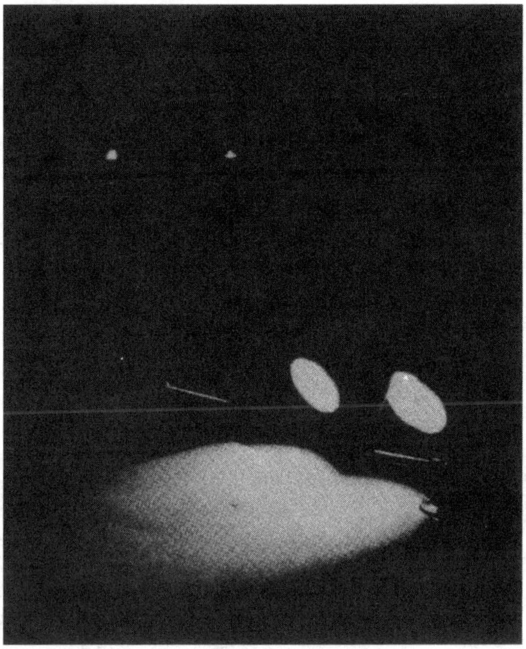

Figure 24.49. Ridges showing up under grazing light.

For the lighting solution of many industrial seeing tasks, one should refer to the *American National Standard Practice for Industrial Lighting* (97).

6 VEILING REFLECTIONS

In offices and similar locations where paper work is done, it has long been recommended that mat paper be used to avoid glossy or mirrorlike reflections that overlay the details to be seen and interfere with visibility. Research since the latter 1950s (102) has revealed that mat paper work continues to exhibit microscopic mirrorlike reflections. Overhead lighting with a heavy downward component of light rays (Fig. 24.51) causes serious loss of visibility because of veiling reflections in the tiny facets of the paper fibers, and the inks or pencil marks used for the printed or handwritten characters reduce the contrast. Such microscopic reflections cause loss of visibility in accordance with Figure 24.52. If one were to work in the shade of a tree (Fig. 24.53[a], 24.53[b]), these glossy reflections would be eliminated and light would come to the work from the sky beyond the tree. Reflections of this light would go away from the eyes of the worker, and the details would be seen ideally. A good compromise has been to reduce the downward components and have the light come from the ceiling and side walls as though they represented the hemisphere of the sky without sun (Fig. 24.53[c]). To account for this loss of visibility in

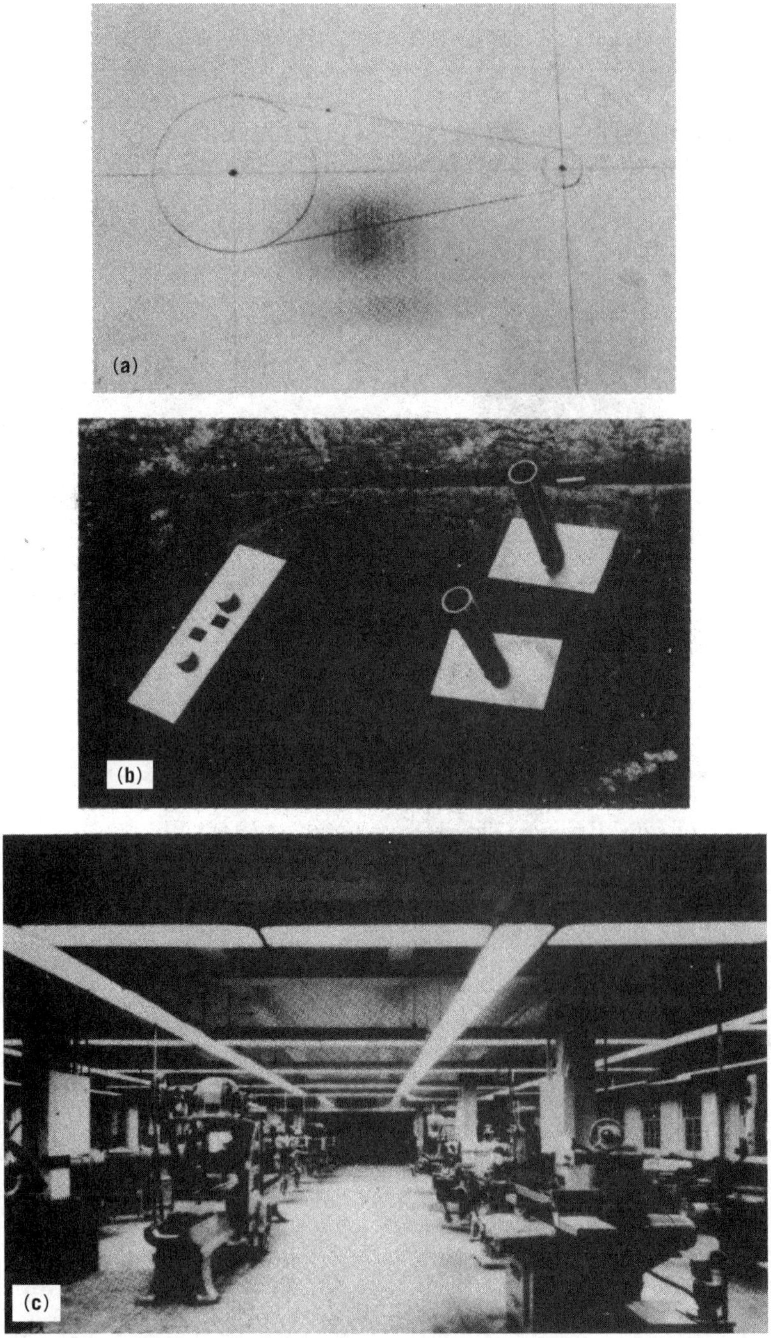

Figure 24.50. (**a**) Appearance of scribed lines on steel surface under large-area, low-brightness lighting equipment. (**b**) Highlight covering metal surfaces reveal contours and details. (**c**) Grid system of fluorescent lighting produces broad highlights on metal surfaces.

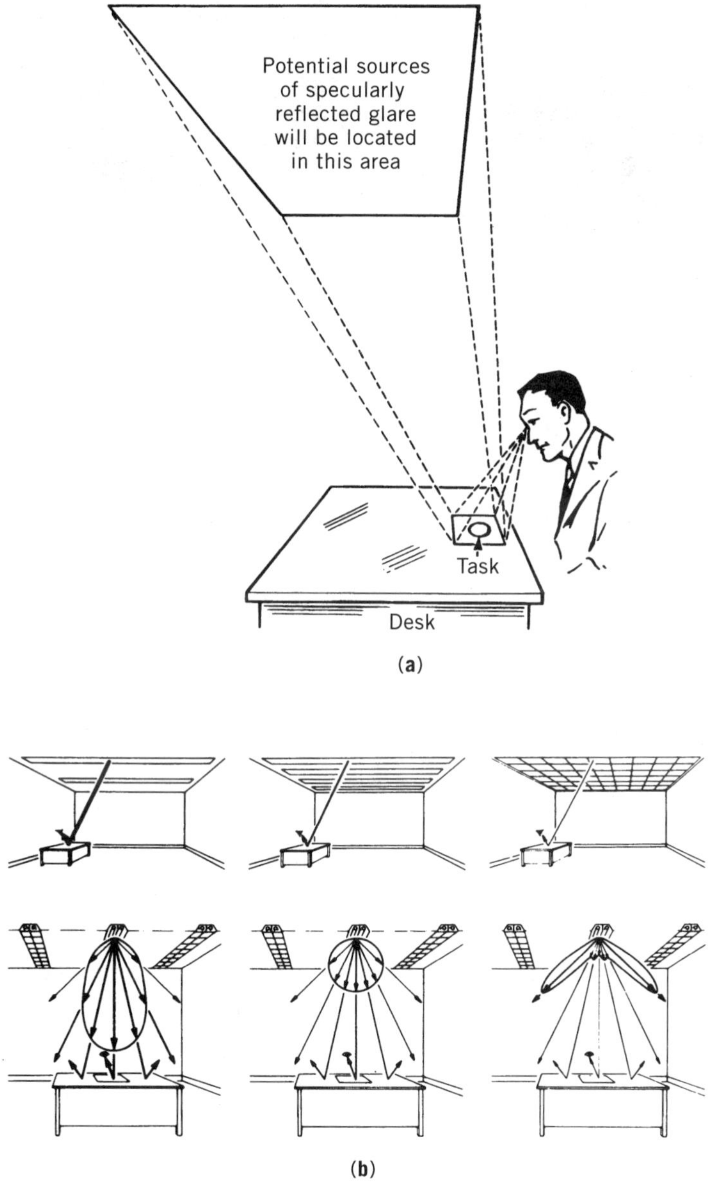

Figure 24.51. (**a**) Downward light from luminaires in this zone will cause veiling reflections and serious loss of visibility of office tasks. (**b**) Different luminaire downward components. The less the downward component in the reflection zone of the desk task, the greater the visibility of the task.

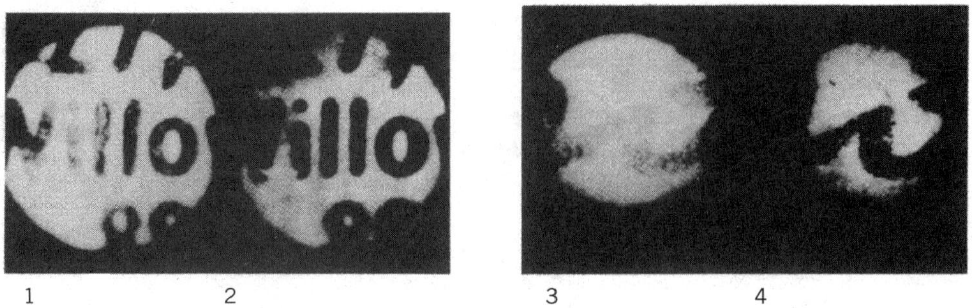

Figure 24.52. Veiling reflections reduce task contrast. (1) Magnified sample of type under a troffer causing veiling reflections. (2) Same sample with veiling reflection shielded out. (3) and (4) Similar situation with letters a, r, and t (parts only shown) handwritten in pencil.

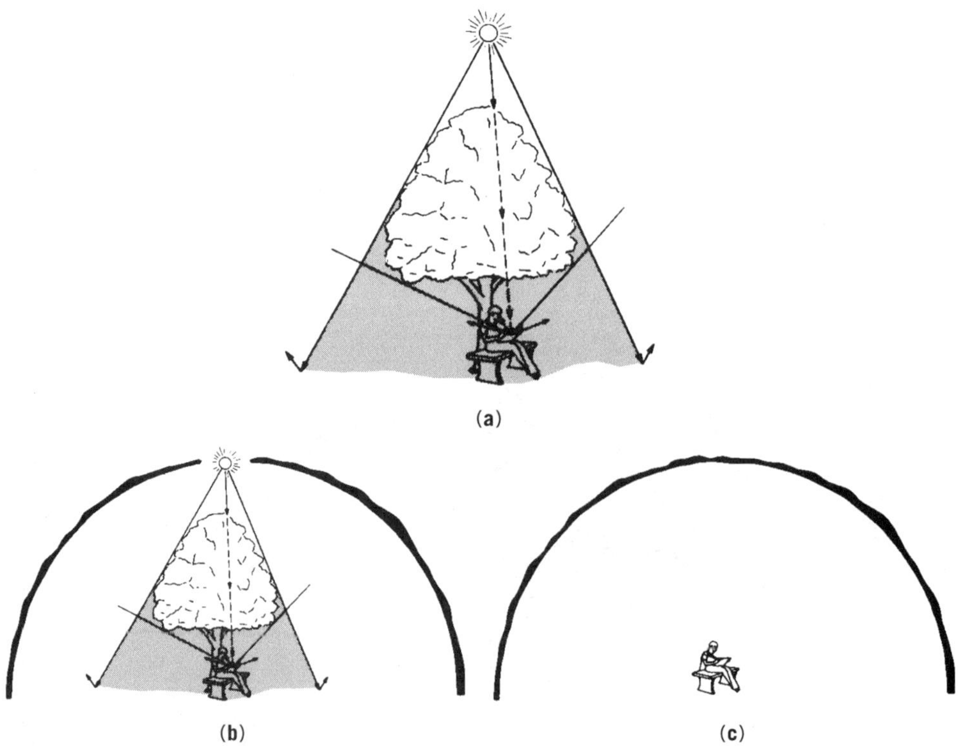

Figure 24.53. (**a**) Tree shields out veiling light. (**b**) Lights to reading task comes from vault of the sky outside veiling zone. (**c**) Vault of sky with no sun produces good results with low component from the veiling zone overcome by majority of light outside the veiling reflection zone.

practice, the concept of contrast rendition factor (CRF), described above under the discussion of the CIE model, was developed. CRF compares the effect on a task's contrast by the distribution of light of a particular lighting system to the contrast of the same task under reference (hemispherical) lighting conditions. The CRF formula is

$$\text{CRF} = \frac{\text{contrast of the task under conditions of interest}}{\text{contrast of task under reference conditions}}$$

where the reference conditions equal complete diffuse, uniform illumination such as occurs in an integrating sphere. The need to consider veiling reflections was shown in a survey of the range of paper-based office tasks (103), which indicated that a large majority were of poor contrast.

A veiling reflection study (104, 105) indicated that commonly used luminaire systems had only 15% to 50% effectiveness in revealing the details of the task. Therefore the lighting system should provide a minimum of downward light immediately over the desks, in comparison with the light coming from wider angles (106, 107), which does not cause glossy reflection toward the eye (Fig. 24.54). Study of the angular reflections of a typical task (pencil handwriting) has resulted in a determination of the effectiveness of light coming toward the task from various angles (108). This effectiveness is represented in Figure 24.55, where the length of the radial lines indicates the relative effectiveness of light approaching the task in revealing the details of the task. This average effectiveness holds

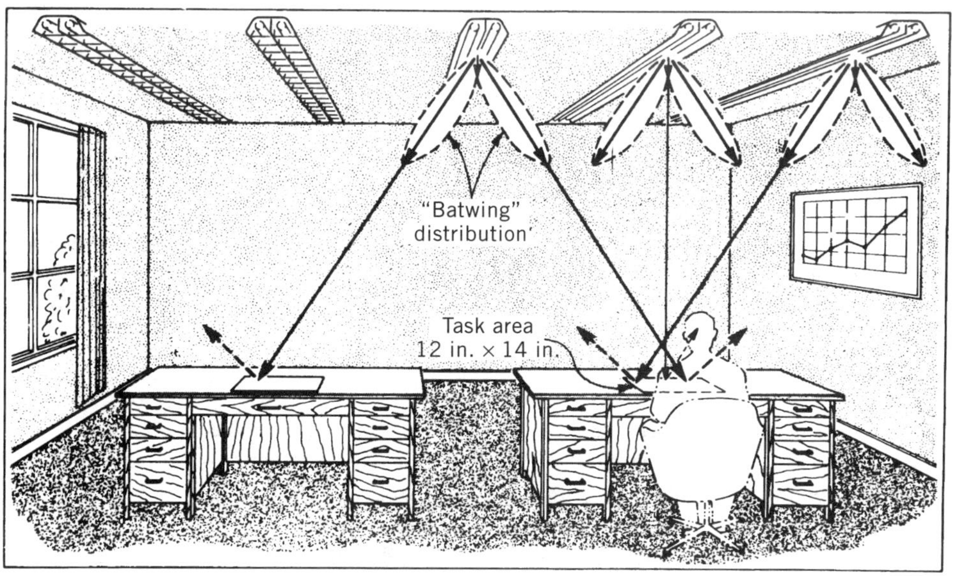

Figure 24.54. Batwing candlepower distribution of light from the luminaire minimizes light in veiling zone, distributes it from other luminaires outside the zone, and reflects it from the walls.

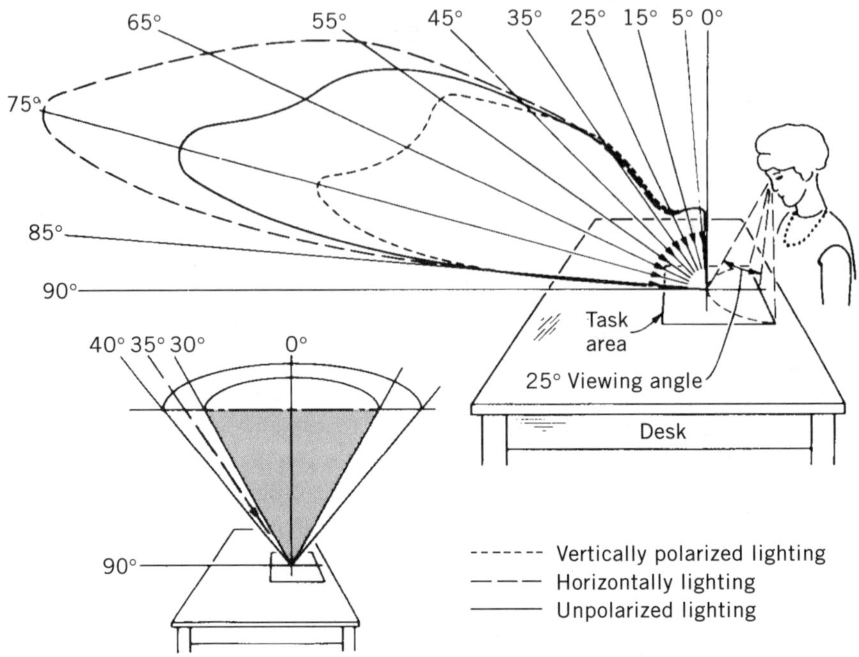

Figure 24.55. Average effectiveness of light approaching the task in different angular cones, of 360° azimuth.

for light approaching from every direction of the compass (i.e., 360° in azimuth including body shadow). It will be noted that the maximum occurs at 75° with the vertical, indicating that the light from walls and windows is very important in rendering more effective visibility of the task. (The reflectance of the walls should not be greater than 60% to avoid glare.) Various measurement techniques are commonly being employed by today's designers to evaluate lighting systems. CRF is being measured using such instruments as the Brüel and Kjaer meter or the Equivalent Sphere Illumination (ESI) meter. These have proved to be helpful evaluation tools. As with all evaluation techniques, there are of course limitations to the CRF method of analysis. Depending on the task, veiling reflections can lead to low or high contrast and thus low or high CRF values. Researchers understand and can control veiling reflections for laboratory experiments; however, in practice, the magnitude of veiling reflections can vary greatly within a space because the incidence of veiling reflections depends on the geometry of the luminaire, the type of task, the observer, and where each of these is located within the space. Because the designer does not always know where people and tasks will be located within a given space, the consequences of veiling reflections on task performance can be readily estimated only for a specific task set at a fixed position and seen from a particular angle (54–57). Even with these limitations taken into consideration, CRF remains the best available measure for the assessment of veiling reflections.

The description of veiling reflections above primarily discusses the effects of lighting distributions on horizontal paper-based tasks. Existing lighting system technology is available that, when properly applied, can adequately handle these veiling reflections. Today however, with the introduction of VDTs into industry and offices, the issue of reducing annoying veiling reflections from the video screen must be addressed. Research is being conducted to determine how best to limit these reflections, as discussed under types of tasks above.

7 EFFECTS OF SURROUNDINGS: TRANSIENT ADAPTATION

The eyes are constantly scanning both the details of the task at hand and the immediate surroundings of the rest of the room. Time-lapse photography of office workers shows that when persons look up from their desk tasks, the peak of the frequency of their angles of viewing is at 90°, or a horizontal line of sight (108). Thus they look around the room and then back at their desk tasks with their immediate surroundings. This involves looking at various surfaces with their different luminances (brightnesses), as in Figure 24.56. These

Figure 24.56. Marked differences in luminance in the field of view cause serious momentary losses of visibility, thereby scanning sensitivity.

sudden changes from one luminance to another cause a serious loss of visual sensitivity or visibility, unless the luminances are kept within a suitable range of each other (109). The additional contrasts necessary to compensate for losses of sensitivity are plotted in Figure 24.57.

The time-lapse studies show that the reaction between the luminance of the background of the details of the task and that of its immediate and more remote surroundings affects the ability to see the task after looking away from it (Fig. 24.58). Furthermore, they demonstrate that the smaller the difference between the brightness of the visual surroundings and that of the visual task, the higher the efficiency of seeing. However, equal brightness in all areas of the room would result in a marked feeling of monotony. Investigation has shown that the immediate surroundings should be less bright than the task, but not less than one-third as bright as the task, and never greater than five times that of the task. These ratios apply to large areas of the surroundings and can be achieved by the use of ordinary lighting equipment and light-colored room surfaces. (Small areas with accent color do not interfere with the transient adaptation.) Every object reflects some portion of the light it receives, and the percentage reflected is known as reflectance. The reflectances of floor and furniture should be in the range of 30% to 50%, the walls 40% to 60%, and the ceiling 70% to 90% (Fig. 24.59). Attempts to quantify the losses of visibility from transitional adaptation are described in the section on the CIE model above and under the discussion of TAF.

8 ELIMINATION OF DISCOMFORT GLARE FROM OVERLY BRIGHT LIGHTING SYSTEMS

For a number of years it had been hypothesized that the physiological basis for the feeling of discomfort is generated in the sphincter muscle controlling the opening of the iris (110).

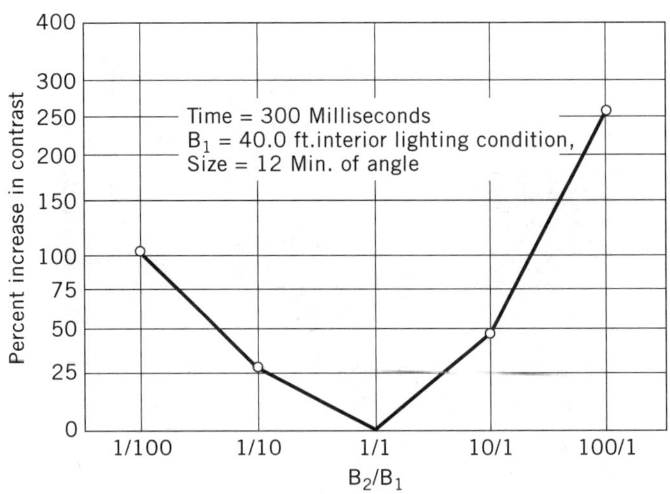

Figure 24.57. Percentage increase in contrast necessary to compensate for loss of visibility due to change from 40 fL to 1/10 or 1/100 or to 10 to 100 times that of luminance.

Figure 24.58. Serious loss of scanning sensitivity occurs when one encounters dark surrounding (or too bright surroundings).

Figure 24.59. Reflectances of room surfaces to favor good scanning sensitivity.

When one looks up from one's work and encounters an overly bright lighting system, the sphincter muscle closes down the iris for protection. If one continues to look near the horizontal, the sphincter muscle first closes down too much, then the system desires more light. It opens up too much, and the luminaire overbrightness causes it to close down again. This alternation continues, and the phenomenon is called *hippus*.

An age effect on the limiting luminance at the borderline of discomfort glare has been determined (112). It appears that on average, the limiting luminance varies inversely with the age of the observer. Discomfort glare sensitivity increases with age. As our age doubles our toleration is halved (112, 113). This may require greater protection for older workers than the current recommendations provide.

Research on the sensitivity of people to peripheral brightness has resulted in a formula for discomfort glare that takes into account the luminances of the lighting units, the position of the lighting units in relation to the line of sight, the additive effect of a number of luminaires in the field of view, and the range of sensitivity of the population (116–118). The result of this research has brought about a standard known as visual comfort probability (VCP) for offices and classrooms. A VCP of 70 means that the visual comfort probability will be such that 70 percent of the occupants at the end or side of the room will be satisfied when the luminances of the luminaires are limited to a given value. From a layman's practical viewpoint, this means that when the lighting system is suddenly exposed to view, 70% of the observers (Fig. 24.60) will obtain no shock of overbrightness. At present VCP is applicable in interiors only for direct lighting systems; research is needed to consider new advances in fixture design. Development of metrics to rate the comfort/discomfort of lighting systems remains a challenge today. Many modern lighting systems are nonuniform in luminance (i.e., low-brightness parabolics) designed for VDT environments. Rating systems are based on uniform luminances and may overestimate the glare rating. Comparison of various discomfort glare metrics has shown that they can rank order lighting systems (119–121). Further research will be required to develop an international standard to consider these modern luminaires (119, 324, 325).

Other applications where discomfort glare is being charted include roadway lighting. A North American approach for rating discomfort glare has been completed and is consistent with similar European work (112). Further research determined the sensitivity to discomfort glare for underground mining populations that showed an increased sensitivity to discomfort glare, having worked underground over a period of time. Formulas to take this into account were developed to assess the comfort of lighting systems to be developed for underground mining equipment (73, 74).

9 DISABILITY GLARE: LOSS OF VISIBILITY

Various researchers (113–115) have found that a high brightness source away from the line of sight has the same effect in reducing visibility as a uniform veil of brightness overlying the details to be seen on the line of sight. This equivalent veiling luminance is expressed as

$$L_V = 10\pi[E1q1 - 2 + E2q2 - 2 + \langle 258\rangle\langle 258\rangle\langle 258\rangle + E_n q_n - 2]$$

where L_v = equivalent veiling luminance (footlamberts)

E = illuminance on a plane through the center of the pupil perpendicular to the line of sight contributed by each glare source (footcandles)

q = angular displacement between the line of sight and each glare source.

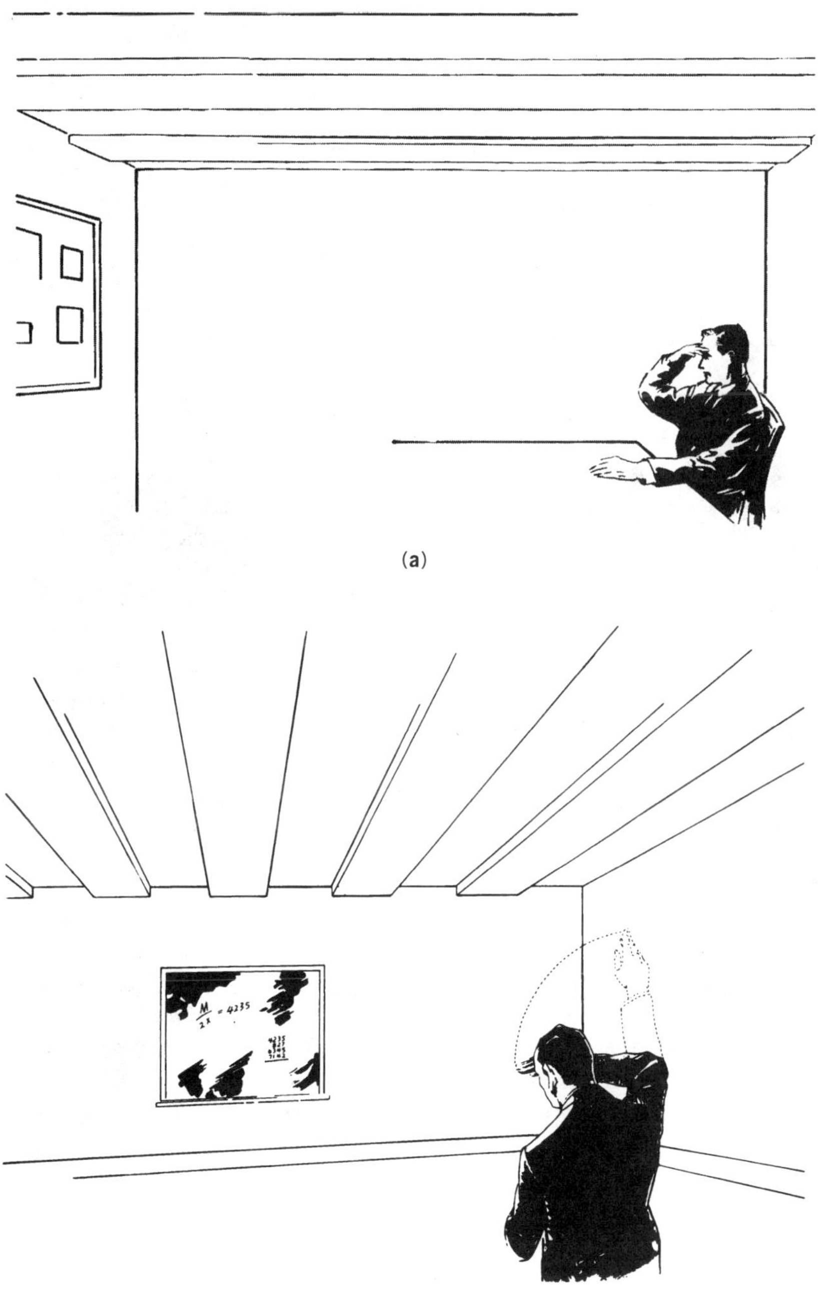

(a)

(b)

Figure 24.60. Practical test for uncomfortable lighting units. (**a**) Stand or sit at one end of the room, look straight ahead. Shade eyes so that no units are seen and hold for a few moments for adaptation. (**b**) Suddenly remove hand; if there is a feeling of overbrightness, the lighting units are too bright for comfortable reading.

If L_v is put in the formula for the resulting contrast of the object to be seen, we arrive at the resulting contrast:

$$C' = \frac{L_b - L_o}{L_b - L_v}$$

where C' = contrast under glare effect
L_b = luminance of the background of detail
L_o = luminance of the detail

The original contrast without glare is expressed as follows:

$$C = \frac{L_b - L_o}{L_b}$$

The effect of surroundings around the immediate background of the details was presented in Figure 24.7, Figure 24.8, and Figure 24.9. When the surroundings become brighter than the background of the task detail, the visibility decreases rather rapidly with this increased luminance. This is a disability glare effect. Even with a uniform and equal surrounding field, there is a 7% loss, but this is better than a dark surrounding field. As an illustration of the disability glare effect even with comfortable surroundings, consider that one cannot see inside a mountain tunnel in bright daylight. As one approaches the mountain with its face illuminated by the sun, the tunnel looks like a dark hole with no details visible. If one could look through a tube at the tunnel opening, the details inside the tunnel could be seen. If one could look out at the sunlit sky from the bottom of a well, one could see the stars still shining. This form of apparently comfortable lighting can be a real danger in dark industrial processes and out on the roadway, where vital cues are missed because of overly bright surroundings.

It has been discovered that age has a significant effect on the veiling luminance, as much as a threefold difference between younger and older observers (123). This has resulted in a factor of age being put in the L_v formula (10, 53–57). Vos (56) presented a formula to estimate the luminance of the scattered light for people of different ages over a wide range of deviations from the line of sight. The CIE model takes account of the disability glare changes on contrast thresholds and the resulting change in visibility. A thorough example of the modifications to present disability glare formulation is that by Adrian (62, 63):

$$Lseq = k \sum_{i=1}^{n} \frac{E_{GLi}}{\theta_i^2} \ (cd/m^2)$$

where the illuminance in lux at the eye from the glare source i
q^2 = the glare angle between the center of the glare source and the target and valid over the range $1.5° < O < 30°$
k = the age-dependent constant (for 20–30 years of age $k = 9.2°$).

For age modifications:

$$k = (0.0752 \text{ Age} - 1.883)2 + 9.2$$

Currently, a modified Fry glare lens that has been used with the Pritchard photometer is being developed for field assessment of disability glare. This can be used as another means to assess the disability glare beyond the factor DGF described in the CIE model and assessed using a modified visual task evaluator (VTE).

10 AGE

Age is a significant factor in the ability to carry out visually guided activity. This is extremely important with regard to safety and productivity. Life span has increased markedly, as Figure 24.61 (326, 327) indicates. As a result of a better quality of life in many nations, people are not only living longer but are also leading active lives both at work and leisure. In the United States alone it was estimated in 1988 that there were 25 million Americans 65 or older, and in the next 25 years this figure will double to 50 million. The effect of aging on requirements for light and vision has been addressed in a number of reviews (18, 126, 127).

Many physiological changes related to the aging eye are fairly well known. In the following discussion of the changes related to the aging process, it would be helpful to refer to Figure 24.62. In considering the effects of age, two aspects of the function of the eye should be borne in mind, its optical function and its physiological makeup. First, the eye functions as an optical receiver, transmitter, and absorber of electromagnetic radiation (the visible portion known as light). As such, light that is processed by the eye obeys basic laws of optics until the optical energy that reaches the retina is transformed into electro-

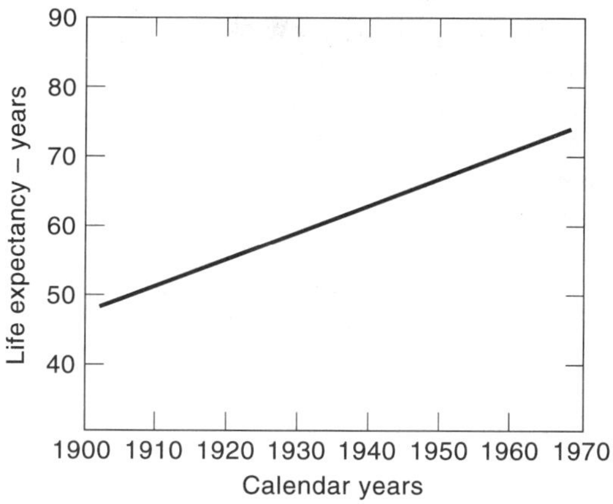

Figure 24.61. Life expectancy since 1900 (327).

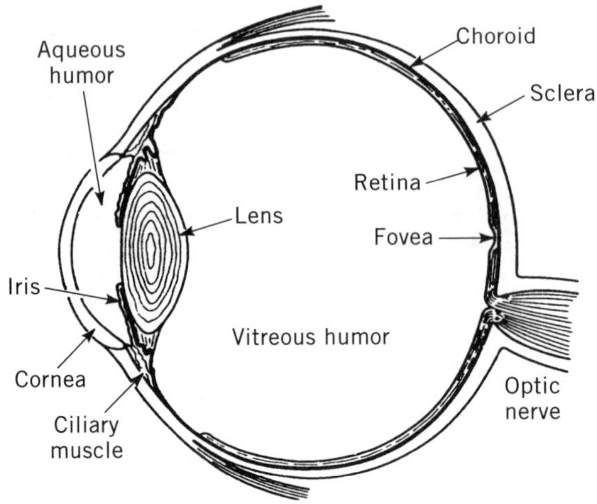

Figure 24.62. Horizontal cross-section of the human eye. Approximate length from cornea to cone layer of retina is 24 mm. Thickness of choroid is about 0.05 mm and the sclera 1.0 mm. Courtesy of *IES Lighting Handbook, Reference Volume*, 1984 (84).

chemical signals. These signals are then processed by the brain for vision and setting of biological rhythms. The second aspect to consider is the makeup of the eye's physical structure. The eye is composed of various living tissues and a fluidlike substance (vitreous humor). These tissues and fluids respond in different ways over time. Some tissues continue to repair or replace themselves over time, but others do not. For those that do not regenerate themselves, understanding the cumulative effects of the absorption of optical energy is extremely important. Understanding the eyes' resilience over time is the key to setting safety standards for living with natural and man-made light sources. Excellent coverage of this topic has been completed by Sliney and Wolbarsht (124, 125). Furthermore, industry standards on safety limits for optical sources have been adopted (128–136).

The quantity of light actually transmitted to the retina is greatly reduced with age (330). As optical radiation moves toward the eye, the first structure it may encounter is the eyelid. During the life of an individual eyelids may become baggy or droopy, occluding some of the light. The cornea—the outer shield of the eye—is initially very transparent. Over its life the cornea gradually builds up fatty deposits that scatter available light before it enters the next structure, the pupil. It has been found that the eyes deteriorate considerably in their ability to adjust the pupil opening in proportion to the light available. Figure 24.63 shows that this eye gate becomes smaller and smaller, even in the dark. The mean difference in pupil size between ages 20 to 29 and 80 to 89 is 2.6 mm in the dark, and 1.7 mm at a level of 1 fL. The reduction in pupillary area with increasing age is equivalent to approximately a 0.3 log unit reduction in luminance. This decrease in the amount of light reaching the retina for image-forming purposes is one of the underlying causes of loss in visual sensitivity with age (137). Critical details to be seen in commerce and industry

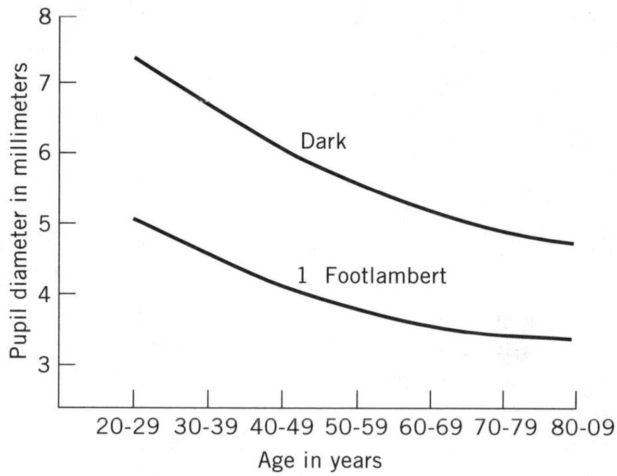

Figure 24.63. The change in pupil diameter with age as measured in the dark and at 1 fL for 222 observers (137).

involve small size and, in many cases, poor contrast. Figure 24.64 demonstrates that the ability to see fine detail is seriously reduced with increasing age. Another closely related factor is the speed in the ability to see small objects, to be able to maintain safety and accurate production. Weston (138) found that this performance was radically reduced with

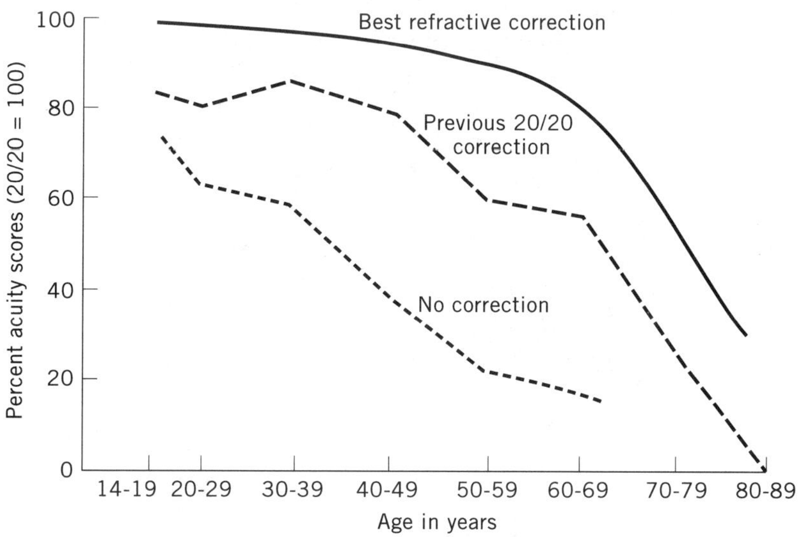

Figure 24.64. The reduction in visual acuity with age (331).

age (Fig. 24.65). One benefit of aging may be the better depth of field experienced from the smaller opening. The lens is the next structure in the path of light to the retina. At a very young age some portion of ultraviolet radiation can enter the eye. This early transmittance of ultraviolet is quickly diminished to a point of total absorption. Over its life the lens of the eye pays a price for guarding the interior of the eye by absorbing ultraviolet radiation. Gradually its transparency is reduced as it becomes more yellow. In some cases opaque cataracts are formed. At later stages it hardens and by age 40 starts to lose its ability to focus different distances flexibly. Finally the lens becomes a source of glare by scattering light in the interior of the eye. The yellowed lens acts as a filter, absorbing more blue and violet. This changes color perception, with white looking more yellow and the fine discrimination of a difference between blues and greens diminished.

What other changes result from aging? A most vital factor in maintaining safety is the ability to see movement of objects representing a potential hazard out of the corner of the eye. Wolf (139) has found that the ability to see flicker (movement) 40° to 80° out from the line of sight is reduced as much as 60% (Fig. 24.66).

A young person sees very keenly in the presence of very little light, but this sensitivity to low brightness is reduced approximately one-half at 80 to 85 years of age, as shown by McFarland et al. (Fig. 24.67). Furthermore, the same investigators found (Fig. 24.68) that it takes 30 min for the 85-year-old to adapt to lower outdoor night brightness. The shape of the curve shows that the older person would be partially blind for a while after having gone outdoors at night.

Most startling of all is the loss of ability to see in the presence of glare sources. Unshielded or poorly shielded light sources with high candlepower directed toward the eyes causes severe losses of visibility, especially to older people (Fig. 24.69); yet inadequately shielded light sources are found in many industrial plants. This glare condition is found especially in night driving and in some sport arenas.

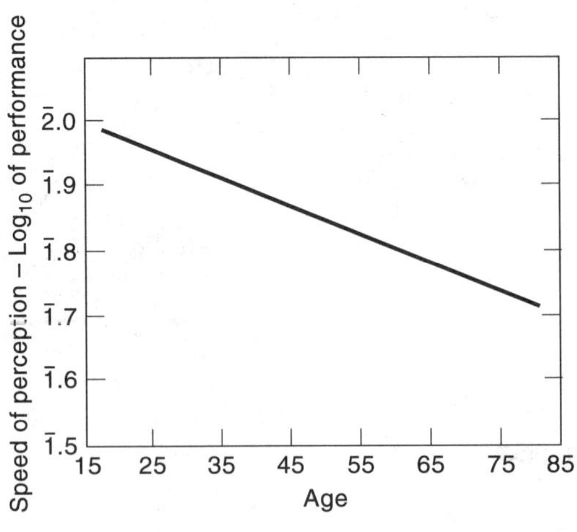

Figure 24.65. Speed of perception is reduced as the age of the eye increases.

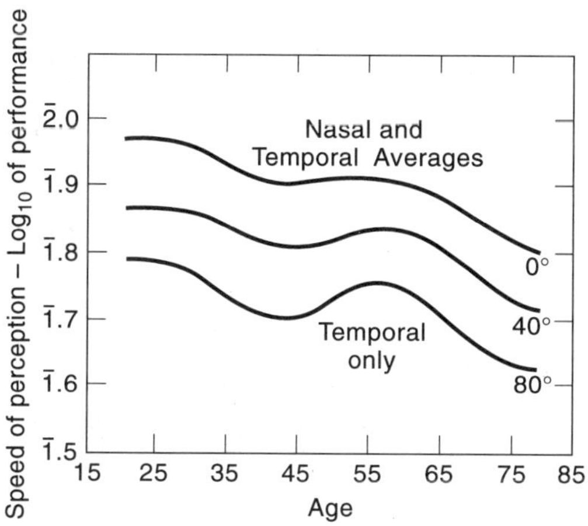

Figure 24.66. The ability to perceive movement at various degrees off center of view exhibits an undulating decline as age increases.

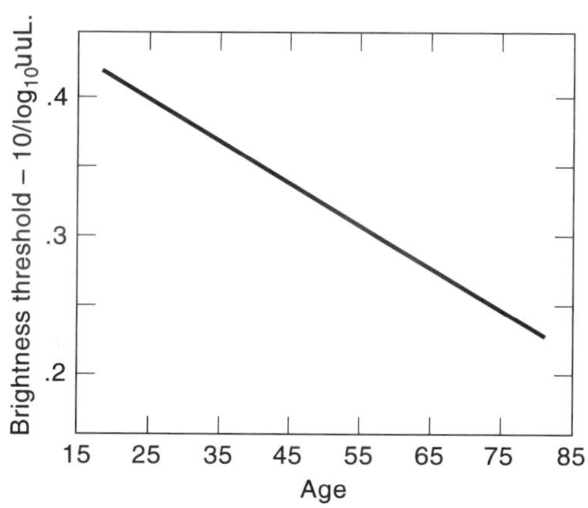

Figure 24.67. In moving from a bright to a dark area, the ability to see is sharply reduced with age (332).

The increased need for more illumination for the older workers has been shown (Figs. 24.27 and 24.29). Studies (140) have indicated that the working curve of contrast versus luminance (Fig. 24.32) is displaced upward for every 10-year interval of age. The displacement is slight for the years 30 to 40, but it is increasingly larger with every succeeding decade of age (Fig. 24.70). These displacements are in terms of static viewing (looking fixedly at detail to be seen), but studies are underway for determining the displacements for dynamic viewing such as the scanning of detail being carried out in commerce and

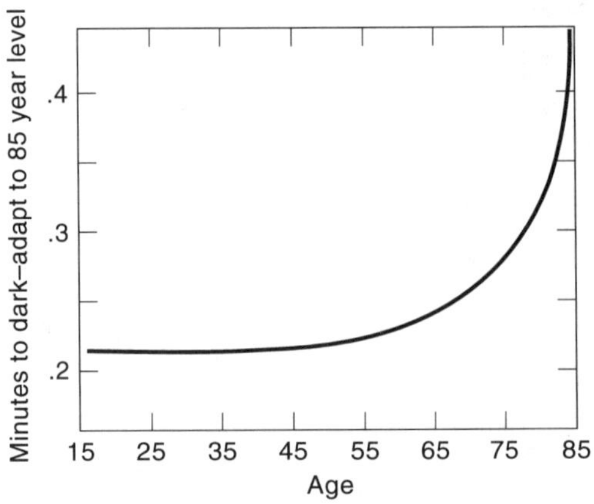

Figure 24.68. Eyes of "young normals" need time to adjust to brightness changes. Eyes aged 20 through 64 years gradually need more time; eyes older than 65 years need a sharp increase in time (332).

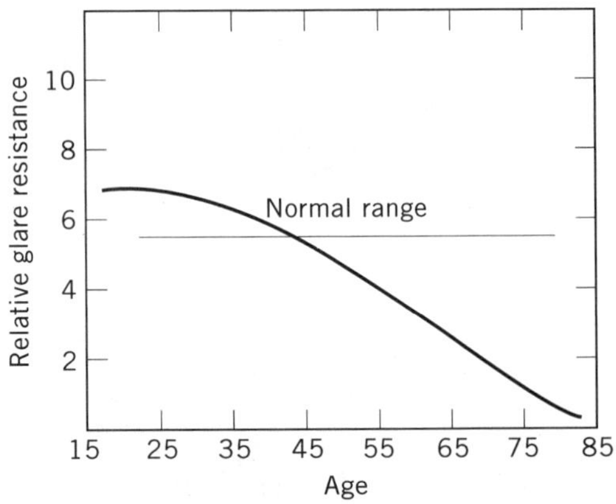

Figure 24.69. The older eye is much less able to resist glare that "washes out" the visual image (333).

industry. If the static viewing is taken as a possible criterion, the additional illumination needed would be as shown in Figure 24.71.

Although the eye decreases in its effectiveness with age, there are a growing number of specialists and specific solutions that can help older people remain productive (18, 126, 127, 141–149, 328, 329, 334, 335). In summary, when attempting to provide a comfortable and productive environment for an aging population, consideration should be given to the following:

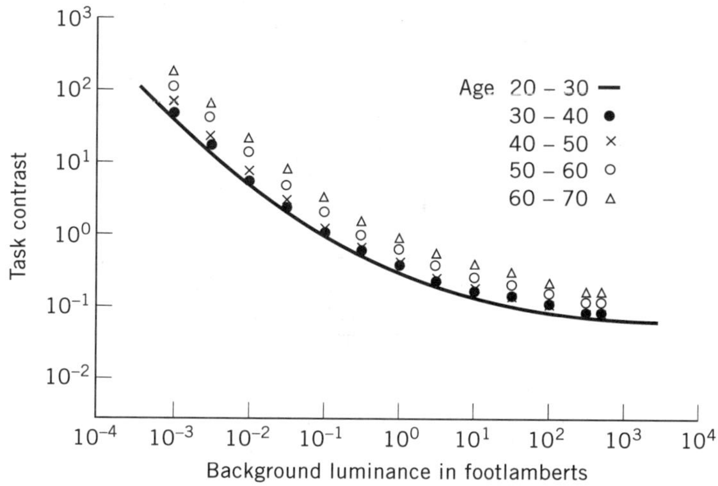

Figure 24.70. Effect of age on required task contrast for task performance.

1. Light and glare can be a problem for older workers. Care must be taken to control light from windows and spotlights. A technique of backlighting of objects may help to reduce glare and focusing problems.
2. Older people have difficulty seeing at low, nighttime levels of light, reading small print, distinguishing similar colors, and coping with glare from VDTs. Consideration should be given to providing stronger, adjustable, task lighting. Other considerations include increasing contrast on stairs and circulation paths, repositioning desks to reduce glare on work and VDTs, and using light sources that would enhance color discrimination lost through yellowed lens.

11 MOTIVATIONAL FACTORS

Since ancient times, architects, designers, and artists have been using light as an expression of grace, beauty, and art. First, because of the lack of suitable man-made light sources, daylight, and particularly the sun, was used to create highlights and shadows and to focus attention. Now a whole artistic palette of electric light sources is available to give highlights, shadow, diffusion, and color in innumerable patterns, depending on the skill and innovation of the designer. Significant effort is being mounted within the Illuminating Engineering Society of North America (IESNA) and the International Association of Lighting Designers (IALD) to develop metrics of quality to quantify what is meant by lighting quality (150–158). People are impressed with and artistically motivated by the beauty of the natural scenes about them, many scenes calling for activity, relaxation, or meditation. In the same way, the moods and behavior of people can be affected by the patterns of electric lighting combined with color and decor of reflecting surfaces. This concept opens

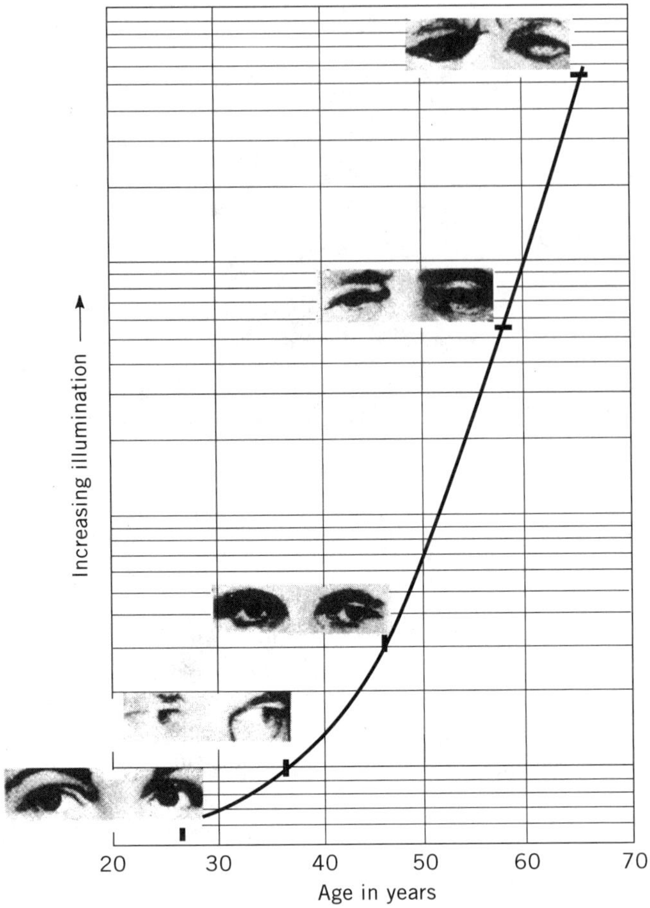

Figure 24.71. For fixed viewing, older eyes need much more light.

up a whole vista of possibilities that seem to defy systematization short of the individual ingenuity of the designer. Some, however, have been inspired to make the attempt. Some of the basic elements of good lighting design (150–156) whether applied by the ancient designer or today's designer would include the following:

- Source and which source is best: daylight, an electric light (and which one), or a combination?
- All space has a clear hierarchy of importance and purpose. How should the lighting be planned to reinforce this hierarchy? The lighting design should

  Reinforce spatial clarity

  Promote productivity

Meet the client's budget
Fully utilize daylight if available
Be readily maintained
Be energy efficient
Have considered all state-of-the-art alternatives

11.1 Color Appearance and Color Preference

Color perception is influenced by direct and indirect aspects of light. Direct aspects include the physics of light as it is absorbed and reflected by surfaces, and the physiology of the eyes' response to the reflected light rays. The physics of light in color perception is discussed in terms of hue, chroma, and value. The physiological aspects are discussed in terms of age, color constancy, and color response. Indirect aspects of color perception are studied in terms of psychological interpretations of these light rays. Subjective rating techniques are used to study the indirect aspects, including color memory, constancy, adaptation, and individual expectations of color.

Lighting plays two major roles in the perception of color. It provides the adaptation luminance needed to trigger the eye's photoreceptors and the spectral composition to trigger color vision. The range of luminances over which the eye operates is shown in Figure 24.72. The rod photoreceptors are active at luminances below about 0.034 cd/m^2 (scotopic

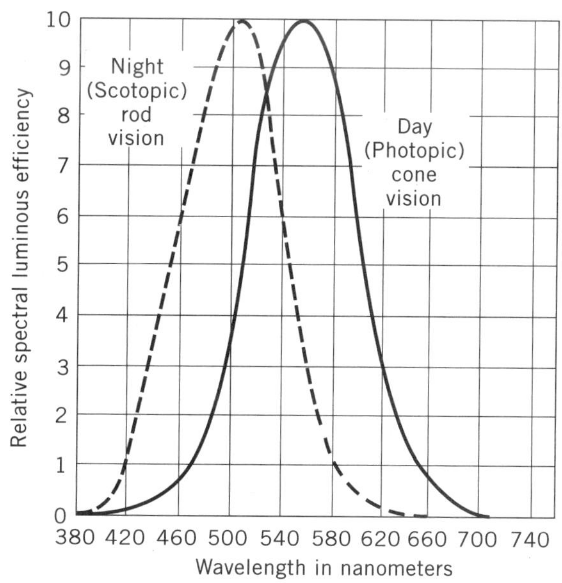

Figure 24.72. The relative spectral sensitivity for photopic (cone) and scotopic (rod) vision. Courtesy of *IES Lighting Handbook 1984 Reference Volume* (84).

vision), and no color vision is possible. The cone photoreceptors, which regulate color vision, are all fully operating at about 3 cd/m^2 and greater. The other factor affecting color perception is the spectral composition of light reaching the cone photoreceptors. Three types of cone photoreceptors are known today, each with different spectral sensitivities. Because different light sources emit different wavelengths, they produce different signals to the photoreceptors. When the output to the three photoreceptors is the same, colors are indistinguishable from each other. Two colors that are indistinguishable from each other may have radically different spectral compositions. Any two colors that have different spectral compositions but are indistinguishable under a given light source are called metamers. Although metamers appear alike under one light source, they may be easily distinguishable under another. For example, in the purchase of pieces of clothing, under the particular store light source there appears to be a color match among the garments. Yet under daylight the colors of these same garments appear radically different. Other factors that affect color perception are the size of the visual field occupied by the color and the context in which the colors are seen. Additionally, a significant minority of the population (8% to 9%) has defective color vision. Factors of aging that influence color perception were covered above.

Research by Berman et al. (160) has shown that the spectral composition of the light can affect performance of tasks that do not involve color judgment. They found that the pupil contracts to different sizes when different light sources (high-pressure sodium and incandescent light) were seen over a large field and balanced for the same photopic luminance. When a large field of view is used, the spectral sensitivity of the mechanism controlling pupil size is dominated by rod photoreceptors. This finding might affect task performance because pupil size is related to visual acuity, retinal illumination, depth of field, and contrast sensitivity.

The resultant changes in task performance due to different light sources are somewhat mixed. Although Piper showed a slight reduction in performance of an achromatic reading task requiring changes in accommodation when lighted by either high-pressure sodium or fluorescent sources, Smith and Rea, on the other hand, found that light sources commonly used in practice have no affect on performance of an achromatic task (161, 162).

The Inter-Society Color Council has been a great aid in isolating and attacking color problems. Its leaders have been of great help to the lighting research community in planning and executing researches in the application of color to the lighting field. As a result, Helson and Judd made studies under average daylight and under various electric lighting sources (163). From this study and previous knowledge, an international system of indexes for rating color rendering of various light sources was developed (164, 165).

Helson and Lansford (166) studied the preferences of a sample of population for various color combinations, using a psychological scale that rated (on a scale of 1 to 10) viewing samples of color space (125 Munsell color cards) against large Munsell-rated backgrounds (25 widely different chromatic samples). The researchers were able to make 156,000 judgments under five different current color emission light sources (average daylight to incandescent to three fluorescent lights having different color temperatures). Rating above 5 indicate increasingly preferred color combinations, and those below 5 increasing lack of preference (166). As a result of his studies and those of others, Judd developed a “flattery index" (167) for light sources to compliment the appearance of human complexions and interior color schemes. This concept has been further developed by Jerome (168).

Today further work is being conducted to augment the early pioneer work of Helson and Judd. Modern researchers in lighting for color judgment work consider the type of light source and the level of illuminance to be the most important aspects. Because most industrial and commercial interiors fall within the illuminance range of 300 to 1000 lx, the choice of light source is the more important of the two.

Practical ramifications of lighting and color perception were considered by Collins (169). She has studied the ability of observers to determine safety colors under seven varying light sources. It was found that the perception of many standardized safety colors was distorted by the spectral composition of the light source used. Collins has suggested alternate safety colors that proved more stable in consistent recognition under a variety of sources, including high-intensity discharge (HID) sources. This emphasizes that choice of light source is critical for color identification.

11.2 Helpful Color Tools

Standardized nomenclatures for color measurement techniques, color specification systems, color temperature specification, colorimetry of light sources, color rendering, and color application have been developed to promote the effective use of color data. These tools are described in the *IES Design Guide and Lighting Handbook* (170, 171). An American Standard System of Specifying Color (D1535) designates the Munsell system as a standard of specifying color in terms of hue, value, and chroma. Collections of carefully standardized color chips in mat or glossy surface are available. A very handy tool for the field is the Munsell Value Scales for Judging Reflectance, which contains a good sampling of color chips from the color space (various hues, value, and chroma), permitting rough judgments of the color specification of unknown color in the field and its reflectance.

A handy guide for the application of color in the lighting field, produced by the Color Committee of the Illuminating Engineering Society, is entitled *Color and the Use of Color by the Illuminating Engineer* (172).

11.3 Subjective Feeling of Interiors

Every interior has a feeling about it, an impression, an impact that each occupant consciously or unconsciously senses. Naturally the senses record this impression, and the visual system shares a major responsibility. Architects and designers are constantly striving to create a stimulating and pleasurable experience, hoping for a reaction that will result in a behavioral pattern suitable to the function of the room. Psychological scaling has long been a method of rating people's feelings about objects or interiors, and a series of reactions is ranked on a numerical scale such as 1 to 10, or 1 to 100. This scaling procedure has now been developed into two approaches: semantic differential scaling (SDS) and multidimensional scaling (MDS). In semantic differential scaling, a series of adjectives, both positive and negative, are used to describe the subjective characteristics of an interior. Such an approach employs a rating sheet with a series of positive adjectives on the left side and the opposing negative adjectives on the right side, with a scale of 6 to 10 divisions in between. The observer, after becoming acclimated to the interior, marks the scale to indicate his or her reaction to each pair of opposite adjectives. For instance, is the interior "neutral," "positive," or "negative"? If it is negative, how negative? By marking the scales in answer to these questions, the observer has recorded his or her impression of the feeling of the room. These answers are analyzed by computer, and the important characteristics are determined out. These now appear to be impressions of (*1*) visual clarity, (*2*) spaciousness, (*3*) complexity (or liveliness), (*4*) personal prominence or anonymity (public vs. private), (*5*) relaxation or tension, and (*6*) pleasantness.

Multidimensional scaling involves trying to determine the psychological dimensions that underlie the judgment of differences or similarities between the impression of different lighted interiors. In this approach one takes a given lighted room or lighted arrangement as representing a base figure such as 10 to 100. Then all other rooms or other lighting arrangements in a given room are rated as bearing a higher or lower value, and computer analysis is set up to determine the two, three, or four psychological dimensions that governed the judgment of the higher or lower values.

Out of such studies (173–179) have come the following three dimensions:

1. The overhead-peripheral mode, which emphasizes vertical surfaces, unlike conventional overhead lighting, which emphasizes horizontal surfaces
2. The uniform-nonuniform mode (sometimes specular-nonspecular) affecting the articulation or modeling of forms or objects in the room
3. The bright-dim mode, affecting the perceived intensity of light on the horizontal activity plane

These aspects are represented in a three-dimensional plot as shown in Figure 24.73(a)–(c).

The color tone of light tends to warp recognition of these dimensions. Light color might also be a fourth multidimensional scaling element if properly manipulated during the experiment. Glare appears to be another modifier of subjective ratings of light settings, but the evidence is ambiguous, and investigation continues.

Both SDS and MDS provide reasonable consistency between spaces and between groups of subjects of similar backgrounds. Putting both methods together provides insight into the ways in which light settings affect judgment.

Beyond subjective ratings of impressions of spaces due to the lighting distribution, behavior appears to be an influence as well. Studies have shown that choice of passageway, selection, and interactions between people all have been influenced by lighting patterns (54–57, 151–156, 159, 180–188).

12 PHYSIOLOGICAL FACTORS

Physiological is defined as characteristic of or appropriate to an organism's healthy or normal functioning. Therefore fatigue and health effects on organic responses, not part of, but related to the visual process, come under this classification. Research reviews (3, 12,

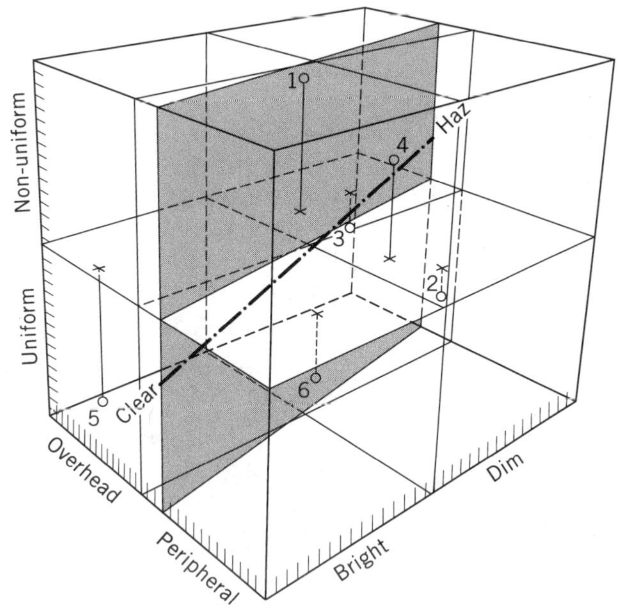

(a)

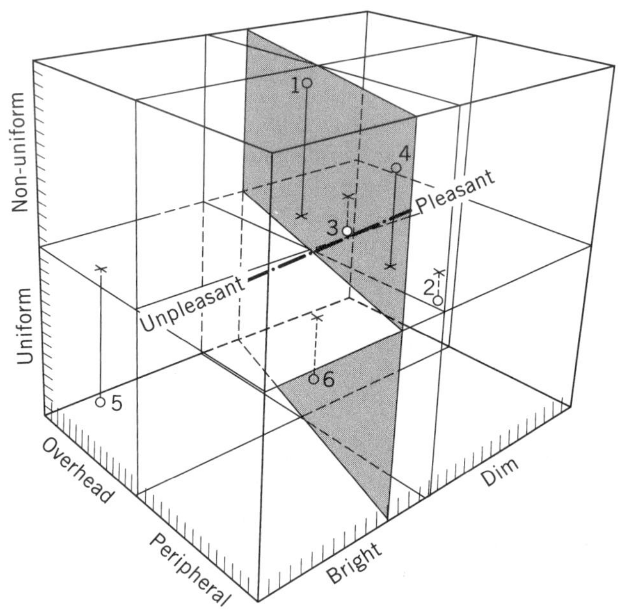

(b)

Figure 24.73. Three-dimensional presentation to indicate lighting design decisions affecting (**a**) perceptual clarity, (**b**) pleasantness, and (**c**) relaxation (and tension) (IERI Project 92).

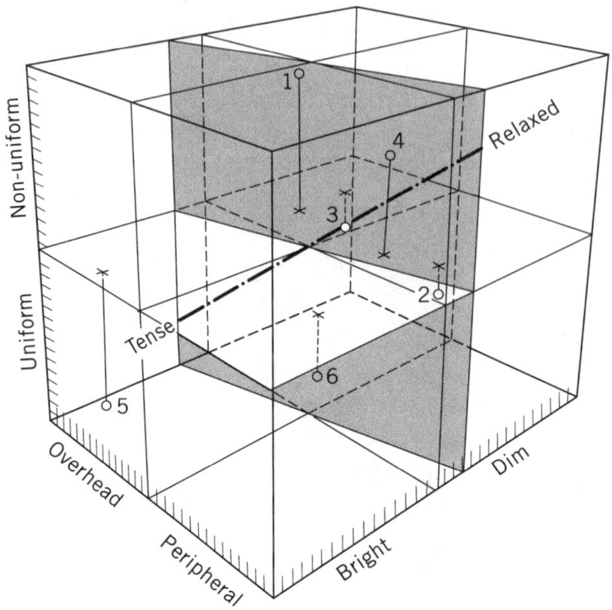

(c)

Figure 24.73. Continued.

16, 189) have broadened the definition of fatigue to include what is necessary to sustain performance. So far, because of the broad adaptability of human beings, no direct correlation between lighting and sustained performance has been determined; however, visual performance measures have behaved in predicted order (189).

12.1 Fatigue and Eyestrain

In early studies (1900–1920) of fatigue (190) dealt with low levels (less than 5 fc) of illumination. Between 1913 and 1924 Ferree and Rand (191–198) made studies of the degree of fatigue from different distributions of light (indirect versus direct), different spectral emission light sources, and different-colored paper backgrounds. During the same early period there was much discussion of eyestrain, which appeared to be related to the low levels and poor quality of electric lighting. Weston (199), conducting a series of studies in industrial plants of different types for the British Industrial Fatigue Research Board, and later for the Industrial Health Research Board, became convinced that fatigue and eyestrain were due to poor lighting, and upon review, Sir Duke Elder, a famous British ophthalmologist, described strain from visual factors, strain from oculomotor factors, and "posteriol" eyestrain. Lancaster (200) described eyestrain from faulty illumination as ocular discomfort with associated symptoms. Ocular discomfort shows up as "sandiness" and tired, aching feelings of the eyeball, orbit, or head. Associated symptoms appear as headache and feelings of fatigue. If eyestrain is bad enough, it may cause indirect symptoms of vertigo and digestive and psychic reactions. However, both Weston and Lancaster addressed levels comparatively lower than those being used in the visual performance studies conducted by Blackwell, Smith, and Rea. As a result there appears to be little or no recent discussion of eyestrain. On the contrary, some ophthalmologists are saying that there is no such thing as eyestrain, which was medically called asthenopia by MacKenzie in 1843 and described in various ophthalmological textbooks throughout the years.

One of the forms of eyestrain mentioned by Weston is eyestrain from oculomotor factors. Hebbard (201) had a similar theory that there might be more oculomotor adjustment activity under lower levels of illumination than under higher levels. He discovered that the amplitude of the microsaccades was significantly greater at 1 fL luminance of the background than at 306 fL (30%). Furthermore he found, under sudden imposition of glare conditions (changing the surroundings from 0 to 306 fL), that the frequency of the microsaccades increased significantly (200%). This would indicate that the oculomotor adjustment system has to work much harder under adverse lighting conditions; therefore, over a working period of one-half to one day's time, there would be fatigue, or under prolonged conditions, eyestrain such as was described in the earlier literature.

12.2 Ease of Seeing

There has always been a dramatic contrast between daylight illumination levels and electric lighting levels. Weston (199) pointed out that lace makers took their tasks involving fine detail and poor contrast out of doors. Campbell et al. (202) described the early weavers' houses in Coventry fitted with tall windows for daylighting. They stated, "The eye is accustomed to the natural illumination provided by sunshine and outdoor daylight, and it is natural for us to seek the best possible daylight for the performance of work which is difficult to see." Sir Duke Elder quoted in *Eyes in Industry* (202, p. 2) wrote:

> In the care of the eyes the most important general principle is that they should have sufficient light whereby to work. It is to be remembered that the eyes of man were evolved to function in daylight; and although there is a popular superstition that artificial light is bad for them, the only fault lies in the fact that artificial light is rarely provided liberally or adequately distributed.

With the dramatic difference, it was natural for a feeling to arise that there was less fatigue (203) or a greater ease of seeing under daylighting and, by inference, under higher illumination levels. Luckiesh and Moss (26), by using a key for recording the completion of each page of reading (actually a pressure-sensitive key), were able to record the pressure exerted on the key during 308 half-hour sessions of 14 observers. The reading sessions were conducted under 1, 10, and 100 fc. The results showed a steady decrease of muscular tension as the level of illumination was increased (Fig. 24.74). Glare was also introduced, and it made the nervous muscular tension greater.

The same authors, in another series of experiments, used the blink rate as an index of mental tension (Fig. 24.75) and found apparent consistent trends related to illumination levels, glare, severe ocular scanning, and simulated refractive errors (204). Others could

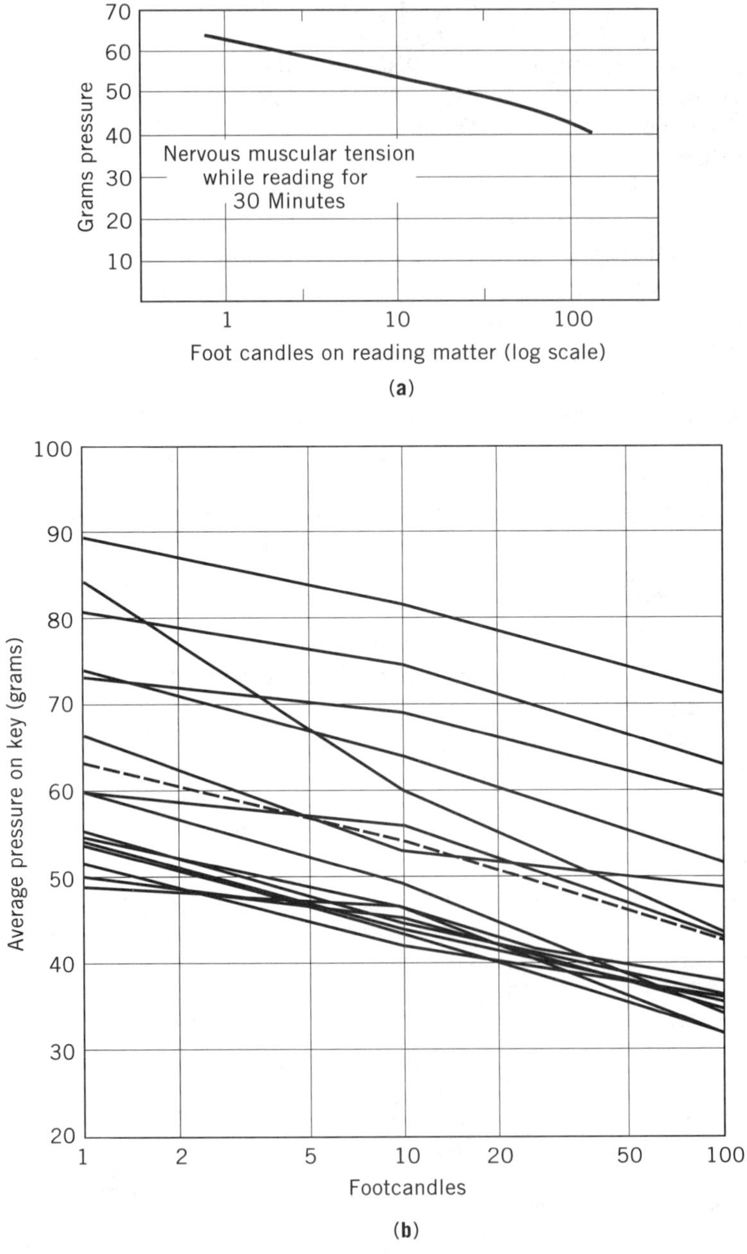

Figure 24.74. (**a**) Mean value of nervous muscular tension under three levels of illumination. (**b**) Relations between indicated nervous muscular tension and level of illumination for each of the 14 subjects; dashed line represents the geometrical mean of the results of the 14 subjects. The probable errors of the average values of nervous muscular tension are less than 1 percent.

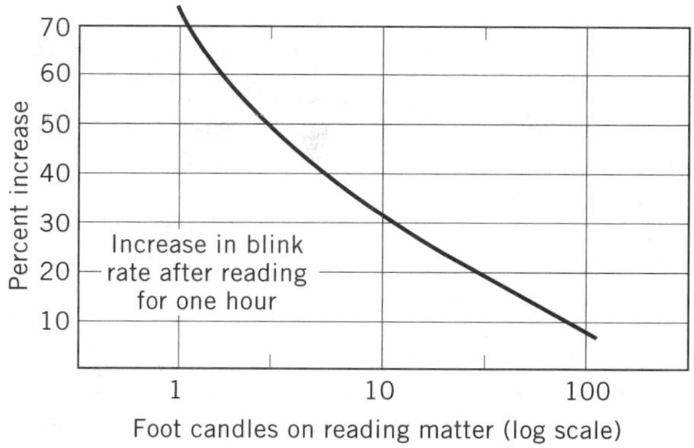

Figure 24.75. Increase of blink rate with lower levels of illumination.

not reproduce the same effects. Much of the work depends on the careful design of the lighting conditions, such as veiling reflections, subtle differences in discomfort glare, and unrealized transient adaptation effects (205). Some of the nonconfirming researchers were not aware of these effects, but knowledge of them has been developed in later years (21). A recent summary of work on fatigue over the past decade concludes that there is sufficient evidence to support the indication that illuminance can affect the occurrence of fatigue (as measured in terms of decreased performance and increased variability), following the prolonged performance of some tasks. However, a widely applicable range of principles on the impact of lighting on fatigue cannot be stated based on the evidence to date. Illustrative of the problem is a study in which a glare source was present during the paced performance of a task. Although the performance remained constant, variability increased with increased work duration. This seemed to indicate that gradually fatiguing subjects were compensating by increased exertion of effort. Therein lies the problem of how to measure both fatigue and possible compensation. This promises to be a fruitful area for further exploration (189).

12.3 Biological Effects

The ability to maintain a proper luminous input for human biological regulation depends on understanding the interaction and role of lighting for tissues (i.e., the eye and skin) and circadian rhythms (i.e., the regulation of body functions based on a light dark cycle). These biological responses (see Fig. 24.76) are based on irradiance, wavelength(s), duration, and time of exposure (336, 339, 346, 358, 359, 365).

Steinmetz (206) stated in 1916 that illuminating engineering embraced not only light for seeing but also knowledge of the effect of radiation from light sources involving the physiologist. Gage (207) in 1930 reviewed the work of physiologists who had discovered in the 1920s that overcoming rickets, the strengthening of bones, the reduction of colds, and the treatment of tuberculosis all are affected by sunshine and simulated sources of sunshine such as mercury lamps. Today with a resurgence in multidrug-resistant strains of

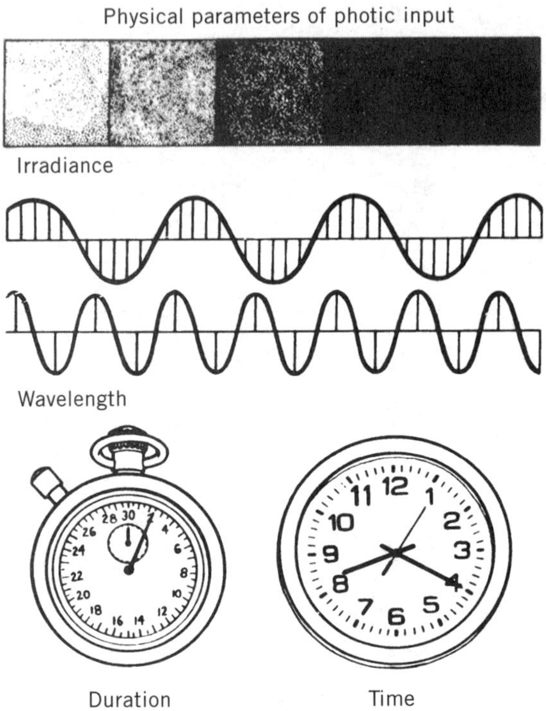

Figure 24.76. Factors affecting biological responses in humans.

tuberculosis, UV air disinfection has once again come to the forefront to combat this disease (208–240). Research in field trials of efficacy to develop predictive models of UV and air mixing are underway and have potential of having a global impact. The World Health Organization released statistics that showed one-third of the world's population have a positive tuberculosis skin test. During the 1930s the lighting industry developed sunlamps S1, S2, RS, and later fluorescent sunlamps, which together with light control luminaires, produced dual-purpose lighting: light for seeing and for ultraviolet radiation in the zone found in sunlight from 280 nm to the visible region. Complaints regarding the use of the emerging fluorescent lamps for general lighting because of suspicions of harmful UV radiation (241) dampened the enthusiasm for dual-purpose lighting. Furthermore, some dermatologists became concerned with the carcinogenic effects of sunlight and urged that all sunlight be excluded. With the rise of skin cancer in the 1980s, dermatologists continue to urge use of protective lotions and avoidance of summer or tropical sun exposure from 11 A.M. to 2 P.M. and issue warnings concerning tanning booths (347). Some claims linked fluorescent lighting to skin cancer; however, further research did not confirm this (242, 243). Light therapy being used to overcome respiratory diseases has served to renew interest in the use of sunlight and electrically simulated daylight (244, 245). Currently light sources such as white or blue fluorescent lamps are being successfully used to overcome jaundice in the newborn, a condition that occurs in 15% to 20% of all premature infants. The portion of the spectrum responsible for photodegradation of bilirubin was found to be in the blue region (Fig. 24.77). Lucey (246), who first reported his work in 1968, led the United States in implementing the discovery by Cremer et al. (247, 248). Precautions need to be taken, however, in the design of the neonatal units so that the infants' visual system is not damaged by exposure to blue light (249).

A study on calcium absorption to avoid osteoporosis was carried out by Neer et al. (249). It was found that the small UV component of currently available fluorescent lamps, when used at a level of 5000 lx, stimulated significantly more calcium absorption. This effect appears to be related to the role of vitamin D3 in curing rickets (250). As people grow older, they receive less outdoor daylight exposure and stay behind UV-absorbing glass; this can result in less calcium being absorbed from their food, with a resulting weakness of the bones (osteomalacia). Mechanical shock may break their bones, and the recovery is very slow or indefinite. Wurtman (244) describes a study by Jean Aaron in England whose autopsies in winter revealed more cases of osteomalacia than those in the summer.

It has been established that light has other curative values for disease. Psoriasis is an affliction characterized by the eruptions of large rough red areas that are very itchy and disfiguring (245). Parrish et al. (251) have developed a successful treatment consisting of ingesting a photoactive drug called psoralen, and after a 2-h time delay, irradiating the area of the psoriasis with UV near the visible spectrum, 320 to 400 nm (UV-A). The photoactive chemical is found in nature in vegetables and fruits such as carrots, parsnips, celery, parsley, limes, lemons, and figs.

Diamond (252–255) has taken an interesting approach for the treatment of cancer. He has used the property of the chemical hematoporphyrin, having been given in a dose, to stay in the malignant cells while clearing out of the normal cells. When hematoporphyrin is excited with a beam of red light, it generates an excited state of the oxygen molecule (oxygen singlet), which destroys the cell in which it is located.

With the increased success of cataract surgery, eye specialists have been studying safety limits for exposure to light from daylight and various man-made sources (128–135, 256–

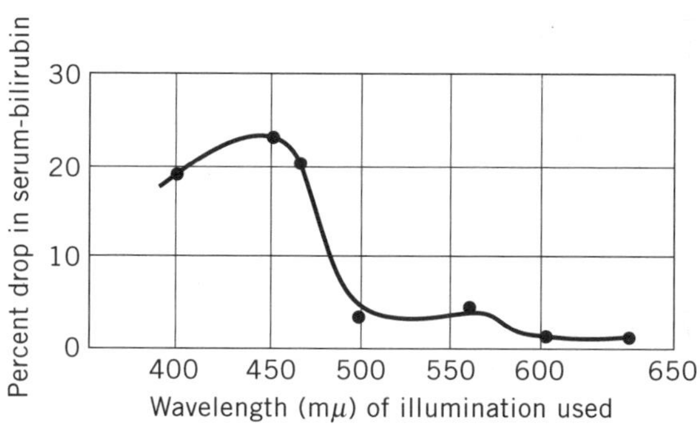

Figure 24.77. Wavelength effectiveness in reducing bilirubin in serum.

258). With the eye's natural protection removed, artificial means must be employed to effectively remove certain wavelengths that damage the visual system. Further, eye specialists are looking at the cumulative effects of absorbed radiation to understand if cataracts and other eye afflictions can be retarded or prevented through screening certain wavelengths throughout life. These findings are being used to set safety standards (128–135).

In the 1970s the whole field of study of photobiology and biological rhythms, tracing the biological influence of light, began to catch researchers' and the public's attention. Wurtman (244) and others (259) described evidence that the duration and color of light affect the hormones, the bodily rhythms, and the sexual activity of animals and humans. Many anecdotes were given widespread coverage but when they were probed for their scientific basis, many of the studies could not be replicated because of lack of adequate controls (260). As a result of confusion in the field, the Lighting Research Institute began its program on the health and biological effects of light. Early basic animal research was conducted by Brainard et al. and Hoffmann et al. (261–268) to establish a baseline for further study of the human's response to light. It was found that if lighting parameters were not carefully controlled, results were inconclusive and unrepeatable. Light was shown clearly to have an impact on various animals' biological systems, activity level, growth, life span, and reproductive responses (264–268). The impact of various light sources to produce biological effects was also studied. Three sources, fluorescent, high-pressure sodium, and fluorescent simulating 5500 K natural daylight in the ultraviolet and visible regions were equalized on the basis of the visual sensitivity curve common to nocturnal animals. Although the animals responded to the intensity of light, there was no significant difference between the results based on the different spectral power distributions. This sets a further standard for future studies. Earlier published studies were flawed because they failed to distinguish between brightness and wavelength in nocturnal animals. Brainard has studied basic physiological mechanisms. He found that light entering through the visual system regulates the production of hormones and affects the metabolism of specific areas of the brain (261–268). In humans, as in some other species, biological functions are regulated on a light–dark cycle. These approximately 24-h oscillations of the physiology of the body are called circadian rhythms. These rhythms must be reset (entrained) every day by a natural or man-made light–dark cycle. Evidence is mounting that many in our 24-h society are suffering from circadian disorders that range from jet lag, shift work accidents, and insomnia, to seasonal affective disorder (SAD) (269–272, 331, 338, 340, 343–346, 348–364). By understanding circadian physiology, new therapeutic approaches are being used to alleviate circadian disorders. Lewy et al. (269) have found that bright morning light (2000 to 2500 lx) is effective in treating SAD patients. This is four to five times current normal office levels. If research establishes that the general population is at risk because of isolation from lighting cues necessary each day to set the circadian clock, then some alteration of daily routine will be needed to ensure daily entrainment. Where light therapy is required, professional guidance should be sought. A society of specialists has recently been organized to disseminate information on light therapy (273). Questions being studied today include how workers' alertness can be maintained during shift work, what is the daily dose of light needed by the general population, and if current lighting standards meet this daily light requirement (340, 343, 348–364). There is a mounting body of knowledge that is making light therapy a recognized medical treatment, and new circadian lighting systems are helping to regulate the alertness/restorative sleep cycle for shift workers as well as regulating the fragile sleep cycles of seniors. New mechanisms, such as the tick/tock gene, are being discovered and are filling out the picture of light being absorbed by the body to regulate the body clock (341, 342). Lighting practitioners are considering how to apply these new findings, and the twenty-first century promises to provide a unique synergy between the design professions and the healing professions (343–346, 348–365).

From all the studies mentioned above it would appear that light can very seriously affect the health and general welfare of humans. Much more knowledge is needed in the establishment, through design, of the role of lighting for good health as well as good vision.

13 APPLICATION INFORMATION

13.1 Lighting Design

With the rapid advance of lighting technology and techniques, the increased concerns of environmental comfort for building occupants, and the need to utilize our energy resources strategically, there is a growing need for lighting professionals who can integrate all these aspects. Traditionally architects and designers have turned to consultants such as electrical engineers to meet their lighting design requirements. Manufacturers and their representatives have also been heavily involved in lighting design. During the late 1960s, in addition to practicing illuminating engineers, a small group of independent lighting consultants evolved, offering lighting design on a fee basis. A professional organization of lighting designers was then established: the International Association of Lighting Designers. New (1997) lighting certification tests are now licensing professionals as Lighting Certified (LC). Since then, lighting design as a profession has steadily grown (274–283). When consideration is being given to building a new structure or remodeling a space, a lighting professional should be involved as part of the design team including the architect, interior designer, and consulting engineer. The lighting professional will be able to establish a program and goals for lighting the space(s) to include integrating natural light if available, visualizing how the space will look, and then producing the desired results through the proper specification and installation of lighting systems to manipulate the color, intensity, direction, and distribution of light. There is a vast array of choices from which to make up a lighting system—from light sources, luminaires (fixtures), and controls, in possible combination with daylight. All of these aspects must address energy efficiency as discussed below.

13.2 Electric Light Sources

The designer has a vast repertoire of available "bulbs and tubes" in various shapes, sizes, wattages, and colors. These vary all the way from a single wavelength of light, such as a low-pressure sodium lamp with its yellow light, to the blue-green of a mercury lamp with its individual four-wavelength emissions, to the white multiwavelength emissions of a metal halide lamp, and the golden light of a high-pressure sodium lamp. They vary in efficacy from 8 lm W (small incandescent) to 180 lm/W (low-pressure sodium). There are two main categories of lamp: incandescent and gaseous discharge. Excellent references on these sources are available from the Illuminating Engineering Society of North America (157, 284, 366). A comparison chart of some aspects of the various sources commonly found in industrial applications is shown in Figure 24.78.

13.3 Incandescent Lamps

Hot tungsten wires that are high enough in temperature to glow with a pale yellowish-white light are the basis of incandescent lights, which are available for general service in pear-shaped bulbs from 10 to 1500 W with the efficacy of 8 lm/W for the 10-W size to

Why use incandescent?	Why use fluorescent?
• Low inital cost • Small size • Excellent color • Directional control • Variety of shapes • Simple circuitry	• Higher efficiency • Long life • Variety of colors • Low brightness • Low operating temperatures • Diffuse source
Drawbacks to incandescents	**Drawbacks to fluorescents**
• Low efficiencies • High heat output • High operating cost • Short life • Glare potential	• Higher initial cost • Sensitive to temperature • Limited optical control • Requires auxiliary equipment

Why use high pressure sodium?	Why use metal halide?
• Very high efficiencies • Long life • Reasonable optical control • Very low operating cost • High Lumen output	• High efficiencies • Long life • Reasonable optical control • Low operating costs • Good color properties
Drawbacks to high pressure sodium	**Drawbacks to metal halide**
• High initial cost • Start-up and restrike times • Relatively poor color properties • Requires auxiliary equipment • Large variations "Ballast Factor" • Glare potential	• High initial cost • Start-up and restrike times • Color consistency • Requires auxiliary equipment • Large variation in "Ballast Factor" • Glare potential

Figure 24.78. Criteria for selecting light sources: color, efficiency, life, size and shape, price.

24 lm/W for the 1000-W size. They arc also available in parabolic and elliptical reflector bulbs. They include low, medium, and high voltages for special applications such as flashlights, bicycles, trains, aircraft, and automobiles. The life of general service lamps varies from 750 to 1000 h.

A longer-life, concentrated incandescent lamp, called the tungsten halogen lamp, has been developed. In this design the filament is put in a small quartz tube with a negative-valence element such as iodine, bromine, chlorine, or fluorine. As this lamp burns, the tungsten gradually evaporates off the filament and normally would collect on the bulb wall as blackening; instead, it combines with the halogen that is in the quartz tube to form a gas. As this gas approaches the filament, it breaks down and deposits tungsten on the filament. These lamps maintain a good output over their life and last up to 3500 h, depending on their particular size and design. They are also available in reflector-type bulbs.

Architects and many designers like to use incandescent lamps because of the complimentary effects on human complexions and on warm colors. They are an extension of candlelight and the light of the fireplace. The emitted light emphasizes the yellows and reds but deemphasizes the blues and greens.

13.4 Fluorescent Lamps

Fluorescent lamps are essentially a mercury gaseous discharge in a long tube, coated inside with phosphor powder. The phosphor powder reacts to the invisible mercury emission spectral line at 253.7 nm and converts the invisible energy into visible light. By using various chemical phosphors usually in a mixture, colors of visible light are emitted, and the emission becomes a continuous spectral output. Aside from particular spectral color emphasis such as blue, green, gold, pink, and red, the phosphor mix most commonly used is such that the emission tends to follow the CIE spectral luminous efficiency curve of the eye for photopic (or day) vision. By an appropriate shift in the mix, the emission emphasis moves toward the warm side (warm white), in having proportionately more yellow and red light, or cool white, with a greater proportion of blue and green. For good portrayal of human complexion and all colors, one could use the deluxe cool white or deluxe warm white, where a proportionately greater red emission is present.

Fluorescent lamps require a ballast to limit the current and in many instances to transform the supply voltage to a suitable level. Lamp performance is influenced by the character of the ballast and luminaire, line voltage, ambient temperature, burning hours per start, and air movement.

The efficacy of fluorescent lamps (157, 284) varies from approximately 30 to 80 lm/W for lamps of standard cool white color, depending on bulb size and shape (lamp length ranges from 6 to 96 inches [0.5 to 244 cm]). This does not include losses in the ballast, which are typically on the order of 20%. Most fluorescent lamps are of conventional tubular design, but there are some special types such as circular, reflectorized, jacketed, and panel shaped. Recent developments in fluorescent technology include improved color rendition, longer life, compact sources, smaller diameters, and reduced wattages, as well as the variety of wattages available. The compact fluorescents are becoming common replacements for incandescent sources. When considering retrofitting an incandescent source with a compact fluorescent source, careful consideration must be given to the selection in order to achieve the same light output and distribution.

13.5 High-Intensity Discharge Lamps

High-intensity discharge (HID) lamps (157, 284) are also electric discharge sources. The basic difference between HID and fluorescent lamps is that the operating pressure of the arc is much higher in HID lamps. HID lamps include the groups commonly known as mercury, metal halide, and high-pressure sodium lamps. Spectral characteristics are different because the higher pressure arc emits a large portion of its visible light. HID lamps produce full light output when full operating pressure has been reached, which generally requires several minutes. Most HID lamps have both an inner and an outer bulb. The inner is made of quartz or polycrystalline aluminum, and the outer bulb is generally made of thermal-shock-resistant glass. Light output is practically unaffected by surrounding temperatures. Like fluorescent lamps, HID lamps also require devices that limit the current. Lewin and Stafford (285–289) have found different combinations of ballasts, lamps, and luminaires can result in up to a 40% reduction in actual light output versus what was predicted. These equipment factors allow more targeted lighting designs. Other recent advances in HID source technology have provided a wider range of available wattages, more compact sizes, improved restrike times, and improved color rendition.

Mercury lamps for general lighting are available with either clear or phosphor-coated bulbs of various sizes and shapes with wattages of 40 to 1000. There is also a 3000-W clear mercury lamp. Typical efficacies range from 30 to 63 lm/W, not including a 6% to 20% power loss in the ballast. Clear mercury lamps produce light that is rich in yellow and green tones and almost lacking in red. Various types of phosphor-coated lamps are available to provide improved color, and they are more popular than clear lamps for industrial lighting. A number of special mercury lamps are available, including semireflector, reflectorized, and self-ballasted.

Metal halide lamps are very similar in construction to the mercury lamp, the major difference being that the arc tube contains various metal halides in addition to mercury. They are available with either clear or phosphor-coated bulbs of various sizes and shapes from 175 to 1500 W. Typical efficacies range from 70 to 100 lm/W, not including power loss in the ballast. The metal halide additives improve the efficacy and color compared to a clear mercury lamp. Further color improvement is achieved with the addition of phosphor coatings. Recent development in metal halide sources have also resulted in better restrike times, improvements in color, an extended range of wattages, and compact sizes.

In the high-pressure sodium (HPS) lamp, light is produced by electricity passing through sodium vapor. The lamps are available in wattages from 250 to 1000. Typical efficacies range from 100 to 130 lm/W, not including power loss in the ballast. This lamp produces a golden-white light. HPS technology has benefited from the availability of compact sizes, a wider range of wattages, improvements in restrike times, and some improvements in color.

In low-pressure sodium (LPS) discharge lamps, the arc is carried through vaporized sodium. The starting gas is neon with small additions of argon, xenon, or helium. To maintain the maximum efficacy of the conversion of the electrical input to the arch discharge into light, the vapor pressure must be at a given level and constant, and the proper operating temperature must be maintained. The arc tube must be enclosed in a vacuum flask or in an outer bulb at high vacuum to ensure that the temperature is maintained. The light produced by a low-pressure sodium arc is almost monochromatic, consisting of a double line in the yellow region of the spectrum at 589 and 589.6 nm. Two types of arc tube construction are used in modern low-pressure sodium lamps: the hairpin, or "U," tube and the linear tube. The efficacy is 180 lm/W, without the ballast.

14 DAYLIGHTING

Over the centuries humans have attempted to bring daylight into indoor visual activities. Central courts open to the sky allowed the light to penetrate into the covered areas. In northern latitudes churches and guild halls had high windows to admit daylight as far as possible into the interior, permitting work to be carried on in spite of inclement weather. All this was necessary because of the weakness of man-made light sources. Since World War II there has been decreasing dependence on daylighting because of availability of far more efficient electric light sources (fluorescent lights and the developing use of high-intensity discharge sources). Architects and designers found they could reduce costs of buildings by lowering the ceilings (reducing the height of the windows). Although they continued to use large areas of glass, workers near the windows became overwhelmed by the high levels of daylight as well as the greater heat loads. Accordingly, low-transmission glass (neutral tinted with 15% to 50% transmission) was developed to overcome these drawbacks. Furthermore, to reduce the amount of dirt accumulated, the windows were sealed and the interiors air conditioned. It was found that the daylighting, and particularly the sunload on the air conditioning, was appreciable. This has resulted in reflecting-type window glazing, where a large part of the heat is reflected back outside, and a low transmission of light is maintained.

Our discussion of transient adaptation pointed out that encountering higher luminances than the task luminance greatly reduces the immediate scanning sensitivity of the visual system (Fig. 24.57). The sky luminances as seen through clear glass windows may range between 10 and 100 times the task luminance; therefore, the scanning sensitivity is severely reduced. This and a significant degree of discomfort glare probability accounted for the trend toward low-transmission glass. In the older industrial plants with large expanses of glass in the walls or the roof, it is common to see the panes covered with paint to reduce the outdoor luminances, with the aim of achieving better balance with interior luminances.

Nevertheless window expanses with appropriate limited luminance do provide a significant illumination to the peripheral areas of a building. For peripheral offices the light that approaches desk tasks at wide angles with the vertical does a most effective job of overcoming veiling reflections and producing high visibility. Illumination from windows is not significant at a distance beyond twice the height of the windows above the floor.

Reflected ground light can make an important contribution by being directed onto a white ceiling as soft indirect illumination. A scheme of having a low-transmission vision strip opposite the eye level, with clear glass above to admit ground light but with the view of the clear glass occluded by a horizontal shield or louvers is helpful (290–292).

Another system is to use silvered louvers that catch the sunlight (in high-percentage sunlight areas) and direct the light out across the ceiling for indirect lighting (293).

Many structural forms such as unilateral, bilateral roof monitor, clerestory, sawtooth, and skylight have been studied as to the effective admission of daylight (294). A *Recommended Practice of Daylighting* (294), available from the Illuminating Engineering Society of North America, has guidelines for designing for daylight. Limitations of outdoor luminances with respect to transient adaptation and discomfort glare should be borne in mind.

During the mid-1970s architects in the United States rediscovered daylighting in the wake of the oil embargo. Today EPA Greenlights and the DOE Federal Energy Management Program are providing programs on daylighting through the national laboratory system. The federal government, through the Department of Energy, began studying how daylight could be used as a force to reduce the energy consumption of electric lighting. Research has been conducted on daylighting control, glare, chilling and heating, ventilation, air-conditioning (HVAC) size, solar gain control, fading of interiors, and so on. Some benefits of a properly integrated daylighting–electric lighting and HVAC design include reduced electric lighting when daylight is available and reduced peak electrical demand for utilities. Sophisticated techniques are being developed in the laboratory including "smart windows" that can change their transmissiveness based on an electric current applied to a specially impregnated window, special daylighting controls that dim the electric lights in response to daylight availability, occupant sensors to turn room lights on or off, and rating systems for various combinations of windows and shading devices (294, 295–305).

15 ENERGY MANAGEMENT

Beginning in 1944 and through 1981, the Illuminating Engineering Research Institute (IERI), a trust affiliate of the Illuminating Engineering Society of North America (IESNA), was concerned with the efficient operation of the visual system for the most efficient use of human resources. This concern is being carried on by the IERI successor, the Lighting Research Institute and its successor the EPRI Lighting Research Office. The pattern of these university-determined results has been reflected in the preceding pages of this chapter. The amount of light on the details of the visual task, the reduction or elimination of veiling reflections for optimal visibility, the proper brightness balance for optimal scanning sensitivity (transient adaptation), the avoidance of discomfort glare and disability glare, the motivation of good subjective feeling, and the provision of good biological health all contribute to optimal ability to perform efficiently.

Having provided the optimal use of human resources, one can begin to plan the strategy of the use of lighting energy (296–306). Research data do not indicate the need for illumination for a given visual performance over all the habitable space, but only on the task area, as long as the surrounding luminances are in balance with task luminance to avoid serious transient adaptation losses, disability glare, or discomfort glare. Therefore, there can be nonuniform lighting with less light in the nontask areas (307, 308). These strategies must be skillfully employed; otherwise, as a postoccupancy evaluation study has shown, a large percentage of occupants might become dissatisfied (308, 367). The IESNA has developed specific recommendations for nonuniform lighting, more efficacious light sources (fluorescent and HID), better and more efficient design of luminaires, combining daylight with electric lighting, providing controls for flexible use of lighting, using lighter finishes on room surfaces, and providing for better maintenance. To aid architects and designers, there is the *IES Recommended Lighting Power Budget Determination Procedure* (EMS-1), supplemented with *An Interim Report Relating the Lighting Design Procedure to Effective Energy Utilization* (EM2), and *Example of the Use of the IES Recommended Lighting Power Budget Determination Procedure* (EM3). Renewed concern for the environment and electrical generation shortages have focussed national attention on employing greater lighting efficiency. This must be done in context of the human requirements stated above. Education is the key to combining the need for energy efficiency with human productivity. New lighting courses and advanced technology guidelines are available and others are being developed that will meet the needs of nonexperts as well as professional practitioners (309–323).

Refer to Appendix A for currently recommended illuminance categories and values for industrial lighting design (target maintained levels).

APPENDIX A. METHOD FOR PRESCRIBING ILLUMINATION*

(This Appendix is not part of the "American National Standard Practice for Industrial Lighting," ANSI/IES RP-7-1983, but is presented as background material for the user of the Standard.)

A1 New illuminance selection procedure. (**a**) Since 1965 this Standard has included single-value illuminance recommendations based on a method established at that time.[1] In recent years it became apparent, through on-going research and design experience, that it was time to move away from the single-value recommendations to a range approach—illuminance ranges accompanied by a weighting-factor guidance system reflecting lighting-performance trends found in research. In 1979, such a new procedure was established.[2]

(**b**) It is intended that this new procedure will accommodate a need for flexibility in determining illuminance levels so that lighting designers can tailor lighting systems to specific needs, especially in an energy conscious era. Such flexibility requires that additional information be available to effectively use the new range approach—a *lighting task* must be considered to be composed of the following:

1. The visual display (details to be seen).
2. The age of the observers.
3. The importance of speed and/or accuracy for visual performance.
4. The reflectance of the task (background on which the details are seen).

(**c**) The visual display is the object being viewed—it will present some inherent visual difficulty. The age of the observer is a predictor of the condition of the observer's visual

*Excerpt from IES RP-7-1983. Used with Permission.

Table A1. Illuminance values, maintained, in lux, for a combination of illuminance categories and user, room and task characteristics

a. General Lighting Throughout Room

Weighting Factors: Average of Occupants Ages	Weighting Factors: Average Room Surface Reflectance (per cent)	Illuminance Categories: A	B	C
Under 40	Over 70	20	50	100
	30–70	20	50	100
	Under 30	20	50	100
40–55	Over 70	20	50	100
	30–70	30	75	150
	Under 30	50	100	200
Over 55	Over 70	30	75	150
	30–70	50	100	200
	Under 30	50	100	200

b. Illuminance on Task

Weighting Factors: Average of Workers Ages	Demand for Speed and/or Accuracy[a]	Task Background Reflectance (per cent)	Illuminance Categories: D	E	F	G[b]	H[b]	I[b]
Under 40	NI	Over 70	200	500	1000	2000	5000	10000
		30–70	200	500	1000	2000	5000	10000
		Under 30	300	750	1500	3000	7500	15000
	I	Over 70	200	500	1000	2000	5000	10000
		30–70	300	750	1500	3000	7500	15000
		Under 30	300	750	1500	3000	7500	15000
	C	Over 70	300	750	1500	3000	7500	15000
		30–70	300	750	1500	3000	7500	15000
		Under 30	300	750	1500	3000	7500	15000
40–55	NI	Over 70	200	500	1000	2000	5000	10000
		30–70	300	750	1500	3000	7500	15000
		Under 30	300	750	1500	3000	7500	15000
	I	Over 70	300	750	1500	3000	7500	15000
		30–70	300	750	1500	3000	7500	15000
		Under 30	300	750	1500	3000	7500	15000
	C	Over 70	300	750	1500	3000	7500	15000
		30–70	300	750	1500	3000	7500	15000
		Under 30	500	1000	2000	5000	10000	20000
Over 55	NI	Over 70	300	750	1500	3000	7500	15000
		30–70	300	750	1500	3000	7500	15000
		Under 30	300	750	1500	3000	7500	15000
	I	Over 70	300	750	1500	3000	7500	15000
		30–70	300	750	1500	3000	7500	15000
		Under 30	500	1000	2000	5000	10000	20000
	C	Over 70	300	750	1500	3000	7500	15000
		30–70	500	1000	2000	5000	10000	20000
		Under 30	500	1000	2000	5000	10000	20000

[a]NI = not important, I = important, and C = critical

[b]Obtained by a combination of general and supplementary lighting.

system. The importance of speed and/or accuracy distinguishes between casual, important and critical seeing requirements. The reflectance will determine the adaptation luminance produced by the illuminance. These characteristics, considered together, determine the appropriate amount of light for the *lighting task.* All four must be considered as comprising the lighting task.

(**d**) In applying the new procedure the first step is to determine a range of illuminances appropriate for the visual difficulty presented by the visual display, the first of the above characteristics (see **Table B1** in Appendix B) and then to determine a target value from that range on the basis of the remaining three characteristics.

(**e**) For a given visual display, a specific value of illuminance can be chosen from the recommended range only if the second, third, and fourth characteristics of the lighting task are known; i.e., observer's age, importance of speed and/or accuracy, and task reflectance. These should be determined at design time, by the designer in consultation with the user.

(**f**) A guide for using the second, third and fourth characteristics of the lighting task, to determine a specific target value of illuminance, takes the form of a table of weighting factors (see Table 3). The designer or user determines the weight of each characteristic. A combined weighting factor then indicates whether the lower, middle or upper value of illuminance in the range is appropriate (see **Table 1**).

(**g**) It can be seen that this procedure is an illuminance *selection* procedure, where consensus-determined recommended ranges combine with user supplied information and judgment. The results is the determination of a specific target value of illuminance appropriate for the lighting task under consideration.

A2 Limitations of the new selection procedure. This illuminance selection procedure is intended for use in interior environment where visual performance is an important consideration. It has been developed from a consideration of experience and research results from visual performance experiments. Its use is then limited to applications where this information can be applied directly. Thus, the illuminance selection procedure[2] is *not* used to determine the appropriate illuminance when:

1. Merchandising is the principal activity in the space and the advantageous display of goods is that purpose of lighting.
2. Advertising, sales promotion or attraction is the purpose of lighting.
3. Lighting is for sensors other than the eye, as in film and television applications.
4. The principal purpose of lighting is to achieve artistic effects.
5. Luminance ratios have a greater importance than adaptation luminance, as when it is desired to achieve a particular psychological or emotional setting rather than provide for visual performance.
6. Minimum illuminances are required for safety.
7. Maximum illuminances are established to prevent nonvisual effects, such as bleaching or deterioration due to ultraviolet and infrared radiation in a museum.
8. Illuminances are part of a test procedure for evaluating equipment, such as for surgical lighting systems.

Table B1. Currently recommended illuminance categories for industrial interiors

Area/Activity	Illuminance Category
Inspection	
Simple	D
Moderately difficult	E
Difficult	F
Very difficult	G
Exacting	H
Iron and steel manufacturing	(see Table B3)[a]
Jewelry and watch manufacturing	G
Laundries	
Washing	D
Flat work ironing, weighing, listing, marking	D
Machine and press finishing, sorting	E
Fine hand ironing	E
Loather manufacturing	
Cleaning, lanning and stretching, vals	D
Cutting, fleshing and stuffing	D
Finishing and scarfing	E
Loather working	
Pressing, winding, glazing	F
Grading, matching, cutting, scarfing, sewing	G
Locker rooms	C
Machine shops	
Rough bench or machine work	D
Medium bench or machine work, ordinary automatic machines, rough grinding, medium butting and polishing	E
Fine bench or machine work, line automatic machines, medium grinding, line butting and polishing	G
Extra-fine bench or machine work, grinding, fine work	H
Materials handling	
Wrapping, packing, labeling	D
Picking slock, classifying	D
Loading, inside truck bodies and freight cars	C
Meat packing	
Slaughtering	D
Cleaning, cutting, cooking, grinding, canning, packing	D
Nuclear power plants (see also Electric generating stations)	
Auxiliary building, uncontrolled access areas	C
Controlled access areas	
Count room	E[c]
Laboratory	E
Health physics office	F
Medical aid room	F
Hot laundry	D
Storage room	C
Engineered safety features equipment	D

Table B1. (continued)

Area/Activity	Illuminance Category
Diesel generator building	D
Fuel handling building	
Operating floor	D
Below operating floor	C
Off gas building	C
Radwaste building	D
Reactor building	
Operating floor	D
Below operating floor	C
Paper manufacturing	
Beaters, grinding, calendering	D
Finishing, cutting, trimming, papermaking machines	E
Hand counting, wet end of paper machine	E
Paper machine reel, paper inspection, and laboratories	F
Rewinder	F
Petroleum and chemical plants	(see Table B3)[a]
Plating	D
Polishing and burnishing (see Machine shops)	
Power plants (see Electric generating stations)	
Poultry industry (see also Farm—dairy)	
Brooding, production, and laying houses	
Feeding, inspection, cleaning	C
Charts and records	D
Thermometers, thermostats, time clocks	D
Hatcheries	
General area and loading platform	C
Inside incubators	D
Dubbing station	F
Sexing	H
Egg handling, packing, and shipping	
General cleanliness	E
Egg quality inspection	E
Loading platform, egg storage area, etc.	C
Egg processing	
General lighting	E
Fowl processing plant	
General (excluding killing and unloading area)	E
Government inspection station and grading stations	E
Unloading and killing area	C
Feed storage	
Grain, lead rations	C
Processing	C
Charts and records	D
Machine storage area (garage and machine shed)	B

Table B1. (continued)

Area/Activity	Illuminance Category
Printing industries	
Type foundries	
Matrix making, dressing type	E
Font assembly—sorting	D
Casting	E
Printing plants	
Color inspection and appraisal	F
Machine composition	E
Composing room	E
Presses	E
Imposing stones	F
Proofreading	F
Electrotyping	
Molding, routing, finishing, leveling molds, trimming	E
Blocking, linning	D
Electroplating, washing, backing	D
Photoengraving	
Etching, staging, blocking	D
Routing, finishing, prooting	E
Tint laying, masking	E

A3 Example of Illuminance Selection. (**a**) A machine shop is to be relighted. The designer in consultation with the foreman has determined the following:

1. The task is diemaking using metal of a reflectance of about 40 percent.
2. The workers are young (under 40).
3. The workers are under pressure to maintain close tolerances on die production. Speed and accuracy are considered to be important, but not critical to production.

(**b**) Using the above step-by-step procedure:

Step 1. The visual task is defined above.

Step 2. Referring to Table B1, an illuminance category of E is found under machine shops, "medium bench or machine work."

Step 3. Referring to Table 1, the illuminance range is found to be 500-750-1000 lux [50-75-100 footcandles].

Step 4. Referring to Table 3 and the above information, the weighting factors selected are: -1 for workers' ages; 0 for speed and/or accuracy; and 0 for reflectance of task background. The algebraic sum is $-1 + 0 + 0 = -1$. Therefore, the illuminance to be selected is the mid-value, i.e., 750 lux [75 footcandles].

(**c**) If the task were considered to be fine bench work where the tolerances were critical, the metal used has a reflectance of 25 percent, and the average of the workers' ages was 50, the illuminance category would become G, the range 2000-3000-5000 lux [200-300-500 footcandles]; and the weighting factor for workers' ages would be 0, for speed and/or accuracy +1, and for reflectance +1. The new algebraic sum of the weighting factors is +0 +1 +2. Therefore, the illuminance to be selected is the high value in the new range, i.e., 5000 lux [500 footcandles].

(**d**) By referring to Table A-1 after Step 2, the illuminance can be selected without referring to Tables 1 or 3.

References

1. IES Committee on Recommendations for Quality and Quantity of Illumination, "RQQ report no. 1—Recommendations for quality and quantity of illumination," *Illum. Eng.*, Vol. 58, No. 5, Aug. 1958, p. 422.
2. IES Committee on Recommendations for Quality and Quantity of Illumination of the IES: "RQQ report no. 6—Selection of illuminance values for interior lighting design," *J. Illum. Eng. Soc.*, Vol. 9, No. 3, Apr. 1980, p. 188.
3. *IES Lighting Handbook, 1931 Reference Volume*, New York: Illuminating Society of North America.
4. IES Lighting Design Practice Committee, "Zonal-cavity method of calculating and using coefficients of utilization," *Illum. Eng.*, Vol. 59, No. 5, May 1964, p. 309.
5. IES Design Practice Committee, "General procedure for calculating maintained illumination," *Illum. Eng.*, Vol. 65, No. 10, Oct. 1970, p. 603.
6. IES Lighting Design Practice Committee, "The determination of illumination at a point in interior spaces," *J. Illum. Eng. Soc.*, Vol. 3, No. 2, Jan. 1974, p. 171.

APPENDIX B. LIGHTING RESOURCE ORGANIZATIONS

CIE: Commission International de l'Eclairage
Kegelgasse 27
A-1030 Vienna, Austria
Email: ciecb@ping.at.
Website: www.cie.co.at/cie

SLTBR: Society for Light Treatment and Biological Rhythms
842 Howard Avenue
New Haven, CT 06519, USA
Fax: (203) 764-4324
Email: sltbr@yale.edu
Website: www.sltbr.org

IESNA: Illuminating Engineering Society of North America
120 Wall Street, 17th Floor
New York, NY 10006-4001
Email: iesna@iesna.org
Website: www.iesna.org

IALD: International Association of Lighting Designers
Suite 11-114-A, The Merchandise Mart
200 World Trade Center
Chicago, IL 60654 USA
Email: IALD@IALD.org
Website: www.iald.org

CIBSE: The Chartered Institution of Building Services Engineers
Delta House
222 Balham High Road
Balham
London
SW12 9BS
Fax No. 0208 675 5449
Tel No. 0208 675 5211
General Email Address: secretary@cibse.org
Website: www.cibse.org

LRC: Lighting Research Center
Rensselaer Polytechnic Institute
110 8th Street
Watervliet Facility
Troy, New York 12180 USA
Email: lrc@rpi.edu
WWW: http://www.lrc.rpi.edu

NCQLP: National Council of Qualification for the Lighting Professions
4401 East-West Highway, Suite 305
Bethesda, MD 20814
301-654-2121
301-654-4273 (fax)
E-Mail: info@ncqlp.org
www: http://www.ncqlp.org/

EPRI: Electric Power Research Institute
Lighting Information Office
Lighting Research Office
EPRI Headquarters
3412 Hillview Avenue
Palo Alto, CA 94303
Email: askepri@epri.com
www: www.epri.com

NIST: National Institute of Standards and Technology
Center for Building Technologies
100 Bureau Drive, Gaithersburg, MD 20899-0001.
Email: inquiries@nist.gov
Website: www.nist.gov

NEMA: National Electrical Manufacturers Association
1300 North 17th Street
Suite 1847
Rosslyn, Virginia 22209
Email: webmaster@nema.org
Website: www.nema.org

NLB: National Lighting Bureau
8811 Colesville Road, Suite G 106
Silver Spring, MD 20910
Email: info@nlb.org
Website: www.nlb.org

Lawrence Berkeley National Lab
1 Cyclotron Road
Berkeley, CA 94720
Website: www.lbl.gov

ASHRAE: American Society of Heating Refrigerating and Air-conditioning Engineers
1791 Tullie Circle, N.E.
Atlanta, GA 30329
Website: www.ashrae.org

ASP: American Society of Photobiology
BioTech Park
1021 15th Street, Suite 9
Augusta, GA 30901-3158 USA
Website: www.photobiology.org

Pacific Northwest National Laboratory
P.O. Box 999
902 Battelle Boulevard
Richland, Washington 99352
E-mail: inquiry@pnl.gov
Website: www.pnl.gov

US EPA: United States Environmental Protection Agency
Green Lights
Energy Star Buildings
Website: www.epa.gov

US DOE: United States Department of Energy
Office of Energy Efficiency and Renewal
Website: www.doe.gov and www.eren.doe.gov/buildings

Legend: Very Important | Important | Somewhat Important | Blank = Not important or not applicable

INDUSTRIAL LOCATIONS AND TASKS[a] / Design Issues	Appearance of Space and Luminaires	Color Appearance (and Color Contrast)	Daylighting Integration and Control	Direct Glare	Flicker (and Strobe)	Intrinsic Material Characteristics	Light Distribution on Surfaces	Light Distribution on Task Plane (Uniformity)	Luminances of Room Surfaces	Modeling of Faces or Objects	Reflected Glare	Shadows	Source/Task/Eye Geometry	Special Considerations	Notes on Special Considerations	Illuminance on Task Plane (b)	Category or Value (lux) (c)	Notes—see end of section		Reference Chapter(s)
Warehousing and storage																				
Inactive	Somewhat Important	Very Important	Somewhat Important	Important			Somewhat Important	Somewhat Important	Important	Somewhat Important	Important	Somewhat Important	Very Important			Somewhat Important	B			
Active: bulky items; large labels	Somewhat Important	Very Important	Important	Important			Somewhat Important	Somewhat Important	Important	Somewhat Important	Important	Somewhat Important	Very Important c			Important	C			
Active: small items; small labels	Somewhat Important	Very Important	Important	Important			Somewhat Important	Somewhat Important	Important	Somewhat Important	Important	Somewhat Important	Very Important			Very Important	D			

Notes:

(a) For details on specific tasks or spaces refer to Chapter 19, Industrial Lighting.

(b) The task may be horizontal, inclined, or vertical.

(c) Low illuminances (less than 30 lux) are given in lux; values greater than 30 lux are given in letter categories.

BIBLIOGRAPHY

1. D. W. Clough, *Vision 2020: Lighting Technology Roadmap, Progress Report*, August 17, 1999, U.S. Department of Energy, Washington, DC, 1999.
2. R. L. Vincent, *Lighting Research Institute Bibliography 1983–1996*, EPRI Lighting Research Office, Hendersonville, NC, 1996.
3. R. L. Vincent, *Lighting Research Institute, Synopsis: Completed and Ongoing Programs*, EPRI Lighting Research Office, Hendersonville, NC, 1996.
4. R. Kwartin and A. Gough, *The EPRI Lighting Research Office: Research Accomplishments and Program Plan*, EPRI Lighting Research Office, Hendersonville, NC, 1996.
5. R. L. Vincent, *LRI Research and Education Program: Status Report*, Lighting Research Institute, New York, 1989.
6. *Research Priorities: 1987, Annual Report*, Lighting Research Institute, New York, 1987.
7. "LRI Research Agenda," *LD+A* **14**, 23–26 (1984).
8. "Illumination Roundtable III: Lighting Research and Education for the Eighties," *LD+A* **14**, 27–34 (1984).
9. W. K. Adrian and R. Gibbons, *CIE Report on the Correlation of Models of Vision and Visual Performance* (CIE Committee TC, 1–19 June 1999) University of Waterloo, Waterloo, Ontario, Canada, 1999.
10. CIE, *Unified Framework of Methods for Evaluating Visual Performance Aspects of Lighting*, CIE Publication 19, Commission Internationale de l'Eclairage, Paris, 1972.
11. CIE, *An Analytical Model for Determining the Influence of Lighting Parameters on Visual Performance, Vol. 1: Technical Foundations, Vol. 2: Summary and Application Guidelines*, CIE Publication 19/2, Commission International de l'Eclairage, Paris, 1981.
12. P. R. Boyce et al., *Lighting and Human Performance: A Review, Report to Lamp Division, National Electrical Manufacturers Association and the Lighting Research Institute*, National Electrical Manufacturers Association, Washington, DC, 1989.
13. W. Adrian, *A Model to Calculate the Visibility of Targets*, Lighting Research and Technology, Chartered Institution of Building Services Engineers, London, 1989.
14. E. W. Bickford, *Human Circadian Rhythms: A Review, Report to Lighting Research Institute*, LRI Project 88 DR NEMA 2, Lighting Research Institute, New York, 1988.
15. *Task Force Report on Light Treatment for Sleep Disorders, J. Biol. Rhythms* (Special Issue) **10**, (1995).
16. G. C. Brainard and C. A. Bernecker, "The Effects of Light on Human Physiology and Behavior," in *Proceedings of the 23rd Session of the CIE*, New Delhi, Nov. 1–8, 1995, CIE 120-1996, Commission Internationale de l'Eclairage, Vienna, 1996, pp. 88–100.
17. U.S. Congress, Office of Technology Assessment, *Biological Rhythms: Implications for the Worker*, OTA-BA-463, U.S. Government Printing Office, Washington, DC, 1991.
18. W. K. Adrian, ed., *LRI Proceedings 3rd International Symposium: Lighting for Aging Vision and Health*, Lighting Research Institute, New York, 1995.
19. C. L. Crouch, "Lighting for Seeing," in G. D. Clayton and F. E. Clayton, eds., *Patty's Industrial Hygiene and Toxicology*, 3rd Rev. Ed., Vol. 1, Wiley, New York, 1978, Chapter 13.
20. G. C. Brainard, "The Effects of Light on Physiology, Mood and Behavior in Humans," in *Psychological Aspects of Architectural Lighting Symposium Proceedings*, University Park, PA, 1990, pp. 90–99.

21. IES, *"Light and Vision,"* in *IES Lighting Handbook*, 5th Ed., Illuminating Engineering Society, New York, 1972, pp. 3–8.
22. S. K. Guth and J. F. McNelis, *Illum. Eng.* **63**, 32 (1968).
23. S. K. Guth and J. F. McNelis, *Illum. Eng.* **64**, 99 (1969).
24. R. J. Lythgoe, *Measurement of Visual Acuity*, Special Report No. 173, Medical Research Council, His Majesty's Stationary Office, London, 1932.
25. H. R. Blackwell, *Illum. Eng.* **54**, 317 (1959).
26. M. Luckiesh, *Light, Vision and Seeing*, 1st Ed., Van Nostrand, New York, 1944, pp. 158–165.
27. S. K. Guth, *J. Am. Ind. Hyg.* **23**, 359 (1962).
28. G. A. Fry, "Blur of the Retinal Image of an Object Illuminated by Low Pressure and High Pressure Sodium Lamps," *J. IES* **5**, 158–164 (1976).
29. S. M. Berman et al., "Lighting Spectral Effect on Landolt C Performance Is Enhanced by Blur and Abolished by Mydriasis," *J. IES* **25**(1), 42–50 (1996).
30. D. Kline, "Visual Requirements for the Aging Driver," in W. Adrian, ed., *Proceedings 3rd International Symposium on Lighting for Aging Vision and Health*, Lighting Research Institute, Hendersonville, NC, 1995, pp. 103–112.
31. W. R. Stevens and C. A. P. Foxell, *Light and Lighting (London)*, **48**, 419 (1955).
32. D. K. Tiller, M. J. Ouellette, and M. S. Rea, "The Effect of Luminous Surroundings on Visual Performance," *J. IES* **19**(2), 45–58 (1990).
33. P. W. Cobb and F. K. Moss, *J. Franklin Inst.* **199**, 507 (1925).
34. P. W. Cobb, *J. Exp. Psychol.* **1**, 540 (1916).
35. C. E. Ferree and G. Rand, *Trans. IES* **22**, 79 (1927).
36. C. E. Ferree and G. Rand, *Trans. IES* **23**, 507 (1928).
37. S. W. Smith, in *1974 Annual Report*, Illuminating Engineering Research Institute, New York, 1974, pp. 10–22.
38. S. W. Smith, *J. IES* **5**, 235 (1976).
39. I. Bailey, R. Clear, and S. Berman, "Size as a Determinant of Reading Speed," *J. IES* **22**(2), 102–117 (1993).
40. S. M. Berman et al, "Luminance-Controlled Pupil Size Affects Word-Reading." *J. IES* **25**(1), 51–59 (1996).
41. I. Lewin, "Road Scholar," *LD+A* **29**(3), 137–142 (1999).
42. A. L. Lewis, "Visual Performance as a Function of Spectral Power Distribution of Light Sources at Luminances Used for General Outdoor Lighting," *J. IES* **28**(1), 37–42 (1999).
43. A. L. Lewis, "Equating Light Sources for Visual Performance at Low Luminances," *J. IES* **27**(1), 80–84 (1998).
44. EPRI LRO, *4th International Lighting Research Symposium: Vision at Low Light Levels*, EPRI Lighting Research Office, Hendersonville, NC, 1998.
45. Y. He et al, "Evaluating Light Source Efficacy under Mesopic Conditions Using Reaction Times," *J. IES* **26**(1), 125–138 (1997).
46. N. H. Eklund, P. R. Boyce, and S. N. Simpson, "Lighting and Sustained Performance," in *IESNA Conference Proceedings, New Orleans August 9–11, 1999*, Illuminating Engineering Society of North America, New York, 1995, pp. 501–524.
47. F. Schieber, "Effects of Visual Aging on Driving Performance," in W. Adrian, ed., *Proceedings: 3rd International Symposium on Lighting for Aging Vision and Health*, Lighting Research Institute, Hendersonville, NC, 1995, pp. 113–127.

48. S. W. Smith and M. S. Rea, "Proofreading under Different Levels of Illumination," *J. IES* **8**(1), 47 (1978).
49. *Annual Report*, Illuminating Engineering Research Institute, New York, 1975.
50. H. C. Weston, *The Relation between Illumination and Industrial Efficiency: (1) The Effect of Size of Work, Joint Report of the Industrial Health Research Board (Medical Research Council) and the Illumination Research Committee (Department of Scientific and Industrial Research)*, Her Majesty's Stationary Office, London, 1935.
51. *The Relation between Illumination and Visual Efficiency: The Effect of Brightness Contrast, Joint Report of the Industrial Health Research Board Medical Research Council) and the Illumination Research Committee (Department of Scientific and Industrial Research)*, Her Majesty's Stationary Office, London, 1944.
52. G. A. Fry, "Assessment of Visual Performance," *Illum. Eng.* **57**, 426–437 (1962).
53. CIE, *An Analytical Model for Determining the Influence of Lighting Parameters on Visual Performance*, Vol. 1: *Technical Foundations*, Vol. 2: *Summary and Application Guidelines*, CIE Publication 19/2, Commission Internationale de l'Eclairage, Paris, 1981.
54. O. M. Blackwell and H. R. Blackwell, "Individual Responses to Lighting Parameters for a Population of 235 Observers of Varying Age," *J. IES* **9**, 205–232 (1980).
55. O. M. Blackwell and H. R. Blackwell, "Recent Investigation of Glare as a Factor in Visibility at Night," in H. Leibowitz and C. A. Johnson, co-chairs, *Night Vision Current Research and Future Directions, Symposium Proceedings, Working Group on Night Vision, Committee on Vision*, National Academy Press, Washington, DC, 1987, pp. 196–206.
56. J. J. Vos, "Disability Glare-A State of the Art Report," *CIE J.* **3**(2), 39–53 (1984).
57. R. E. Greule, "Contrast Thresholds at Transient Adaptation," *CIE J.* **8**(1), 25–26 (1989).
58. W. K. Adrian, "Die Unterschiedsempfindlichkeit des Augs und Die Möglichkeit Ihrer Berechnung," *Lichtechnik* **21**, 118 (1969).
59. B. Indistsky, H. W. Bodmann, and H. J. Fleck, "Visual Performance-Contrast Metric-Visibility Lobes-Eye Movements," *Lighting Res. Technol.* **14**, 218 (1982).
60. M. S. Rea, "Toward a Model of Visual Performance: Foundation and Data," *J. IES*, **15**, 41 (1986).
61. S. W. Smith and M. S. Rea, "Check Value Verification under Different Levels of Illumination," *J. IES* **16**(1), 21–38 (1987).
62. W. Adrian, "Visibility Levels under Night-time Driving Conditions," *J. IES* **16**, 3–12 (1987).
63. W. Adrian, "A Model to Calculate the Visibility of Targets," *Lighting Res. Technol.* **21**(4), 181–188 (1989).
64. *IESNA: Roadway Lighting*. RP-8-98, Illuminating Engineering Society of North America, New York, 1999.
65. M. E. Keck and R. E. Stark, "Evaluation of Visibility Models in the Roadway Situation," in *Proceedings of the CIE 21st Session*, Venezia, June 17–25, 1987, CIE Publication No. 71, Vol. 1, Commission International de l'Eclairage, Vienna, 1987, pp. 282–285.
66. P. L. Vrable, C. A. Bernecker, and R. G. Mistrick, "Visual Performance and Visual Clarity under Electric Light Sources, Part II: Visual Clarity, *J. IES* **27**(1), 29–41 (1998).
67. P. R. Boyce, "Illuminance Selection Based on Visual Performance—and Other Fairy Stories, *J. IES* **25**(2), 41–49 (1996).
68. E. J. Rinalducci, *Models of Visual Performance, Final Report to Illuminating Engineering Society of North America*, New York, 1995.

69. W. K. Adrian, R. Gibbons, "Visual Performance and Its Metric," *Light Eng.* **2**(3), 1–34 (1994).
70. P. L. Vrable, C. A. Bernecker, R. G. Mistrick, "Visual Performance and Visual Clarity under Electric Light Sources, Part I: Visual Performance, *J. IES* **24**(1), 69–80 (1995).
71. *1979 Annual Report*, Illuminating Engineering Research Institute, New York, 1979, pp. 5–22.
72. H. R. Blackwell and O. M. Blackwell, *Final Report IERI Project 30B, Report to the IERI*, Illuminating Engineering Research Institute, New York, 1984.
73. C. L. Crouch and R. L. Vincent, *Survey of Illumination Needs for Coal Processing Plant Areas both Interior and Exterior, Report to Bituminous Coal Research Institute by the IERI*, Illuminating Engineering Research Institute, New York, 1977.
74. C. L. Crouch and R. L. Vincent, *Survey of Illumination Needs of Tasks for Safe Work Performance on Walkways and Work Areas of Mobile Surface Mining Machinery, Report to Mine Safety Appliances Company by the IERI*, Illuminating Engineering Research Institute, New York, 1980.
75. C. L. Crouch and R. L. Vincent, *Survey of Illumination Needs for Safe and Productive Performance for Shipboard Tasks in the Underway Replenishment Area, the Destroyer Refueling Area, the Jet Engine Test Cell Area, and the Hangar Bay Area, Report to the Naval Air Systems Command, Dept. of the U.S. Navy, by the IERI*, Illuminating Engineering Research Institute, New York, 1980.
76. L. A. Jones, *Phil. Mag.* **39**, 96 (1920).
77. C. Dunbar, *Trans. IES (London)* **5**, 33 (1940).
78. C. L. Cottrell, *Illum. Eng.* **46**, 95 (1951).
79. A. E. Simmons and D. M. Finch, *Illum. Eng.* **48**, 517 (1953).
80. A. A. Eastman, *Illum. Eng.* **63**, 37 (1968).
81. H. R. Blackwell, *Illum. Eng.* **65**, 267 (1970).
82. C. L. Crouch and R. L. Vincent, "Disability and Discomfort Glare Field Studies under Low Luminance Conditions" in *Proceedings of the CIE 20th Session*, Amsterdam, 1983, CIE Publication No. 56, Commission International de l'Eclairage, Vienna, 1983.
83. C. L. Crouch and R. L. Vincent, *Survey of Illumination Needs for Safe and Productive Performance of Flight Deck Tasks and Measurements of Disability Glare, Report to the Naval Air Systems Command, Dept. of the U.S. Navy, by the IERI*, Illuminating Engineering Research Institute, New York, 1981.
84. J. E. Kaufman and J. F. Christensen, eds., *IES Lighting Handbook, Application Volume*, Illuminating Engineering Society of North America, New York, 1987.
85. J. E. Kaufman and J. F. Christensen, eds., *IES Lighting Handbook, Reference Volume*, Illuminating Engineering Society of North America, New York, 1984.
86. B. W. Tansley, K. W. Houser, and I. Pasini, "The IQCam Digital Image Photometer System: Principles of Operation and Comparative Performance," *J. IES* **28**(1), 182–42 (1999).
87. I. Lewin, R. Laird, J. Young, "Video Photometry for Quality Control," *LD+A*, **22**(1), 16–20 (1992).
88. M. S. Rea, ed., *IESNA Lighting Handbook: Illuminance Selection*, 8th Ed., Illuminating Engineering Society of North America, New York, 1993, Chapter 11.
89. M. S. Rea, ed., *IESNA Lighting Handbook: IESNA Lighting Design Guide*, 9th Ed., Illuminating Engineering Society of North America, New York, 2000, Chapter 10.
90. R. Clear and S. Berman, "Economics and Lighting Level Recommendations," *J. IES* **22**(2), 77–86 (1993).

91. M. Kohn, "Lighting Offices Containing VDTs," *LD+A* **28**, 9–11 (1988).
92. C. J. Lloyd, M. Mizukami, and P. R. Boyce, "A Preliminary Model of Lighting-Display Interaction," *J. IES* **25**(2), 59–69 (1996).
93. C. A. Bernecker, P. A. Sanders, and R. G. Mistrick, "Reflected Luminance Patterns in Visual Display Units," *J. IES* **24**(2), 8–18 (1995).
94. IESNA, *American National Standard Practice for Office Lighting*, ANSI/IESNA-1-1993, Illuminating Engineering Society of North America, New York, 1993.
95. J. M. Wolfe and K. M. O'Connell, "Adaptation of the Resting States of Accommodation: Dark and Light Field Measures," *Invest. Ophthalmol. Vis. Sci.* **28**(6), 992–996 (1987).
96. W. K. Adrian and M. L. Fleming, *Review of the Literature on Vision and Visual Display Terminals (VDT) and Lighting Interactions, Report to the Lighting Research Institute*, University of Waterloo, Waterloo, Ontario, Canada, New York, 1989.
97. *American National Standard Practice for Industrial Lighting (RP-7)*, Illuminating Engineering Society, New York, 1991.
98. J. F. Chorlton and H. F. Davidson, *Illum. Eng.* **54**, 474 (1959).
99. H. R. Blackwell, *Illum. Eng.* **58**, 161 (1963).
100. W. Allphin, *Illum. Eng.* **58**, 244 (1963).
101. C. L. Crouch and J. E. Kaufman, *Illum. Eng.* **58**, 277 (1963).
102. R. L. Henderson, J. F. McNelis, and H. G. Williams, *J. IES*, **4**, 150 (1975).
103. F. K. Sampson, *Contrast Rendition in School Lighting*, Technical Report No. 4, Educational Facilities Laboratories, New York, 1970.
104. I. Lewin and J. W. Griffith, *Illum. Eng.* **65**, 594 (1970).
105. N. S. Florence and S. B. Glickman, *Illum. Eng.* **66**, 149 (1971).
106. I. Goodbar et al., *LD+A*, **4**, 45 (1974).
107. I. Goodbar, in *1973 Annual Report*, Illuminating Engineering Research Institute, New York, 1973, pp. 22–29.
108. L. J. Buttolph, in *1973 Annual Report*, Illuminating Engineering Research Institute, New York, 1973, p. 34.
109. R. M. Boynton, E. J. Rinalducci, and C. Sternheim, *Illum. Eng.* **64**, 217 (1969).
110. G. A. Fry and V. M. King, *J. IES*, **4**, 307 (1975).
111. T. Liu, *Discomfort Glare: Luminaire Light Distribution, Vehicle Speed and Background in a Dynamic Roadway Lighting Simulation*, Masters Thesis, LRI Project 84 SP 7, Kansas State University, Manhattan, KS, 1989.
112. C. A. Bennett, *Discomfort Glare: Demographic Variables*, Special Report No. 118, Kansas Engineering Experimental Station, Kansas State University, Manhattan, KS, 1976.
113. L. L. Holladay, *J. Opt. Soc. Am.* **12**, 279 (1926).
114. W. S. Stiles, *Proceedings of the 7th Session of the CIE*, Saranac, Commission International de l'Eclairage, Paris, 1928.
115. G. A. Fry, *Illum. Eng.* **50**, 31 (1955).
116. "Outline of a Standard Procedure for Computing Visual Comfort Ratings for Interior Lighting," *J. IES* **2**(3), 328–344 (1973).
117. S. K. Guth, *Illum. Eng.* **58**, 351 (1963).
118. "An Alternate Simplified Method of Determining Acceptability of a Luminaire from the VCP Standpoint for Use in Large Rooms," *J. IES* **1**, 256–260 (1972).

119. D. S. Sliney, ed., *Symposium Notes: 1st International Symposium on Glare*, Lighting Research Institute, Hendersonville, NC, 1991.

120. S. M. Berman et al., "The Influence of Spectral Composition on Discomfort Glare for Large-Size Sources." *J. IES* **25**(1), 34–41 (1996).

121. J. J. Vos, "Age Dependence of Glare," in W. Adrian, ed., *Proceedings: 3rd International Symposium on Lighting for Aging Vision and Health*, Lighting Research Institute, Hendersonville, NC, 1995, pp. 11–25.

122. P. Ngai, and P. R. Boyce, "The Effect of Overhead Glare on Visual Discomfort," in *IESNA Conference Proceedings, New Orleans, August 9–11, 1999*, Illuminating Engineering Society of North America, New York, 1999, pp. 485–500.

123. A. J. Fisher and A. W. Christie, *Vision Res.* **5**, 565 (1965).

124. D. Sliney and M. Wolbarsht, *Safety with Lasers and Other Optical Sources*, Plenum Press, New York, 1980.

125. D. Sliney, "Ocular Injury Due to Light Toxicity," *Int. Ophthalmol. Clinics* **28**(3), (1988).

126. IESNA, *Lighting and the Visual Environment for Senior Living*, IES RP 29-98, Illuminating Engineering Society of North America, New York, 1998.

127. W. K. Adrian, *The Aging Eye and Adaptation of Lighting Techniques.*, University of Waterloo, Waterloo, Ontario, Canada, Final Report LRI Project 92 SP 1, Lighting Research Institute, New York, 1995.

128. Measurements Subcommittee, IESNA Photobiology Committee, *Recommended Practice for Photobiological Safety for Lamps and Lamp Systems—Measurement Techniques*, ANSI/IESNA RP-27.2-98, Illuminating Engineering Society of North America, New York, 1998.

129. IESNA Photobiology Committee, *Recommended Practice for Photobiological Safety for Lamps and Lamp Systems—General Requirements*, ANSI/IESNA RP-27.1-96, Illuminating Engineering Society of North America, New York, 1996.

130. IESNA Photobiology Committee, *Recommended Practice for Photobiological Safety for Lamps and Lamp Systems—Risk Group Classification and Labeling*, ANSI/IESNA RP-27.3-96, Illuminating Engineering Society of North America, New York, 1996.

131. IESNA Testing Procedures Committee, *IESNA Guide for the Measurement of Ultraviolet Radiation from Sources*, IESNA LM-55-96, Illuminating Engineering Society of North America, New York, 1996.

132. C. M. H. Driscoll, "Measurement Methods for Ultraviolet Radiation." in *Proceedings of the 24th Session of the CIE, Warsaw, June 24–30, 1999*, CIE Publication No. 133, Vol. 1, Part 2, Commission Internationale de l'Eclairage, Vienna, 1999, pp. 287–290.

133. ACGIH, *Threshold Limit Values for Chemical Substances and Physical Agents, Biological Exposure Indices*, American Conference of Governmental Industrial Hygienists, Cincinnati, OH, 1999.

134. R. E. Levin, "Photobiological Safety and Risk—ANSI/IESNA RP 27 Series," *J. IES* **27**(1), 136–143 (1998).

135. R. Matthes and D. Sliney, eds., *Measurement of Optical Radiation Hazards, A Reference Book Based on Presentations Given by Health and Safety Experts on Optical Radiation Hazards*, International Commission on Non-Ionizing Radiation Protection, IRCNIRP 6/98, Oberscheißheim, Germany, and Commission Internationale de l'Eclairage, x-016-1998 Vienna, Austria, 1998.

136. A. D. Ryer, *IL 1400A Radiometer/Photometer: Instruction Manual*, Rev. 920210, International Light, Inc., Newberryport, MA.

137. J. E. Birren, R. C. Casperson, and J. Botwinick, "Age Changes in Pupil Size," *J. Gerontol.* **5**, 216 (1950).

138. H. C. Weston, "The Effects of Age and Illumination upon Visual Performance with Close Sights," *Br. J. Ophthalmol.* **32**, 645 (1948).

139. E. Wolf, *Effects of Age on Peripheral Vision*, Highway Res. Board Bull. No. 336, 1962, p. 26.

140. O. M. Blackwell and H. R. Blackwell, "Visual Performance Data for 146 Normal Observers of Various Ages," *J. IES* **1**, 3–13 (1971).

141. H. Bouma, "Aging Effects on Vision," in *Proceedings of the 24th Session of the CIE, Warsaw, June 24–30, 1999*, CIE Publication No. 133, Vol. 1, Part 2, Commission Internationale de l'Eclairage, Vienna, 1999, pp. 368–370.

142. RPI-LRC, *Delta Portfolio: South Mall Tower—Lighting residences for the elderly and disabled*, Lighting Research Center, School of Architecture, Rensselaer Polytechnic Institute, Troy, NY, 2000.

143. CIE, *Low Vision, Lighting Needs for the Partially Sighted*, CIE 123-1997, Commission Internationale de l'Eclairage, Vienna, 1997.

144. H. Brandston, A Vision for Lighting the Lighthouse, *LD+A*, **27**(5), 54–57 (1997).

145. T. H. Monk et al., "Subjective Alertness Rhythms in Elderly People," *J. Biol. Rhythms* **11**(3), 268–276 (1996).

146. S. Sörensen and G. Brunnström, "Quality of Light and Quality of Life: An Intervention Study among Older People," *Lighting Res. Technol.* 27(2), 113–118 (1995).

147. E. E. Faye and C. S. Stuen, eds., *The Aging Eye and Low Vision: A Study Guide for Physicians*, 2nd Ed., Lighthouse National Center for Vision and Aging, The Lighthouse, Inc., New York, 1995.

148. J. M. Halloin, "Application of Current Knowledge in Lighting to the Needs of the Homebound Aged," *J. Long-Term Home Health Care PRIDE Inst. J.* **14**(4), 32–39 (1995).

149. C. M. Singer, "Non-Visual Effects of Lighting," in W. K. Adrian, ed., *Proceedings of the 3rd International Symposium: Lighting for Aging Vision and Health*, Lighting Research Institute, New York, 1995, pp 143–148.

150. H. Brandston, "Lighting Design—A Practitioner's Problem," *Int. Lighting Rev.* **3**, 67–68 (1983).

151. T. McGowan and N. Miller, "What Is Lighting Quality and How Do We Apply It to Lighting Design? in *Proceedings of the 24th Session of the CIE, Warsaw, June 24–30, 1999*, CIE Publication NO. 133, Vol. 1, Part 1, Commission Internationale de l'Eclairage, Vienna, 1999, pp. 11–16.

152. J. A. Veitch and G. R. Newsham, "Preferred Luminous Conditions in Open-Plan Offices: Implications for Lighting Quality Recommendations," in *Proceedings of the 24th Session of the CIE, Warsaw, June 24–30, 1999*, CIE Publication No. 133, Vol. 1, Part 2, Commission Internationale de l'Eclairage, Vienna, 1999, pp. 4–6.

153. J. A. Veitch, ed., *Proceedings of the 1st CIE Symposium on Lighting Quality, Ottawa, Canada, May 9–10, 1998*, Publication CIE X015-1998, Commission Internationale de l'Eclairage, Vienna, 1998.

154. J. A. Veitch and G. R. Newsham, "Determinants of Lighting Quality, I: State of Science," *J. IES* **27**(1), 92–106 (1998).

155. J. A. Veitch and G. R. Newsham, "Lighting Quality and Energy—Efficiency Effects on Task Performance, Mood, Health, Satisfaction, and Comfort," *J. IES* **27**(1), 107–129 (1998).

156. T. McGowan, "Measures for Lighting Quality in Interiors," in *Proceedings of the 23rd Session of the CIE, New Delhi, November 1–8, 1995*, CIE 120-1996, Vol. 2, Commission Internationale de l'Eclairage, Vienna, 1995, pp. 73–74.

157. P. R. Boyce and N. H. Eklund, "The Development of a Reliable, Valid, and Simple Office Lighting Survey," *J. IES* **25**(2), 25–40 (1996).

158. B. L. Collins and P. A. Sander, Post-Occupancy Evaluation of the Forrestal Building," *J. IES* **25**(2), 89–103 (1996).

159. I. Lewin and T. Cheknis, "An Index of Lamp Hiding Power for Lensed Fluorescent Luminaires," *J. IES* **18**(1), 9–15 (1989).

160. S. M. Berman et al., "Pupillary Size Difference under Incandescent and High Pressure Sodium Lamps," *J. IES* **16**, 3–20 (1987).

161. H. A. Piper, "The Effect of HPS Light on Performance of a Multiple Refocus Task," *LD+A*, **11**, 36 (1981).

162. S. W. Smith and M. S. Rea, "Relationship between Office Task Performance and Ratings of Feelings and Task Evaluation under Different Light Sources and Level," in *Proceedings of the CIE 19th Session*, Kyoto, 1979, Publication No. 50, Commission Internationale de l'Eclairage, Paris, 1980, pp. 207–211.

163. H. Helson, D. B. Judd, and M. H. Warren, "Object-Color Changes from Daylight to Incandescent Filament Illumination," *Illum. Eng.* **47**, 221–233 (1952).

164. D. Nickerson and C. W. Jerome, "Color Rendering of Light Sources: CIE Method of Specification and Its Application," *Illum. Eng.* **50**, 262 (1965).

165. CIE, *Methods of Measuring and Specifying Color Rendering Properties of Light Sources*, CIE Publication 13.2, Commission Internationale de l'Eclairage, Paris, 1974.

166. H. Helson and T. Lansford, "The Role of Spectral Energy of Source and Background Color in the Pleasantness of Object Colors," *Appl. Opt.* **9**, 1513 (1970).

167. D. B. Judd, "A Flattery Index for Artificial Illuminants," *Illum. Eng.* **52**, 593 (1967).

168. C. W. Jerome, "The Flattery Index," *J. IES*, **2**, 351 (1973).

169. B. L. Collins, "Safety Color Appearance under Different Illuminants," *J. IES* **16**(1), 21–38 (1987).

170. IESNA, *Color and Illumination, DG-1-90*, Illuminating Engineering Society of North America, New York, 1990.

171. M. S. Rea, ed., *IESNA Lighting Handbook: Color*, 8th ed., Illuminating Engineering Society of North America, New York, 1993, Chapter 4.

172. "Color and the Use of Color by the Illuminating Engineer," *Illum. Eng.* **57**, 764 (1962).

173. J. E. Flynn, "A Study of Lighting as a System of Spatial Cues," Environmental Design Research Association Workshop on "The Psychological Potential of Illumination," University of Kansas, April, 1975.

174. J. E. Flynn et al., "The Influence of Spatial Light on Human Judgment," in *Proceedings of the CIE 18th Session, London, 1975*, Publication CIE No. 36, Commission Internationale de l'Eclairage, Paris, 1976, pp. 39–46.

175. J. E. Flynn, "A Study of Subjective Responses to Low-Energy and Non-Uniform Lighting Systems," *LD+A* 7(2), 6–15 (1977).

176. J. E. Flynn et al., "A Guide to Methodology Procedures For Measuring Subjective Impression in Lighting," *J. IES* **8**, 95–110.

177. J. E. Flynn and T. J. Spencer, "The Effects of Light Source Color on User Impressions and Satisfaction," *J. IES* **6**, 167–179 (1977).

178. J. E. Flynn, A. W. Segal, and G. R. Steffy, *Architectural Interior Systems: Lighting, Acoustic, Air Conditioning*, 2nd Ed., Van Nostrand Reinhold, New York, 1988.

179. R. J. Hawkes, D. L. Loe, and E. Rowlands, "A Note towards the Understanding of Lighting Quality," *J. IES* **8**, 111–120 (1979).

180. B. L. Colins, *Evaluation of Subjective Response to Lighting Distributions: A Literature Review*, NISTIR 5119, U.S. Department of Commerce, National Institute of Standards and Technology, Gaithersburg, MD, 1993.

181. C. A. Bernecker and P. L. Vrabel, *IALD Annotated Bibliography—For Studies in Luminance and Luminance Distribution. Report to Metric of Quality Committee*, International Association of Lighting Designers, New York, 1993.

182. B. L. Collins et al., "Second Level Post Occupancy Evaluation Analysis," *J. IES* **19**(2), 21–44, (1990).

183. B. L. Collins, "Evaluation of the Role of Luminance Distributions in Occupant Response to Lighting," in *Proceedings, Cambridge, England: National Lighting Conference*, Chartered Institution of Building Services Engineers, London, 1990, pp. 1–10.

184. B. L. Collins et al., *Evaluating Office Lighting Environments: Second Level Analysis*, NISTIR 89-406, National Institute of Standards and Technology, Gaithersburg, MD, 1989.

185. B. L. Collins et al., *Post-Occupancy Evaluation of Several U.S. Government Buildings*, NISTIR 89-4175, National Institute of Standards and Technology, Gaithersburg, MD, 1989.

186. G. Gillette, *Evaluating Office Lighting Environments: Reference Lighting Power Density Data*, NBSIR 88-3691, NBS, Gaithersburg, MD, 1988.

187. G. Gillette and M. Brown, *Occupant Evaluation of Commercial Office Lighting: Volume III, Data Archive and Database Management System*, ORNL/TM-10264/V2, Oak Ridge National Laboratory, Oak Ridge, TN, 1987.

188. G. Gillette and M. Brown, *Occupant Evaluation of Commercial Office Lighting: Volume 1, Methodology and Bibliography*, ORNL/TM-10264/V1, Oak Ridge National Laboratory, Oak Ridge, TN, 1986.

189. N. H. Eklund, P. R. Boyce, and S. N. Simpson, "Lighting and Sustained Performance," in *IESNA Conference Proceedings, New Orleans, August 9–11, 1999*, Illuminating Engineering Society of North America, New York, 1999, pp. 501–524.

190. L. T. Troland, "An Analysis of the Literature Concerning the Dependency of Visual Functions upon Illumination Intensity," *Trans. IES* **26**, 107 (1931).

191. C. E. Ferree and G. Rand, "Tests for the Efficiency of the Eye under Different Systems of Illumination and a Preliminary Study of the Causes of Discomfort," *Trans. IES* **8**, 40 (1913).

192. J. R. Cravath, "Some Experiments with the Ferree Test for Eye Fatigue," *Trans. IES* **9**, 1033 (1914).

193. C. E. Ferree and G. Rand, "The Efficiency of the Eye under Different Conditions of Lighting: The Effects of Varying the Distribution of Factors and Intensity," *Trans. IES* **10**, 407 (1915).

194. C. E. Ferree and G. Rand, "Some Experiments of the Eye with Different Illuminants, Part I," *Trans. IES* **13**, 50–82 (1918).

195. C. E. Ferree and G. Rand, "Some Experiments of the Eye with Different Illuminants, Part II," *Trans. IES* **14**, 107–132 (1919).

196. C. E. Ferree and G. Rand, "The Effect of Variations in Intensity of Illumination on Functions of the Importance to the Working Eye," *Trans. IES* **15**, 769–801 (1920).

197. C. E. Ferree and G. Rand, "The Effect of Variations of Visual Angle, Intensity and Composition of Light on Important Ocular Functions," *Trans. IES* **17**, 69–102 (1922).
198. C. E. Ferree and G. Rand, "Further Studies on the Effect of Composition of Light on Important Ocular Functions," *Trans. IES* **19**, 424–455 (1924).
199. H. C. Weston, *Sight, Light and Efficiency*, H. K. Lewis, London, 1949.
200. W. G. Lancaster, "Light and Lighting," in C. Berens, ed., *The Eye and Its Diseases*, 2nd Ed., Saunders, Philadelphia, 1950, p. 81.
201. F. W. Hebbard, "Micro Eye Movements: Effects of Target Illumination and Contrast," *Illum. Eng.* **64**, 199 (1969).
202. D. A. Campbell, W. J. B. Riddell, and A. S. MacNalty, *Eyes in Industry*, Longmans, Green, New York, 1951.
203. M. Poser, "Eye Fatigue in Industry," *Trans. IES* **16**, 431 (1921).
204. M. Luckiesh and F. K. Moss, *The Science of Seeing*, Van Nostrand, New York, 1937.
205. C. L. Crouch, "Discussion," *Illum. Eng.* **47**, 344 (1952).
206. C. P. Steinmetz, "The Scope of Illuminating Engineering," *Trans. IES* **11**, 625 (1916).
207. H. P. Gage, "Hygienic Effects of Ultraviolet Radiation in Daylight," *Trans. IES* **25**, 377 (1930).
208. Centers for Disease Control and Prevention, "Tuberculosis Mortality—United States, 1992," *MMWR Morb. Mortal. Wkly Rep.* **42**, 696–697, 703–704 (1992).
209. T. R. Frieden et al., "The Emergence of Drug-Resistant Tuberculosis in New York City," *N. Engl. J. Med.* **328**, 521–526 (1993).
210. Centers for Disease Control. Nosocomial Transmission of Multidrug-Resistant Tuberculosis among HIV-Infected Persons—Florida and New York, 1988–1991," *MMWR Morb. Mortal. Wkly Rep.* **40**, 585–591 (1991).
211. Centers for Disease Control. "National Action Plan to Combat Multidrug-Resistant Tuberculosis," *MMWR Morb. Mortal. Wkly Rep.* **41**, 5–8 (1992).
212. J. M. McAdam et al., "The Spectrum of Tuberculosis in a New York City Men's Shelter Clinic (1982–1988)," *Chest* **97**, 798–805 (1990).
213. P. W. Brickner, L. L. Scharer, J. M. McAdam, "Tuberculosis in Homeless Populations," in L. B. Reichman and E. S. Herschfield, eds., *Tuberculosis: A Comprehensive International Approach*, Marcel Dekker, New York, 1993, pp. 433–453.
214. T. Dumyahn, M. First, "Characterization of Upper-room Ultraviolet Air Disinfection Devices," *AIHA J.* **60**, 219–227 (1999).
215. CIE, *Ultraviolet (UV) Air Disinfection: Present State of the Art*, Report CIE TC 6–35, Commission Internationale de l'Eclairage, Vienna (in press).
216. M. W. First et al., "Guidelines for the Application of Upper-Room Germicidal Irradiation for Preventing the Transmission of Airborne Contagion, Part I: Basic Principles," *ASHRAE Trans.* **105**, 869–876 (1999).
217. M. W. First et al., "Guidelines for the Application of Upper-Room Germicidal Irradiation for Preventing the Transmission of Airborne Contagion, Part II: Design and Operation Guidance," *ASHRAE Trans.* **105**, 877–887 (1999).
218. R. L. Vincent, "Airborne Disease Control: Measurement of Ultraviolet Germicidal Irradiation (UVR) in High Risk Environments," in R. Matthes and D. Sliney, eds., *ICNIRP/CIE: Measurement of Optical Radiation Hazards, A Reference Book Based on Presentations Given by Health and Safety Experts on Optical Radiation Hazards, Gaithersburg, Maryland, USAm September 1–3, 1998*, IRCNIRP 6/98 and CIE x016-1998, International Commission on Non-

Ionizing Radiation Protection, Oberscheißheim, Germany, and Commission Internationale de l'Eclairage, Vienna, Austria. 1998, pp. 369–386.

219. E. A. Nardell, "Interrupting Transmission from Patients with Unsuspected Tuberculosis: A Unique Role for Upper-Room Ultraviolet Air Disinfection," *Am. J. Infect. Control* **23**, 156–164 (1995).
220. Centers for Disease Control, "Guidelines for Preventing the Transmission of Mycobacterium Tuberculosis in Health Care Facilities," *MMWR* **43** (RR-13), 1–132 (Oct. 28, 1994).
221. R. L. Riley, "Ultraviolet Air Disinfection: Rationale for Whole Building Irradiation," *Infect. Control Hosp. Epidemiol.* **15**, 324–328 (1994).
222. E. A. Nardell, "Environmental Control of Tuberculosis." *Med. Clin. N. Am.* **77**, 1315–1334 (1993).
223. R. L. Riley and E. A. Nardell, "Controlling Transmission of Transmission of Tuberculosis in Health Care Facilities: Ventilation, Filtration, and Ultraviolet Air Disinfection," *Plant Technol. Safety Manage. Ser.* **1**, 25–31 (1993).
224. M. S. Rea, ed., *IESNA Lighting Handbook*, 8th Ed., Illuminating Engineering Society of North America, New York, 1993, pp. 151–155.
225. E. A. Nardell and R. L. Riley, "A New Ultraviolet Germicidal Irradiation (UVGI) Fixture Design for Upper Room Air Disinfection with Low Ceilings," in *Program and Abstracts of the World Congress on Tuberculosis*, Bethesda, MD, 1992, p. 38.
226. Philips Lighting, *Disinfection by UV-Radiation*, Eindhoven, Holland, August, 1992.
227. D. H. Sliney, "Ultraviolet Radiation and the Eye," in M. Granddolfo, A. Rindi, D. H. Sliney, eds., *Lights, Lasers, and Synchronous Radiation: A Health Assessment*, Plenum, New York, 1991 pp. 237–245.
228. I. L. Shechmeister, "Sterilization by Ultraviolet Irradiation," in S. S. Block, *Disinfection, Sterilization, and Preservation*, 4th Ed., Lea and Febiger, Philadelphia, pp. 553–565.
229. E. A. Nardell et al., "Airborne Infection: The Theoretical Limits of Protection Achievable by Building Ventilation," *Am. Rev. Resp. Dis.* **144**, 302–306 (1991).
230. R. L. Riley and E. A. Nardell, "Clearing the Air—The Theory and Application of Ultraviolet Air Disinfection," *Am. Rev. Resp. Dis.* **139**, 1286–1294 (1989).
231. National Institute for Occupational Safety and Health, *Criteria for a Recommended Standard for Occupational Exposure to Ultraviolet Radiation*, National Institute for Occupational Safety and Health, Cincinnati, OH, 1972.
232. R. L. Riley, M. Knight, G. Middlebrook, "Ultraviolet Susceptibility of BCG and Virulent Tubercle Bacilli," *Am. Rev. Resp. Dis.* **113**, 413–418 (1976).
233. R. L. Riley, S. Permutt, J. E. Kaufman, "Convection, Air Mixing and Ultraviolet Air Mixing in Rooms," *Arch. Environ. Health* **222**, 200 (1971).
234. R. L. Riley, S. Permutt, "Room Air Disinfection by Ultraviolet Air Disinfection of Room Air-Air Mixing and Germicidal Effectiveness," *Arch. Environ. Health* **22**, 208 (1971).
235. R. L. Riley, S. Permutt, and J. E. Kaufman, "Room Air Disinfection by Ultraviolet Irradiation by Ultraviolet Irradiation of Upper Room Air. *Arch. Environ. Health* **23**, 35 (1971).
236. R. L. Riley et al., "Infectiousness of Air from a Tuberculosis Ward—Ultraviolet Irradiation of Infected Air: Comparative Infectiousness of Different Patients," *Am. Rev. Respir. Dis.* **84**, 511–525 (1962).
237. W. F. Wells, *Airborne Contagion and Air Hygiene*, Harvard University Press, Cambridge, Massachusetts, 1955.

238. L. J. Buttolph and H. Haynes, *Ultraviolet Air Sanitation*, General Electric Lamp Division, Cleveland Ohio, December 1953.

239. L. J. Buttolph, "Practical Application and Sources of Ultraviolet Energy," in A. Hollaender et al., eds., *Radiation Biology*, Vol. II, McGraw-Hill, New York, 1952, pp. 41–93.

240. M. Luckiesh, *Application of Germicidal, Erythemal, and Infrared Energy*, Van Nostrand, New York, 1946.

241. C. Berens and C. L. Crouch, "Is Fluorescent Lighting Injurious to the Eyes?" *Am. Ophthalmol.* **45** , 47 (1958).

242. International Non-Ionizing Radiation Committee of the International Protection Association, "Fluorescent Lighting and Malignant Melanoma," *Health Phys. J.* **58**(1), 111–112 (1990).

243. F. Urbach et al., *Artificial Lighting: Influence on the Production of Solar UVR-Induced Skin Tumors in Mice, Report to Lighting Research Institute*, Project 84 DR NEMA 1, Lighting Research Institute, New York, April 1987.

244. R. J. Wurtman, "The Effects of Light on the Human Body," *Sci. Am.* **233**, 68 (1975).

245. T. P. Vogl, "Photomedicine," *Opt. News* (Spring), 6 (1976).

246. J. R. Lucey, "The Effects of Light on the Newly Born Infant," *J. Perinat. Med.* **1**, 1 (1973).

247. R. J. Cremer, P. W. Perryman, and D. H. Richards, "Influence of Light on the Hyperbilirubinemia of Infants," *Lancet* **1**, 1094 (1958).

248. J. Bullough and M. S. Rea, "Lighting for Neonatal Intensive Care Units: Some Critical Information for Design, *Lighting Res. Technol.* **28**(4), 189–198 (1996).

249. R. M. Neer et al., "Stimulation by Artificial Lighting on Calcium Absorption in Elderly Human Subjects," *Nature* **229**, 255–257 (1971).

250. W. F. Loomis, "Rickets," *Sci. Am.* **233**, 77 (1970).

251. J. A. Parrish et al., *New Engl. J. Med.* **291**, 1207 (1974).

252. I. Diamond et al., *Lancet* **15**, 1175 (1972).

253. S. G. Granelli et al., *Cancer Res.* **35**, 2567 (1975).

254. T. J. Dougherty et al., *J. Natl. Cancer Inst.* **55**, 115 (1975).

255. J. F. Kelly, M. E. Snell, and M. C. Berenbaum, *Br. J. Cancer* **31**, 237 (1975).

256. D. G. Pitts, J. P. G. Bergtmanson, and L. W.-F. Chu, "Rabbit Eye Exposure to Broad-Spectrum Fluorescent Lighting," *Acta Ophthalmol.* (Suppl.) **159**, 1–54 (1983)

257. S. Zigman and R. J. Collier, "The Grey Squirrel Lens Protects the Retina from Near-UV Radiation Damage," in J. G. Hollyfield, R. E. Anderson, and M. M. LaVail, eds., *Degenerative Retinal Disorders: Clinical and Laboratory Investigations*, Alan R. Liss, New York, 1987, pp. 571–586.

258. S. Zigman and R. J. Collier, "Comparison of Retinal Photochemical Lesion after Exposure to Near-UV or Short-wavelength Visible Radiation," in M. M. LaVail, R. E. Anderson, and J. G. Hollyfield, eds., *Inherited and Environmentally Induced Retinal Degenerations*, Alan R. Liss, New York, 1989, pp. 569–575.

259. L. R. Ronchi, "An Annotated Bibliography on Some Basic Effects of Optical Radiation on Humans," *CIE J.* **7**(2), 42–48 (1988).

260. D. L. Jewett et al., "The Lack of Effect on Human Muscle Strength of Light Spectrum and Low Frequency Electromagnetic Radiation in Interior Lighting," *J. IES* **15**, 19 (1986).

261. G. C. Brainard et al., "Circadian Rhythms of Melatonin in Primate (*Cynomologus macaque*) Cerebrospinal Fluid," *Anat. Rec.* **221**(3), 25A (1985).

262. G. C. Brainard et al., "The Influence of Light Irradiance and Wavelength on Pineal Physiology of Mammals," in P. J. O'Brien and D. Klein, eds., *Pineal and Retinal Relationships*, Academic Press, 1986, pp. 253–277.

263. G. C. Brainard and F. M. Barker, "Regulation of the Pineal-Reproductive Axis by Near Ultraviolet Radiation (UV-A) in Three Rodent Species," in G. P. Trentini, C. DeGaetani, and P. Pevet, eds., *Fundamentals and Clinics in Pineal Research*, Raven Press, New York, 1987, pp. 207–210.

264. R. A. Hoffman, L. B. Johnson, and R. J. Reiter, "Harderian Glands of Golden Hamsters: Temporal and Sexual Differences in Immunoreactive Melatonin," *J. Pineal Res.* **2**, 161–168 (1985).

265. R. A. Hoffman and L. B. Johnson, "Effects of Photic History and Illuminance Levels on Male Golden Hamster," *J. Pineal Res.* **2**, 209–215 (1985).

266. R. A. Hoffman, L. B. Johnson, and G. Brown, "Growth and Development of Golden Hamsters: Influence on the Pineal Gland, Melatonin and Photic Input," in G. Brown and S. D. Winwright, eds., *The Pineal Gland, Endocrine Aspects, Proceedings of a Symposium Held in Conjunction with the Seventh International Endocrinology Congress, Canada*, Pergamon, 1984, Chapter 42.

267. R. A. Hoffman, L. B. Johnson, and R. Corth. "The Effects of Spectral Power Distribution and Illuminance Levels on Key Parameters in the Male Golden Hamster and Rat with Preliminary Observations on the Effects of Pinealectomy," *J. Pineal Res.* **2**, 217–233 (1985).

268. R. A. Hoffman and L. B. Johnson "Interactions of Diet and Photoperiod on Growth and Reproduction in Male Golden Hamsters Growth," *Growth*, **49**, 380–399 (1985).

269. A. J. Lewy et al., "Antidepressant and Circadian Phase-Shifting Effects of Light," *Science*, **16**(5), 11–15 (1986).

270. E. W. Bickford, *Human Circadian Rhythms: A Review, Report to Lighting Research Institute*, Project 88 DR, *National Electrical Manufacturers Association*, New York, December 1988.

271. A. L. Lewis, "Light Affects Health," *LD+A* **17**, 32–36 (1987).

272. C. A. Czeisler et al., "Bright Light Induction of Strong (Type O) Resetting of the Human Circadian Pacemaker," *Science*, **244**, 1328–1333 (1989).

273. Society for Light Treatment and Biological Rhythms, Inc., New Haven, CT, *www.sltbr.org*

274. M. S. Rea, ed., *IESNA Lighting Handbook: Lighting Design Process*, 8th ed., Illuminating Engineering Society of North America, New York, 1993, Chapter 10.

275. *Illuminating Engineers, Chronicle of Guidance Publication*, Monrovia, NY, 1988.

276. J. R. Benya, "The Lighting Design Professional, Making a Living at Lighting," *Architectural Lighting*, **40**, 42 (1989).

277. "The Profession of Lighting Design: A Roundtable Discussion," *Lightview International*, Vol. 2, No. 1, International Association of Lighting Designers, New York, 1989.

278. W. J. Jankowski, "The Origins of Lighting Design," *Lightview International*, Vol. 2, No. 1, International Association of Lighting Designers, New York, 1989.

279. "What to Consider When Looking for a Lighting Designer and Who to Chose" *Lightview International*, Vol. 2, No. 1, International Association of Lighting Designers, New York, 1989.

280. J. R. Benya, "Essay by Invitation" (Lighting Certified), *LD + A* **28**(7), 12–13, (1998).

281. G. Gordon and J. Nuckolls, eds., *Interior Lighting for Designers*, 3rd Ed., John Wiley & Sons, New York, 1995.

282. Eley Associates, *Advanced Lighting Guidelines: 1993 (Revision 1), Report for EPRI, California Energy Commission, U.S. Department of Energy*, EPRI TR-101022, R1, Electric Power Research Institute, Palo Alto, CA, 1993.

283. *National Product Information Program (NLPIP)*, Lighting Research Center, Rensselaer Polytechnic Institute, Troy, NY, 1991–1999.

284. B. L. Collins and P. A. Sander, "Post-Occupancy Evaluation of the Forrestal Building," *J. IES* **25**(2), 89–103 (1996).

285. I. Lewin and L. Stafford, "Photometric and Field Performance of High Pressure Sodium Luminaires," *J. IES* **16**(2), 106–114 (1987).

286. I. Lewin and L. Stafford, "Real World Use of Photometric Test Reports," *LD + A* **18**(1), 18–24 (1988).

287. I. Lewin and L. Stafford, "Photometric and Field Performance of Metal Halide Luminaires," *J. IES* **17**(1), 15–19 (1988).

288. P. R. Boyce, *The Emergency Egress Roundtable: 1991*, Lighting Research Center/RPI, Lighting Research Institute and National Electrical Manufacturers Association, New York, 1991.

289. W. E. Brackett and R. E. Levin, "Zero-Distance In-Place Luminaire Flux Measurements," *J. IES* **25**(2), 173–179 (1996).

290. M. S. Rea, ed., *IESNA Lighting Handbook: Daylighting*, 8th ed., Illuminating Engineering Society of North America, New York, 1993, Chapter 8.

291. IESNA, *Daylighting*, RP5-99, Illuminating Engineering Society of North America, New York, 1999.

292. EPRI, *Daylighting Design Smart and Simple*, TR-109720, Electric Power Research Institute, Palo Alto, CA, 1997.

293. D. Rosenfeld, *Efficient Use of Energy in Buildings, Report on the 1975 Berkeley Summer Study*, LBL 4411, Lawrence Berkeley Laboratory, University of California, Berkeley, 1976.

294. IESNA, *Recommended Practice of Daylighting*, RP-5, Illuminating Engineering Society of North America, New York, 1979.

295. IES Committee on Calculation Procedures, *IES Recommended Practice for the Lumen Method of Daylight Calculations*, RP-23, Illuminating Engineering Society of North America, New York, 1989.

296. S. Selkowitz, "Smart Windows," *Glass Mag.* (August) 86–91 (1986).

297. S. Selkowitz and R. McClunney, *Commercial Building Fenestration Performance Indices Project; Phase I: Development of Methodology*, prepared by Windows and Lighting Program, Applied Science Division, Lawrence Berkeley Laboratory, and the Florida Solar Energy Center for the Lighting Research Institute, under contract to the Electric Power Research Institute, Palo Alto, CA, December 1988.

298. K. Papamichael and F. Winkelmann, "Solar-Optical Properties of Multilayer Fenestration Systems," 1986 International Daylighting Conference Proceedings II, American Society of Heating, Refrigerating, and Air-conditioning Engineers, 1988.

299. S. Selkowitz, K. Papamichael, and J. Klems, "Determination and Application of Bidirectional Solar-Optical Properties of Fenestration Systems," Proceedings, 13th National Passive Solar Conference, June 19–24, 1988, Massachusetts Institute of Technology Cambridge, MA, 1988.

300. S. Selkowitz et al., "Determining Daylight Illuminance in Rooms Having Complex Fenestration Systems," 1986 International Daylighting Conference Proceedings II, American Society of Heating, Refrigerating, and Air-conditioning Engineers, 1988.

301. R. Sullivan et al., "An Approach for Evaluating the Thermal Comfort Effects of Nonresidential Building Fenestration Systems," presented at the International Symposium on Advanced Comfort Systems for the Work Environment, Troy, NY, May 1–3, 1988, published in the *Proceedings*, LBL-25060, BS-249.

302. R. Sullivan et al., "An Indices Approach for Evaluating the Performance of Fenestration Systems in Nonresidential Buildings," *ASHRAE Trans.* **94**(2), (1988).

303. K. M. Papamichael and S. Selkowitz, "Simulating the Luminous and Thermal Performance of Fenestration Systems," *LD + A* **17**, 37–45 (1987).

304. M. D. Parent and J. B. Murdoch, "The Expansion of the Zonal Cavity Method of Interior Lighting Design to Include Skylights," *J. IES* **17**, 141–173 (1988).

305. M. D. Parent and J. B. Murdoch, "Analysis and Comparison of Skylights Based on Their Intensity Distribution Curves," *Lighting Res. Technol.* **21**(3), 111 (1989).

306. C. L. Crouch, *Effective Seeing in an Era of Energy Conservation*, Illuminating Engineering Research Institute, New York, 1973.

307. C. L. Crouch, *Effective Seeing and Conservation Too*, Illuminating Engineering Research Institute, New York, 1974.

308. J. E. Kaufman, "Optimizing The Use of Energy for Lighting," *LD + A* **4**, 8 (1973).

309. IES Energy Management Committee, *IES Recommended Procedure for Lighting Power Limit Determination*, IES LEM-1-1982, Illuminating Engineering Society of North America, New York, 1983.

310. IES Energy Management Committee, *IES Recommended Procedure for Lighting Energy Limit Determination*, IES LEM-2-1984, Illuminating Engineering Society of North America, New York, 1985.

311. IES Energy Management Committee, *IES Design Considerations Effective Building Lighting Energy Utilization*, IES LEM-3, Illuminating Engineering Society of North America, New York, 1987.

312. IES Energy Management Committee, *IES Recommended Procedure for Energy Analysis of Building Designs and Installations*, IES LEM-4-1984, Illuminating Engineering Society of North America, New York, 1985.

313. A. B. Gough, "Progress Report on Setting a National Agenda for Lighting Research and Education," Keynote Speech IESNA Conference, Baltimore, MD, August 30, 1990, Lighting Research Institute, New York.

314. USDOE, *National Energy Strategy: Interim Report* (April 1990), U.S. Department of Energy, Washington, DC, 1990.

315. EPRI, *Efficient Electrical Use: Estimates of Maximum Energy Savings*, EPRI CUG 746 Project 2788, Electric Power Research Institute, Palo Alto, CA, March 1990.

316. IES Education Committee, *Introductory Lighting Education, ED 100 Series*, Illuminating Engineering Society of Lighting Guidelines (Pub. No. 400-90-014), California Energy Commission, Sacramento, CA, March 1990.

317. EPRI, *Lighting Retrofit Manual*, prepared by Eley Associates, TR-107130-R1, Electric Power Research Institute, Palo Alto, CA, 1998.

318. NCQLP, *Lighting Certification: Candidate Handbook, 1997*, National Council on Qualifications for the Lighting Professions, Bethesda, MD, 1997.

319. EPRI, *Lighting Controls: Patterns for Design*, prepared by Rundquist Associates, TR-107230, Electric Power Research Institute, Palo Alto, CA, 1996.

320. IESNA, *Recommended Practice for the Economic Analysis of Lighting*, RP-31-96, Illuminating Engineering Society of North America, New York, 1995.
321. P. A. Bleasby, *Energy and Environmental Aspects of Lighting*, CIE Technical Reporter, Division 7: 1995–1999 Quadrennial Report, Commission Internationale de l'Eclairage, Vienna, 1999.
322. E. Mills, "Energy and Lighting," in *Proceedings of the 23rd Session of the CIE, New Delhi, November 1–8, 1995*, CIE 120-1996, Vol. 2, Commission Internationale de l'Eclairage, Vienna, 1996, pp. 76–78.
323. A. I. Slater, "Lighting and Energy in Buildings," in *Proceedings of the 23rd Session of the CIE, New Delhi, November 1–8, 1995*, CIE 120-1996, Vol. 2, Commission Internationale de l'Eclairage, Vienna, 1996, pp. 81–87.
324. R. G. Mistrick, and A. Choi, "A Comparison of Visual Comfort Probability and Unified Glare Rating Systems," *J. IES* **28**(2), 94–101 (1999).
325. C. E. Waters, R. G. Mistrick, and C. A. Bernecker, "Discomfort Glare from Sources of Nonuniform Luminance," *J. IES* **24**(2), 73–85 (1995).
326. P. C. Box, "Public Lighting Needs," *Illum. Eng.* **61**, 585 (1966).
327. *1964 World Almanac*, New York, 1964.
328. R. Cowen, *Eyes on the Workplace, Report for the Committee on Vision, Commission on Behavioral and Social Sciences and Education, National Research Council*, National Academy Press, Washington, DC, 1988.
329. R. Cowen, *Work, Aging, and Vision, Report for the Working Group on Aging Workers and Visual Impairment, Committee on Vision, Commission on Behavioral and Social Sciences and Education, National Research Council*, National Academy Press, Washington, DC, 1987.
330. R. A. Weale, "Do Years or Quanta Age the Retina," *Photochem. Photobiol.* **50**(3), 429–438 (1989).
331. L. B. Zerbe and H. W. Hofstetter, "Prevalence of 20/20 with Best Previous and No Lens Correction," *J. Am. Optical Assoc.* **29**, 772 (1957–1958).
332. R. A. McFarland et al., *Dark Adaptation as a Function of Age and Tinted Windshield Glass*, Highway Res. Board Bull. No. 255, 1960, p. 51.
333. E. D. Fletcher, "An Investigation of Glare Resistance and Its Relationship to Age," *Motorists' Vision* **5**, 1 (1952).
334. W. K. Adrian, "Visual Performance, Acuity and Age," in *Proceedings of LUX Europa, the 7th European Lighting Conference*, Heriot-Watt University, Edinburgh, UK, (April 4–7, 1993), Chartered Institution of Building Services Engineers, London, 1993, pp. 278–290.
335. W. Adrian, "The Aging Eye, and Its Effect on Vision and Visual Performance," in *Technical Papers IESNA Annual Conference*, Houston, Texas, August 8–12, 1993, Illuminating Engineering Society of North America, New York, 1993, pp. 455–478.
336. S. Shankman, "Light in the Laboratory," *LD + A*, **16**(5), 11–15 (1986).
337. R. Küller, *The Influence of Daylight and Artificial Light on Diurnal and Seasonal Variations in Humans, a Synergistic Computer-Based Bibliography, Report of CIE TC 6016: Psychobiological Effects of Lighting*, Commission Internationale de l'Eclairage, Vienna, 1999.
338. R. Küller et al., "Shortness of Daylight as a Reason for Fatigue and Sadness: A Cross Cultural Comparison," in *Proceedings of the 24th Session of the CIE, Warsaw, June 24–30, 1999*, CIE Publication No. 133, Vol. 1, Part 2, Commission Internationale de l'Eclairage, Vienna, 1999, pp. 291–294.
339. G. C. Brainard, J. M. Greeson, and J. P. Hanifin, "Action spectra for Circadian and Neuroendocrine Regulation in Mammals," in R. Matthes and D. Sliney, eds., *ICNIRP/CIE: Measure-*

ment of Optical Radiation Hazards, A Reference Book Based on Presentations Given by Health and Safety Experts on Optical Radiation Hazards, Gaithersburg, Maryland, USA, September 1–3, 1998, IRCNIRP 6/98, and CIE x016-1998, International Commission on Non-Ionizing Radiation Protection, Oberscheißheim, Germany, and Commission Internationale de l'Eclairage, Vienna, 1998, pp. 131–142.

340. L. Lamberg, "Dawn's Early Light to Twilight's Last Gleaming," *JAMA* **280**, 1556–1558 (1998).
341. S. S. Campbell and P. J. Murphy, "Extraocular Circadian Phototransduction in Humans," *Science* **279**(5349), 396 (1998).
342. S. Blakeslee, "Biologists Close in on the 'Tick-Tock' Genes." *New York Times, Science Section F3*, Tuesday, December 15, 1998.
343. *Philips Forum: Non-Visual Effects of Lighting*, Lighting Research Center, Rensselaer Polytechnic Institute, Troy, NY, 1998, *http://www.lrc.edu/Philips/index.html*.
344. P. R. Boycc, "Light, Sight and Photobiology: Light, Sight—Is That All There Is?" in M. E. Fitzgerald, ed., *Lighting Futures*, Vol. 2, No. 3, Lighting Research Center, Rensselaer Polytechnic Institute, Troy, NY, 1997, pp. 1, 3–6.
345. P. R. Boyce et al., "Lighting the Graveyard Shift: The Influence of a Daylight-Simulating Skylight on the Task Performance and Mood of Night-Shift Workers, *Lighting Res. Technol.* **29**(3), 105–134 (1997).
346. G. C. Brainard, M. D. Rollag, and J. P. Hanifin, "Photic Regulation of Melatonin in Humans: Ocular and Neural Signal Transduction," *J. Biol. Rhythms* **12**(6), 537–546 (1997).
347. J. P. Cèsarini, "UVA Radiation and Health Effects: New Issues with Regard to UVA," in *Proceedings of the 23rd Session of the CIE, New Delhi, November 1–8, 1995*, CIE 120-1996, Vol. 2, Commission Internationale de l'Eclairage, Vienna, 1996, pp. 67–72.
348. "Task Force Report on Light Treatment for Sleep Disorders," *Special Issue, J. Biol. Rhythms* **10**(2), 105–176 (1995).
349. T. L. Baker, "Use of Circadian Lighting System to Improve Night Shift Alertness and Performance at the USNRC's Headquarters Operations Center," *Proceedings, Safety of Operation Reactors: An American Nuclear Society International Topical Meeting, Seattle (Bellevue), Washington, September 17–20, 1995.*
350. W. Ehrenstein, "Circadian Lighting Systems," *Int. Lighting Rev.* **2**, 64–67 (1995).
351. C. A. Bernecker et al., "Biological Effects of Architectural Lighting and Their Associated Energy Utilization," *J. IES* **23**(2), 31–39 (1994).
352. G. C. Brainard et al., "Architectural Lighting: Balancing Biological Effects with Utility Costs," in M. F. Holick and E. G. Jung, eds., *The Biologic Effects of Light*, Walter de Gruyter & Co., New York, 1994, pp. 169–185.
353. EPRI, *Investigating Innovative Concepts 1993: Identifying Lighting Impacts on Human Performance*, No. 9, Electric Power Research Institute, Office of Exploratory and Applied Research, Palo Alto, CA, May, 1993.
354. M. Shepherd and T. Schneider, "Lighting and the Human Condition," *EPRI J.* **11**, 6012 (1986).
355. T. Moore, "Human Performance in the Spotlight," *EPRI J.* **18**(6), 14–21 (1993).
356. T. Schneider, "Architectural Lighting for Improved Human Performance," *EPRI Explor. Res. Lett.* **14**, 2 (1993).
357. P. Hannon et al., "Bright Light Suppresses Melatonin and Improves Cognitive Performance during Nighttime Hours in Humans," in *22nd Annual Meeting of the Society for Neuroscience, Anaheim, CA*, **18**, 1 (1992).

358. U.S. Congress, Office of Technology Assessment, *Biological Rhythms: Implications for the Worker*, OTA-BA-463, U.S. Government Printing Office, Washington, DC, September 1991.

359. G. C. Brainard and C. A. Bernecker, "Biological and Behavioral Effects of Light in Humans," in *Commission Internationale de l'Eclairage, Proceedings of the 22rd Session of the CIE, Melbourne, Australia.* Vol. 1, 1991, pp. 29–30.

360. G. C. Brainard et al., "The Influence of Bright Illumination on Plasma Melatonin, Prolactin and Cortisol Rhythms in Normal Subjects during Sustained Wakefulness," *Sleep Res.* **20**, 444 (1991).

361. S. S. Campbell and D. Dawson, "Timed Exposure to Bright Light Improves Sleep and Alertness during Simulated Night Shifts," *Sleep* **14**(6), 511–516 (1991).

362. S. S. Campbell and D. Dawson, "Enhancement of Nighttime Alertness and Performance with Bright Ambient Light," *Physiol. Behav.* **48**, 317–320 (1990).

363. C. A. Czeisler et al., "Exposure to Bright Light and Darkness to Treat Physiologic Maladaptation to Night Work," *New Engl. J. Med.* **322**, 1253–1259 (1990).

364. J. French, P. Hannon, and G. C. Brainard, "Effects of Bright Illuminance on Body Temperature and Human Performance," *Annu. Rev. Chronopharmacol.* **7**, 37–40 (1990).

365. G. C. Brainard et al., "Dose-Response Relationship between Light Irradiance and the Suppression of Melatonin in Human Volunteers," *Brain Res.* **454**, 212–218 (1988).

366. IESNA, *Choosing Light Sources in General Lighting*, DG-10-98, Illuminating Engineering Society of North America, New York, 1996.

367. B. L. Collins, W. S. Fisher, and R. W. Marans, "Second-Level Post-Occupancy Evaluation (POE) Analysis," *J. IES*, **19**(2), 21–44 (1990).

CHAPTER TWENTY-FIVE

Safety and Health Effects of Video Display Terminals

Marvin J. Dainoff, Ph.D., CPE

1 INTRODUCTION

1.1 Historical Context

Since the last appearance of this chapter (1), there has been a great deal of attention and research interest in health and safety concerns in the office. The prime focus of concern has been on work with visual display terminals (VDTs) as they tend to be referred to in the United States, or visual display units (VDUs) as they tend to be referred to abroad. At the time of writing of the Briggs chapter, the major sources of collected information on this topic had been an international conference entitled Work with Display Units (WWDU) (2), and a study panel on visual problems of VDTs organized by the National Academy of Science (3). [A literature review by present author (4) was part of the materials reviewed by the study panel.] In addition, the Human Factors Society (now the Human Factors and Ergonomics Society) had just issued the first ergonomic standard on VDT workplaces (5). At that time, the primary concerns associated with VDT work were associated with the visual problems, levels of radiation, and concerns about pregnancy problems among female VDT operators. (The latter was presumably related to radiation issues.)

Since that time, the WWDU conferences have become regular biannual events, along with a second series of conferences on human-computer interaction (HCI International). VDT-related research has comprised significant fractions of the regular scientific meetings of associated professional organizations such as the Human Factors and Ergonomics Society and the International Ergonomics Society. In addition, there have been numerous special purpose conferences devoted to VDT issues at both national and international

Patty's Industrial Hygiene, Fifth Edition, Volume 2. Edited by Robert L. Harris.
ISBN 0-471-29754-2

levels. In the U.S., two industry-funded groups (The Office Ergonomics Research Committee, and the Johns Hopkins University Center for VDT and Health Research) have sponsored research and issued reports on health impacts of working with VDTs. The ANSI/HFES 100 standard is in the process of being revised, while a parallel international standard has been issued by the International Standards Organization as ISO 9241: Ergonomic requirements for office work with visual display terminals (VDTs). ISO 9241 has multiple parts, each with its own reference number and date of approval.

During this interim period, concerns regarding radiation and pregnancy issues seem to have faded. At the WWDU 94 conference (6), 87 papers were selected for the published proceedings; only three related to electromagnetic radiation or pregnancy outcomes. Visual concerns, while still present, appear to have receded somewhat. Presumably this can be related to the significant increases in display quality of VDTs since 1989. However, these concerns have been replaced by the issues of work-related musculoskeletal disorders.

Work-related musculoskeletal disorders (WMSDs) are, at this writing, the source of a great deal of attention and political controversy. Part of the controversy is related to labels, which have diagnostic and insurance implications: these disorders have also been referred to as Repetitive Strain Injuries (RSIs), Repetitive Motion Injuries (RMIs), and, more recently, Cumulative Trauma Disorders (CTDs). Although these disorders obviously encompass work environments other than offices, the number of office workers complaining of such disorders is large, and the potential risk factors associated with such disorders in office environments form a major portion of this chapter. In making this statement, however, it is important to also emphasize that WMSDs cannot be considered in isolation without also considering associated organizational, psychosocial, and visual factors. Hence, an integrated systems approach is essential.

In the U.S., there are two governmental agencies which have, as their prime concern, occupational safety and health. These are the National Institute for Occupational Safety and Health (NIOSH), which is part of the Centers for Disease Control under the Department of Health and Human Services, and the Occupational Safety and Health Administration (OSHA), which is under the Department of Labor. NIOSH researchers early identified musculoskeletal problems, along with visual, and psychosocial problems, associated with intense work with VDTs (7). OSHA's initial major concern was with WMSDs in meatpacking plants; resulting in an OSHA ergonomic guideline (8). This *guideline* has formed the basis for attempts by OSHA to develop an ergonomic *regulation* applicable to all workplaces, including offices. In fact, OSHA has issued an ergonomic guideline specific to office work (9).

This regulatory direction has generated a great deal of political resistance in Congress and from trade associations. One result has been the appointment, in 1998, of a special study panel by the National Academy of Science/National Institute of Medicine to review the evidence related to WMSDs in general, and to answer specific questions emerging from congressional testimony. The published report of this panel concludes: "There is a higher incidence of reported pain, injury, loss of work, and disability among individuals who are employed in occupations where there is a high level of exposure to physical loading than for those employed in occupations with lower levels of exposure" (10). An excellent brief overview of the OHSA regulation and its background as of this writing can be found in Albin (11).

OSHA is not the only governmental agency concerned with ergonomics. The state of California has passed its own ergonomic regulation, other states may be following suit. In addition, the European Community has its own ergonomic regulation relating to work with VDTs already in place (EEC/90/270). For a review of the relationship between this regulation and ISO 9241 (which also relates to VDTs), see Stewart (12).

Finally, the National Safety Council has initiated, through the ANSI process, a standard entitled: Z365—Prevention and Control of Cumulative Trauma Disorders (CTDs). The initial focus of this standard was only on disorders of the upper extremity. This follows the somewhat confusing practice of restricting the definition of CTD to upper extremity disorders only. The broader category of WMSD includes both upper extremities and back disorders.

1.2 Work-Related Musculoskeletal Disorders (WMSDs)

Issues of work-related musculoskeletal disorders are central to the subject matter of this chapter. Accordingly, this section discusses WMSDs in general; the following section focuses on the specific case of WMSDs in the office.

An "epidemic" of muscular disorders among telecommunication operators, then described as Repetitive Strain Injuries (RSI), brought WMSDs to public attention in the early 1980s. However, a number of inconsistencies about this event led to questions as to whether such disorders could, in fact, be attributed to strain and/or trauma to musculoskeletal systems (13). The alternative hypothesis, that of individual or psychosocial causation, has important implication for prevention/intervention programs. In particular, Hocking (14) has argued that there is little evidence ergonomic interventions were effective in the subsequent reduction in incidence of disorders in that particular case. Hadler (15) has discussed the possibility of media attention and secondary gains as implicated in the rise in reported symptoms. With regard to the latter, these can occur both for individual workers via compensation systems, but also for medical service providers and consultants.

Nevertheless, the incidence of WMSDs has appeared to rise to such an extent that it is predicted to be one of the major occupational disorders of the 21st century (16), with estimates of direct costs (lost days and compensation claims) ranging from $13 to $20 billion annually (10). It is likely that associated indirect costs (lost productivity, need to recruit and hire additional employees) would increase these costs by several times. These issues form the context for the OSHA activity and subsequent critical reaction described above.

Several efforts at clarification of the question of work-relatedness of musculoskeletal disorders have been attempted. The National Safety Council initiated the development of an ANSI standard (Z365) Prevention and Control of Cumulative Trauma Disorders, whose task was to review the literature critically and to develop recommendations for prevention and control. Several preliminary drafts of this document have been released for public comment. A parallel effort in Canada, undertaken by the Qué Research Institute on Occupational Health and Safety (Institut de recherche en santé et en sécurité du travail du Québec) resulted in a published reference book (16). In the U.S., the National Institute for Occupational Safety and Health conducted its own review of epidemiological evidence (17). This report, along with other evidence, was reviewed by the congressionally mandated National Academy of Science expert panel on WMSDs. The NAS report (10) constitutes, as of this writing, the most recent and comprehensive review of WMSDs and will, therefore, be discussed in some detail in the following sections.

2 NAS CONCEPTUAL FRAMEWORK

The concept of *work-relatedness* is based on World Health Organization terminology (18). Disorders are work-related when one or a combination of multiple risk factors (physical, organizational, psychosocial, sociological) contribute significantly to the cause of the disorder. Figure 25.1 (19) reflects the multifactorial nature of potential WMSD causes.

The underlying physiological mechanisms are seen in the central portion of Figure 25.1. If there are external physical loads on soft tissues (e.g., muscles, tendons, joints, nerves) of sufficient force or duration (from repeated activity), the response of such soft tissues may be damage, inflammation, or fatigue. These external loads are generated by combinations of physical risk factors (e.g., repetition, awkward posture, excess force, duration of effort, temperature, vibration) which may be associated with work procedures, equipment, and environments. Examples include work tasks in which the work is located overhead requiring the workers to raise their hands over their heads, or work located on the floor which requires workers to bend or stoop.

These tissue responses can lead to experienced *symptoms* (inflammation, swelling, pain, fatigue) and/or *adaptations* in which alternative patterns of muscle use are developed. These patterns may in turn result in additional problematic tissue reactions (e.g., inflammation leads to release of additional amount of joint lubricants, which leads to additional swelling). If the duration and severity of symptoms and adaptations is severe, the result

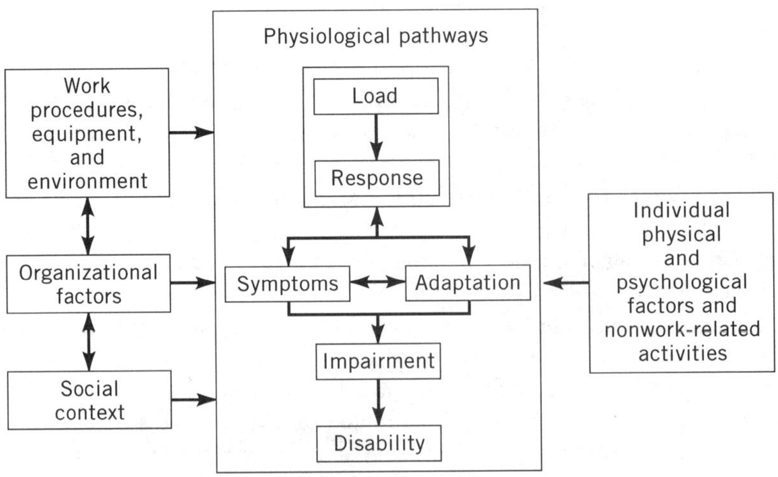

Figure 25.1. Conceptual framework of physiological pathways and factors that potentially contribute to musculoskeletal disorders. From Ref. 10.

can be *functional impairment*. Work capability might be reduced, and this could be reported as work-related disorder. Finally, if the impairment is serious enough for lost work to occur, it becomes a *disability*.

As can be seen by the left and right hand portions of Figure 25.1, there are a variety of organizational, social, and individual factors which can affect this process. The left-hand portion of Figure 25.1 depicts environmental factors which might play a mediating role in the tissue adaptation and the potential transition to impairment and disability. These include organizational factors and social context factors. An example of the former might be that time pressure to meet deadlines produce organizational pressures encouraging workers to ignore safe work practices. An example of the latter might be lack of social support from family and friends to cope with the psychological stress associated with physical impairment.

The right-hand portion of Figure 25.1 represents individual physical and psychological factors (both work-related and non work-related) which might affect the progression of the disorder. An example of psychological factors might include willingness to claim that an impairment is a disability. Examples of physical factors might include previously existing disease states, such as arthritis, age, or weight.

3 NAS REVIEW OF EVIDENCE FOR WORK-RELATEDNESS

Based on the conceptual framework depicted in Figure 25.1, the NAS panel reviewed the existing literature relating to: soft tissue responses to physical stressors, work factors and biomechanics; epidemiological evidence that physical, and nonbiomechanical factors can cause musculoskeletal disorders; and effectiveness of interventions. It is worth quoting directly the critical portion of their conclusions (10)

> . . . while there are many points about which we like to know more, there is little to shake our confidence in the thrust of our conclusions, which draw on converging results from many disciplines using many methods:
>
> - There is a higher incidence of reported pain, injury, loss of work, and disability among individuals who are employed in occupations where there is a high level of exposure to physical load than for those employed in occupations with lower levels of exposure.
> - There is a strong biological plausibility to the relationship between the incidence of musculoskeletal disorders and the causative exposure factors in high-exposure occupational settings.
> - Research clearly demonstrates that specific interventions can reduce the reported rate of musculoskeletal disorders for worker who perform high-risk tasks. No known single intervention is universally effective. Successful interventions require attention to individual, organizational, and job characteristics, tailoring the corrective actions to those characteristics.

In addition to these general conclusions, there are two specific issues commented upon by the panel that are relevant to the discussion. The first is the question of malingering. The report indicates that, while there are clearly individuals who abuse medical systems for their own gain, there has been no evidence that malingering is a large contributor to overall levels of WMSDs. Secondly, while there is evidence for the causative role of individual factors (age, prior medical condition, body mass, and gender), these factors appear to account for a relatively small proportion of the overall variance.

4 MUSCULOSKELETAL DISORDERS IN VDT OPERATORS

At the time of the initial concerns regarding health effects associated with VDTs, these devices were newly emerging in the workplace. Initial concerns centered around isolating those aspects of VDT work which differentiated them from comparable jobs performed without use of VDTs (7). In addition to the obvious differences in visual demand, VDT work seemed to be characterized by clusters of physical and psychosocial factors which might plausibly be related to musculoskeletal disorders. VDTs were often placed on standard office desks at keying heights which were too high, thereby leading to awkward working postures. Workload levels tended to be higher than for non-VDT work.

However, in the intervening years, the rapid diffusion of computer technology led to the necessity of asking different kinds of questions. The computer has become so ubiquitous that it is quite difficult to even find comparable groups of people doing similar group with and without VDTs. Thus, within the conceptual framework discussed above, the question now might be: are there risk factors associated with VDT use which consistently lead to WMSDs?

This question is not as straightforward as it might be. VDT studies were certainly included within the scope of the NAS review. However, VDT work does not typically involve the levels of exertion seen in high exposure industrial work settings such as automotive assembly and meatpacking. Thus, the causal chains leading from levels of exposure to specific risk factors to disorder have not been easy to establish.

Since the VDT entered the workplace, there have been a number of published reports that associate work with VDTs with WMSDs. (e.g., Refs. 7, 20, 21). However, many results from these studies are inconsistent, and fail to show specific links between exposure and disorder outcomes. Gerr, Marcus, and Ortiz (22) have critically reviewed much of this literature. They indicate several possible sources of inconsistency. There tend to be two types of studies comprising this literature. One type involves laboratory-based human factors–ergonomics studies. These studies allow relatively good control of exposure conditions (workstation equipment and duration of work) but tend to be limited in terms of numbers of subjects and amount of exposure. In particular, Gerr et al. point out that it is difficult to utilize information from experiments lasting only a few hours to predict long term disability effects. On the other hand, the existing epidemiological studies have different kinds of problems. Exposure and outcome measures are often not clearly defined. Most of the studies have been cross-sectional and are therefore subject to selection bias (employees most effected negatively by VDTs may have left their jobs and are not, therefore, included in the exposure group.) Finally, there are often confounding variables (individual and psychosocial factors) which have not been controlled for.

At the same time, there have also appeared a number of published intervention studies with positive outcomes. Several of these are documented by Sauter, Dainoff, and Smith (23). A study conducted by the U.S. Army Corps of Engineers (24), conducted a workplace redesign project in a government agency. After nine months, productivity and satisfaction were improved as compared with two control groups; one receiving new but traditional furniture, the second maintaining their present working conditions. Ong (25) described two case studies in Singapore in which data entry performance increased and health complaints decreased after ergonomic changes (furniture, lighting, rest breaks) were made. Sullivan (26) and Westin (27) described implementation of successful large-scale ergonomic programs at Aetna and Federal Express, respectively.

Dainoff (28) conducted a controlled laboratory study in which the same operators alternated, on successive days, between an optimal and a "typical" VDT workstation. The results indicated higher keystroke rates and lower musculoskeletal complaints on the days operators worked on the optimal workstation. Each subject in this study participated for five days, eight hours per day. Thus, the study meets at least some of the objections about the short duration of controlled studies voiced by Gerr et al. (22) as discussed above.

More recently, the U.S. Congress commissioned an investigation by the Government Accounting Office as to the effectiveness of ergonomic programs in reducing WMSDs. The ergonomic programs at five facilities were examined. One of the locations (American Express Financial Advisors) involved predominantly VDT operators. The conclusions of the report are that: "Officials at all the facilities we visited believed their ergonomics programs yielded benefits, including reductions in workers compensation costs associated with MSDs. . . . also . . . reductions in overall injuries and illnesses." (29). There were also indications of improved productivity, employee morale, and product quality. Finally, the investigators report that costs associated with programs were lower than has been reported in the literature.

Finally, a series of studies showing improvements among telecommunication workers (30) was one of those explicitly cited by the NAS panel as providing relatively convincing evidence of effectiveness of ergonomic interventions. (See also Refs. 31 and 32 for evidence of highly significant [850% return on investment] productivity improvements resulting from ergonomic interventions).

4.1 Visual Disorders

While visual problems associated with VDT work have not had the degree of recent attention as WMSDs, they remain a continuing concern despite the technical improvements in display quality since the first version of this chapter. Currently, the primary focus of attention in VDT work is *aesthenopia*, which is a clinical term for complaints of eye strain or eye fatigue. An extensive investigation of VDT effects on visual function was carried out by a multidisciplinary, multisite team in Italy during the last decade. A special issue of "Bollettino di Oculistica" (33) was devoted to this study, which involved ocular examinations of over 31,000 subjects. An update of this continuing effort (now involving over 70,000 subjects) was presented by a number of researchers at the 1994 Work With Display Units conference in Milan. Piccoli (34, 35), in reviewing this work, states: " . . . [this work] . . . supports quite clearly the existence of a positive correlation between the intensity and the prevalence of aesthenopic disturbances and the 'characteristics' of the display." On the other hand, a collaborative group of researchers (36) failed to find that VDT use is related to permanent organic ocular damage.

As in the case of WMSDs, the issue is to relate specific components of the VDT work environment which might be causally linked to symptoms. Unfortunately, the underlying causal mechanisms relating to aesthenopia are poorly understood perhaps even less than for musculoskeletal problems (3, 4, 37). Nevertheless, there are some general principles which can be stated, and certain interventions (typically multifactorial) have been successful. Piccoli (34) identifies three classes of variables that might potentially be linked to aesthenopia: (a) *Subject variables*—current ophthalmic deficiencies (e.g., lack of proper eyeglasses for prolonged near vision); (b) *Environmental variables*—(i) presence of irritants to the ocular surface (e.g., indoor air quality problems and/or microclimate problems), (ii) presence of disturbances to visual process (e.g., presence of glare sources); (c) *Visual workload variables*—short viewing distance, duration of viewing, image quality.

4.2 Are Health Effects Associated with VDT Work?

It may be useful, at this point, to review the issues. In the years since the first version of this chapter, the question of comparing VDT with non-VDT work seems to have effectively vanished. The VDT is now ubiquitous in the office workplace, and a significant presence in nonoffice workplaces as well. It is, therefore, more useful to attempt to identify those elements of the VDT work environment that might be causally linked to specific symptoms and/or disease. This was, in fact, the underlying motivation of the Dainoff (28) studies mentioned above. In these studies, the task (VDT data entry) and the operators were kept constant, while elements of the workstation and work environment were systematically manipulated. In the first study, operators alternated, on successive days, between optimal and "typical" workstations. The optimal workstation consisted of well-designed task lighting and antiglare filters, and adjustable worktables, adjustable chairs with lumbar support, and auxiliary equipment (wrist rest and copyholder). The typical workstation simulated office environments characteristic of the time of the study (1981); poor lighting resulting in glare, unadjustable worksurfaces and chairs, and lack of copyholder and wrist rest. In the second study, lighting and glare conditions were equated, so that the only difference between the two workstations was furniture and auxiliary equipment. In both studies, work performance was higher and MSD symptoms lower in the optimal workstation. However, the average difference in work performance rates (keystrokes minus errors) between optimal and typical workstations reflected the reduced ergonomic differences between workstations in the second study (furniture only) as compared with the first (furniture plus lighting). The work performance differences were, respectively, 24% and 19%.

The individual ergonomic elements chosen to differentiate between these workstations reflected an underlying conceptual model of working posture and visual comfort. By providing an adjustable chair, worksurfaces, copyholder and wrist rest, the operator is afforded the possibility of avoiding awkward working postures. By providing glare control and the design of appropriate lighting, the operator is afforded a higher degree of visual comfort. A fundamental hypothesis falling out of this model is that requiring the operator to work under conditions of awkward posture and visual discomfort will result in increased levels of load on postural and visual systems. These loads cause work performance inefficiencies, and, if maintained for sustained periods of time, can result in symptoms of postural and visual disorders (e.g., WMSDs and aesthenopia). While the laboratory and field evidence cited above support the performance component of the load hypothesis, direct epidemiological evidence linking short term symptoms associated with eventual disorders is lacking, as was discussed above (e.g., Ref. 22). However, the success of ergonomic program interventions described above makes these links at least plausible.

At the same time, the question of task duration is critical. The work task requirements of participants in the Dainoff studies involved continuous interaction with the VDT for the entire workday. Several authorities [e.g, Hettinger, (38) Greico, (39)] have argued that working for prolonged periods in fixed postures constitutes a risk factor in itself. Indeed, using German industrial insurance records, Hettinger (38) showed that office workers who were required to maintain fixed postures yielded levels of muscular disorders which exceeded all other industry groups except construction, and construction materials workers.

Although the role of physical ergonomics appears reasonably clear from the above discussion, the conceptual framework depicted in Figure 25.1 is a reminder that that individual, organizational, and social context factors also must be considered. (Although this figure refers to WMSDs, exactly the same logic can be applied to aesthenopi). With regard to individual factors, there are two issues to be discussed. First, with regard to the role of individual differences in physical and psychological factors, such differences will not be the subject of a separate discussion but will, instead, be considered at points where appropriate. Secondly, the question of the role of non work-related activities will not be considered part of the scope of this chapter. The reasoning is as follows: if an individual employee spends time at an activity not connected with his/her primary job (perhaps even a second job) where ergonomic conditions are such that excessive loads are present, the question of how much that load contributes to a particular symptom or disorder must be determined (typically for insurance purposes) on an individual basis by an occupational health specialist. However, the relationship between load and symptoms, in and of itself, is already part of the discussion in this chapter.

With regard to the question of organizational and social factors, a NIOSH sponsored conference on psychosocial aspect of MSDs in office work (40) resulted in a number of important contributions to this discussion. From that conference, Smith and Carayon (41) have outlined several mechanisms by which social and organizational factors can causally effect the outcome of WMSDs. Many of these mechanisms are intimately related to the concept of the *stress response* as originally elucidated by Selye (42). The stress response consists of three stages. In the first or *alarm* stage, an external threat results in the mobilization of bodily resources. If the threat persists, a second or *adaptation* stage occurs in which bodily functions appear to return to normal homeostatic balance, but at a higher than normal expenditure of resources. Finally, a third stage of *exhaustion* results when bodily resources fail to be able to maintain homestasis. Disability or death is the result.

Smith and Carayon (41) point out that increasing levels of occupational stress have paralleled the rise of WMSDs, and that this is perhaps not an accident. There is a large literature indicating linkages between work organizational factors and stress-related diseases such as cardiovascular disease and mental disturbance. For example, Karasek (43) has shown high workloads combined with inability to control the rate or flow of work renders employees particularly susceptible to cardiovascular disease.

Smith and Carayon (41) have indicated two groups of mechanisms by which organizationally determined stress can impact MSDs. First, there are known psychobiological mechanisms by which the stress response can modulate effects of physical strain and/or limit the capacity to repair musculoskeletal microtrauma. For example, the alarm stage of the stress response results in reduction in blood flow to peripheral organs, including muscles and tendons. The secretion of norepinepherine, which accompanies the alarm stage, causes hypersensitivity of neuromuscular junctions; the result can be an increased sensitivity to pain. Thus, if a combination of physical risk factors (force, posture, repetition) results in an elevated physical load, and the individual is also under stress, already overworked muscles are denied blood, while the accompanying pain is exacerbated. In addition, the higher blood pressure associated with hypertension can result in higher carpal tunnel pressure, and the increased corticosteroid secretion associated with the stress response can produce fluid retention in peripheral body tissues which can exacerbate nerve compression. Moreover, the known effects of stress on the immune system can hamper the repair of minor tissue damage associated with prolonged physical load. Finally, there is evidence for direct pathways from the emotional centers of the brain to muscle fibers. Hence, strong emotions associated with the stress response (e.g., fear, anger) can result in increases in muscle tension.

The second group of mechanisms described by Smith and Carayon (41) relates to *behavioral* mechanisms by which the stress response may produce behaviors which increase an individual's exposure to physical risk factors, reduce the motivation to seek help, or generally increase the sensitivity to pain. For example, a physical load which might otherwise be part of a normal adaptation to work may be perceived as more significant in an individual simultaneously subjected to stressful organizational factors. (It must be emphasized that this is *not* the same as malingering, which is a conscious attempt to deceive. It may be recalled that the National Academy of Science panel did not feel that malingering accounted for more than small proportion of WMSD cases.) Alternatively, pain symptoms themselves may be perceived as stressors, resulting in increased sensitivity causing an upwardly spiraling positive feedback loop. Finally, the negative mood states associated with the stress response (anger, fear, apathy) might result in lack of motivation to carry out those very policies and procedure which have been put in place to reduce or control risk factors. This is particularly critical in the case of office workplaces, since avoidance of risk factors requires proper use of adjustable ergonomic equipment.

This chapter is not the place to go into detail on organizational and social context issues, but a few comments by Smith and Carayon (41) relating to conditions likely to be considered stressful might be appropriate. In some cases, VDT work has been associated with *deskilling*, which entails lowered levels of responsibility, initiative, and loss of control over the pace and scheduling of details of daily work tasks. [See also the work of Karasek (43) mentioned above.] Even for those for whom the VDT might represent a valuable tool, significant stressors are present in what might be called *clumsy technology*; software systems which are difficult to use, not transparent in their structure, and fail often. (It seems to be the case that hard disks crash or systems fail at the moment that they are needed the most!)

Finally, organizational factors and social context are critical in the extent to which management is willing to recognize the nature of the problem and establish programs to reduce and control risk factors. See Kuorinka and Forcier (44) and current drafts of the NSC Z365 Control and Prevention of CTDs.

4.3 Indoor Air Quality and Work with VDTs

The issue of indoor air quality has already been raised in Piccoli's (34) discussion above regarding the impact of air contaminants on ocular function. There has also been some discussion regarding skin complaints to electrical fields associated with VDTs (45). However, recent work by Hedge and Erickson (46) documenting a more global relationship between indoor air quality and ergonomic considerations ought to be of particular interest to readers of this volume. The focus of their investigation was Sick Building Syndrome (SBS), which they define as a cluster of symptoms (headache, lethargy, eye, nose, throat, and skin irritation) which may have an unusually high prevalence in some buildings, and which disappear when the people reporting the symptoms leave the building. The authors review literature suggesting that, contrary to expectations, SBS is more characteristic of air-conditioned buildings than building with natural ventilation, and that correlations of symptoms with levels of gaseous indoor air pollutants have not been consistent. Two alternative (but not mutally exclusive) hypotheses have been proposed. The first states that SBS symptoms are associated with ergonomic factors. The second states that SBS symptoms are mediated by psychosocial and organizational factors. The study was designed to examine these alternative possibilities.

Four multistoried office buildings were studied. A number of indoor air quality (IAQ) measures were taken on a given day at multiple sites within each building. These included CO, CO_2, illuminance, temperature, relative humidity (RH), formaldehyde, respirable dust mass, and particulate counts. In addition, dust samples were taken from carpet and seat pans and assayed for dust mite allergens, and measures of noise and infrasound were taken. On the day of the study, self-report questionnaires were distributed to all employees in the building. These included questions on work equipment, office furniture, job uncertainty, postural discomfort, and musculoskeletal symptoms. A number of psychological measures included scales of affect, stress, chemical sensitivity and seasonal sensitivity.

Multiple-correlation analyses indicated that the IAQ variables were not related to SBS symptoms, nor were the psychosocial variables. Thus, the authors conclude that neither the gaseous environment nor malingering could be considered important determinants of symptomatology. However, musculoskeletal symptoms showed a highly significant relationship with SBS and yielded the highest beta weights in the multiple regression equation. Thus, the authors conclude that MSDs play an important role in the etiology of SBS. The mechanisms for this are unclear but emphasize the importance of a multifactorial interdisciplinary approach in examining health effects of VDT workers.

4.4 Complexity and the Inappropriateness of the Industrial Hygiene Model

The previous review has indicated that VDT workers can experience groups of symptoms which fall into overlapping areas of professional expertise (WMSDs, aesthenopia, SBS). Can some clarity be brought to this situation? Moon (47) has contributed a particularly thoughtful discussion of the complexity issue. While his discussion focussed specifically on CTDs (that is, upper extremity MSDs), his arguments are quite relevant to the broader questions of health problems related to VDT use.

Moon's analysis of the question of definition of exposure and of the inappropriateness of the industrial hygiene model to this problem is worth quoting directly from Ref. 48:

> Although all occupational exposure occurs through an interface between the environment and the worker, the ergonomic interface differs significantly from a typical industrial hygiene model. Hazardous exposure to chemical substances is a function of quantitative levels in the work environment. Ergonomic exposure occurs across an interface largely *defined by human activity*. Physical exertion, contact, movement, or static positioning typically create the physical exposure . . . factors other than the physical work environment modify physical actions and may affect the resulting perception of sensory feedback relating to those actions. . . . These and other factors may influence the individual's capacity, opportunity, or motivation to alter the exposure interface in response to such feedback.

The problem, therefore, in defining exposure, is that any definition of exposure must somehow account for human actions occurring in direct and continuous interaction with workstation equipment and work procedures. This is what Moon calls the ergonomic interface. However, since human action does not occur in a motivational vacuum, psychosocial variables necessarily have a direct effect on any exposure measure.

Moon then goes on to describe the dilemma this situation poses for the medical practitioner. On the one hand, if a practitioner observes, for example, that stress levels from inappropriate management styles are affecting MSD symptom rates, can he or she intervene by providing medical advice on what is a management issue? On the other hand, the practitioner may find that he or she is in conflict with an ergonomic approach. That is, an ergonomic approach to a situation in which an operator complains of pain or discomfort might be to adjust the work environment until the pain or discomfort is reduced. This approach assumes an underlying model of *efficiency*. If the complaint is plausible given current understanding of physical risk factors, reducing such risk factors can increase the operator's work efficiency. However, that same complaint can become a *symptom* within the medical models, thus triggering a different set of events (OSHA reporting, worker's compensation) which focuses on the "patient" rather than the work environment. Moon warns that inappropriate *medicalization* of the situation may, in fact, be counterproductive for both organization and individual.

The basic question, then, is how can these complexities be examined at an appropriate level of analysis. If "exposure" represents human action affected by human motivation interacting with a physical interface, a descriptive framework is needed which matches the complexity of the problem. It is the belief of the author that an emergent discipline called "ecological ergonomics" does contain analytic tools which are sufficiently rich to deal with the interactive factors described above. The remainder of this chapter explicates this approach.

5 AN ECOLOGICAL FRAMEWORK

5.1 Recap: The Office as a VDT Work Environment

It is necessary to answer a fundamental question: what is an office? In the modern electronic workplace, the answer is not straightforward. *The Compact Edition of the Oxford English Dictionary*, 1971, defines an office as. . . . "a place for the transaction of private or public business." With the profusion of portable computers and communications equipment, almost any location can fit that definition: an airport waiting room, a kitchen table, an automobile, even a park bench. For purposes of this chapter, the focuses on workplaces whose primary purpose is some aspect of information processing and transformation, and whose occupants are expected to remain in place for extended periods of time (e.g., several hours). Furthermore, the ubiquitous role of the computer terminal—or Video Display Terminal (VDT)—in such workplaces necessitates that VDT considerations will play a central role in this discussion. Having established this necessarily limited scope however, it should be indicated that the general principles to be discussed herein, can be usefully applied to more temporary venues (e.g., setting up of a temporary workspace with a laptop and modem in a hotel room).

There are compelling reasons to pay attention to office ergonomics. As has already been discussed, there is a significant body of published interventions which indicates that investments in ergonomic improvements in an office environment are cost-effective (24). Other examples may be found in MacLeod (48). The essence of these arguments is well captured in Van Wely's (49) conception of ergonomics as: "a kind of eternal triangle linking efficiency, comfort, and health." Much office work with VDTs tends be characterized by prolonged periods of static posture coupled with high visual demands. In such postures, movement occurs in only a few muscle groups such as wrists and fingers. To the extent that poorly designed workstations require the operator to take up working postures which are inefficient ("awkward"), the onset of fatigue—with consequent discomfort and/or pain—will be relatively rapid. Conversely, the application of ergonomic principles to workplace design will delay the onset of fatigue, resulting in improved work efficiency and increased feelings of well-being (28). In the longer term, ergonomic improvements should reduce the incidence of those musculoskeletal disorders which might be linked to poor working postures.

It is possible to give a few simple rules of thumb which will improve office workplaces, and many of these have appeared in the popular press. However, a systematic analysis of the scientific principles underlying such ergonomic recommendations indicates that the relationships among such recommendations are anything but simple. Optimizing individual workstation components (i.e., seat height, gaze angle) without taking into account the interactive effects of such components on each other can easily result in an overall outcome which is suboptimal. What is required is a systems perspective which considers the characteristics of the human operator *in relationship to* the physical and psychosocial work environment within which he or she acts. The so-called ecological approach to ergonomics (e.g., Refs. 50 and 51) provides such a perspective and will be employed as the conceptual approach to this chapter.

6 FUNDAMENTALS

6.1 An Ecological Approach to Ergonomics

The seminal research of James J. Gibson (see Ref. 52) has formed the basis of the ecological movement within the field of perceptual psychology. This ecological approach has, in turn, provided an important set of conceptual tools for applied concerns within the field of human factors and ergonomics (50). Rasmussen, in particular, has utilized Gibson's work as a major theoretical underpinning of his approach to Cognitive Systems Engineering (53, 54). These are the theoretical foundations upon which this chapter will be built.

Ergonomics can be described as the fit between people and the elements of the physical environment with which they use/interact (55). As such, ergonomics is inherently relationship oriented in that absolute dimensions and physical characteristics of objects in the work environment must be defined relative to the relevant characteristics of the user. Gibson's work provides a principled approach to conceptualizing such relationships. A key element in this approach involves the terms *affordance* and *effectivity*. Affordances refer to the characteristics of the physical environment measured with respect to the individual or "actor." Affordances thus represent, in physical terms, the potential for action allowed (afforded) by the environment for a particular individual. Effectivities are complementary in that they refer to action capabilities of individuals measured with respect to the physical environment.

The usefulness of this pair of concepts can be illustrated in the following example. Two parents wish to sit down to dinner with their one year old child. The physical characteristics of their "work environment" consist of four chairs and a table. The chairs will afford sitting at the table only for the parents, since the relevant effectivity (leg height) of the child is too small for the action capability of sitting to be performed. To accomplish the desired action of having the child eating with her parents during dinner, a different type of chair is provided (a high chair), into which the child must be lifted. Now the chair affords an eating posture for the child which is approximately the same as that of the parents. This example is perhaps rather trivial, but becomes less so if one now assumes that the parents have some physical disability (effectivity) which prevents either of them from lifting the child into the high chair. Some other (ergonomic) solution must be found to accomplish the overall system goals (dining together with one's child.) In a real sense, then, affordances represent an abstract way of conceptualizing the things which a designer must provide for his/her client by taking into account that client's effectivities.

6.2 Means-End Abstraction Hierarchies

In order to carry out goal-directed actions in the world, the individual actor must be able to perceive the affordances of that world. (A 250-lb adult will most likely not choose to sit in a chair made for a two-year old!) However, most real-world work environments consist of collections of affordances, some of which may be nested within others. To meet the challenge of characterizing such environments, and to allow perception of complex sets of affordances by the user/actor, Vicente and Rasmussen (56) have utilized a conceptual tool called the *means-end abstraction hierarchy*. Central to this concept is the notion that groups of affordances can be hierarchically organized by function. At any point in the hierarchy, the affordances at one level act as the ends or goals with respect to the affordances (means) at the next level down. Rasmussen et al. (53) has effectively demonstrated the use of means-end abstraction hierarchies as tools for representing functional properties of complex real-world environments within which human information-processing, control, and decision-making must occur and which will allow us to characterize the possibilities for user action/decision making in such complex problem spaces.

The remainder of this chapter attempts to solve the practical problem of discussing the ergonomics of a VDT-based office workplace by specifying the affordances (possibilities for action) of that workplace through the use of a means-end abstraction hierarchy while, at the same time, considering the relevant effectivities (user characteristics which affect their action capabilities). In this description, the primary focus must be on the individual actor who is actually using the furniture/VDT system to accomplish work. At the same time, an understanding of this discussion ought to provide valuable information for both equipment designers as well as system integrators/facilities managers who must assemble the components (collection of affordances) for their users/customers.

7 OVERVIEW: STRUCTURE OF THE MEANS-END ABSTRACTION HIERARCHY

This section will present a very brief overview of the approach of Rasumussen et al. (53) to the means-end abstraction hierarchy. Five levels are utilized (see Table 25.1).

At the lowest end of the hierarchy, *physical form* refers to the representation of relevant components of the physical structure of the elements comprising the system. For example, consider a worktable which is height-adjustable. The range of adjustability in physical units (centimeters or inches) is represented at level 5. Moving upward, level 4 represents *physical function*, the ends for which level 5 is the means. In this case, physical function refers to the *capability* of height adjustability. At level 3, a *general function* called *clearance*, which refers to the area under the worksurface, is an end for which height adjustability is a means. Clearance itself is subsumed under the broader concept of person-environment fit, which is a level 2 *abstract function* characterization. Finally, the overall system *functional purpose or goal* (good ergonomic design) is found at the highest level of the hierarchy.

In Rasmussen's approach, the complete characterization of a work space should include (along a second, orthogonal *part-whole* dimension) a description of components and subcomponents within the system. Since means-end and part-whole entries are usually highly correlated, it is possible, for the purposes of this chapter, to disregard the part-whole description.

What follows will be an elaboration of the means-end abstraction hierarchy as a characterization of ergonomic problem space of an office environment. Table 25.2 depicts an overview of this approach.

Table 25.1. Example of a Means–End Hierarchy[a]

Level 1	Functional purpose	Good ergonomic design
Level 2	Abstract function	Person–work environment fit
Level 3	General function	Clearance
Level 4	Physical function	Worksurface height adjustment capability
Level 5	Physical form	Worksurface height adjustment of 25 cms

[a]Ref. 53, p. 38.

Table 25.2. Means–End Hierarchy of an Office Environment

Level 1	Functional purpose		Comfort	Health	Productivity			
Level 2	Abstract function: Efficiency	Biomechanical	Perceptual	Cognitive	Resource allocation			
Level 3	General function	Postural efficiency	Postural change	Reach	Clearance	Usability	Image quality	Illumination and glare
Level 4	Physical function: Interaction zones	User-chair	Below work surface	Shoulder, arms, hands	Eyes, head, neck			

7.1 Level 1: Functional Purpose (System Goals and Values)

At the top of the abstraction hierarchy lies the overall purpose of the system: a specification of the *ends* to be achieved embodied in goals and values along with appropriate system constraints. Van Wely's (49) "eternal triangle" relating comfort, health, and productivity is quite useful as a way of concretizing the goals and values of good ergonomic design. Van Wely's ends can also be considered organizational as well as individual values in the sense that reducing health costs and increasing productivity represent pathways to organizational success, with individual comfort representing the working conditions which allow the two previous goals to be achieved.

7.2 Level 2: Abstract Function (Flow and Distribution)

The second level in the abstraction hierarchy represents *abstract functions and priorities* (means) by which the higher order ends are to be achieved. Rasmussen et al. (53) characterizes level two in terms of relevant underlying causal structures specified in abstract terms such as mass, energy, information, and value. For example, Vicente and Rasmussen (56) utilized means-end abstraction hierarchy analysis to design an "ecological interface" for operator control of a power plant. In this analysis, the abstract function level described the flow of mass and energy throughout the plant in terms of the basic laws of thermodynamics and the principle of conservation of energy.

In the case of ergonomic analyses, the underlying causal structures have not been described at the same level of theoretical detail. Instead, a series of conceptual abstractions will be proposed as representative of the underlying assumptions characterizing the existing body of ergonomic theory and practice. Four such entities are defined: *biomechanical efficiency*, *perceptual efficiency*, *cognitive efficiency*, and *resource allocation efficiency*. The first three of these entities represent different aspects of the fit or compatibility (57) between human attributes and elements of work environment. The fourth, resource allocation efficiency, represents a somewhat broader operational context within which the person–workplace system operates. For each of these conceptual entities, the use of "efficiency" as a descriptor attempts to capture Rasmussen's emphasis on flow, accumulation, and conservation of resources as characteristic of this level of the hierarchy.

7.2.1 Biomechanical Efficiency

Biomechanical efficiency reflects the underlying assumption that comfort, health and productivity can be achieved through postural adjustments which, at some level, represent combinations of muscle, tendon, joint and metabolic function which strive toward some degree of optimality at a given point in time under given conditions. (See, for example, Ref. 58.) Although these optimal states can be conceptualized in principle, their precise specification in practice may be quite problematic (59). Nevertheless, some underlying concept of optimal posture which generates the most efficient use of metabolic resources seems to underlie most postural analyses. See Ref. 60 for a general discussion on motor control and efficiency.

A further assumption that may be made is that work activities involving continual and marked departures from optimal efficiency may result in symptoms of musculoskeletal disorder. The transition from inefficiency to disorder has, of course, been extensively discussed in previous sections of this chapter. For the remainder of this chapter, the perspective taken will be that reduction in biomechanical inefficiency is a worthwhile goal in itself, and that the links between such inefficiencies and subsequent disease states is at least plausible.

7.2.2 Perceptual (Visual) Efficiency

Perceptual efficiency reflects a parallel set of underlying assumptions regarding the conditions under which the pickup of perceptual information can be optimized. For purposes of this chapter, concerns are restricted to visual perception; focusing on the spatial and temporal quality of visually displayed information on VDTs and paper copy; and associated issues regarding ambient light and glare. There is a large body of research evidence on visual perception and psychophysics which can be used as a basic for design decisions. See, for example, Ref. 61.

7.2.3 Cognitive Efficiency

Cognitive efficiency reflects the *usability* of elements present in the work environment. It is possible (and all too frequent) for displays and workplace furniture to be constructed such that biomechanical and perceptual efficiency are theoretically achievable but so difficult to use that they are, in fact, inefficient.

Cognitive efficiency is related to the degrees of freedom problem (57, 62). The great variety in anthropometric and sensory capacities to be found among users of office workplaces dictates that some degree of adjustability in elements of these workplaces is required in order to achieve biomechanical and perceptual efficiency. Each such adjustment can be considered to be a single degree of freedom within the overall system. If each degree of freedom is independently adjustable, the outcome, as the number of degrees of freedom increases, is likely to be cognitive overload. Imagine trying to operate a motor vehicle in which each of the four wheels had a separate steering mechanism. The solution to this dilemma is to attempt to *constrain* or link multiple degrees of freedom. In the office environment, this can be done mechanically (as with chair controls) or by training of operators.

It should be noted that the description of cognitive efficiency in this chapter is deliberately limited. There is a much larger area of application for this concept in, for example, the field of software or cognitive ergonomics.

7.2.4 Resource Allocation Efficiency

Resource allocation efficiency deals with the broader context of the workplace. This is the point at which task demands must be considered. What resources does the organization provide for accomplishing the task; what is the nature of the task demands (including work flow)? These contextual demands act as constraints on the first three components. Specifically, task demands greatly affect the resulting work postures as well as the capability to interrupt work for rest breaks. Financial resources dictate the extent of any ergonomic improvements, including operator training. This is the point at which the psychosocial mechanisms discussed by Smith and Carayon (41) referred to previously would play a role, either positively in terms of enhancing resources, or negatively in terms of diminishing them.

Within this level, consideration of financial resources is appropriate. Explicit cost-benefit analyses of effectiveness of ergonomic interventions are available (32, 63). Muckler (64) provides an excellent overview of issues surrounding the measurement of productivity in an ergonomic context. He reminds us that removing inefficiencies (e.g., reducing overhead costs such as medical and retraining expenses) are also significant contributors to overall productivity. Discussions of the employment of ergonomic programs as preventative measures in controlling musculoskeletal disorders are contained in Kuorinka and Forcier (16).

7.3 Level 3: Generalized Function

Level 3 contains a list of functional requirements for office workplaces specifically in service of the abstract efficiency goals described at level 2. It should be emphasized that the mapping between levels is not necessarily one to one. Several of these requirements could be derived from more than one efficiency goal.

7.3.1 Postural Efficiency

Postural efficiency, which is derived from biomechanical efficiency, refers to the requirement to avoid awkward postures while working. "Awkwardness" can be defined in terms of postures in which the orientation of bones are near the limits of the range of motion of their corresponding joints. Functionally, this can be achieved by organizing the workspace so as achieve working postures in which joints are maintained close to the middle of their range of motion (65). This minimizes biomechanical load.

7.3.2 Postural Change

Maintaining any posture for prolonged periods of time (static loading) impairs blood circulation and increases muscle fatigue (39, 66). Therefore, working conditions should be organized so as to allow for changes in posture and provide for adequate rest breaks (67).

7.3.3 Reach

The requirement to be able to grasp and/or operate work elements during work operation is fundamental to ergonomic design of the workplace (68). Design of so-called reach envelopes must take into account anthropometric variability of the user population together with task constraints.

7.3.4 Clearance

A second fundamental requirement of workspace design is that adequate room must be provided for legs, thighs, elbows, etc. under and adjacent to the worksurface (68).

7.3.5 Image Quality

The size, spacing, clarity, temporal and spatial resolution of displayed characters and graphics—both on the display screen and on printed material—should afford adequate visual performance (69, 70). This requirement is derived from the higher level goal of sensory efficiency.

7.3.6 Illumination and Glare

The level of illumination falling on elements in the work space should be adequate for comfortable viewing. At the same time, sources of illumination should not act as glare sources which are uncomfortable and/or lead to decrements in visual performance (71).

7.3.7 Usability

The requirement for usability is derived primary from the goal of cognitive efficiency. Usability in this context refers to means available to the user to carry out adjustments (manipulate degrees of freedom). Users should be able to easily detect the function of such controls, as well as determine whether the intended outcome of operating the control has actually been realized (72). In the office workplace, this requirement entails a combination of equipment design and user education and training.

7.4 Level 4: Physical Function

The functional requirements described at level three must now be translated into specific hardware requirements for office furniture and displays. These requirements may be grouped into four zones; each zone refers to a region of interaction between the user and the work environment. Zone descriptions are simply listed here; further elaboration of requirements are found in the following section.

Zone 1: Interactions between user and chair.

Zone 2: Interactions below the work surface.

Zone 3: Interactions involving shoulder, arm and hand.

Zone 4: Interactions involving eyes, head, and neck.

7.5 Level 5: Physical Form

This is the lowest level of the means-end abstraction hierarchy. At this point would be found physical descriptions (using appropriate metric dimensions), of office furniture and VDT equipment which would accomplish the functional requirements specified at level 4. This is the level at which, for example, the Design Specifications found in ANSI-HFES 100/1988, The American National Standard for Human Factors Engineering of Video Display Workstations, would be found.

8 APPLICATIONS AND EXAMPLES

The conceptual structure provided by the means-end abstraction hierarchy is now utilized to present a principled examination of ergonomic requirements of the office workplace. The bulk of the discussion will take place at the levels of Physical Form and Physical Function (levels 4 and 5); however, each requirement should be traceable upward through increasing levels of abstraction and generality. The intention is to minimize the sense of arbitrariness and lack of prioritization which all too often accompanies lists of ergonomic requirements and specification.

8.1 Anthropometry and Equipment Dimensions

Some comments and explanations are necessary regarding specific workstation dimensions that appear in this discussion. The process of fit between user and workplace ultimately comes down to a discussion of flexibility and adjustability. The basic problem is to achieve the goal of biomechanical efficiency while accommodating a population of users who come in a variety of sizes and shapes.

The solution to this problem is technological. Typically, an anthropometric database is selected which most closely corresponds to the group of users to be accommodated. A decision is made as to what proportion of the target population is to be accommodated. Usually, the range of accommodation is from the 5th percentile female to the 95th percentile male. Specific anthropometric dimensions from the database are then selected. These can be considered *effectivities*, which presumably play some role in determining the corresponding furniture dimensions (affordances) which allow the desired range of individuals to be accommodated (achieve biomechanical efficiency).

In practice, there are many technical issues in this process that have yet to be resolved. In the original (1988) version of the ANSI-HFES 100 standard, the anthropometric database utilized was derived from a survey of U.S. civilians conducted in 1962 (73). However, ANSI-HFES 100 is, at this writing, in the process of being revised. In this revision, a more recent anthropometric survey of U.S. Army personnel (74) was utilized. Because body dimensions change over time, and because the ethnic mix of the U.S. population is considerably different now than it was in 1966, it was assumed that the Army database would be a better representation of the current civilian population. However, the specific workstation dimensions derived from the Army database are not yet publicly available, since the revision process is not yet complete.

In addition, there are recent methodological developments that will likely affect the process of determining equipment dimension specifications. Particularly in the area of seated workstations, it is frequently necessary to construct geometrical simplifications or models to estimate workstation dimensions. Such models typically require combinations of more than one anthropometric measure. In the past, designers were limited to combining dimensions of the same percentile. Thus, for example, to determine clearance under the workstation, the appropriate data base dimension was thigh thickness. However, this is measured only with respect to the seat pan. An additional dimension, popliteal height, which is the distance from the floor to the back of the knee, is needed to determine the height of the seat. What has typically been done is to simply add the values of both dimensions at the level of the fifth percentile female to construct a model of the fifth percentile female. However, it is well known that such procedures are statistically inappropriate (68). Because such dimensions are rarely highly correlated, the error resulting from making decisions based on such models can be considerable. Alternative methods using, for example, multivariate approaches, are becoming available (75). Recent improvements in computer processing capacity have made it possible to now work directly with original raw data sets using personal computers. Moreover, a major new effort is presently underway to update the U.S. civilian anthropometric database. This effort, called the CAESAR project, is a joint venture between the U.S. Air Force and the Society of Automotive Engineers.

Accordingly, the actual dimensional recommendations contained in this chapter must be considered tentative. Instead, the goal will be to illustrate the underlying logic by which such dimensions are obtained.

8.2 Zone 1: Interactions between User and Chair

The interaction between the operator and the chair forms the foundation for examining the seated workstation. A primary consideration is support of thighs and buttocks; this is accomplished by the surface of the seat or seatpan. In addition, the trunk can be supported by a backrest, and the arms by arm rests. These components, while important, are not considered fundamental, since some seated work postures do not require backrests or armrests.

In analysis of seated work, it is convenient to begin with a basic posture in which the seatpan is horizontal and the users legs are vertical with the feet firmly on the floor. Figure 25.2 from Ref. 76, provides a reference of critical variables. To design a chair which provides the means to achieve this efficient (level 3 end) posture, the height of the seatpan should correspond to the distance from the floor to the back of the user's knee. In the anthropometric databases, this dimension is called *popliteal height*. Therefore, a first approximation to a design specification for ergonomic chairs can be achieved by requiring the seatpan height to correspond to the user's popliteal height plus a correction for heels. In the Army database (74) the 5th percentile female popliteal height is 351.3 mm; the corresponding value for 95th percentile males is 476.3 mm. Rounding, and adding a 25 mm correction for heel height, the resulting design range for seatpan heights should be 376 to 501 mm (14.8–19.7 in.) so as accommodate the desired range of users.

However, a second level 3 requirement is for movement. This requirement has been recognized by chair designers, and horizontal seatpans have given way to seats which have variable angles, both forward and backward. Newer ergonomic chairs now afford a range of working postures from backward leaning (in which both seatpan and backrest are inclined backward) to forward tilting (in which seatpan is inclined forward while the backrest is vertical); the latter case, however, requires that the seatpan height be increased if the leg is to remain vertical. As a rule of thumb, an increase of 10 mm in height is needed for each 5° of forward inclination.

Furthermore, field studies have shown that the actual postural adjustments carried out in the work environment can not be predicted by anthropometric variables alone. The data

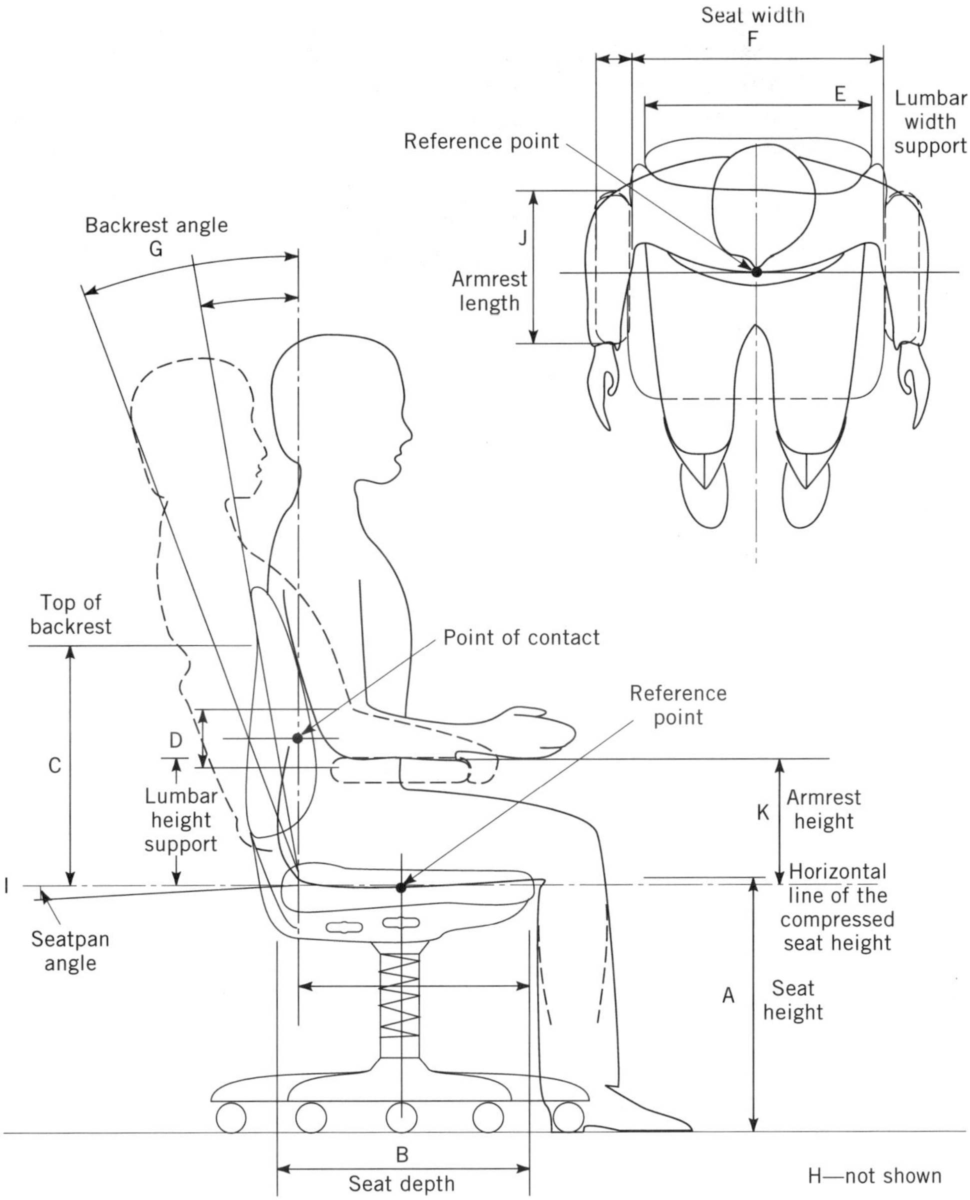

Figure 25.2. Reference variables for seated posture. From Ref. 76.

presented by Cornell and Kokut (77) indicate that seatpan heights may need to be as high as 560 mm (22 in.) to satisfy observed user preferences.

The results from such field studies remind us that all of the functional requirements laid out in level 3 of the means–end abstraction hierarchy may be operative in the apparently simple act of chair adjustment. The user may raise the seatpan higher than his/her popliteal height in order to trade off postural efficiency with respect to leg support for a more efficient keying posture (zone 3) with respect to elbows, hands and fingers or a more efficient viewing distance (zone 4). Alternatively, the chair controls may be too difficult to use (cognitive efficiency), or the user may be too distracted by work, or simply have not been informed of the adjustment capabilities.

8.3 Zone 2: Interactions Below the Worksurface

8.3.1 Vertical Clearance Below the Worksurface

Applications in zone 2 are primarily concerned with the general (level 3) function of clearance. In a seated workstation, it is necessary to allow adequate room under the worksurface for the user's thighs, legs, and feet. The term worksurface here refers to the lowest component of the workstation with which the user interacts. If there is a separate keyboard support shelf mounted under a fixed height work table, this shelf must be considered as the relevant worksurface. See Figure 25.3.

The problem will first be analyzed by assuming that the surface of the seat (seatpan) is flat. As such, the clearance requirement can be satisfied by ensuring that the height of the underside of the worksurface is greater than the thickness of the thigh as measured from the seat surface. In general, if clearance requirements can be met for the large (95th percentile) male, all other users will be accommodated. In the Army database, the 95th percentile male thigh thickness is 190 mm (7.48 in.). This value must be added to seat height.

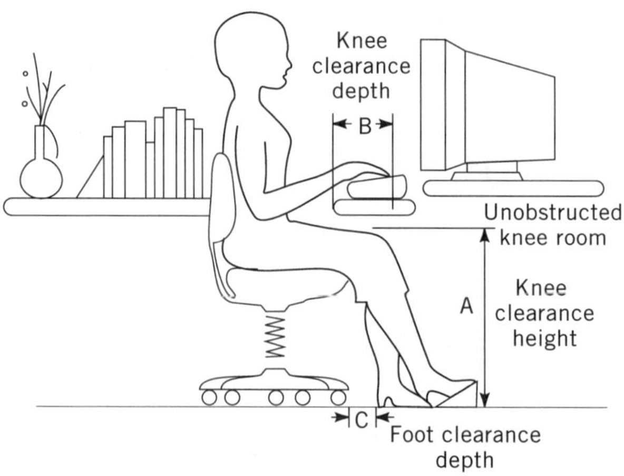

Figure 25.3. Clearance dimensions for seated users. From Ref. 76.

In the previous section, it was shown that a seatpan height of 501 mm would accommodate a 95th percentile male popliteal height plus heel correction. Adding the thigh thickness would yield a value of 691 mm (27.2 in.), which defines the distance from the floor to the underside of the worksurface. Hence, a standard fixed height work table which measures 686 mm (28 in.) to the top of the worksurface and is 25 mm (1 in.) thick will only afford a clearance of 661 mm (686 − 25), which is 30 mm less than is required. Presumably, the large male user will have to place his seatpan below popliteal height to fit under the table; this will mean that his legs can no longer be vertical, and he will be unable to work with his feet flat and supported. Postural efficiency is thereby diminished.

There are at least two reasons why this logic leads to a less than exact solution. First, as discussed previously, it is most unlikely than a male who is at the 95th percentile of thigh thickness will also be at the 95th percentile of popliteal height. In fact, according to Roebuck (78), the correlation between these two variables is close to zero. Secondly, reliance on a military database, by necessity, excludes those individuals who are too large or small to meet military criteria for admission. The strict definition of 95th percentile is that only 5% of the population are larger on any given dimension. Nevertheless, simple observation of workplaces with large numbers of employees usually will yield considerably more than 5% of the workforce who are "95th percentile males." The first of these problems can be dealt with by more sophisticated statistical procedures as discussed previously. The second will require new anthropometric surveys of civilian populations (79).

8.3.2 Horizontal Clearances

Thus far, the discussion has dealt with vertical clearances. Two horizontal clearance issues need to be considered as well. The first relates to the clearance under the worksurface. The work surface should be deep enough and wide enough so that adequate clearance is provided at the levels of the knee and the foot. In particular, adjustment controls and keyboard trays must not protrude into this space. The appropriate depth at the level of the knee can be estimated by subtracting two anthropometric dimensions: the distance between the knee and buttock minus the thickness of the abdomen (assuming the user's abdomen is directly adjacent to the edge of work surface). For the 95th percentile male, this clearance value is 376 mm or 14.8 in. However, the same caveats apply as discussed above. Nemeth and Dainoff (80) have shown that this particular approach of subtracting two 95th percentile dimensions actually underestimates the clearance requirement considerably.

The clearance requirement for width under the table can be determined by the breadth of the user's hips. This can also be used to determine the spacing of arm rests on the chair. In this case, the hip breadth of the 95th percentile female (432 mm) is used since it is larger than that of the 95th percentile male (412 mm). However, this value is likely to be too low; it is measured with the two thighs parallel to one another, but this is an unrealistically rigid posture.

8.4 Zone 3: Interactions Involving Shoulder, Arm, and Hand

8.4.1 Working Height of Shoulder, Arm, and Hand

The level 3 requirement for postural efficiency can be broadly translated into postural requirements which state that the design of the worksurface should allow the major joints of the upper limbs to be in neutral posture. As such, the forearms should be parallel to the floor, the wrists should be flat (minimizing the angles between hands and forearm) and the upper arms should hang vertically. To accomplish this exactly for an anthropometrically diverse population would require a combination of seatpan height adjustability together with worksurface/keyboard support adjustability. In practice, some degree of deviation from neutral posture can be tolerated. ANSI-HFES 100 (81) allows the lower arm to be located within a range of 70–135° with respect to the upper arm (where 90° is neutral posture); the upper arm is permitted to bend forward (flexion) up to 20°. Recent research suggests that 15° might be more conservative (82). Determining the appropriate working height analytically is complicated by the necessity of allowing users to work with backrests inclined backwards as much as 30°. While a fully adjustable worksurface would be preferred, it might be possible to accommodate some users by attaching an adjustable tray to a fixed height worksurface. However, as will be discussed below, this solution moves the user away from the desk, thereby making much of the worksurface inaccessible without stretching. In addition, such trays may well constrain the user to a single keying posture which is not necessarily optimum. For example, for some applications, it might be more efficient to angle the monitor and keyboard off to the side; this would not be possible with most keyboard trays. Finally, a particularly problematic aspect of many currently available keyboard trays is that they do not provide space for input devices such as mice or trackballs. Hence, if frequent use of such devices is required, users must alternate hand position between keyboard tray and table top.

8.4.2 Use of Keyboards and Other Input Devices

While the overall working posture of hands and arms is constrained by the chair and support surface height, what might be called the microposture of the hands is constrained by the configuration of the keyboard or other input device (mouse, trackball, etc.). Consider the keyboard first. If the naturally curved positions of the hands as they rest on the lap can be considered neutral posture, a traditional flat keyboard will force the hand out of neutral posture as it first must bend outward (ulnar deviation) and then rotate downward (pronation). If the keyboard height is different from the operational height of the elbow, the hand may have to bend upward or downward (extension/flexion).

A variety of alternative keyboard designs are available or under development which attempt to allow users to work in postures which are closer to neutral. Work by Remple and his colleagues (83) have demonstrated that: (a) keyboards which are split and angled resulted in hand postures which more closely approximated neutral posture than did traditional keyboards; and (b) measured minimum carpal tunnel pressures tended to correspond to neutral hand posture. This was significant since elevated carpal pressures have been associated with wrist disorder (84). The carpal tunnel is an anatomical location near the wrist through which pass tendons and nerves servicing a large portion of the hand.

An alternative approach to approximating neutral posture consists of slightly tilting the keyboard support surface away from the user so that the rear of the keyboard is lower than the front. There is some supporting evidence for this approach (85).

The level 2 goal of biomechanical efficiency can also be approximated by support of the forearms while keying or operating alternative keyboard devices. This can be achieved by supporting the forearms on a worksurface (86) or a wrist support (87).

The previous discussion has focused on the postural impacts on biomechanical efficiency (level 2 goal). However, there is also the possibility that excessive force exerted in keying might also result in reduced efficiency. What constitutes "excess" necessitates an understanding of the relationship between physical design characteristics of the key action (affordances) and the muscular response capability of the operator (effectivities). There is some evidence that keyboard operators use significantly more force than is required (Ref. 88; but see Ref. 89). In addition, Armstrong et al. (88) argue that force requirements in using a mouse—particularly in dragging operations—could be great enough to be considered excessive, as discussed above.

Technical specifications for keyboard and other input devices will be found in ANSI-HFES 100, and are discussed in Cushman and Rosenberg (90) and Smith, Cohen, and Grandjean (91). Many of these specifications are, however, based on industry practice rather than empirical research evidence. A solid analysis of the biomechanical and psychophysiological fundamentals underlying input device design remains to be accomplished. However, recent work by Aarås and Ro have shown reductions in trapezius loads (92) using a joystick-like vertical mouse, which affords a more neutral posture for the hand.

8.4.3 Work Layout and Reach Envelopes

The computerization of office work environments has appropriately focused much attention on ergonomic requirements necessary to support such equipment. An unfortunate byproduct of this attention is the impression that paper documents and other, more traditional, components of the office, are less important. In fact, computer equipment does not necessarily replace these older elements, but competes with them for working space. Accordingly, efficient work layout becomes a critical issue.

A starting place for discussion is with the reach envelope (see Fig. 25.4). For purposes of this chapter, this can be defined as a region in space where objects can be placed such that they could be used/operated by a seated operator without needing to bend his/her trunk. From military data described in Kroemer et al. (59), horizontal reach envelopes from 54 to 85 cm (measured from the intersection between seat and backrest) would accommodate a range of users from a 5th percentile female to a 95th percentile male. Recent data by Mark et al. (93), however, indicates that users tend to begin to change posture when the target object is located at a point about 85% of their anthropometrically defined maximum reach.

The extent to which such bending is a problem depends on the overall situation, and, in effect, reflects the need to consider interacting level 3 functional requirements. If the target object is used frequently—such as a document lying flat on an adjacent worksurface which must be consulted—the bending act becomes an awkward posture resulting in a decrease in biomechanical efficiency. If, in this situation, a 25-cm deep keyboard tray is employed, a small female would have virtually no worksurface space available without bending. On the other hand, if items with lower frequency of use are located outside of the work envelope, requiring the operator to bend or even leave the chair to retrieve them, this could be considered a beneficial postural change. Consequently, spatial organization

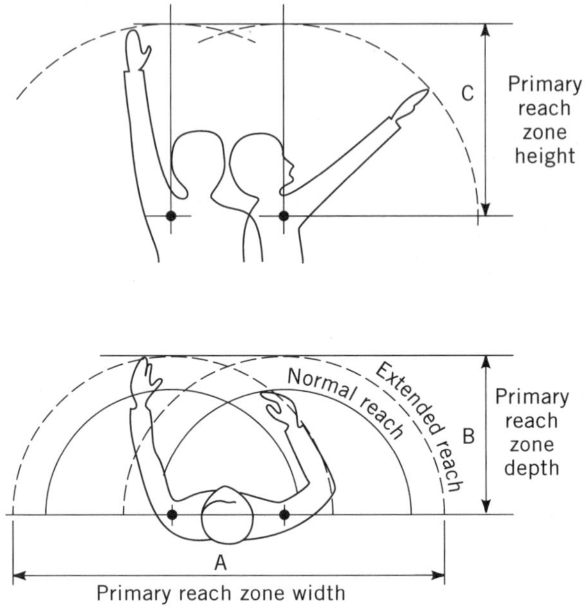

Figure 25.4. Primary reach zones for the user. From Ref. 76.

of major elements of the work environment, including the operator reach envelope, must be integrated with temporal considerations such as importance and frequency of use.

Achieving the level 3 postural change goal can be accomplished by either shifting postures while working or by taking rest pauses. However, due to what might be called "work momentum", operators attempting to maintain cognitive efficiency (level 2) often forget to take breaks. Hence, the opportunity to make significant changes in posture while maintaining work momentum can be achieved by designing a workstation which affords alternating between standing and sitting (94). Several studies have shown the benefit of sit-stand workstations (94–98). However, since frequent alternation is unfamiliar to most office workers, appropriate training is essential.

8.5 Zone 4: Interactions Involving Eyes, Head, and Neck

The final zone of interaction between user and work environment to be considered is centered around the visual interaction with video display unit monitor, paper documents, and other work objects. The following level 4 components are presented as means for satisfying level 3 functional requirements. In discussing these components and their interactions, it should be understood that a large volume of technical information and specifications are available but not presented here. Appropriate references have been provided for those interested.

8.5.1 Visual Display Characteristics

The focus of this discussion is on video display monitors; however, the basic principles can also be applied to paper documents or any other visual display. Image quality, as it affects visual comfort and performance, can usefully be conceptualized in terms of three fundamental characteristics: spatial resolution, temporal resolution, and intensity (99). These requirements can be found in ANSI/HFES-100 and are discussed in detail in Ref. 62.

With regard to the spatial component of image quality, the size of the visual object must be large enough to be clearly visible, while its shape must be clearly formed (not blurry) and interpretable. Size is specified in terms of visual angle: a minimum character height of 16 minutes of arc is a minimum requirement; 20–22 minutes is preferred. Good form can be achieved by monitor characteristics such as an appropriate number of scan lines or pixels.

The issue of temporal resolution is a byproduct of the electronic generation of images by painting images on a screen over time. The rate at which this is done by the display technology (refresh rate) must be fast enough so that elements in the display do not appear to flicker; this is disturbing to many users. The perception of flicker is related to the brightness of the screen; the brighter the screen, the more likely that flicker will be seen. For example, an older display screen with a refresh rate of 60 Hz and positive polarity (dark characters on a bright background) might produce discernible flicker, whereas the same screen in negative polarity (light characters on a dark background) might be flicker-free.

The sharpness or crispness of a visual image is affected by the relationship between the intensity of light used to form the image and the intensity of the background. This relationship is called contrast. Contrast ratios of 3:1 between the brighter and dimmer portions of the screen are considered minimal for both positive and negative polarity displays. However, these ratios must take into account the presence of *unwanted light* (diffuse glare or reflected images) appearing on the face of the screen. To the extent that such unwanted light is present, the contrast ratio is reduced and image quality affected. Here there is a trade off between positive and negative polarity screens. Older positive polarity screens are more susceptible to flicker, but they are less susceptible to reflected glare. (The dark background of a negative polarity display effectively becomes a kind of mirror for the surrounding environment.) Glare problems can be dealt with by various antiglare treatments at the face of the display screen or they can be dealt with by controlling the location and intensity of surrounding light sources.

8.5.2 Lighting Characteristics

In traditional offices, solutions to lighting problems were usually straightforward: more lighting was better. Computer technology complicates the problem and requires more thoughtful solutions. Rea (71) presents a thorough discussion; see also Ref. 100. The basic problem for the VDT operator is to have enough light falling on the work area so that paper documents (and other similar items) are clearly visible while, at the same time, unwanted light, as just discussed, is minimized.

The problem of unwanted light can be divided into two categories: reflection and balance. The reflection issue, as already discussed, becomes a problem to the extent that the mirrorlike surface of a display produces reflected images. These reflections lower the effective contrast on the screen, thereby degrading image quality. The upright orientation of the monitor makes available, for reflection, all objects in front of the monitor. Bright overhead lights are particularly likely to be reflected. The issue of balance refers to the field of view in front of, rather than behind, the operator. Visual comfort and performance may be impaired if there are large differences (i.e., 10:1 or greater) in light levels within the immediate field of view. Thus, for example, viewing a negative polarity (dark background) monitor located in front of a sunlit window might be quite disturbing to most users.

The solution of these problems requires consideration of the relative placement of monitor and lighting fixtures, decisions as to the type of lighting fixtures (i.e., direct vs. indirect; ambient vs. task), and understanding of the monitor characteristics (certain types of flat panel displays actually require reflected light to function). However, placement of the monitor introduces a new set of issues.

8.5.3 Monitor and Document Location

Monitor location issues include viewing distance, height, and tilt angle. Unless the task is completely paperless, parallel considerations apply to paper documents. Level 3 functional requirements of postural efficiency, postural change, illumination and glare, and image quality are all affected by the location of monitor and document. The minimum visual angle requirement of 16 minutes of arc discussed previously is operationally determined by the relationship between actual image height and viewing distance. For angles less than 10 degrees of arc, an accurate approximation of visual angle is 3438* (character height/viewing distance) i.e., (3438)(0.125/16) = 26.9 minutes of arc. The 1988 version of ANSI/HFES 100 specified a minimum viewing distance of 20–30 cm for monitors, but more recent published research (101) has indicated that even 50 cm is considered too close; viewing distances as large as 100 cm were preferred. There is additional research evidence which suggests that users prefer the monitor to be as much as 60 degrees below eye level and to be tilted backward (102, 103). With regard to document location, Hamilton (104) found the poorest levels of postural efficiency (assessed by head rotation and muscle tension) when documents were laid flat on the worksurface, and the best efficiency when documents were on a copyholder directly in front of the operator. The latter could be achieved with a copyholder below the monitor or by moving the monitor to the side if the primary task is paper-based. Copyholders placed to the side were, however, more effective than documents lying flat.

In attempting to integrate the previous sets of requirements, certain trade-offs may be required. Tilting the monitor backward to a preferred condition may exacerbate a glare situation by increasing the number of reflected lighting fixtures. Larger monitors require larger desks to achieve the desired viewing distances, and may be impossible to position sufficiently below eye level. Locating a copyholder below the monitor conflicts with the requirements for lower monitors. Laptop computers may be particularly difficult to place optimally. These were originally meant to be used for short-term purposes, but many individuals are now being asked to utilize such devices as their primary computers.

9 SUMMARY

This chapter has attempted to present a systematic, structured approach to the ergonomics of office workplaces. The means-end hierarchy, by allowing individual component recommendations to be traced to higher-order goals, provides a mechanism for prioritization of such recommendations. A basic assumption in the preceding material has been that overall work efficiency, as defined in terms of the means-end hierarchy, will be improved to the extent that lower level requirements are met. If, however, the equipment design and layout requirements cannot be met, it should be realized that the resulting increased loads on the operators can be compensated for through decreased work demand, typically in the form of longer and more frequent breaks. This solution, of course, leads to different kinds of inefficiencies.

The treatment of the various recommendations and requirements in the preceding section has been deliberately uneven. This discussion has been aimed at a nonspecialist who might need to integrate many components into a functioning office system. More technical information, and deeper scientific justification for such information, may be found in the appropriate references.

BIBLIOGRAPHY

1. R. Briggs, "Safety and Health Effects of Visual Display Terminals," in G. D. and F. E. Clayton eds. *Patty's Industrial Hygiene and Toxicology*, 4th ed. Vol. 1, Pt B, Wiley, New York, 1991.
2. B. Knave and P-G. Wideback, eds., *Work with Display Units 86*, Elsevier (North-Holland), Amsterdam, 1987.
3. National Research Council *Video Displays, Work and Vision*, Washington, D.C. National Academy Press, Washington DC, 1983.
4. M. J. Dainoff, "Occupational Stress Factors in VDT Operation: A Review of Empirical Literature," *Behavior and Information Technology* **1**, 141–176 (1982).
5. ANSI/HFS 100, *American National Standard for Human Factors Engineering of Video Display Terminal Workstations*, Human Factors and Ergonomics Society, Santa Monica, CA, 1988.
6. A. Grieco, G. Molteni, B. Piccoli, and E. Occhipinti, eds. *Work with Display Units 94*. Elsevier Science B.V., Amsterdam, 1995.
7. M. J. Smith, B. Cohen, L. Stammerjohn, and A. Happ, "An Investigation of Health Complaints and Job Stress in Video Display Operations," *Human Factors* **23**, 387–400 (1981).
8. *OSHA 3123, Ergonomics Program Management Guidelines for Meatpacking Plants*, Occupational Safety and Health Administration, U.S. Department of Labor, Washington, DC, 1993.
9. *OSHA 3092, Working Safely with Video Display Terminals*, Occupational Safety and Health Administration, U.S. Department of Labor, Washington, DC, 1997.
10. National Research Council *Work-Related Musculoskeletal Disorders: A Review of the Evidence*, National Academy Press, Washington, DC, 1998.
11. T. Albin, "OSHA Draft Ergonomics Regulation Released," *Human Factors and Ergonomics Society Bulletin*, March 1–6, 1999.
12. T. F. M. Stewart, "Round Table: An Unholy Alliance—Experience of the Development and Implementation of Standards and Legislation on Display Screen Equipment in Europe—The

CEN/ISO Experience," in A. Grieco, G. Molteni, B. Piccoli, and E. Occhipinti, eds., *Work with Display Units 94*, Elsevier Science B.V., Amsterdam, 1995.

13. B. Hocking, "Some Social and Cultural Anthropologic Aspects of Musculoskeletal Disorders as exemplified by the Telecom Australia RSI Epidemic," in S. D. Moon and S. L. Sauter, eds. *Beyond Biomechanics: Psychosocial Aspects of Musculoskeletal Disorders in Office Work.* Taylor and Francis, London, 1996.
14. Ref. 13, p. 49.
15. N. M. Hadler, "Cumulative Trauma Disorders: An Iotrogenic Concept," *J. Occupat. Med.* **32**, 38–41 (1990).
16. I. Kuorinka and L. Forcier, *Work-related Musculoskeletal Disorders (WMSDS): A Reference Book for Prevention*, Taylor and Francis, London, 1995.
17. B. P. Bernard, Ed. *Musculoskeletal Disorders and Workplace Factors: A Critical Review of Epidemiological Evidence for Work-Related Musculoskeletal Disorders of the Neck, Upper Extremity, and Low Back*, U.S. Department of Health and Human Services, Cincinnati, OH, 1997.
18. *Identification and Control of work-related Diseases.* 9th Revision. World Health Organization, Geneva, Switzerland, 1985.
19. Ref. 10, p. 3.
20. T. R. Hales, S. L. Sauter, M. Peterson, L. J. Fine, V. Putz-Anderson, L. R. Schleifer, T. T. Ochs, and B. P. Bernard, "Musculoskeletal Disorders among Visual Display Terminal Users in a Telecommunications Company," *Ergonomics* **37**, 1603–1621 (1994).
21. U. Bergqvuist, E. Wolgast, B. Nilsson, and M. Voss, "Musculoskeletal Disorders among Visual Display Terminal Workers: Individual, Ergonomic, and Work Organizational Factors," *Ergonomics*, **38**, 763–776 (1995).
22. F. Gerr, M. Marcus, and D. Ortiz, "Methodological Limitations in the Study of Video Display Terminal Use and Upper Extremity Musculoskeletal Disorders," *Am. J. Ind. Med.* **29**, 649–656 (1996).
23. S. L. Sauter, M. J. Dainoff, and M. J. Smith, eds. *Promoting Health and Productivity in the Computerized Office: Models of Successful Ergonomic Intervention*, Taylor & Francis, London, 1990.
24. J. Francis and D. L. Dressel, "Workspace Influence on Worker Performance and Satisfaction: An Experimental Field Study," in S. L. Sauter, M. J. Dainoff, and M. J. Smith, eds., *Promoting Health and Productivity in the Computerized Office: Models of Successful Ergonomic Intervention.* Taylor & Francis, London, 1990.
25. C-N. Ong, "Ergonomic Intervention for Better Health and Productivity: Two Case Studies," in S. L. Sauter, M. J. Dainoff, and M. J. Smith, eds., *Promoting Health and Productivity in the Computerized Office: Models of Successful Ergonomic Intervention*, Taylor & Francis, London, 1990.
26. C. Sullivan, Employee Comfort, Satisfaction and Productivity: Recent Efforts at Aetna, in S. L. Sauter, M. J. Dainoff, and M. J. Smith, eds., *Promoting Health and Productivity in the Computerized Office: Models of Successful Ergonomic Intervention*, Taylor & Francis, London, 1990.
27. A. Westin, "Organizational Culture and VDT Policies: A Case Study of the Federal Express Corporation," in S. L. Sauter, M. J. Dainoff, and M. J. Smith, eds., *Promoting Health and Productivity in the Computerized Office: Models of Successful Ergonomic Intervention.* Taylor & Francis, London, 1990.

28. M. Dainoff, "Ergonomic Improvements in VDT Workstations: Health and Performance Effects," in S. L. Sauter, M. J. Dainoff, and M. J. Smith, eds., *Promoting Health and Productivity in the Computerized Office: Models of Successful Ergonomic Intervention*, Taylor & Francis, London, 1990.

29. C. Jeszeck, L. Rectanus, and J. B. Werth, "GAO Report Highlights Ergonomic Benefits," *Human Factors and Ergonomics Society Bulletin*, October 1–5, 1997.

30. A. Aarås, "The Impact of Ergonomic Intervention on Individual Health and Corporate Prosperity in a Telecommunications Environment," *Ergonomics* **37**, 1679–1696 (1994).

31. A. Aarås, R. Westgaard, E. Stranden, and S. Larsen, "Postural Load and the Incidence of Musculoskeletal Illness," in S. L. Sauter, M. J. Dainoff, and M. J. Smith, eds., *Promoting Health and Productivity in the Computerized Office: Models of Successful Ergonomic Intervention*. Taylor & Francis, London, 1990.

32. S. Spilling, J. Eitrheim, and A. Aaras, "Cost Benefit Analysis of Work Environment Investment at STK's Telephone Plant at Kongsvinger," in: N. Corlett, J. Wilson, and I. Manenica, eds., *The Ergonomics of Working Postures*, Taylor and Francis, London, 1986.

33. G. Rubino and F. Grignolo. "Epidemiological Survey of Ocular Disorder among VDT Operators: an Italian Multicentric Research on 31,570 Subjects—Research Promoted by ASSILS," *Bollettino di oculistica*, **68**, Suppl. N7, (1989).

34. B. Piccoli, "The Definition and Assessment of Quality of Images on VDUs. Part I: Relationships with Aesthenopia," in: A. Grieco, G. Molteni, B, Piccoli, and E. Occhipinti, eds., *Work with Display Units 94*: Elsevier Science B.V., Amsterdam, 1995, p. 161.

35. Ref. 35 p. 179.

36. ASSILS (Voluntary Mutual Assistance Association of Telecommunication Workers), "Eye Disorders in Employees Working at VDU: Longitudinal Study," in A. Griece, G. Moltini, B. Piccoli, and E. Occhipinti, eds., *Work with Display Units 94*. Elsevier Science B. V. Amsterdam, 1995, p. 161.

37. L. Stark, "Visual Fatigue and the VDT Workplace," in J. Bennett, ed., *Video Display Terminals: Usability Issues and Health Concerns*, Prentice-Hall, Englewood Cliffs, NJ, 1984.

38. T. Hettinger, "Statistics on Diseases in the Federal Republic of Germany with Particular Reference to Diseases of the Skeletal System," *Ergonomics* **28**, 17–20 (1985).

39. A. Greico, "Sitting Posture: An Old problem and a New One," *Ergonomics* **29**, 345–362 (1986).

40. S. D. Moon and S. L. Sauter, eds., *Beyond Biomechanics: Psychosocial Aspects of Musculoskeletal Disorders in Office Work* Taylor and Francis, London, 1996.

41. M. J. Smith and P. Carayon, Work Organization, Stress, and Cumulative Trauma Disorders. In S.D. Moon and S.L. Saunter, eds., and S. L. Sauter, eds., *Beyond Biomechanics: Psychosocial Aspects of Musculoskeletal Disorders in Office Work*, Taylor and Francis, London, 1996.

42. H. Selye, *The Stress of Life* McGraw-Hill, New York, 1956.

43. R. A. Karasek, "Job Decision Latitude, Job Design, and Coronary Heart Disease," in G. Salvendy, and M. J. Smith, eds. *Machine Pacing and Occupational Stress*. Taylor and Francis, London, 1981.

44. Ref. 16, Chapts. 4–9.

45. B. Stenberg, Skin Symptoms and VDT Work—Reports from Three Continents, in A. Grieco, G. Molteni, B. Piccoli, and E. Occhipinti, eds., *Work with Display Units 94*, Elsevier Science B.V., Amsterdam, 1995.

46. A. Hedge and W. A. Erickson, Office Environment Syndrome Study: Part I, in *Proceedings of the Human Factors and Ergonomics Society 42nd Annual Meeting*, Santa Monica, CA, Human Factors and Ergonomics Society, 1998, pp. 634–638.
47. S. D. Moon, "A Psychosocial View of Cumulative Trauma Disorders: Implications for Occupational Health and Prevention," in S. D. Moon and S. L. Sauter, eds. *Beyond Biomechanics: Psychosocial Aspects of Musculoskeletal Disorders in Office Work*, Taylor and Francis, London, 1996.
48. D. MacLeod, *The Ergonomics Edge*, Van Nostrand Reinhold, New York, 1995.
49. P. Van Wely, "Design and Disease," *Applied Ergonomics* **5**, 262–269 (1970).
50. J. Flach, "The Ecology of Human-Machine Systems: A Personal History," in J. Flach, P. Hancock, J. Caird, and K. Vicente, eds., *Global Perspectives on the Ecology of Human-Machine Systems*, Lawrence Earlbaum Associates, Hillsdale, NJ, 1995.
51. K. Vicente, "A Few Implications of an Ecological Approach to Human Factors," *Human Factors Soc. Bull.* **33**(11), 1–2 (1990).
52. J. J. Gibson, *The Senses Considered as Perceptual Systems*, Houghton Mifflin, Boston, 1966.
53. J. Rasmussen, M. Pejtersen, and L. Goodstein, *Cognitive Systems Engineering*, New York, Wiley, 1994.
54. K. Vicente, *Cognitive Work Analysis: Towards Safe, Productive, and Healthy Computer-based Work*, Lawrence Erlbaum Associates, Hillsdale, NJ, 1999.
55. M. J. Dainoff and M. H. Dainoff, *People and Productivity: A Manager's Guide to Ergonomics in the Electronic Office*, MDA Books, Cincinnati, OH, 1986.
56. K. J. Vicente and J. Rasmussen, "The Ecology of Human-Machine Systems II: Mediating "Direct Perception" in Complex Work Domains," *Ecological Psychology* **2**, 207–249 (1990).
57. W. Karwowski, "Complexity, Fuzziness, and Ergonomic Incompatibility Issues in the Control of Dynamic Work Environments," *Ergonomics*, **34**, 671–686 (1991).
58. Y. Nubar and R. Contini, "A Minimal Principle in Biomechanics," *Bulletin of Mathematical Biophysics* **23**, 377–391 (1961).
59. K. H. E. Kroemer, H. B. Kroemer, and K. E. Kroemer-Ebert, *Ergonomics*, Prentice-Hall, Englewood Cliffs, NJ. 1994.
60. D. A. Rosenbaum, *Human Motor Control*. Wiley, San Diego, 1991, pp. 7–8.
61. K. R. Boff and J. E. Lincoln, *Engineering Data Compendium: Human Perception and Performance*, AAMRL, Wright-Patterson AFB, OH, 1988, Vol I, Image quality 1.211–1.221; Vol III. Image quality 11.8.
62. L. S. Mark, M. J. Dainoff, R. Moritz, and D. Vogele, "An Ecological Framework for Ergonomic Research and Design," in R. R. Hoffman and D. A. Palermo, eds., *Cognition and the Symbolic Processes*, Vol. III, Erlbaum Associates, Hillsdale, NJ, 1991.
63. M. J. Dainoff and L. S. Mark, "Use of a Means-end Abstraction Hierarchy to Conceptualize the Ergonomic Design of Workplaces," in J. Flach, P. Hancock, J. Caird, and K. Vicente. *Global Perspectives on the Ecology of Human-Machine Systems*, Lawrence Earlbaum Associates, Hillsdale, NJ, 1995.
64. F. A. Muckler, "Evaluating Productivity," in M. D. Dunnette and E. A. Fleishman, eds., *Human Performance and Productivity*, Vol. I, *Human Capacity Assessment*, Lawrence Erlbaum Associates, Hillsdale NJ, 1982, pp. 13–49.
65. J. Dul and B. Weerdmeester, *Ergonomics for Beginners*, Taylor and Francis, London, 1993.
66. D. B. Chaffin and G. Andersson, *Occupational Biomechanics*, 2nd ed. Wiley-Interscience. New York, 1991.

67. N. Swanson, S. Sauter, and R. Chapman, "The Design of Rest Breaks for Video Display Terminal Work: A Review of the Relevant Literature," in A. Mital, ed. *Advances in Industrial Ergonomics and Safety 1*, Taylor and Francis, London, 1989.
68. S. Pheasant, *Bodyspace: Anthropometry, Ergonomics and Design*, Taylor and Francis, London, 1986.
69. H. L. Snyder, "Toward the Determination of Electronic Display Image Quality," *Advances in Man-Machine Systems Research* **4**, 1–68 (1988).
70. S. Orsini and L. Milanesi, "The Definition and Assessment of Quality of Images on VDUs. Part II: Physical Aspects," in A. Grieco, G. Molteni, B. Piccoli and E. Occhipinti, eds., *Work with Display Units 94*, Elsevier Science B.V., Amsterdam, 1995.
71. M. S. Rea, "Technics: Solving the Problem of VDT Reflections," *Progressive Architecture*. **72**, 35–40 (1991).
72. D. Norman, *The Psychology of Everyday Things*. Basic Books, New York, 1988.
73. Webb Associates, *Anthropometric Sourcebook* (3 volumes), NASA Reference Publication 1024, National Aeronautics and Space Administration, Houston, 1978.
74. C. C. Gordon, T. Churchill, T. E. Clauser, B. Bradtmiller, J. T. McConville, I. Tebbetts, and R. A. Walker, *1988 Anthropometric Survey of U.S. Army Personnel: Summary Statistics Interim Report* (Technical Report NATICK/TR-89-027), United States Army Natick Research, Development and Engineering Center, Natick, MA, 1989.
75. G. F. Zehner, R.S. Meindl and J.A. Hudson, *A Multivariate Anthropometric Method for Crew Station Design: Abridged.* (Tech. Report AL-TR-1992-0164), Air Force Material Command, Armstrong Laboratory. Wright-Patterson Air Force Base, Ohio: 1993.
76. B. Rutter and M. J. Dainoff, *The Ergonomic Office (revised)*, Metaphase, St. Louis, MO, 1996.
77. P. Cornell and D. Kokot, "Use of Adjustable VDT Stands in an Office Setting," in R. Lueder and K. Noro, eds., *Hards Facts About Soft Machines: The Ergonomics of Seating*, Taylor and Francis, London, 1994, pp. 47–58.
78. J. A. Roebuck, *Anthropometric Methods*. Human Factors and Ergonomics Society, Santa Monica, CA, 1995.
79. Ref. 60, p. 29.
80. K. J. Nemeth and M. J. Dainoff, "Inappropriately Applying Anthropometric Methods for Ergonomic Design Testing," *Proceedings of the International Ergonomics Association 13th Triennial Congress*, Finland, Tampere, 1997.
81. Ref. 5, p. 47.
82. Ref. 33, pp. 68–93.
83. M. Honan, E. Serina, R. Tal, and D. Rempel, "Wrist Postures while Typing on a Standard and Split Keyboard," *Proceedings of the Human Factors and Ergonomics Society, Santa Monica, California, 39th Annual Meeting*, Human Factors and Ergonomics Society, Santa Monica, CA; 1995, p. 366.
84. N. D. Weiss, L. Gordon, T. Bloom, Y. So, and D. M. Remple, "Position of the Wrist Associated with the Lowest Carpal-Tunnel Pressure: Implications for Split Design," *J. Bone Joint Surgery* **77-A**, 1695–1699 (1995).
85. A. Hedge, D. McCrobie, B. Land, S. Morimoto, and S. Rodriguez, "Healthy Keyboarding: Effects of Wrist Rests, Keyboard Trays, and a Preset Tiltdown System on Wrist Posture, seated posture, and musculoskeletal discomfort," in *Proceedings of the Human Factors and Ergonomics Society 39th Annual Meeting*, (pp. 630–634). Human Factors and Ergonomics Society, Santa Monica, CA 1995, pp. 630–634.

86. A. Aarås, K. I. Fostervold, O. Ro, M. Thoresen, and S. Larsen, "Postural Load during VDU Work: A Comparison between Various Work Postures," *Ergonomics* **40**, 1255–1268 (1997).

87. E. A. Damann and K. H. E. Kroemer, "Wrist Posture during Computer Mouse Usage," *Proceedings of the Human Factors and Ergonomics Society 39th Annual Meeting*, HFES, Santa Monica, CA, 1995, pp. 625–629.

88. T. J. Armstrong, J. A. Foulke, B. J. Martin, and D. Rempel, "An Investigation of Applied Forces in Alphanumeric Keyboard Work," *Am. Ind. Hyg. Assoc. J.* **55**, 30–35 (1994).

89. C. S. Pan and L. M. Schleifer, "An Exploratory Study of the Relationship Between Biomechanical Factors and Right-arm Musculoskeletal Discomfort and Fatigue in a VDT Data-entry Task," *Applied Ergonomics* **27**, 195–200 (1996).

90. W. H. Cushman and D. J. Rosenberg, *Human Factors in Product Design*, Elsevier, Amsterdam, 1991, pp. 149–204.

91. M. J. Smith, W. J. Cohen, and E. Grandjean, "Design of Computer Terminal Workstations," in G. Salvendy, ed., *Handbook of Human Factors and Ergonomics*, 2nd. Ed., John Wiley & Sons, Inc., New York, 1997.

92. A. Aarås, and O. Ro, Workload when Using a Mouse as an Input Device. *Intern. J. Human-Computer Interaction* **9**, 105–118 (1998).

93. L. S. Mark, K. Nemeth, D. Gardner, J. Paasche, M. J. Dainoff, M. Duffy, and K. Grandt, "Postural Dynamics and the Preferred Critical Boundary for Visually-Guided Reaching," *J. Exper. Psych. Human Perception and Performance*, 1997.

94. M. J. Dainoff, J. Paasche, K. Simons, K. Terlaga, "Periodic Standing as a Relief from Fatigue in Office Work," *Applied Ergonomics Case Studies*, **2**, 116–125 (1999).

95. J. Winkel and M. Oxenburgh. "Towards Optimizing Physical Activity in VDT/Office Work," in S. L. Sauter, M. J. Dainoff, and M.J. Smith, eds., *Promoting Health and Productivity in the Computerized Office: Models of Successful Ergonomic Intervention*. Taylor & Francis, London, 1990.

96. H. L. Nerhood and Thompson, "Adjustable Sit–Stand Workstations in the Office," in *Proceedings of the Human Factors and Ergonomics Society 38th Annual Meeting*, Human Factors and Ergonomics Society, Santa Monica, CA, 1994, p. 668.

97. R. D. Paul, "Effects of Office Layout and Sit–Stand Adjustable Furniture: A Field Study" in *Proceedings of the Human Factors and Ergonomics Society 42nd Annual Meeting*, Human Factors and Ergonomics Society, Santa Monica, CA, 1995, p. 422.

98. R. D. Paul and M. G. Helander, Effect of Sit-Stand Schedule on Spinal Shrinkage in VDT Operators, in *Proceedings of the Human Factors and Ergonomics Society 42nd Annual Meeting*, Human Factors and Ergonomics Society, Santa Monica, CA, 1995, p. 563.

99. Beaton, personal communication.

100. Illuminating Engineering Society Committee on Office Lighting, *American National Standard Practice for Office Lighting, ANSI/IES RP-1-1993*, Illuminating Engineering Society of North America, New York, 1993.

101. W. Jaschinski-Kruza, "Eyestrain in VDU Users; Viewing Distance and the Resting Position of Ocular Muscles," *Human Factors* **33**, 59–83 (1991).

102. Ref. 60, p. 444.

103. D. R. Ankrum, E. E. Hansen, and K. J. Nemeth, "The Vertical Horopter and Viewing Distances at Computer Workstations," in Y. Anzai, K, Ogawa, and H. Mori, eds., *Symbiosis of Human and Artifact*, Elsevier, Amsterdam 1995.

104. N. Hamilton, "Source Document Position as it Affects Head Position and Neck Muscle Tension," *Ergonomics* **39**, 593–610 (1996).

GENERAL REFERENCES

ANSI-HFES 100: the American National Standard on Human Factors Engineering of Video Display Terminal Workstations: now in the process of revision, can be consulted for more detailed information regarding technical specifications for workstation furniture, keyboards, and displays, as well as their integration into a functioning system.

ANSI-NSC Z365: the American National Standard on Prevention and Control of Cumulative Trauma Disorders, focuses on ergonomics as related to musculoskeletal disorders.

ANSI-HFES 300: The American National Standard for Anthropometry and Biomechanics, will provide methods for resolving some of the anthropometric issues discussed above.

Ergonomics in Design, The Human Factors and Ergonomics Society, Santa Monica, CA, provides articles on office ergonomics which are aimed at a general readership.

More technical research reports may be found in the following journals: *Human Factors, Ergonomics, Applied Ergonomics, International Journal of Human-Computer Interaction, International Journal of Industrial Ergonomics, Behaviour and Information Technology.*

Refs. 49, 56, and 66 provide introductions to ergonomics aimed at the general reader. More advanced reviews can be found in textbooks such as Ref. 60.

CHAPTER TWENTY-SIX

Biohazards and Occupational Disease

Patricia A. Heinsohn, Ph.D., MPH, CIH

1 INTRODUCTION

Microbes inhabit every environmental niche on the earth, and some have adapted to survival under extremely harsh conditions. Most do not cause disease in humans or animals. Indeed, they have many beneficial uses. Fungi can deteriorate organic materials, including, textiles, leather, wood, foods, paper, and paint. Molds synthesize chemical reagents and medicines and are used in food fermentation, pest control, and other industrial processes. This chapter deals only with microorganisms that can cause occupational disease.

2 BIOHAZARDOUS AGENTS

Biohazardous agents include microorganisms and their toxins, arthropods, allergens and toxins derived from higher plants, and protein allergens from vertebrate animals. Arthropods (crustaceans, arachnids, and insects) may bite or sting, causing skin inflammation, systemic intoxication, or transmission of infectious agents (e.g., mosquitoes transmit). Inhalation may also result in disease. For example, the inhalation of fecal dusts and dust mite bodies may result in allergic disease. Allergens and toxins derived from higher plants may cause dermatitis on skin contact (e.g., poison ivy) or rhinitis or asthma by inhalation. Protein allergens from vertebrate animals, present in the urine, feces, saliva, hair, dander, and feathers, may become airborne in agricultural settings, during care of laboratory ani-

Patty's Industrial Hygiene, Fifth Edition, Volume 2. Edited by Robert L. Harris.
ISBN 0-471-29754-2

mals, or in other occupational settings. This chapter is limited to infectious biohazards and occupational allergens.

An understanding of basic microbiology is necessary to understand the nature and importance of infectious biohazards. Microorganisms vary in size (see Table 26-1) and include viruses, bacteria, fungi, rickettsiae, chlamydiae, and mycoplasmas. Microbes, like all other organisms, are given a binomial Latin name. The first is the genus name, capitalized, and the second, lowercase, is the species name. Both are italicized, for example, *Herpes simiae* (a virus), *Mycobacterium tuberculosis* (a bacterium), *Aspergillus fumigatus* (a fungus), *Coxiella burnetii* (a rickettsia), and *Chlamydia psittaci* (a chlamydia).

Viruses are obligate intracellular pathogens, meaning that they cannot metabolize or reproduce outside of a living cell. Their survival is limited to a few hours to a few weeks outside of a living host. In their simplest form they consist of a nucleic acid, either DNA or RNA, encapsulated in a protein coat, or capsid, and in some cases surrounded by a membrane. The capsid protects the nucleic acid from enzymes and physical agents. The capsid is also responsible for entry of the viral nucleic acid into a host cell. Once inside, the host's replication machinery is altered to produce more viral nucleic acids and proteins. The fate of virus-infected cells depends on the specific virus. Cell lysis may occur either quickly or only after a long period of growth. Viruses are responsible for the majority of human diseases, especially those transmitted by inhalation.

Bacteria are unicellular, prokaryotic microorganisms. The nucleus in prokaryotes contains DNA that is in direct contact with the cytoplasm, that is, it is not membrane-bound. They have a single chromosome and do not undergo mitosis. Instead, they reproduce by fission. Some can reproduce in less than 30 min. Bacteria are characterized by their staining reaction (e.g., gram stain, acid-fast stain); morphology (size; shape; and presence of flagellae, spore, and capsule formation); use of carbohydrates; and others.

Fungi represent a separate kingdom of life forms, the Eumycota, and include yeasts, fleshy mushrooms, and molds. They are neither plant nor animal. Fungi are heterotropic and absorb their food from other organisms, living or dead. Fungi are eukaryotic (i.e., they possess a nuclear membrane). Yeasts are unicellular and reproduce by budding. Molds are multicellular and filamentous. They have tubular, branched, and diffuse bodies and reproduce by means of spores. Most molds produce sexual and asexual spores. The structure, arrangement, and location of the developing spores provide taxonomic information for identification. Although mold spores vary in size from 2 μm to >200 μm, most are 2 to 20 μm, that is, small enough remain airborne for long periods. They are able to survive in a dormant state for years, even decades, maintaining the ability to germinate. Most are

Table 26.1. Relative Size Ranges of Microorganisms

Microorganism	Size Range
Mycoplasmas	0.12–0.25 μm
Viruses	0.02–0.3 μm
Rickettsiae	0.3–0.7 × 1.5–2.0 μm
Bacteria	0.5–1.0 × 2–5 μm
Fungi	2–>200 μm (largest aspect)

disseminated through the air, especially during dry weather. Germination can occur indoors or outdoors as long as they are provided with sufficient nutrients for growth. Depending on environmental conditions, fungi can exist as either yeasts or molds. More than 60,000 species of fungi have been described. Relatively few of them cause infectious disease in man. Fungi are associated with allergic disease.

Rickettsiae are obligate intracellular parasites that are closely related to gram-negative bacteria. With the exception of *Coxiella burnetii*, all rickettsial diseases are transmitted from animal to animal by the bite of an infected arthropod vector. Chlamydiae comprise a unique group of microorganisms due to their complex method of reproduction. Mycoplasmas are the smallest known organisms capable of growth and reproduction outside living host cells. They differ from other prokaryotes in that they lack a cell wall.

3 INFECTIOUS DISEASE TRANSMISSION

3.1 Infection and Infectious Disease

Infection is not synonymous with infectious disease. Infection is the entry and development or multiplication of an infectious agent in the body of humans or animals. Infection may be either inapparent (e.g., tuberculin skin test positive only) or manifest (e.g., tuberculosis). Some body cavities (e.g., nose, mouth, and gut) and skin have naturally occurring commensal microorganisms called normal flora. Zoonoses are infections or infectious diseases not normally occurring in humans but transmitted from vertebrate animals to humans under natural (rather than experimental) conditions. Hantavirus pulmonary syndrome is an example of a zoonosis. Hantavirus normally infects deer mice but can infect exposed, susceptible people.

3.2 Reservoirs and Hosts

A reservoir of an infectious agent is any person, animal, arthropod, plant, soil, or substance (or combination of these) in which an infectious agent normally lives and multiplies. The infectious agent primarily depends on its reservoir for survival. Reservoirs for some infectious agents associated with occupational disease are listed in section 3.1. Hosts are people or living animals, including birds and arthropods, that provide subsistence or lodging to an infectious agent under natural conditions. Some protozoa and helminths pass successive stages in alternate hosts of different species. Hosts are further differentiated as primary or definitive, secondary or intermediate, or transport hosts. For example, the reservoir for legionellae is primarily water and humans, a host for infection.

3.3 Modes of Transmission and Infectious Dose

Routes of exposure to a variety of agents are typically described in industrial hygiene courses as ingestion, dermal absorption, and inhalation. Public health terminology describes mechanisms of infection as direct transmission; indirect transmission, including vehicle-borne and vector-borne transmission; and airborne transmission. Ingestion is a type of direct transmission and inhalation is equivalent to airborne transmission involving droplet nuclei (e.g., tubercle bacilli) or dusts (e.g., aerosolized *C. burnetii* rickettsiae or spores of *A. fumigatus* fungi) containing microorganisms.

Dose is directly related to infectivity; a sufficient dose must be received at a target site in the body of a susceptible individual to initiate infection. Experimentally, the infectious dose is the minimum number of microorganisms required to establish infection in 50% of a group of hosts of the same species. This quantity, called the ID_{50}, is a function of the infectious agent, the route of administration, the source of the agent, and host susceptibility factors. The ID_{50} is determined in test animals and provides a means of ranking infectious agents. Viral infectious doses are expressed similarly but with cells grown in culture as the "host." A $TCID_{50}$ (tissue culture infectious dose) is the end point of a quantal titration of virus that infects 50% of the inoculated cell culture. Limited quantitative human infectivity data exists.

3.4 Pathogenicity and Virulence

Pathogenicity is the capability of an infectious agent to cause disease in a susceptible host. Virulence is the degree of pathogenicity of an infectious agent, indicated by case fatality rates, or its ability to invade and damage host tissues, or both. The route of transmission affects pathogenicity and virulence, that is, the same microorganism may be pathogenic by one route of exposure but not by another. An opportunistic pathogen is one that does not normally infect healthy people but may cause disease in people who are immunocompromised.

3.5 Host Susceptibility

It is generally thought that people who contract infection or develop an infectious disease cannot mount sufficient resistance against a particular pathogen when exposed to it. Risk factors for susceptibility to infectious disease include age. For example, both the young (under 3 years old) and the elderly are more susceptible to tuberculosis. The risk of developing tuberculosis is highest in children under 3 years old and reactivations of long-latent infection account for a large proportion of cases of tuberculosis in older persons. Another risk factor for tuberculosis is co-infection with the human immunodeficiency virus (HIV). Other risk factors include underweight and malnourished condition, silicosis, and diabetes.

3.6 Chain of Infection

Infection control is a term used by medical professionals to denote techniques aimed at preventing nosocomial infection in hospitalized patients and staff. In other settings, biosafety professionals also practice a form of infection control. The goal of infection control is to break the chain of infection and reduce the occurrence of infection and disease in the workplace. Any intervention that can break the chain of infection will prevent infection. The following are offered as examples that biosafety professionals use.

- Biological safety cabinets are engineering controls that can prevent inhalation exposure to a wide of variety of infectious agents in the handling of human body fluids that can generate aerosols. The reader should refer to section 2.3.1 in Chapter 27, Control of Biohazards, for discussion of the design, use, and maintenance of biological safety cabinets.
- Personal protective equipment, including gloves, gowns or laboratory coats, masks, and protective eyewear or face shields, is critical in preventing mucous membrane and non–intact skin exposure to blood-borne pathogens such as hepatitis B virus (HBV) and human immunodeficiency virus (HIV). The reader should refer to section 2.4 in Chapter 27, Control of Biohazards, for a discussion of personal protective equipment.
- Administrative controls such as the proper disposal of sharps and work practices such as not recapping or shearing needles are important control methods for prevention of parenteral inoculation of blood-borne pathogens.
- Immunization and vaccination can confer resistance to disease and are used to prevent many occupational diseases, for example, vaccination to prevent HBV infection in health care employees. The reader should refer to section 2.2.3 in Chapter 27, Control of Biohazards, for additional information.

3.7 Tables of Specific Biohazardous Agents and Occupational Diseases With Which They Are Associated

The following tables list the type of microorganism or allergen of occupational significance and disease. Table 26.2 contains viruses and occupational viral diseases; Table 26.3 contains bacteria, rickettsiae, and chlamydiae and occupational bacterial, rickettsial, and chlamydial diseases; Table 26.4 contains fungi and occupational fungal diseases; and Table 26.5 contains occupational allergens and disease. The following information is given in each table:

- Description of the disease
- Symptoms
- The etiologic biohazardous agent (i.e., microorganism)
- Occupations at risk
- Mode of transmission and whether person-to-person transmission is possible
- The incubation period and period of communicability
- Host resistance and susceptibility
- Some control measures (Note: The reader should consult Chapter 27, Control of Biohazards for additional information.)

4 CURRENT TOPICS IN BIOSAFETY

Several biohazards have been selected for more detailed discussion because they are common workplace risks. Whereas Chapter 27 discusses the concepts of biosafety, a detailed

Table 26.2. Occupational Infectious Diseases Associated With Viruses

Disease and Symptoms	Agent and Reservoir	Occupations at Risk	Mode of Transmission Person-to-Person Transmission Possible: Y/N	Incubation Period Period of Communicability	Resistance and Susceptibility
AIDS (acquired immunodeficiency syndrome), AIDS-related syndrome	Human immunodeficiency virus (HIV-1 and HIV-2)	Health care workers who are exposed to blood products and other potentially infectious material	Contact with blood, semen, or other infected body fluids via (*1*) percutaneous exposure, (*2*) mucous membrane exposure, (*3*) exposure to abraded or otherwise nonintact skin, (*4*) breastfeeding, (*5*) perinatal routes from an infected mother, (*6*) sexual activity.	Variable. Generally 1–3 months for detectable antibodies, but varies from less than 1 year to more than 10 years from infection to onset of AIDS.	Unknown, but susceptibility presumed to be general. Presence of other STDs, specifically those with ulcerations, and penile foreskin may increase susceptibility. No recovered cases documented. Degree of immunity is unknown.
Virus targets CD4 immune cells, resulting in susceptibility to various malignancies and opportunistic infections. Within several weeks to months after infection, many people develop an acute self-limited mononucleosis-like condition lasting a week or two. Infected people may then remain unaffected for several months to years before opportunistic infections and constitutional and neurological symptoms appear.	Humans		Y	Unknown, presumed to begin early after onset of HIV infection and extend throughout life. Infectiousness may increase with increasing immune deficiency, clinical symptoms, and presence of other STDs.	
CMV mononucleosis	Cytomegalovirus (CMV)	Child-care workers	Intimate contact with secretions or body fluids of an infected individual	3–8 weeks following transfusion of infected blood	Infection is nearly universal; fetuses, patients with debilitating diseases, patients on immunosuppressive drugs, specifically organ allograft recipients and patients with AIDS, are more susceptible.
Infection later in life is usually inapparent but may cause a condition similar to mononucleosis.	Humans		Y	Virus excreted in urine and saliva for many months and can persist for several years following primary infection	
Hemorrhagic fever with renal syndrome, nephropathy epidemic, hantavirus pulmonary syndrome	Hantaan, Seoul, Puumala, Belgrade, and Muerto Canyon viruses	Outdoor workers in endemic areas, rodent-control workers, laboratory workers.	Direct contact with infected rodent secretions, urine, and feces and inhalation of airborne infected dust particles either by direct aerosolization of infected rodent urine (or other potentially infected body fluid) or by secondary aerosolization of dried rodent excreta	As short as a few days or as long as 2 months, but usually 2–4 weeks	Those who do not possess serologic evidence of past exposure are uniformly susceptible. Inapparent infections can occur and second attacks have not been observed.
Five clinical phases: Febrile, hypotensive, oliguric, diuretic, and convalescent Febrile: High fever, headache, malaise, and anorexia followed by severe abdominal or lower back pain, often accompanied by nausea and vomiting, which lasts 3–7 days Hypotensive: Several hours to 3 days—defervescence and abrupt onset of hypotension may progress to shock and hemorrhagic manifestations Oliguric: Blood pressure returns to normal or is high, which lasts 3–7 days; nausea and vomiting may persist, severe hemorrhage may occur, and urinary output falls dramatically Convalescence: Takes weeks to months Case fatality rates ~5% in Asia, higher in Balkans, <1% caused by the Puumala virus in Europe.	Field rodents; humans are accidental hosts		N		

Hepatitis C, non-A-non-B hepatitis	Hepatitis C virus (HCV)	Health care workers	Parenteral, sexual possible but infrequent	From 2 weeks to 6 months; most commonly 6–9 weeks	General susceptibility. Degree of immunity after infection is unknown. In experiments, chimpanzees infected repeatedly.
Onset is insidious with anorexia, vague abdominal discomfort, nausea, and vomiting, progressing to jaundice less frequently than HBV. Severity ranges from inapparent in 75% of cases to rare fatal cases.	Humans		Y	From one or more weeks before onset of symptoms and may persist indefinitely	
Herpes B virus or monkey B virus	*Herpes simiae*	Zoo attendants, nonhuman primate animal handlers, veterinarians, researchers	Bites and scratches, parenteral	Within 1 month of exposure	Unknown
Vesicular skin lesions at or near site of inoculation, neurologic symptoms. Myalgias, headaches, dizziness. Symptoms may progress rapidly and half are fatal. Majority of survivors have had significant neurologic impairment.	Monkeys		Y	Unknown	
Herpes simplex virus (HSV) Common names: Cold sores, fever blisters, genital herpes, herpetic whitlow	Herpes simplex virus (HSV) *Herpes viridae*	Health care workers, particularly dental, nursing, and respiratory care workers	Direct contact with infected secretions. Virus enters through mucosal surfaces or abraded skin.	2–12 days.	Probably universally susceptible
HSV1—illness of varying severity, marked by fever and malaise lasting a week or more; reactivation commonly results in fever blisters and cold sores.	Humans		Y	Secretion of virus in saliva has been reported for as long as 7 weeks after recovery from stomatitis.	
Infectious hepatitis	Hepatitis A virus (HAV)	Foreign travelers to areas of high endemnicity, day care workers, institutional and some military settings, health care, primate handling, dentistry	Fecal-oral route almost exclusively; by saliva or blood transfusion is possible but uncommon.	15–50 days depending on dose; average 28–30 days	General susceptibility. Low incidence of manifest disease in infants and preschool children suggests that anicteric and mild infections are common. Immunity after infection probably lasts a lifetime.
Abrupt onset of fever, malaise, anorexia, nausea, and abdominal discomfort followed by jaundice. Disease can vary in severity from a mild illness lasting 1–2 weeks to a rare severely disabling disease lasting several months.	Humans and rarely captive chimpanzees. Less frequently, other nonhuman primates.		Y	Maximum infectivity during the latter half of incubation period, continuing few days after onset of jaundice. Most cases noninfectious after first week.	
Influenza, flu, grippe	Influenza virus	Workers in hospitals, nursing homes, and child care facilities	Transmitted predominantly by aerosols (mean diameter <10 μm) generated by coughing, sneezing, or talking by infected persons	Short, usually 1–3 days	When new subtype appears, all children and adults are susceptible. Infection produces immunity, but duration and strength depend on degree of antigenic drift and number of previous infections.
Acute viral disease of respiratory tract characterized by fever, headache, malaise, myalgia, prostration, coryza, sore throat, and cough. Recovery usually occurs in 2–7 days. In major epidemics, fatalities can occur among elderly and debilitated.	Humans. Mammalian reservoirs such as swine and birds are likely sources on new human subtypes.		Y	Probably 3–5 days from clinical onset in adults, up to 7 days in children	
Measles, red measles, rubeola	Measles virus	Employees in health care settings (particularly pediatrics workers), child care workers, and school employees	Transmitted through respiratory route through aerosol droplets. It is one of the most communicable of infectious diseases.	About 10 days, varying from 7–18 days from exposure to onset of fever, usually 14 days until rash appears	All persons who have not had the disease or been successfully immunized are susceptible.
Acute, highly communicable disease with prodromal fever, conjunctivitis, coryza, cough, and Koplik spots on buccal muscosa.	Humans		Y	From slightly before the prodromal period to 4 days after appearance of rash, minimal after second day of rash	

Table 26.2. (continued)

Disease and Symptoms	Agent and Reservoir	Occupations at Risk	Mode of Transmission Person-to-Person Transmission Possible: Y/N	Incubation Period Period of Communicability	Resistance and Susceptibility
Mumps	Mumps virus	Persons working in child care, schools, research, and laboratories	Transmitted through direct contact with saliva or recently contaminated materials	About 12–25 days, commonly 18 days	Immunity is usually lifelong and develops after inapparent as well as clinical infections.
Acute viral disease characterized by fever and swelling and tenderness of one or more salivary glands, usually the parotid and sometimes the sublingual or submaxillary glands	Humans	Y	Transmitted through droplet nuclei, and entry is by nose or mouth. Substantial contact is needed.	Virus has been isolated from saliva from 6–7 days before overt parotitis to 9 days after. Exposed nonimmune people should be considered infectious from 12th to 25th day after exposure. Maximum infectiousness occurs hours before onset of illness.	
			Y		
Rabies	Rabies virus family rhadoviridae	Veterinarians, animal handlers, laboratory workers, foreign travelers to endemic areas, recreational hunters, animal-control officers, outdoor workers. Health care workers dealing with patients with rabies.	Mostly transmitted to humans through animal bite with virus-laden saliva. Rarely, nonbite transmission can occur from affected saliva and other infectious materials such as brain tissue through abraded skin, open wounds, and mucous membranes. Rare cases of aerosol exposure in rabies laboratories.	Usually 3–8 weeks, rarely as short as 9 days or as long as 7 years depending on severity and site of wound in relation to nerve supply and distance to the brain	All mammals susceptible to varying degrees. Humans are more resistant than several other species.
Fatal disease begins with malaise, fatigue, headache, anorexia, and fever. In ~50% of patients, pain or paresthesia occurs at the site of exposure. Sore throat, cough, abdominal pain, nausea and vomiting, diarrhea, or chills occur in prodrome phase lasting 2–10 days. Neurologic phase follows and includes hyperactivity, hallucinations, disorientation, bizarre behavior, seizures, nuchal rigidity, and paralysis and lasts 2–7 days, ending in death without treatment. With treatment, the duration of illness averages 25 days.	Wild and domestic animals, including dogs, foxes, coyotes, and wolves. Also skunks, raccoons, mongooses, and other biting animals.			In dogs and cats, usually 3–7 days before onset of clinical signs, and throughout disease	
			N (theoretically possible, but not documented)		
Respiratory syncytial virus pneumonia	Respiratory syncytial virus (RSV)	Hospital and child care workers	Through respiratory route; large droplet is primary mode. Small aerosols can cause infection but are less contagious. Also through contact with contaminated fomites and skin.	Usually 2–8 days (average 5 days)	Secondary infection is less severe in healthy adults.
Infection in healthy adults results in upper respiratory tract illness with nasal congestion and cough. Tends to be more prolonged than common cold with symptoms averaging 9 days. Incapacitation occurs in about 50% of adults with RSV. Some experience pulmonary function change and airway hyperactivity after 8 weeks following illness.	Humans			Viral shedding in adults occurs for 3–6 days on average.	Elderly or debilitated individuals may require supportive therapy.
			Y		

Serum hepatitis Less than 30–50% of adults with acute HBV infection will have icteric disease. Onset of clinical symptoms is usually insidious: anorexia, vague abdominal discomfort, nausea and vomiting, sometimes arthralgias and rash, often progressing to jaundice. Fever may be absent or mild.	Hepatitis B virus (HBV) Humans. Chimpanzees are susceptible, but an animal reservoir in nature has not been discovered.	Health care workers who are exposed to blood or blood products	Parenteral, sexual, perinatal. Permucosal, from fomites. Y	Usually 45–180 days, average 60–90 days. As short as 2 weeks for appearance of Ab and rarely as long as 6–9 months. HBsAg positive people are potentially infectious. In experiments, blood has been shown to be infective weeks before onset of disease and all through acute clinical course of disease.	General susceptibility. Usually disease is milder and often anicteric in children and asymptomatic in children. Persons with Down's syndrome, lymphoproliferative disease, or HIV and those on hemodialysis are more susceptible.
Simian AIDS Causes a chronic disease similar to AIDS. Opportunistic infections, diarrhea, thymic atrophy, decrease in CD4 cell numbers, lymphomas, and encephalitis.	Simian immunodeficiency virus (SIV) Nonhuman primates	Veterinarians, nonhuman primate handlers, animal caretakers, and virus researchers	Via needle sticks, possibly bites and scratches, and mucous membrane contact with certain nonhuman primate body fluids	Seroreactivity developed within 3 months in one needle stick case. Unknown	Unknown
Vaccinia Healthy, nonimmune individuals experience mild fever and sometimes lymphadenopathy. Generalized vaccinia is widespread but usually mild dissemination of vaccinia lesions occur in patients without preexisting skin disease.	Vaccinia virus Virus is found mainly in research laboratories.	Molecular biology research laboratory workers	Through intentional immunization. The pustules contain live virus and may be transferred from pustule to wound or via mucous membranes. Laboratory uses of large quantities of virus can expose workers via mucous membranes or wounds or perhaps by inhalation. Y	Papule forms after 3–5 days from exposure. Live virus lives in pustules for several weeks.	Most healthy, nonimmune individuals experience mild symptoms after vaccination. Immunocompromised individuals can experience systemic vaccinia that leads to destruction of skin, subcutaneous tissues, and even viscera. For those with agammaglobulinemia or T-cell deficiencies and those receiving immunosuppressive therapy, the disease is usually fatal.
Viral hemorrhagic fever Sudden onset of fever, malaise, myalgia, and headache followed by pharyngitis, vomiting, diarrhea, and maculopapular rash. Hepatic damage, renal failure, and CNS involvement and terminal shock with multiorgan dysfunction can also occur.	Filoviruses (Ebola and Marburg viruses) Natural reservoir is unclear, but both are associated with green monkeys.	Researchers, primate handlers, health care workers treating infected patients	Direct contact with blood and other body fluids Y	3–9 days for Marburg; 2–21 days for Ebola virus disease As long as blood and secretions contain virus; Ebola virus was isolated from seminal fluid on 61st but not 76th day in one case.	All ages are susceptible.

Refs. 1, 2.

Table 26.3. Occupational Infectious Diseases Associated With Bacteria

Disease or Bacterium and Symptoms	Agent and Reservoir	Occupations at Risk	Mode of Transmission Person-to-person Transmission Possible: Y/N	Incubation Period Period of Communicability	Resistance and Susceptibility
Acinetobacter Gram-negative bacteria Occupational asthma can result from exposure to aerosols in a contaminated environment. Endotoxin is the agent that is implicated in bronchospasm and has been associated with headache, nausea, cough, nasal and eye irritation, chest tightness, phlegm, fever, diarrhea, and fatigue.	Acinetobacter (especially *A. calcoaceticus*) Soil	Workers in swine-containment buildings, sewage treatment plants, composting plants, poultry processing plants, foundries, cotton mills. Workers who use industrial microscopes are at higher risk for ocular infections.	Inhalation N	Unknown	Unknown
Anthrax, woolsorter's disease (inhalation anthrax) *Cutaneous* form involves toxins that cause tissue necrosis. *Inhalational* anthrax from inhalation of aerosols causes upper respiratory infection leading to severe respiratory distress and death, usually within 24 hours. It can be associated with anthrax hemorrhagic meningitis, after hematogenous spread to the meninges. *Intestinal* is rare and tends to occur in explosive outbreaks of the food-poisoning type.	*Bacillus anthracis* Domestic herbivores, especially cattle but also sheep, goats, horses, and mules	Humans who have contact with animals (agricultural) or animal products (industrial anthrax)	Via direct contact with contaminated animal products or infected animals, by indirect contact with a contaminated environment, or by contact with airborne *B. anthracis* N	A few hours to 7 days; most cases occur within 48 hours of exposure. Transmission person to person is rare. Contaminated articles and soil may remain infective for decades.	Uncertain. There is some evidence of inapparent infection among people with frequent contact with infectious agent. Second attacks are rare.

Brucellosis, Bang's disease, Mediterranean fever, undulant fever, Neapolitan fever, Malta fever, Gibraltar fever	Brucella species *Brucella abortus*, biovars 1–6 and 9; *B. melitensis*, biovars 1–3; *B. suis*, biovars 1–5, and *B. canis*	Workers who slaughter and milk animals or handle bull semen, aborted fetal tissue, and placentas; livestock handlers and slaughterhouse workers. Other occupations with significant risk are veterinarians, government meat inspectors, farmers, dairy workers, and microbiology laboratory workers.	Inhalation of infectious aerosols. The bacteria can also enter the body by penetrating the mucosa of the mouth or throat, or through minor breaks in skin. Ingestion is likely route of food-borne cases.	Highly variable and difficult to determine, usually 5–60 days, 1–2 months common; occasionally several months	Severity and duration vary widely. Duration of acquired immunity is uncertain.
Systemic bacterial disease with acute or insidious onset. Fever of variable duration, headache, weakness, profuse sweating, chills, arthralgia, depression, weight loss, and generalized aching. Localized suppurative infection of organs, including liver and spleen. Disease may last several days, months, or occasionally a year without proper treatment. Osteoarticular complications occur in 20–60% of cases, with sacroiliitis the most frequent joint manifestation. Genitourinary involvement in 2–20% of cases, with orchitis and epididymtis most common. Recovery is usual but disability is often pronounced, with <2% case-mortality of untreated cases.	Cattle, swine, goats, and sheep. May occur in bison, elk, caribou, and some species of deer. *B. canis* is occasional problem in laboratory dog colonies and kennels. Coyotes found to be infected.		N		
Campylobacteriosis	Campylobacter species *Campylobacter jejuni* and less commonly *C. coli*. *C. laridis* an *C. fetus* also associated with diarrhea.	Abattoir workers, butchers, veterinarians, farmers, laboratory animal workers, and animal handlers	Ingestion of bacteria are responsible for the majority of infections. Contact with feces or gastrointestinal contents can result in ingestion.	2–5 days with a range of 1–10 days, depending on dose ingested.	Not fully understood, by lasting immunity to serologically related strains follows infection. In developing countries, most people develop immunity in first 2 years of life.
Acute bacterial enteric disease of variable severity involving diarrhea, abdominal pain, malaise, fever, nausea, and vomiting. Illness is frequently over in 2–5 days and usually does not last more than 10 days.	Animals, particularly poultry and cattle. Puppies, kittens, other pets, swine, sheep, rodents, and birds.		Y	Throughout the course of infection, usually from several days to several weeks. Untreated individuals may excrete organisms for as long as 2–7 weeks.	

Table 26.3. (continued)

Disease or Bacterium and Symptoms	Agent and Reservoir	Occupations at Risk	Mode of Transmission Person-to-person Transmission Possible: Y/N	Incubation Period Period of Communicability	Resistance and Susceptibility
Chancroid, soft chancre	*Haemophilus ducreyi*	Occurs in military and civilian populations. Contact with prostitutes is high risk factor.	Sexually transmitted. Direct sexual contact with discharges from open lesions and pus from buboes. Self autoinoculation to other sites can occur.	From 3–5 days up to 14 days	Susceptibility is general. Uncircumcised are higher risk than circumcised. No evidence of natural resistance.
Causes genital and perianal ulcerative lesions. Lesions begin as tender erythematous papule that becomes pustular and then ulcerates. Bleeding and pain are common, with progression to abscess and spontaneous draining.	Humans		Y	Until healed and as long as infectious agent persists in lesion or discharging regional lymph nodes. Can last several weeks or months without antibiotic treatment.	
Erysipeloid, erysipeloid of Rosenbach, fish poisoning, seal finger	*Erysipelothrix rhusiopathiae*	Workers handling meat and fish products—fishermen, abattoir workers, and butchers. Veterinarians, farmers, and animal handlers are at risk through contact with infected animals. Less commonly, kitchen workers, furriers, fertilizer producers, soapmakers, sewer workers, and bacteriology laboratory workers.	Cutaneous inoculation of bacteria from contaminated source (i.e., fish scales or bone fragments that puncture skin surface). Also, consumption of undercooked infected meat.	2–7 days	Unknown
Harbored in pharynx and excreted in feces. Most common form is erysipeloid, a localized skin infection at site of contact. Area is demarcated and edematous, with violaceous lesion that fades centrally as it spreads peripherally. Accompanied by burning, itching, or throbbing pain that can be severe. Fever and arthralgia occur in ~10% of cases and lymphangitis and lymphadenopathy can be associated. Diffuse cutaneous form: Multiple skin lesions occur in generalized pattern. Third form is severe systemic illness consisting of sepsis and endocarditis that is rare.	Swine are primary reservoir. Also found in freshwater fish or shellfish, turkeys, and rarely sheep, cattle, chickens, and pheasants.		N		

Glanders	*Pseudomonas mallei*	Laboratory workers, veterinarians, equine butchers, pathologists, and other workers who handle horses	Via broken skin, mucous membranes, and possibly inhalation.	Unknown	Unknown
Cellulitis with necrosis, abscess formation, and granuloma formation. After cutaneous inoculation, subcutaneous nodules and regional lymphadentitis appear. After inhalation, draining mucosal ulcers, necrotizing lobar pneumonia or bronchopneumonia, nodular lung infiltrates, and lung abscess may occur. Either site of infection can lead to systemic infection with pustular rash. This form is uniformly fatal in 7–10 days. Chronic infection can cause subcutaneous and intramuscular abscesses on mostly arms and legs and affects liver, spleen, lung, eye, and central nervous system. Case fatality rate without treatment is high.	Horses, mules, and donkeys		N, not readily		
Gonorrhea, clap	*Neisseria gonorrheae*	Outside of prostitution, very few occupations affected. Seafarers appear to be at greater risk. Cutaneous infection of laboratory workers.	Direct contact of mucous membranes with organism, usually during sexual intercourse. Parenteral inoculation through cut skin has occurred.	Usually 2–7 days, sometimes longer when symptoms occur	Susceptibility is general. Gonococcal strains are heterogeneous, so reinfection is often. Women using intrauterine device have higher risks of salpingitis during first 3 months from insertion. Those deficient in complement components susceptible to bacteremia. Vaginal epithelium resistant, but prepubertal epithelium susceptible.
Urethra, endocervix, anal canal, pharynx, and conjunctiva are infected directly by contact. Penetrates mucosal epithelium and causes local inflammatory response. Men—urethritis, with local extension in urogenital tract without treatment. Women—vaginitis, cervicitis, and urethritis; untreated can cause endometritis, salpingitis, and pelvic inflammatory disease. If left untreated becomes systemic and causes bacteremia, arthritis, tenoxynovitis, endocarditis, meningitis, or disseminated rash.	Human		Y	May be months without treatment. Effective therapy ends communicability within hours.	

Table 26.3. (continued)

Disease or Bacterium and Symptoms	Agent and Reservoir	Occupations at Risk	Mode of Transmission Person-to-person Transmission Possible: Y/N	Incubation Period Period of Communicability	Resistance and Susceptibility
Impetigo, erysipelas, strep throat, scarlet fever, rheumatic fever	Streptococcal species (other than *S. suis*)	Construction and farm workers, meat packers, mortuary technician, occasionally pig handlers.	Ubiquitous human pathogens. Skin infection occurs through breaks in skin. Infection more common in surgical incisions or lacerations. Respiratory exposure through inhalation of droplets. Direct contact with patients or carriers. Food-borne outbreaks from milk and milk products or egg products.	Short, usually 1–3 days, rarely longer	Susceptibility to sore throat and scarlet fever is general. Immunity develops only for that specific type of Group A streptococcus and may last years. Antibiotic therapy may interfere with development of specific immunity.
Group A streptococcus is the most important human pathogen causing skin infections, including sipelas, a rapidly progressive skin infection producing fever and sometimes bacteremia; pyoderma, a purulent skin infection; and cellulitis. Pharyngitis is common followed by fever and sore throat. Complications include local head and neck infections. Scarlet fever is caused by toxins from some strains. Acute rheumatic fever and acute poststreptococcal glomerulonephritis are two delayed complications. Acute rheumatic fever is systemic illness involving connective tissue and heart, joints, and blood vessels. Acute poststreptococcal glomerulonephritis can follow pharyngitis or pyoderma.	Humans		Y	Untreated uncomplicated cases, 10–20 days. Untreated complicated cases, weeks or months. With treatment, within 24 hours	
Impetigo, toxic shock syndrome, scalded skin syndrome, staphylococcal food poisoning	Staphylococcal species. *S. aureus* most common pathogen in occupational environment.	Occupations where skin trauma can occur: Agricultural workers, construction workers, meat packers. Outdoor workers such as river rafting guides. Health care workers. Food poisoning outbreaks caused by food handlers.	Hands of and objects touched by nasal carriers with draining lesions or any purulent discharge. Needle sticks and sharps in hospitals.	Variable and indefinite, commonly 4–10 days	Compromised immune systems, prior viral infections, indwelling foreign bodies, prior antibiotic treatment, and preexisting illness can predispose people to infection. Use of steroids and antimetabolites can also increase susceptibility.
Facultative anaerobe that colonizes human skin. Bacteria produce a number of enzymes and toxins. *Superficial infections*: Folliculitis, furunculosis, carbuncles, impetigo, abscesses, and infected lacerations. *Systemic infections*: Superficial scalded skin syndrome that can lead to desquamation of entire skin and toxic shock syndrome. *Organ infections*: Endocarditis, pericarditis, pneumonia, osteomyelitis, septic arthritis, septic bursitis, and pyomyositis. Bacteremia has occurred from needle sticks. Food poisoning caused by enterotoxin.	Humans, rarely animals		Y	As long as purulent lesions continue to drain or carrier state persists. Autoinfection may continue for period of nasal colonization or duration of active lesions.	

Legionellosis, Legionnaires' disease, Pontiac fever	Legionella species, mostly *Legionella pneumophila*	Many occupational environments where cleaning operations in cooling towers and steam turbine have caused outbreaks. Dental workers may be exposed through high-speed drills and sprays.	Via inhalation of aerosols from contaminated water	Pontiac fever 5–66 hours, most often 24–48 hours	Age is a risk factor (most of those affected are at least 50 years of age), especially in patients who smoke or have diabetes mellitus, chronic lung disease, or renal disease or malignancy and in the immunocompromised. Male-to-female ratio is 2.5:1.
Pontiac fever: Flulike with attack rate of >90%, mild with rapid recovery. Fever, myalgias, headache, chills, and fatigue. Cough is minimal, with clear chest radiograph. Legionnaires' disease—low attack rate 0.1–5%—pneumonia. Begins like Pontiac fever with nonproductive cough and evidence of consolidation on chest radiograph. Chest pain and hemoptysis, diarrhea, and gastrointestinal complaints in 40%; neurologic abnormalities may occur. Case fatality rate may be as high as 39% in hospitalized cases.	Primarily aqueous. Hot-water systems (showers), cooling towers, evaporative condensers, humidifiers, whirlpool saunas, etc.		N	Legionnaires' disease 2–10 days, most often 5–6 days Person-to-person transmission not documented.	
Leptospirosis, Weil syndrome, canicola fever, hemorrhagic jaundice, mud fever, swineherd disease	*Leptospira interrogans, L. biflexa*	Rice and sugarcane workers, farmers (especially dairy farmers), miners, sanitation and sewage workers (from rat infections), rodent-control workers, zoo workers, veterinarians, slaughterhouse and other meat workers, and fishing industry workers	From contaminated fluids, tissues, or waters through breaks in the skin; mucous membranes; and possibly intact skin. Urinary shedding of organisms is most important source, and meat handling is also important source of exposure.	Usually 10 days, with range of 4–19 days	General susceptibility of humans. Immunity to specific serovar follows infection or (occasionally) immunization but may not protect against infection with different serovar.
Sudden onset of fever, headache, chills, severe myalgia (calves and thighs), and conjunctival suffusion. Other manifestations: diphasic fever, meningitis, rash, hemolytic anemia, hemorrhage into skin and mucous membranes, hepatorenal failure, jaundice, mental confusion/depression, myocarditis, and pulmonary involvement with/without hemoptysis.	Wild and domestic animals—rats, swine, cattle, dogs, and raccoons		Y, rare	Person-to-person rare, but leptospires may be excreted in urine for 1 month and as long as 11 months after acute illness.	
Lyme disease, Lyme borreliosis	*Borrelia burgdorferi*	Outdoor workers	Transmitted to humans by tick vectors. In experimental animals, transmission does not occur until the tick has been attached for more than 24 hours. This may be true in humans.	For EM, from 3–32 days after tick exposure. The early stages may be asymptomatic and patient may present later with manifestations.	All persons are probably susceptible. Resistance has occurred by treating individuals with antibiotics for early disease.
Distinctive rash, erythema migrans (EM), develops in 60–90% of patients within 3–30 days. EM may be accompanied by malaise, fatigue, fever, headache, stiff neck, myalgia, migratory arthralgias, and/or lymphadenopathy, possibly lasting several weeks or more in untreated patients. Within weeks or months of EM lesion, early neurologic and cardiac abnormalities may occur. Weeks to years after onset, swelling and pain in large joints may occur. Chronic neurologic problems may develop.	Spirochete distributed by tick: *Ixodes scapluaris* in eastern and midwestern U.S. *I. pacificus* in western U.S., *I. ricinus* in Europe, and *I. persulcatus* in Asia Deer are important mammalian hosts for vector-tick species.		N	No evidence of natural transmission from person to person. Rare case reports of congenital transmission, but epidemiologic studies show no link between maternal Lyme disease and adverse outcomes of pregnancy.	

Table 26.3. (continued)

Disease or Bacterium and Symptoms	Agent and Reservoir	Occupations at Risk	Mode of Transmission Person-to-person Transmission Possible: Y/N	Incubation Period Period of Communicability	Resistance and Susceptibility
Meningococcal meningitis, meningococcemia, cerebrospinal fever Sudden onset of fever, intense headache, nausea, vomiting, stiff neck, and often petechial rash. Delirium and coma are often observed. Case mortality formerly 50%, but with early diagnosis and modern therapy, 5–15%. 5–10% of populations with endemic disease asymptomatic carriers in the nasopharynx. Small minority of people proceed to invasive disease, which may include bacteremia, sepsis, meningitis, or pneumonia.	*Neisseria meningitidis* Humans	Military recruits, day care workers, prison workers, chronic care facility workers, and others who work in close contact and crowded areas	Direct contact with and inhalation of respiratory droplets from nose and throat of those infected with or asymptotically carrying bacteria. Y	From 2–10 days, commonly 3–4 days Until meningococci are no longer present in nose and mouth. Usually 24 hours after treatment.	Susceptibility is low and decreases with age. More carriers than cases. Those deficient in complement proteins are susceptible to recurrent disease. Splenoctomized persons are susceptible to bacteremic illness. Group-specific immunity of unknown duration follows even subclinical infections.
Pasteurellosis Present in GI or respiratory tracts of many animals. Many animals are carriers. 70–90% of oral and nasal secretions in cats shown to carry this bacteria. After bite or scratch, local erythema, warmth, tenderness, and purulent drainage in ~40% of cases. Pulmonary pasteurellosis is primarily a problem with those with underlying lung disease. Systemic infections can primarily develop in patients with underlying medical disease.	*Pasteurella multocida* Cats, dogs, rodents, and larger domestic and wild animals ranging from cattle to lions.	Animal laboratory workers, veterinarians, animal breeders, zoo workers, abattoir workers, and agricultural workers. Work involving contact with animals.	Usually through animal bites or scratches, possibly animal secretions. Pulmonary pasteurellosis—inhalation of aerosolized secretions from animals. N	Within 24 hours	Those with underlying medical disease more susceptible.

Pertussis, whooping cough	*Bordetella pertussis*	Health care and child care workers, especially in parts of world where vaccination is not widespread.	Transmitted by inhalation of infectious airborne respiratory secretions.	Commonly 6–20 days	Susceptibility is universal in nonimmunized individuals. Incidence rates highest in children under 5 years of age. Waning immunity in previously immunized adults can result in second attacks.
First stage is marked by 1–2 weeks of nonspecific upper respiratory infection symptoms. Second stage is characterized by increasing bouts of severe coughing as well as "whoop" caused by inhaling air through a narrowed glottis, cyanosis, engorged neck veins, bulging eyes, and protruding tongue. There is no fever. Second stage lasts 2–4 weeks. Third stage is convalescent stage.	Humans		Y	Highly communicable in early catarrhal stage before paroxysmal cough stage. Communicability decreases and becomes negligible in about 3 weeks for nonhousehold contacts, despite persistent spasmodic cough with whoop. When treated with erythromycin, period of infectiousness is 5 days or less after onset of therapy.	
Psittacosis, ornithosis, parrot fever	*Chlamydia psittaci*	Bird handlers, pet shop workers, zoo attendants, poultry workers, laboratory workers, and veterinarians	Inhalation of agents from desiccated droppings; secretions and dust from feathers of infected birds	1–4 weeks	Susceptibility is general; immunity following infection is incomplete and transitory. Older adults may be more severely affected.
Once inhaled, rapidly enters bloodstream and is transported to reticuloendothelial cells of the liver and spleen. It replicates in sites, then invades the lungs and other organs by hematogenous spread. The incubation period ranges from 4–15 days. Often begins abruptly with chills and fever. Headache, malaise, anorexia, myalgias, and arthralgias are common. A persistent hacking cough is prominent. Respiratory symptoms are often mild compared to X-ray. Hepatosplenomegaly may occur and myocarditis and encephalitis rarely occur.	Principally parakeets, parrots, and lovebirds; less often poultry, pigeons, canaries, and seabirds		N	Diseased and seemingly healthy birds may shed the agent intermittently and sometimes continuously for weeks or months.	

Table 26.3. (continued)

Disease or Bacterium and Symptoms	Agent and Reservoir	Occupations at Risk	Mode of Transmission Person-to-person Transmission Possible: Y/N	Incubation Period Period of Communicability	Resistance and Susceptibility
Q fever	*Coxiella burnetti*	Abattoir and livestock workers, dairy workers, veterinarians, veterinary and laboratory technicians, laboratory animal handlers, farmers, ranchers, and hide and wool handlers. Farmers exposed to cattle, sheep, and goats and who assist with birthing of lambs are exposed to birth by-products and are at high risk.	Inhalation of particles or direct contact with infected animals, primarily cattle, sheep, and goats. Placentas from infected sheep are extremely infectious.	Depends on size of the infecting dose, usually 2–3 weeks	Susceptibility is general. Immunity following recovery from illness is probably lifelong with cell-mediated immunity lasting longer than humoral. Workers with prosthetic heart valves and liver disease and particularly at risk.
Presents as mild respiratory illness and described as atypical pneumonia. Resolves in 2–4 weeks. Most cases are mild and self-limiting, but prolonged course can lead to granulomatous hepatitis with jaundice. Chronic form associated with infective endocarditis can develop as late as 1–20 years following acute illness (most of time associated with preexisting valvular disease or prosthetic valve). Rarely acute pleuronecrosis, hemolytic anemia, bone marrow necrosis, inflammatory pseudotumor of the lung meningitis, optic neuritis, arthritis, cerebella ataxia, myocarditis, thrombophlebitis, nephritis, cystitis, and epididymitis.	Sheep, cattle, goats, cats, dogs, some wild animals (bandicoots and some feral rodents), birds and ticks		N	Direct transmission from person to person occurs rarely, if ever.	
Rat-bite fevers, streptobacillary fever, Haverhill fever, sodoku	*Streptobacillus moniliformis*	Animal laboratory workers, veterinarians, animal breeders, agriculture workers, and other workers with exposure to rodents	By urine or secretions of mouth, nose, and conjunctival sac. Bites from wild or laboratory rats, which may have colonization of oral cavities and upper respiratory tracts. Bites or scratches from other animal carriers. Haverhill fever may be from ingesting food contaminated with rat feces containing the organism. Blood from experimental animals has been known to transmit the disease.	Streptobacillus moniliformis, 3–10 days, rarely longer. Spirillum minor, 1–6 weeks.	No information.
Streptobacillus moniliformis—rats excrete bacteria in saliva and urine. Causes Haverhill fever in the U.S. Illness consists of prodrome of fever, chills, headache, vomiting, myalgias, and arthralgia. The initial wound appears healed, but within 2–4 days rash develops over extremities, followed by polyarthritis. Complications include endocarditis, pneumonia, septic arthritis, and localized abscess in organs. Spirillum minor—causes sodoku in Asia. Symptoms similar to *S. moniliformis*, except ulceration at site of bite. Associated lymphadenopathy and lymphangitis and arthritis are unusual.	Rats; rarely other animals such as squirrels, weasels, gerbils		N		

Rocky Mountain spotted fever, North American tick typhus, New World spotted fever, tick-borne typhus fever, Sao Paulo fever Sudden onset of moderate to high fever that persists for 2–3 weeks if untreated, significant malaise, deep muscle pain, severe headache, chills, and conjunctival infection. In about half of cases, maculopapular rash appears and spreads rapidly to rest of body; petechiae and hemorrhages are common. Case fatality rate is 13–25% untreated, but with treatment death is uncommon.	*Rickettsia rickettsii* Ticks by transovarian and transstadial passage	Persons in outdoor occupations, including farming, forestry, logging, construction, and telephone line work. Also environmental technicians and some laboratory workers.	Bite of an infected tick. At least 4–6 hours of attachment and feeding of the tick are needed before the rickettsiae become reactivated and infectious for people. Contamination of breaks in skin or mucous membranes with crushed tissues or feces of the tick can lead to infection. N	3–14 days Not directly transmitted from person to person, but tick stays infective for life, commonly up to 18 months.	Susceptibility is general. One attack probably confers lasting immunity.
Shigellosis, dysentery, bacillary dysentery Only small inoculum needed to cause disease. After ingestion and passage through stomach, invades the mucosa of colon. Produces several toxins, primarily leads to local cell invasion and destruction. Mucosal layer in colon ulcerates and sloughs. Fever; abdominal cramping; watery diarrhea; and some blood, pus, mucus, and tenesmus. If left untreated, dehydration may result. Mild cases resolve. More severe cases can lead to fatal colonic perforation or bacteremia. Less common are hemolytic-uremic syndrome, seizures, and arthritis.	Shigella species Humans, but prolonged outbreaks in primates have occurred.	Workers who travel to endemic areas (i.e., military personnel). Seafarers, day care and nursery workers, health care workers, medical laboratory workers, and other workers who handle contaminated materials and wastes.	Fecal-oral; involves ingestion of the organisms; occurs most often after direct person-to-person contact. Less commonly, may be water- or food-borne. Y	12–96 hours (usually 1–3 days; up to 1 week) During acute infection and until agent no longer present in feces, usually within 4 weeks. Asymptomatic carriers may rarely be infectious, lasting months or longer. Appropriate treatment reduces duration of carriage to a few days.	Susceptibility is general. More severe in young children than adults who may be asymptomatic. Elderly, debilitated, and malnourished of all ages susceptible to severe disease and death. Breastfeeding is protective.
Streptococcus suis Group R beta-hemolytic species. Cause of meningitis/sepsis in both pigs and humans. Found in blood and cerebrospinal fluid of infected individuals. Symptoms include flulike prodrome followed by fever with mengismus and sometimes polyserositis. Permanent hearing loss and vestibular dysfunction uveitus, endophthalmitis, and arthritis are possible complications.	*Streptococcus suis* Pigs	Pig farmers and handlers of raw pork, such as meat packers and butchers	Mainly through skin breaks, although nasopharyngeal transmission can occur.	Unknown	Unknown

Table 26.3. (continued)

Disease or Bacterium and Symptoms	Agent and Reservoir	Occupations at Risk	Mode of Transmission Person-to-person Transmission Possible: Y/N	Incubation Period Period of Communicability	Resistance and Susceptibility
Syphilis	*Treponema pallidum*	Prostitutes, workers with enforced separation from families (i.e., seafarers and migrant laborers), possibly laboratory workers handling infected animals.	Direct contact of mucous membrane or skin with an infectious lesion like chancres, skin rashes, mucous patches, or condylomata lata; almost always involves sexual contact. Also acquired congenitally and through contact with human blood.	10 days to 3 months, usually 3 weeks.	Susceptibility is universal, but only 30% of exposures result in infection. Infection leads to immunity and sometimes to heterologous treponemes. Treatment during primary and secondary stages may stop immunity from developing. Concurrent HIV infection may change normal host response.
Primary lesion develops at site of infection. Chancre of primary syphilis is typically painless and usually heals within 2 months. Secondary syphilis is systemic and occurs usually 6–8 weeks after primary lesion has healed. It affects skin; mucous membranes; and lymphatic, renal, GI, and skeletal systems. Late syphilis occurs years later and can include neurosyphilis (symptomatic and asymptomatic), and cardiovascular disease (aortitis and valvular disease). Gummas may occur in skin and many organ systems.	Humans		Y	Variable and indefinite during primary and secondary stages and also in recurrences during the first 4 yr of latency. Unsure of communicability during latent period. Inapparent lesions may be potentially infectious. Adequate penicillin treatment usually ends infectivity within 24–48 hr.	
Travelers' diarrhea, turista, food poisoning	*Escherichia coli* Various virulent strains	Day care and chronic care facility workers; workers who travel to less-developed countries, animal handlers.	ETEC—ingestion of contaminated food or water. Direct contact transmission by fecally contaminated hands is rare. EPEC—by contaminated infant formulas and weaning food and by fomites and by contaminated hands by fecal-oral route. EIEC—scanty evidence suggests transmission by contaminated food. EHEC—ingestion of inadequately cooked hamburgers, unpasteurized milk.	ETEC—1–3 days EPEC—2–6 days EIEC—2–3 days EHEC—3–5 days	ETEC—serotype-specific immunity is acquired following infection. EPEC—susceptibility mainly in young infants in nature EIEC—little is known. EHEC—infectious dose is very low. Little known about susceptibility and immunity. Old age is risk factor.
4 major categories: 1. Enterotoxigenic *E. coli* (ETEC)—travelers' diarrhea after ingestion of food. Resolves without treatment. 2. Enteropathogenic *E. coli* (EPEC)—major cause of childhood diarrhea, leads to mucosal lesions. Watery diarrhea for 1–3 weeks. 3. Enteroinvasive *E. coli* (EIEC)—invades intestinal cells, leading to inflammatory response that destroys intestinal mucosa. Fever and bloody diarrhea with numerous fecal leukocytes, lasting 1–2 weeks. 4. Enterohemorrhagic *E. coli* (EHEC) produces Shigalike toxin that causes 7–10 days of bloody diarrhea without inflammatory response, fever, or leukocytes.	ETEC—humans EPEC—humans EHEC—cattle, but humans may serve as reservoirs for person-to-person transmission. EIEC—humans		Y	ETEC—duration of excretion EPEC—duration of excretion or prolonged EIEC—duration of excretion EHEC—duration of excretion, 1 week in adults	

Tuberculosis, consumption	*Mycobacterium tuberculosis*	Workers in health care; clinical or research laboratories; funeral homes; drug treatment centers; and facilities for homeless, alcoholics, and persons with AIDS. Primate caretakers. Maintenance workers for biological safety cabinets and other ventilation system parts are also at risk.	Contracted after inhalational exposure of droplet nuclei 1–5 um in size that remain airborne for long periods. Is not contracted by skin contact with surfaces.	From infection to primary legion or significant tuberculin reaction, 4–12 weeks. Subsequent risk of progressive pulmonary or extrapulmonary TB greatest within first year or two; latent infection may be for lifetime.	Directly related to degree of exposure. Most hazardous period for development of clinical disease is first 6–12 months after infection. Risk is highest in children under 3 years old, lowest in later childhood, and high again in adolescent, young adults, very old, and immunosuppressed. HIV, malnourishment, drug use, and debilitating diseases increase susceptibility.
May be insidious at onset, causing anorexia, weight loss, low-grade fever, fatigue, and cough. Pulmonary and pleural TB most common acute manifestation—pleuritic chest pain, coughing of bloody sputum, high fever, and profuse sweating. Multidrug-resistant strains have high case fatality rate. Other organs can be affected. Genitourinary system, lymph nodes, and vertebral bodies can be affected. Meningeal TB and tuberculous peritonitis can occur. TB can affect pericardium, causing heart failure, and larynx, causing hoarseness. Addison's disease and skin infiltration can also occur. If infiltration is overwhelming and disseminated, can mimic leukemia.	Primarily humans, rarely primates. In some areas, diseased cattle, swine, and other mammals.		Y	Theoretically, as long as bacilli are in sputum. Some untreated individuals may be sputum-positive intermittently for years. Effective chemotherapy usually stops communicability within a few weeks, at least in households.	
Tularemia, rabbit fever, deer-fly fever, Ohara disease, Francis disease	*Francisella tularensis*	Muskrat trappers, outdoor workers (especially military personnel), those exposed to insect and arthropod vectors, and farmers	Direct handling of infected animal carcasses, hides, or fur, through nonintact skin; via bites from infected ticks and biting flies; or bites by infected animals. Inhalation exposure is important for laboratory workers handling dead animals.	Depends on virulence of infecting strain and of size of inoculum; 1–14 days, usually 3–5 days	All ages are susceptible. Long-term recovery follows recovery. Reinfection has occurred in laboratories.
High fever, headaches, rigors. Meningitis, rhabdomyolysis (especially septicemia), disseminated intravascular coagulopathy, shock, and death may follow. Skin inoculation may lead to ulceroglandular tularemia cutaneous ulcer and lymphadenopathy. Mortality <5%. Infection of eyes can occur if contact is made with infectious material. Inhalation of infected aerosols can lead to septicemia and typhoidal form associated with fever, prostration, and weight loss without adenopathy. With appropriate therapy, <1% mortality.	Numerous wild animals, especially rabbits, hares, muskrats, beavers, some domestic animals, and various hard ticks. Rodent-mosquito cycle described in Russia and Sweden.		N	Not directly transmitted from person to person. In blood in first 2 weeks and in lesions for a month. Flies infective for 14 days; ticks for lifetime (2 yr).	

Table 26.3. (continued)

Disease or Bacterium and Symptoms	Agent and Reservoir	Occupations at Risk	Mode of Transmission Person-to-person Transmission Possible: Y/N	Incubation Period Period of Communicability	Resistance and Susceptibility
Yersiniosis, intestinal yersiniosis, extraintestinal yersiniosis Acute febrile diarrhea with abdominal pain that can be and can mimic acute appendicitis. Septicemia is less common and usually associated with preexisting illness. Some complications can be erythema nodosum (in about 10% of adults, mostly women), postinfectious arthritis, and systemic infection. Joint pain is reported in half of infected adults.	*Yersinia pseudotuberculosis* and *Y. enterocolitica* Animals. *Y. enterocolitica*—pig. *Y. pseudotuberculosis*—avian and mammalian, particularly rodents and other small mammals.	Abattoir workers, workers who come in contact with refrigerated meat after slaughter (i.e., butchers, meat handlers, and meat packers)	Fecal-oral route by ingestion of contaminated food or water, direct contact with infected animals or patients. Nosocomial transmission reported and transmission through blood transfusion. Y	Mostly 3–7 days; usually under 10 days Secondary transmission is rare, but fecal shedding at least as long as symptoms exist, usually 2–3 weeks, possible up to 2–3 months	Diarrhea is more severe in children; postinfectious arthritis is more severe in adolescents and older adults. *Y. pseudotuberculosis* favors male adolescents whereas *Y. enterocolitica* attacks both genders equally. Reactive arthritis and Reiter's syndrome more often affect HLA-B27 genetic type. Septicemia occurs among people with excess iron or those with immunosuppressive illness or therapy.

Refs. 1, 2.

discussion of biohazard control for legionellae and Creutzfeldt-Jakob prion is presented. Microbial indoor air quality is also discussed.

4.1 Legionellae and Legionellosis

4.1.1 Clinical Syndromes and Epidemiology

Infection with legionellae causes two distinct clinical syndromes with distinct epidemiological features. Legionnaires' disease, discovered after a 1976 outbreak among American Legion convention attendees (3), presents clinically as pneumonia occasionally accompanied by multiorgan dysfunction. It may occur as both outbreak-related epidemics of pneumonia and nosocomial infection. Pontiac fever is an acute febrile, self-limiting illness, without pneumonia. Legionnaires' disease has a low attack rate but a high fatality rate (15–39%) if untreated. The fatality rate is generally higher among the immunocompromised. Pontiac fever has a high attack rate but a low fatality rate. It is believed to be a reaction to inhaled bacterial antigen rather than true infection.

Initially, both diseases present with fever, chills, anorexia, malaise, myalgia, and headache. The incubation period is generally 2 to 10 days for Legionnaires' disease and 48 to 72 hours for Pontiac fever. Diarrhea, abdominal pain, and a nonproductive cough are common. In legionella pneumonia, there are patchy or focal areas of lung consolidation that may progress to bilateral involvement and respiratory failure. Diagnosis depends on isolation of legionellae, direct immunofluorescence stain in involved tissue or respiratory secretions, detection of antigens of *L. pneumophila* serogroup 1 in urine by radioimmunoassay, or a fourfold or greater rise in levels of convalescent serum antibodies. So-called fingerprinting techniques of multilocus enzyme assay (MEA), ribotyping, restriction fragment length polymorphism (RFLP), random amplified polymorphic DNA (RAPD), and pulsed-field-gel electrophoresis (PFGE) can be used to correlate the specific *Legionella* strain clinically isolated to that isolated from a suspected environmental source. Antibodies to serogroup-specific antigens are useful for diagnostic or epidemiological studies, but they do not appear to confer immunity.

Although more than 60 serogroups within at least 39 species of legionellae have been identified, 85 to 90% of all cases of legionellosis are caused by a single species, *L. pneumophila*. Most (65–70%) of these involve one of 15 serogroups, namely *L. pneumophila* serogroup 1. *L. micdadei, L. bozemanii, L. longbeachae*, and *L. dumofii* have also been isolated, primarily from immunocompromised patients with pneumonia.

Legionnaires' disease is underreported. Approximately 10,000 to 100,000 of the 2.4 million cases of pneumonia occurring annually in the United States are legionella pneumonia. However, only 1,000 to 3,000 of these are actually reported. Buffered charcoal yeast extract agar, which contains the necessary L-cysteine cofactor, is required for successful cultivation of legionella species. Because many laboratories may not have access to such media, physicians often may not perform appropriate diagnostic tests in patients with pneumonia (4).

4.1.2 Transmission of Legionellae

Legionellosis is transmitted primarily by inhalation. Some investigators believe that the disease may also be acquired by drinking legionellae-contaminated water, especially if

Table 26.4. Occupational Infectious Diseases Associated With Fungi

Disease or Fungus and Symptoms	Agent and Reservoir	Occupations at Risk	Mode of Transmission Person-to-Person Transmission Possible: Y/N	Incubation Period Period of Communicability	Resistance and Susceptibility
Coccidioidomycosis, valley fever, desert rheumatism, valley bumps, California disease, Posada's mycosis	*Coccidioides immitis*	Agricultural workers, construction workers, telephone post diggers, archaeologists, and laboratory workers	Primary route is inhalation of spores. Skin inoculation and transplacental infection from mothers with disseminated disease are rare.	10–16 days	Although no racial, gender, or age differences in susceptibility to primary coccidioidomycosis exist, dark-skinned races are more prone to severe primary illness and disseminated disease. Pregnancy also seems to increase risk of dissemination.
Usually is mild respiratory infection or asymptomatic. Symptoms are fever; pleuritic or dull chest pain; cough; white or blood-streaked sputum; and malaise, headache, anorexia, and myalgia possibly followed by skin rash, arthralgias, and mild conjunctivitis. Acute coccidioidal pneumonia may disseminate rapidly and be fatal. Some experience chronic progressive form of pulmonary coccidioidomycosis that mimics TB. Skin and subcutaneous lesions are most common. Bone lesions and meningitis are possible complications.	Soil		N	Cases of person-to-person transmission are rare.	
Cryptococcosis, torulosis, European blastomycosis	*Cryptococcus neoformans*	Pigeon breeders and other workers in areas contaminated with pigeon and other bird droppings	Inhalation of fungal spores most likely found in old pigeon droppings	Unknown. Pulmonary disease may precede brain infection by months or years.	Usually only immunocompromised individuals are affected: persons with AIDS, sarcoidosis, or lymphoma, and those requiring chronic steroids.
Deep mycosis usually subacute or chronic meningitis; infection of lungs, kidneys, prostate, and bone may occur. Acneiform lesions, ulcers, or subcutaneous tumorlike masses may appear on skin. May act as endobronchial saprophyte in patients with lung disease. Untreated disease may end in fatality within weeks to months.	Found in pigeon droppings, and occasionally in droppings of other birds		N		

Histoplasmosis, Ohio Valley disease, Cave disease, Tingo Maria fever, Darling's disease, reticuloendotheliosis Five clinical forms: 1. Asymptomatic 2. Acute benign respiratory form with cough, pleuritic chest pain, fever, chills, myalgias, malaise, and nausea. Calcifications of lung, hilar lymph nodes, and spleen in late findings. 3. Acute disseminated histoplasmosis 4. Chronic disseminated disease with low-grade fever, weight loss, weakness, hepatosplenomegaly, mild hematologic abnormalities, focal disease. Progression over 10–11 months and usually fatal if left untreated. 5. Chronic pulmonary, resembles chronic pulmonary tuberculosis mostly in elderly and middle-aged men with lung disease	*Histoplasma capsulatum* Soil with high organic content and undisturbed bird droppings	Working on or under structures that are habitats for birds or bats. Activities such as exploring bat-infested caves, clearing bird roosts, or cleaning chicken houses. Laboratory and hospital workers from cutaneous inoculation.	Primarily from inhalation of airborne conidia. Cutaneous inoculation reported. N	3–17 days after exposure, but may be shorter with heavy exposure, usually 10 days	Susceptibility is general. Inapparent infections are common in endemic areas and usually result in resistance to infection. May be opportunistic infection with immunocompromised individuals.
Humidifier lung suberosis (*P. frequentans*), cheese washer's disease (*P. caseii, P. roquefortii*), cheese worker's lung, woodman's disease, allergic bronchopulmonary mycosis (ABPM) Hypersensitivity disease, most commonly hypersensitivity pneumonitis but also asthma. Hypersensitivity pneumonitis symptoms range from severe and acute to subtle and chronic. Acute illness: Fever, chills, cough, dyspnea, abnormal chest radiograph, leukocytosis with improvement within few days of removal from exposure. Chronic illness: Insidious onset of cough and progressive dyspnea on exertion. Rarely, allergic bronchopulmonary penicilliosis causing intermittent airway obstruction, transient pulmonary infiltrates, blood and sputum eosinophilia, and positive dual skin test (Types 1 and 3), and precipitating antibodies in serum. Proximal sacular bronchiectasis is often found in the segment of lung containing infiltrate	*Pencillium* species Ubiquitous in nature.	Cork workers, cheese workers, laboratory workers, farmers, tree cutters, sawmill workers, and other handlers of contaminated wood. Contaminated humidifier water and moldy HVAC systems also occupational hazard.	Inhalation of airborne spore. N	Unknown	Unknown

Table 26.4. (continued)

Disease or Fungus and Symptoms	Agent and Reservoir	Occupations at Risk	Mode of Transmission Person-to-Person Transmission Possible: Y/N	Incubation Period Period of Communicability	Resistance and Susceptibility
Hypersensitivity diseases: Extrinsic allergic alveolitis, farmer's lung, malt worker's lung, allergic bronchopulmonary aspergillosis (ABPA), allergic Aspergillus sinusitis, allergic fungal sinusitis, baker's asthma, mushroom grower's asthma. Three major types of disease: 1. Allergic aspergillosis—asthma, hypersensitivity pneumonitis, allergic sinusitis 2. Colonizing aspergillosis—otomycosis, fungus ball of paranasal sinuses and lung, endobronchial colonization 3. Invasive aspergillosis—uncommon but most serious. Fatal invasive microgranulomatous aspergillosis. Disease presents as pneumonia, but necrotizing pneumonia that develops is usually fulminant and death occurs in 1–2 weeks.	*Aspergillus* species, especially *A. fumigatus* Ubiquitous in nature, particularly in decaying vegetation. Commonly present in indoor and outdoor air during all seasons.	Farmers, sawmill workers, mushroom workers, cane mill workers, greenhouse workers, brewery workers, and bird hobbyists. Workers who deal with compost piles, decomposing haystacks, moldy grains. Bakers have been known to become allergic to Aspergillus-derived enzymes.	Airborne conidia in chains 2–5 μm in diameter are airborne and readily inhalable. Most diseases caused by inhalation of spores but can also enter tissue exposed during surgery. Other routes include through eyes, paranasal sinuses, skin, GI tract, and via intravenous drug use. N	Probably a few days to weeks.	Ubiquity of fungal species and occurrence of disease as secondary infection suggest high resistance in healthy people. Allergic fungal sinusitis and invasive aspergillosis usually occurs in immunocompromised patients. Invasive disease primarily seen in those with prolonged neutropenia. Patients with HIV or chronic granulomatous disease of childhood also susceptible.
Maple bark disease, maple bark stripper's disease Rare disease that can affect both trees and humans. Sensitizing asthma and hypersensitivity pneumonitis. Lung histology has demonstrated scattered fibrosis and granulomas.	*Cryptostroma corticale* Wood	Wood and sawmill workers who de-bark logs before cutting	Inhalation of respirable spores that can cause sensitization and occupational lung disease	Unknown	Unknown
Mushroom spore asthma, hypersensitivity pneumonitis, extrinsic allergic alveolitis, mushroom worker's lung, mushroom compost worker's lung, lycoperdonosis	Basiodiomycetes including *Merulius lacrymans*. Lycoperdon, and mushrooms	Mushroom workers	Inhalation of basidiospores N	First few months of employment, but sensitization has been described years after exposure.	Unknown

Merulius lacrymans is found in buildings in cool temperate climates and is associated with asthma and hypersensitivity pneumonitis. Inhalation of dried powder from Lycoperdon (puffball) associated with lycoperdonosis, a form of hypersensitivity pneumonitis that develops following treatment of epitaxis. One report of 8 workers exposed to mushroom dust and developing asthma and rhinoconjunctivitis. Dermatitis can also occur.	Agents are rusts, puffballs, common rusts, smuts, mushrooms, and certain yeasts.				
Sporotrichosis	*Sporothrix schenckii*	Gardeners, farmers, masonry workers, outdoor workers, floral workers, miners, nursery workers, forestry workers, laboratory workers, and others with exposure to contaminated soil or vegetation such as sphagnum moss, prairie hay, salt marsh hay, or roses	Traumatic implantation (cutaneous infection), but inhalation of fungal spores is occasionally associated with pulmonary infection.	Unknown	Unknown
Cutaneous disease arises from minor trauma and begins as small, erythematous papule that grows over days or weeks, usually remaining painless but may secrete clear discharge. Usually, discrete nodular lesions spread along lymphatic channels. Skin lesions may result from hematogenous dissemination, which may lead to osteoarticular sporotrichosis, a disease that is manifested by stiffness and pain in a large joint, specifically the knee, elbow, ankle, or wrist. May spread to additional joints and lead to pain, limitation of movement, and nerve entrapment. Endopthalmitis, brain abscess, and chronic meningitis are rare. Pulmonary sporotrichosis typically involves cough, low-grade fever, weight loss, hemoptysis, and upper lobe single cavitary lesion with or without parenchymal infiltrate. Pleural effusion, hilar adenopathy, and calcification are rare. Without treatment, lung lesion leads to death.			N		
Wood pulp workers' disease	*Alternaria* species	Woodworkers	Inhalation of spores	Unknown	Unknown
Hypersensitivity pneumonitis, asthma, and allergic rhinitis due to inhalation exposure to Alternaria spores.	Wood		N		

Refs. 1, 2.

Table 26.5. Occupational Diseases Associated With Allergens

Disease or Allergen and Symptoms	Agent and Reservoir	Occupations at Risk	Mode of Transmission Person-to-Person Transmission Possible: Y/N	Incubation Period Period of Communicability	Resistance and Susceptibility
Bakers' asthma, wheat flour asthma IgE-mediated, immediate-type allergic response. Bakers' asthma is associated with intensity and duration of exposure. Rhinitis and conjunctivitis precede development of asthma. Many allergens associated with flour can cause asthma: molds, fungi, insects, storage mites, pollens, and bacteria.	Wheat flour	Bakers; also flour mill workers, grain handlers, pastry and cereal factory workers, and animal feed factory workers	Flour and grain dust become airborne and inhaled. N	After months or years of exposure	Persons with atopic characteristics are at increased risk.
Crustacean, crab, lobster, shellfish, prawn IgE-mediated immediate hypersensitivity with watery, itchy eyes and nose; sneezing; less frequently cough and wheezing and shortness of breath. Cross-reactivity between shrimp, crab, lobster, and crayfish antigens reported.	Shellfish and other marine invertebrates	Food-processing facilities that handle prawns, lobster, crabs, and other shellfish. Oyster farming, any setting where marine animals are handled. Exposure to horseshoe crab–derived antigen may occur in laboratory settings where assays for bacterial endotoxins are performed.	Inhalation of allergenic components of shellfish may result in respiratory sensitization. Allergens become airborne in food-processing plants by aerosolization or by being carried along with steam and water vapor. Dust from processing-dried products also may be inhaled. N	Unknown	Unknown

Detergents, digestive acids, dough improvers, papain, others Enzymes are proteins used to catalyze chemical reactions. They are potent respiratory sensitizers and can cause immediate, delayed, or dual allergy consisting of rhinoconjunctivitis and several patterns of asthma. Also, skin allergies have been described from enzyme exposure. Skin disorders can be irritant, digestive, or allergic. Urticaria and contact dermatitis may occur. Allergic symptoms may include rhinorrhea, conjunctivitis, often shortness of breath, cough, chest tightness, and wheezing that occurs minutes to hours following exposure. Anaphylaxis has occurred by ingestion of papain in sensitized individuals. Cellulase, xylanase, papain, and bromelain cross-react, but conflicting evidence exists on alcalase, sarinase, and esperase.	Enzymes Respiratory allergy: Cellulase, xylanase, papain, flaviastase, trypsin, bromelain, pepsin, alpha-amylase, and others Urticaria: Papain, cellulase, xylanase Contact dermatis: Xylanase, cellulase, and combination with alcalase	Enzymes are used in the chemical, pharmaceutical, cosmetic, textile, medical, detergent, and food and beverage industries.	Enzyme powders tend to be fine and easily airborne. Exposure may be through inhalation or skin contact with liquid enzyme preparations. Dust may impact the upper airways or be carried directly into the lungs. N	Allergy may develop after months to years of exposure.	Asthmatic illness occurs usually in sensitized individuals.
Poison ivy, eastern poison oak, western poison oak, poison sumac, coffee beans, tobacco, psyllium, tea leaves, ipecac, colophony, natural rubber latex, others Generalized urticaria or contact dermatitis can occur. Repeated allergen exposure can cause thickening and lichenification of the skin. Although rare, cutaneous reactions have been reported of olive oil causing contact allergy and two cases of hand eczema in pedicurists. Allergens in latex gloves have caused urticaria, contact dermatitis, rhinoconjunctivitis, angioedema, asthma, and anaphylaxis. Asthma can occur from inhalation of green tobacco leaves, green coffee, pharmaceutical products, and other plants.	Plants	Agricultural and processing plant workers, tobacco and garlic farmers, coffee bean processors, tea blenders, spice processors, paper and rubber workers, florists, beekeepers, forest rangers, surveyors, utility company workers, and pharmaceutical and health care workers	Dermal or respiratory exposure. For those substances that cause skin allergy, exposure occurs via direct contact between plant material and exposed skin. Contamination of objects with antigen that then contacts skin. Inhalation of plant-derived allergens for respiratory sensitization. N	One to several days.	A history of spina bifida and early surgeries may predispose an individual to latex allergies.

Table 26.5. (continued)

Disease or Allergen and Symptoms	Agent and Reservoir	Occupations at Risk	Mode of Transmission Person-to-Person Transmission Possible: Y/N	Incubation Period Period of Communicability	Resistance and Susceptibility
Rats, mice, guinea pigs, rabbits, hamsters, monkeys Prevalence of allergy among laboratory animal handlers is ~15–30%. About 10% develop asthma. Allergy typically involves rhinitis and conjunctivitis, which progresses into asthma in a minority of patients. Symptoms usually occur within 5–30 minutes after exposure. Asthma may occur immediately or may exhibit dual or delayed patterns. Skin reactions include contact urticaria from direct contact or maculopapular rash in association with airborne exposure or respiratory symptoms. The antigens come mainly from urine, dander, saliva, and serum.	Laboratory animals	Laboratory workers who work with animals	Laboratory animal allergy can result from skin exposure or respiratory exposure. Percutaneous exposures may result from bites or allergen contamination of wounds. May result from direct contact as well as by indirect contact through airborne dust. Allergens, such as from rodent urine or in fur, may become airborne and inhaled. N	It has been suggested that 50% of those who develop laboratory animal allergies do so in first 2 years of working, although reports exist of more than 20 years after initial exposure.	Presence of several general allergic symptoms and several historical indicators of atopy is moderately predictive of new onset of laboratory animal allergy. Development of asthma from laboratory animals may cause other hypersensitivities.
Rye grass, soybeans, buckwheat, oat grass, barley Potential allergens range from insects, bacteria, and fungi to components in grains themselves like soybean flour and buckwheat flour. Wheat, rye, and triticale grasses show cross-antigenicity. Rye, barley, and oat show some cross-reactivity. Allergic individuals may be allergic to more than one antigen. Urticaria has been associated with buckwheat and cellulase. Community asthma has been associated with soybean loading and unloading. Nonallergic respiratory symptoms include cough, shortness of breath, and decrease in lung function. Nonallergic acute grain-related respiratory disorders include an asthmalike acute syndrome and grain fever. Severity of symptoms depends on duration of exposure.	Grain dust	Any occupation where grains are handled. Farmers, cereal manufacturers, grain loaders and unloaders.	Inhalation to respirable dust. Some grain associated with allergens such as cellulase and buckwheat may also cause skin reactions. N	Unknown	Unknown

Refs. 1, 2.

legionellae are aspirated into the lungs. Colonization of the oropharynx followed by aspiration into the lungs may be possible but has not been proven conclusively (5).

It has recently been proposed that airborne transmission may involve inhalation of vesicles containing legionellae released from amoebae (6). These vesicles measure 2.1 to 6.4 μm in diameter, a size range that can penetrate and deposit deep in the lungs. Each vesicle may contain several hundred legionellae. Interestingly, *L. pneumophila* is more invasive after growth with amoebae (7). This theory is consistent with the epidemiology and aerobiology of legionella infections; cooling towers as far as 1.2 km away have been documented as sources. Moreover, bacteria-containing vesicles may survive air transport better than bacterial aerosols, which are prone to dessication.

4.1.3 Environmental Biology of Legionella

Legionellae are poorly staining, gram-negative bacteria found primarily in aquatic environments. Water environments epidemiologically linked to Legionnaires' disease include domestic water systems, cooling towers and evaporative condensers, respiratory care equipment, showers, faucets, whirlpool baths, humidifiers, decorative fountains, air-perfused footbaths (8), and a grocery store mist machine. Legionellae have also been isolated from hot- and cold-water taps and showers, hot tubs, and creeks and ponds and surrounding soil.

Many natural environments contain insufficient nutrients to support bacterial growth (9). Bacteria that do not form spores may adopt specific starvation responses to survive (10). A fraction of the bacterial population can enter and remain in a viable, starved state until favorable growth conditions are restored (10). There is some evidence that legionellae can remain viable (and possibly infectious) but not culturable for up to 16 months in the absence of nutrients. A fresh culture of *L. pneumophila* had an LD_{50} of 10 cfu for embryonated yolk sacs (11). When suspended in tap water for 14 days at 4°C and 37°C, the LD_{50} values decreased to 0.5 and 0.25 cfu respectively. The investigators concluded that these results were due to the cells going into a viable but nonculturable state. This phenomenon has been observed for *Vibrio cholerae* (12) and *Campylobacter jejuni*, both bacteria found in fecally contaminated water. Steinert et al. (13) have demonstrated the successful resuscitation of a strain of nonculturable but viable *L. pneumophila*.

There are several other problems with using culture as a method for detecting legionellae bacteria. As with other bacteria, legionellae can form biofilms. Not only do legionellae colonize various metal and nonmetal surfaces, but they also are found preferentially on surfaces, that is, adherent rather than free-living (14, 15). Assessing only the free-living populations, such as is done with typical swab sampling, would therefore underestimate total populations. Moreover, although standard methods of isolation have been developed, legionellae are difficult to culture on synthetic media and require at least 3 to 7 days at 37°C. Not all laboratories have access to appropriate growth media.

Many microorganisms can either support or inhibit the growth of legionella. Microorganisms that support the growth of legionellae include algae such as Fischeriella (16, 17), green algae (18), cyanobacteria (19), amoebae (20–22), ciliated protozoa (22–26) and nonlegionellae heterotropic bacteria (27). Growth of legionellae on synthetic media is inhibited by numerous heterotrophic bacteria commonly present in the same environmental sample (28–30). Other investigators have documented the inhibition of legionellae by bacterial strains isolated from water samples (31) and by bacterial strains isolated from heterotrophic plate counts of drinking water (32). Inhibition by other bacteria is important in that a water sample containing both legionellae and bacterial strains that inhibit their growth will show a reduced legionellae plate count. Although heat and acid pretreatment of environmental samples is recommended (33) to kill undesirable nonlegionellae bacteria, such treatment can potentially affect survival of legionellae and reduce plate counts.

Intracellular growth of legionellae cells has been demonstrated in ciliated protozoa and amoebae. The ciliated protozoa *Tetrahymena* spp. (34) and *Cyclidum* spp. have been shown to support intracellular multiplication of *L. pneumophila*. Legionellae can grow in trophozoites of amoebae and become incorporated into their cysts. Species of *Acanthamoeba, Hartmanella, Valkampfia*, and *Naegleria* can all be infected and, depending on conditions, may coexist for prolonged or indefinite periods (20–23, 35). Tens or hundreds of legionellae cells may exist within the vesicles of amoebae. At low temperature (20°C), only limited growth of legionellae occurs, but at 35°C they rapidly multiply intracellularly (23, 36). The protozoa feed on legionellae but cannot digest virulent strains. Instead of being metabolized, virulent strains multiply within protozoa just as they do in human phagocytic cells (37, 38). Multiplication within certain protozoa has been linked to pathogenicity (25, 36).

Sodium hypochlorite is often used as a biocidal control strategy (see section 4.1.4). However, legionellae may remain viable but nonculturable at hypochlorite concentrations up to 100 ppm (39). Amoebae and protozoa are also more resistant to hypochlorite than free-living legionellae, implying that water disinfected with hypochlorite can be reinoculated when amoebae and protozoa release legionellae within them.

In summary, negative culture or low cfu counts, particularly of cooling tower water samples, may be obtained for any of the following reasons:

1. Legionellae are difficult to grow and require appropriate L-cysteine-containing media.
2. Legionellae can remain viable but nonculturable and may therefore be more resistant to standard biocides.
3. Other microorganisms present in environmental samples may inhibit the growth of legionellae on agar plates. Pretreatment of such samples may also kill legionellae.
4. Legionellae living within amoebae cysts and protozoa are protected from disinfection and are then free to recolonize cooling tower water and surfaces. This provides a mechanism whereby viable legionellae can be released into cooling tower waters unpredictably.
5. Free-living legionellae are more effectively killed by biocidal treatment than adherent populations, which remain to colonize surfaces.

All of these findings suggest that hypochlorite may have reduced biocidal effectiveness as a biocide in cooling towers and water-distribution systems, and culture analysis may be an insensitive indicator of adequate maintenance and contamination.

4.1.4 Legionella Control Strategies

Several strategies may be useful in the control of Legionella Infections. Common techniques used to control the growth of legionellae are cooling tower maintenance, increased heat and heat shock treatments, and the use of oxidizing and nonoxidizing chemical biocides.

The American Society of Heating, Refrigerating, and Air-conditioning Engineers (ASHRAE) has emphasized the importance of preventive maintenance of cooling towers and other evaporative condensers. Appropriate cooling tower maintenance procedures include water treatment to control biologic activity, prevention of scaling, periodic cleaning, maintenance of a written record of procedures performed, and periodic inspections (40, 41). ASHRAE has recently updated its recommendations to include having effective drift eliminators and maintaining mechanical components in working order (40). Although maintenance does not control legionellae growth with certainty (42), lack of adequate preventive maintenance leads to inefficient system operation and conditions optimal for growth and transmission of legionellae. ASHRAE maintains that more emphasis should be placed on clean equipment in excellent repair than on periodic testing with concurrent system neglect (40).

Although legionellae are capable of growth and survival under the range of physical and chemical conditions commonly encountered in most human-made water systems, they prefer warm water. Cooling tower water counts of legionellae have been found to correlate positively with water temperature, pH, and protozoan counts (43). Heat shock treatment and maintenance of water-distribution systems above a minimum temperature have both been used as techniques to control legionellae growth in potable water due to their well-known dependence on temperature for optimal growth. The effectiveness of increasing the water temperature to 50 to 70°C has been challenged since *L. pneumophila* has been isolated in two domestic hot water systems that delivered water at temperatures of >50°C, and in one instance, at 70°C (44). One possible explanation for this finding is that the temperature is not homogenous throughout the system (45). In addition, in one report, strains isolated after heat shock were not the same as those detected in the original water (46). These data spawned the theory that *L. pneumophila* may possess heat shock genes analogous to those in *Escherichia coli* and that heat shock can alter monoclonal expression and virulence.

4.1.5 Cooling Towers

Cooling towers play an important role as environmental sources of Legionnaires' disease (47). Cooling towers are mass heat-transfer devices; their basic function is that of evaporative cooling and exchange of sensible heat. Water is lost via discharge to the atmosphere as hot moist water vapor and aerosol mist, resulting in a cooling effect of the remaining water. *Drift* refers to the whole water droplets released to the atmosphere and includes water, minerals, and presumably microorganisms.

Cooling towers are of two basic types: direct and indirect contact. In the former, air and water to be cooled are in direct contact with each other; in the latter, they are kept separate. Only direct contact between the ambient air and legionellae-containing water is of interest as an environmental exposure source for community-acquired legionellosis and for occupational exposure during cooling tower maintenance. Workers should wear respiratory protection during cooling tower maintenance that involves scraping and high-pressure steam cleaning.

4.1.6 Cooling Tower Maintenance

Both inspection and water treatment are critical aspects of an effective cooling tower maintenance program. Water treatment is generally aimed at controlling deposition of scale and biomass. Scale is the deposition of inorganic compounds in the tower. Biofouling results from the growth of algae and the production of slime by bacteria and fungi. Both scale and slime interfere with the proper functioning of the cooling tower by reducing heat transfer. Interestingly, a water-treatment program may be effective in controlling scaling and biofouling but may not control legionellae. Water-treatment programs that are ineffective against scaling and biofouling will definitely be ineffective in controlling legionellae.

Although the level of all aquatic bacteria in cooling towers (including legionellae) should be controlled, it is not feasible to attempt to eradicate them completely. Because cooling towers are an environmental niche for legionellae, it is inevitable that they will release legionellae to the ambient air. The systematic use of biocides is generally coordinated with all other maintenance activities such as mechanical cleaning of the cooling tower. Continuous feed of biocides is often preferred to slug or intermittent treatment as it is more likely to provide a consistent biocide concentration. *Legionella* control strategies should also include control of blue-green algae, which both enhance growth of legionellae in aquatic reservoirs and improve their survival in aerosols.

The goal of disinfection is to control growth of bacteria, algae, and protozoa that may contribute to the growth of legionellae. Of the oxidizing biocides, chlorine is the most effective in controlling legionellae, but only if the appropriate concentration and pH are maintained (41). Free chlorine at 0.4 mg/L kills *L. pneumophila* (48). Unfortunately, corrosion and pH maintenance make use of chlorine difficult. Legionellae have been shown to be inhibited or killed by a wide range of oxidizing and nonoxidizing biocides (14), even at low concentrations (49). Chlorination of cooling towers at 15 mg/L for routine cleaning and 50 mg/L for decontamination has been recommended (50). The Wisconsin Division of Health has recommended a free-residual chlorine concentration of 10 ppm for maintenance and 50 ppm in an emergency protocol (51). However, many biocides that showed promising results in the laboratory have been less efficacious *in situ* (52, 53). Thus, it is not surprising that ASHRAE has recommended that biocidal efficacy be determined in the field under actual working conditions rather than relying entirely on laboratory testing under controlled conditions (54).

Three proposed mechanisms underlie the decreased efficacy of chemical biocidal treatment: (*1*) decreased biocide susceptibility of legionellae within protozoa, (*2*) decreased biocide susceptibility of legionellae within biofilm, and (*3*) a biocidal stimulatory effect on amoebae. Protozoa that support intracellular growth of legionellae are resistant to normal levels of biocide. *L. pneumophila* has been recovered from *Acanthamoeba polyphaga* cysts after exposure to 50 mg/L free chlorine for 18 h at 25°C (50). Trophozoites of *Acanthamoeba* were sensitive to 1.25 mg/L free chlorine, but cysts can survive concentrations of up to 50 mg/L (50). Encysted legionellae may therefore be able to survive cooling tower disinfection and recolonize the system. Although biocides are relatively effective against free-living *L. pneumophila* over a short time (minimum of 9–12 h), substantially longer periods (at least 48 h) were required to reduce the number of culturable bacteria to below detectable levels in the population that had colonized cooling tower construction materials (15). Cooling tower amoeba isolates of *A. hatchetti* and a *Cochliopodium* species resembling *C. minus* may both be more resistant to biocides than laboratory-grown species and actually be stimulated by exposure to recommended concentrations of biocides (55). These results suggest that cooling tower amoebae adapt to biocidal treatment and that the use of biocides may actually enhance populations of amoeba host organisms for pathogenic legionellae. Finally, legionellae within vesicles of *Acanthamoeba* species remained viable after amoebicidal biocide exposure (6). Assessing only the number of legionellae with source waters where amoebae may be present may greatly underestimate the number of legionellae within vesicles.

Chemical treatment involving the addition of ionic copper and silver to free chlorine has been used to enhance inactivation of *L. pneumophila* (56, 57). Water-disinfection systems utilizing electrolytically generated copper and silver ions and low levels of free chlorine were evaluated at room temperature and elevated temperature for their ability to kill *L. pneumophila*. Significantly increased inactivation rates were observed at 0.4 mg/L free chlorine when 400 and 40 mg/L copper and silver were added, regardless of temperature (56, 57). Copper and silver ions are believed to interfere with enzymes involved in cellular respiration and bind specifically to DNA (58–60).

4.1.7 Domestic Water System Design

Domestic water systems have also been sources of human infections with legionellae. General control strategies for domestic water systems should prevent stagnation of water, use materials that do not provide nutrients for growth of legionellae, and address conditions that allow for buildup of scale and sediment.

As in cooling towers, heat shock is an appropriate general control strategy for domestic water systems. However, hot water tanks in hot water generators may have cool zones since only about 65% of the water is maintained at the set temperature and because hot water is not constantly demanded from the system. Efforts to minimize such cool zones include relocating the supply and return taps and installing a top-to-bottom circulating pump. Sediment and warm temperatures at the bottom of a tank beneath the heating element are ideal conditions for legionellae growth.

High-temperature semi-instantaneous steam heaters, available in newer hot water tanks, are advantageous in that they are not conducive to legionellae growth. If there is insufficient steam capacity, consideration should be given to a tank that keeps a high percentage of water at the set temperature as well as one that can produce water at 70°C for heat shock disinfection.

4.1.8 Domestic Water System Operation and Maintenance

Water temperatures >55 to 60°C in tanks and 50°C at all taps will control legionellae in the absence of scale and biofilm. Domestic cold water should be maintained below 25°C.

Because the complete absence of biofilm is unlikely, temperature control only minimizes but does not eradicate colonization or amplification of legionellae.

The temperature of water drawn from several sources, including the drainage valve, water returning to the tank, and water taken from the drain of cold-water storage tanks, should be monitored on a regular basis. The hot-water tap should read 50°C within 1 min of full flow and the cold-water tap should read ≤20°C after running for 2 min, even at the most distal locations. Points farthest from the point of distribution should be routinely checked to assess the water quality of the entire system.

Recommended water temperatures for controlling the growth of legionellae in tanks and at taps vary by geographic locations. Hospitals in the United Kingdom are required to keep water at or above 50°C. In contrast, regulations in the United States are based on a maximum temperature, usually about 60°C, that will not result in scalding. Cold-water storage and distribution should be ≦20°C. The appropriate choice and use of thermostatic mixing values and antiscald valves are essential in achieving optimum temperatures.

4.1.9 Environmental Sampling

There is considerable debate in the United States as to whether domestic water systems and cooling tower water should be routinely sampled (61, 62). If the decision to sample is made, the sampling objectives must be clearly defined *a priori* and the limitations of resultant data must be clearly understood. The Centers for Disease Control (CDC) has stated that "the relationship between the results of water cultures and the risk of legionellosis remains undefined" and that "[sampling] data are insufficient to assign a level of risk of disease." ASHRAE's current position is consistent with that of CDC (53). However, even in the absence of a fully defined risk relationship, routine sampling may accomplish particular objectives. For example, *in situ* efficacy of biocides can be determined only by taking routine samples. In doing so, the sampling strategy should be planned so that (*1*) a sufficient number of sites are sampled to assess an entire system, (*2*) data from one sample location can be compared to samples taken at the same location over time, and (*3*) the data are examined as a whole, for example, what percentage of samples are positive. Objective data can be useful in determining whether facility operations are meeting the goal of its maintenance responsibilities and not necessarily whether hazard exists.

Bulk water and swab samples are most useful. Air sampling, discussed later, is less useful. Bulk water samples best address free-living populations of legionellae, which probably comprise the major inhalational exposure. Swab samples best address sessile populations residing in biofilms, such as in a faucet. Swab samples generally yield higher counts than bulk water samples (63), an example of the potential for unpredictable release of legionellae from amoebae into the water.

Water sample bottles should be polypropylene, sterile, of 250- to 1,000-mL capacity, leakproof, and autoclavable for reuse. Aseptic technique should always be used. The required sample volume varies by laboratory according to the desired limit of detection. Sterile swabs should be within the expiration date. Because differences in temperature greatly affect growth, the water temperature should be consistently measured either before or after a sample is taken. The thermometer used should be disinfected and dried before use. It may be necessary to neutralize residual chlorine by having the laboratory add sodium thiosulfate to the sample bottle. One approach is to determine if residual chlorine is present before adding sodium thiosulfate. While in theory residual chlorine can adversely affect the concentration of culturable legionellae, one commercial laboratory has not found that residual chlorine decreases the concentration of several isolates (verbal communication, George Morris, Ph.D., PathCon Laboratories, Norcross, GA, US). Both American Society for Testing and Materials (ASTM) (62) and American Public Health Association (APHA) (33) describe rinsing sample bottles with the subject water before taking the actual sample. Rinsing is unnecessary if the sample bottles are truly clean and sterile; it should not be done if sterile sample bottles contain sodium thiosulfate. Samples should be shipped at ambient temperatures in an insulated container on the date of collection by overnight courier to the laboratory; analysis should begin within 24 h of the time of collection. If sample analysis cannot begin within 24 h of collection, samples should be maintained between 3 and 30°C (62). All samples should be identified with unique identification numbers and handled following standard chain-of-custody procedures.

Samples of sludge and sediment are taken to determine the presence of culturable legionellae in any biofilm present. If the sludge and sediment are taken from areas in the absence of water, the data are not particularly meaningful. Such samples are also difficult to process, and aliquots taken for culture may not be representative of the entire sample.

Culturable air sampling has been used in some outbreak investigations. An epidemiologic study with culturable air sampling and monoclonal antibody subtyping demonstrated that the source of one Legionnaires' disease outbreak was aerosolized water from an evaporative condenser (64). Air sampling results also corroborated monoclonal antibody subtype patterns found in patient and environmental samples in another outbreak investigation (65). Though culturable air sampling can be a useful adjunct to epidemiologic studies, this author does not recommend it due to the potential for false negatives resulting from environmental factors on survival, loss of viability during sampling stress, and the fact that not all *Legionella* aerosol generators are continuous-point sources (e.g., grocery misters).

Nonculture analysis of bulk samples is done using direct fluorescent antibody (DFA) tests or polymerase chain reaction (PCR), a molecular biology technique. However, the standard or primary method of detection and quantitation of legionellae is culture. Only competent laboratories with experience isolating legionellae from environmental samples using APHA's methods (33) should be used to culture legionellae.

4.1.10 Interpretation of Data

This author is unaware of the existence of published standards or regulatory guidelines for interpretation of analytical data regarding legionellae concentrations (e.g., the implication of 1 cfu/mL of water). Commercial laboratories, such as PathCon Laboratories, Inc. (Norcross, GA, US), have developed such guidelines.

4.1.11 Outbreak Investigation

Investigation of a legionellosis outbreak starts with the identification of two or more infected individuals. The goal of the investigation is to identify the likely environmental source of those cases. Culture and DFA have been used to determine the presence of legionellae in environmental sources and patient specimens. An enzyme-linked immunosorbent assay (ELISA) has also been used to assess the antibody titer in acute and convalescent clinical serum samples. Co-cultivation with amoebae has been proposed as particularly useful when indirect fluorescence antibody (IFA) results are positive and culture results are negative for environmental water samples (66). An epidemiological link between the suspected source and case can be strengthened when the environmental and clinical isolates are of identical monoclonal antibody subtypes.

4.2 Creutzfeldt-Jakob Prion and Disease

A blood-borne pathogen that deserves separate discussion is the Creutzfeldt-Jakob disease (CJD) prion. CJD is a chronic, rapidly progressive dementia syndrome in humans characterized by spongiform encephalopathy. CJD prions gained worldwide attention when an epidemic of bovine spongiform encephalopathy (BSE), or "mad cow disease," was reported in the United Kingdom. Of particular interest was whether the epidemic of BSE might be related to the increase in the incidence of sporadic CJD and the high incidence of CJD among dairy farmers. One author concluded that they might be related to an unusual neurological profile in a group of young people (67).

Prions were originally thought to be filterable, self-replicating agents, or "slow viruses." They are more accurately defined as a host-encoded protein that is apparently infectious and resists inactivation by processes that modify nucleic acids. Importantly, CJD prion is not inactivated by normal laboratory disinfection and sterilization procedures, and it is uniformly fatal.

Three epidemiological forms of CJD are recognized: sporadic (the most common), familial (i.e., inherited), and infectious/iatrogenic. Only the infectious exposure and its relationship to biohazards and biosafety are discussed here. However, it is of particular note that the incubation period for iatrogenic transmission is greatly shortened from years to months (68).

Considerable attention has been focused on CJD prion because it is a known blood-borne pathogen. As such, there are tremendous implications with regard to blood product collection and processing. However, in a review of currently available animal experimentation and human data, one author concluded that if transmission by blood in humans occurs, it is likely to be rare (68).

CJD is not communicable or contagious by standard routes of infection. There is no evidence of contact or aerosol transmission of prions from one human to another. However, infection has occurred in the following ways:

- Ritualistic cannibalism (ingestion) in New Guinea, causing kuru
- The administration of prion-contaminated growth hormone, causing iatrogenic CJD
- Transplantation of prion-contaminated dura mater grafts
- Corneal transplants
- Through prion-contaminated EEG leads
- Consumption of prion-contaminated fertility drugs derived from human sources

CJD agent is extremely resistant to disinfection and inactivation procedures, including the following: alcohol, ethylene oxide, formaldehyde, glutaraldehyde, hydrogen peroxide, iodine, ionizing radiation, phenolics, quaternary ammonium compounds, steam sterilization at 121°C, and urea concentrations of 6 to 8 mol/L (69). Prions are inactivated by 1N NaOH, 4M guanidinium hydrochloride or isocyanate, or sodium hypochlorite ($\geq$2% free available chlorine) and steam autoclaving at 132°C for 4.5 h (70). Formaldehyde-fixed and paraffin-embedded tissues, especially from the brain, remain infectious.

The highest concentrations of prions are found in the central nervous system and its coverings. It is likely that high concentrations are found in the spleen, thymus, lymph nodes, and lung. Unfixed samples of brain, spinal cord, and other tissues containing human prions should be processed with extreme care at biosafety level (BSL)-3. Eye protection and gloves should be worn. All manipulations should be conducted in biosafety cabinets (BSCs). Extreme care should be taken to avoid aerosol production and cuts and punctures of the skin. Tissue processors should not be used because of the problems of inactivation. Jars and beakers should be used instead. Specific precautions are required for the following: standard precautions for autopsies of patients with suspected prion disease, autopsy suite decontamination procedures, brain cutting procedures, and tissue preparation. The reader should consult the fourth edition of *Biosafety in Microbiological and Biomedical Laboratories* (70) for detailed information.

4.3 Microbial Indoor Air Quality

Microbial indoor air quality is a broad term that describes the quality of indoor air in any building with respect to microbial growth or contamination. For example, microbial indoor air quality investigations might be done to assess the airborne potential for dust mite allergens Der f1 and Der p1, or fungal growth in building materials as a result of water intrusion. The knowledge necessary to understand the microbial aspect of this field comes under the rubric of biosafety only in the sense that an in-depth knowledge of the microorganisms of interest and health outcomes associated with exposure to them is critical in evaluating microbial indoor air quality and assessing any potential health risks involved. The previous chapter contains further information on microorganisms implicated in poor microbial indoor air quality. Because much of the literature on health effects associated with airborne fungal exposure falls outside of the realm of industrial hygiene, it is critical that other sources be consulted, such as the following:

- Allergy/immunology texts, for example, *Medical Immunology* (Stites 1997), ninth edition, edited by Daniel P. Stites, Abba I. Terr, and Tristram G. Parslow, Appleton & Lange, 1997 (71)
- Publication on the specific topic, for example, *Health Implications of Indoor Environments* (72), *Field Guide for the Determination of Biological Contaminants in Environmental Samples* (73), *Bioaerosols-Assessment and Control*, (74)

With regard to fungal growth in or on building materials that might degrade indoor air quality, we do not know if there is an increased risk above background exposure to ambient fungal aerosols. It is therefore critical that microbial indoor air quality practitioners stay abreast of the current scientific peer-reviewed literature to modify their risk assessments accordingly.

In addition to knowledge of the microorganisms of interest and health effects associated with their exposure, special knowledge of the ecology of microorganisms in indoor environments is required, as well as the following rubrics of industrial hygiene: aerosol technology, biostatistics, toxicology, and epidemiology. One must also understand building construction and building performance in order to conduct effective microbial indoor air quality investigations.

For a more in-depth discussion of indoor air quality, the reader should consult Chapter 65, Indoor Air Quality in Nonindustrial Occupational Environments in this edition.

5 ASSESSING THE RISK OF BIOHAZARDS

5.1 Sampling Methods to Evaluate the Potential for Biohazard Exposure

Industrial hygienists are accustomed to using standardized, validated sampling methodologies in investigations and then comparing the sample results to either permissible exposure limits (PELs) promulgated by the Occupational Safety and Health Administration, or threshold limit values (TLVs) established by the American Conference of Governmental Industrial Hygienists (ACGIH), or both of these. Far fewer exposure limits exist for the bioaerosols or biologically derived airborne contaminants. ACGIH defines bioaerosols as those airborne particles that are living or originate from living organisms and biologically derived airborne contaminants to include bioaerosols as well as gases and vapors that living organisms produce (74). TLVs exist for the following bioaerosols: cellulose, some wood, cotton, and grain dusts; nicotine; pyrethrum; starch; subtilisins; sucrose; and vegetable oil mist (75). The reader should consult the most current year TLV and BEI (Biological Exposure Indices) book for the justification of why TLVs cannot be established for a far greater number of these contaminants. For example, ACGIH cites the lack of human dose–response data for infectious bioaerosols as the reason that there are no TLVs for infectious agents.

Nonetheless, environmental assessments are possible for a number of bioaerosols and biologically derived airborne contaminants. Such assessments are also useful in identifying surface growth of microorganisms or contamination and bulk liquids supporting microbial growth. If well-designed sampling strategies are executed with clear objectives, risk assessments can often be done. Industrial hygienists must always use professional judgment, but it is even more important in the absence of critical issues such as dose-response relationships for bioaerosols, biologically derived airborne contaminants, and surface or bulk samples.

Special expertise is required to interpret such data, especially as it applies to allegedly water-damaged building materials. Please refer to section 4.3, "Microbial Indoor Air Quality," and the discussion there of such expertise. It cannot be overemphasized that determining the objectives of the investigation is critical to the overall sampling design strategy and interpretation of resulting data. The sampling design strategy must allow for the following:

- Maximizing the probability of a true positive and minimizing the possibility of a false negative.
- Including sample locations such as worst-case sample locations, control and comparison sample locations, and sample locations chosen randomly. A significant number of samples and multiple rounds should be taken.
- Including a variety of sampling methods that, when combined, offer the most useful information possible to assess exposure or the potential for exposure
- Including a variety of analytical methods that, when combined, offer the most useful information possible to assess exposure or the potential for exposure
- Ensuring that equipment is properly selected, calibrated within acceptable limits, and competently operated
- Ensuring that proper chain-of-custody procedures are followed, samples are shipped promptly, and sample integrity is maintained
- Using only highly qualified and reputable laboratories for analysis
- Critically reviewing the reliability of the data
- Using only recognized, defensible criteria for interpretation of the sample data

As previously stated, microbes are ubiquitous in our environment. Thus, finding biological contamination does not in and of itself constitute hazard. The reader is referred to the sources such as those cited in section 4.3 for more in-depth information on interpretation of sample data, as well as to Chapter 65, Indoor Air Quality in Nonindustrial Occupational Environments. The following discussion will be limited to sampling and analytical methods used to assess biohazard exposure or the potential for exposure.

5.1.1 Bulk Sampling and Analytical Methods

The choice of analytical method dictates what bulk sampling method or methods are used. Culture analysis of surface contact plate samples, swipe (swab) samples, swab-rinse samples, and rinse samples (76) is performed. Culture analysis can also be performed on bulk material samples and vacuumed-settled dust samples. Only microscopic analysis can be performed on tape lift samples, and these are useful only in identifying morphological features of fungi that are multicellular. Culture provides more information for the microbiologist or mycologist than microscopic analysis for identification of microorganisms and is therefore more reliable. On the other hand, identification of fungal growth structures, for example, by microscopy is indicative of past or present fungal surface growth. Obtaining both kinds of samples and analyses provides more complete and accurate data on which to base a decision. Direct microscopic examination can also be done on a bulk material sample that is subsequently analyzed by culture.

Culture involves multiple steps. First, the sample, such as a swipe sample, is inoculated onto preselected synthetic growth media in petri dishes in the laboratory. Then the plates are incubated for enough time to allow for slow-growing organisms to be identified, generally 7 to 10 days. Subculturing onto subsequent plates may be necessary for identification. The choice of media and incubation temperature is based on the objective of the investigation. Specific growth requirements of microorganisms dictate the use of media optimal for their isolation, for example, buffered yeast charcoal extract agar for legionellae and MacConkey's media for gram-negative bacteria. Optimal growth of microorganisms occurs within temperature ranges that depend on the specific organism. Thus, the investigator must specify the incubation temperature desired.

Culture on synthetic growth media involves the involves the development of a colony, or a colony forming unit (cfu). Theoretically, one organism is the minimum number of organisms required for development of a colony. A single organism is not visible to the naked eye (see Table 26.1), but a colony is. Colonies are identified as to the organism growing and are then counted. Microorganisms that are incapable of growth outside of living cells, such as viruses, are grown in cell or tissue culture. The visible units of growth are called plaque forming unit (PFU).

The potential number of other types of chemical, immunological analyses that might be useful in an investigation is large. Some examples include but are not limited to the following: ergosterol, endotoxin, mycotoxins, antigens, volatile organic compounds, and (1 → 3)-β-D-glucan. The reader should refer to the *Field Guide* cited in section 4.3 for many of these methodologies.

Bulk samples also include sampling liquids and fluids that are submitted for culture analysis. Such samples might include those from potable water systems and cooling towers for culture of legionellae or industrial fluids such as water-based metal-working fluids for culture of gram-negative bacteria or assay of endotoxin. Sterile containers must be used for sample collection. Examples of devices available for sampling from large bodies of water such as lagoons and ponds include the Kemmerer sampler, the vanDorn sampler, and the Nasen bottle. These allow for sample collection at desired depths (76). Aliquots of the liquid or fluid are plated out on the desired media to initiate culture analysis.

5.1.2 Air Sampling and Analytical Methods

Just as the choice of analytical method dictates what bulk sampling method or methods are used, the same is true of air-sampling devices. Bioaerosol samplers are generally placed into three categories: inertial impactors, filtration devices, and gravity samplers, or settling plates. Settling plates are not recommended as they are biased toward the collection of larger culturable particles and provide only quantitative information at best. Inertial impaction devices include the following:

- All glass impingers: AGI-4, AGI-30
- Slit samplers
- Multiple-hole impaction samplers

All of these samplers are used for culture analysis of the collected aerosol. In addition, sporetrap samplers, which are also slit samplers, are used to collect bioaerosols to be microscopically identified and counted. Sporetrap samples represent a collection of both culturable and nonculturable bioaerosols in the same sample without distinction. A filtration technique known as CAMNEA (collection of airborne microorganisms on nucleopore filter estimation and analysis) provides for collection of both culturable and nonculturable bioaerosols in the sample with distinction through the use of a vital stain. The reader is encouraged to consult texts that discuss bioaerosol sampling in-depth, such as the most current edition of *Air Sampling Instruments for Evaluation of Atmospheric Contaminants*, published by ACGIH.

5.2 Risk Assessment of Biohazards

In general, risk assessment of biohazards involves a careful review of all available information so that a biosafety level of containment can be assigned. It involves assessing all of the factors that determine infectious disease transmission as presented in section 3. Those include the mode(s) of transmission of disease, the infectious dose, the pathogenicity and virulence, and host susceptibility factors. The CDC offers additional factors to be considered, including agent stability, concentration of the agent, the origin of potentially infectious material, the availability of prophylaxis, and an evaluation of the experience and skill level of the at-risk personnel (70).

Agent stability affects aerosol infectivity because stability in the environment is directly related to viability of the organism. Of course, the airborne concentration of viable organisms is directly related to the exposure concentration, which is related to the actual dose received at the target site, which either constitutes the required infectious dose or does not. The origin of the potentially infectious material may refer to a geographic location, the nature of the source, or the infected host (70). The availability of prophylaxis such as vaccination is part of a medical surveillance program.

The CDC also suggests that infectious agents whose risk is often evaluated will fall into the following categories (70):

- *Materials containing known infectious agents.* A risk assessment has already been done for all of the agents for which there is an agent summary statement. A excellent source of information for others is *Control of Communicable Diseases Manual* (2), published by APHA, which was used to prepare the tables in section 3.
- *Materials containing unknown infectious agents.* The CDC suggests that answering the following questions may reveal the level of containment needed: Why is an infectious agent suspected? What epidemiological data are available? What route of transmission is indicated? What is the morbidity and mortality rate associated with exposure to the agent? What medical data are available?
- *Materials containing recombinant DNA molecules.* The CDC suggests that some of the points to consider include the following, combined with guidance offered through the most current *Guidelines for Research Involving Recombinant DNA Molecules* published by the National Institutes of Health in the *Federal Register*: Does the inserted gene encode a known toxin or relatively uncharacterized toxin? Does the modification have the potential to alter the hose range or cell tropism of the virus? Does the modification have the potential to increase the replication capacity of the virus? Does the inserted gene encode a known oncogene? Does the inserted gene have the potential for altering the cell cycle? Does the viral DNA integrate into the host genome? What is the probability of generating replication-competent viruses?

- *Materials that may or may not contain unknown infectious agents.* In the absence of information that suggests an infectious agent, CDC recommends use of universal precautions.
- *Animal studies.*

6 SUMMARY

The industrial hygiene principles of anticipation, recognition, evaluation, and control of occupational hazards are as applicable to biohazards as to other occupational hazards. Many occupational diseases and their etiologic biohazards have been presented, accompanied by pertinent information pertaining to the disease, exposure, and annotated control methods. This chapter addressed anticipation, recognition, and evaluation of occupational biohazards across many occupational settings by presenting occupational biohazards, the occupations in which they can occur, methods to evaluate them, and risk assessment. Chapter 27, presents an in-depth discussion of control of biohazards.

ACKNOWLEDGMENTS

The author acknowledges the invaluable contribution of Mona Shum, M.S. for the construction of the tables that constitute a major section of this chapter.

BIBLIOGRAPHY

1. P. H. Wald and G. M. Stave, *Physical and Biological Hazards of the Workplace*, Van Nostrand Reinhold, New York, 1994.
2. A. S. Benenson, *Control of Communicable Diseases Manual*, 16th ed., American Public Health Association, Washington, DC, 1995.
3. D. W. Fraser et al., *New Engl. J. Med.* **297**, 1189–1197 (1977).
4. H. M. May, T. S. Harrison, and B. D. Harrison, *Respir. Med.* **88**, 693–696 (1994).
5. V. L. Yu et al., *Ann. Intern. Med.* **97**, 724–726 (1982).
6. S. G. Berk et al., *Appl. Environ. Microbiol.* **64**, 279–286 (1998).
7. J. D. Cirillo, S. Falkow, and L. S. Tompkins, *Infect. Immun.* **62**, 3254–3261 (1994).
8. J. W. Den Boer et al., *Lancet* **351**, 114 (1998).
9. R. Y. Morita, *Can. J. Microbiol.* **34**, 436–441 (1988).
10. H. M. Lappin-Scott and J. W. Costerton, *Experientia* **46**, 807–813 (1990).
11. D. Hussong et al., *Biotechnology* **5** (1987).
12. H. Xu et al., *Microb. Ecol.* **8**, 313–323 (1982).
13. M. Steinert et al., *Appl. Environ. Microbiol.* **63**, 2047–2053 (1997).
14. J. V. Lee and A. A. West, *Soc. Appl. Bacteriol. Symp. Ser.* **20**, 121S–129S (1991).
15. J. B. Wright, I. Ruseska, and J. W. Costerton, *J. Appl. Bacteriol.* **71**, 531–538 (1991).

16. G. A. Bohach and I. S. Snyder, *Infect. Immun.* **42**, 318–325 (1983a).
17. G. A. Bohach and I. S. Snyder, *Appl. Environ. Microbiol.* **46**, 528–531 (1983b).
18. D. H. Pope et al., *Curr. Microbiol.* **7**, 319–322 (1982).
19. D. L. Tison et al., *Appl. Environ. Microbiol.* **39**, 456–459 (1980).
20. T. J. Rowbotham, *J. Clin. Pathol.* **33**, 1179–1183 (1980).
21. R. L. Tyndall and E. L. Domingue, *Appl. Environ. Microbiol.* **44**, 954–959 (1982).
22. R. M. Wadowsky et al., *Appl. Environ. Microbiol.* **54**, 2677–2682 (1988).
23. C. M. Anand et al., *J. Hyg. (Lond).* **91**, 167–78 (1983).
24. B. S. Fields et al., *Appl. Environ. Microbiol.* **47**, 467–471 (1984).
25. B. S. Fields et al., *Infect. Immun.* **53**, 553–559 (1986).
26. B. S. Fields et al., *Curr. Microbiol.* **18**, 131–137 (1989).
27. J. E. Stout, V. L. Yu, and M. G. Best, *Appl. Environ. Microbiol.* **49**, 221–228 (1985).
28. J. E. Stout, M. G. Best, V. L. Yu, and J. D. Rihs, *J. Appl. Bacteriol.* **60**, 297–299 (1986).
29. R. M. Wadowsky and R. B. Yee, *Appl. Environ. Microbiol.* **46**, 1447–1449 (1983).
30. R. M. Wadowsky and R. B. Yee, *Appl. Environ. Microbiol.* **49**, 1206–1210 (1985).
31. T. Makin, *Med. Lab. Sci.* **43**, 854 (1986).
32. S. Toze et al., *Curr. Microbiol.* **21**, 139–143 (1990).
33. American Public Health Association (APHA-American Public Health Association; AWWA-American Water Works Association; WEF-Water Environment Federation), *Standard Methods for the Examination of Water and Wastewater*, APHA, Washington, DC, 1992.
34. K. Nahapetian et al., *Res. Microbiol.* **142**, 677–685 (1991).
35. T. J. Rowbotham, *Isr. J. Med. Sci.* **22**, 678–689 (1986).
36. T. J. Rowbotham, "*Legionellae and Amoebae.*" *Legionella, Proceedings of the 2nd International Symposium*, Washington, DC, 1984, pp. 325–327.
37. B. S. Fields et al., *Infect. Immun.* **58**, 3139–3142 (1990).
38. M. A. Horwitz and S. C. Silverstein, *J. Exp. Med.* **153**, 398–406 (1981).
39. A. K. Bej, M. H. Mahbubani, and R. M. Atlas, *Appl. Environ. Microbiol.* **57**, 597–600 (1991).
40. American Society of Heating, Refrigerating, and Air-conditioning Engineers (ASHRAE), "Cooling Towers," in *Handbook—Equipment*, ASHRAE, Atlanta, 1988, pp. 20.1–20.18.
41. J. M. Barbaree, *ASHRAE Journal* **33**, 38–42 (1991).
42. R. A. Swann, M. M. Peel, and G. J. Rouch, *Med. J. Aust* **2**, 226–228 (1982).
43. H. Yamamoto et al., *Appl. Environ. Microbiol.* **58**, 1394–1397 (1992).
44. S. Cameron et al., *Aust. N. Z. J. Med.* **21**, 65–70 (1991).
45. J. S. Colbourne et al., *Lancet* **1**, 210–213 (1984).
46. J. S. Colbourne et al., *Water Sci. Technol.* **20**, 5–10 (1988).
47. P. L. Garbe et al., *JAMA* **254**, 521–524 (1985).
48. E. L. Domingue et al., *Appl. Environ. Microbiol.* **54**, 741–7 (1988).
49. W. L. Wang et al., *Ann. Intern. Med.* **90**, 614–618 (1979).
50. S. Kilvington and J. Price, *J. Appl. Bacteriol.* **68**, 519–525 (1990).
51. Wisconsin Division of Health, *Control of Legionella in Cooling Towers. Summary Guidelines*, Wisconsin Department of Health and Social Services, Division of Health, Bureau of Community Health and Prevention, Section of Acute and Communicable Disease Epidemiology, 1987, Appendix C. Madison, WI. Gov. doc #: HE.6/2:L 43/1987.

52. A. C. England et al., *Appl. Environ. Microbiol.* **43**, 240–244 (1982).
53. C. B. Fliermans and R. S. Harvey, *Appl. Environ. Microbiol.* **47**, 1307–1310 (1984).
54. American Society of Heating, Refrigerating, and Air-conditioning Engineers (ASHRAE), *Legionellosis: Position Statement*, ASHRAE, Meitz, A. K., Atlanta, 1998.
55. S. Srikanth, and S. G. Berk, *Appl. Environ. Microbiol.* **59**, 3245–9 (1993).
56. L. K. Landeen, M. T. Yahya, and C. P. Gerba, *Appl. Environ. Microbiol.* **55**, 3045–3050 (1989a).
57. L. K. Landeen, M. T. Yahya, and S. M. Kutz, *Water Sci. Technol.* **21**, 267–270 (1989b).
58. D. C. Liebe and J. E. Stuehr, *Biopolymers* **11**, 145–166 (1972a).
59. D. C. Liebe and J. E. Stuehr, *Biopolymers* **11**, 167–184 (1972b).
60. K. Shinohara et al., *J. Nutr. Sci. Vitaminol. (Tokyo)* **29**, 671–678 (1983).
61. M. R. Freije and J. M. Barbaree, *Legionellae Control in Health Care Facilities: A Guide for Minimizing Risk*, HC Information Resources, Indianapolis, 1996.
62. American Society for Testing and Materials (ASTM), *Standard Guide for Inspecting Water Systems for Legionellae and Investigating Possible Outbreaks of Legionellosis (Legionnaires' disease or Pontiac Fever)*, ASTM, West Conshohocken, PA, 1996, pp. 5952–5996.
63. A. C. Ta et al., *J. Clin. Microbiol.* **33**, 2118–2123 (1995).
64. R. F. Breiman et al., *J. Infect. Dis.* **161**, 1257–1261 (1990).
65. J. A. Zingeser et al., *JAMA* **264**, 2625–2626 (1990).
66. R. J. Fallon and T. J. Rowbotham, *J. Clin. Pathol.* **43**, 479–483 (1990).
67. S. N. Cousens et al., *Br. Med. J.* **315**, 389–395, (1997).
68. M. N. Ricketts et al., *Is Creutzfeldt-Jakob Disease Transmitted in Blood*? Laboratory Centre for Disease Control, Health Canada, Ottawa, Ontario, Canada and Montreal Neurological Institute, Montreal, Canada, 1997.
69. M. Fishman, G. F. Fort, and D. J. Mikolich, *Am. J. Infect. Control.* **26**, 74–79 (1998).
70. Centers for Disease Control and Prevention and National Institute of Health, *Biosafety in Microbiological and Biomedical Laboratories*, U.S. Government Printing Office, Washington, DC, 1999.
71. D. P. Stites, A. I. Terr, and T. G. Parslow, *Medical Immunology*, Appleton & Lange, Stanford, 1997.
72. R. A. Samson et al., *Health Implications of Fungi in Indoor Environments*. Elsevier Science B.V., Amsterdam, 1994.
73. H. K. Dillon, P. A. Heinsohn, and J. D. Miller, *Field Guide for the Determination of Biological Contaminants in Environmental Samples*, AIHA Publications, Fairfax, Virginia, 1996.
74. J. Macher, *Bioaerosols: Assessment and Control*, American Conference of Government Industrial Hygienists, Cincinnati, 1999.
75. American Conference of Governmental Industrial Hygienists, (ACGIH), *Threshold Limit Values and Biological Exposure Indices for 1995–1996*, ACGIH, Cincinnati, 1972.
76. American Industrial Hygiene Association (AIHA). *Biosafety Reference Manual, 2d ed.*, AIHA Biosafety Committee, Fairfax, VA, 1995.

CHAPTER TWENTY-SEVEN

Control of Biohazards

Patricia A. Heinsohn, Ph.D, MPH, CIH

1 HISTORY OF OCCUPATIONAL EXPOSURE LIMITS

Before any biohazardous exposure can be controlled in the workplace, it must be recognized and evaluated. The extensive tables in the preceding chapter were assembled specifically to help the reader recognize occupational biohazards. Those tables list many of the known occupational biohazards, the occupational settings where they can be found, and the diseases with which they are associated. It is also critical to gain a thorough understanding of the specific process or activities whereby a biohazardous exposure might occur. However, the typical industrial hygiene approach to evaluating an airborne chemical hazard by air sampling for the contaminant during an industrial process is not applicable to evaluating a potentially infectious biohazard (see Chapter 26). Despite this, risk assessment can be done (see also Chapter 26) and effective control measures implemented.

The term biosafety describes a complete program of administrative controls, medical surveillance, vaccination (when appropriate), and containment strategies for reducing the risk of disease among laboratory workers and other persons with potential exposure to biohazards (1). Interest in biosafety has waxed and waned over the past six decades. Biosafety was critically important during research with biological warfare agents in the 1940s. Concern over the genetic manipulation of microorganisms was brought about by the birth of biotechnology in the 1970s. Biosafety regulation began with the Occupational Safety and Health Administration's (OSHA's) bloodborne pathogen rule (29 CFR 1910.1030) and continues with the proposed rule on occupational exposure to tuberculosis (2). But potential biohazard exposure and occupational disease are not unique to research laboratories, biotechnology research and manufacturing, and the health care industries. An estimated several hundred million workers worldwide, workers in workplace settings other

Patty's Industrial Hygiene, Fifth Edition, Volume 2. Edited by Robert L. Harris.
ISBN 0-471-29754-2

than those of historical note, are at risk from occupational biohazards (3). Though the biohazard control strategies discussed in depth here are for the laboratory setting, the same principles are applicable to many workplace settings.

2 CONTAINMENT OF BIOHAZARDS IN THE LABORATORY

In general, the kinds of controls used to prevent biohazard exposure are no different from the traditional industrial control measures of administrative controls, engineering controls, and the use of personal protective equipment. Working safely with biohazards is perhaps more highly dependent on work practices than in some other occupational hazards, particularly in the laboratory. The overall goal of administrative controls, engineering controls, and the use of personal protective equipment is to provide containment of biohazards. Containment is essential in preventing or minimizing personnel exposure and the release of biohazards to the environment beyond the area of use. In the laboratory, employee protection, as well as prevention of biohazard release beyond the workstation, is accomplished through the use of standard microbiological practices and techniques and the effective use of appropriate safety equipment. Facility design and construction contribute to this primary containment but also provide a secondary barrier so that biohazards are not released to the environment outside the laboratory to the community.

2.1 Biosafety Levels of Containment

The concept of containment is best explained by describing biohazardous agents as classes of microorganisms based on degree of hazard, and levels of containment required to contain those microorganisms. As described in the preceding chapter, microorganisms were identified as a diverse group of microscopic organisms including bacteria, viruses, rickettsiae, chlamydiae, fungi, algae, and protozoa. Based on hazard, microorganisms are assigned a hazard category in one of four classes, namely Class 1 through Class 4. Other countries may have similar hazard-ranking categories using a different nomenclature. Numerous documents list biological agents according to hazard rating, but the parent of them all is the Centers for Disease Control (CDC) document *Classification of Etiologic Agents on the Basis of Hazard* (4). (Any such list should be reviewed on a regular basis to account for changes in taxonomy and information pertaining to the virulence between strains.) Working safely with these microorganisms in laboratories involves four containment levels, namely Biosafety Level (BSL)-1 through BSL-4, as described in publications by the CDC and the National Institutes of Health (NIH) for Biosafety in Microbiological and Biomedical Laboratories (BMBL) (5) and in laboratories doing recombinant DNA (rDNA) work (6).

In general, working safely with one class of agents requires that the corresponding containment level be used. For example, working with Class 1 agents requires containment at BSL-1, Class 2 agents require containment at BSL-2, and so on. However, this generalization does not always hold, and the appropriate BSL should never be arbitrarily assumed based solely on hazard classification (see Chapter 26). One or more factors involved in the risk assessment can dictate a higher or lower level of containment than the hazard classification of the agent. For example, the manipulation of an agent in the laboratory may necessitate that the required BSL be higher than the agent classification. For example, BSL-2 containment is required for non-aerosol-producing manipulations of sputum to make acid-fast smears, a diagnostic test for tuberculosis, with all aerosol-producing activities conducted in a Class I or II biological safety cabinet (BSC). However, BSL-3 containment is required for laboratory activities in the propagation and manipulation of cultures of *Mycobacterium tuberculosis* (MTB) or *M. bovis* (5). In contrast, the 20MP-12 vaccine strain of rift valley fever virus, a Class 3 agent, may be handled at BSL-2 (5). Furthermore, Class 1 agents that are not ordinarily associated with disease in humans can pose a risk of infection. For example, opportunistic infection can result in susceptible individuals due to age or compromised immune status.

The CDC and NIH have developed so-called agent summary statements for biological agents that have been responsible for laboratory-acquired infections (5). Each agent summary statement constitutes a risk assessment in that all available information was taken into account to arrive at the stated required level(s) of containment. Unfortunately, the number of existing agent summary statements is small (fewer than 100) compared to the lists categorizing agents according to hazard. For this reason, the biosafety professional must conduct his or her own risk assessment to determine the level of containment required to safely handle agents other than those with agent summary statements. One of the first steps in that risk assessment is to ascertain the hazard classification of the agent. (See the preceding chapter for an in-depth discussion of risk assessment.)

BSLs 1 through 4 consist of combinations of laboratory practices and techniques, safety equipment, and facility design. BSL-1 practices, safety equipment, and facilities are appropriate for undergraduate and secondary educational training and teaching laboratories, and for other facilities in which work is done with defined and characterized strains of viable microorganisms not known to consistently cause disease in healthy adults. BSL-2 practices, equipment, and facilities are applicable to clinical, diagnostic, teaching, and other facilities in which work is done with a broad spectrum of indigenous moderate-risk agents that are present in the community and are associated with human disease of varying severity. BSL-3 practices, equipment, and facilities are applicable to clinical, diagnostic, teaching, research, or production facilities in which work is done with indigenous or exotic agents with a potential for transmission by inhalation and that may cause serious and potentially lethal infection. BSL-4 practices, equipment, and facilities are applicable for work with dangerous and exotic agents that pose a high risk of life-threatening disease, that may be transmitted via the aerosol route, and for which there is no available vaccine or efficacious therapy. Table 27.1 summarizes the recommended BSL requirements for infectious agents. It should be noted that BSLs were developed to prevent or minimize the potential for infectious disease in healthy adults as a result of exposure to living microorganisms and are not appropriate for protection against biogenic allergens (e.g., Der f1 dust mite allergen).

The reader should consult BMBL (5) for additional explanation of the requirements for each BSL.

The work practices, equipment, and facility requirements for each BSL include those of the lower levels.

Although BSLs 1 through 4 are described for work in laboratories and animal care facilities, these concepts of containment can be applied to any occupational setting.

Table 27.1. Summary of Recommended Biosafety Levels for Working With Infectious Agents

BSL	Agents	Work Practices	Safety Equipment and Personal Protective Equipment (PPE)	Facilities
1	Not known to consistently cause disease in healthy adults	Standard microbiological practices	None required	Open bench work permitted
2	Associated with human disease via percutaneous injury, ingestion, or mucous membrane exposure routes	BSL-1 practices plus: • Limited access • Biohazard warning signs • Sharps precautions • Biosafety manual	Primary barrier: Class I or II BSCs or other physical containment devices used for all manipulations of agents that cause splashes or aerosols of infectious materials PPE: Laboratory coats, gloves, face protection as needed	BSL-1 plus: • Autoclave must be available
3	Indigenous or exotic agents with potential for aerosol transmission to cause disease with serious or lethal consequences	BSL-2 plus: • Controlled access • Decontamination of all waste • Decontamination of laboratory clothing before laundering • Serum banking	Primary barrier: Class I or II BSCs or other physical containment devices used for all open manipulations of agents PPE: Laboratory coats, gloves, respiratory protective equipment (RPE) as needed	BSL-2 plus: • Physical separation from access corridors • Self-closing, double-door access • No recirculation of exhausted air • Airflow into the laboratory
4	Dangerous or exotic agents that pose high risk of life-threatening disease, aerosol-transmitted lab infections, or related agents with unknown risk of transmission	BSL-3 practices plus: • Clothing change before entering • Shower on exit • All material decontaminated on exit from facility	Primary barriers: All procedures conducted in Class III BSCs or Class I or II BSCs in combination with full-body, air-supplied, positive pressure personnel suit	BSL-3 plus: • Separate building or isolated zone • Dedicated supply and exhaust, vacuum, and decontamination systems • Other requirements (see BMBL)

Adapted from BMBL (5).

2.1.1 Biosafety Levels of Containment for Large-Scale Work

Large-scale work describes work with production volumes of $\geq$10 L of potentially infectious material. BSL-1LS, BSL-2LS, BSL-3LS, and BSL-4LS are appropriate containment levels for large-scale work that could be handled safely for agents at BSL-1, BSL-2, and BSL-3 on a small scale. Detailed descriptions of containment requirements for each level can be found in appendices of the NIH document *Guidelines for Research Involving Recombinant DNA Molecules* (7).

2.1.2 Animal Biosafety Levels

BSL containment requirements for *in vivo* work with vertebrate laboratory animals is similar to *in vitro* work with infectious agents described in section 1.1.1. Biohazardous exposure in medical or microbiological laboratories is the result of work practices or the equipment used when doing *in vitro* work with infectious agents. Animals themselves can be an additional source of exposure. Animals can generate potentially infectious aerosols or can infect personnel through scratching or biting.

Laboratory animal facilities must in general meet all applicable guidelines and regulations (e.g., *Guide for the Care and Use of Laboratory Animals* [8] and *Laboratory Animal Welfare Regulations* [9]).

Four levels of biological containment are specified for animal facilities, similar to laboratories involved with nonvertebrate animal infectious work. These levels are designated Animal Biosafety Level (ABSL)-1 through ABSL-4 and describe the combination of standard and special work practices, use of safety equipment, and facility design that constitute minimal standards for activities involving laboratory animals infected experimentally or naturally with agents assigned to BSL-1 through BSL-4. Please refer to the most recent edition of BMBL (5) for standard work practices and all other requirements. Table 27.2 summarizes these requirements.

The reader should consult BMBL (5) for additional explanation of the requirements for each BSL.

2.1.3 Clinical Laboratories

Perhaps the most significant biohazard in clinical laboratories is blood-borne pathogens in blood and other body fluids (see also section 3.1.) Diagnostic work, such as chemical assays using serum, in clinical and private laboratories may be done at BSL-2. Gloves should be worn when handling specimens. Stoppered tubes of blood should be grasped using gauze, and stoppers should be lifted away from the operator so that splashing does not occur. Additional control measures are specifically discussed in section 3.1, entitled "OSHA Occupational Exposure to Blood-borne Pathogens," and in section 2.2, "Work Practices and Techniques."

2.1.4 Biotechnology

With the discovery of rDNA technology and advances in molecular biology, there has been considerable concern over possible effects of recombinant products in the environment. However, occupational biosafety in the biotechnology industry was once of much greater

Table 27.2. Summary of Biosafety Levels for Activities Involving Naturally or Experimentally Infected Vertebrate Animals

ABSL	Agents	Work Practices	Safety Equipment and Personal Protective Equipment (PPE)	Facilities
1	Not known to consistently cause disease in healthy adults	Standard animal care and management practices	As required for normal care of each species	Standard animal facility • No recirculation of exhaust air • Directional airflow recommended • Handwashing sink recommended
2	Associated with human disease via percutaneous injury, ingestion, or mucous membrane exposure routes	ABSL-1 practices plus: • Limited access • Biohazard warning signs • Sharps precautions • Biosafety manual • Decontamination of all infectious wastes and of animal cages prior to washing	ABSL-1 equipment plus primary barriers: Containment equipment appropriate for animal species PPE: Laboratory coats, gloves, face and RPE as needed	ABSL-1 facility plus: • Autoclave available • Handwashing sink available in the animal room • Mechanical cage washer used

3	Indigenous or exotic agents with potential for aerosol transmission to cause disease with serious or lethal consequences	ABSL-2 practices plus: • Controlled access • Decontamination of clothing before laundering • Cages decontaminated before bedding is removed • Disinfectant footbath as needed	ABSL-2 equipment plus: • Containment equipment for housing animals and cage dumping activities • Class I or II BSCs available for manipulative procedures (inoculation, necropsy) that may create infectious aerosols PPE: Appropriate RPE	ABSL-2 facility plus: • Physical separation from access corridors • Self-closing, double-door access • Sealed penetrations • Sealed windows • Autoclave available in facility
4	Dangerous or exotic agents that pose high risk of life-threatening disease, aerosol-transmitted lab infections, or related agents with unknown risk of transmission	ABSL-3 practices plus: • Entrance through change room where personal clothing is removed and laboratory clothing is put on; shower on exiting • All wastes are decontaminated before removal from the facility	ABSL-3 equipment plus: • Maximum containment equipment (i.e., Class III BSCs or partial containment equipment in combination with full body, air-supplied positive-pressure personnel suit) used for all procedures and activities	ABSL-3 facility plus: • Separate building or isolated zone • Dedicated supply and exhaust, vacuum, and decontamination systems • Other requirements as outlined in BMBL

BSC: Biological safety cabinet.
PPE: Personal protective equipment.
RPE: Respiratory protective equipment.
Adapted from BMBL (5).

concern than it is today, ostensibly for two reasons. First, the microorganisms most often used in manufacturing operations are those requiring a minimum of containment. These include nonpathogenic, genetically engineered bacteria (e.g., *Escherichia coli* K-12), molds, yeasts, and plant and animal cells. Second, production operations are usually closed systems with redundant design features to minimize release of any potential biohazard. Even inactivation of waste materials and contaminated by-products must be validated. The NIH publication *Guidelines for Research Involving Recombinant DNA Molecules* should be consulted for guidance in establishing the required BSL and the requirements for safely working with recombinant microorganisms at that level. The reader is urged to consult other sources for in-depth discussions of biosafety of genetically engineered microorganisms such as *Genetically Modified Organisms: A Guide to Biosafety* (10).

2.2 Work Practices and Techniques

Work practices are the most critical element in preventing or minimizing occupational exposure to biohazards. *In vivo* or *in vitro* work with biohazards requires appropriate training and experience. The background necessary for working with infectious microorganisms includes, but is not limited to, the following: an understanding of the mechanism of infectious disease transmission, infectious dose, routes of exposure, host susceptibility and resistance, work activities that could result in exposure, how to use engineering controls and personal protective equipment, and how to proceed if exposure occurs. While the ultimate responsibility of biosafety rests with the laboratory's principal investigator (PI), the employee accepts an individual responsibility that only he or she can assume by using safe practices identified in the previously performed risk assessment. Biosafety training to ensure employees have been given the necessary background should be mandatory.

The cornerstone of biohazard containment in the laboratory is standard microbiological practice, which includes the following:

- Limited or restricted access to the laboratory when experiments or work with cultures and specimens are in progress
- Handwashing after handling viable materials, after removing gloves, and before leaving the laboratory
- Prohibition of eating, drinking, smoking, handling contact lenses, applying cosmetics, and storage of food for human consumption in work areas
- Prohibition of mouth pipetting
- Safe handling of sharps
- Prevention or minimization of splashing and aerosolization
- Decontamination of work surfaces on a daily basis and after spills
- Posting of a biohazard sign at the entrance to the laboratory with pertinent information (identity of the agent in use; name and phone number, both on- and off-site, of the investigator)
- Implementation of an integrated pest management program
- Decontamination of cultures, stocks, and other regulated wastes before disposal

- Specific precautions for materials to be transported outside the laboratory before decontamination

The National Research Council (NRC) has developed seven basic rules of biosafety similar to the standard microbiological practices listed here (11). The NRC elaborates on the use of sharps and advocates wearing a laboratory coat and gloves. In addition, the NRC restricts use of needles and syringes to procedures for which there are no alternatives.

Handwashing results in a significant reduction in the carriage of potential pathogens on the hands. This is true concerning both occupational biohazardous exposure and nosocomial infection control. Issues facing biosafety professionals and infection control staff alike are which antiseptic agents to use and when to use them. The reader is encouraged to consult a recent review article published by the Association for Professionals Infection Control and Epidemiology, Inc. for an in-depth discussion of handwashing (12).

2.2.1 Laboratory Procedures that Emit Bioaerosols and Work Practices to Control Exposure

Aerosols, including infectious ones, can be created any time energy is imparted to a liquid or solid material. Laboratory procedures with a high potential for creating infectious aerosols include the following:

- Pipetting
- Mixing
- Shaking
- Centrifuging
- Grinding
- Blending
- Flaming bacteriologic loops
- Vigorous shaking or mixing (such as by vortexing)
- Sonic disruption
- Opening lypholized cultures
- Entering or opening vessels and withdrawing liquid under pressure
- Breaking used needles

Note that mouth pipetting is strictly prohibited due to the large number of documented laboratory-acquired infections associated with it. Prohibition of mouth pipetting eliminates the potential for aspiration of infectious material into the mouth and hand-to-mouth exposure when a contaminated finger is placed on the mouthpiece and then into the mouth. However, an aerosol exposure can still occur using mechanical devices to draw up and expel fluids in transfer pipettes. To minimize aerosolization as well as ensure transfer of liquids of accurate volumes, pipettes should be drained gently with the tip against the inner wall of the receiving vessel rather than allowed to run and drip from the tip freely into the vessel.

Other safe work practices involving the use of pipettes include the following:

- Pipettes should have cotton plugs to reduce contamination of pipetting devices.
- Air should never be forcibly blown through a liquid containing infectious agents.
- Infectious agents should not be mixed by alternate aspiration and expulsion of air through the pipette into the liquid.
- Contaminated (used) pipettes or automated pipette tips should be completely submerged in a suitable disinfectant, contained in an unbreakable container, and allowed sufficient contact time for disinfection.
- This unbreakable container for pipettes, pipette tips, or both should be placed within a BSC rather than outside it.

Splatter is also of concern because as adjacent work surfaces are contaminated, the potential for a percutaneous or ingestion hazard is created. Splatter can also result in exposure to mucosal surfaces of the eye, nose, or mouth. Also, because satellite aerosols are generated whenever splatter occurs, splatter creates the potential for an inhalation exposure. Aerosolization of dusts from animal cages, such as by dry sweeping or release of fungal spores when opening cultures in petri dishes, are examples of aerosolization of solid materials. Thus, it is critical that appropriate steps be taken when these laboratory procedures are used. For example, placing the sonicator in a BSC, wearing a face shield, capping all tubes containing potentially infectious material before mixing by vortexing, and opening cultures within a BSC are all methods of controlling exposure.

Improper flaming of inoculating loops can also result in aerosolization. Heating the shaft until the sample has been heat-dried before flaming the loop itself can prevent spatter and release of aerosols. Using a microincinerator can also control such aerosolization. The potential hazard from flaming can be completely eliminated by using sterile, disposable loops.

Proper use of a BSC is as important as its design and proper functioning. First, excess storage in the BSC can affect the air-distribution pattern and may result in reflux of contaminants into the laboratory. Only the reagents and supplies used in the experiment in progress should be stored in the cabinet, and such items should be placed in the rear of the cabinet behind the actual working area. Work should be laid out in the BSC in an ordered progression from one side to the other. Small temporary waste receptacles can be stored in the cabinet rather than outside the cabinet to allow for fewer hand and arm movements through the face of the cabinet, which break the air curtain. The fan should be run for at least 5 min after completion of work in the cabinet. A Bunsen burner should not be used in the cabinet because the heat produced will distort the airflow pattern. A microincinerator is permissible, but use of disposable loops is preferred.

The centrifuge is an indispensable piece of equipment in many laboratories. It can also be the source of potentially infectious aerosols if not properly used or equipped. Tabletop placement of centrifuges is acceptable if all operators in the laboratory can see into the bowl to place trunnions and buckets correctly, as well as load and unload tubes. Centrifuge rotors and buckets should be inspected routinely to detect signs of corrosion or cracks. Buckets and trunnions should be paired by weight to maintain proper balance. Simply placing centrifuges into Class I or Class II BSCs does not ensure containment of infectious aerosols that could be generated during centrifuging because particle velocity outstrips the velocity of air in the face of the cabinet. Thus, it is necessary to use securely capped tubes and sealable centrifuge buckets or safety cups to contain spillage and aerosols. Centrifuges with sealed rotor heads should be used in lieu of safety cups when centrifuging Class 3 or Class 4 agents. Centrifuge buckets (like homogenizers, shakers, and sonicators) should be loaded, sealed, and opened in a BSC. Centrifuge buckets, rotors, and bowls should be regularly decontaminated.

2.2.2 *Medical Surveillance*

Medical surveillance is a system for the identification and management of individual cases of illness as well as for the early recognition of latent disease (screening) and for the provision of health-promotion activities. It is also the systematic collection, analysis, and evaluation of health data in the workplace to identify cases, patterns, or trends suggesting an adverse effect on workers' health. Biological monitoring is often a useful tool. Medical surveillance of employees working with biohazards is particularly important because the traditional exposure-assessment techniques for hazard evaluation commonly performed by industrial hygienists for nonbiohazards are not applicable to biohazards. Most often the infectious dose for infection or consequence for noninfectious biohazard exposure is unknown or hypothetical. The primary purpose of any occupational medical surveillance program is to detect occupational disease early, before the employee would have sought medical care independently, and at which time treatment is more successful at eradicating disease. Thus, it focuses on secondary prevention of disease.

A medical surveillance program must be tailored to the specific risks and biohazards of the work setting. The medical surveillance program must also be designed to deal with accidental exposure and prophylaxis, such as following needle sticks involving blood or other potentially infectious material. Components of a medical surveillance program might include the following:

- Preplacement examination
- Periodic monitoring evaluation
- Treatment and documentation of exposure
- Prophylactic treatment regimes
- Procedures for massive contamination
- Contact information for infectious disease physicians
- Information from CDC
- Educational information re early symptoms
- Immunization
- Postemployment evaluation
- Occupational epidemiology

2.2.2.1 Preplacement Examination and Periodic Monitoring Examination. The goal of a preplacement examination for a prospective employee is to identify any medical condition that might put that individual or future coworkers at increased risk in the assigned job. Preplacement examination should specifically address job activities and associated biohazards actually involved. Preplacement examination also establishes a baseline health status and ensures that the employee has received adequate immunization for future work with infectious biohazards. At the time of hire, employees should be given questionnaires including a detailed medical and occupational history and a vaccination record. These may be useful in evaluating future medical complaints and for interpreting the significance of nonspecific early manifestations of illness (e.g., frequency of coughing).

Some employees may require a more complete evaluation requiring a physical examination or laboratory testing. For example, preplacement of employees in a sheep-containment facility could require an antibody test for antibodies to the rickettsia *Coxiella burnetti*, which is normally found in parturient fluids of sheep and causes Q fever. Employees who are seronegative may be required to wear respirators while working in the sheep-containment facility. Spirometry testing should be performed. Individuals with valvular heart disease should not work with *C. burnetii* because it can cause endocarditis. The placement evaluation should attempt to identify such individuals.

Serum banking can be another important aspect of medical surveillance. Serum banking is invaluable in establishing the occurrence and timing of a laboratory-acquired infection. Serum banking should be considered only when infection causes a change in the level of assay-detectable serum antibodies. At time of preplacement, a baseline sample should be taken and frozen without analysis. At least two serum samples, called acute and convalescent, are necessary to ascertain the relevance of occupational exposure in a specific incident. At the time of infection, both samples are analyzed. A typical example of serum banking is collection of sera from health care professionals who are at higher risk of infection with human immunodeficiency virus (HIV). Note that CDC and NIH require that baseline sera be taken from employees working at BSL-3 and BSL-4 (5).

There is an important difference between biological monitoring and serum banking. Biological monitoring involves continual surveillance of biological indicators of infection (such as antibody production) and is thus proactive. A great advantage of biological monitoring is that it allows for evaluation of control strategies. Detection of infection by serum banking is obviously reactive and is used mostly for diagnostic purposes.

Monitoring evaluations should be performed at regular intervals, for example, annually on the employee's birthday. This enables occupational health staff to detect changes in health status since the last evaluation. The nature of the evaluation, whether it includes physical examination, biological monitoring, and so forth, should be decided on a case-by-case basis.

2.2.2.2 Treatment and Documentation of Exposure. All accidents, such as needle sticks or spills of potentially infectious materials, should be reported immediately to the laboratory supervisor, who then refers the individual for medical treatment. Medical treatment could include administration of pentavalent toxoid (as in the case of exposure to the toxin of *C. botulinum*), AZT as chemical prophylaxis for HIV infection, or INH for prophylaxis of infection with MTB. Testing, if relevant, should precede treatment. The incident should then be reported to the biosafety professional, who should determine the cause of the accident and make recommendations for preventing reoccurrence and train employees accordingly. Such training could include training key personnel as "first responders." Legal requirements for treatment and documentation of exposure are as follows:

- 29 CFR 1904 Recording and Reporting Occupational Illnesses and Injuries
- 29 CFR 1910.20 Medical Records
- 29 CFR 1910.1030 Occupational Exposure to Bloodborne Pathogens

2.2.2.3 Immunization. The availability and effectiveness of immunization for employees should be evaluated by the biosafety professional in conjunction with occupational health staff. The biosafety professional may best know what biohazards exist in the workplace, potential routes of exposure based on activities, and what employees are at risk of exposure. The occupational health staff should decide which licensed immunizations are recommended to specific employees and those that are required for specific job assignments. The relative benefits of immunization should be weighed against the potential for rare but severe vaccine-related adverse events.

There are a number of different types of immunobiologics, including live attenuated vaccines, inactivated vaccines, toxoids, immune serum globulin, recombinant vaccines, and lyophilized live bacterial or viral vaccines. Vaccines are suspensions of live (usually attenuated) or inactivated microorganisms (or fractions thereof) administered to induce immunity. Examples of live attenuated vaccines include those for measles, rubella, the pox viruses, yellow fever, tuberculosis, and tularemia. Examples of inactivated vaccines include those for anthrax, cholera, eastern and western equine encephalomyelitis, influenza, plague, poliomyelitis (IPV), Q fever, and typhoid. Toxoids are modified bacterial toxins that have been made nontoxic but that retain the ability to stimulate the formation of antitoxin. Examples include those for botulism and diphtheria-tetanus. Immune serum globulins contain antibodies from human blood and are used to confer passive immunity in certain immunodeficient persons and for passive immunization against measles and hepatitis A. Live bacterial or viral vaccines convey immunity rather than disease by specifically infecting recipients with live but nonpathogenic organisms. One such example is the vaccinia virus lyophilized vaccine.

Careful consideration should be given to use of licensed vaccines that do not provide clear protection or are associated with high rates of local or systemic reactions, and unlicensed vaccines given under investigational new drug (IND) protocols. Examples of IND vaccines include a phase IQ fever vaccine for protection against *C. burnetii*. This vaccine is available from the Special Immunizations Program, U.S. Army Medical Research Institute for Infectious Diseases, located in Fort Detrick, Maryland.

Another example of prevention of infection disease using an IND is the use of a pentavalent botulism toxoid to prevent the clinical syndrome of botulism, caused by the toxin of the bacteria *Clostridium botulinum*. Note that botulism is caused by the toxin of the bacteria *C. botulinum*. The toxin may be absorbed through the skin; after ingestion; or following contact with the skin, eyes, or mucous membranes, including those of the respiratory tract; and via accidental parenteral inoculation. The toxoid is recommended for employees working with cultures of *C. botulinum* or its toxins. This IND is available through the Centers for Disease Control and Prevention. The reader is urged to consult the current recommendations of the Public Health Service Advisory Committee on Immunization Practices (ACIP) to find out which vaccines are available, how to obtain them, and who should receive them (13). Note that the toxoid does not prevent infection; rather, it limits symptoms upon infection or exposure to toxin.

2.2.2.4 Postemployment Evaluation. The purpose of an exit examination is to allow for a comparison to preemployment health status. If a serum sample was banked at the time of hire, it may be appropriate to obtain a final sample at exit.

2.2.3 Biosafety Programs and Manuals

As previously defined, biosafety describes a complete program of administrative controls, medical surveillance, vaccination, and containment strategies for reducing the risk of disease in employees with potential exposure to biohazards. A biosafety manual codifies all of these program elements into one written document specifically tailored to the institution, and in some cases to individual laboratories. At a minimum, individual laboratories should have standard operating procedures (SOPs) for work with infectious agents. The laboratory director should write SOPs. Most often biosafety manuals are written by the institution's biosafety officer and approved by the institutional biosafety committee (IBC). While the institution's biosafety manual should be a stand-alone document, it should cite all other applicable programs in place at the institution that interface with it, for example, the institution's blood-borne pathogen program.

2.2.3.1 Biosafety Manual Content. Typical sections of a biosafety manual include the following:

- Purpose
- Scope
- Definitions
- Requirements
- Required records
- Appendixes

Various biosafety manuals can be downloaded from the Internet. CDC's biosafety information is part of its general laboratory health and safety manual and can be downloaded from *www.cdc.gov.* Biosafety manuals of educational institutions can also be found by searching the Internet. (Note: Reference to this does not necessarily constitute endorsement of said manuals.) Manuals specific to other institutions should be used only as rough guidelines for development of an institution-specific biosafety manual. Every effort should be made to make the manual a document that will be useful and not merely a "paper program." Careful consideration should be given to what information goes into the manual as well as how it should be organized and made available, distributed, reviewed, and updated.

2.2.3.2 Responsibilities. All employees have a role in handling biohazards safely. Though policies may vary among institutions, the following describes typical responsibilities. An individual employee is responsible for using standard and special microbiological practices, using safety equipment properly, and complying with all other aspects of the biosafety program. Employees must report all accidents resulting in personal injury, illness, or both and the circumstances associated with the accident. Employees must also report any symptoms associated with exposure. The success or failure of a biosafety program in the most practical sense rests with the employee for whom the program was developed.

The biosafety officer (BSO) is the biosafety professional typically responsible for developing the biosafety program. The BSO must be aware of all the biohazard work done at the institution, including the location, agents involved, activities or manipulations involved, adequacy of the laboratory SOPs, and so on. The BSO participates in policy decisions concerning biohazardous exposure. As such, the BSO is an integral part of the IBC and provides consultation services. The BSO is the interface between the IBC and employees. Other BSO responsibilities include the following:

- Organize and implement biosafety compliance programs and audit their effectiveness
- Prepare and routinely review the biosafety manual
- Review biohazard user authorization (BUA) applications and make recommendations to the PI and IBC
- Interact with occupational health staff
- Institute, conduct, and document biosafety training of employees
- Develop and implement an infectious and medical waste management program
- Provide technical information and advice on products affecting biological safety
- Investigate accidents and recommend control strategies
- Develop and manage a BSC certification program

It is appropriate to discuss the background necessary to qualify as a biosafety professional. Industrial hygiene is the anticipation, recognition, evaluation, and control of occupational and environmental health hazards despite the nature of the hazard, whether chemical, biological, or physical. Obviously, biosafety comes under the purview of the industrial hygienist. Because the field of industrial hygiene is so large and because many industrial hygienists do not have specific training or expertise in biosafety, an institution's industrial hygienist may not be qualified to handle biosafety program responsibilities. Often industrial hygienists are asked or expected to handle biosafety in addition to all other health and safety issues at the institution without the specific background necessary. No one who is not qualified through education and training should take on biosafety responsibilities such as those of the BSO. This author has known very competent biosafety professionals with varied backgrounds, such as industrial hygiene, microbiology, medical technology, molecular biology, and others, as well as a combination of these.

Rigorous certification of industrial hygienists has been in practice since 1963 as administered by the American Board of Industrial Hygiene (telephone 517-321-2638). Sitting for core and comprehensive certification (certified industrial hygienist, or CIH) examination requires extensive relevant academic training and work experience. There are over 5,000 CIHs. One of AIHA's technical committees is the Biosafety Committee, which has been extremely active in all manner of biosafety issues, including training members in basic biosafety, publishing biosafety texts directed toward industrial hygienists, and providing testimony to OSHA on the proposed tuberculosis rule.

Just as there has been increased interest in protecting the practice of industrial hygiene within the American Industrial Hygiene Association (AIHA) through licensure in a number of states, there has been a dedicated effort to "protect turf" within the American Biological Safety Association (ABSA). A certification exam recently made available for biosafety professionals is administered by the National Registry of Microbiology (202-737-3600). Passage earns the title of Certified Biological Safety Professional (CBSP). There are currently approximately 70 CBSPs. Like the ABIH CIH exam, the CBSP exam must be maintained by accumulating points. This author feels strongly that what matters is the competent practice of biosafety by qualified biosafety professionals whatever their background, whether or not they are certified as a CBSP or CIH. In this regard, many references are made to "biosafety professionals" throughout this chapter.

The PI is ultimately responsible for the overall safety of his or her employees. Typically, the PI applies for institutional approval to work with biohazardous agents or BUA. The application requires the PI to describe how infectious agents will be manipulated in the laboratory or animal facility, the work practices and safety equipment to be used, the personnel that will be involved, where experiments will be done, the method of waste treatment and disposal, and so on.

The first IBCs were required by the NIH at institutions conducting recombinant DNA work (14). Today's IBCs also oversee work involving infectious agents. The IBC issues institutional approval to PIs to do recombinant DNA work and work with biohazardous agents. The NIH required that IBCs have no fewer than five members selected for their expertise in recombinant DNA technology and their capability to assess the environmental impact of such work. At least two members must not be employed at the institution and must represent the public's interests rather than those of the institution. The IBC must work effectively with many other institutional committees, such as the radiation safety committee, the animal care and use committee, chemical safety committees, and perhaps others.

2.2.3.3 Assessing Biosafety Program Compliance. All safety programs, including the biosafety program, should be assessed for compliance. Only through compliance audits can it be determined whether the program was successfully implemented and is effective. An audit tool such as a checklist can easily be tailored to a specific institution or site when the required biosafety containment level has been predetermined. Assessing compliance in nonlaboratory work settings is simply based on knowledge of the elements required to prevent exposure.

A blood-borne pathogen program, if applicable, should be part of an overall biosafety program. A blood-borne pathogen program must meet all of the requirements of the OSHA blood-borne pathogen rule, 29 CFR 1910.1030, and the applicable regulations in states with state-approved OSHA plans. Many of the required elements are similar. Implementation and auditing of a blood-borne pathogen program is relatively simple under the broad umbrella of an overall biosafety program. The reader is encouraged to consult the OSHA instruction document (OSHA Instruction CPL 2-2.44C) used by OSHA compliance officers to determine compliance with the blood-borne pathogen rule.

2.3 Engineering Controls

2.3.1 Biological Safety Cabinets

Biological safety cabinets (BSCs) are designed to prevent exposure to workers and the environment. Some of these cabinets also prevent contamination of the material or product being handled in the cabinet. Such protection is quantifiable. All BSCs allow the employee to access potentially infectious material to be manipulated. Class I and Class II BSCs have an opening through which the employee can carry out manipulations inside the cabinet. Class III BSCs have gloves that are mechanically attached to the cabinet, thus providing a totally enclosed environment for the manipulation of infectious material. Room air is supplied to the cabinet through the opening and exhausted continuously from the Class I BSC at a rate sufficient to minimize the escape of aerosol generated in the cabinet. Room air is also supplied to the cabinet through the opening in the Class II BSC, but a portion of that air is recirculated after passing through a high-efficiency particulate air (HEPA) filter and passed down the front of the cabinet in a laminar fashion. Laminar flow inside the cabinet minimizes the escape of aerosol generated in the cabinet. There are several types of Class II BSCs, namely types A, B1, B2, and B3. Differences among these BSC types will be further discussed. Only HEPA-filtered air is supplied to the Class III BSC, and it is entirely exhausted. This cabinet offers the maximum amount of protection to personnel and environment.

2.3.1.1 Class I BSCs. The minimum face velocity at the work opening is 75 linear feet/min (lfpm). All of the air from the cabinet is filtered either into the laboratory or to the outside. This cabinet, as depicted in Figure 27.1, is designed for use with low (Class 2) and moderate (Class 3) risk agents but not toxic or radioactive materials. It is also not appropriate for handling materials that are vulnerable to airborne contamination, since the room air supplied to the cabinet is unfiltered and can carry contaminants into the cabinet. If the Class I BSC is retrofitted with a front closure panel with gloves and ducting to the outside, then it is safe to use with toxic or radioactive materials.

2.3.1.2 Class II BSCs. The Class II BSC is designed to provide personnel, environmental, and product protection by providing an inward face velocity of 75–100 lfpm and HEPA-filtered exhaust. Class II BSCs are suitable for use of low to moderate biohazards. The differences between types A, B1, B2, and B3 derive from varying construction, face velocities and patterns, and exhaust systems, as apparent in Figures 27.2–27.5. Like Class I BSCs, type A Class II cabinets cannot be used with volatile chemicals or radioactive materials because air is recirculated in them. Radioactive materials and volatile chemicals can be used in all three type B cabinets. The airflow pattern of each type B cabinet is summarized in Table 27.3.

Only Class II BSCs that meet the standards for materials, design, construction, and performance published by the National Sanitation Foundation (NSF) (NSF Standard 49)

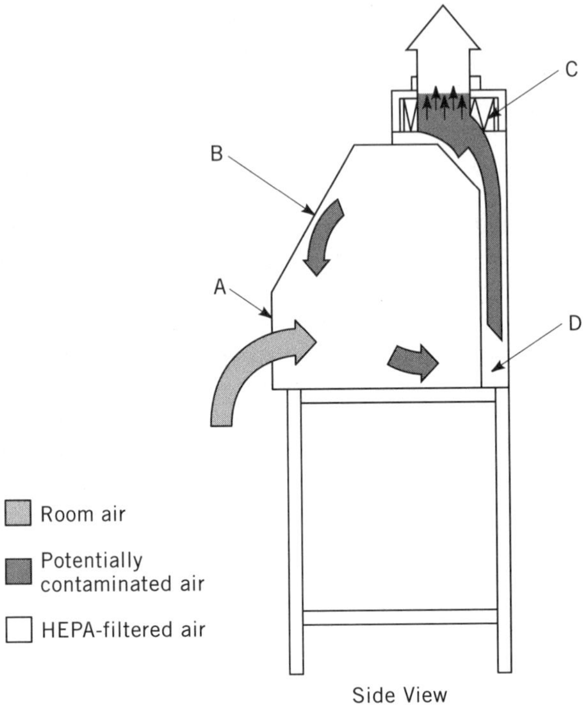

Figure 27.1. Class I Biological Safety Cabinet. A, Front opening; B, sash; C, exhaust HEPA filter; and D, exhaust plenum (CDC, 1999).

should be purchased. A list of products that meet NSF standards is available from National Sanitation Foundation International, Ann Arbor, Michigan (15). The effectiveness of these BSCs depends on the integrity of the HEPA filters and sealed construction as well as the ability of the air curtain at the face to prevent the escape of bioaerosols into the laboratory. To ensure adequate performance, they must be tested and certified.

2.3.1.3 Testing and Certification of Class II BSCs. Class II BSCs are tested and certified according to the protocols described in NSF Standard 49 (16), which test performance to provide reliable function; durability; and personnel, product, and environmental protection. The standard also defines testing requirements for cleanability, noise, illumination, temperature rise, and vibration. Class II BSCs must be tested and certified in the laboratory after initial installation, whenever the BSC is moved, and on an annual basis to ensure that it meets performance standards developed by the NSF. Individuals who themselves test and certify Class II BSCs must now be certified in the NSF Biohazard Cabinet Field Certifier Accreditation Program. A listing of all certified individuals is also available from the NSF (15).

Academic institutions and private research companies usually have many Class II BSCs and typically negotiate a single contract with a vendor to test and certify BSCs. This

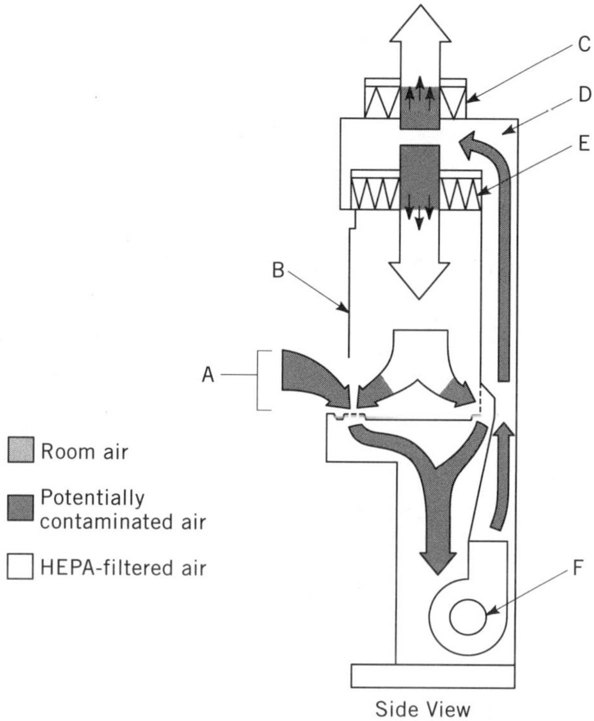

Figure 27.2. Class II, Type A Biological Safety Cabinet. A, Front opening; B, sash; C, exhaust HEPA filter; D, rear plenum; E, supply HEPA filter; and F, blower (CDC, 1999).

author strongly recommends that the biosafety professional work with the purchasing department to provide input early in the contract negotiation to make vendor requirements specification-oriented. In addition, a database should be implemented and maintained to manage all BSCs in inventory at the institution by serial number. Typical entries into this database include the PI's name and extension, the room location, the type of BSC, the serial number, the last testing and certification date, and the biohazards handled in the cabinet. An arrangement for testing can be made with the vendor such that it is their responsibility to make the necessary arrangements for access with laboratory personnel. Obviously, testing must be done when the BSC is not in use. If test reports are sent directly to the person who manages the BSC certification program, he or she can update the database and approve payment to the vendor. This process ensures that the institution's BSCs are tested in a timely manner. Whether or not the cabinet is certified, that is, meets performance criteria, is essential in auditing laboratories and, in many cases, meeting requirements imposed by some research funding organizations, such as NIH.

Several test methods are now used in the United Kingdom to address operator protection factors that quantitatively characterize the performance of open-front cabinets. An open protection factor is defined as the ratio of the exposure to airborne contamination generated

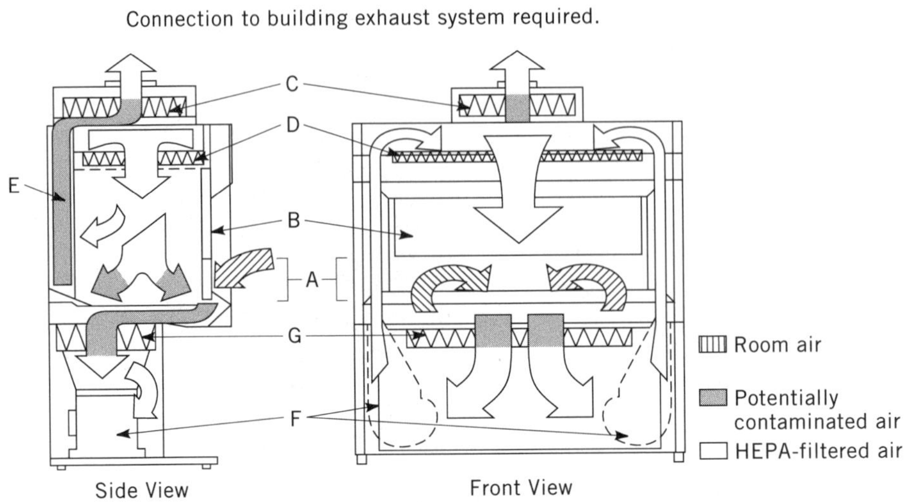

Figure 27.3. Class II, Type B1 Biological Safety Cabinet (classic design). A, Front opening; B, sash; C, exhaust HEPA filter; D, supply HEPA filter; E, negative pressure exhaust plenum; F, blower; and G, additional HEPA filter for supply air. Note: The cabinet needs to be connected to the building exhaust system (CDC, 1999).

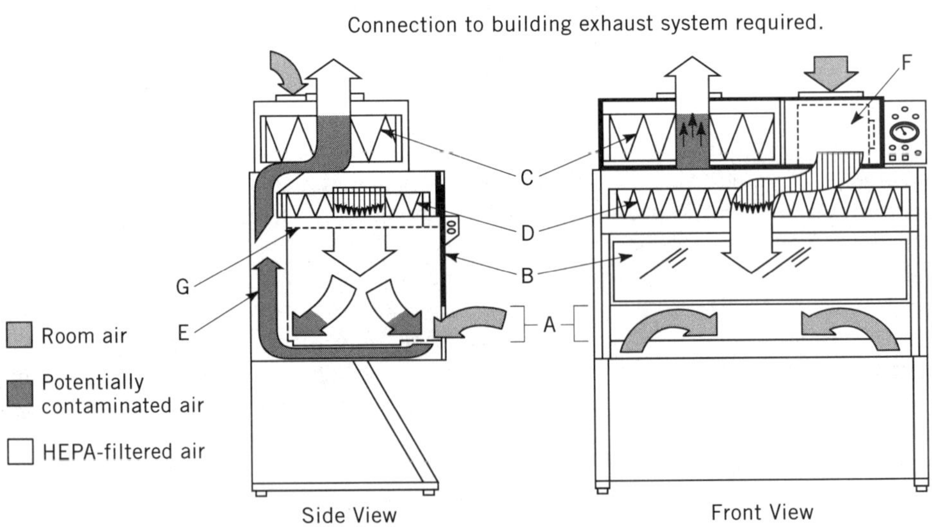

Figure 27.4. Class II: Type B2 Biological Safety Cabinet. A, Front opening; B, sash; C, exhaust HEPA filter; D, supply HEPA filter; E, negative pressure exhaust plenum; and F, filter screen. Note: The carbon filter in the building exhaust system is not shown. The cabinet exhaust needs to be connected to the building exhaust system (CDC, 1999).

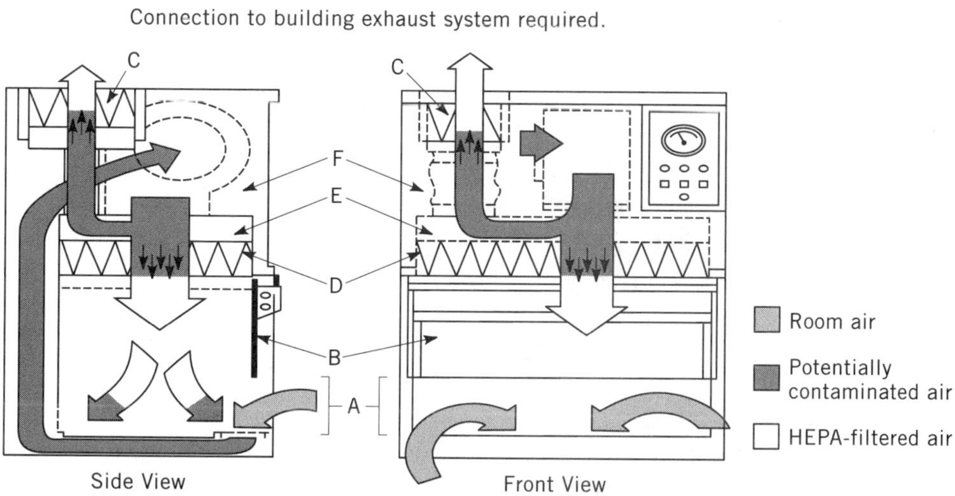

Figure 27.5. Tabletop Model of a Class II: Type B3 BSC. A, Front opening; B, sash; C, exhaust HEPA filter; D, supply HEPA filter; E, positive pressure plenum; and F, negative pressure plenum. Note: The cabinet exhaust needs to be connected to the building exhaust plenum (CDC, 1999).

on the open bench to the exposure resulting from the same dispersal within the cabinet (17). Three types of tests have been used in the United Kingdom, including generation of viable bioaerosols (e.g., *Serratia marcescens* and *Bacillus subtilis* var. niger spores), potassium iodide aerosol (the so-called KI-Discus method), and tracer gas. This author is not aware of any testing in the United States to determine operator protection factors, but they are mentioned here for completeness. A strong word of caution is advised with regard to the generation of either *S. marcescens* or *B. subtilis* and inadvertent human exposure in an improperly controlled operator protection factor test. Normally nonpathogenic, *S. marcescens* is a rare cause of severe infection in humans. Also, whereas the KI-Discus and tracer gas methods can be used *in situ*, the viable bioaerosol method cannot practically be performed *in situ*.

2.3.1.4 Class III BSCs. Maximum personnel protection can be achieved in one of two ways: either enclose the biohazard or enclose the personnel. Class III BSCs offer the ultimate personnel, environmental, and material protection by enclosing the biohazard (see Figure 27.6). All manipulations of the potentially infectious material are done using arm-length rubber gloves of half-suits. Class III cabinets are gas tight and are usually set up in series to allow for containment of all required equipment such as incubators, refrigerators, and centrifuges. The cabinet is attached to a double-doored autoclave, chemical disinfection tanks, or both to allow for safe exit of materials leaving the cabinet as well as entry of supplies.

2.3.1.5 Laminar Flow Clean Benches. Laminar flow clean benches are not BSCs. They are designed so that room air is filtered through HEPA filters before being passed over the

Table 27.3. Summary of Biosafety Cabinet Features

Type	Face Velocity (lfpm)	Source of Air Supply/% Recirculated	Use of Radioactive Materials/Volatile Chemicals	Use with Agent Hazard Class	Product Protection
Class I, open front	75	Unfiltered room air/None	No	2, 3	No
Class I with glove panels	150	Unfiltered room air/None	Yes	2, 3	No
Class II, Type A	75	70% recirculated through HEPA/ Exhaust through HEPA; hard-ducted	Yes (low levels/volatility)	2, 3	Yes
Class II, Type B1	100	30% recirculated through HEPA/ Exhaust through HEPA; hard-ducted	Yes	2, 3	Yes
Class II, Type B2	100	No recirculation/Total exhaust through HEPA; hard ducted	Yes	2, 3	Yes
Class II, Type B3	100	Same as IIA, but plena under negative pressure to room and exhaust air is ducted	Yes	2, 3	Yes
Class III	NA	Supply air inlets and exhaust through two HEPA filters	Yes	3, 4	Yes

Adapted from BMBL (5).

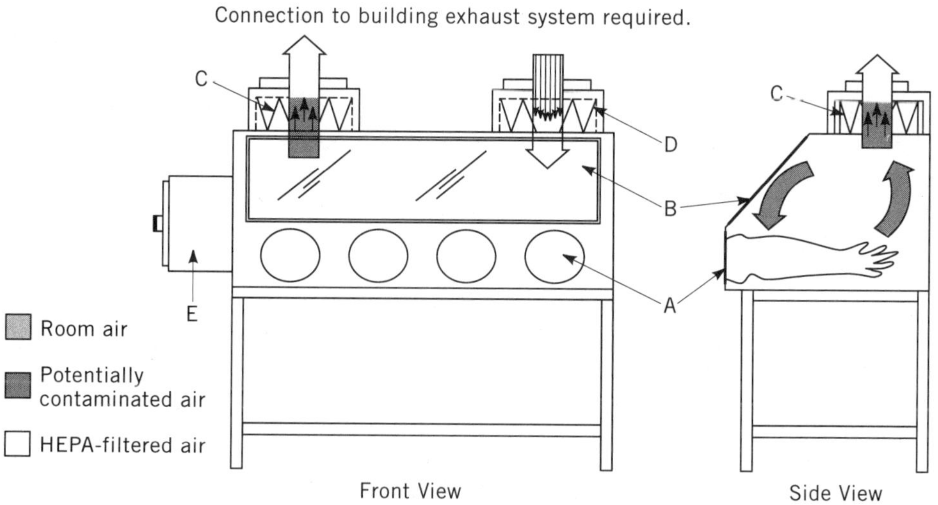

Figure 27.6. Class III Biological Safety Cabinet. A, Glove ports with O-ring for attaching arm-length gloves to cabinet; B, sash; C, exhaust HEPA filter; D, supply HEPA filter; and E, double-ended autoclave or pass-through box. Note: A chemical dunk tank may be installed that would be located beneath the work surface of the BSC with access from above. The cabinet exhaust needs to be connected to an independent building exhaust system (CDC, 1999).

interior work surface and exiting out toward the employee. Such a hood design is meant to protect only the product. As such, its only use in a microbiology or biomedical laboratory is for making sterile growth media. They are not to be confused with BSCs and should never be used in work with any infectious materials.

2.3.1.6 Placement of Biological Safety Cabinets. Proper selection of BSCs as well as testing and certification of Class II BSCs have been discussed. Also of importance is the actual placement of the BSC in the laboratory and how it is used. Air drafts from supply vents should be controlled because they can interfere with BSC performance. DiBerardinis (18) recommended that air velocity at the occupant's level not exceed 50 lfpm. Heating, ventilating, and air-conditioning (HVAC) design employing multiple supply vents can alleviate drafts exceeding this velocity. Placement of the BSC near the door to the laboratory can create drafts sufficient to reflux aerosols generated in the cabinet into the laboratory. In general, foot traffic behind the operator should be minimized. It is also critical that sufficient make-up air is supplied to the laboratory if the BSC's exhaust air is not recirculated within the laboratory, as is the case with Class II B1 and Class II B2 cabinets. All of these factors should be taken into consideration when designing the laboratory and placing the BSC within the laboratory.

2.3.1.7 Decontamination of Biological Safety Cabinets. Decontamination of BSCs is often necessary, for example, before filter replacement, before some repairs are performed, sometimes after accidental spills, and after discontinuing use of the cabinet. Decontamination of BSCs is typically done using formaldehyde gas by heating flakes of paraformaldehyde in the presence of high humidity after completely sealing off the cabinet. This procedure should be done only when personnel are not present to minimize the potential for exposure should the cabinet not be completely sealed during decontamination. The procedures are described in depth in an appendix of NSF Standard 49. Some Class III cabinet designs have connection ports to formaldehyde-generating equipment that allow for a more controlled delivery of gas than is achieved through the use of the previously described "frying pans."

Use of formaldehyde gas poses the risk of inhalational exposure. It is a suspected human carcinogen and known irritant. The OSHA permissible exposure level (PEL) and ACGIH threshold limit value (TLV) of 0.3 ppm should not be exceeded during any part of a working exposure. In addition to the potential exposure issue, the efficacy of formaldehyde decontamination has also been questioned (19). Biological indicators can be used to determine the efficacy of BSC decontamination. Recently, a comparative study of methods to validate formaldehyde decontamination of BSCs found that of three biological indicators manufactured by 3M Company, only one (Attest 1291) was useful in detecting gross decontamination failure of the cabinet (20).

2.3.2 Positive-Pressure Personnel Suits

Positive pressure suits are one-piece suits that are ventilated by a life-support system protected by HEPA filtration. These specially designed suits are intended to provide personnel protection that is equivalent to that provided by Class III BSCs and are used when working with Class 3 or 4 agents. The attached life support system includes redundant breathing air compressors, alarms, and emergency backup breathing air tanks. The air pressure within the suit is positive to the surrounding laboratory. When the employee exits the laboratory, he or she decontaminates the outside of the suit in a chemical shower area of the facility. All penetrations into the internal shell of the suit area, chemical shower, and airlocks are sealed.

2.3.3 Sharps Containers Handling

Sharps containers are an engineering control, but their proper use is work-practice dependent. Sharps containers are intended for safe placement and storage of sharps until the container is disposed of. Sharps include anything that can cause percutaneous (including parenteral) exposure. Examples include the sharp or broken edges of glass microscope slides, glass capillary tubes, test tubes, glass cuvettes, pipettes, needles, syringes, laboratory glassware, scalpels, and razor blades.

There are two ways to prevent percutaneous or parenteral exposure via sharps: limit or eliminate their use, or properly collect and dispose of them. As previously stated, needles and syringes can be restricted to procedures for which there are no alternatives. Options include the following: syringes that resheathe the needle, needleless systems, needle-locking syringes, and disposable needle-syringe units (i.e., the needle is integral to the syringe).

When no alternatives exist, safe handling is the key to prevention of percutaneous or parenteral exposure. No needle should be bent, sheared, broken, recapped, removed from disposable syringes, or otherwise manually handled before disposal. Disposable sharps should be disposed of only in puncture-resistant containers known as sharps containers. Common sense dictates that sharps containers be located conveniently in the work area so that employees can access them quickly and easily. Nondisposable sharps must be placed in a hard-walled container for transport to a processing area for decontamination.

The design of a sharps container is also important. There are several manufacturers, and thus different designs of sharps containers are available. In general, sharps containers load from either the front or the top. Needle sticks are more likely in those that load from the side because the needle end, the lighter end, is prone to tip up, especially when the container is almost full or full. By design, the needle is already facing down when loading into top-loading sharps containers. To prevent overloading and possible needle sticks, sharps containers should be disposed of before they are full. For example, an institution's biosafety program could stipulate that sharps containers must be disposed of when two-thirds full to avoid overloading. Many sharps containers lock, making it more difficult to remove used needles for possible reuse or resale by intravenous drug users or others.

In the laboratory setting, policies for the safe handling of sharps are required as a standard microbiological practice (see section 2.2, "Work Practices and Techniques"). The blood-borne pathogen rule in health care settings (see Section 3.1) mandates the following sharps requirements:

- Prohibition of recapping or needle removal unless there is no demonstrable, feasible alternative or unless such action is required by a specific medical procedure
- If done, recapping or needle removal must be accomplished through the use of a mechanical device or a one-handed technique.
- Immediately or as soon as possible after use, contaminated reusable sharps must be placed in appropriate containers until properly reprocessed.
- Appropriate containers must be puncture resistant, labeled or color-coded, and leak-proof on the sides and bottom.

Although safe sharps handling has always been critical in the prevention of occupationally acquired hepatitis B virus (HBV) infection, its importance was heightened by the AIDS epidemic. Though an effective vaccine exists for HBV, one does not presently exist for either HIV or hepatitis C virus (HCV) infection. Nor is there an effective postexposure prophylaxis for HCV needle stick exposure (21). There is effective prophylaxis for HIV (22).

Cleanup of broken glass and other sharps should never be done using bare or gloved hands. Only forceps or tongs should be used to pick up pieces of broken glass, and the pieces should be placed in a dedicated broken-glass waste container or sharps container. Another safe technique involves sweeping pieces into a dustpan using a whisk broom.

2.3.4 Engineering Controls for Prevention of Transmission of Tuberculosis

Tuberculosis can be occupationally acquired. OSHA has issued a proposed rule designed to prevent occupational exposure to MTB. Thus, control of this biohazard warrants individual discussion both here and in section 2.4.2, "Respiratory Protective Equipment."

The most important public health intervention in controlling transmission of tuberculosis is early diagnosis to identify infectious individuals prior to exposure to other susceptible individuals. The most important industrial hygiene controls are the use of respiratory protective equipment and engineering controls. The reader is urged to consult additional sources for discussions of dilution ventilation and air mixing as well as specifications for available equipment (23, 24).

The principal engineering controls used to prevent transmission of tuberculosis to health care employees or other patients involve filtration and ultraviolet irradiation of air suspected of containing bacilli of mycobacteria. These are discussed next. Both sputum induction booths and isolation tents are discussed separately, although they both involve filtration.

2.3.4.1 Air Filtration and Filtration Devices. Filtration has long been used as an effective means of cleaning air. Air filters remove particles by five known mechanisms, including interception, inertial impaction, diffusion, gravitational settling, and electrostatic attraction. Filtration efficiency, and its reciprocal penetration, are dependent on both the size of particles in the airstream passing through and the air velocity. Particles of size 0.3 μm penetrate most filters maximally (25, 26). Particle sizes larger and smaller than 0.3 μm will be removed with greater efficiency.

There are several types of filters. HEPA filters are expendable, extended-media, dry-type filters in rigid frames. HEPA filters have a minimum particle collection efficiency of 99.97% for 0.3 μm thermally generated dioctyl phthalate (DOP) particles or specified alternative challenge aerosol and a maximum clean-filter pressure drop of 1 in. water column when tested at rated airflow capacity (27). This requirement takes into account variables affecting testing results, including particle size, method of particle generation, airflow, and pressure drop. HEPA filtration is endorsed by the CDC as a method of air cleaning that supplements dilution ventilation (28).

HEPA filters are categorized into five performance classes and two construction classes. Together, these class combinations provide a range in construction, performance, and cost suitable for different application requirements. Performance classes A through E are based on DOP penetration and resistance to airflow (28). Type A filters meet the minimum performance level or removal efficiency that satisfies the basic definition of a HEPA filter. Type B filters meet the minimum efficiency of Type A filters but are certified to be free of pinhole leaks that would cause penetration at low flow rates. Type C filters meet the minimum efficiency of Type A filters but are tested using aerosols at 80–100 lfpm face velocity with an upstream DOP smoke concentration of approximately 10–20 μg/L. They are fully scanned to detect and eliminate all significant leakage streams having concentrations greater than 0.01% of the upstream DOP smoke concentration at these concentrations. Type D filters must meet the requirements of Type C filters but must also yield a penetration of ≤0.001% at rated flow. Type D HEPA filters can suffice when ultralow penetration is required. Type E filters are designed, constructed, and tested in accordance with two U.S. military standards for use in air-cleaning systems involving toxic chemical, carcinogenic, radioactive, or hazardous biological particles. HEPA filters are also categorized into two construction classes, Grades 1 and 2 (28). Grade 1 construction class is fire resistant, whereas Grade 2 is semicombustible.

Additional performance and construction considerations should be taken into account when selecting HEPA filters:

- Single filters having a depth of ≤¼ of the minimum side dimension may not meet pressure-loading requirements in standardized test methods. Consideration should be given to the desired separator direction to avoid media failure. Where filter faces are mounted vertically, pleats and separators should be vertical. Sagging of the pack may occur in filters installed with horizontal separators, increasing the risk for media failure. Installation of lower efficiency prefilter(s) upstream of the HEPA filter can extend the life of a HEPA filter up to 25%. Similarly, following the lower-efficiency filter with a 90% extended surface filter can extend the life of the HEPA filter by nearly 900% (29). Both used prefilters and HEPA filters should be handled and disposed of as if contaminated with droplet nuclei.
- A manometer or other pressure-sensing device should be installed in the filter system to monitor the need for filter replacement. The manufacturer should be able to describe the pressure drop characteristics of the filter as it loads particles. The filter's installation design should allow for maintenance that will not contaminate the ventilation system or the area it serves.
- A quantitative leak test should be performed when the filter is initially installed, when it is changed or moved, and every 6 months. There are several test procedures, including the leak test, the scan test, the two-flow test, and a laser test (26).
- Proper filter installation should prevent leakage between filter segments and between the filter mat and its casing.
- Manufacturers of equipment containing HEPA filters provide a maintenance schedule and recommended procedures for installation, removal, and disposal of filter elements. Only properly trained personnel should do such maintenance.

Improper design, installation, or maintenance of HEPA filters or HEPA filtration units can result in exposure to MTB droplet nuclei.

2.3.5 Filter Configuration

HEPA filters can be used in numerous configurations, including the following: in booths or enclosures recirculating air back into a room, within a room either in freestanding (portable) air-cleaning units or in wall-mounted recirculation units, in HVAC ductwork prior to recirculation to other areas, or in exhaust ducts prior to exhaust to the outside.

2.3.5.1 In-Room Filtration. Whether HEPA filters effectively remove MTB from the air has not been studied. However, CDC guidelines state they are likely to be effective since they have been shown to remove spores of the fungus *Aspergillus*, which range in size from 1.5 μm to 6.0 μm (30, 31). Unfortunately, the volumetric flowrate conditions under which these studies were conducted were not well described. The ability of a HEPA filter to remove particles of a certain size (e.g., 1.5–6.0 μm) addresses only the issue of filter efficiency. Any properly manufactured HEPA filter should also perform *in situ* consistent with its performance and construction rating when installed properly. More importantly, the ability of HEPA filters, either in freestanding or in-room mounted units, to effectively remove particles of a given size from an occupied space should be documented because variables unique to the room may reduce the unit's effectiveness. Such variables include ventilation configuration, air-mixing patterns in the room, placement of the unit, unit design, proximity of health care workers to the infectious patient, and the duration and intensity of exposure.

Miller-Leiden et al. (32) evaluated the effectiveness of freestanding HEPA room air units to exhaust aerosol within a mock hospital room. Three commercially available freestanding HEPA filtration units tested performed equally well. As expected, the highest concentration of particles occurred closest to the source and placement within the room, and the flow geometry in the room significantly impacted effectiveness. Unfortunately, because caregiving demands close proximity to the infectious patient (where particle concentration is highest), it is likely that exposure will occur despite adequate air changes per hour (ACH) provided by the ventilation system, the presence of a freestanding HEPA filtration unit, and, for that matter, the presence of ultraviolet germicidal irradiation (UVGI). Clearly, adequate respiratory protection is still required when caring for patients with clinically active TB.

Miller-Leiden et al. (33) also investigated the effectiveness of ceiling-mounted filtration units. The experimental protocol involved simultaneous aerosol generation and sampling for 2 hours and determination of air concentrations as a function of space and time. Three units were evaluated using various combinations of ventilation and equivalent ACH filtration (24). Ceiling-mounted filtration units appeared to work well. The units provided 6 equivalent ACH and reduced the average room concentration to a value less than that achieved by 6 ACH with general ventilation only. However, there was substantial intraunit and interunit variation. Moreover, the concentration nearest the source was always higher than the average room concentration. Increasing the airflow rate beyond 12 equivalent ACH resulted in little change in air concentration.

CDC's preferred method for recirculating HEPA-filtered air within a room is a standalone system. Air is exhausted from the room into a duct separate from the HVAC ductwork, passes through a HEPA filter in the duct, and returns to the room. Such a design requires placement of ductwork in the ceiling cavity of the room. The reasoning behind CDC's preference for this system (over portable or wall-mounted units) is not readily apparent, as installation and maintenance in this configuration are more difficult.

2.3.5.2 Filtration in HVAC Ductwork. Although HEPA filters can also be placed within HVAC ductwork, this practice is not appropriate. Air from tuberculosis isolation and treatment rooms should be exhausted to the outside in accordance with applicable federal, state, and local regulation. Consequently, there is no reason to filter the air exhausted from such rooms for recirculation purposes. Notably, if air exhausted to the outside can reenter the building ventilation system supply, then HEPA filtration is needed. Only in existing facilities where recirculation of the air into the general ventilation system from such rooms is unavoidable should HEPA filters be installed in the exhaust duct leading to the general ventilation system. Note that HEPA filtration within the HVAC system does not lower the airborne concentration of droplet nuclei *within* the room from which it is being exhausted.

2.3.5.3 Sputum Induction Booths. A sputum induction booth, used for obtaining sputum samples from patients who cannot produce sufficient sputum spontaneously, typifies an enclosing device. It is a miniature, freestanding room that should be designed to pull room air past the breathing zone of a patient seated in the booth into a prefilter and HEPA filter. The booth design should not allow room air to be pulled directly from the air-supply inlet to the exhaust outlet, nor should it allow exfiltration, that is, air leakage outward through cracks and interstices or through ceilings, floor, and walls of a space or building. Air should be exhausted only through the HEPA filter into the room or outdoors. The integrity of the system can be checked using a smoke tube at all exfiltration points while the unit is running.

2.3.5.4 Isolation Tents. Isolation tents are commonly used enclosing devices that create an isolated environment within a room (e.g., a negative pressure bed enclosure for infectious patients) for aerosolized drug therapy (such as for ribavirin treatment in a tented crib) or for isolation of an infected patient being transported. The typical design draws air upward from the area around the bottom of the skirt through HEPA filters prior to recirculation into the room.

Notably, if air exhausted to the outside can reenter the building ventilation system supply, then HEPA filtration is needed.

2.3.5.5 Ultraviolet Irradiation. Ultraviolet (UV) radiation is non-ionizing radiation in the electromagnetic spectrum within three wavelength bands, viz. UV-C (100–280 nm), UV-B (280–315 nm), and UV-A (315–400 nm). Commercially available UV lamps used for germicidal purposes are low-pressure mercury vapor lamps that emit radiant energy in the UV-C range, predominantly at 253.7 nm.

UV lights can be mounted on a wall or ceiling of a space to irradiate the upper portion of the air, in HVAC ductwork to irradiate air passing through ducts, in freestanding UVGI units or combination UVGI/HEPA units, or in self-contained units mounted in ceiling cavities. CDC recommends the use of UVGI lamps in general-use areas in which the risk of transmission is relatively high as an adjunct to ventilation for reducing the airborne concentration of MTB. However, the effectiveness of this practice has not been adequately evaluated (28).

UV lights mounted on a wall or ceiling of a space are passive; they do not contain air-moving devices that bring MTB droplet nuclei to the UV beam for a germicidal dose. Ultraviolet lamps mounted in HVAC ductwork may be more effective because the air-handling unit(s) of the HVAC system draw contaminated air from an occupied space past the lamps mounted on the interior duct surface. Freestanding UVGI units or combination HEPA/UVGI units contain pumps to move contaminated room air past UV lamps. Actively moving large volumes of room air past UV lamps may result in more effective MTB control than passive UVGI, assuming sufficient residence time in the zone of irradiation. UV lamps must be installed properly and maintained according to the manufacturer's instructions. Only specially trained employees should handle and maintain UV lamps.

2.3.6 Effectiveness of Ultraviolet Germicidal Irradiation

The biocidal properties of UV light have been studied. *M. tuberculosis, M. phlei*, and BCG are less susceptible to loss of viability than *Staphylococcus albus*, *Listeria monocytogenes*, *E. coli*, *Pseudomonas aeruginosa*, *Salmonella enteritidis*, and *Serratia marcescens* (34).

These data contrast with those of Riley et al. (35), who found that virulent tubercle bacilli and BCG were equally susceptible to UV irradiation but that *M. phlei* was 10 times more resistant. It is important to note that these studies involved the artificial generation of aerosols in test rooms. It has not been demonstrated that the rates of inactivation observed in test rooms can be attained in occupied rooms in mechanically ventilated buildings.

Chang et al. (36) measured the biocidal dose response of UV-C to other human pathogens such as bacteria, viruses, spores, and amoebal cysts. Greater doses of UV-C are required to kill cysts and spores than viruses (36). Several strains of virus were effectively inactivated with high rates of kill, 99.6 to 100%, by UV exposure in an experimental chamber (37).

Biocidal effectiveness of UVGI is a function of relative humidity. Under experimental conditions, a sharp decrease in the number of culturable *S. marcescens* recovered was observed above 60–70% relative humidity (38). The practical implication of this finding is that relative humidity, an environmental condition that is often not tightly controlled, may reduce effectiveness of UVGI.

Total room ventilation and air exchange between portions of a room are also important considerations. One study found that upper air irradiation with a passive UVGI device in the waiting room of an outpatient HIV clinic reduced the concentration of ambient airborne bacteria by 15 to 30% (39). However, the authors subsequently indicated that the total ventilation to the room may have been underestimated, resulting in an overestimate of the reduction attributable to UVGI. Air currents of approximately 20 lfpm are sufficient to waft MTB droplet nuclei into an upper zone of irradiation. However, the amount of air exchanged between upper and lower room zones is not known. Moreover, the effectiveness of passive upper air UVGI in reducing MTB exposure due to close proximity to infectious patients has not been documented.

Ultimately, research performed to date has not determined the value of UV-C in real-world applications such as hospitals and other institutions, where variables such as temperature, relative humidity, and volumetric flow rate cannot be or are not controlled. Although UVGI has been shown to be effective in preventing the transmission of measles and influenza viruses in classrooms (40), subsequent placement in schools was unsuccessful (41). Additional studies are needed to accurately determine the extent of UVGI-attributable droplet nuclei reduction and, more importantly, to evaluate whether the resulting reduction is sufficient to reduce infection. It is unclear whether the health risks inherent to exposure to UV light emitted by UVGI devices (reviewed next) are justified by their efficacy in killing mycobacteria. It is likely that passive UVGI reduces both the concentration at the source and the average room concentration much more slowly and to a lesser degree than HEPA filtration units or UVGI units equipped with an air-moving device.

2.3.7 Comparison of UVGI Designs

Ultraviolet lamps are sometimes mounted in HVAC ductwork; however, in this location, they only disinfect (to an unknown degree) the air downstream of the lamps. Thus, CDC recommends that UVGI devices not be used as substitutes for HEPA filters in ducts that discharge air from TB isolation rooms into the general ventilation system (28).

HVAC duct-mounted, freestanding UVGI, HEPA/UVGI units over passive wall-mounted UV lamps, and units recessed in a ceiling cavity offer several distinct advantages. First, they minimize room occupant exposure to UV radiation. Second, recessed units do not distort room air mixing by taking up room space. Third, permanently mounted systems probably have greater biocidal effectiveness than freestanding floor units, whose effectiveness is subject to floor placement within the room. Portability of freestanding UVGI or HEPA/UVGI units is advantageous only if TB isolation rooms are not dedicated, permanent rooms. Freestanding UVGI or combination HEPA/UVGI and recessed (ceiling cavity) UVGI all provide some level of "equivalent" ACH of ventilation. Strictly speaking, ventilation implies supplying outdoor air. Because these units contain an air-moving device, they in effect provide ACH in addition to those of the mechanical ventilation system because the air is presumed disinfected and, therefore, in theory, as fresh as outdoor air.

Critics of combination HEPA/UVGI units claim that there is no need for both HEPA filtration and UVGI capability within the same device. If both were equally effective, that claim would be true. However, combination units are built primarily as HEPA filtration units and typically provide only a fraction of the UV intensity of dedicated UVGI units. If the filter fails, the effectiveness of the UVGI alone may be minimal. Although combined units offer the advantage of redundancy (if one technology fails, the other is still present), their incremental effectiveness is probably low.

2.3.8 Health Effects of Ultraviolet Radiation Exposure

The adverse effects of human exposure to UV light have been well documented. Prolonged exposure to UV radiation causes several dermatologic reactions. Prolonged exposures cause a delayed (by 2–10 h) erythema, similar to a sunburn. Some subjects may experience photosensitivity reactions, abnormal skin reactions to UV light in the presence of certain chemical agents. Direct phototoxins—substances directly activated in the skin by exposure to UV light—include coal tar, colognes, lipsticks, hair preparations, psoralens, and sunscreens. Photoallergic reactions, resulting from immune responses to radiation-induced chemical reactions in the skin, can occur with some salicylanilides and antibiotics, hexachlorophene, cosmetics, and colognes. Urticaria (hives) and eczema (scaly skin reactions) have also been reported.

UV-C and UV-B of wavelengths less than 300 nm are highly absorbed by the eye's cornea; those of wavelengths greater than 300 nm are increasingly absorbed by the lens. The result is dose-dependent photokeratitis (inflammation of the cornea) and conjunctivitis (inflammation of the outer epithelial lining of the eye). Signs and symptoms include pain, scratchy eyes, blepharospasm (uncontrollable blinking), tearing, congestion of the conjunctiva, photophobia, blurry vision, and occasionally blepharitis (red, swollen eyelids). Although these effects may be incapacitating for up to 2 days, permanent damage is rare.

Due to these and other adverse health effects of UV exposure, the American Conference of Governmental Industrial Hygienists (42), NIOSH (43) and the International Radiation Protection Association (44) recommend occupational exposure limits for wavelengths between 180 and 400 nm. ACGIH TLVs are given as a measure of irradiance and radiant exposure. Irradiance is the level of radiation arriving at a point in space, the quotient of the radiant power and the cross-sectional area, or time-averaged energy flow. Radiant exposure is the change of the incremental radiant energy in a given cross-sectional area, the energy dose of radiant energy. Irradiance is measured in watts per square meter (W/m^2), and radiant exposure in joules per square meter (J/m^2). The ACGIH TLVs for radiant exposure at 254 nm are 60 J/m^2 (6.0 mJ/cm^2). The product of a TLV radiant exposure at a specified wavelength and its relative spectral effectiveness (RSE) is the limiting dose. RSE is a hazard-weighting function indicating the biologic efficacy of that wavelength. The use of RSEs allows irradiance data to be weighted to the limiting TLV at 270 nm of 30 J/m^2, which is based on the potential for photokeratitis.

The NIOSH recommended exposure limit (REL) requires that the measured irradiance level be multiplied by the RSE at the specific wavelength to obtain the effective irradiance. The maximum permissible exposure time can then be calculated (43). Neither the REL nor the TLV exposure limits should be used to determine acceptable exposure of photosensitive individuals or regarded as protective for individuals lacking a crystalline lens.

Qualified, experienced personnel should measure irradiance levels using a radiometer. Before measurement, personnel should determine source characteristics, operator-source interaction, and the location of the source relative to reflective objects such as aluminum surfaces. Should the measurement data warrant use of PPE for the eyes, qualified, experienced personnel can determine which protective eyewear is compatible with the spectral distribution and intensity of the UV light source.

2.4 Personal Protective Equipment

The use of PPE received unprecedented attention with the advent of the AIDS epidemic and increased incidence of tuberculosis in the 1980s. Attention was focused on proper selection and adequacy of challenge test methods for biological hazards. This was particularly true of barrier clothing, including gloves to prevent nonintact skin exposure to blood-borne pathogens and RPE to prevent inhalation exposure to MTB.

2.4.1 Personal Protective Clothing

There is no single comprehensive federal regulation governing the use of PPE for biohazard protection. Numerous OSHA regulations set minimum standards for PPE usage (e.g., 29 CFR 1910.133, eye and face protection) that might be required to prevent mucous membrane exposure to a biohazard (such as blood that is potentially infected with HIV). The fact that biohazards often coexist with physical and chemical hazards must be taken into account to properly select PPE. In addition, there are guidelines pertinent to the use of PPE in specific workplaces, such as the CDC/NIH guidelines in biomedical and microbiological laboratories summarized in Tables 27.1 and 27.2. Additional examples include guidelines applicable to recombinant DNA laboratories, clean-room facilities, dental facilities, emergency medical operations, and hospitals and ambulatory care facilities.

The purpose of this section is to acquaint the reader with specific issues regarding the use of PPE and clothing used for protection against biohazards, which have important performance characteristics that distinguish them from other forms of PPE and clothing. The intended purpose of the PPE dictates the manufacturer's design criteria. Examples of design criteria are as follows:

- Interruption of transmission of infectious agents, for example, PPE worn by health care veterinary employees and public safety workers who may have direct contact with infectious tissues or bodily fluids
- Prevention of contamination of special environments or products from endogenous biohazardous agents and clothing lint, for example, food service, pharmaceutical, biotechnology, and industries involving clean-room technology
- Attenuation of the effects of exposure if a biohazard penetrates the primary barrier, for example, protective undergarment apparel, glove liners, or fabric treatments that may reduce infection during war, such as those applied to certain military uniform components
- Preservation of the human body or inhibition of microbial growth, for example, outer space garments or barrier materials with surface treatments to reduce microbial growth or odor
- Repellent capability against arthropods, for example, pesticide surface treatment of barrier materials used by outdoor sportsmen and soldiers (1).

Laboratory and health care employees typically use various disposable or reusable fabric options to provide general body protection from daily soiling and splattering. Various fabrics (e.g., woven, nonwoven) and surface treatments (e.g., for lipid resistance and antistatic qualities) are available. The efficiency of these fabrics in preventing liquid penetration or "strike-thru," is variable. In general, woven materials and single-layer materials provide the least amount of strike-thru protection as opposed to reinforced or layered materials (1). Reinforced fabrics with film or polymer coatings provide the highest amount of strike-thru protection but do not allow for vapor and heat transfer.

There are standard test methods for determining the permeability resistance of clothing materials, such as air permeability, water repellency, and water resistance. Blood more readily penetrates fabrics because it has a lower surface tension than water. Thus, because test methods involving water do not accurately portray the strike-thru involving pathogenic body fluids, biopenetration test methods have been developed. There are numerous test methods for glove performance, including bacterial and viral challenge tests. For an in-depth discussion of these methods, the reader should consult "Glove Performance Measurements and Microbial Permeation and Penetration of Gloves" in the AIHA *Biosafety Reference Manual* (1).

2.4.2 Respiratory Protective Equipment

Industrial hygienists often use RPE to control occupational exposure. Industrial hygienists especially understand that surgical masks and respiratory protective equipment are fundamentally different, a fact sometimes misunderstood by health care professionals. Surgical masks provide a barrier only for splashes and splatter. The mask material filters large particles (>5 μm) fairly well. However, because of their generally poor fit to the face, surgical masks provide little protection against MTB. Small particles can easily travel with the airstream around the nonsealing periphery of the mask into the breathing zone. For example, under experimental conditions, Tuomi (45) showed that 5% and 45% of 5.5-μm particles passed through sealed and unsealed masks, respectively. Notably, 94% and 98%, respectively, of 1-μm particles passed through sealed and unsealed masks (45). Mask wearers, especially those with poorly fitting masks, are thus at substantial risk of exposure to 1- to 5-μm MTB droplet nuclei.

All respirators are assigned a protection factor (assigned protection factor, APF) depending on the type of respirator and respiratory inlet covering. APF is the expected workplace level of respiratory protection that would be provided by a properly functioning respirator or a class of respirators to properly fitted and trained users (45). APF is calculated as $C_I = C_A/APF$, where C_I is the contaminant concentration in inspired (respirator) air and C_A is the contaminant concentration in ambient air. APF is essentially the magnitude of reduction of airborne contaminant concentration provided by respirators; the higher the APF, the higher the protection of the wearer. All respirators of the same class are given the same APF value; for example, the APF for an approved disposable air-purifying half mask HEPA respirator is 10. The APF is also 10 for a nondisposable elastomeric half mask air-purifying respirator equipped with dust/mist filters or with HEPA filters. Such respirators would be ineffective against an MTB airborne concentration of $200/m^3$ because the concentration of MTB in respired air would be $20/m^3$ or 2 to 10 times the infectious dose.

Health care providers, CDC, NIOSH, and health and safety professionals have engaged in intense debate since CDC issued its recommendation for a disposable particulate respirator (46). That recommendation was based on the particle size distribution of MTB aerosols and the filtration efficiency of air-purifying respirators. Both the 1994 CDC guidelines and the OSHA proposed rule (2) require negative pressure air-purifying or more protective respirators capable of the following:

HEPA or N95 filtration

1. The ability to be qualitatively fit tested in a reliable way to obtain a face seal leakage of no more than 10% for most workers
2. The ability to be fit checked by the wearer each time he or she puts on the respirator

The proposed OSHA rule for occupational exposure to tuberculosis does not specifically stipulate any of the preceding requirements.

Implied in these recommendations is that a total leakage of 15% (5% filter leakage and 10% face seal leakage) is acceptable. However, when considering the potency of MTB infection, this level of leakage is most likely unsafe. If the ambient airborne concentration were 100 MTB/m^3, then the concentration in inhaled air would be 100/6.7 or 15 MTB/m^3, which is clearly unacceptable because the infectious dose will be exceeded in less than 1 h. The most obvious flaw is the assumption that a respirator with a low APF value (6.7) is suitable in a wide variety of situations involving MTB exposure. The recommendations also imply that APFs of 50, 100, and 1,000 are not needed, an assumption that is not warranted.

The debate over the minimum required level of respiratory protection was so intense that NIOSH responded with revised 42 CFR Part 84 regulations on testing and certification of respirators to account for the CDC criteria just described. NIOSH plans to revise its regulations in three modules, including APFs, which will be addressed at a later date.

In the revised 42 CFR Part 84, NIOSH created three series of particulate filters (Types N, R, and P) with three filter efficiencies for each series (95%, 99%, and 100%). Type N filters may be used against solid particles in workplaces free of oil aerosols, whereas Type R and Type P filters may be used against both solid and liquid (including oil) aerosols. Type N filters are tested against a sodium (NaCl) aerosol, whereas Type R and P filters are tested against a DOP aerosol. The Type R and P filters differ in that the latter filters are more rigorously tested. Both the NaCl and DOP challenge aerosols have a mass median aerodynamic diameter of approximately 0.3 μm. Efficiencies of 95%, 99%, and 100% indicate that the filters are able to remove at least, 95%, 99%, and 99.7%, respectively, of the challenge aerosol particles. A labeled filter efficiency of 100% is misleading because the filter need not remove every particle to pass the test. In its introduction to the new regulations, NIOSH stated that all nine classes of particulate filter respirators meet or exceed the respirator-performance criteria in the CDC guidelines for preventing TB transmission. Interestingly, NIOSH did not require quantitative determinations of the amount of face seal leakage as part of the respirator certification process; instead, NIOSH presumed that if a health care worker wearing a Type N95 respirator passed a qualitative fit test, the worker would experience less than 10% face seal leakage. At present, it appears that many health care institutions may use disposable, filtering-respirator Type N95 respirators against MTB aerosols. A filtering respirator is a one-piece disposable device similar in appearance to a surgical mask. This author's experience is that disposable respirators cannot be reliably fit tested, resulting in poor protection against MTB.

A subcommittee of the American National Standards Institute (ANSI) Z88 committee on respiratory protection (Z88.12) has been charged with the task of developing a standard for the selection and use of respirators for protection against infectious aerosols. ANSI standards are consensus documents that involve due process requirements of openness and balance, ensuring that the process is not dominated by any single interest category.

The ANSI subcommittee has used a risk-based approach despite the fact that air-sampling data as well as a legally defined acceptable risk level are unavailable. This risk-based approach relies in part on a modification of the traditional Poisson probability model for airborne MTB exposure and infection, also known as the Wells/Riley equation (47). The user employer can consult tables based on infectious dose, anticipated exposure concentration, and a personally acceptable level of risk and choose a class of respirators (represented by APF) based on those parameters. Although it may take several years to finalize this standard, such a standard may take less time than resorting to air-sampling data currently considered infeasible. The legal determination of an "acceptable risk level" may take even longer still because OSHA has yet to address the issue. CDC has also avoided the issue of acceptable risk, stating only that the infection-control staff must individually estimate risk and may, based on this assessment or best judgment, determine that respiratory protection exceeding their criteria may be considered appropriate (28).

A common misconception is that respirators should be used to the exclusion of other controls, including engineering controls. In other words, a risk manager may choose a single strategy (e.g., dilution ventilation, UVGI, HEPA filtration, or RPE) when none of these controls used singly will likely prevent transmission of occupational TB. The limitations of each method described in this chapter argue that a combination of controls must be used.

With the widespread use of the infection-control practices of Universal Precautions and Body Substance Isolation, the health care industry has become accustomed to the use of disposable supplies. Disposable negative pressure air-purifying respirators may therefore be appealing to the health care industry and others. Disposable respirators are advantageous in that they are less expensive than elastomeric half-mask respirators, and they do not require a scheduled maintenance program as do other respirators. However, disposable respirators have several disadvantages. Most disposable respirators are reused until the development of excessive air resistance or physical damage. The number of times these devices may be safely reused has not been determined. At issue are the sanitary condition of the respirator and maintenance of the respirator's physical integrity. Disposable respirators cannot be chemically sanitized or disinfected like nondisposable respirators. Thus, even for an individual user, good hygiene cannot be maintained. All respirators, including disposable ones, must be stored in a manner that prevents distortion, contamination, and damage due to physical agents such as vibration, electric shocks, sunlight, cold, and moisture. Though such storage is possible with a disposable respirator, it is likely that disposable respirators would be handled with less care.

2.5 Infectious Waste Management

2.5.1 Overview

The assumption that biohazardous waste is infectious is just that, an assumption. However, proving otherwise is not feasible. Therefore, "infectious" waste management follows three basic steps: identify biohazard waste streams, treat them, and dispose of them. Identifying waste streams is the first step (see section 2.5.2). Every effort should be made to minimize biohazardous waste to prevent potential occupational exposure during handling, to control costs, and to protect the environment. Segregation of waste streams will serve to minimize waste and maintain the separate integrity of each waste stream. On-site treatment (i.e., decontamination) before disposal is much preferred and is typically accomplished through autoclaving or chemical disinfection (see section 2.5.3). Spills of biohazardous material are a special case requiring decontamination (see section 2.5.4). It is necessary to collect and store waste properly. For example, pathological waste should be refrigerated before disposal. Another example is the "red bag" requirements in the OSHA blood-borne pathogen rule that govern containment and temporary storage of blood and other potentially infectious material as defined by the standard. Following on-site treatment, formerly biohazardous waste can usually enter normal waste streams for liquids (sewer) or solids (landfill or incineration). If on-site treatment is not permissible due to local ordinances for various institutions (e.g., health care facilities), then it must go to a landfill or be incinerated. The resulting costs are much higher than for nonbiohazardous waste, and compliance with interstate transportation regulations must be met when applicable (see section 2.5.5). Disposal of biohazardous waste is currently mandated through individual state and local regulations.

The primary responsibility for biohazardous waste resides with the generator. Management of biohazardous waste must be done in a systematic manner by dedicated staff. An effective biohazardous waste management program includes the following elements:

- Waste stream identification
- Segregation

- Packaging
- Storage
- Transport and handling
- Treatment techniques
- Disposal
- Contingency planning
- Staff training

2.5.2 Identification of Biohazardous Waste

This author recommends that the combined EPA recommended infectious waste and optional infectious waste categories be used as working categories of biohazardous waste due to the assumption of infectiousness (48):

- Isolation wastes
- Cultures and stocks of infectious agents and associated biologicals (e.g., specimens from medical and pathology laboratories; cultures and stocks of infectious agents from clinical, research, and industrial laboratories; disposable culture dishes; and devices used to transfer, inoculate, and mix cultures)
- Human blood and blood products (e.g., waste blood, serum, plasma, and blood products)
- Pathological wastes (e.g., tissues, organs, body parts, blood, and body fluids removed during surgery, autopsy, and biopsy)
- Sharps (e.g., contaminated hypodermic needles, syringes, scalpel blades, Pasteur pipettes, and broken glass)
- Animal carcasses, body parts, and animal bedding
- Wastes from surgery and autopsy (e.g., soiled dressings, sponges, drapes, lavage tubes, drainage sets, underpads, and surgical gloves)
- Miscellaneous laboratory wastes (e.g., specimen containers, slides and cover slips, disposable gloves, lab coats, and aprons)
- Dialysis unit wastes (e.g., tubing, filters, disposable sheets, towels, gloves, aprons, and lab coats)

2.5.3 Decontamination

Decontamination is the treatment process whereby the number of microorganisms is reduced to an acceptable level. Chemical decontamination is critical in health care settings to prevent nosocomial infection and in many occupational settings to prevent employee infection. Antimicrobial or biocidal activity is complicated by the fact that we do not thoroughly understand all of the mechanisms of action (e.g., against fungi, viruses, and protozoa) or the means by which microorganisms become resistant (49). Antimicrobial activity is also affected by other factors, including, but not limited to, formulation effects, presence of organic material (e.g., blood), temperature, and dilution.

Chemical sanitizers, disinfectants, and sterilants are used to obtain different levels of reduction of microorganisms. They are classified by their active ingredient. Disinfection inactivates virtually all recognized pathogenic microorganisms but not necessarily all microbial forms (e.g., bacterial endospores) on inanimate objects. The differences between disinfectants and antiseptics is that antiseptics inhibit or destroy microorganisms on or in living tissue. Sterilization is the use of a physical or chemical procedure to destroy all microbial life, including highly resistant bacterial endospores.

The effectiveness of a disinfectant depends on the following factors:

- The kind and number of contaminating microorganisms
- The concentration of the chemical
- The length of exposure to the chemical
- The amount of organic matter (e.g., soil, feces, blood)
- The type and condition of the material to be disinfected

A wide variety of disinfectants is available. Table 27.4 describes the degree of decontamination and target microorganisms of some groups of chemicals used.

Among the most common disinfectants used in occupational settings are sodium hypochlorite (bleach), isopropyl and ethyl alcohol, iodophors (e.g., Wescodyne®), and phenolics (e.g., Lysol® and Amphyl®). The OSHA blood-borne pathogen rule, which applies to health care settings, requires only that "an appropriate disinfectant" be used to disinfect against blood contamination. There are many blood-borne pathogens, including HIV, the Creutzfeldt-Jakob prion, and HBV; HBV is more resistant to disinfection than HIV, and Creutzfeldt-Jakob prion requires extreme decontamination procedures (see below). There is no single disinfectant appropriate for all potential bloodborne pathogens (see also section 3.1).

The OSHA compliance document for blood-borne pathogens requires that an EPA-registered disinfectant be used. Presumably efficacious disinfectants might not be used because manufacturers had not sought registration. EPA regulates and registers antimicrobials under the Federal Insecticide, Fungicide, and Rodenticide Act (FIFRA). Lists of EPA-registered antimicrobials are available as follows:

- List A: products registered as sterilants
- List B: products labeled as tuberculocides
- List C: products labeled for use against HIV-1.
- List D: products labeled as effective against HBV and HIV

Registration does not constitute approval, although the manufacturer is required to submit efficacy data using standardized test methods to support their product claims, and EPA reviews these data. These lists should be reviewed to determine whether the disinfectants an institution currently uses are EPA registered. For example, isopropyl and ethyl alcohol are widely used to disinfect bench tops in laboratories such as those in which blood might be used. Neither is an EPA-registered disinfectant. Viruhol®, which is 70% isopropyl alcohol, is registered by EPA. It contains an excipient that reduces the evaporation time

Table 27.4. Effectiveness of Decontamination Chemicals

Chemical(s)	Degree of Decontamination	Target Microorganisms
Alcohols	Antisepsis and disinfection, especially of hard surfaces	Vegetative bacteria (including MTB [49]), lipid viruses and hydrophilic viruses to a lesser extent, fungi (49), HIV and HBV to a lesser extent
Glutaraldehyde	Disinfection (especially low temperature) and sterilization	Vegetative bacteria (including MTB) and their spores, lipid and hydrophilic viruses, HIV, HBV, fungi (49)
Formaldehyde	Disinfection and sterilization in liquid or in combination with low-temperature steam	Vegetative bacteria (including MTB) and their spores to a lesser extent, lipid and hydrophilic viruses, HIV, HBV
Halogen-releasing agents: (*1*) Sodium hypochlorite, (*2*) Iodine and iodophors	(*1*) Disinfection of hard surfaces, (*2*) disinfection and antisepsis	(*1*) Bacteria, spores, viruses; (*2*) vegetative bacteria (including MTB), bacterial spores to a lesser extent, fungi, lipid viruses and hydrophilic viruses to a lesser extent, HIV and HBV to a lesser extent
Phenols	Antisepsis, disinfection	Vegetative bacteria (including MTB), lipid viruses, HIV and HBV to a lesser extent
Quaternary ammonium compounds	Antisepsis, disinfection	Vegetative bacteria, lipid-enveloped viruses, HIV
Vapor-phase sterilant: ethylene oxide	Sterilization	Vegetative bacteria (including MTB) and their spores, lipid and hydrophilic viruses, HIV, HBV

HBV: Hepatitis B virus
HIV: Human immunodeficiency virus
MTB: Mycobacterium tuberculosis

of the alcohol and significantly increases product efficacy. Clorox® is the only sodium hypochlorite (bleach) product that is an EPA-registered disinfectant. The reader can obtain a copy of the EPA lists from the National Antimicrobial Information Network at http://ace.orst.edu/info/nain for a nominal charge.

Disinfectants and sterilants are themselves chemical hazards with their own properties, including lower and upper explosive limits and toxic effects. OSHA substance-specific regulation exists for formaldehyde, 29 CFR 1910.1048, and for ethylene oxide, CFR 1910.1047.

Dry heat and moist heat can also be used to decontaminate materials. Dry heat at 171°C for 1 h, 160°C for 2 h, or 121°C for 16 h or longer is required for sterilization (50). Steam autoclaving (moist heat) at 121°C at 15 psi for ≤60 min for gravity-displacement autoclaves or at 132°C at 27 psi for ≤10 min for vacuum-type autoclaves is required for sterilization. Autoclaving is the preferred method of decontaminating cultures, laboratory glassware, pipettes, syringes, and other small items contaminated with infectious agents. Animal carcasses, human body parts, and large volumes of contaminated clothing cannot be effectively decontaminated by autoclaving. Rather, incineration is the preferred method. When autoclaving is used, strict adherence to proper loading and operation procedures is necessary to ensure decontamination. In addition, at least one state (California) has legal requirements applying to the use of autoclaves.

Note that validation of measuring devices (e.g., thermometers, pressure gauges, and timers) is required. The accuracy of these must be certified independently when the autoclave is initially installed and before waste is treated, and annually thereafter. Careful attention must be paid to loading autoclaves as it can affect efficacy of decontamination. Loading should be done in such a manner as to allow for maximum steam penetration. The California regulation cited earlier requires that spore strips (e.g., *Bacillus stearothermophilus*) be used to determine whether in fact decontamination is achieved. The spore strip should be placed at the center of the largest pack in the largest load or in an area of the load that is least likely to reach sterilization temperature. Chemical indicators such as heat-sensitive tape or indicator inks on waste packaging materials provide a color change when exposed to the design temperature. These provide a visual indication that materials reached an indicated temperature only and do not provide verification of decontamination.

For more information on decontamination, the reader is urged to consult the latest edition of *Disinfection, Sterilization and Preservation*, edited by S. S. Block.

2.5.4 Spill Management

Despite the use of safe work practices, safety equipment, engineering controls, and protective clothing, accidental spills can occur, resulting in employee infection. Spills constitute a potential skin contact and, more importantly, an inhalation hazard. An important part of a biosafety program is having established spill cleanup procedures for infectious materials (i.e., those agents requiring BSL-2 or higher containment) to prevent exposure. The following is a recommended stepwise procedure:

- Evacuate the room immediately
- Close the doors
- Remove and decontaminate clothing
- Notify the BSO and PI
- Allow sufficient time for sedimentation and/or removal of aerosols generated during the spill by the mechanical ventilation system (discussed further later in this section)
- Reenter wearing appropriate PPE, including RPE
- Clean up the spill (discussed further later in this section)
- Wash or shower

The amount of time required for settling of aerosols produced during spills is dependent on the particle size, the amount of aerosol generated, the degree of air mixing within the room, and the number of room ACH supplied by the mechanical HVAC system design and operation. The amount of aerosol generated will probably not be known. The degree of mixing in the room may not be known. To understand the principle of air mixing and its effect on airborne concentration, the reader is encouraged to consult an in-depth discussion of air mixing and removal of MTB from the air by dilution ventilation (23). The recommendation of waiting 30 min for settling to occur (51) appears somewhat arbitrary.

Spill kits are commercially available specifically for biohazards. Whether an institution purchases preassembled spill kits or assembles them on-site, they must be located in convenient locations, clearly evident by signage, and continuously stocked. The spill should be surrounded with absorbent material and cleaned from the outer edge inward to prevent further spreading beyond the original size of the spill. An appropriate disinfectant should then be applied after absorption and disposal of the absorbent material. Disinfectant should not be poured directly into the spill before absorption because doing so will result in aerosolization of the biohazard.

Spills within BSCs are already contained within a HEPA exhaust filtered cabinet. A disinfectant appropriate for the agent should be applied to clean up the spill while the BSC is running. Vapors from disinfectants that are organic solvents can reach high concentrations within the BSC. This is an important consideration when the BSC is either a Class IIA or Class IIB1 cabinet with recirculation of cabinet air.

2.5.5 Shipping and Transportation of Biological Materials

Many regulations apply to the shipment and transportation of biohazardous materials in order to anticipate and prevent accidents during transport and prevent harm to individuals and the environment (52). All biohazardous materials leaving facilities must meet all applicable requirements.

To ship biological agents safely, the following steps must be followed:

- Biohazardous materials must be identified and categorized correctly.
- Proper packaging of the agents must be completed.
- Packages must include warning labels and proper documentation.
- Anyone involved in the shipping of biohazardous materials must be trained properly (52).

Numerous regulatory agencies regulate the proper transportation of biohazardous materials. The U.S. Public Health Service (PHS) and Department of Transportation (DOT) govern interstate transportation of biohazardous materials by rail, air, vessel, and public highway. The regulations of the International Air Transport Association (IATA) and the International Civil Aviation Organization (ICAO) also apply when biohazardous materials are shipped by air. An employee who uses a commercial carrier or personally carries a package in his or her own car, a company vehicle, a taxi, or a bus is responsible for ensuring that all legal requirements are met for packaging, labeling, and documentation. The type of packaging, labeling, and documentation required depends on the type of biohazardous material being shipped and how it is to be shipped (i.e., the mode of transportation and the temperature requirements of the materials). Specific DOT, PHS, and IATA/ICAO packaging and labeling requirements are not provided in this chapter. DOT and PHS regulations can be found in the DOT Hazardous Materials Regulations (49 CFR Parts 171–180) and Interstate Shipment of Etiologic Agents (42 CFR Part 72), respectively. IATA regulations can be found in the IATA Dangerous Goods Regulations manual. The United States Postal Service also regulates the transport of biological agents. Importation permits must be obtained from the CDC for any infectious agents affecting humans that are imported into the United States. Certain animal pathogens require permits from the U.S. Department of Agriculture for importation and interstate shipment. Shipping carriers should also be consulted and qualified by the waste generator before shipping any biological materials.

It is important to understand when various regulations or agency requirements apply. Infectious substances are biohazardous, whereas diagnostic specimens that are sent to a laboratory for analysis are potentially biohazardous. Consequently, the requirements for their shipping and transportation are different.

The reader should also be aware that there are specific legal requirements for biological products as well. These should not be confused with those for infectious substances or diagnostic specimens. A biological product is a material prepared and manufactured in accordance with the provisions of the following regulations and that may be shipped in interstate commerce:

9 CFR Part 102: Licenses for biological products

9 CFR Part 103: Experimental production, distribution, and evaluation of biological products prior to licensing

9 CFR Part 104: Permits for biological products

21 CFR Part 312: Investigational new drug application

21 CFR Parts 600–680: Biological products

2.6 Laboratory Design

The facility-design requirements described by CDC are more performance oriented than prescription oriented. In general, those requirements are given in Table 27.1for BSL-1 through BSL-4 laboratories. The overall traffic pattern for employees and materials must be considered. Of critical importance is ventilation to allow for directional airflow and no recirculation when required, and sufficient supply air for BSCs. Building exhaust should not be placed upwind of air intakes. The need for proper maintenance of the ventilation system and monitoring of ventilation characteristics (e.g., smoke tube for directional airflow) cannot be overstated. The following needs must also be addressed in the design phase: electrical, sewage, vacuum, waste storage and handling, safety equipment (e.g., BSCs, autoclaves), laundry, and other storage areas. More in-depth information is beyond the scope of this text. The reader is urged to consult other references. One such reference is *Anthology of Biosafety—I. Perspectives on Laboratory Design*, published by the American Biological Safety Association (53).

3 REGULATIONS

OSHA has promulgated one rule to protect employees against a biohazard and has issued a second proposed rule. Both are discussed next.

3.1 OSHA Occupational Exposure to Blood-borne Pathogens

The OSHA blood-borne pathogens rule (29 CFR 1910.1030) became effective in 1992. The rule is principally intended to protect health care employees from exposure to blood-borne pathogens. However, the rule applies to all employees with exposure or potential exposure, including those in clinical laboratories and HIV or HBV production facilities. Blood-borne pathogens include, but are not limited to, HBV and HIV. As defined by the standard, blood-borne pathogens are pathogenic microorganisms that are present in human blood. Other blood-borne pathogens recognized by biosafety professionals as occupational biohazards are often not readily recognized as requiring blood-borne pathogen program elements outside of the health care industry. One such example is Creutzfeldt-Jakob disease prion (see section 4.2 of Chapter 26 for an in-depth discussion). Others include:

- *Treponema pallidum*: This spirochete is the etiological agent of syphilis. Cases of transmission by needle stick, by tattooing instruments, and by blood transfusion have been documented.
- *Plasmodium*: This genus includes the etiologic agents of malaria. It has been transmitted by needle stick and transfusion.
- *Babesia microti*: Transmission by transfusion of fresh blood from asymptomatic donors has been reported.
- *Brucella* species: Etiologic agents of brucellosis. Transmission by blood transfusion has been reported.
- *Leptospira interrogans*: No cases of transmission by blood have been reported.
- Human T-lymphotropic virus type I (HTLV-1): There is some evidence that neurologic disease may be associated with blood transfusion.
- Hemorrhagic fever viruses (e.g., Lassa, Marburg, Ebola, Crimean-Congo): These viruses are not indigenous to the U.S. Blood and other body fluids of infected people are considered infectious.

The principal element of the blood-borne pathogen rule is the written exposure control plan. This plan must contain the following: the schedule and method of implementation of engineering and work-practice controls that will achieve compliance, hepatitis B vaccination, hazard communication to employees, and record keeping. The hallmark of work-practice controls is the required use of infection-control procedures known as universal precautions, the underlying assumption of which is that all human blood and most human body fluids are treated as if known to be infectious for HIV, HBV, and other blood-borne pathogens. Universal precautions are intended to prevent parenteral, mucous membrane, and nonintact skin exposures.

3.2 OSHA-Proposed Tuberculosis Rule

OSHA published the *Occupational Exposure to Tuberculosis: Proposed Rule* in the Federal Register in October 1997. Rule-making public hearings have subsequently been held. The proposed rule applies to occupational MTB exposure occurring in hospitals, long-term-care facilities, correctional facilities, hospices, homeless shelters, drug abuse treatment facilities, and laboratories that handle specimens potentially containing MTB or perform activities potentially resulting in MTB aerosolization. Key elements of the rule include the following requirements: employee exposure determination, written exposure-control plans, work-practice and engineering controls, clinical and research laboratories, respiratory protection, medical surveillance, training, labeling, and record keeping. Issues still being discussed include requirements for two-step versus one-step skin testing, laboratory biosafety, respirator fit testing, and monitoring pressure relationships in isolation rooms. Until the final rule is promulgated, OSHA can issue citations under existing OSHA rules, including the respiratory protection rule (29 CFR 1910.134), accident-prevention signs and tags (29 CFR 1920.145), access to employee and medical records (29 CFR 1910.20), and the OSHA log and summary of occupational injuries and illnesses (29 CFR 1904).

The Tuberculosis Task Force of AIHA has provided OSHA hearing testimony and posthearing comments supporting all of these requirements. AIHA's white paper and position statement is available on the Internet at http://www.aiha.org.

Many recognize that the CDC (13) guidelines are more stringent than the OSHA proposed rule. CDC guidelines are guidelines only and are not legally enforceable. At least one study demonstrated that voluntary compliance with CDC guidelines is not widespread (54). In this study, 28% (7/25) of isolation rooms were under positive pressure; air moved toward, rather than away from, likely worker locations in almost all (26/27) rooms tested, and none of the hospitals regularly checked the performance of engineering controls. Although 83% (20/24) of the isolation rooms tested had six or more nominal ACH, supply air did not mix rapidly. Thus, the nominal ACH overestimated the effective ACH and the protection provided. Voluntary compliance is clearly lacking.

Inspections are, however, done and citations given in the absence of a promulgated standard. Inspections are conducted for occupational exposure to TB only in response to employee complaints or as a part of industrial hygiene inspections in five types of facilities identified by CDC as having higher incidence of TB than the general population. Presently, health care settings, one of the facility types identified, do not include private physicians' offices and dental offices. Citations have been issued under the general duty clause of the 1970 Occupational Health and Safety Act where employees had potential exposure to the exhaled air of a confirmed or suspected case of TB or were exposed during a high-hazard procedure performed on an individual who may have had TB. High-hazard procedures include aerosolized medication treatment, bronchoscopy, sputum induction, endotracheal intubation, suctioning procedures, and autopsies. The presence of the hazard is the basis of a general duty clause citation. A violation of the general duty clause requires proof that the employer failed to keep the workplace free of a hazard to which his or her employees were exposed, that the hazard was recognized, that the hazard was causing or likely to cause death or serious physical harm, and that a feasible and useful method to abate the hazard existed.

The following is intended to guide the biosafety professional to prepare for inspection. An employer's TB-control program (or generic infection-control program) can serve as evidence of knowledge and recognition of the general duty clause. Feasible and useful TB control methods include the following:

1. *A protocol for the early identification of individuals with active TB*: Anyone with symptoms of TB (including productive cough, coughing up blood, weight loss, loss of appetite, lethargy/weakness, night sweats, or fevers) should have diagnostic testing and close medical follow-up.
2. *Training and information*: Employees must ensure employee knowledge of the hazard of TB transmission, its signs and symptoms, medical surveillance, therapy, and site-specific protocols including the purpose and proper use of exposure controls. In addition, the training requirement of the OSHA standard for the selection and use of RPE (29 CFR 1910.134) can be cited for absent or inadequate respirator training.
3. *Medical surveillance at no cost to the employee*: Employers should offer preplacement evaluation, administration and interpretation of skin tests, as well as periodic employee testing. All employees should undergo a baseline skin test. Workers potentially exposed to TB due to the performance of high-risk procedures should have biannual tests. Employees who are definitely exposed to an individual with infectious TB should be managed according to CDC recommendations (13).
4. *Postexposure evaluation and follow-up*: Employees with a positive skin test or a history of positive skin tests who are exhibiting symptoms of TB must be evaluated and managed. Proper management includes work restrictions for infectious employees.
5. *Respiratory isolation rooms and special procedure areas*: Individuals with suspected or confirmed TB must be placed in respiratory isolation rooms in the hospital. Such rooms or areas where high-hazard procedures are performed must be maintained under negative pressure and be exhausted either directly to the outside away from intake vents or through properly designed, installed, and maintained HEPA filters. Despite not knowing the efficacy of available engineering controls, OSHA instructs compliance officers to encourage the employer to follow the CDC guidelines with respect to local exhaust and UV lights.

4 SUMMARY

The need for industrial hygienists and other health and safety professionals to recognize specific occupational biohazards across a variety of work settings is coupled with the need to effectively control occupational exposure to biohazards. Because many of the guidelines for biosafety apply to laboratory and health care settings, control techniques in these settings were presented in this chapter. The key to occupational biosafety is containment of biological hazards. Containment is achieved through the use of work practices and techniques (including administrative controls), primary barriers (including engineering controls and personal protective equipment), and facility design.

Biosafety is a specialized practice area of industrial hygiene. It combines knowledge from industrial hygiene, microbiology, infectious disease, and other disciplines.

BIBLIOGRAPHY

1. American Industrial Hygiene Association (AIHA), *Biosafety Reference Manual*, 2nd ed., AIHA Biosafety Committee, Fairfax, VA, 1995.

2. "OSHA Occupational Exposure to Tuberculosis: Proposed Rule 62:54159–54309," *Federal Register*, (1997), Vol. 62 no. 201 (Oct. 17, 1997).
3. J. Dutkiewicz, L. Jablonski, and S. A. Olenchock, "Occupational Biohazards: A Review," *Am. J. Ind. Med.* **14**, 605–623 (1988).
4. Centers for Disease Control (CDC), *Classification of Etiologic Agents on the Basis of Hazard*, 4th ed., CDC Office of Biosafety, U.S. Department of Health, Education and Welfare, Public Health Service, 1974, Atlanta, GA.
5. CDC, *Biosafety in Microbiological and Biomedical Laboratories*, 4th ed., U.S. Department of Health and Human Services, National Institutes of Health, 017-040-00547-4, 1999, Washington, DC.
6. "Guidelines for Research Involving Recombinant DNA Molecules," *Federal Register* **62**, no. 49 (March 1996), Vol. 62 No. 49 (March 1996).
7. "Guidelines for Research Involving Recombinant DNA Molecules, 51:16958–16968." *Federal Register* (1986), Vol. 51, No. 88 (May 7, 1986).
8. NIH, *Guide for the Care and Use of Laboratory Animals*, DHHS Pub. No. NIH 86-23, U.S. Government Printing Office, Washington, DC, 1985.
9. Code of Federal Regulations, "Laboratory Animal Welfare Regulations (9 CFR)," Title 9, Parts 1–3, Subchapter A, 1992.
10. G. T. Tzotzos, ed., *Genetically Modified Organisms. A Guide to Biosafety*, United Nations Industrial Development Organization, 1995, Wallingford: CAB International.
11. National Research Council (NRC), *Biosafety in the Laboratory. Prudent Practices for the Handling and Disposal of Infectious Materials*, Committee on Hazardous Biological Substances in the Laboratory, Board on Chemical Sciences and Technology, Commission on Physical Sciences, Mathematics, and Resources, National Research Council, National Academy Press, Washington, DC, 1989.
12. E. L. Larson, "APIC Guideline for Handwashing and Hand Antisepsis in Health Care Settings," *Am. J. Infect. Control* **23**, 251–269 (1995).
13. CDC, *General Recommendations on Immunization: Recommendations of the Advisory Committee on Immunization Practices (ACIP)*, MMWR 1994:43(No. RR-1), U.S. Department of Health and Human Services, Centers for Disease Control and Prevention, Atlanta, GA, 1994.
14. "Recombinant DNA Research Guidelines, 41:27902–27943." *Federal Register* (1976), Vol. 41 No. 131 (July 7, 1976).
15. National Sanitation Foundation (NSF), *Listings of Biohazard Cabinetry Class II Cabinet Certification Field Certifier Accreditation*, NSF International, Ann Arbor, MI, 1996.
16. NSF, *Standard 49 Class II (Laminar Flow) Biohazard Cabinetry*, NSF, Ann Arbor, MI, 1992.
17. R. P. Clark, "Safety Cabinets, Fume Cupboards and Other Containment Systems," in *Bioaerosols Handbook*, ed. Christopher S. Cox and Christopher M. Wathes, CRC Lewis, 1995.
18. L. DiBerardinis et al., *Guidelines for Laboratory Design: Health and Safety Considerations*, 2nd ed., John Wiley & Sons, Inc., New York, 1993.
19. L. A. Taylor, M. S. Barbeito, and G. G. Gremillion, "Paraformaldehyde for Surface Sterilization and Detoxification," *Appl. Microbiol.* **17** (4), 614–618 (1969).
20. K. Munro, J. Lanser, and R. Flower, "A Comparative Study of Methods to Validate Formaldehyde Decontamination of Biological Safety Cabinets," *Appl. Environ. Microbiol.* **65**, 873–876 (1998).
21. S. S. Hamid et al., "Risk of Transmission and Features of Hepatitis C After Needlestick Injuries," *Infect. Control Hosp. Epidemiol.* **20**, 63–64 (1999).

22. P. S. Lurie et al., "Postexposure Prophylaxis After Nonoccupational HIV Exposure: Clinical, Ethical, and Policy Considerations," *JAMA* **280**, 1769–1773 (1998).
23. P. Heinsohn, *Tuberculosis Resource Guide*, prepared for Electric Power Research Institute, Palo Alto, CA, Heinsohn Consultation Services, Pacifica, CA, 1996.
24. P. A. Heinsohn, "Respiratory Tract Infections—Bacteria" in *Micro-organisms in Home and Indoor Work Environments*, Brian Flannigan and Robert Samson, eds., Harwood Academic Publishers GMBH, in press, Amsterdam, The Netherlands.
25. W. C. Hinds, *Aerosol Technology: Properties, Behavior, and Measurement of Airborne Particles*, J. Wiley, New York, 1982.
26. Cambridge Filter Corporation, *Absolute Filter Testing Bulletin 12-106*, Cambridge Filter Corporation, Syracuse, NY, 1982.
27. Institute of Environmental Sciences (IES), *HEPA Filters* IES-RP-CC-001-86, Institute of Environmental Sciences, Mount Prospect, IL, 1986.
28. CDC, "Guidelines for Preventing the Transmission of Mycobacterium Tuberculosis in Health-Care Facilities," *MMWR Morb Mortal Wkly Rep*, **43**, 1–132 (1994).
29. American Conference of Government Industrial Hygienists (ACGIH), *Industrial Ventilation: A Manual of Recommended Practice*, ACGIH, Cincinnati, OH, 1992.
30. S. M. Opal et al., "Efficacy of Infection Control Measures During a Nosocomial Outbreak of Disseminated Aspergillosis Associated With Hospital Construction," *J. Infect. Dis.* **153**, 634–637 (1986).
31. R. J. Sherertz et al., "Impact of Air Filtration on Nosocomial Aspergillus Infections. Unique Risk of Bone Marrow Transplant Recipients," *Am. J. Med.* **83**, 709–718 (1987).
32. S. Miller-Leiden et al., "Effectiveness of HEPA-filter Units in Controlling Airborne Tuberculosis Transmission," *Health Care Initiative Information Seminar*, New Orleans, LA, 1995.
33. S. Miller-Leiden and W. W. Nazaroff, "Evaluation of Ceiling-Mounted Air Filters for reducing tuberculosis transmission," *JAWMA*, 1995.
34. Collins, F. M. "Relative Susceptability of acid-fast and non-acid-fast bacteria to ultraviolet light." *Appl. Microbiol.* 1971. Mar 21(3): 411–413.
35. R. L. Riley, M. Knight, and G. Middlebrook, "Ultraviolet Susceptibility of BCG and Virulent Tubercle Bacilli," *Am. Rev. Respir. Dis.* **113**, 413–418 (1976).
36. J. C. Chang et al., "UV Inactivation of Pathogenic and Indicator Microorganisms," *Appl. Environ. Microbiol.* **49**, 1361–1365 (1985).
37. M. M. Jensen, "Inactivation of Virus Aerosols by Ultraviolet Light in a Helical Baffle Chamber," *Proc. First Intern. Symp. Aerobiol.* Oakland, CA, 1963. Editor: R. L. Dimmick. Publisher: Naval Biological Laboratory.
38. R. L. Riley and J. E. Kaufman, "Effect of Relative Humidity on the Inactivation of Airborne *Serratia marcescens* by Ultraviolet Radiation," *Appl. Microbiol.* **23**, 1113–1120 (1972).
39. J. M. Macher et al., "Effect of Ultraviolet Germicidal Lamps on Airborne Microorganisms in an Outpatient Waiting Room," *Appl. Occup. Environ. Hyg.* **7**, 505–513 (1992).
40. W. F. Wells, "Aerodynamics of Droplet Nuclei," In *Airborne Contagion and Air Hygiene*, Harvard University Press, Cambridge, MA, 1955, pp. 13–19.
41. J. M. Macher, "The Use of Germicidal Lamps to Control Tuberculosis in Healthcare Facilities," *Infect. Control Hosp. Epidemiol.* **14**, 723–729 (1993).
42. ACGIH, *Threshold Limit Values and Biological Exposure Indices for 1995–1996*, ACGIH, Cincinnati, OH, 1995.
43. National Institute for Occupational Safety and Health (NIOSH), *Occupational Exposure to Ultraviolet Radiation (HSM 73-11009)*, NIOSH, Cincinnati, OH, 1972.

44. International Radiation Protection Association (IRPA), "Guidelines on Limits of Exposure to Ultraviolet Radiation of Wavelengths Between 180 nm and 400 nm (Incoherent Optical Radiation)," *Health Phys.* **49**, 331–340 (1985).
45. T. Tuomi, "Face Seal Leakage of Half Masks and Surgical Masks," *Am. Ind. Hyg. Assoc. J.* **46**, 308–312 (1985).
46. S. W. Dooley, Jr. et al., "Guidelines for Preventing the Transmission of Tuberculosis in Health-Care Settings, with Special Focus on HIV-Related Issues," *MMWR Morb. Mortal. Wkly Rep.* **39**, 1–29 (1990).
47. M. Nicas, "Refining a Risk Model for Occupational Tuberculosis Transmission," *Am. Ind. Hyg. Assoc. J.* **57**, 16–22 (1996).
48. U.S. EPA, *EPA (Environmental Protection Agency) Guide for Infectious Waste Management*, U.S. Environmental Protection Agency, Office of Solid Waste, Washington, DC, 1986.
49. G. McDonnell and A. D. Russell, "Antiseptics and Disinfectants: Activity, Action,and Resistance," *Clin. Microbiol. Rev.* **12**, 147–179 (1999).
50. M. S. Favero, "Sterilization, Disinfection, and Antisepsis in the Hospital," in Edwin H. Lennette, Albert Balows, William J. Hausler, Jr., and H. Jean Shadomy, eds., *Manual of Clinical Microbiology*, 4th ed., American Society of Microbiology, 1985, Washington, DC.
51. NRC. *Biosafety in the Laboratory: Prudent Practices for the Handling and Disposal of Infectious Materials*, NRC, 1989, p. 65, Washington, DC.
52. S. R. DiNardi, *The Occupational Environment—Its Evaluation and Control*, American Industrial Hygiene Association, Fairfax, VA, 1998.
53. J. Y. Richmond, ed., *Anthology of Biosafety—I. Perspectives on Laboratory Design*, American Biological Safety Association, Mundelein, IL, 1999.
54. P. M. Sutton et al., "Evaluating the Control of Tuberculosis among Healthcare Workers: Adherence to CDC Guidelines of Three Urban Hospitals in California," *Infect. Control Hosp. Epidemiol.* **19**, 487–493 (1998).

CHAPTER TWENTY-EIGHT

Potential Exposures in the Manufacturing Industry—Their Recognition and Control

William A. Burgess, CIH (Ret.)

1 INTRODUCTION

The conventional approach in an introduction of the occupational health practitioner to the health hazards in industry is to provide an industry by industry check list of the major chemical and physical exposures and the available control methods. This approach will not be used in this chapter. Rather, a limited number of operations will be presented in some detail with the topics weighted toward metal, chemical, and rubber products. Although the selection represents only a fraction of the major industrial workplaces, the issues that are identified can be used to evaluate health risks in many other work settings. Where possible, comments are made on the health status of today's workforce, and if exposure data are available, summary information is noted. Although the details of control approaches cannot be presented in this brief review, the proven control techniques are cited.

2 ABRASIVE BLASTING

Abrasive blasting has widespread application in a range of manufacturing, construction and service activities. It is estimated that there are 100,000 U.S. workers with the job title

Patty's Industrial Hygiene, Fifth Edition, Volume 2. Edited by Robert L. Harris.
ISBN 0-471-29754-2

of abrasive blaster and there are two to three times that number who do some abrasive blasting as a part of their job function.

In the metal working industry, abrasive blasting is widely used to remove surface coatings, scale, rust, and fused sand in preparation for subsequent finishing operations. No other method is as effective and economical for this purpose. The process is also used to remove flashing, tooling marks, or burns from cast, welded or machined metal fabrications and to provide a matte finish to metal parts. In addition, chilled-iron and steel shot are used to peen metal surfaces to improve resistance to fatigue stress, reduce surface porosity, and increase the wearability of parts.

In addition to its widespread use to prepare steel structures for painting, abrasive blasting is now widely used for cleaning wooden and brick buildings. In quarry operations it offers an alternative surfacing technique for architectural stone. Abrasive blasting is also applied in lettering stone monuments, etching of glass and in stone and metal art work.

2.1 Abrasives

For decades the terms abrasive blasting and sandblasting were used interchangeably. Although sand is still widely used in the United States, there are a large number of other abrasives available for specific applications. In evaluating worker exposure it is important that the abrasive in use be precisely identified to assist in predicting the health hazards.

The common heavy duty abrasives for metal surfaces are silica sand, metal shot and grit, coal and metallurgical slags, and synthetic abrasives such as silicon carbide and aluminum oxide. Sand is by far the most commonly used abrasive with metal grit and shot the next most common. For light duty cleaning of plastic and metal parts where one does not wish to erode the workpiece, a number of ground organic products based on corn, oat, and fruit pits are available in addition to glass beads, solid carbon dioxide, plastic chips, and sodium bicarbonate.

2.2 Abrasive Blasting Processes

The three blasting methods of projecting the abrasive at high velocity against the work surface are compressed air, hydroblast, and the centrifugal impeller. The most common compressed-air blasting equipment is the so-called direct pressure method (Fig. 28.1). In this system the abrasive receiver or pot is pressurized and the abrasive is forced into the throat of the abrasive blasting nozzle. In an alternative system the pot is open to the atmosphere and the abrasive is siphoned into the throat of the nozzle. In both system the compressed air fed to the nozzle carries the abrasive to the work surface at velocities in excess of 200 feet per second. This equipment is commonly used in both outdoor blasting and in abrasive blasting enclosures inside the plant. In addition to the person doing the blasting, the crew includes the pot man who adds abrasive to the pot and services the air compressor. The pot man also has responsibility for cleaning up debris from the blasting operation. Unfortunately cleaning is frequently done with an air hose resulting in high concentrations of airborne dust.

Hydroblasting uses a high pressure stream of water to convey the abrasive to the work surface. High pressures of 1,000 to 10,000 psig and ultra high pressure in excess of 20,000

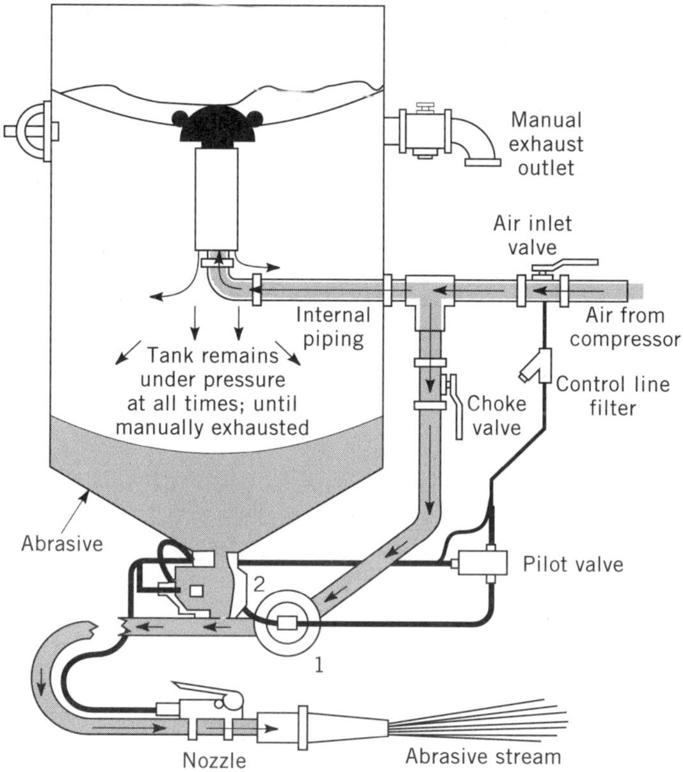

Figure 28.1. Schematic of abrasive pot. High pressure air from the compressor enters from the right and splits into two main lines immediately after air inlet valve. One line pressurizes the tank containing the abrasive; the second line picks up the abrasive at the apex of the tank cone and directs the abrasive under pressure to the hand held nozzle.

psig are currently in use. This operation, which also requires a minimum of two workers, is applied principally on outdoor work.

In the centrifugal wheel system, the abrasive flows to the center of the wheel and is discharged tangentially at high velocity. This unit is normally a component of an abrasive blasting enclosure as are discussed later in this chapter.

2.3 Applications

2.3.1 In-Plant, Containment With Ventilation

In metal fabrication, abrasive blasting is conducted in specially designed enclosures with integral systems for recirculation of abrasive, exhaust ventilation, and air cleaning. The enclosure design, application of the unit, ventilation design and other controls are discussed in this section.

2.3.1.1 Cabinets. A number of manufacturers provide small bench top or free-standing cabinets for abrasive blasting of small parts in repair or job shops. The units are glove boxes with a number of special features. The cabinet is of heavy steel construction with either a roll-up front panel or a side door for loading the workpiece to be cleaned. The abrasive blasting hose usually enters the top of the cabinet through a rubber star fitting which provides a partial dust seal. The nozzle is manipulated by the operator using heavy duty rubber gloves while viewing the workpiece through a sealed window. A dust-tight light fixture is installed so that the cabinet can be used for organic abrasives without the fear of a dust explosion. The cabinet is exhausted at a minimum air flow rate of four air changes per minute. The dust laden air passes through a cyclone collector and then to a bag house before release to outdoors.

The common abrasives for cabinets of this type are silicon carbide and aluminum oxide. Although exhaust ventilation will normally provide good dust control, it is important that an abrasive with low toxicity be chosen since lapses in maintenance of the equipment may result in worker exposure. Silica sand should not be used in this type of equipment. Periodic inspection of gloves, star gasket, door seals, and the structural integrity of the enclosure are important since the exhaust ventilation cannot control the dust projected through a major opening. The operator need not wear a respiratory protective device when working with a properly designed and maintained cabinet.

2.3.1.2 Tumble or Barrel Blasting. The abrasive blasting cabinets described above are not efficient for large lots of small parts. In this case the parts are processed in a tumble blasting barrel. The horizontal barrel, supported on trunions and rotated at low speed, tumbles the parts as they are air-blasted by silicon carbide or aluminum oxide abrasive. Dust released from the attrition of the abrasive or the workpiece residue is collected by a local exhaust ventilation system with sweep air designed to provide a velocity of 500 fpm though all openings.

2.3.1.3 Rotary Blasting Table. This equipment is widely used in ferrous and nonferrous foundries to remove silica sand fused on castings. The parts are placed on a large swing table attached to the main access door. After the parts are loaded, the door is closed and the parts table retracts into the tight enclosure. Metal shot or grit is projected to the surface of the workpiece positioned on the rotating table using a centrifugal impeller. Control of this dust is accomplished by local exhaust ventilation on the main enclosure and the associated abrasive transport equipment including the skip hoist. The ventilation is designed to provide a velocity of 500 fpm through all openings.

2.3.1.4 Abrasive Blasting Room. Large workpieces such as welded assemblies, forgings, and castings are cleaned in large rooms (Fig. 28.2) where the workpiece is transported to the blasting position by a fork lift truck or a flat bed rail car. After the part is positioned the access doors are closed and the blasting is done manually by an operator wearing special protective clothing and respiratory protective equipment.

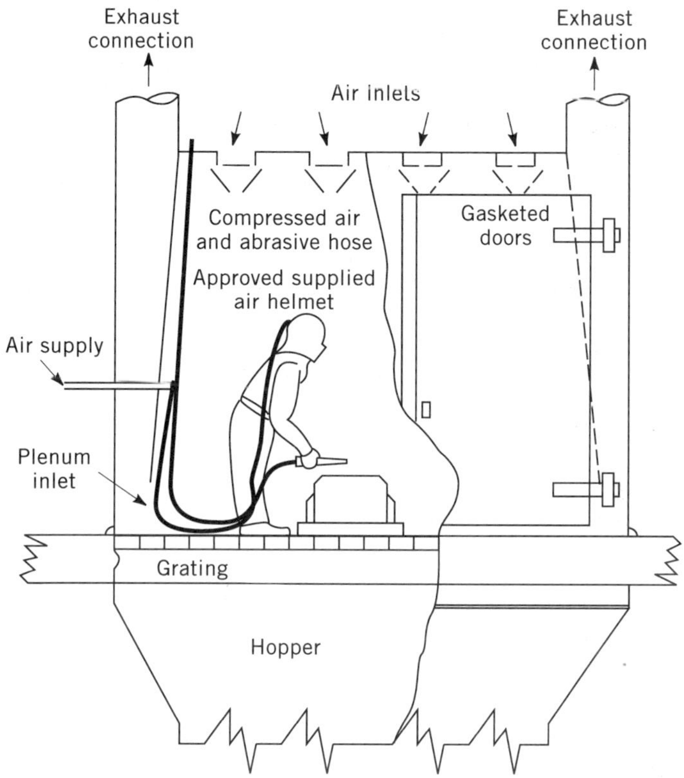

Figure 28.2. Abrasive blasting room. Access door is shown open to illustrate the nozzle.

The abrasive blasting room is serviced by a local exhaust ventilation system with airflow from ports in the ceiling downward past the work position to either a floor grate or exhaust slots along the base of the wall. The exhaust design is based on a cross-sectional velocity of 100 fpm. As with other systems, the air passes through a settling chamber and to a fabric filter before being exhausted to outdoors.

The primary hazard in abrasive blasting rooms is the high dust concentration (10–100 mg per m^3.) from the abrasive and the surface debris on the workpiece. The down draft ventilation permits visibility but it cannot control the dust to a hygenic level. The abrasive blasting helmet is the principal form of protection. As is discussed later in this section, this device may not be adequate to prevent lung disease.

The hand held nozzle which projects abrasive at a high velocity presents a vibration hazard to the worker. Isolation gloves may be of some value. The nozzle is equipped with a dead man switch; if the nozzle is dropped, the air supply shuts off. If this precaution were not taken the whipping hose could injure the worker. The noise level in the enclosure exceeds 110 dBA and a hearing conservation program is necessary (1).

2.3.2 Open Air Blasting

Abrasive blasting of large transport equipment such as ships, trucks and trains and structures including bridges, towers, and water tanks must be done outdoors (Fig. 28.3). This activity presents a major challenge to protect the worker and the environment from both the abrasive dust and the surface coating on the structure.

The preparation of bridge structures for repainting is a major challenge. In most cases previously applied lead-based paint must be removed and the surface blasted to what is known as a white metal condition. To prevent general environmental contamination, this work is done with temporary enclosures with or without ventilation. This approach minimizes dust release to the community but frequently results in the exposure of blasters not only to abrasive dust but to high concentrations of lead dust. Again one must rely on the use of respiratory protective devices to protect the worker. The structural steel painting industry has given the protection of the environment and the general population considerable study (2).

2.4 Health Hazards

Historically the principal hazard associated with abrasive blasting in the United States is attributable to the widespread use of silica sand as the abrasive. The popularity of this

Figure 28.3. Open blasting of a bridge structure. Abrasive blasting of a bridge painted with lead-based paint requires enclosure of the blasting area and exhaust of the dust laden air to a mobile cleaner. The blasting personnel must wear appropriate protective clothing.

abrasive is due to its availability, effectiveness, and low cost. Unfortunately, many of the abrasive blasting sands have very high concentrations of free silica which places the worker at risk of silicosis. In the U.S. the disease pattern in this industry was clearly identified by Public Health Service studies conducted in the 1930s.

Studies of the health status of blasters in the United States in the 1970s showed that workers blasting in confined spaces without adequate respiratory protective devices had a working lifetime of 10 years. In one study population of 100 workers, 28 men died from silicosis at an average age of 45 years (3). Another startling result was that eight of the workers who died from silicosis had worked in abrasive blasting less than three years. In an alert published in 1992 the National Institute for Occupational Safety and Health identified a registry of 99 cases of silicosis in abrasive blasting with 14 either dead or severely impaired (4). In 1993 the frustration of public health professionals with this issue was voiced by the Director of NIOSH who proclaimed the prevalence of silicosis in abrasive blasting to be a national disgrace.

In the United Kingdom silicosis in sandblasters was acknowledged in 1949 and the use of sand was prohibited for in-plant abrasive blasting. A ban on the use of sand for both in-plant and outside applications has been adopted by the European Community. In contrast, in the United States, in the late 1970s OSHA requested comments on the banning of sand for abrasive blasting. As a result of the information obtained it was decided not to ban the use of sand but to require close control of abrasive blasting operations. A draft standard was written but hearings were not held and the proposed standard was withdrawn.

In addition to the health hazards associated with the abrasive, one must consider the debris removed from the workplace. If the workpiece is an alloy containing chromium, nickel or a highly toxic material such as berryllium, the airborne metal dust may present a hazard. If the workpiece is mild steel and only rust and scale are being removed the hazard from the surface debris is minimal. However, the removal of toxic coatings such as lead paint on bridge structures and organic mercury fungicidal paints from ship hulls must be closely controlled.

2.5 Control

The most effective control of the hazard from sandblasting continues to be substitution of low toxicity abrasives. One of the widely used substitutes for sand is a series of metal and coal slags. Animal studies have demonstrated fibrotic changes in rats exposed to coal slags; such effects were not seen in comparable studies of copper slags (5). Concerns have also been directed to the presence of heavy metals and trace carcinogens in certain slags. A major concern is directed to the variability of the chemical composition of the slags that occurs with changes in source material. Although this knowledge is disquieting there is still a strong opinion that slag-based abrasives are an appropriate substitute for silica sands.

Other substitutes for sand include selected silicate minerals which contain low percentages of silica. One manufacturer provides a mineral sand consisting primarily of ferrous aluminum silicate that has been conditioned to reduce the concentration of free silica to less than 1% (6). This abrasive, which has seen application in the steel industry for removal of mill scale and rust, offers low dustiness and requires a low abrasive to air ratio.

No matter what abrasive is in use, the smaller the abrasive grain the greater the potential for high dust concentrations in the respirable size. To minimize this problem many authorities have specified that the abrasive grain size must be such that prior to blasting not more than 1% by weight of the abrasive may pass a #70 U.S. sieve and after use not more than 1.8% of the particles will be less than 5 microns.

Ventilation plays an important role in dust control from blasting in enclosures and accepted standards are available (7). In open blasting, however, ventilation control is difficult to achieve. The one approach that has been tried commercially is a high velocity-low volume capture system. This technique requires close attention to the seal between the nylon brush and the work surface and as result it is much slower than conventional techniques.

Attempts to utilize wet blasting to minimize dust generation on open blasting has found favor on certain operations. A common criticism is that its use before painting results in surface moisture which interferes with the paint bond. This criticism is countered by the manufacturer of the equipment who states that a conditioning wash can be used after the blasting that will protect the surface until the painting is completed.

An early study found great variability in the effectiveness of respiratory protective programs operated by abrasive blasting companies (8). The equipment in use ranged from simple paper bag hoods to approved Type C supplied-air respirators with continuous air flow supply and equipped with full facepiece masks, hoods, or helmets. The protection factors ranged from 1.9 to 3 750. The principal failing was the lack of a written respiratory protective device program and poor selection, fitting and maintenance of respirators. In other studies in this industry, air supplied devices were used with hazardous air supplies and inadequate air flow to the helmet.

As previously noted, respiratory protective device are not required on properly designed and externally operated abrasive blasting enclosures. It is also obvious that Type C air supplied abrasive blasting helmets are necessary to protect personnel doing open blasting and working in abrasive blasting rooms. Less obvious is the selection of proper respiratory protective devices for personnel who assist in open blasting or for the other trades who work in the area. In most cases it is prudent to require the proper wearing of a half or full face respirator equipped with a high efficiency filter.

The release of air at high velocity by the blasting nozzle generates a wide band noise that will easily exceed 110 dBA. This noise is greatly attenuated in a properly designed, externally-operated enclosure, but this alternative is not available for open blasting. Engineering control by nozzle redesign has not been possible to date and a muffler is not feasible. The only satisfactory control is the implementation of a Hearing Conservation Program including the use of hearing protectors. Frequently this requires the use of both insert plugs and muffs.

3 DEGREASING

Before discussing current and emerging processes and materials, the evolution of degreasing technology during the past four decades should be understood. In the 1940s the normal practice of degreasing metals was cold degreasing in immersion tanks using a range of petroleum hydrocarbons. These materials did a reasonable job of removing oils and greases, however, they had limited solvency, were slow drying, and left a residual film on the workpiece. Since these solvents were flammable (flash point below 100°F) they required rigorous in-plant fire protection controls. In addition, most of these solvents were photochemically reactive and produced smog. As a result, these volatile organic chemicals (VOCs) were designated as hazardous air pollutants under the Clean Air Act and were subject to both federal and local air pollution district regulations. For these reasons there was a gradual shift away from cold immersion cleaning with petroleum hydrocarbons to vapor phase degreasing using trichloroethylene, perchloroethylene, and methylene chloride. Of these three common vapor degreasing solvents only trichloroethylene was photochemically reactive. Vapor degreasers became universally acceptable in most sophisticated plant settings and more than 20,000 units were installed in the United States by 1960.

Toxicology studies of the the major degreasing solvents conducted in the 1960s and 1970s identified trichloroethylene, perchloroethylene, and methylene chloride as having significant systemic toxicity and trichloroethylene and perchloroethylene were considered suspect carcinogens. Two additional solvents, 1,1,1-trichloroethane and chlorofluorocarbons 113, were introduced in the late 1950s. They were widely used in the 1970s and 1980s because of their superior performance, low toxicity, and nonphotochemical reactivity. By 1990 over 180 thousand metric tons of chlorofluorocarbons were produced each year.

In the 1970s scientists became concerned that the stable chlorofluorocarbons would migrate to the upper atmosphere where halogen-catalyzed chemistry would result in the loss of stratospheric ozone. This concern was justified by a series of studies identifying a hole in the ozone layer in the Antarctic polar stratosphere. Depletion of the ozone layer had a number of major consequences including increased levels of ultraviolet at the earth's surface with a resulting increase in malignant melanoma in the worlds' population. By the 1980s there was definite evidence that 1,1,1-trichloroethane and a range of chlorofluorocarbons including chlorofluorocarbon 11 (CFC 11) and chlorofluorocarbon 113 (CFC 113) were major contributors to depletion of the stratospheric ozone layer at both polar atmospheres.

As a consequence of the public health importance of this issue there was a worldwide consensus that the manufacture and use of these materials must be curtailed on an emergency basis and that acceptable substitutes be developed. The case against CFCs was concluded with the 1987 Montreal Protocol, an international agreement on the production and use of chlorofluorocarbons, 1,1,1-trichloroethane, Halons (fire-suppression agents), and carbon tetrachloride. Insofar as the degreasing solvents were concerned, the agreement required a production and consumption freeze in all developed countries with a complete phase-out of chlorofluorocarbons by the year 2000 and 1,1,1-trichloroethane by 2005. An additional 10-year delay was permitted for the developing countries. In addition a phase-out of hydrochlorofluorocarbons was planned. The Clean Air Act of 1990 also contained a comprehensive regulation for chlorofluorocarbons, hydrofluorocarbons, and hydrochlorofluorocarbons. The initial Protocol was modified twice with the present agreement calling for phase-out of chlorofluorocarbons by January 1996 1,1,1-trichloroethane to 50% of the 1989 production level by January 1, 1996, and total phase-out by January 1, 2002. Hydrochlorofluorocarbons production will be frozen on January 1, 2015 and subsequent use will be limited to refrigeration equipment. Complete curtailment of hydrochlorofluorocarbon production will be effective on January 1, 2030. Pressure to eliminate the application of these materials was given additional force by an EPA regulation that all products manufactured with a process including the use of chlorofluorocarbons or 1,1,1-trichloroethane-based cleaning agents must be so labeled. In addition, in 1993, a heavy tax was levied on chlorofluorocarbons and 1,1,1-trichloroethane.

3.1 Cold Degreasing

The conventional cold degreasing practiced in maintenance shops and automotive garages in the early 1900s was frequently based on the use of pails of kerosene and gasoline in an uncontrolled fashion. By the 1940s mineral spirits, Stoddard solvent, and naphtha were commonly used for degreasing in many industries. Currently a range of aliphatic and chlorinated hydrocarbons are used in immersion tanks of the type shown in Figure 28.4 with a vertical agitated perforated tray and a spray nozzle. The immersion tanks are provided with the necessary fire protection controls including a fusible link-actuated cover. Dermatitis is the principal health hazard in this operation. Skin contact should be avoided

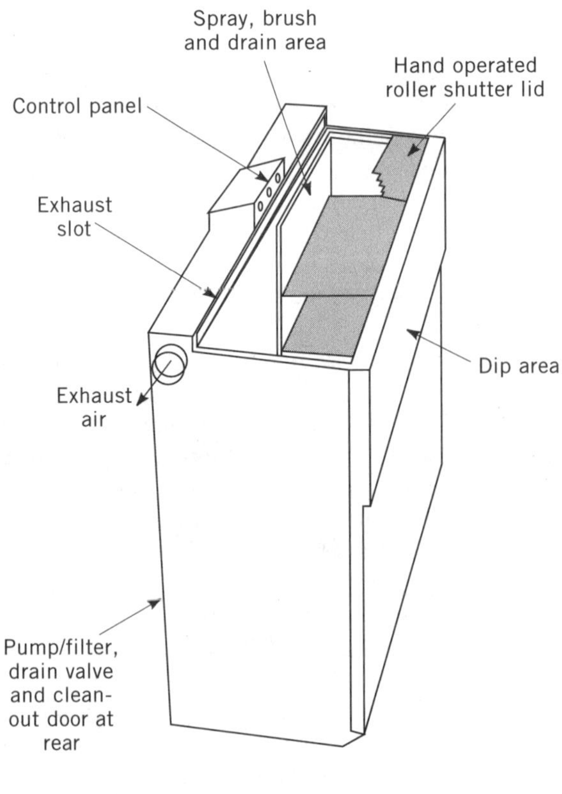

Figure 28.4. Cold degreasing tank.

and heavy duty protective gloves are mandatory. Glasses and a face shield should be used to protect the eyes and face from accidental splashing. In some cases, a work table with an exhaust is necessary for unloading and inspecting parts.

Although immersion tanks and spray washing are widespread, the most common cold degreasing technique is the use of a small amount of solvent in a container that is taken into the plant for on-site use. In the case of the conventional petroleum hydrocarbons described above, the principal health hazards are dermatitis from skin contact and the inhalation hazard posed when these materials are used in confined spaces.

A major shift in the hazard profile of cold degreasing occurred during the 1960 and 1970s when greater solvency power was required and, as described earlier, many plants converted to chlorinated hydrocarbon solvents such as trichloroethylene, perchloroethylene, and methylene chloride. Subsequently the advantages of 1,1,1-trichloroethane were recognized and this solvent became very popular for cold degreasing. The widespread use of such high vapor pressure solvents for general purpose cold degreasing in the confined space applications without adequate controls resulted in many deaths during this period.

3.2 Vapor Phase Degreasing

A straight vapor-phase degreaser is a steel tank containing a quantity of solvent that is heated to its boiling point by steam or electricity (Fig. 28.5). The solvent vapor rises and fills the tank to an elevation determined by the location of a condenser which in its simplest form is a cold water coil or jacket. The vapor condenses and returns to the liquid sump. Units are designed for a specific solvent based on its boiling point, vapor density, and latent heat of vaporization.

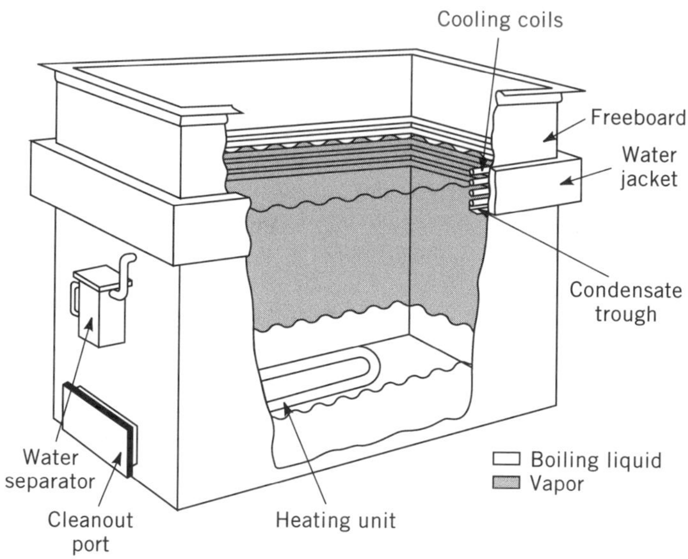

Figure 28.5. Conventional vapor-phase degreaser.

Individual workpiece or racks of small parts to be cleaned are lowered into the hot vapor, the vapor condenses on the cold parts and the surface oils and greases are dissolved by the solvent. This oily condensate drops back into the liquid solvent sump at the base of the tank. The solvent is evaporated continuously to replace the condensed vapor and maintain the vapor blanket. Since the oils are not vaporized, they form a sludge in the bottom of the tank which must be removed periodically. The scrubbing action of the condensing vapor on the workpiece continues until the temperature of the part reaches the temperature of the vapor [82°C (182°F) for trichloroethylene]. At this time condensation stops, the part appears dry, and it is removed from the degreaser. The time required to reach this point depends on the solvent, the temperature of the vapor, the weight of the workpiece, and its specific heat. In addition to straight cleaning, the vapor phase degreaser does an excellent job of drying parts after aqueous cleaning and before plating; it is frequently used for this purpose in the jewelry industry.

The straight vapor-cycle degreaser may not be effective on small, light work since the part will reach the temperature of the vapor quickly and the condensing action stops before the soil is removed from the workpiece. For such applications the vapor-spray cycle degreaser is frequently used. The part to be cleaned is first placed in the vapor zone for initial cleaning as in the straight vapor cycle degreaser. A portion of the vapor is condensed by a cooling coil and fills a liquid solvent reservoir. This liquid solvent is pumped to a spray nozzle that can be used to spray the solvent on the part, washing off surface oils and cooling the part, thereby permitting vapor condensation to provide the final cleaning.

A third common degreaser design has a compartment with warm liquid solvent and a second compartment with the conventional vapor zone. This degreaser is used for heavily soiled parts or cleaning baskets of small parts that nest together and cannot be effectively cleaned by vapor alone. Other designs have both boiling and warm liquid compartments. The boiling liquid compartment maintains the vapor phase when parts of high heat capacity are processed through the unit. In addition to the open top degreasers described above which are designed only for batch operations, conveyorized in-line enclosed units are available for high production levels.

As described above, the popular solvents used with vapor phase degreasers until the 1990s included trichloroethylene, perchloroethylene, methylene chloride, 1,1,1-trichloroethane, and trichlorotrifluoroethane. The solvents contain a small amount of stabilizer to neutralize any free acid that might result from oxidation of the degreasing liquid in the presence of air, hydrolysis in the presence of water, or pyrolysis under the influence of high temperatures.

Although 1,1,1 trichloroethane and trichlorotrifluoroethane are being banned, trichloroethylene, perchloroethylene, and methylene chloride will continue to see wide utilization in degreasing although there will be continued pressure to provide more effective control of solvent vapor for both worker and environmental health reasons. In certain cases equipment designed for chlorofluorocarbons or 1,1,1-trichloroethane will be converted for use with trichloroethylene, perchloroethylene, or methylene chloride.

3.3 Controls

The ability to operate vapor degreasers with minimum release of solvent vapors to the work place depends on the careful integration of the individual control design features shown in Figure 28.6 and summarized in Table 28.1. The controls are designed to minimize diffusion losses and disturbance losses due to rapid air movement across the top of the degreaser, movement of the load, convective velocities in the freeboard zone and overloading with subsequent vapor zone collapse. The various controls noted below are discussed in detail in a comprehensive review by the Center for Emissions Control (9).

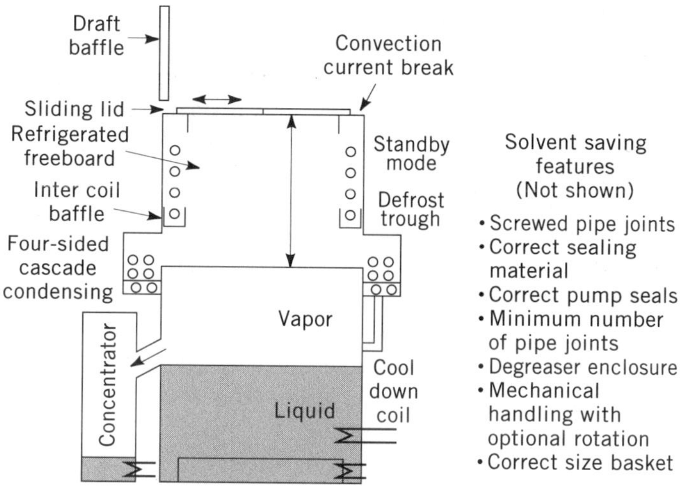

Figure 28.6. Controls on a vapor-phase degreaser.

Table 28.1. Effectiveness of Various Control Techniques for an Open-Top Vapor Degreaser[a]

Control Technique	Control Efficiency (%)	
	Idling[b]	Working
Manual cover	40	—
Freeboard ratio		
0.75 → 1.0	20	20
1.0 → 1.25	10	10
Reduced condenser temperature	40	40
Reduced room draft (100 → 50 fpm)	50	50
Refrigerator freeboard		
above freezing	40	40
below freezing	40	40
Hoist	—	35
Dwell	—	30
Automated cover	40	40

[a]Ref. 9, Operating Schedule—6 hr working; 2 hr idle; 16 hr downtime.
[b]Breakdown of total solvent loss: working—69%; downtime—22%; idling—9%.

The installation design features noted above must be integrated with acceptable work practices. Vapor phase degreasers should be cleaned periodically to prevent the accumulation of sludge and metal chips. The solvent should be distilled off until the heating surface or element is nearly, but not quite exposed, or until the solvent vapors fail to rise to the collecting trough. After cooling, the oil and solvent should be drained off and the sludge removed. It is important that the solvent be cooled before draining since removing hot solvent causes serious air contamination and frequently requires the evacuation of plant personnel from the building. A fire hazard may exist during the cleaning of machines heated by electricity because the flash point of the residual oil may be reached. Trichloroethylene is flammable at elevated temperatures. After sludge and solvent removal, the degreaser must be ventilated mechanically before any maintenance work involving flames or welding is undertaken. A person should not be permitted to enter a degreaser or place his head in one until all controls for entry into a confined space have been put in place. Anyone entering a degreaser should wear a respirator suitable for conditions immediately dangerous to life or health; a lifeline should be held by an attendant. In such circumstances, anesthetic concentrations of vapor may be encountered and the oxygen concentration may be insufficient. Such an atmosphere may cause unconsciousness with little or no warning; a number of deaths have occurred because of failure to observe the foregoing precautions.

The EPA National Air Pollution Control Techniques Advisory Committee is preparing a national emission standard for vapor degreasers. This standard will include operating records, work practices, and solvent consumption records.

3.4 Alternative Degreasing Systems

A principal precision cleaning activity in the early 1990s was the development and evaluation of various alternative cleaning techniques for electronic equipment to replace the banned chlorofluorocarbons. The options included a range of terpenes, aqueous and semi-aqueous cleaners and advanced hydrocarbon fluids. One manufacturer devised a printed circuit board test bed to evaluate the ability of candidate materials to handle the critical cleaning demands. Sophisticated vapor recovery systems were also been introduced in the 1990s. One system, the Reverse Brayton cycle, uses a hydrophobic adsorbent that has twice the capacity of conventional carbon adsorbents. In addition it will not differentially strip the stabilizer from the degreaser.

In the early 1990s a number of semi-aqueous cleaning processes were developed in response to the banning of chlorofluorocarbons and 1,1,1-trichloroethane. A representative process using such a semi-aqueous cleaner includes the initial immersion of parts in the hydrocarbons wash tank followed by a series of rinse tanks containing progressively cleaner solvent (Fig. 28.7). A cascade system provides cleaner make-up solvent to the last tank with recycling of the overflow of dirty solvent from the initial immersion tank. Drying of the workpiece, an energy intensive step, depends on the specific requirements of the operation. Adequate drying may be achieved by spinning, wiping, or an air blow-off. If totally dry parts are required, a Class A hot air dryer must be used.

The new semi-aqueous processes have a number of advantages over previous applications of chlorofluorocarbons and 1,1,1-trichloroethane solvents while providing comparable cleaning for a range of surface soils encountered in general metalworking. Cleaning

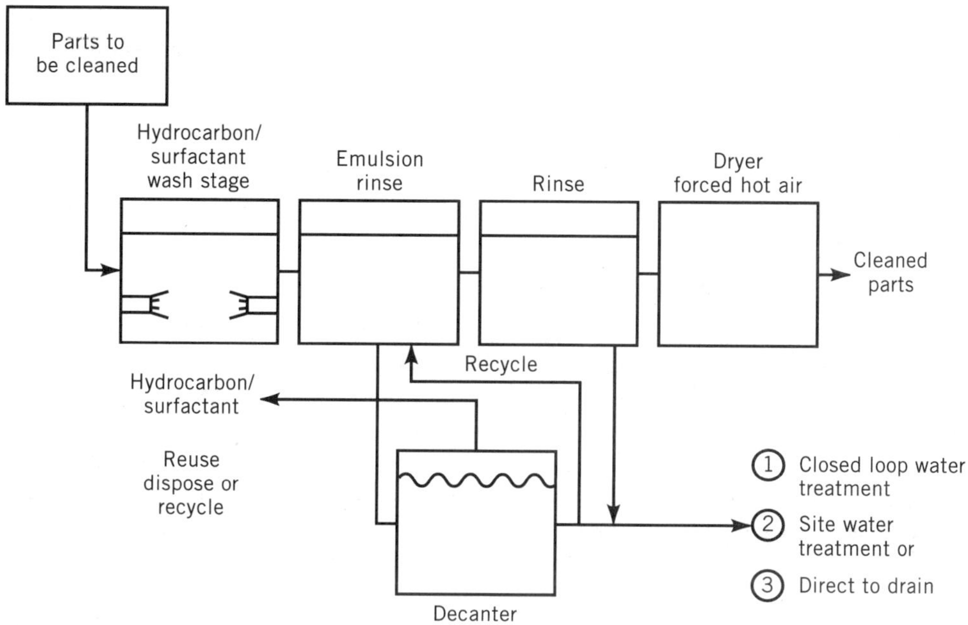

Figure 28.7. Semi-aqueous cleaning process using an immiscible hydrocarbon solvent.

systems are now designed for specific applications such as washing printed circuit boards after soldering to remove flux. Degreasing equipment previously used with chlorofluorocarbons and 1,1,1-trichloroethane may be retrofitted for these semi-aqueous systems by removing electrical heaters, condensing coils, and adding ventilation and drying components. The most prudent approach, however, is to purchase equipment designed specifically for this work.

The advantages cited by the manufacturers for semi-aqueous processes are elimination of ozone depleting solvents, low toxicity, high flash point and low vapor pressure with resulting minimum worker exposure to solvent vapors. Another important feature is the ability to easily separate the solvent and water thereby minimizing the need for extensive water treatment. On the negative side, the active component may be declared non-exempt by the EPA, a water rinse may be required, and a drying operation is necessary.

4 ELECTROPLATING

Metal, plastic, and rubber parts are plated to prevent rusting and corrosion, for appearance, to reduce electrical contact resistance, as a base for soldering operations, to provide electrical insulation, and to improve wearability. The common plating metals include cadmium, chromium, copper, gold, nickel, silver, and their alloys. Prior to electroplating, the parts must be cleaned and the surfaces treated to ensure good bonding and appearance.

There are approximately 160,000 electroplating workers in the United States. Of that number probably 40,000 work in independent job shops and the balance are employed in large captive shops. The potential health hazards in electroplating operations are numerous; however, a comprehensive epidemiologic study of this industry in the United States has not been completed so that the true impact of the health hazards on the workers is not known. A limited number of studies have identified a series of health effects in electroplaters ranging from dermatitis to elevated mortality for selected cancers.

4.1 Electroplating Techniques

A schematic of a conventional electroplating system is shown in Figure 28.8. The electrolyte consists of a salt of the metal to be applied dissolved in water. Two electrodes powered by a low voltage DC-power supply (4 to 12 V DC) are immersed in the electrolyte. The cathode is the workpiece to be plated and the anode is either an inert electrode or most frequently, a slab or a basket of spheres of the plating metal to maintain the bath metal ion concentration. If the anode is an inert electrode, the metal ion concentration in the bath is maintained by periodically adding metallic salts to the bath. When power is applied, the positively charged metal ions deposit out of the bath on the cathode or workpiece. Water is dissociated releasing hydrogen at the cathode and oxygen at the anode. These gases are released as bubbles at the liquid surface. The current density expressed in amperes per ft^2 of workpiece surface, varies depending on the operation; the higher the current density the greater the gas evolution rate. In addition to the salt containing the metallic ion, the plating bath may contain chemicals to adjust the electrical conductivity of the bath, additives that determine the type of plating deposit, and a buffer for pH control.

The electroplating tank shown in Figure 28.9 is constructed of rigid plastic or steel with a plastic, rubber, or lead lining. In many common processes the anodes and cathodes, hung from copper rods and immersed in the electrolyte, are electrically connected to the DC-power supply with heavy copper bus bars. When required the tank may be fitted with a local exhaust ventilation hood (7).

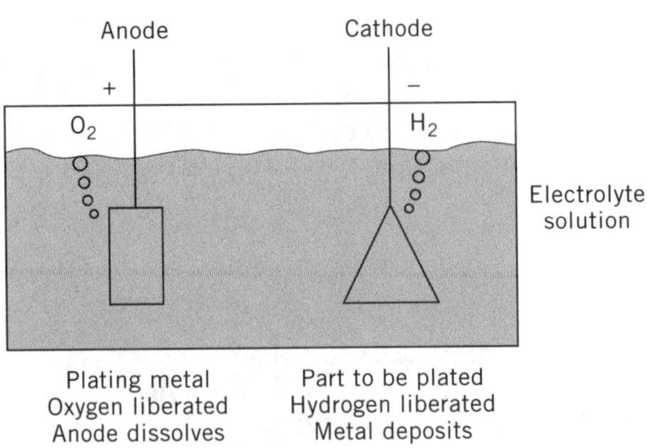

Figure 28.8. Schematic of conventional electroplating system.

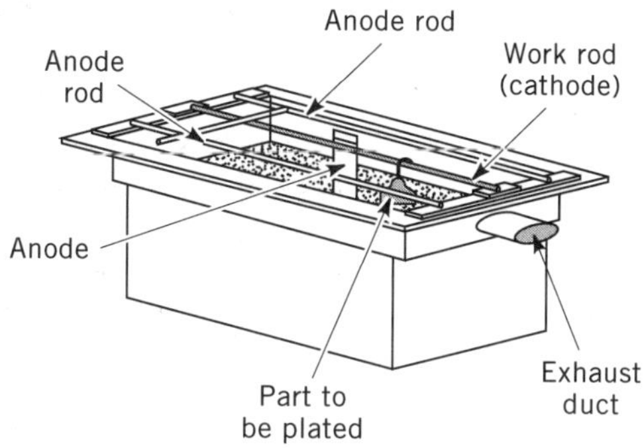

Figure 28.9. Principal components of an electroplating tank showing only one anode plate and one workpiece for clarity. The tank is exhausted with a perimeter exhaust.

In chromium electroplating, a plating process with significant health hazards, the plating bath contains chromic acid and the anode is chemically inert. In decorative chrome plating, the part is first nickel plated and then a thin coating of chromium is plated on this base. Decorative chrome is widely used for plumbing fixtures and automobile components. In a second chromium plating process, hard chrome, used to plate tools and gauges and to build up worn parts, a thick layer of chromium is applied directly to the base metal. This process results in higher exposure to chromic acid mist than decorative chrome.

In anodizing, a common surface treatment for decoration, corrosion resistance, and electrical insulation on such metals as magnesium, aluminum, and titanium, the polarity of the system is the reverse of the electroplating process. The workpiece is the anode, and the cathode is a lead slab. The oxygen released at the workpiece causes a controlled surface oxidation that is desirable for many applications. Anodizing is usually conducted in a chromic acid bath with high current density; since its efficiency is quite low, the rate of misting is high.

In the last decade there has been a great increase in the electroplating of plastic parts due to the ability of the industry to match the quality of plating on metal parts. Electroless plating of a base coat on the plastic is a necessary first step in the process. Electroless plating does not require an external DC-power supply, but rather the metal is deposited on the workpiece by a catalytic reaction of the bath. A number of plastics, including nylon and polycarbonate, are suitable for electroless plating. Nickel, copper, cobalt, gold, and palladium can be deposited on the plastic workpiece. First the surface of the plastic part is etched to provide a microscopically rough surface that will permit a mechanical bond between the plastic and the metal. As an example, an acrylonitrile–butadiene–styrene part can be etched with a chromic and sulfuric acid solution to produce a pitted surface for mechanical bonding of the molecules of the plating metal to the plastic. The etching process also forms the basis for a chemical bond. After alkaline neutralization, the part is treated with a hydrochloric acid solution of stannous and palladium salts. The tin–palladium hydrosol which is formed is treated with an accelerator to remove the stannous ion resulting in a palladium-rich surface on the plastic. This surface is the base for subsequent electroless nickel plating from a bath of a nickel salt and a reducing agent. This coating can then be plated by any conventional electrode system.

In production plating operations, individual parts or racks of small parts are hung manually from the cathode bar. If many small pieces are to be plated, the parts are placed in a perforated plastic barrel in electrical contact with the cathode bar and the barrel is immersed in the bath. The parts are tumbled to achieve a uniform plating.

In small job-shop operations, the parts are transferred manually from tank to tank as dictated by the type of plating operation. The complexity of this operation is illustrated by the steps in decorative chromium plating (Fig. 28.10). The extensive pretreatment in this sequence includes degreasing to remove grease and oil, pickling to remove heavy scale, alkaline anodic cleaning to remove soils, and finally an acid dip to remove light scale and activate the surface for plating. In high production shops an automatic transfer unit is programmed to cycle the parts from tank to tank; the worker is only required to load and unload the racks or baskets. This automatic plating operation permits the facilities engineer to design enclosing exhaust hoods on the tanks, to achieve more effective control of air contaminants. Exposure to air contaminants is also limited in automatic plating lines since the worker is stationed at either a loading or unloading position remote from the process and therefore is not directly exposed to air contaminants released at the tanks.

4.2 Air Contaminants

The principal source of air contamination in electroplating operations is the release of the bath electrolyte to the air by the formation of bubbles in the bath. As mentioned above,

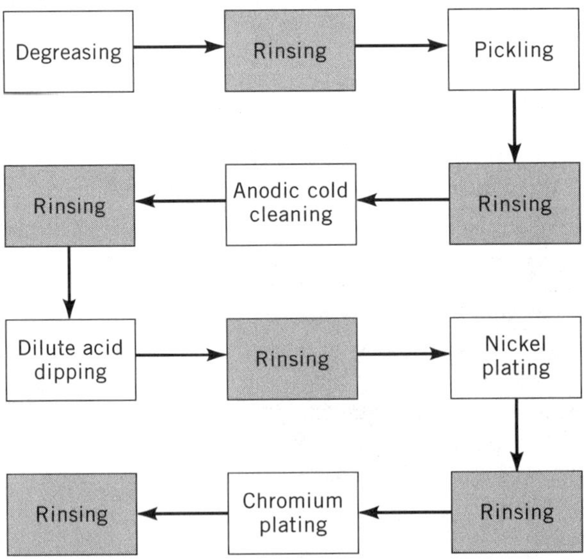

Figure 28.10. Steps involved in decorative chromium plating of steel.

the bath operates as an electrolytic cell; thus water is dissociated and hydrogen is released at the cathode and oxygen at the anode. When the film cap of the gas bubble ruptures, a small sized droplet is formed as shown in Figure 28.11. The liquid jet rising from the crater left by the bubble is then the source of a second droplet that is an order of magnitude larger than the original droplet (10). It is this droplet that represents the principal air contaminant, a respirable mist of the liquid bath. The presence of this mist can be detected by placing a clean piece of paper approximately 2 cm above the surface of the bath. Staining of the paper indicates that mist is released from the surface of the bath.

The mist generation rate of electroplating tanks depends on the efficiency of the process. As shown in Table 28.2 for copper plating, the efficiency of the sulfate plating bath is nearly 100%; that is, essentially all the energy goes into the plating operation and little goes into the dissociation of water and the release of mist. Mist generation in nickel plating baths is minimal since these baths operate at 95% efficiency, that is, only 5% of the energy is directed to the dissociation of water and gasing. However, chromium plating operations are quite inefficient and up to 90% of the total energy may be devoted to dissociation of the water with resulting severe gassing and potential exposure of the operator to chromic acid mist (11).

Although the contaminant generation rate of the bath is governed principally by the inherent efficiency of the process, it also varies with the metallic ion concentration, the current density, the bath additives, and bath temperature. Air or mechanical agitation of the bath, used to improve plating quality, may also result in release of mist (12).

The health significance of the mist generated by electroplating processes depends, of course, on the contents of the bath. In considering the nature of the airborne contaminants released from the bath it is convenient to class the bath electrolyte as alkaline or acidic. A majority of alkaline baths, based on cyanide salt solutions, are used for cadmium, copper,

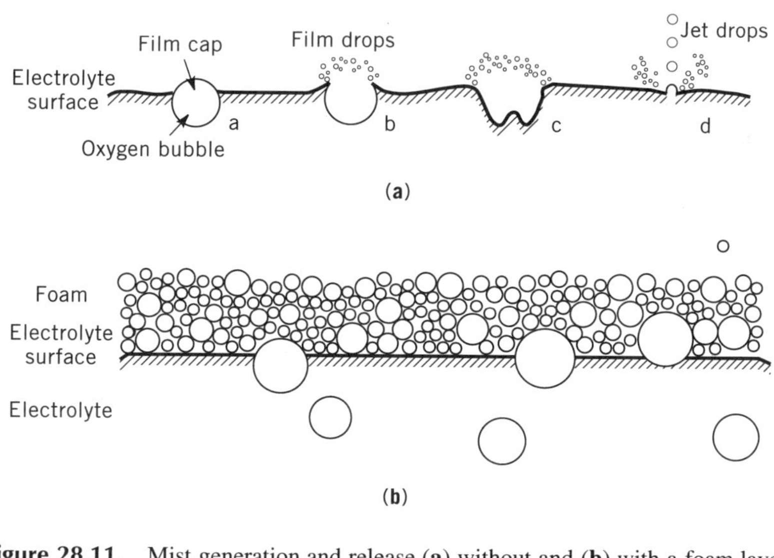

Figure 28.11. Mist generation and release (**a**) without and (**b**) with a foam layer.

Table 28.2. Electroplating Bath Operating Conditions

Deposited Metal	Application	Bath Composition	Temperature (°C)	Cathode Current Density (A/m^2)	Cathode Current Efficiency (%)
Cadmium	Decorative	Cyanide	30	5–50	90–95
	Protective				
Chromium	Hard plate	Dilute CrO_3	45–55	50–200	10–20
	Decorative	Conc. CrO_3	40–50	80–120	10–20
Copper	Strike	Cyanide	40–60	10–40	30–50
	Heavy plate	Sulfate	20–50	20–100	95–100
Gold	Electrical	Cyanide	60–70	1–10	95
	Decorative				
Nickel	Protective	Watts	55	10–60	95
		Chloride	55	50–100	95
		Fluoborate	55	50–100	95
Silver	Strike	Cyanide strike	20	30	99
	Cyanide heavy	Cyanide heavy	20–50	10–150	99
Tin	Tin plate	Acid sulfate (1)	25–30	10–60	95
		(2)	40–50	100–400	95
		Alkaline (1)	80	10–60	95
		(2)	85	100–400	95
Zinc	Brightener	Cyanide	30–50	10–50	75–95
	Brightener	Sulfate	25–35	10–30	95

[a]Ref. 11.

silver, brass, zinc, and bronze plating. Chromium, copper, nickel, and tin are commonly plated from acidic baths. The exact composition of the baths can be obtained from an electroplaters handbook or from the supplier.

An inventory of the chemicals in the common electroplating baths, the form in which they are released to the air, and the rate of gassing, has evolved over the past decades drawing heavily on the experience of the state industrial hygiene programs in New York and Michigan (7). In addition to the release of electrolyte mist, the acidic action of carbon dioxide in air (350–400 ppm) may cause small quantities of hydrogen cyanide to be released from cyanide-based baths. This issue has not been clearly identified in the literature.

4.3 Control

Proprietary bath additives are available to reduce the surface tension of the electrolyte and, therefore, to reduce misting. The minimum effective viscosity is 35 dynes/cm but one should operate at a viscosity of 25 dynes/cm. Additives are available to provide a thick foam that traps the mist released from the bath. These agents are best used for tanks that operate continuously. A layer of plastic chips, beads, or balls on the surface of the bath will also trap the mist and permit it to drain back into the bath.

The three exhaust hoods for open surface tanks shown in Figure 28.12 are commonly installed in plating shops. All are identified as lateral exhaust hoods since they provide a horizontal sweep of air across the surface of the tank to the hood opening. As noted in the discussion of automatic plating, the application of partial and total enclosures in conjunction with these hoods is frequently possible with excellent control of air contaminants at modest flow rates.

To assist the ventilation designer and the industrial hygienist in the design of local exhaust ventilation for plating tanks, the New York Division of Industrial Hygiene published a set of tables on various electroplating operations which describe the bath components which may be released from each process, the physical form of the contaminant and the alphanumeric code which becomes the basis for the ventilation design (7). This design procedure permits one to determine a minimum control velocity based on the hazard potential of the bath and the rate of contaminant generation. The exhaust volume is calculated based on this control velocity and the tank geometry. A detailed presentation of the design ventilation procedure is available elsewhere (7, 13). In addition to the proper design and installation of good local exhaust ventilation, one must provide adequate replacement air, backflow dampers on combustion devices to prevent carbon monoxide contamination of the workplace, and suitable air cleaning.

The low voltage DC-power supply in use in electroplating does not present an electrical hazard. A fire and explosion risk may result if solvent degreasing and spray painting are conducted in areas contiguous with the plating shop. The major chemical safety hazards result from the handling of concentrated acids and alkalies when preparing the baths and the accidental mixing of acids with cyanides or sulfides during plating, bath preparation, and waste disposal with the formation of hydrogen cyanide or hydrogen sulfide. The

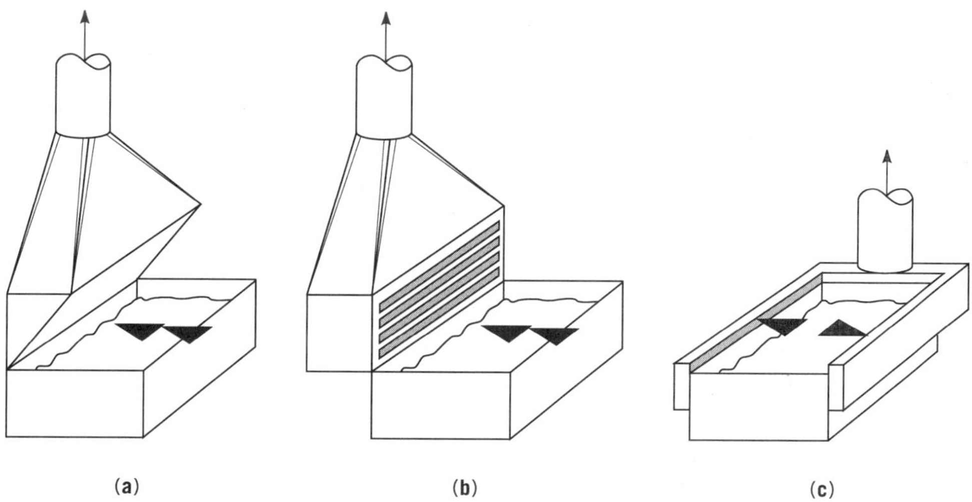

Figure 28.12. Conventional open surface tank exhaust hoods: (**a**) open fishtail, (**b**) slotted plenum fishtail and (**c**) lateral slot.

educated use of protective equipment by electroplaters is extremely important in preventing contact with the various sensitizers and corrosive materials encountered in the plating shop. The minimum protective clothing should include rubber gloves, aprons, boots, chemical handler's goggles, and face shields. Aprons should come below the top of the boots. All personnel should have a change of clothing available at the workplace. A shower and eye-wash station serviced with tempered water should be available. The wide range of chemicals handled in an open fashion does present a major dermatitis hazard to the plater, and skin contact must be avoided. Nickel is a skin sensitizer and may cause nickel itch developing into a rash with skin ulcerations. If solutions are splashed on the work clothing it should be removed, the skin washed, and the worker should change to clean garments. Hands shoulds be washed frequently and eating, drinking, and smoking should be prohibited in the work area.

A summary of the health hazards encountered in electroplating shops and the appropriate control are presented in Table 28.3.

Table 28.3. Summary of Major Electroplating Health Hazards

	Inhalation
Mist, gases and vapors	
Hydrogen cyanide	Accidental mixing of cyanide solutions and acids
Chromic acid	Released as a mist during chrome plating and anodizing
Hydrogen sulfide	Accidental mixing of sulfide solutions and acids
Nitrogen oxides	Released from pickling baths containing nitric acid
Dust	
Released during weighing and transfer of various solid bath materials including cyanides and cadmium salts	
Fumes	
Generated during on-site repair of lead-lined tanks using torch-burning techniques	
	Ingestion
Accidental ingestion of particulate contamination during smoking and eating at workplace	
	Skin Contact
Absorption of cyanide compounds through the skin	
Defatting by solvents	
Primary irritants	
Sensitizers	
	Control Technology
Local exhaust ventilation	
Mist reduction	
Reduce surface tension	
Coat surface	
Tank covers	
Isolation of stored chemicals	

5 PAINTING

Paint products are used widely in industry to provide a surface coating for protection against corrosion, for appearance, as electrical insulation, for fire retardation, and for other special purposes. The widespread application of this technology, from small job shops to automated painting of automobiles, includes more than half a million workers in the United States. Of this number 200,000 workers are employed in automobile body repair shops.

Although the health status of painters has not been well defined, the studies that have been completed suggests acute and chronic central nervous system effects, hematologic disorders, and excess mortality from cancer. In addition, respiratory sensitization may occur from two-part urethane systems, and skin sensitization may be caused by the amine catalyst in epoxy paint systems. In the last decade the market for traditional solvent-based industrial coatings has been declining at the rate of 5% per year due to the conversion to high solids and non solvent paints. During this time the manufacturing sector was lagging behind the construction sector in the rate of conversion to the newer systems. Recently this situation has changed with manufacturing now showing a steep increase in the conversion from high solvent-based paints.

5.1 Paint Systems

The term paint is commonly used to identify a range of liquid organic coatings including paints, varnishes, enamels, and lacquers. Conventional paint is an inorganic pigment dispersed in a vehicle consisting of a binder and a solvent with selected fillers and additives. Varnish is a nonpigmented product based on oil and natural resin in a solvent that dries first by the evaporation of the solvent and then by the oxidation of the resin binder. Varnishes are also available based on synthetic resins. A pigmented varnish is called an enamel. Lacquers are coatings in which the solvent evaporates leaving a film that can be redissolved in the original solvent.

The second major paint system, powder coating, utilizes a fine powder which is a blend of the resin, pigment, and additives. The powder is conveyed from a reservoir to the spray gun where a charge is imparted to the individual paint particles. This dry powder is gently blown on the electrically grounded workpiece and after coating, the parts are baked to fuse the powder film into a continuous coating (14). This coating technique provides a high-quality job while eliminating solvent exposure to the worker. Thermosetting resins, used for decorative and protective coatings, are based on epoxy, polyester, and acrylic resins. Thermoplastic resins are applied in thick coatings for critical applications.

5.1.1 Liquid Paint Systems

5.1.1.1 Solvent-Based Paints. The conventional solvent-based paints consist of the vehicle, filler, and additives. The vehicle represents the total liquid content of the paint and includes the binder and solvent. The binder, which is the film-forming ingredient, may be a naturally occurring oil or resin including linseed oil and oleoresinous materials or a synthetic material such as alkyd resins. The solvent systems are varied and complex. The most common organic solvents include aliphatic and aromatic hydrocarbons, ketones, alcohols, glycols, and glycol ether/esters. These solvents have high vapor pressure and represent the critical worker exposure component in most painting techniques. High-solids paint systems (low solvent) represent a major contribution to reduced solvent exposure.

The fillers, pigments, and extenders have historically presented a major hazard in painting. The common white pigments include bentonite and kaolin clay, talc, titanium dioxide, and zinc oxide. Mineral dusts used as extenders to control viscosity, texture, and gloss include talc, clay, calcium carbonate, barite, and both crystalline and amorphous silica. The pigments and extenders do not represent an exposure during brush or roller application, but during spray painting and preparation of the surface by sanding or abrasive blasting prior to repainting, these materials are released to the air and may present significant exposures. Pigments based on lead, cadmium, and chromium compounds present a critical exposure during spray application and surface preparation.

Additives to hasten drying, reduce skin formation in the can, and control fungus are added to paints. The fungicides warrant special attention because the ingredients, including copper and zinc naphthenate, copper oxide, and tributyltin oxide, are biologically active.

5.1.1.2 Water-Based Paints. Water-based paints now represent 30 to 40% of construction and industry paints, and expanded use is anticipated. The present systems use binders based on copolymers of several monomers including acrylic acid, butyl acrylate, ethyl hexyl acrylate, and styrene. The other major ingredients in a conventional water-based paint include pigments, coalescing solvents, surfactants, biocides, plasticizers, and driers. The pigments include the conventional materials mentioned under solvent-based paints. Coalescing solvents, including hydrocarbons, alcohols, esters, glycol, and glycol ethers and esters, although present in low concentrations, may present an inhalation hazard. The proprietary biocides frequently result in exposure to significant airborne formaldehyde concentrations. The surfactants include known skin irritants and sensitizers.

5.1.1.3 Two Component Paints. Special attention must be given two-component epoxy and urethane paint systems. The urethanes consist of a polyurethane prepolymer containing a reactive isocyanate; the second component is the polyester. Mixing of the two materials initiates a reaction with the formation of a chemically resistant coating. In the early formulations, the unreacted toluene diisocyanate (TDI) resulted in significant airborne exposures to TDI and respiratory sensitization of workers. Formulations have been changed to include TDI adducts and derivatives that are stated to minimize the respiratory hazard from these systems. The urethane/polyester system includes a blocked isocyanate which is not activated until an oven temperature of 250°C is reached. In this system, the unreacted monomer content is less than 0.01%.

Epoxy paint systems offer excellent adhesive properties, resistance to abrasion and chemicals, and stability at high temperatures. The conventional epoxy system is a two-component system consisting of a resin based on the reaction products of bisphenol A and epichlorohydrin. The resin may be modified by reactive diluents such as glycidyl ethers. The second component, the hardener or curing agent, was initially based on highly reactive amines. The use of low molecular weight aliphatic amines such as diethylenetetramine and triethylenetetramine, both strong skin irritants and sensitizers, presented major health hazards during the early use of epoxy systems. This has been overcome to a degree by the use of low-volatility amine adducts and high molecular weight amine curing agents. The epoxy powder system includes an amine catalyst which is activated at an oven temperature of 350°C. Insofar as dermatitis is concerned, the epoxy resins can be divided into three grades. The solid grades are felt to be innocuous; however, skin irritation may occur from solvents used to take up the resin. The liquid grades are mild to moderate skin irritants. The low-viscosity glycidyl ether modifiers are skin irritants and sensitizers and have systemic toxicity.

5.1.2 Liquid Paint Application Methods

In the industrial setting, paints can be applied to parts by a myriad of processes including brush, roller, dip, flow, curtain, tumbling, spray, and powder coating. Spray painting by air atomization is the most common application method encountered in industry and presents the principal hazards. To a large degree the hazard in a given spray paint operation depends on the transfer efficiency of the paint application method. Transfer efficiency denotes the percentage of the paint that is deposited on the workpiece; the balance is lost by overspray and rebound (Fig. 28.13).

The quantity of paint applied per unit time and therefore the potential amount of solvent released during the operation is impressive: a 6-in, wide brush may use 7 gal/day, a 9-in. roller may use 14 gal/day, and air spray use varies from 10 to 70 gal/day. The operator is exposed to the solvent vapor in processes in which the paint is flowed on, as in brushing and dipping, and during drying of the parts. However, during atomization techniques, the exposure is to both the solvent and the paint mist. The worker's exposure is dependent on

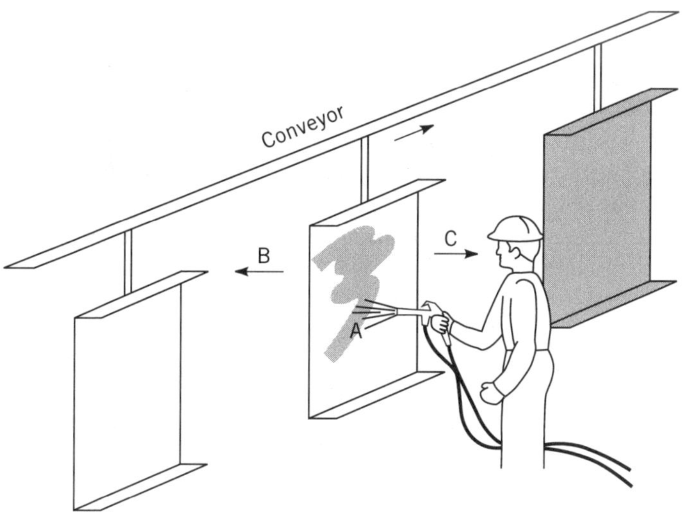

Figure 28.13. Spray painting showing the fate of paint released from the spray gun; A, Deposition on the workpiece; B, overspray, and C, rebound. Transfer efficiency is the percentage of the total paint released from the gun that deposits on the workpiece.

the overspray and rebound that occurs during spraying. The principal spray painting methods used in industry are described below.

5.1.2.1 Air Atomization. In this painting system the paint is conveyed from a paint reservoir by either siphon pick up created by the airflow, or in the case of heavy, viscous or high solids paints, the reservoir is pressurized and the paint flows to the spray gun (Fig. 28.14). In either case, compressed air atomizes the paint at the nozzle to form droplets or mist, releases the droplet cloud from the gun and conveys it to the workpiece. The atomization can take place external to the gun in the zone between the horns of the nozzle tip or inside the gun itself. The system commonly used in industrial finishing painting is an external mist generating system with pressure feed. This system requires high airflow, generates fine particles (fog) and has a low transfer efficiency in the range of 25–30%. If the paint is heated to lower its viscosity, the pressure used in the system can be reduced with a resulting reduction in over-spray and rebound and an increase in transfer efficiency.

5.1.2.2 Low Pressure–Low Volume. This system utilizes a special air-atomizing gun with low pressure air jets impinging on a primary stream of paint to provide the atomization mechanism. The paint fluid stream exits separately from the gun and is introduced into a secondary tip zone. The solid elliptical-shaped paint stream is atomized externally by two small air jets. The atomization is not as fine as conventional air atomization however rebound is reduced and transfer efficiencies in the range of 60–80% are achieved. Low pressure–low volume spraying has seen wide application in the furniture, automobile and general metal finishing industries for low to medium viscosity paints.

5.1.2.3 High Volume–Low Pressure. The high volume–low pressure system uses compressed air at 3–6 psig and air flows of approximately 20 cfm to atomize the paint. The paint mist size is in the range of 1–50 μm. Since the air pressure is lower than conventional air atomization, there is a reduction in overspray and rebound with resulting transfer efficiencies in excess of 65%. A number of environmental regulatory agencies have recognized the advantages of this system and have restricted spray painting operations to either high volume–low pressure or electrostatic methods. Frequently this system is chosen over electrostatic painting for water-based paints since electrostatic systems used with water-based paint require special isolation of all paint reservoirs. In addition to its high transfer efficiency and resulting environmental advantages, this system produces a high quality finish with minimal loss of paint. In conventional air atomization, two gallons of paint are lost by overspray and rebound for every 3 gallons sprayed. In the high volume-low pressure system, this loss is reduced to one pint of paint for each gallon sprayed.

5.1.2.4 Electrostatic. This system (Fig. 28.15) places a negative charge on the paint particle by passing the paint mist through a corona discharge field. The charged particles are then attracted to the grounded workpiece. The early systems for fixed nozzle work utilized an external high voltage grid between the spray gun and the workpiece. In the present designs the high voltage corona is established in the gun; the system can be used on air, airless, and air-assisted airless guns with transfer efficiencies in excess of 90%. The hand-held guns operate at voltages in the range of 30 to 60 KV and a current of

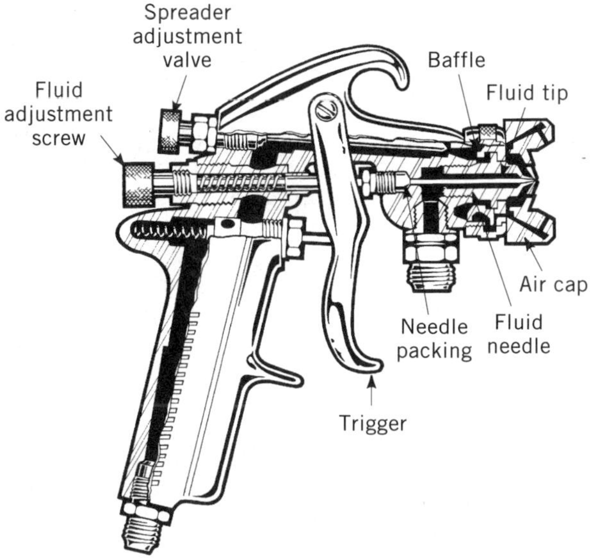

(a)

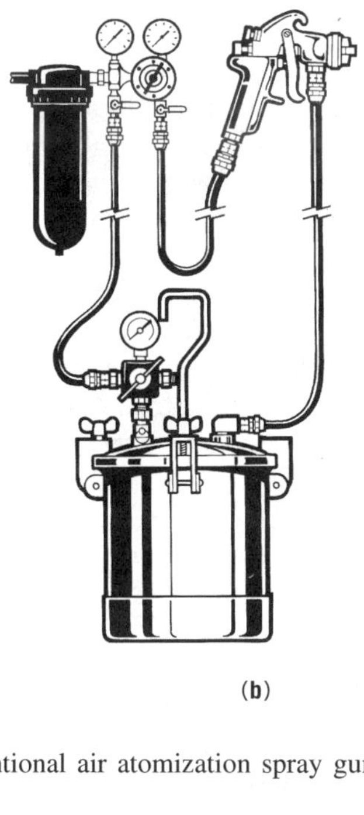

(b)

Figure 28.14. (**a**) Conventional air atomization spray gun and (**b**) pressure feed tank hook up. Courtesy of DeVilbiss Co.

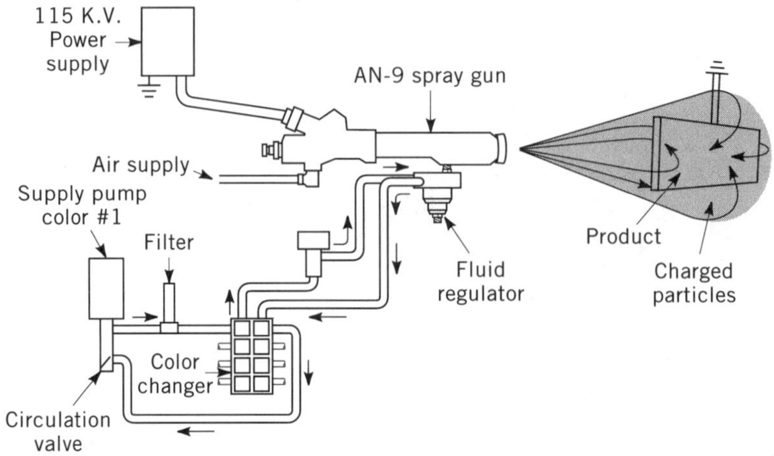

Figure 28.15. Automatic electrostatic spray gun. Courtesy of Nordson Corp.

approximately 100 μA; the guns do not present an electrical hazard to the worker because of the low current.

5.1.4 Powder Paint System

During the past ten years the application of powder coating has increased dramatically. By the mid-1990s, 20% of the appliances manufactured in the United States were finished with powder; it is now used on auto primer coats and clear coats. A major impetus in its acceptance has been the introduction of restrictive environmental air quality regulations. Powder coating blends productivity improvements with minimal adverse environmental impact. In this system a layer of powder incorporating all paint components is deposited on a workpiece. The workpiece is then processed through an oven which melts the powder converting it to a continuous paint film.

The composition of the powder mimics the composition of liquid paint with the exception that it contains no solvent. Each particle contains the resin and hardener (50–60%), pigments and fillers (40–50%), and various additives (1–2%). The common powder systems include epoxy, epoxy–polyester, polyurethane, and various enamels. In the standard powder manufacturing process, the various components are dry-blended in a mixer and processed through an extruder to form a thick, viscous putty. This extrudate is then milled to form a ribbon or strip which is ground and size-classified. The final powder has a geometric mean diameter in the range of 20 to 30 μm.

The parts to be painted undergo a rigorous preparation process which may involve as few as 6 or as many as 24 cleaning steps. The bulk powder is fluidized to provide a particle cloud. In early systems the parts were immersed and coated directly in the fluidized powder cloud but now the fluidized powder is usually applied with a gun dispenser. After the workpiece is coated it, passes through an oven operating at temperatures of 160 to 250°C where the powder melts to form a continuous film and hardening takes place. Enamels based on metallic oxide frits are fired at temperatures in the range of 780 to 830°C.

5.1.5 Powder Coating Application Methods

The fluidized powder is conveyed through a corona discharge (60–100KV) where the powder particles pick up a negative charge; they are then directed by the electrostatic field to the grounded workpiece and deposit as a uniform coating. If the workpiece geometry is complex, there will be poor deposition of the powder at angles and recesses due to the Faraday cage effect. In this case a different powder gun is used, one in which the particles obtain a positive charge by tribo-electrification. The powder laden airstream passes through. Teflon tubing where electrons are stripped off. The positive-charged particles are deposited on the workpiece with high efficiency and achieve better deposition in difficult areas than the negatively charged particles. The transfer efficiency of the powder system is in the high 90% range so little powder is lost from the hood. However, no air sampling data are available to demonstrate the dust exposure of the operator.

The powder dispensing gun may be (*1*) operated manually, (*2*) in a fixed position with automatic triggering to activate coating only when the conveyorized workpiece passes the gun station, or (*3*) it may be operated by an active robotic system. The automatic application head is best illustrated by the tunnel operation in Figure 28.16. In both cases, the inward air flow through the open face of the partially enclosing hood collects any fugitive powder losses for reuse. The powder can be recovered in a number of different ways: it

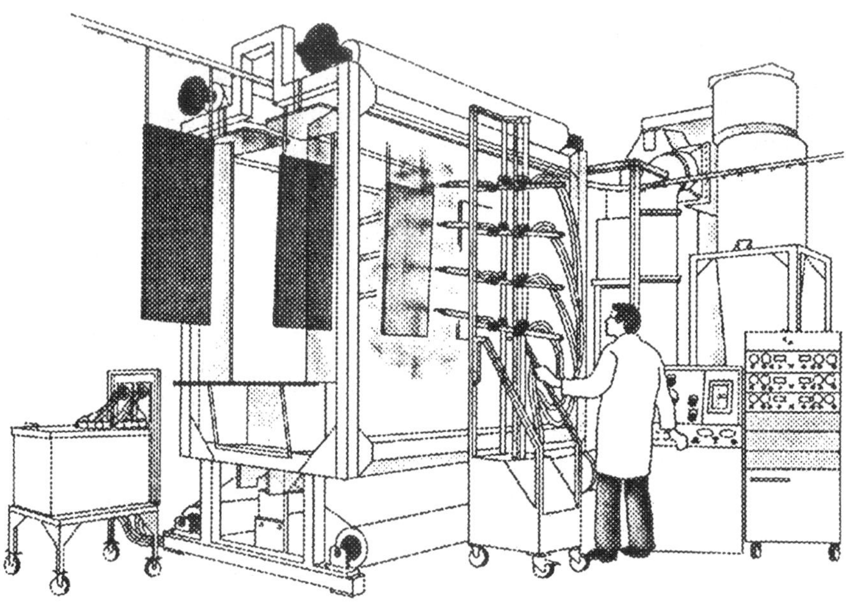

Figure 28.16. Automatic powder spraying in tunnel booth. Courtesy of Gema Volstatic

can settle onto a slow moving belt on the hood floor and be continuously vacuumed off for reuse; an integral two stage cyclone-filter system may be installed to collect the material; or a bank of cartridge filters collects the powder and it is recovered by reverse jet cleaning. The manufacturers of powder coating equipment state that powder utilization approaches 99%. The cartridge filter system is applicable on operations using a number of different colored powders. The system described above represents an application approach that is modeled after a wet paint system, that is the part is fabricated first and then painted. A new approach involves painting the base stock before the components are fabricated minor touch-up painting is completed after assembly. The production advantages of this approach are many and include its application to "just in time" manufacture. The precoating technique may present occupational health exposures if the coating is thermally stressed in the fabrication steps resulting in the release of thermal degradation products of the paint. (See Welding for a more detailed discussion of this topic.)

5.2 Worker Exposure Profile

5.2.1 Solvent and Mist Concentrations

In a review of solvent exposures in Danish manufacturing and service industries, a series of 72 air sampling measurements were conducted during spray painting (15). Thirty-eight percent of the mixed solvent exposures exceeded 0.25 of the ACGIH Threshold Limit Value, Time Weighted Average (TLV–TWA), 29% were between 0.25 and 1.0 of the ACGIH TLV–TWA and 33% exceeded the combined ACGIH TLV–TWA.

An extensive air sampling program was completed on a product line of paints which includes an isocyanurate resin with a residual monomer content of less than 0.3% and a polyisocyanate resin whose free monomer content is 0.7% at the time of manufacture but rises to a maximum of 1.6% during storage (16). Over an eight-year period, 28 spray painting operations were studied in addition to selected brush and roller applications; the latter data were grouped for discussion with a limited number of samples from paint mixing and preparation operations. The air sampling programs covered a wide application range including in-plant manufacturing, transportation and construction sectors. The authors sampled for the hexamethylene diisocyanate (HDI)-based monomer with an assigned ACGIH TLV–TWA of 5 ppb and a ceiling limit of 20 ppb proposed by the manufacturer. There is no TLV for the HDI polyisocyanate, however, the manufacturer established a guideline limit of 1.0 mg/m^3 and a eight-hour TWA guideline of 0.5 mg/m^3.

Since this is the most comprehensive data in the industrial hygiene literature on isocyanate spray painting exposures, the air sampling results are important. In the nonspray data the HDI polyisocyanurate results were all less than 3.5% of the manufacturer's guidance value of 1.0 mg per m^3 and the geometric mean value of the HDI concentrations were all less than one-third of the ACGIH TLV–TWA of 5 ppb. The spray painting data indicates the manufacturer's guidance level of 1.0 mg per m^3 was exceeded in several industrial segments; in some cases the value exceeded the guidance level by a factor of 10. Exposures to HDI in excess of the ACGIH TLV–TWA also occurred. In all cases, the air samples were taken outside of any respiratory protective device in use at the time.

A survey of 46 automobile shops involving 64 workers who did spray painting intermittently, with the balance of the shift devoted to preparation of the body work, was conducted in Australia (17). Thirty percent of the shops used only an acrylic-based paint, 20% used only a polyurethane paint and the balance used both paint systems. Two-thirds of the shops had local exhaust control and the balance used general exhaust. Reusable, combination paint mist and vapor air-purifying respirators were available in all shops, however, approximately one-half of the workers used nuisance-type respirators. Airline respirators were available in one-tenth of the shops but used only occasionally. The air concentrations of solvent vapors were low with a mean concentration of one-fifth of the occupational exposure standard. The authors state that the conditions noted in the study are probably representative of small repair shops where work is intermittent and of short duration.

A Danish study continues to be the most comprehensive review of the potential occupational health problems associated with the new generation of water-based paint (18). Although the work was conducted in architectural settings, it does offer insight into the health hazards in industrial applications. The study was prompted by the fact that as early as the mid-1980s, approximately 90% of all construction paints in Denmark were water-based. The study included 34 water-based products including both high quality acrylic latex paints formed with acrylic copolymer binders for indoor/outdoor work and lower quality indoor paint formed with acrylic co-polymers. The contents of the raw materials in the water-based paints are summarized in Table 28.4. Since the latex paints do not contain pigments, plasticizers, and driers, a number of these categories show zero concentration. The binders in the water-based paints are based on the monomers listed in Table 28.5. The latices are produced by emulsion polymerization of the monomer with droplets dispersed in water. The latex binders may be copolymers based on two to five monomers including butyl acrylate, acrylic acid, and styrene. The high molecular weight polymers (approximately 1,000,000) have negligible toxicity. The latices are not solvent-free; solvents are added in low concentration for proper film formation during application of the

Table 28.4. Minimum and Maximum Concentration of Raw Materials in Water-Based Paints[a]

Chemical Compound	Content (% w/w)
Water	26–86
Biocides	0–1.1
Surfactants	0.7–5
Pigments/extender pigments	0–54
Binders	
Latex (dry weight)	6–27
Alkyds (dry weight)	6–14
Coalescing agents/cosolvents	0–15
Amines	0–0.5
Plasticizers	0–2.2
Driers	0–0.6
Others	0–4

[a]Ref. 18. Courtesy of *Scandinavian Journal of Work, Environment and Health.*

Table 28.5. Monomers for Latex Production in Water-Based Paints[a]

Chemical Compound	Typical amount (% w/w)
Acryamide	
Acrylonitrile	0.001
Acrylic acid	
Butyl acrylate	0.02–0.06
Butyl methacrylate	
2-Ethylhexyl acrylate	0.01
Methacrylic acid	
Methyl methacrylate	0.01
Styrene	0.01–0.04
Vinyl acetate	
Vinyl ester of Versatic 10® (VeoVa 10®)[b]	

[a]Ref. 18. Courtesy of *Scandinavian Journal of Work, Environment and Health.*
[b]Versatic 10® is a mixture of branched acids with nine carbon atoms.

paint. The solvent may be used alone or as a complex solvent mix. The maximum solvent concentrations present in the 34 paints in the Danish study ranged from 0.01 to 10%. The residual monomer concentration was in the hundreds of parts per million range.

Biocides are used to maintain stability of water-based paint in the can. The paint manufacturer may purchase a proprietary biocide which may contain one of a dozen individual compounds. The common pigments in white water-based paints in the study include clays, mica, talc, zinc oxide, and titanium dioxide. The pH of the paint is adjusted with ammonia. A series of driers includes cobalt naphthenate and barium, cobalt, and zirconium octoates.

An extension of the study by Hansen (18) included short period air sampling conducted for periods of 20 min. at fifteen work sites to define worker exposure. The concentration of most of the volatile components were low except that formaldehyde, ammonia, mineral spirits, and ethylene glycol butyl ether were in the range of 0.25 to 1.0 of the occupational exposure limit. Sensitization from the biocides is not considered a real problem since the chemical biocide concentrations are too low to initiate sensitization. Skin irritation may occur from the surfactants. The residual monomers were not considered a problem due to their low concentration.

In developing an air sampling strategy for reactive elements in an epoxy paint mist, one must acknowledge that the resin molecules in the mist are undergoing cross-linking with curing agents; this reaction starts at the time of mixing and continues while the material is airborne (19). The author reminds the industrial hygienist that spray painting mist is a complex mixture of organic binder, organic pigment, solid inorganic pigment, curing agent, and residual solvent (20).

5.2.2 Particle Size of Paint Mist

The conversion of an automobile paint system from a lacquer to a high solids paint prompted an investigation of the particle size of the paint mist (21). Basecoat and clearcoat application by conventional air atomizing manual and robotic systems operating at 200 to

1356 psig had a mass median aerodynamic diameter (MMAD) of 4 to 12 μm depending on the air pressure. The clearcoat application by a rotary atomizer with rotational speeds of 17,000 to 40,000 rpm generated particles of 20 to 35 μm.

In another study, the MMAD of lacquer mist was characterized as 6.4 μm with a standard deviation of 3.4; an enamel had a MMAD of 5.7 with a standard deviation of 2.0. In both cases, spray painting was conducted by conventional air atomization (22).

5.3 Control Technology

The two major determinants of exposure in the painting workplace are the toxic components of the paint and the application technique. The ideal joining of these two issues would result in a paint without toxic components applied with a technique that deposits all of the paint on the workpiece. This is an ideal which cannot be attained, however, a number of advances permit a more close approach to that ideal.

5.3.1 Paint Systems and Application Techniques

The form of a given paint may be modified to provide the same performance while reducing the potential exposure to workers. As an example, the following actions have been taken over the past two decades in the design of isocyanate paint systems. In order to reduce the concentration of the monomeric isocyanates, chains of the monomers (oligimers) have replaced the straight monomers in many paint systems (23). As noted earlier, a number of systems may be used to block the isocyanate group until the reaction is initiated by heating. A major occupational health problem with epoxy paint systems due to the amine curing agent can be controlled by the substitution of less toxic polyamide resins (24).

The major study of water-based paints and worker exposure cited earlier (18) confirms that these systems represent an improvement over solvent-based systems and should be the paint of choice if they provide the necessary performance. Although the water-based paints are usually preferable there may be hazardous compounds in water-based systems, albeit in low concentration. The authors of this study recommend that water-based paint formulations be designed to eliminate formaldehyde, minimize the unreacted monomer content in the various polymers, eliminate ethylene glycol ethers because of their reproductive hazard, and minimize the ammonia content.

Although attention has been given to the elimination of chromium in paint systems these toxic compounds may still be present as a pigment in chrome green, chrome yellow, and molybdenum orange or as a drier in oleoresinous and alkyd paints. Also, the elimination of asbestos and crystalline silica in paint systems is necessary to eliminate the hazard during subsequent refinishing work.

The choice of the paint application system is frequently governed by production quantities and quality demands. There is a clear difference in contaminant release rates between passive coating systems such as brush, roller, dip, and flow compared to spray painting. In most industrial operations, however, spray painting will be the production choice. Within the spray painting techniques there are a number of options. The best single index for workplace contamination is transfer efficiency. All other factors being equal, one should pick the spray painting system with the highest transfer efficiency.

An environmental problem frequently occurs in paint shops when conveyor hangers, racks, and hooks become heavily coated with paint. The encrustation is usually removed with chemical paint strippers that present an exposure to the worker. Small furnaces are now available which destroy the coating by pyrolysis at 800°C to form an ash which can easily be removed.

5.3.2 Workplace Design, Work Practice, and Medical Surveillance

A number of straightforward control measures have been proposed to control fugitive emissions in paint mixing areas (paint kitchens) and storage areas (25). All storage and mixing vessels should be provided with close fitting covers designed with access ports. It should be normal design practice to equip such tanks with integral agitators since clamp-on agitators contribute to spills and resulting elevated air concentrations. All dispensing stations should be provided with collection trays and safety cans. Transfer of solvent should be done by closed-pump systems and not by open pouring.

A common exposure to organic vapors occurs when spray operators place the freshly sprayed parts on a rack directly behind them. The air movement to the spray booth sweeps over the drying parts and past the breathing zone of the operator, resulting in an exposure to solvent vapors.

Controls in the application of two-component urethane and epoxy paint systems must include excellent housekeeping, effective ventilation control, and protective clothing; moreover, in applications not effectively controlled by ventilation, the operators should wear air-supplied respirators. Adequate washing facilities should be available, and eating, drinking, and smoking should be prohibited in the work area. Dermatitis due to primary irritation and defatting from solvents or thinners as well as sensitization from epoxy systems is common. Skin contact must be minimized, rigorous personal cleanliness encouraged, and suitable protective equipment used by the operator.

5.3.3 Ventilation

Most spray painting operations utilizing solvent-based paints require exhaust ventilation for control of solvent vapors at the point of application and during drying and baking operations. Water-based paints may require ventilation only when spray application is utilized. Flow application of solvent-based paints may require local exhaust ventilation depending on the application technique. A number of references are available that provide the necessary design criteria (7, 13).

There are two main types of hoods available for in-plant use. The most common is designed to provide a lateral flow of clean air past the breathing zone of the operator with a uniform velocity distribution across the hood face, thereby minimizing worker exposure to air contaminants. Downdraft hoods are installed in heavy industry for painting large, irregular shaped objects such as castings; since these hoods are of special design and their use is limited they will not be discussed here.

The conventional spray booth is a heavy gauge sheet metal enclosure with one open face (Fig. 28.17). The side panels and roof may have openings to allow a conveyor to bring the parts directly to the spray area. This feature is especially important in automated painting operations. The choice of hood size is primarily dependent on the size of the parts

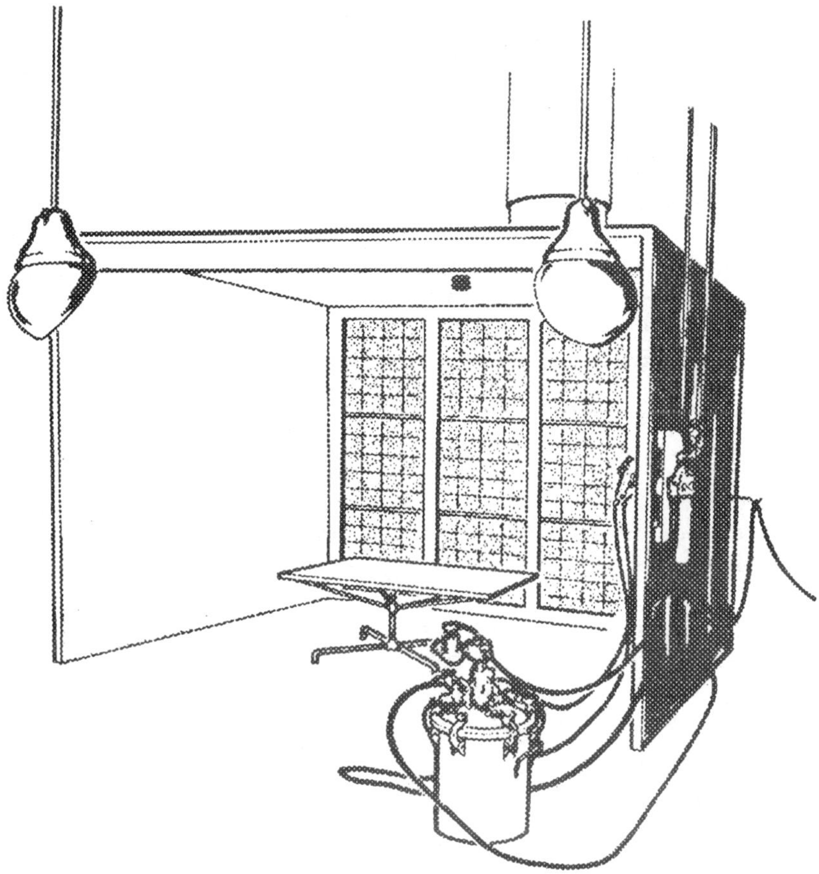

Figure 28.17. Liquid paint spray booth.

to be painted. It is important that the face area of the hood be large enough so the part does not block airflow through the hood face, which will result in poor capture efficiency. Geometry of the parts to be painted frequently requires a holding fixture or a turntable for complete spray coverage; the booth must be sized to accommodate this equipment.

Booth depth is critical to good performance since it helps to ensure a flat velocity profile at the hood face with resulting effective capture of toxic air contaminants. Rebound from flat surfaces sprayed in a shallow booth may escape capture; a deep booth assists in the recovery of such contaminants. It is important to maintain good airflow across the entire face of the paint spray booth. This is conventionally accomplished by the use of metal baffles, dry filter pads defining the rear plenum or by the use of a water-wash curtain.

The booths are commonly equipped with some type of air cleaning component to provide paint mist removal varying from a simple water curtain or a throwaway dry filter to a more sophisticated scrubber. The efficiency of these devices against paint mist has not been thoroughly evaluated. There is little information on the performance of water wash paint spray painting booths. One investigator found the efficiency of the scrubber to be size dependent with efficiencies as low as 64% for paint mist particles 0.6 μm in size (26). The performance could be improved by increasing the pressure drop across the scrubber element or more impressively by reducing the size of the fine overspray.

An extensive study of the control efficacy of paint spray booths was conducted in three companies with a total of seven plants (26). A total of 34 paint spray booths with a normal face opening of 2 m by 2.5 m and a mean face velocity of 0.55 m/s were used for conventional, airless and high solids spray paint applications. The authors classed the paints as either low or high aromatic paints although they did not indicate the basis for this classification. The study demonstrated the high performance of properly designed paint spray booths; in almost all cases the air concentration were less than one-tenth the ACGIH TLV.

5.3.4 Respiratory Protection

In general, the minimum respirator for all paint applications should be a combination mist-organic vapor air-purifying device. Higher levels of protection including air-supplied hoods or helmets may be necessary on certain paint systems such as the spray application of isocyanate-based paint. In the use of the air line systems, caution is required to ensure that the air supply meets minimum requirements. In a series of fatalities due to the use of inert gas supplies on air line equipment, five of the eleven fatalities were painters (27).

6 METAL MACHINING

The fabrication of complex metal parts from solid stock is done with a variety of machine tools the most common of these are the lathe, milling machine, shaper, planer, drill press, and surface grinder. The occupational health hazards from these operations are similar, so they will are grouped together for discussion. Five special machining techniques, electrochemical machining, electrical discharge machining, hard metal tool grinding, laser cutting and waterjet cutting are seeing expanded use, but will not be covered in this brief treatment of the subject.

The major occupational health issues associated with metal machining operations are attributable to dermal contact and inhalation exposures to the machining fluids mists generated during machining. Dermatoses from skin contact with machining fluids is, of course, the most common occupational disease of the industry and each year will affect one in three machinists. There is also convincing epidemiologic evidence that respiratory disease is associated with exposure to airborne machining fluids. With the exception of highly toxic metals, such as berryllium, the airborne metal particles released during the operation do not represent a health risk.

6.1 Conventional Metal Machining Practice

All dimensional machining operations are done by forcing the point of a cutting tool against the metal workpiece in a given direction with a controlled speed and force (Fig. 28.18).

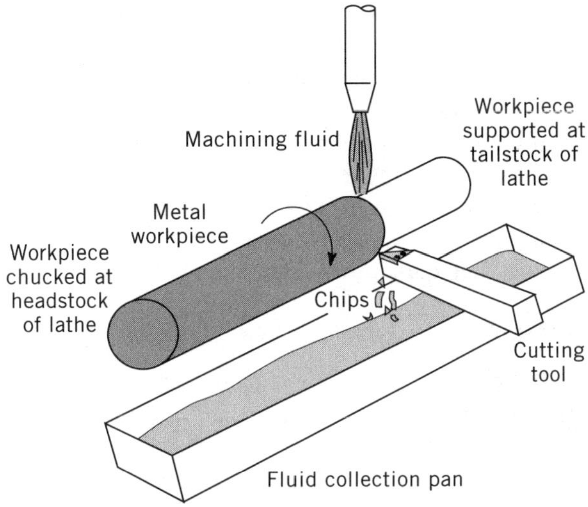

Figure 28.18. Components of a machining operation.

The cutting element removes the metal by a shearing or abrading action. This material is released as a fine continuous wire-like chip, a heavy short chip, or as individual particles.

The four elements in a conventional machining operation shown in Figure 28.18 are the machine tool, cutting tool, machining fluids, and workpiece metal. The health hazards these elements may present and controls for these hazards are discussed below.

6.1.1 Machine Tools

A large number of machine tools have been designed to accomplish dimensional cutting on a range of metals and plastics. The cutting tool may be fixed and the moving workpiece fed past the tool as in a lathe or the part may be fixed and the cutting tool moves past the stock as in a milling machine. In a grinding operation both the tool, in this case the grinding wheel, and the workpiece are moving at high speeds. The workpiece may be held by an adjustable collet or chuck on a lathe, mechanically clamped on the bed of a milling machine or held in place with a magnet on the workbed of a surface grinder. The cutting tool position, the workpiece speed, and the cutting tool speed may be controlled manually by the machinist or by numerical control (NC) or computer-assisted numerical control (CNC) systems. In a small job shop doing custom machining it may be "one worker-one tool"; in a large production shop with semiautomatic machine tools, each machinist may "tend" a large number of machines located by functional or group layout.

6.1.2 Cutting Tools

Special alloys have been developed for use in the manufacture of cutting tools. These materials include: (*1*) high carbon steels with alloying elements of vanadium, chromium, and manganese; (*2*) high speed steels containing manganese and tungsten; (*3*) special cobalt steels; (*4*) cast alloys of tungsten, chromium, and cobalt; and (*5*) tungsten carbide. In conventional machining the loss of mass from the cutting tool is insignificant and therefore does not represent a potential air contamination problem. On grinding operations, the cutting tools are individual grains of abrasive (aluminum oxide, silicon carbide, diamonds or cubic boron nitride) embedded in a matrix. Although air contamination arising from the cutting tool is insignificant during its use, in the manufacture of the cutting tools a significant exposure to toxic metal dusts occurs during grinding and dressing operations; these operations should be provided with local exhaust ventilation. Special hazards are also associated with the manufacture of tungsten carbide tools.

6.1.3 Workpiece Metal

The common view held by most in the occupational health community is that machining of mild steel and its alloys with cutting tools does not result in significant airborne metal concentrations. This was confirmed in work under a UAW–GM sponsored study on machining fluids by Woskie et al. (28) who found that less than 10% of the aerosol generated from grinding operations was non-oil; on conventional cutting operations such as turning and milling less than 2–3% was non-oil. Since all metal particles are found in the non-oil based fraction of the air sample it is obvious that the bulk of the airborne material originates from the machining fluids and minimal amounts of the metal workpiece becomes airborne. Machining of highly toxic metals such as beryllium do present significant potential exposures and require evaluation. Machining operations on magnesium and titanium may generate explosive concentrations of dust, and these operations must be conducted with suitable ventilation control and air cleaning.

Recent studies have shown that machining fluids may become contaminated with trace quantities of the base metal. The metals are then contacted by the worker from splashing or immersion of hands in the fluid in the course of the work. As is discussed below, this contact route may be responsible for the sensitization to chromium, nickel, and cobalt that occurs in this industry.

6.1.4 Machining Fluids

Extremes of temperature and pressure occur at the interface between the cutting tool and the work. To cool this point, provide an interface lubricant, and help flush away the chips formed by the cutting tool, a machining fluid is directed to the cutting point in a solid stream (flood) or a spray. The terms applied to these machining fluids are shown in Table 28.6.

6.1.4.1 Composition. These fluids are available in the form of: (*1*) mineral oils which contain greater than 80% mineral oil with the balance consisting of various additives; (*2*) emulsified oils consisting of 3–10% mineral oil, emulsifiers and other additives and water; and (*3*) synthetic machining fluids usually consisting of active chemical ingredients in aqueous solution. The synthetic machining fluids are based on a complex group of water soluble chemicals that result in a fluid that is basically a noncorrosive fluid with cooling and lubricating properties. A fourth class of machining fluids, semisynthetic, are based on emulsified oils with the addition of the chemicals found in synthetic oils.

Table 28.6. Machining Fluid Nomenclature

Preferred	Widely Used Synonyms
Straight oil	Mineral oil
	Petroleum oil
	Cutting oil
	Neat oil
Emulsified oil	Soluble oil
	Water-based fluid
	Suds
	Emulsions
Synthetic fluid	Chemical fluid
Machining fluid	Cutting fluid
	Coolants
	Metal-working fluids

The mineral oil-based machining fluids are usually blends of paraffinic and naphthenic hydrocarbons from refinery streams. Since the 1970s, solvent refining or deep hydrotreating has been used to reduce the polycyclic aromatic hydrocarbon content of these oils and therefore their carcinogenic potential. Aromatics from the vacuum distillation of paraffinic and naphthenic oils may be part of the blend.

Mineral oil-based machining fluids have been in extensive use for decades with the peak utilization probably in the 1960s. Since that time emulsified oil and synthetic machining fluids have seen a great surge in use. The composition of mineral oil machining fluids has changed with the addition of additives to improve the range of application and safety (Table 28.7).

6.2 Airborne Contaminants

Data are now available on the concentration of machining fluid aerosols and the particle size distribution for various metals and machining operations. Much of this information is available from a series of studies on the health status of machinists sponsored by the UAW-GM National Joint Committee on Health and Safety. In the phase of the study directed to retrospective occupational exposure assessment of machining fluid aerosols, Hallock et al. (30) defined past exposures in three large machining facilities. The arithmetic mean concentrations for all samples taken prior to 1970 was 5.42 mg/m^3; from 1970 to 1989 the arithmetic mean concentration dropped to 1.82 mg/m^3. This reduction in air concentrations of machining fluid aerosols is attributed to the implementation of plant control measures such as local exhaust ventilation, general exhaust ventilation, air cleaning, enclosure of the cutting tools, and increased use of water-based machining fluids.

In a Swedish study the average air concentration of machining fluid aerosol was approximately 2 mg/m^3 (31). In a general machine shop operation reviewed by NIOSH the concentration of machining fluids in a series of 20 personal air samples ranged from 0.11

Table 28.7. Additives to Mineral Oil Machining Fluids[a]

Lubricity additives
Extreme pressure additives
Antioxidants
Viscosity index improvers
Pour point depressants
Antimist additives
Tackiness additives
Corrosion inhibitors
Metal passivators
Scourability additives
Dyes, scents, reodorants
Antifoam additives

[a]Ref. 29

to 1.57 mg/m^3 with an arithmetic mean concentration of 0.38 mg/m^3 (32). A variety of machine tools and all three types of machining fluids were in use. In a NIOSH Health Hazard Evaluation of a shop machining aluminum, the air concentrations of soluble oil mist ranged from 0.14 to 1.08 mg/m^3 (33).

One of the few studies of the properties of machining fluids in plant use for an extended work period was conducted on severely refined mineral oil containing a sulfurized-fat extreme pressure additive (34). A centralized machining fluids system with a capacity of 20,000 gallons serviced a large number of automatic lathes machining mild steel. A principal goal of the study was to note changes in 4–6 ring polycyclic aromatic hydrocarbons (PCA) over the test period. The maintenance of the machining fluids system included periodic "topping-off" with machining fluids as required and an annual "laundering" which included the removal of water, sterilization at 110°C, removal of solids by pretreatment with subsequent plate and frame filtration and the adjustment of viscosity with mineral oil. Approximately 38% of the volumetric capacity of the system was replaced annually by tramp lubricating oil leaking into the system and the addition of new oil. The authors found that the PCA build up was not important as illustrated by the B(a)P content of the oil. The PCA content of the oil mist mirrored the content of PCAs in the bulk oil with the exception that the more volatile PCAs, i.e., fluoranthene and pyrene had a significantly higher content in the mist than in the bulk oil. Parallel studies of the sensitizing potential, irritancy, and skin carcinogenicity of the oil were negative over the five-year period.

In a study of machining fluids aerosols in an automotive transmission plant, Chan et al. (35) noted a bimodal distribution with large droplets greater than 8 μm arising from direct mist application of the machining fluids to the cutting tool and submicron particles generated by the centrifuges used to separate chips from the fluid. The total concentrations ranged from 0.60 to 2.99 mg/m^3 and the respirable fraction of the total from 20 to 50% of the total mass. The investigators noted that bacterium were present in the aerosol size fraction greater than 2 μm while bacterium were not noted in the less than 1 μm size fraction generated at the cutting tool.

6.3 Health Effects

Three major sources of information are available to the health professional on the occupational health risks from exposure to machining fluid aerosol. A symposium on this subject sponsored by the Institute of Occupational Health at the University of Birmingham in 1989 is extensively cited in this section (36). In the US a symposium sponsored by the American Automobile Manufacturers Association in 1995 with papers from government, industry, and university investigators covered the toxicology, epidemiology, assessment of exposure and control of machining fluids (37). Finally, the NIOSH Criteria For A Recommended Standard For Occupational Exposure to Metal Working Fluids provides a comprehensive summary of the health risks to workers (38).

6.3.1 Dermatitis

The majority of occupational dermatitis in machinists is due to water-based machining fluids, that is both emulsified oil and synthetic machining fluid. Although there is agreement on that point, there is divided opinion on the identity of the causal agent. Rycroft (39) states that the irritant emulsifiers are the principal cause. Ruane (40), however, believes that the high pH of machining fluids is the principal cause. In addition to these two mechanisms which together do account for the majority of cases (70–80%), sensitization may account for the balance (20–30%). The principal causal agents for sensitization include formaldehyde and the formaldehyde-containing release agents which are frequently the active agents in the biocide. In addition, rosin used as a base for the emulsifier and leachates of chromium, nickel, and cobalt from alloy steels may cause sensitization.

The rare case of dermatitis from mineral oil-based machining fluids is usually folliculitis or oil acne which is easily resolved by hygienic measures. Occasionally, sensitization is noted in machinists working with oil-based materials. Rycroft (39) cites sensitization in a worker exposed to dipene in a honing oil formulation and in a second case the sensitizer was an epoxide introduced in the machining fluid formulation to scavange chloride ions released from chloroparaffins, an extreme pressure additive in the machining fluid.

Rycroft (39) ranks the three major types of machining fluids, in descending order of potential for dermatitis as synthetics, emulsified oil, and finally mineral oil; for skin cancer the ranking is mineral oil, emulsified oil, and synthetic machining fluid.

6.3.2 Respiratory Disease

In the 1970s and 1980s, studies of respiratory cancer in workers with exposure to machining fluids showed inconsistent results. Eisen et al. (41) demonstrated a two-fold excess of cancer of the larynx in machinists in the automotive industry. Although a few cases of lipoid pneumonia have been attributed to machining fluids exposure, Burge (42) does not believe it is a problem at the current exposure range of 1–2 mg/m^3. Burge does feel that machinists are at risk for bronchitis and that occupational asthma is initiated by sensitization to the machining fluids aerosol. Finally, it is generally accepted that exposure to mineral oil-based machining fluids mist does not cause fibrotic changes in the lung.

Other malignancies attributed to machining fluids include cancers of the digestive tract (43–45). Bronchitis and asthma are also associated with inhalation exposures to machining fluid aerosols and a before and after work shift decrement in pulmonary function has been noted by Kennedy et al. (46).

6.3.3 Microbial Contamination

Emulsified oil and synthetic machining fluids provide the necessary water, nutrients, and temperature conditions for the growth of microorganisms. The organisms include aerobic bacteria which cause a drop in the pH of the machining fluids, anaerobic bacteria including sulfate-reducing bacteria which produce sulfides including hydrogen sulfide, yeast and fungus (47). The benchmark concentrations of microorganisms in machining fluids cited by Hill include concentrations of 50×10^3 colony forming units (CFU) per mL for systems with exceptional control programs, 10^6 CFU per mL for a reasonably achievable control program, and with a poor control program the concentrations may be a high as 10^9 CFU per mL (48).

6.4 Control Of Exposures To Machining Fluids

6.4.1 Machine Design

Mechanical splash guards are designed to keep the chips and oil inside machine enclosures. The general approach is the use of overlapping, drip-proof shields. To maintain their effectiveness, the guards must be robust and rigidly supported. If access is required, the guards should be designed to slide; removable guards are not recommended. If visual monitoring of the machine function is required, a clear view shield is appropriate.

The generation of machining fluid mist can be minimized by maintaining low temperatures at the cutting point by increasing the machining fluid flow rates. In an example of this control cited by Lee (49), the mist concentration was reduced from 67 to 0.4 mg/m^3 by this simple change. The machine tool and the associated machining fluid system must be designed to handle these high flow rates. The machine tool should also be designed for ease of access to the sump for cleaning. The use of "through the tool" machining fluid systems should be considered in tool design. If a chip conveyor is installed on high production machine tools, it should be designed for ease of cleaning.

6.4.2 Machining Fluid Design

The amount of emulsifier in water-based machining fluid formulations should be minimized since these agents are common skin irritants. The formulations should also be designed to control alkalinity; systems that run at high pH will cause defatting and respiratory irritation.

The ultimate control for MWF mist is the elimination of the metal working fluid itself. Currently some shops conduct dry machining on cast iron, nodular iron, and steel. Machine tool cutting should be conducted at the lowest possible speed compatible with production requirements to reduce mist generation.

6.4.3 Personal Protective Equipment

The following recommendations by Rycroft (39) reflect the wide experience and common sense of this occupational dermatologist.

"With regard to protective clothing, plastic over-sleeves can sometimes be helpful. Their materials and design could probably be improved. Gloves are often unusable for safety reasons but, when they can be used, their materials and design also requires attention. Barrier cream does not seem to help a great deal, probably because water-based cutting fluids tend to remove them. Sensitization by lanolin in barrier cream very rarely occurs. After-work skin conditioning cream does seem to help but achieving usage of after-work creams is often difficult. The choice of skin cleansers needs to be a balance between effectiveness and their adverse effect on the skin."

6.4.4 Work Practice

The ultimate approach for mist control is the elimination of the machining fluid. Currently some shops conduct dry machining on cast iron, nodular iron, and steel. Machine tool manufacturers are now considering this technique for other metals. Also machine tool cutting and grinding operations should be conducted at the minimum speed compatible with production and quality to minimize the generation of mist. The higher the speed, the greater the mist generation, especially in grinding. A written machining fluids management program is necessary to ensure properly designed fluids are used in the manner prescribed by the manufacturer (29).

6.4.5 Ventilation

The exhaust hoods commonly used to capture toxic metal dusts from machining include conventional exterior hoods, high velocity-low volume capture hoods, and enclosures. It is widely understood that all machining operations on beryllium metal require local exhaust ventilation to control airborne dust. Although it is widely accepted that grinding of all beryllium alloys does require local exhaust ventilation. The level of control required on machining of copper–beryllium alloys is less clear and the operations should be evaluated to determine the required ventilation control.

The application of machining fluids under conditions of high load may result in airborne machining fluids mist concentrations that exceed 1 mg/m^3. In this case the application of general or local exhaust ventilation should be considered. In high production shops general exhaust is commonly used with air cleaning and recirculation of air. This approach requires monitoring of the system to ensure conditions are acceptable for recirculation. The most effective approach to control machining fluid aerosols is local exhaust based on partial or full enclosing hoods.

7 WELDING

Welding is a process for joining metals in which coalescence is produced by heating the metals to a suitable temperature. In the United States it is estimated that there are 185,000 welders, brazers, and cutters with an additional 700,000 workers who do some welding as a part of their job. Although current data on the distribution of welding processes are not available in the United States, Ulfvarson (50) states the distribution in Sweden is 58% shielded metal arc welding, 20% metal inert gas welding and metal active gas, 10% tungsten inert gas welding, 9% gas and 3% submerged welding. The distribution by metal welded in Sweden is 64% mild steel, 22% stainless steel, 9% aluminum, and all other metals, 5%. In the United States there is probably less inert gas shielded arc welding and stainless steel work than in Sweden. Stern (51) states that shielded metal arc welding, metal inert gas welding, and tungsten inert gas welding on mild steel and stainless steel coupled with metal inert gas welding on aluminum accounts for 80% of all welding done in the world.

Welding processes can be classed as pressure, nonpressure, or brazing. The nonpressure welding techniques warrant major attention because they have a higher contaminant generation rate than the other processes. The American Welding Society (AWS) lists eighty welding processes in general industrial use. Only two welding techniques are chosen for discussion in this brief coverage—shielded metal arc welding and metal inert gas welding, which together, represent more than three quarters of the welding done in the United States today. A large number of processes including tungsten inert gas, gas, compression, ultrasonic, friction, percussion, thermit, electron beam, and explosive welding are not covered either because they present limited potential for exposure or the hazards they present are similar to the processes that are described.

In the nonpressure welding techniques, metal is vaporized in the arc and then condenses to form a fume in the 0.01 to 0.1 μm particle size range that agglomerates rapidly. The source of this respirable metal fume is the base metal, the metal coating on the workpiece, the electrode, and the fluxing agents associated with the particular welding system. A range of gases and vapors including ozone and nitrogen dioxide may be generated depending on the welding process. The wavelength and intensity of the electromagnetic radiation emitted depend on the welding procedure, inert gas, and base metal.

The International Agency for Research on Cancer has classed welding fume in Group 2A, that is, possible carcinogenic to humans (52). A number of studies have shown pulmonary disease in welders including small airway disease, chronic bronchitis, and X-ray abnormalities. Reduced sperm quality in welders has also been found. The most common acute conditions seen in welders include pulmonary edema from gas welding in confined spaces, severe photokeratitis from UV exposure from viewing the welding arc without eye protection, and metal fume fever from welding and cutting copper- and zinc-coated metals.

The common welding techniques are reviewed assuming that the welding is done on mild steel. The nature of the base metal is important in evaluating the metal fume exposure and occasionally it has impact in other areas. When such impact is important, it is discussed.

7.1 Shielded Metal Arc Welding

7.1.1 Process

This widely used fusion welding process is also commonly called manual metal arc welding, stick or electrode welding. An electric arc is drawn between a welding rod and the workpiece, melting the metal along a seam or a surface (Fig. 28.19**a**). The molten metal from the workpiece and the electrode form a common puddle and cool to form the weld and its slag cover (Fig. 28.19**b**). The slag formed over the weldment, made up of flux and

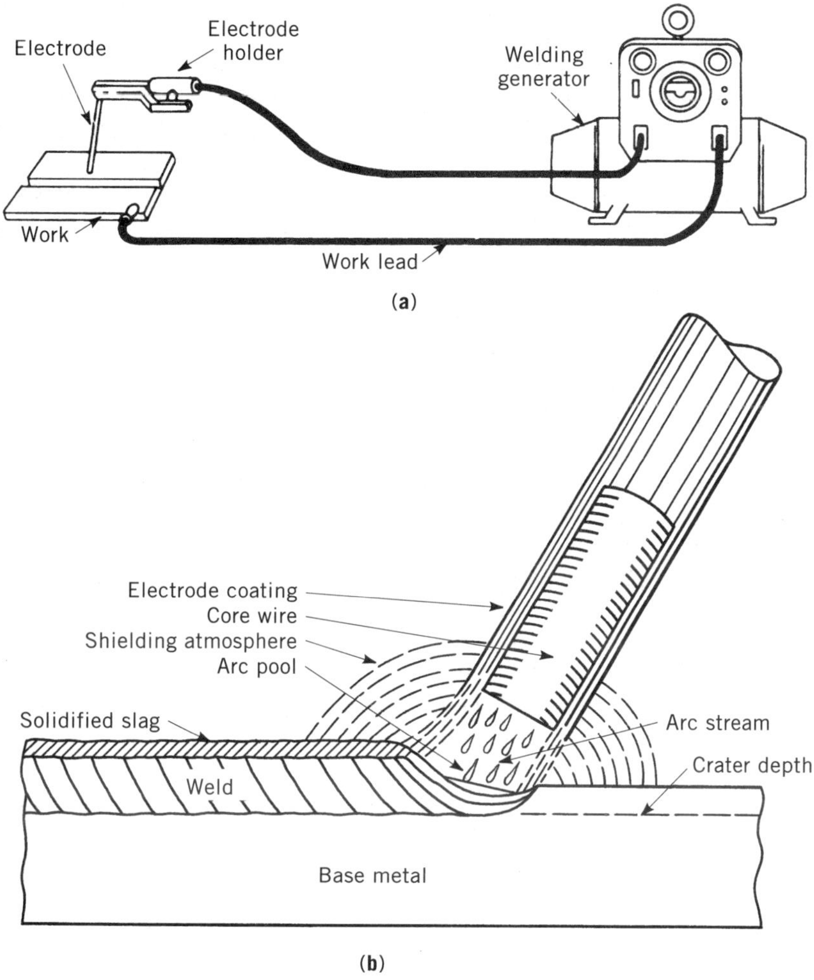

Figure 28.19. (**a**) Shielded metal arc welding (SMA) (**b**) Cross section of the arc area. Courtesy Hobart Brothers, Inc.

impurities, is removed with a manual or pneumatic chipper. Power, DC or AC, is used in either straight (electrode negative, work positive) or reverse polarity. The most common technique involves DC voltages of 10 to 50 volts and a wide range of current up to 2000 A. Although operating voltages are low, under certain conditions an electrical hazard may exist.

The welding rod or electrode may have significant occupational health implications. The electrodes contains the same or similar metal alloy as the parent metal. The three principal electrode types are cellulosic (TiO_2, sand, and magnesium silicate), rutile (TiO_2, $CaCO_3$ plus some cellulose) and basic (high content of calcium carbonate or fluoride).

Initially, a bare electrode was used to establish the arc and act as filler metal. Now the electrode covering may contain 20 to 30 organic and inorganic compounds and perform several functions. The bonding agent that holds the extruded material on the electrodes is water glass. The principal function of the electrode coating is to release a shielding gas such as carbon dioxide to insure that air does not enter the arc puddle and thereby cause failure of the weld. The carbonates in the electrodes coating are degraded thermally in the arc to form carbon dioxide. In addition, the covering stabilizes the arc, provides a flux and slag producer to remove oxygen from the weldment, adds alloying metal, and controls the viscosity of the metal.

A complete occupational health survey of shielded metal arc welding requires the identification of the rod and its covering. The range of compounds in the fume from coated electrodes is shown in Table 28.8. The composition of the electrode can be obtained from the AWS classification number stamped on the electrode as shown in Figure 28.20.

The percent of total electrode mass that is released as airborne fume ranges from 0.5 to 5%. Pantucek (54) states that the "fume" generated from coated electrodes may contain iron oxides, manganese oxide, fluorides, silicon dioxide, and compounds of titanium, nickel, chromium, molybdenum, vanadium, tungsten, copper, cobalt, lead, and zinc. The manganese content of welding fume ranges from 3 to 10%. In a short communication, Moreton and Jenkins (55) state the content of barium in welding fume from self-shielded, flux-coated electrodes made by one U.K. manufacturer ranges from 0.1 to 34%. Silicon dioxide is routinely reported as present in welding fume, apparently originating from the various silicon compounds in the electrode coating. Buck and Dessler (56) state that the silicon dioxide is present in an amorphous form or as silicates, not in the crystalline form.

One electrode type of great occupational health importance is the low hydrogen electrode used with conventional shielded metal arc welding systems to maintain a hydrogen-free arc environment for critical welding tasks on certain steels. The electrode coating is

Table 28.8. Welding Fume Concentration Range for Various Processes[a,b]

Contaminant	Shielded Metal	Gas Metal	Gas Metal Flux Core	Gas
		Fumes (mg/m^3)		
Total fume	1–25	1–30	1–25	2–15
Iron oxide	0.2–10	1–10	0.25–15	1–10
Copper	0.01–0.15	0.01–0.5	0.01–0.2	0.01–0.05
Manganese	0.01–1	0.01–0.05	0.02–2	0.01–0.05
Fluoride	0.01–3	0.01	0.02–5	0.01
		Gases (ppm)		
Nitrogen dioxide	0.1–3	0.1–5	—	0.01–1.5(60)
Carbon monoxide	1–15	1–20	—	5–10
Ozone	0.01	0.01–0.4	—	—

[a]Ref. 53.

[b]These data are based on over 3000 welding investigations in industry, representing a wide range of ventilation and degree of enclosure

Fourth digit	Type of coating	Welding current
1	Cellulose potassium	AC or DC reverse or straight
2	Titania sodium	AC or DC straight
3	Titania potassium	AC or DC straight or reverse
4	Iron powder titania	AC or DC straight or reverse
5	Low hydrogen sodium	DC reverse
6	Low hydrogen potassium	AC or DC reverse
7	Iron powder iron oxide	AC or DC
8	Iron powder low hydrogen	DC reverse or straight or AC

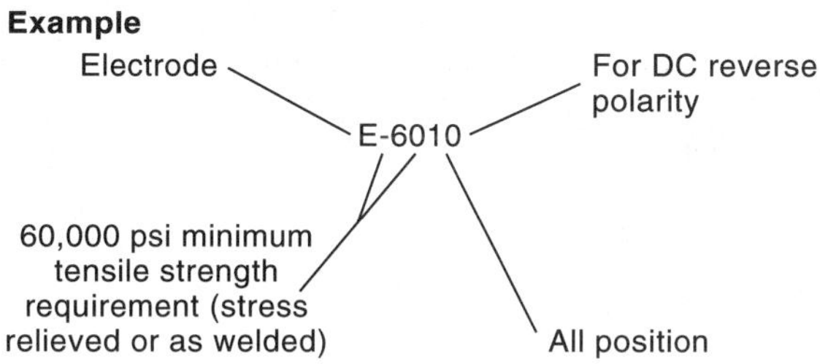

Figure 28.20. American Welding Society shielded electrode classification.

a calcium carbonate–calcium fluoride system with various deoxidizers and alloying elements such as carbon, manganese, silicon, chromium, nickel, molybdenum, and vanadium. A large part of this coating and of all electrode coatings becomes airborne during welding. Early studies showed that 9% of the fumes from low hydrogen electrodes may be fluorides; of that quantity, 10 to 22% is in the form of soluble fluorides. Fluorides may comprise up to 20% of the total fume concentration depending on the coating thickness and electrode core diameter.

It has been suggested that the calcium fluoride is degraded in the arc to silicon hexafluoride, which then forms hydrogen fluoride in the presence of water vapor. Although hydrogen fluoride is found in the low hydrogen welding environment, it is not known if the hypothesized mechanism is correct.

7.1.2 Exposure Profile

7.1.2.1 Metal Fumes. The potential health hazards from exposure to metal fume during shielded metal arc welding obviously depend on the metal being welded and the composition of the welding electrode. The principal component of the fume generated from mild steel is iron oxide. The hazard from exposure to iron oxide fume appears to be limited. The deposition of iron oxide particles in the lung does cause a benign pneumoconiosis known as siderosis. There is no functional impairment of the lung nor is there fibrous tissue proliferation. In a comprehensive review of conflicting data, Stokinger (57) has concluded that iron oxide is not carcinogenic to humans.

The concentration of metal fume to which the welder is exposed depends not only on the composition of the parent metal and the electrodes but on the welding process including the current density (amperes per unit area of electrode), wire feed rate, the arc time (which may vary from 10 to 30%), the power configuration, that is, DC or AC supply, and straight or reverse polarity. The work environment also defines the level of exposure to welding fume and includes the type and quality of exhaust ventilation and whether the welding is done outdoors, in an open building workstation, or in a confined space. Much of the data on metal fume exposures have been generated from shipyard studies; in these studies the air concentration of welding fume ranged from less than 5 to over 100 mg per m^3 depending on the welding process, ventilation, and the degree of enclosure.

Fume Generation Rate. Studies by the American Welding Society in the 1970s provide extensive data on the fume generation rate from various welding processes (58). Development of standard methods to define fume generation rate has continued in both the U.K. and the Scandinavian countries and additional data on rates have been published. Ulfvarson (59) states that all welding processes have similar generation rates with the exception of tungsten inert gas welding which is significantly lower than the others. Observed generation rates cited by Ulfvarson range from 20 mg per min. for shielded metal arc welding to 80 mg per min. for metal inert gas welding. The authors state that differences in fume exposures noted in workplace studies are mainly due to differences in duty cycle, i.e., the amount of time the arc is struck.

Field Studies. Exposure data continue to emerge from shipyard studies. The health status of over 300 welders in 65 Singapore shipyards working with shielded metal arc welding on mild steel was studied by Phoon and Tan (60). Samples of total fume were taken outside the welders helmet for periods of one hour. Twenty-three percent of the total fume samples exceeded 5 mg per m^3 in open welding spaces while 68% exceeded that concentration in confined space work. The bulk of the fume samples contained oxides of iron, manganese, and zinc. Occasionally, the air samples contained lead in excess of 0.2 mg per m^3.

Exposure to fumes from low hydrogen welding under conditions of poor ventilation may prompt complaints of nose and throat irritation and chronic nosebleeds. There has been no evidence of systemic fluorosis from this exposure. Monitoring can be accomplished by both air sampling and urinary fluoride measurements.

A comprehensive study was completed by Smith (61) of metal fume exposure during low hydrogen welding in various industrial settings classified as confined, enclosed, and open conditions. In this study 75% of the samples taken in confined spaces exceed 10 mg per m^3; 50% of the enclosed samples and 27% of the open samples exceeded this condition. The current TLV for fluorides of 2.5 mg per m^3 was exceeded in 37% of the samples taken in confined spaces and in 13% of those collected in enclosed and open locations.

Dawada and Twano (62) have shown that low hydrogen welding on low and high alloy steels produces oxides of nickel, chromium, molybdenum, and copper in addition to the normally observed oxides of iron, silicon, and manganese. Airborne fluorides are also present in significant concentrations. In other studies, there did not appear to be a direct correlation between the percent of fluorides in the electrode coating and in the welding fume. The exposure to various gases including nitrogen dioxide and ozone during low hydrogen welding is apparently the same as that during conventional electrode welding.

Sjogren et al. (63) have demonstrated a linear correlation of both personal breathing zone total fume concentrations and fluoride concentrations in air with post-shift urinary fluoride in welders employing basic electrodes containing 18–20% fluoride. Air samples were taken inside the welding helmet for a minimum of one-half shift. The maximum total fume concentration was 13 mg per m^3 and the air concentration of fluoride was approximately 1.2 mg per m^3.

7.1.2.2 Gases and Vapors. Shielded metal arc welding has the potential to fix atmospheric nitrogen as nitrogen oxides at temperatures above 600°C. Concentrations are not a problem in open shop welding. In over 100 samples of shielded metal arc welding, the author has not identified an exposure to nitrogen dioxide in excess of 0.5 ppm under a wide range of operating conditions. Oxygen is also fixed as ozone by the arc but again, it is not a significant contaminant in shielded metal arc welding operations. Carbon monoxide and carbon dioxide are produced from the electrode covering, but air concentrations are usually minimal.

In two studies of shipboard welding, the maximum concentrations of various gases noted were nitrogen dioxide (212 ppm), carbon monoxide (110 ppm), acrolein (0.08 ppm), and ozone (0.6 ppm) (64, 65). These high concentrations were observed in enclosed space welding with limited exhaust ventilation.

7.1.2.3 Radiation. The radiation generated by shielded metal arc welding covers the spectrum from the IR-C range of wavelengths to the UV-C range. To date, there has been no evidence of eye damage due to infrared radiation from arc welding. The acute condition known to the welder as "arc eye," "sand in the eye," or "flash burn" is due to exposures in the UV-B range. The radiation in this range is completely absorbed in the corneal epithelium of the eye and causes a severe photokeratitis. Severe pain occurs 5 to 6 hr after exposure to the arc and the condition clears within 24 hr. Welders likely experience this condition only once and then protect themselves against a recurrence by the use of a welding helmet with a proper filter, in addition to tinted safety goggles. Skin erythema or reddening may also be induced by exposure to UV-C and UV-B.

Of the major processes, metal inert gas welding produces the most ultraviolet radiation with shielded metal arc welding the next most effective and finally tungsten inert gas welding (52). The UV flux at one meter from the arc is in shielded metal arc welding is 0.7–2.5 W per m^2 (66).

The typical magnetic flux density at the 60 Hz welding generator is 2–200 μT (67). Current pulses of up to 100 000 A produce magnetic fields of up to 10 000 μT at 0.2 to 1.0 m from the welding transformer cables (68).

7.2 Gas Metal Arc Welding

7.2.1 Process

In the 1940s a consumable wire electrode with an annular inert gas blanket was developed to replace the nonconsumable tungsten electrode used in the gas tungsten arc welding system. Originally developed to weld thick, thermally conductive plate, this gas metal arc welding process now has widespread application on aluminum, copper, magnesium, nickel alloys, titanium, as well as steel alloys.

In this system (Fig. 28.21) the welding torch has a center consumable wire fed from a reel or spool that maintains the arc as it melts into the weld puddle. Around this electrode

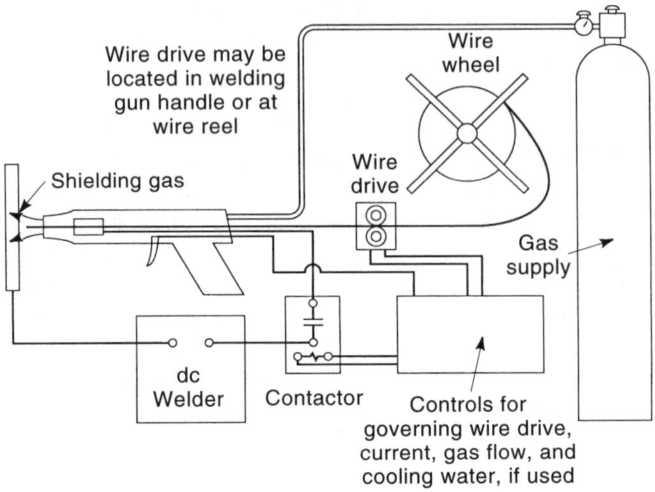

Figure 28.21. Equipment for gas metal arc welding.

is an annular passage for the flow of helium, argon, carbon dioxide, nitrogen, or a blend of these gases. The wire usually has the same or similar composition to the base metal with a flash coating of copper to ensure electrical contact in the gun and to prevent rusting. If filler wire is used, it contains deoxidizers or scavengers. Manganese, silicon, and aluminum are used as deoxidizers for steel filler wires; titanium and silicon, for nickel alloy wire; and titanium, silicon, and phosphorus, for copper alloy wires.

An improvement in gas metal arc welding is the use of a flux-cored consumable electrode. The electrode is a hollow wire with the core filled with various deoxidizers, fluxing agents, and metal powders. The arc may still be shielded with carbon dioxide or the inert gas shield may be generated by the flux core alone. In the latter case the process is called self shielding. The fume will include oxides of the filler wire as well as copper fume if the wire is copper clad.

Initial attempts to use CO_2 as the gas were not completely successful. The introduction of wire with an antioxidant made its use possible. The process using CO_2 is termed metal active gas welding or MAG. The duty cycle of metal inert gas welding is much higher than shielded metal arc welding since there is no slag to remove.

7.2.2 Exposure Profile

Iron oxide fume concentrations with gas metal arc welding are generally lower than shielded metal arc welding. The nitrogen dioxide concentrations are of the same order of magnitude; however, ozone concentrations are higher with the gas metal arc welding technique. Two separate investigations indicate that gas metal arc welding produces higher concentrations of ozone than gas tungsen arc welding (69, 70). The ozone generation rate increases with an increase in current density, but plateaus rapidly. The arc length and the inert gas flow rate do not have a significant impact on the ozone generation rate.

Carbon dioxide is widely used on gas metal arc welding due to its attractive price; argon and helium cost approximately 15 times as much as carbon dioxide. The carbon dioxide process is similar to other inert gas arc welding shielding techniques and one encounters the usual problem of metal fume, ozone, oxides of nitrogen, decomposition of chlorinated hydrocarbons solvents, and UV radiation. In addition, the carbon dioxide gas is reduced to form significant concentrations of carbon monoxide. The generation rate of carbon monoxide depends on current density, gas flow rate, and the base metal being welded. Although the concentrations of carbon monoxide may exceed 100 ppm in the fume cone, the concentration drops off rapidly with distance and with reasonable ventilation; hazardous concentrations should not exist at the breathing zone.

The intensity of the radiation emitted from the arc is an order of magnitude greater than that noted with shielded metal arc welding. The impact of such a rich radiation source in the UV-B and UV-C wavelengths has been covered in the discussion of shielded metal arc welding. Both procedures present an additional problem due to the decomposition of trichloroethylene and other chlorinated hydrocarbon vapors by UV emission from the arc forming chlorine, hydrogen chloride, and phosgene.

Although it has generally been accepted that nickel carbonyl can be formed during welding on nickel alloys with a CO_2 shielded arc, there has not a comprehensive assessment of this problem. Wiseman and Chapman (71) have found that nickel carbonyl is formed at concentrations just exceeding the detection level. It would appear from this study that the risk from this exposure is nil.

Ulfvarson (50) states that metal inert gas welding and tungsten inert gas welding generate significant ozone concentrations. The highest exposures occur with metal inert gas welding on aluminum–silica alloys; the rate of ozone generation in this case is 5 to 20 times that for metal inert gas welding on stainless steel and exposures may approach 1 ppm.

In the study of 45 welding shops in Sweden by Ulfvarson (59), 36% of the total fume samples exceeded 5 mg per m^3 and 16% of the ozone samples exceeded 0.1 ppm. The highest nitrogen dioxide concentration of 3 ppm was found on metal inert gas arc welding on aluminum. The highest ozone concentrations were noted while welding aluminum–silicon alloys.

NIOSH investigators evaluated a welding shop that built and repaired long wall mining equipment built of mild steel using metal inert gas welding (72). Although marginal dilution ventilation was the major control technique, the contaminant concentrations were generally below the occupational exposure limits. Concentrations of iron fume were 0.09 to 0.93 mg per m^3, the TWA concentrations for nitrogen dioxide ranged from 0.08 to 0.38 ppm, carbon monoxide concentration were 1.4 to 14 ppm and short term ozone concentrations were all less than 0.05 ppm.

7.3 Control of Exposure

7.3.1 The Welding Process and Equipment

Over the past decade Hewitt and co-workers in the U.K. have investigated the mechanics of fume generation in welding in order to better understand the operative mechanisms with the hope that modifications to the welding process can be made to reduce welders exposure to air contaminants. Hewitt and Hirst (73) have proposed that metal fume from metal inert gas welding is generated by four mechanisms: (*1*) vaporization of the metal from the homogeneous pool; (*2*) vaporization of metal from small diameter hot spatter particles of great total surface area; (*3*) vaporization of metal from the incompletely mixed molten pool with surface depletion occurring; and (*4*) initial fume undergoing a reaction which changes its vapor pressure and therefore its rate of vaporization. The authors state that mechanism (*1*) will generate a fractionated fume whose fraction composition is unlike the parent alloy. As an example, the authors state that as an alloy material, manganese with a high vapor pressure would result in a disproportionately high concentration in the fume. Mechanisms (2) and (3) result in a fume with the same composition as the parent alloy.

In stainless steel welding the content of the alloy is not important in defining the welder's exposure; the significant contributor to the fume is the electrode. Shielded metal arc welding on stainless steel results in higher hexavalent chromium exposure than metal inert gas welding; alkaline coated electrode generates more chromium than other electrodes. A change in the binder from potassium to sodium silicate significantly reduces the generation rates of soluble fluorides in low hydrogen welding.

Welding blankets and mats are widely used to protect equipment from spatter and to insulate workpieces thermally that are heated for strain relief during welding. Until the 1980s, these blankets were frequently made of asbestos; since that time there has been a conversion to glass and ceramic-based materials. Waterman et al (74) investigated the ability of these materials to stand the physical abuse of the welding workplace and to consider the concentration of fibers released from the mats during active work. Three types of mats identified in the United Kingdom as Type A Brazing Pad, Type B Glass Cloth, and Type C Ceramic Heat Resistent Mat, all generated fiber concentrations in the work area of less than 0.01 fibers per mL.

7.3.2 Preparation of The Parent Metal and Surface Coatings For Welding

In the discussions of the individual processes earlier in this chapter it was assumed that the workpiece was mild steel and, therefore, that the predominant metal fume contaminant was iron oxide. One must also consider the fume exposures that occur while welding steel alloys, nonferrous metal, and copper alloys. Metal coatings and the welding electrodes also contribute to the metal fume exposure. Table 28.8 shows the wide range of metal fume concentrations encountered in welding (53). Recent interest in the possible carcinogenicity of chromium and nickel fume has prompted special attention to these exposures during welding. Lead has been used as an alloy in steels to improve its machinability, and welding on such material requires rigorous control. This is also the case with the use of manganese in steel alloys to improve metallurgical properties. Beryllium, probably the most toxic alloying metal, is added to copper and these alloys warrant close control during welding and brazing operations.

Welding or cutting on workpieces that have metallic coatings may be especially hazardous. Lead-based paints have been used commonly on marine and structural members. Welding on these surfaces during repair and shipbreaking generates high concentrations of lead fume. In the author's experience, cutting and welding on structural steel covered with lead paint resulted in concentrations exceeding 1.0 mg per m^3 in well ventilated conditions out-of-doors; concentrations in excess of 10 mg per m^3 may occur in poorly ventilated confined spaces.

Steel is galvanized by dipping in a hot zinc dip. Air concentrations of zinc during the welding of galvanized steel and steel painted with zinc silicate range from 3 to 12 times the ACGIH TLV under conditions of poor ventilation. In general, the air concentrations are lower on oxygen–acetylene torch work than arc cutting.

Studies on weldable paints have shown that significant concentrations of respiratory irritants occur from the thermal degradation products of the paint. Where possible, an ethyl silicate shop primer should be used since it has a negligible release potential for such irritants (75). If not, the surface coating on metal to be welded should be removed for approximately 1 in. and for burning 2 to 4 in.

7.3.3 Ventilation

Shielded metal arc welding on ferrous metals in open areas can usually be performed safely with dilution ventilation; however, welding in confined spaces will always require local exhaust. Reliance on general exhaust ventilation alone for control of welding fume may not be appropriate at all times. Ulfvarson (59) notes the reduced effectiveness of general exhaust ventilation in the heating season resulted in an increase in air concentrations 2–3 fold that of the non-heating season.

Local exhaust ventilation is also needed when using gas tungsten arc welding or gas metal arc welding techniques on stainless steel, high alloy steels, nickel alloys, copper alloys, or when the base metal is coated with toxic metal. Guidelines for ventilation control by dilution and conventional local exhaust are available (7, 12, 76). A high velocity-low air volume exhaust system has been evaluated for several welding applications with mixed reviews.

The most difficult control problem is welding in confined spaces such as on shipboard. In the 1940s, a portable exhaust system unit was developed for shipboard welding use with an integral particulate filter for removal of metal fume with direct recirculation. A similar system with both a particulate filter and an adsorption system for the removal of gases has been introduced recently in Denmark and may have application on operations when one cannot utilize conventional local exhaust systems.

Since ozone is formed in the UV-C field (most efficiently by wavelengths less than 200 nm), shielding of the arc may be effective. As mentioned earlier, in many cases mechanical shielding of the arc is not possible and the only control is exhaust ventilation. Hazardous concentrations of nitrogen dioxide may be generated in enclosed spaces in short periods of time and, therefore, require effective exhaust ventilation.

The effectiveness of local exhaust booths on Arcair operations in a foundry were evaluated in a NIOSH-sponsored study (76). Several conclusions were drawn from this valuable study. The authors found that there was no significant difference between lapel and in-helmet samples and that individual work practice had a significant impact on the exposure to metal fume. In conventional welding, air concentrations inside the helmet may be as low as one-fifth to one-tenth the ambient air concentrations. Despite rather good ventilation control, TLVs for total welding fume or iron oxide fume were exceeded routinely. If time weighted concentrations were calculated based on full shift operations, only two out of four of the workers exceeded the TLV for total and iron oxide fume. The air jet caused violent dispersal of the metal fume, which made it difficult to achieve efficient collection by the hood. The authors concluded that control could be established with booth ventilation but that individual work practice is an important consideration.

7.3.4 *Confined Space Work*

In evaluating the performance of a direct reading aerosol photometer for measuring the concentration of fume from burning and welding operations in a shipyard, Glinsmann and Rosenthal (77) developed a small data set of short period samples varying from 10 to 90 min. in duration. In their review of the data, the authors categorized the samples by the degree of confinement and the quality of exhaust ventilation. The ventilation was graded by the location of the local exhaust ventilation hood relative to the welding point. The strong dependence of the fume concentration on both the degree of confinement and the quality of ventilation is shown in Figure 28.22 with the degree of confinement having the greatest effect.

7.3.5 *Radiant Energy*

Eye protection from exposure to UV-B and UV-C wavelengths is obtained with filter glasses in the welding helmet of the correct shade, as recommended by the American Welding Society (78). As noted, within a given process, e.g., metal inert gas welding, there may be a choice of shade depending on the operating current. The attenuation afforded by tinted lens is available in NBS reports and the quality of welding lens has been evaluated by NIOSH (79). Glass safety goggles will provide some protection from UV as do certain plastic safety glasses. The manufacturer will provide this information on request.

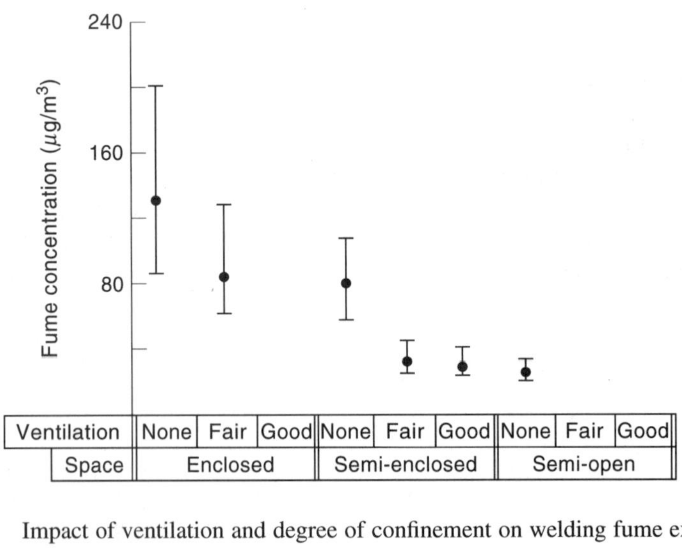

Figure 28.22. Impact of ventilation and degree of confinement on welding fume exposure. From Glinsmann and Rosenthal (77).

Eye protection must also be afforded other workers in the area. To minimize the hazard to nonwelders in the area, flash screens or barriers should be installed. New plastic "see-through" screens are effective in minimizing UV radiation. Plano goggles or lightly tinted safety glasses may be adequate if one is moderately distant from the operation. If one is 9 to 12 m (30 to 40 ft) away, eye protection is probably not needed for conventional welding; however, high current density gas tungsen arc welding or gas metal arc welding may require that the individual be 30.5 m (100 ft) away before direct viewing is possible without eye injury. On gas welding and cutting operations, IR wavelengths must be attenuated by proper eye protection for worker health and comfort.

The ultraviolet radiation from inert gas shielded operations will cause skin erythema or reddening, and therefore, the welder must adequately protect his or her face, neck, and arms. Heavy chrome leather vest armlets and gloves must be used with such high energy arc operations.

7.3.6 Decomposition of Chlorinated Hydrocarbon Solvent Vapor

The decomposition of chlorinated hydrocarbon solvent vapors occurs in the UV field around the arc and not in the arc itself. The most positive control is to prevent the solvent vapors from entering the welding area. Merely maintaining the concentration of solvent below the ACGIH TLV is not, in itself, satisfactory. If vapors cannot be excluded from the workplace, the UV field should be reduced to a minimum by shielding the arc. Pyrex® glass is an effective shield that permits the welder to view his or her work. Rigorous shielding of the arc is frequently possible at fixed station work locations, but it may not be feasible in field welding operations.

7.3.6 Personal Protective Equipment

The welding helmet is designed to protect the welders from hot sparks and spatter from the weldment and the installed plate provides eye protection from UV radiation. The helmet must be worn in conjunction with side shield eye glasses for protection when the helmet is in the up position. In addition to its primary function, the helmet also reduces the air concentration of contaminants. A series of studies in the late 1960s indicated that the welder's helmet reduced the fume exposure by a factor of 3 to 15. In a more recent study by Goller and Paik (80) on welders constructing locomotive frames by shielded metal arc welding, arc-air gouging and torch cutting, the helmets reduced the outside concentration by 36 to 71%.

As one component of a Health Hazard Evaluation, NIOSH investigators evaluated the respiratory protection program at a plant making stainless steel autoclaves (81). The company had provided supplied air respirators for inside tank work where high ozone concentration were encountered. The results of an interview on the acceptance of the devices is enlightening. In the group of 13 welders, 8% wore the respirators at all times, 23% wore them only when welding inside tanks, 46% occasionally wore them on inside tank work, and 23% never wore them. The low level of acceptance was due to complaints of inconvenience, dryness of throat and nose and concern for their safety, i.e., the quality of the respirable air.

The design of the welder's protective clothing reflects the hazard from hot spatter. The clothing is designed with a minimum number of pockets, cuffs, and indeed any spot which would catch a spark. The clothing should provide continuous area coverage and be worn so sleeves overlap gloves and pants overlap shoes. Flame resistent gloves, capes, sleeves, and leggings must be constructed of heavy material reflecting their use (82).

8 CHEMICAL PROCESSING

There are several thousand chemical processing plants in the United States employing over 200,000 technicians and engineers. The chemical processes used in these plants derive from the simple equipment and techniques of the student chemistry laboratory differing only in size and complexity. The design of the production plant and the process equipment will vary greatly depending on whether it produces bulk chemicals, fine chemicals, or specialty products such as pharmaceuticals. At one end of the spectrum is a facility in which a company manufactures a household cleaner by diluting a purchased concentrate with an alcohol and adding an odorant. In a major chemical company the manufacturing history of a new product is far more complex. The research group is assigned the job of inventing, e.g., a new antioxidant for rubber products. If the bench research leads to a new molecule with the desired features, a few grams are made for initial review. If the results are positive the work is scaled up to pilot plant level for the manufacture of several kilograms to be used for definition of chemical and physical properties, product performance, and initial toxicity studies. Successful completion of this phase may result in a decision to undertake full scale production of the chemical. Manufacture may take place at a company plant using existing equipment or it may require the design and construction of a new facility. In either case the company must take certain initiatives including the submission of a Pre-Manufacturing Notification (PMN) to the Environmental Protection Agency before the manufacture of the chemical for sale. The synthesis and purification of the new chemical may be assigned to an existing plant to be completed in a batch operation using open or semi-enclosed equipment, or if the market potential exists and the process chemistry permits, the work is done in a closed, continuous process in a highly automated plant.

It is beyond the scope of this brief review to identify the health issues faced by each of the many different types of chemical plants. However one industry, fine chemical manufacture, includes the principal operations and equipment common to all chemical processing. One such hypothetical plant shown in Figure 28.23, designed for batch production of organic dyes with open or semi-enclosed processes, will be used to illustrate the basic concepts of exposure and control in this industry.

8.1 Raw Materials

This plant requires procurement and on-site storage of one hundred bulk chemicals in quantities from 100 to 100,000 kg. The materials are either granular or liquid; general tank and warehouse storage is appropriate for most but some highly reactive chemicals may require special environmental conditions and isolation in limited quantities.

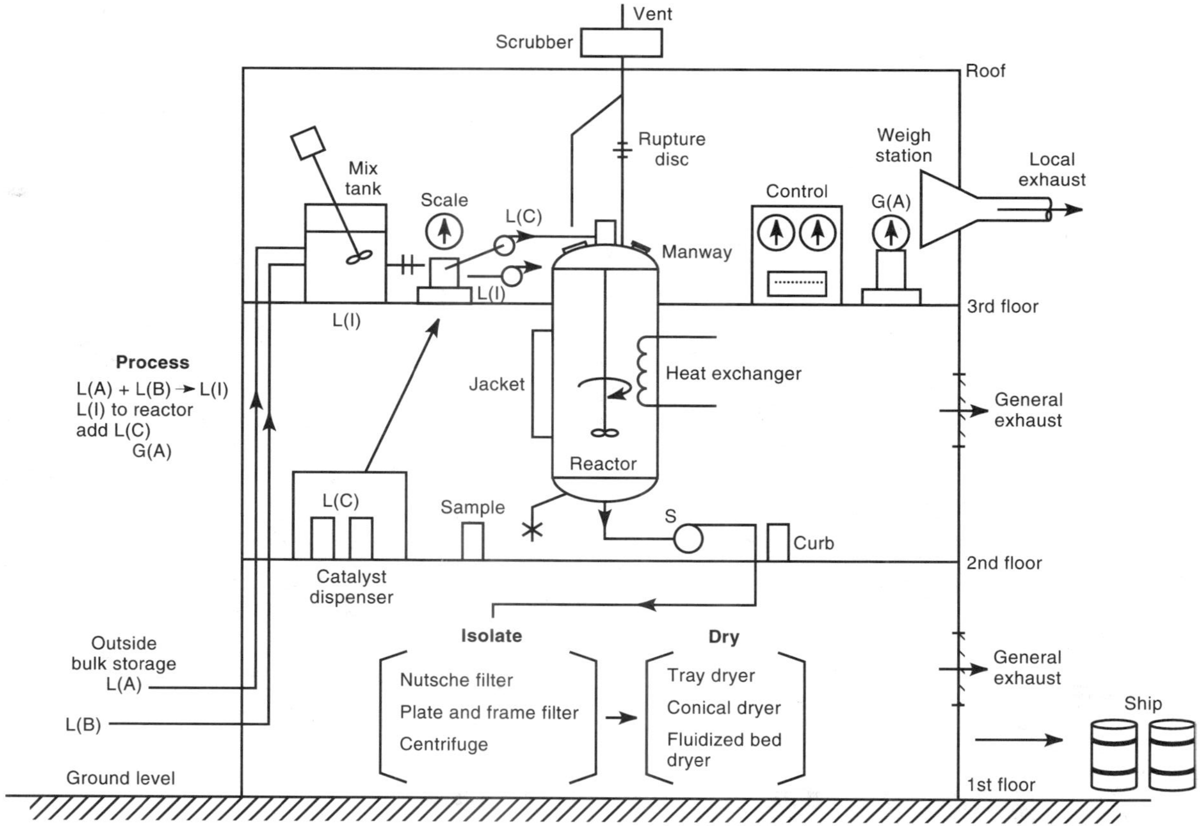

Figure 28.23. A fine chemical plant. Liquid A [L(A)] is mixed with liquid B [L(B)] in tank 104 to make liquid 1 [L(1)]. Liquid 1 is pumped to reactor A 682 and a liquid chemical C [L(C)] and a granular chemical A [G(A)] are added. The reactor is held at temperature and contents are mixed while the reaction runs. Samples are taken periodically and when the reaction is complete the reactor contents are pumped to one of three separation units for filtration and subsequent drying. L = liquid, G = granular, S = suspension.

8.1.1 Granular Materials And Powder

The volume of solids will range from catalysts used in small quantities to large volumes of polymers. The small quantity materials will be shipped in appropriate drums or bags. Intermediate quantities of granular chemicals are effectively transported to the plant in bins or tote containers. Flexible tote containers of polypropylene textile are available for transport of up to 4000 lb of dry and semi-moist chemical. Rigid aluminum and stainless steel bins are also available. An advantage of both the flexible and metal containers is that they are equipped with top filling and bottom dispensing fittings so that sealed connections can be made to minimize dust release during material transfer. Since the containers can be transported directly to a transfer point, such as a reactor charge port, the manual lifting and associated dust exposures in multiple drum dumping is eliminated. The supplier of the bulk chemical will also deliver preweighed quantities for a specific batch size thereby eliminating dusty weighing procedures at the plant. If the bulk material is combustible, classed as dangerous, or is water sensitive, the rigid bins can be hermetically sealed and if needed, inerted with nitrogen.

The large volume granular chemicals delivered to the plant by truck or rail are pneumatically conveyed to storage silos; if the process scale permits the material is pneumatically conveyed to the process vessel. The anticipated material transport of bulk granular materials to process locations should be given early attention since not only will handling costs be reflected in a poor layout but dust release to the workplace will be excessive and require elaborate local exhaust ventilation. In the more common batch process shown in Figure 28.23 using moderate quantities of granular chemical, the chemical is added manually to the vessel through a charging port or manway.

Quantities of granular material in the range of a few drums can be dumped directly into the reactor or mix vessel using a drum dumping device. If the material is highly reactive or if the reactor environment cannot be opened to air, a charging lock is used.

Dumping the granular material into a reactor already charged with solvent induces air into the vessel with the resulting displacement of particles and vapor-laden air back to the workplace. The charging port is therefore an important exposure point and must be controlled by local exhaust ventilation as is discussed later.

8.1.2 Liquids

The fine chemical plant in Figure 28.23 uses a number of solvents including acetone, toluene, methylene chloride, isopropyl alcohol, and methyl cellosolve. The solvents are transported to the plant by rail or truck tanker and delivered by closed line to individual storage tanks each holding 10,000 gallons. The exposure of workers to chemical vapors at large dock receiving facilities has been evaluated in both the petrochemical and chemical processing industry, and although the operations are usually conducted in open air, the exposure to vapors are high and control ventilation may be required. Closed system unloading is the best method to control exposures. The intermittent exposures at the facility in Figure 28.23 would not be significant and special controls are not required.

Both above ground storage tanks and underground storage tanks are encountered in chemical processing facilities. Since the enactment of federal regulations on underground storage tanks in the late 1980s, almost all new tank installations are above ground. The above ground storage tanks offer many advantages. Of major importance is the ability to inspect the tank routinely and repair defects before a leak occurs. In addition, a double-walled above ground storage tank is more economical to install than a comparable-sized underground storage tank and it can be moved if the plant layout is changed in the future. Four types of above ground storage tank options are available (*1*) a double-walled above ground storage tank with built-in monitoring well, (*2*) a single-walled tank enclosed with 6 inches of concrete, (*3*) a single-walled tank in a dike, and (*4*) a single-walled tank in a concrete vault.

In the small batch operation of Figure 28.23 using multi-use equipment, all eight solvents are transferred to every reactor using a piping manifold. Pipe lines are run from each above ground storage tanks to the production facility where they terminate in a manifold. If solvent is to be transferred from a storage tank to a specific reactor or mix tank, hard piping or a flexible line is made up to a fitting on the manifold and connected to the vessel. The solvent is then pumped into the vessel through a flow meter. If flexible hose is used to make this connection, spills will occur during line drainage resulting in exposure to operating personnel. This problem is alleviated in a large single product facility by hard-piping directly from the storage tanks to the vessel. Leakage which occurs when lines are disconnected can be eliminated by first pressurizing the lines with nitrogen to drive the solvent either into the vessel or back to the storage tank.

If the batch process requires liquid chemicals be introduced into the reactor at a specific point in the process, it can be done in several ways depending on the volume, chemical properties and the toxicity of the material. Small quantities can be transferred by drum pump. This widely used pump has a simple configuration with a dip tube which is immersed in the drum and a direct-attached, motor-pump which transfers fluids at the rate of 5 to 20 gpm. The worker is exposed to the vapor in the headspace of the drum and evaporation products from the wetted tube as it is removed from the drum. Control of this exposure can be achieved with a bung hole exhaust and a ventilated scabbard for the wand (83).

If greater quantities are to be transferred from drum or holding tank to the reactor, a wide range of pumps are available including gear and double-diaphraghm pumps. In a batch operation, the pumps may be mounted on wheeled dollies so that they can be moved throughout the facility to serve temporary connections as described earlier. The pump is usually paired with a cartridge filter which must be changed routinely. This operation results in some spillage and exposure to the worker. Highly toxic or reactive fluids can be transferred to a reactor without worker exposure by connecting the chemical drum to a reactor fitting and "pulling" a vacuum on the reactor with a steam ejector to draw over the chemical.

8.2 Facility Design

A building complex designed for chemicals production must provide on-site chemical storage, well-defined and preferably separate building locations for synthesis, separation, drying, packaging, and shipping. The facility can be either a single story or multiple story building; in either case the building must be designed for easy access by roof or wall sections to "drop-in" major pieces of equipment. Various fire prevention and electrical codes must be observed which require explosion relief venting, minimum general and local exhaust ventilation, and explosion-rated electrical components. Flooring and drains must be designed for containment and treatment of all spills. Quick-acting floor drain blank-offs are available to ensure spills do not enter the drains. Air and gas cleaning facilities must be available for the contaminant-laden process streams and local exhaust streams. The health and safety considerations for the design and construction of chemical processing facilities are myriad. A comprehensive checklist for the occupational health practitioner has been proposed (84).

8.3 Major Process Equipment

The laboratory bench-top organic synthesis equipment used in undergraduate chemistry laboratories is the prototype for the production facility. In that laboratory equipment the reagents are added to a round bottom flask placed in a heating mantle. The flask is equipped with a thermometer and a magnetic stirrer. The reaction proceeds to completion while monitored by the student. The solid reaction products are separated from the reaction fluid by pouring the fluid with the suspended solids through a filter paper supported in a Buchner funnel. The recovered material is then placed in a convection drying oven. The equipment in a chemical processing production facility consists of the same components although scaled-up in both size and complexity as described below.

8.3.1 Reactor

The reactor in Figure 28.24, modelled after the laboratory round bottomed flask, is the heart of the fine chemical facility. The reactor is a pressure rated vessel (10–150 psig) with glass lining and ranges in size from a few gallons for pilot plant work to ten thousand gallons for large scale production. The major design features of the reactor can be divided into those necessary for production purposes and those installed for safety. The production features include the glass lining which ensures that trace metals from the reactor shell will not interfere with the reaction. The lining can be rigorously cleaned to prevent product cross-contamination between runs. The access ports on the reactor include a charge port (manway) designed primarily for charging or loading the vessel with granular chemical. The operator may also take samples from the reaction fluid through the manway and it is large enough to permit workers to enter the vessel for cleaning and inspection. A number of flanged pipe fittings are available to connect permanent or temporary transfer lines for pumping liquid into the reactor. As described earlier, these fittings can be used to connect the solvent manifold to the reactor. The suction leg from the steam ejector, installed on the top dome of the reactor, is used to strip off volatile materials or pull in liquid reagent. The reactor is equipped with a heat-exchange jacket serviced with both cooling and heating fluids. This feature, combined with an internal heat exchange coil, allows the operator to achieve the optimum temperature for the reaction. The large reaction mass in the reactor is mixed with a top-mounted impeller driven by a large motor mounted on top of the reactor. A reflux condenser is mounted on the reactor dome with a final condenser for stripping off material.

Although reactor designs include many safety features as listed below, a number of serious accidents have occurred resulting from overpressure and rupture of reactors with

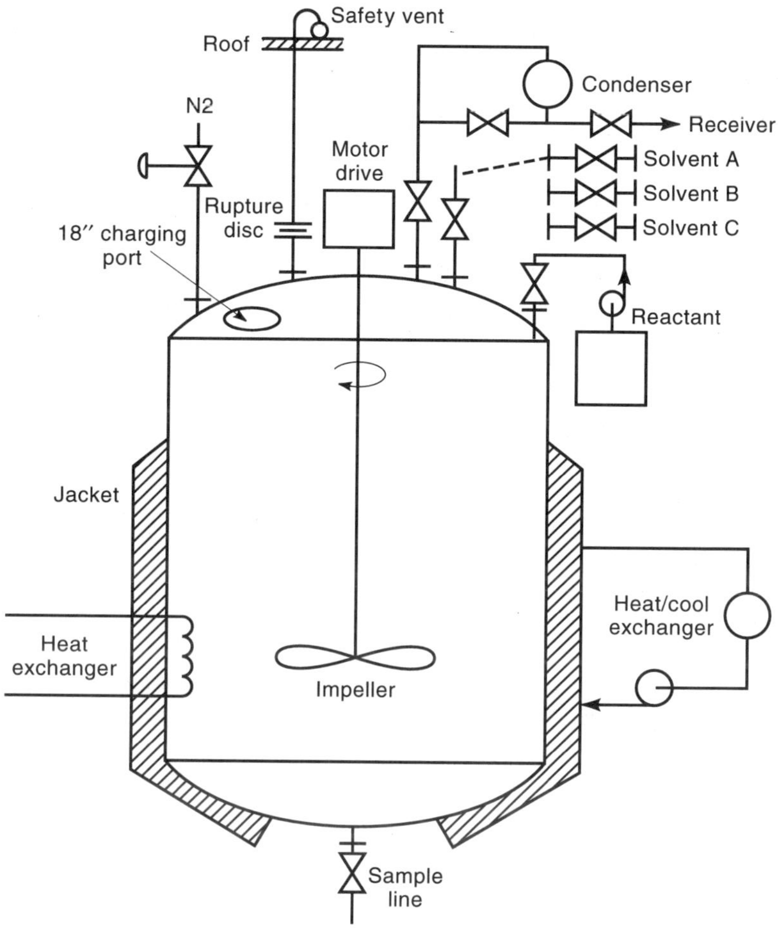

Figure 28.24. Reactor. —— hard piped; - - - - flexible hose; +, flange; —▷◁—, valve.

the release of large quantities of toxic chemicals and resulting in deaths and serious environmental damage. The overall mechanical design identifies the reactor as a pressure vessel. The head and bottom of the vessel are dished, access port hatches are fastened with heavy J bolts, and all fittings are massive. In addition to the basic mechanical design to handle overpressure, the reactor is equipped with a pressure relief system. A resin impregnated graphite rupture disc is mounted in a large diameter pipe leading from the top of the vessel. Discs are available for burst pressures of 10 to 150 psig. Graphite is chosen as the construction material since it is nonfatiguing, corrosion resistant, and can be machined to provide burst pressures within 5% of rated pressure. This performance and reliability cannot be achieved with a metal disc or a conventional pressure relief valve. The vent stack from the rupture disc should be connected to a containment tank and air cleaner as shown in Figure 28.23.

For fire protection purposes it is common practice to inert the headspace of the reactor with nitrogen or carbon dioxide at flow rates of 10–30 Lpm. Rigorous confined space entry procedures must be followed in entering spaces that have been inerted. The availability of piped inert gases throughout the plant has serious implications. A number of deaths have occurred in other industries when air line respirators have been inadvertently connected to such inert gas supply lines (27). Although identified earlier as a production feature, the heat-exchange jacket on the reactor fulfills an important safety function since it is used to control reaction exotherms that might result in overpressure and release of the contents of the reactor. As noted earlier, the manway or charge port positioned at the top front section of the reactor is opened to load or "charge" bulk chemicals. The dumping of granular material into the reactor induces air to flow into the reactor with the resulting displacement of particles and vapor-laden air into the workplace (85). The primary control of this air contaminant is achieved by charge port ventilation using an exterior hood with exhaust flow rates of 1,000 to 12,000 cfm to achieve collection efficiencies in excess of 90%.

If small quantities of toxic or reactive liquid chemicals are to be added to the reactor during processing it is best done in a closed system. Drummed material is positioned close to the reactor on the main floor or in an ancillary exhaust booth. In both cases the chemical can be pumped to the reactor or pulled over by vacuum.

Periodically, samples must be taken of the reactor contents to check on the status of the reaction or as a final product quality check. This is frequently done by opening the charge port on the top of the reactor and using a dipstick with a sampling cup. Although there is potential exposure from the headspace contaminants in the reactor and the vapor from the sample, exposures are minimal when the reactor is equipped with charge port ventilation. Headspace exposure during sampling can also be controlled by placing the reactor under vacuum with the steam ejector. This action causes an inward flow of air through the charge port, thereby minimizing worker exposure. A second sampling technique involves the use of a special sampling loop positioned at the bottom of the reactor. This sampling location should be equipped with a small exhausted enclosure (87).

8.3.2 Liquid–Solid Separation

The reaction product to be isolated in this hypothetical plant is the solid dye suspension formed in the reaction. In batch operations this separation frequently represents the principal exposure point in the process since it is done in an open system with an opportunity for close contact with the chemicals. The three common separation techniques are the nutsche, plate and frame filter press, and centrifuge.

8.3.2.1 Nutsche. In the chemistry laboratory set-up the synthesized crystal was isolated with a filter paper supported in a Buchner funnel. The gravity nutsche is the production version of this technique. Nutsches, available in a range of sizes in round or box configuration, are constructed of stainless steel or rigid plastic. The geometry is simple, providing a filter media mounting support and a bottom drain fitting for removal of the filtrate. The reactor contents are discharged to the nutsche in batches and filtered by gravity or with vacuum or pressure assist. The particles are collected on the filter media and the filtrate or mother liquor is pumped to a holding tank for recovery and reuse. This filtration process is completed on the plant floor and presents a large wetted surface area for evaporation of carrier solvents. If uncontrolled the operation will result in high concentrations of solvent vapor. The most effective control is local exhaust ventilation with an enclosing hood (87). The worker exposures, however, may still be high during product recovery. The conventional nutsche is a relatively slow operation with percent solids of 30–40% in the collected product usually achieved.

8.3.2.2 Plate and Frame Filter Press. The application of the simple filter media for separation of product can be extended in many ways. The available techniques include belt filters, leaf filters, cartridge filters, and the plate and frame filter press. The plate and frame is probably the most widely used separation technique in the chemical industry with relatively high capacity and the ability to provide a product cake in the 40–50% solids range. The initial equipment cost is low, however the operation is labor intensive. In this equipment design, a series of polypropylene support plates with recessed channels are arranged in a parallel configuration, usually with separating plates to retain the solids. The synthetic fibrous filter media is placed over the support plates with solid–liquid separation occuring at the face of the filter. Commonly the plate and frame filter press contains 10 to 30 plates presenting a total filtration area of 1 to 100 m^2 (11 to 1080 ft^2). After the filters are positioned, a hydraulic ram compresses the plates to obtain an interplate seal. The feed slurry is then pumped from the reactor to the plate and frame filter press under pressures up to 250 psig. A cake 1/2 to 2 in. thick is built up and monitored by the system pressure drop. The filtrate from the plate and frame filter press is delivered to a storage tank for recovery. At a given cake build up the feed from the reactor is stopped and the cake is usually washed in place with a solvent. At the completion of washing, the plate sealing pressure is released and the plates are separated either manually or semi-automatically. The solvent-wet cake is scraped from the filter by the operator using a plastic spade and drops into a bin. The air concentrations of solvent vapor are high at this point due to the large surface area of the solvent-wet cake and the confined working space between plates. This condition is further complicated by the extended time period required for cake removal.

Control of airborne chemical vapors on the plate and frame filter press is difficult to achieve. In certain plants this operation can be automated thereby minimizing worker exposure, however, in most cases the product cake must be removed manually. The entire plate and frame filter press can be enclosed with rigid sliding panels or flexible plastic screens and the enclosure exhausted to ensure inflow through the principal work location (13, 87). This approach minimizes general plant contamination, however, the operator working between the two plates is not protected. In such a situation the worker must be provided with high quality respiratory protection such as a full face air-purifying respirator or an air line respirator.

Occasionally it is possible to strip the carrier solvent from the cake before the plates are opened for cake removal. In this scenario, if toluene were the carrier solvent, the cake would be washed in place with an alcohol that strips the toluene from the cake. The alcohol is then stripped by washing with water. The procedure requires that the product will not be degraded by water. If this can be accomplished, the worker is faced with a water wet product cake when it is stripped from the filter.

8.3.2.3 ***Centrifuge.*** The basket centrifuge is a technique for liquid–solid separation in the 60–65% solids range. These devices, as large as 4 ft in diameter and operating at speeds up to 2000 rpm, can be operated in a batch or in a semicontinuous mode with automatic plow-down and product discharge. The perforated basket is lined with a support or backing screen on which a synthetic fabric filter is placed. The feed liquor flows from the reactor to the centrifuge fill pipe and is sprayed on the inside surface of the basket. The particles are collected on the filter face; the filtrate flows to the base of the shell (curb) of the centrifuge and is pumped to a holding tank for later treatment and reuse. When a suitable cake is built up on the basket, the feed from the reactor is slowed and the cake is removed with a plow or peeler knife. This technique will remove up to 90% of the product cake and the balance must be removed manually. The hinged access cover on the top of the centrifuge is raised and the operator leans into the centrifuge to scrape off the balance of the cake with a plastic spade. This work results in exposures to high concentrations of the vehicle or solvent since the vapor concentrations inside the centrifuge quickly reach saturation.

There are a number of control options available to minimize this exposure. Before ploughing, the cake can be washed with a less toxic material, or one with a lower vapor pressure. If acceptable from a production standpoint and if it does not present a fire and explosion problem, the filtrate discharge line can be adapted to pull air through the open access cover to the exhaust system. The inflow of air through the access cover will minimize the worker's exposure. A third approach is the installation of two winged slotted exhaust hoods at the open cover. The last and probably most effective control is the provision of a glove port system (Figure 28.25). During active centrifuging the gloves and scraper are mounted inside the centrifuge housing; they are deployed when the centrifuge is stopped and the access cover is opened.

The operation of the centrifuge with a feed liquor which contains material that present a risk of fire and explosion requires special controls. If a material such as toluene is used

Figure 28.25. Glove port system on centrifuge.

which presents a static charge problem in bulk transfer, a number of standard procedures are followed in the plant including the inerting of the centrifuge with nitrogen or carbon dioxide. This effective control presents a risk of asphyxiation if the operator puts his head in the inert atmosphere during spading or inspection.

8.3.3 Drying and Packaging

The above filtration techniques provide a cake that contains 30–70% solvent depending on the specific system. If the isolated material is the final product it is dried for packaging and shipping. If the material is an intermediate and is the starting material for the next synthesis step, it may or may not require drying. Several drying processes are widely used in chemical processing. The choice of process is dependent primarily on production requirements although the chemistry of the compound may present limitations on the candidate techniques. For example, if the product degrades at moderate temperatures, a number of high temperature drying techniques must be excluded. Although both batch and continuous drying processes are available, emphasis in this discussion will continue to be on batch operations since they represent the greatest potential risk for exposure. One general rule regarding exposure applies to all drying operations. The transfer of the wet cake from the separation device to the dryer presents the potential for exposure to solvent vapor while the unloading of the dryer and packaging of the product presents a dust exposure.

The most widely used drying technique is a tray dryer using a straight convection dryer or a vacuum oven dryer. This process is labor intensive and involves intimate contact with the product. The loading of the wet cake on trays and the removal of the dry product is difficult without significant exposure of the chemical processor to both solvent vapor and respirable dust. The installation of local exhaust ventilation at the tray loading/unloading station is usually a necessary control measure. Another common drying technique is the double-walled conical dryer. In this device, drying is accelerated by tumbling the product in a wall-heated cone while pulling a vacuum on the chamber to strip off the solvent. The concentration of the solvent in the effluent stream will be high and require air cleaning before discharging to the environment. The charging station can be controlled with local exhaust ventilation; the packaging of the dried material is again a difficult problem although a dust tight loading chute may be possible.

Fluidized beds provide a high quality drying technique applied to fine chemicals and pharmaceuticals. A heated air or gas stream lifts the product from the support and the volatile vehicle is stripped off. The geometry of the system lends itself to effective drumming without loss of dusty product to the workplace.

A number of other drying techniques including spray and flash drying are widely used in the industry for continuous, high volume operations.

8.3.4 Transport

A review of the above operations indicate the need for transport of raw materials, byproducts, product, waste streams, intermediates and ancillary/service fluids. These materials may be liquids, syrups, pastes, powders, granules and encapsulated chemicals. If possible, all transfer should be by permanent, hard piped lines to minimize exposure potential. However, the batch operations encountered in many chemical processing operations do not permit dedicated lines. Therefore, each time a temporary flexible line is set up, the possibility exists for leakage and worker exposure.

A second important source of contaminant are fugitive losses from pipe flanges, pump seals, agitator seals, valve stem seals, and process instrument lines. This leakage may seem trivial when an individual piece of equipment is considered but studies of such fugitive losses in the chemical processing and petroleum industry indicate it is a major air pollution problem and may be a worker exposure issue. A program should be implemented to identify and control losses (88). High pressure leaks from flanges and other fittings on pipe lines carrying corrosive and toxic liquids may result in serious accidents. Although primary control is maintenance, the use of piping connection control shields is prudent. The shields are treated synthetic textile jackets that cover the fitting and dissipate a pressure stream originating at the leak.

8.3.5 Work Practice

As indicated in the introduction, the manufacture of the chemicals may be a one-step process requiring the mixing of two chemicals to effect a simple reaction or it may involve multiple reactions with the isolation of several intermediates. Such complex synthesis may require several dozen individual steps and require a month to complete; the cost of the product may be as high as $10,000 per kilogram. It is obvious that sophisticated chemical syntheses must be conducted in a precise way to optimize yield and ensure the batch is not lost. At the per kilogram cost cited for a product requiring an involved synthesis, the loss of a batch toward the end of the procedure could easily reach one-half million dollars. For these reasons, in addition to the potential for major accidents involving workers, the neighborhood and property, all operations must be conducted under carefully developed procedures. These detailed operating procedures include the chemical stock list, the process equipment and operating conditions, specific health and safety precautions and personal protective equipment. The identification of potential hazards that may arise due to impurities in materials, exotherms that may occur and the generation of toxic by-products must also be an integral part of this process instruction (89).

8.4 Workplace Exposure Information

Limited data are available on workers' exposure to air contaminants in batch chemical operations. The air concentrations noted during charging, sampling, filtering, drying and packaging operations vary widely between synthesis steps and even between identical batches. It would be extremely helpful if a chemical exposure data bank were available for this industry. The EPA working group (90) recommended NIOSH and OSHA undertake a joint effort to establish ranges of contaminant air concentrations for each unit operation under various control configurations. This initiative has not been undertaken, nor has a second recommendation from that same group, that workplace dispersion modelling be developed to permit estimation of worker exposure.

One company has developed a job exposure profile which compiles estimated exposures to each chemical and physical agent based on job class at a specific work site (91). The estimation of exposure is a difficult element to generate. The company wisely choose to rank degree of exposure in three exposure categories based on the task or operation and its potential for exposure (Table 28.9). A second key element in establishing a job exposure profile is the duration of exposure; individual work elements in batch operations vary from a few minutes to one-half shift. Most of the specific work tasks take less than ten minutes; however some of the filtration tasks are one hour to half-shift operations depending on the batch size. Identifying the type and level of exposure for a wide range of chemicals is certainly an arduous task, and in some instances, impossible with current sampling techniques. If done rigorously, the exposure profile may be represented by a low concentration of solvent vapor (fraction of a ppm) from fugitive leaks with an occasional significant spike during a specific operation.

8.5 Control Approaches

8.5.1 Containment

One of the principal factors defining the potential for exposure in chemical operations is the degree of containment afforded by the plant design. This issue was clearly identified in a workshop devoted to the prediction of workplace exposure to new chemicals processed under the the PMN (90):

> "The degree of containment of a process facility was suggested as the basis for predicting worker exposure when controls are used. Standard control measures may often be categorized for each process: a sealed and isolated system where airborne contaminant levels are very low (ppb range for vapors); a substantially closed system where airborne contaminant levels are still relatively low (fraction of ppm or low ppm range for vapors); a semiclosed system which is typical of nondedicated equipment used in job shop chemical processing facilities; and an open system."

Table 28.9. Job Exposure Profile[a]

	Degree 1
Minor exposure	Operations conducted in closed processing systems or carried out in a laboratory hood.
	Degree 2
Moderate exposure	Operations conducted in open systems usually provided with local exhaust ventilation.
	Degree 3
Major exposure	Operations conducted without local exhaust ventilation or activities such as cleaning, maintenance, and repair.

[a]Ref. 91. Reprinted with permission of American Industrial Hygiene Association.

The approach used by Holzner et al. (91) in developing the job exposure profile mentioned earlier involves an estimate of potential for exposure which reflects the type of operation in which the chemical is used. The simple numerical code reinforces the importance of containment. The specific tasks assigned to each category are shown in Table 28.9.

The correctness of the containment approach certainly seems obvious to the occupational health practitioner who has experience in this industry. However, open processes are frequently necessary for separation and isolation of final product and intermediates. In addition to presenting opportunity for exposure, these batch operations are time consuming and costly. A recent example of conversion from open batch operations to a closed system in a specialty chemical company confirms the virtues of containment. In the original plant the separation process was conducted using plate and frame and centrifuge equipment. Although local exhaust ventilation was applied to the equipment, exposures to the carrier solvents continued to exceed the exposure guidelines and respiratory protective equipment was mandatory.

In a radical departure from this series of batch operations, the plant introduced a new filter dryer technology which integrated several of the batch processes in a single vessel. This equipment complements the reactor for distillation, extraction, and crystallization processes. The technique improves utilization of process equipment, reduces the number of interequipment transfers and thereby greatly reduces the potential for worker exposure and environmental releases. The work is completed in a closed system without exposure. Sequential isolation of intermediates is not necessary and time consuming step processing through separation and drying is eliminated.

8.5.2 Exhaust Ventilation

Although the use of closed systems as a primary control is certainly desirable, the fact remains that for many reasons, including equipment cost, this approach is not available to all facilities and semi-enclosed operations using batch techniques will continue to be common. In this case, control must rely on a range of local exhaust ventilation techniques shown in Figure 28.26. Specific design data are available from several sources (7, 13, 87).

8.5.3 Personal Protective Equipment

The designer of a respiratory protection program for a chemical processing plant, of the type described in this chapter, is faced with a difficult task. First, a large number of chemicals are in use. The principal exposure will be to carrier solvent vapors and the volatile or dusty chemicals encountered in open or semi-enclosed operations such as separation, drying, and packaging. The concentrations will vary widely ranging from one-tenth the exposure guideline to ten times the guideline, usually for very short periods of time.

If plant operations warrant the use of respirators the lowest level of protection to be used should be a full face air purifying respirator with chin canister for particles, acid gases and organic vapors. The full facepiece is appropriate since it provides splash protection and eye protection in addition to affording a reasonable face seal. The multi-sorbent chin canister is recommended since several air contaminants may be present. In practice

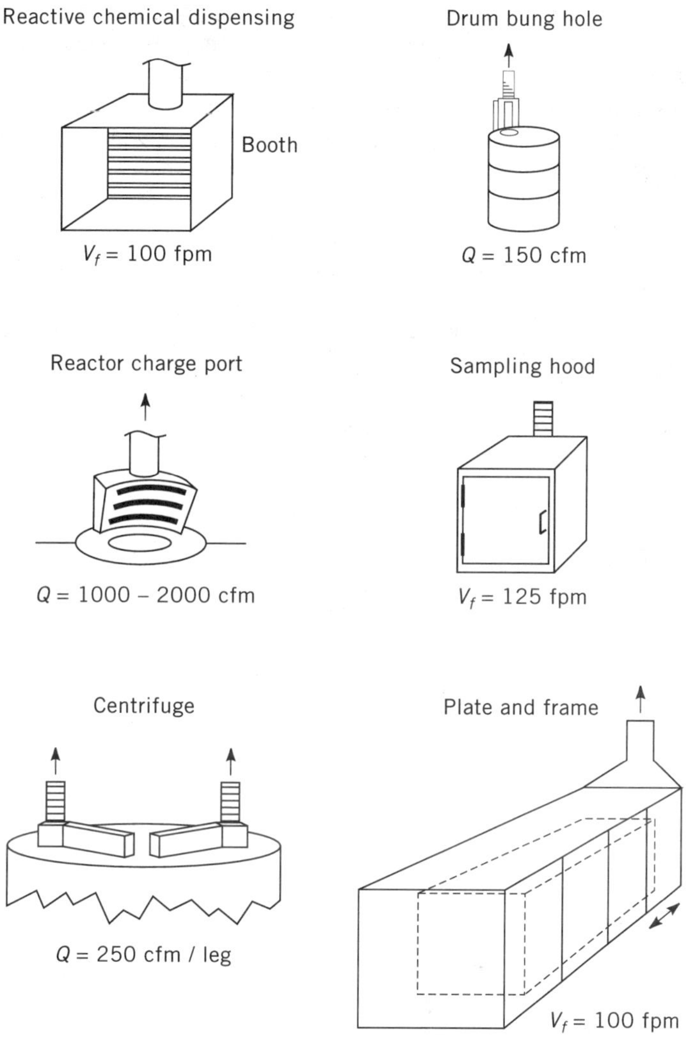

Figure 28.26. Ventilation details.

it is difficult to stock and effectively utilize a number of different cartridges or canisters. Since exposures are usually brief, the service life of the chin canister will be acceptable while reasonable in size and weight.

In those operations where the full facepiece air purifying respirator is not acceptable, due to the nature of the chemical(s) and/or their concentration, an airline respirator should be considered. This equipment may be appropriate at the reactor charge station and on plate and frame operations where the work area is well defined and air concentrations are high. The normal restrictions on the application of airline equipment must be observed.

Although self contained breathing apparatus is occasionally encountered in routine use, due to its bulk and weight it is usually only appropriate for emergency and certain maintenance operations.

The choice of protective clothing is also difficult to specify due to the range of chemicals in use. The availability of selection data based on standardized testing does permit rational selection. Disposable work garments, based on coated, nonwoven fabric, have been widely accepted by the chemical industry. Glove and clothing selection has been enhanced by the availability of permeation data and the introduction of new materials (92). The choice of clothing should also be based on static electricity control where it is a major factor in the fire protection program.

Although the primary operations discussed earlier do not appear to present a noise hazard, the chemical plant is noisy and many facilities require a hearing, conservation program. The source of noise is both from primary processing equipment such as the motor drive on the reactor impeller and the ancillary equipment including pumps, air driven equipment and cooling fans.

9 RUBBER PRODUCTS

The first rigid polymer based on natural rubber has been joined by at least two dozen synthetic rubber polymers for the manufacture of a variety of industrial products including vehicle tires, belting, hose, footwear, rainwear, and chemical protective clothing including gloves, aprons, and respirator facepieces. These products are made by coating techniques, extrusion, calendering, and compression molding.

The rubber workers in the United States, numbering over 200,000 in approximately 1,500 plants, have been the subject of a number of epidemiologic studies. In a series of epidemiological studies supported by the United Rubber Workers and the major United States tire companies, excess stomach cancer was identified in compounding workers, excess lung cancer in curing operators, and leukemia in tire builders (93). In a follow-up, study the authors identified excess pancreatic, prostate and colon cancer in rubber workers (94). In a third paper, the investigators noted excess deaths from bladder cancer, leukemia in white workers, in addition to esophageal cancer, bilary and liver cancer, lymphoma and multiple myeloma (95). Fine and Peters (96) noted increased respiratory morbidity and pulmonary function decrement in processing workers in tire plants. European studies have also defined excess disease in rubber workers.

9.1 Materials

9.1.1 Rubber

More than three million metric tons of natural and synthetic rubber are used annually in the manufacture of products used for shock absorption, sealing, corrosion protection, electrical and thermal insulation, load bearing, tires, and household goods. The natural rubber is harvested from the tropical rubber tree by recovery of the milk-like serum (latex) from the inner bark of the tree. In this century, a series of synthetic rubbers have been developed including Thiokol rubber (polysulfide) in the 1920s, neoprene rubber (chloroprene), and Buna rubber (styrene–butadiene) in the 1930s. The synthetic rubbers are frequently blended with natural rubber to meet specific requirements. Styrene–butadiene rubber represents one-half the total consumption of all rubbers in the United States and of that amount, one-half goes into auto tires.

Both synthetic rubbers and natural rubber have limitations for certain applications. These qualities, however, can be usually be corrected by various additives. The addition of various chemicals to the rubber is called compounding and, as will be discussed later in this chapter, it is the process that offers opportunity for exposure to a range of toxic chemicals in powder form. In the compounding operation, the first material to be added is an accelerator, then the plasticizer, reinforcing agents, antioxidant, fillers, and colorants. This bulk material is then processed through a mixer and stored as sheet stock. In this condition, it is known as master batch material. When required in the process it is processed through a warm-up mill, the vulcanizing agent is added, and the material is used in the product.

9.1.2 Additives

Various rubber processing chemicals are presently in use to permit fabrication into finished product and to ensure specific properties for the product. Usually organic materials, these chemicals are added in relatively small quantities to the rubber stock formula. A normal tread stock recipe (Table 28.10) gives an idea of the quantities of materials involved. The chemicals and production facilities employed in processing these rubber materials are similar and can be discussed in a general fashion.

Sulfur is still the most important vulcanizing agent used either as elemental sulfur or in one of many organic forms. The vulcanizing agent may be added in the initial preparation of the batch or to the masterbatch material just prior to the final fabrication and curing processes. Since the early 1900s, chemical accelerators have been added to rubber systems to hasten vulcanization. Initially, inorganic lead compounds were used, then aniline compounds and now the principal accelerators in use are a series of organic com-

Table 28.10. Natural Rubber Tire Tread Compounds

Material	Parts
Natural rubber	100
Sulfur	3
Accelerator	1
Zinc oxide	5
Stearic acid	1
Antioxidant	1
Softener	5
Pigment	50
Total	166

pounds. In conjunction with the vulcanizer and the accelerator, chemical compounds called activators are added to the batch to achieve a given property. The common activators include zinc oxide, fatty acids, litharge, magnesium oxide, amines, and amine soaps.

Stabilizers are designed to protect the polymer during extended storage before its end use in manufacturing. Antioxidants are added to the batch to protect the finished product; the type and concentration of the antioxidant will vary with the location and usage of the product. The important antioxidants include arylamines and phenols. An antioxidant containing beta-napthylamine as a contaminant in a concentration of 0.25% was identified as the agent that caused bladder cancer in rubber workers in England in the 1950s. This material was banned and excess bladder cancer is no longer seen in the U.K. rubber worker population (97). The principal antiozonants are symmetrical *p*-phenylenediamine, unsymmetrical *p*-phenylenediamine, dihydroquinolines, and dithiocarbonate metal salts. Paraffin and microcrystalline waxes also are used in rubber components. The antiozonant formulation is very important in tires made for the urban markets with high environmental ozone levels.

The viscosity and therefore the workability of the rubbers can be improved by adding organic lubricants or physical softeners to the rubber. These additives may include coal tar, petroleum oils, ester plasticizers, liquid rubbers, fats and oils, and synthetic resins. The final additive, is the pigment which principally contributes addition color, but is also applied as a reinforcement, filler, or extender. The common pigments are carbon black, zinc oxide, clay, and silicates.

In a comprehensive review of the mutagenicity of rubber compounding additives (98), 15 of the 46 rubber additives tested were found positive by a bacterial test system.

9.2 Manufacturing Processes

9.2.1 Compounding and Mixing

In the 1970s and early 1980s, the routine operations involved preparation of master batches of rubber with Banbury mixers. The three major raw materials included rubber stock, carbon black, and chemical additives. A schematic of the production pathways for master batch material is shown in Figure 28.27. The formula or rubber recipe for a given product is chosen by the rubber chemist. In the compounding process, the natural or synthetic rubber is received at the workstation in bale form and is cut into weighable pieces with a guillotine cutter as shown in Figure 28.27. The various rubber chemicals specified in the formula are then weighed out for a specific batch size. This operation involves opening bagged and barreled material and placing it in hoppers for weighing. Since most of the compounds are in powder or granular form this usually is a dusty operation, and weigh stations must be exhausted. Where possible, small quantities of chemicals should be prepackaged in plastic bags that can be placed directly in the batch, thereby eliminating a dust exposure in emptying bag contents. The chemical additives are weighed out in plastic bags and placed on the charging conveyor with the rubber. An automatic weigh system delivers a given amount of carbon black directly to the mixer from either a remote silo, a rigid container located near the Banbury, or a flexible bulk container. Such automatic, closed-pipe delivery systems have a major impact in reducing the exposure of workers to

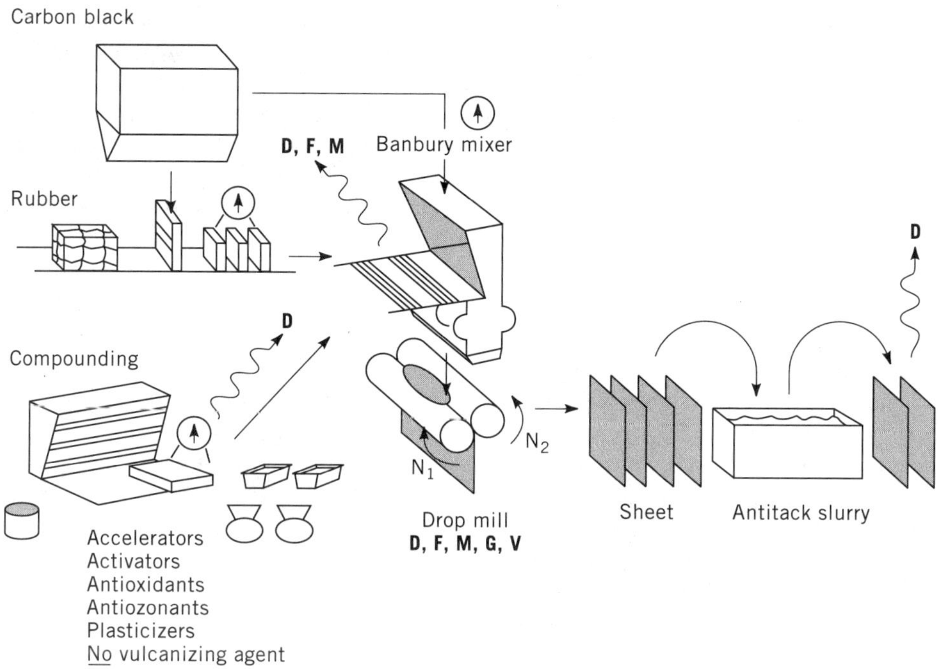

Figure 28.27. Schematic of compounding and mixing processes for the preparation of a rubber master batch with arrows showing the dust (D), fume (F), mist (M), and solvent vapor (V) release points.

carbon black. Only the very small job shop will now manually dump bags of carbon black into the mixer with its resulting dust exposure. The carbon black, preweighed rubber, and additives on the conveyor are then charged to the Banbury, the charge door closes, and the mass falls into the mixer chamber where it is homogenized to ensure uniform dispersion of all ingredients. This system requires a large powerful motor drive, reduction gears and a drive system and as a result the nearby work areas exceed the OSHA PEL of 85 dBA.

Reviewing the compounding operations shown in Figure 28.27, the major exposures occur during the preweighing or compounding work where the chemical additives are scooped from the bulk bags or drums and weighed into plastic bags. As noted in extensive data from a study of tire plants in the United States, Williams et al. (99) cite a median concentration of 0.69 mg/m^3 with a range of 0.39 to 1.57 mg/m^3 during compounding; the Banbury operators were exposed to about the same concentration as in compounding. A survey of plants in the Netherlands (100) reveal geometric mean concentrations of about 2.00 mg/m^3 in compounding. These exposures can be minimized by proper ventilation and work practice controls as will be discussed later. The most effective control of the dust exposure in compounding is a computer-controlled weighing system which delivers the required amount of material to the mixer without manual intervention.

As noted in Figure 28.27, when the Banbury door is opened and the charge is dumped, an exposure occurs to the dust of the chemical additives, condensed fume, and oil mist. This exposure cloud is hard to characterize and equally difficult to control. The most effective control is a partial enclosing hood. In normal practice the exposure of the mixer operator is in the range of 1–2 mg/m^3 depending on the workload and the efficacy of the ventilation control. The installation of well-designed ventilation on this process is frequently neutralized by the disruptive effects of drafts from pedestal fans positioned to cool the operator.

After the mixing operation is completed, the Banbury mixer charge is discharged to a drop mill located directly under the Banbury mixer on a lower floor. The two roll drop mill has rolls of 2–3 ft in diameter which rotate in different directions at different speeds. This process generates considerable heat and requires the rolls be cooled with water. During this transfer the hot stock may release volatile fractions of oils, high vapor pressure organic components, or possibly degradation products of the worked material. If biocidal additives are in the formula, they may be released. The mills are usually provided with canopy type hoods to take advantage of the hot, buoyant, rising air, however, the operator is exposed to particulate exposures with a median concentration of 0.5 to 1.0 mg/m^3. At the drop mill the stock is blended, sheeted, cooled, and cut for racking.

When the drop mill processing is completed, the rubber is slit and sheeted off to a conveyor. This work requires that the operator be positioned directly in front of the mill and leaning over the rolls. Accidents have occurred when clothing or hair is caught in the nip of the mill pulling in the worker and causing serious injury. For this reason the mills are provided with an emergency bar or cable to shut down the mill. All workers are trained in emergency response for such an incident.

The sheeted rubber stock is cut into slabs and the slabs are dipped in an antitack slurry as shown in Figure 28.27. The antitack material is frequently based on a clay, soapstone, or talc and is designed to prevent sticking. If talc or soapstone is used, it should be analyzed to ensure that it does not contain asbestiform material and that it has negligible free silica. The wet slurry process is obviously the most desirable way to add the antitack material, since the level of dustiness is much lower than for dry dusting. Exposure to the antitack occurs when the solution is prepared, when make-up is added, and from the dried residual which is resuspended from the slabs.

The rubber material which has been prepared is called master batch and has all the necessary ingredients except the vulcanizing agent. Master batches are normally made as an interim material and stored on pallets until needed for a production run or for sale to other processors. At that time the slab rubber is worked on a Banbury, breakdown, or feed mill, and the sulfur-bearing vulcanizing agent is added and homogenized with the master batch.

9.2.2 Processing

The rubber processing steps in tire production shown in Figure 28.28 involve all of the processes encountered in general rubber manufacturing except molding. It therefore provides a convenient way to identify the major occupational health issues in the fabrication of a range of rubber products by following the process flow for tires. In the process flow

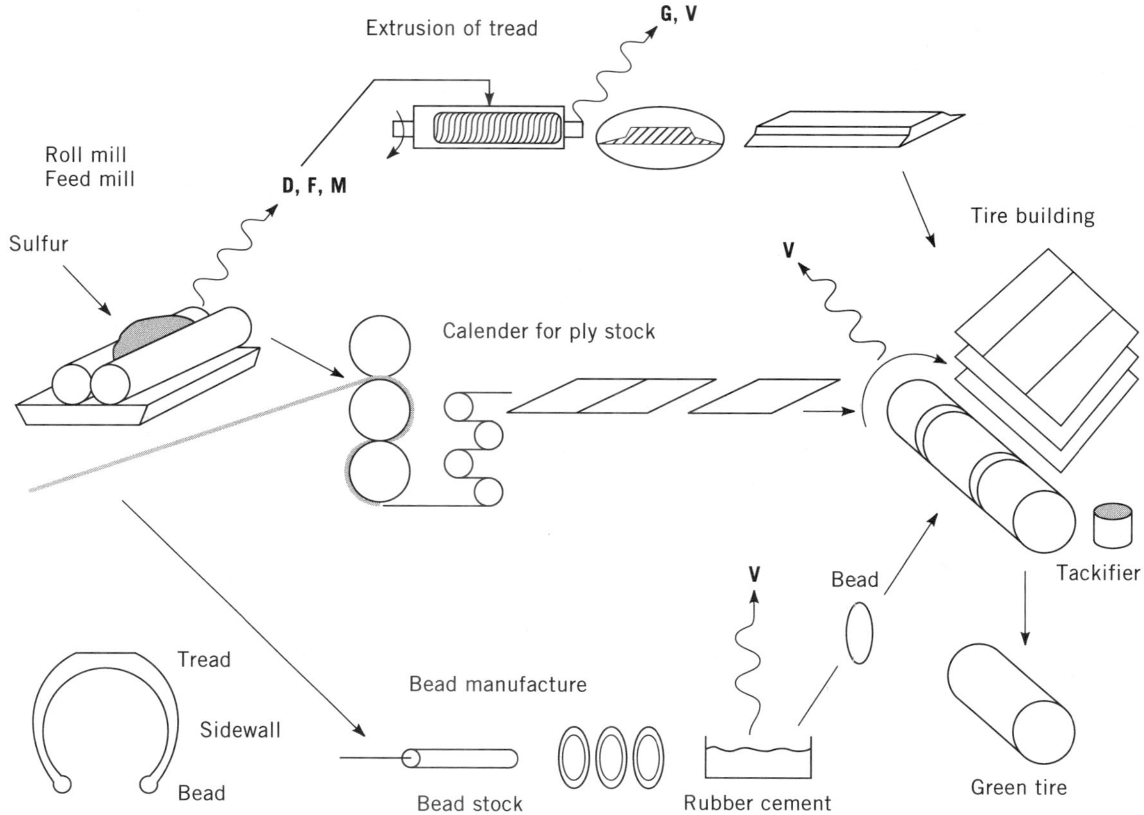

Figure 28.28. Schematic of processing of rubber materials to form the various piece parts for a radial automobile tire including the bead, tread, and plystock. The tire is built up on the machine form. The arrows show the dust (D), fume (F), mist (M), and solvent vapor (V) release points.

for the manufacture of an automobile tire, three piece parts must be fabricated: (*1*) the tire bead, which is a rubber-coated steel wire ring that holds the tire on the wheel rim, (*2*) plystock which is rubber coated fabric used in multiple layers to build the oval structure of the tire and (*3*) the solid rubber tread which is the road bearing element. Although many research programs have been directed to injection molding of a tire from plastic using automated methods, manual tire building with rubber, using a built-up concept, continues to be the method of choice.

The general process flow for the manufacture of these three main components of a tire is shown in Figure 28.28. In this schematic, the rubber is processed from one feed mill to the three processes as a convenience in presentation. This is not the case in the manufacturing plant where a special compounded rubber would be used for each part.

In the manufacture of the tire bead shown in the lowest processing line in Figure 28.28 the shredded rubber is mixed with solvent to make a viscous material which is used to coat the wire ring. The cement is prepared in a dedicated room called the cement house designed for the mixing of flammable solvents and constructed to conform to fire protection codes. Benzene was used in cements decades ago, the normal solvent now is hexane, heptane or white gasoline. The observed air concentrations to hexane and heptane in a tire plant range from 5 to 20 ppm.

The manufacture of the plystock (ply) is completed on the calender as shown in the middle flow pattern in Figure 28.28. The strength and flexibility of the tire is based on a radial structural belt consisting of several layers of a reinforced fabric made by filling and coating the fabric with rubber. The synthetic fiber fabric is woven with a fine steel wire reinforcing structure. It is fed through the calender and led to a roll up station or directly to a cutting bench. At the entrance nip to the calender, a softened lump of rubber is pulled through with the fabric, smearing the fabric and filling it with rubber. The calender is designed to provide close tolerance on thickness of the fabric-rubber material. The continuous web is cooled and cut to length depending on the tire size. The calender operator is exposed to low solvent vapor concentrations and fume from condensed high boiling hydrocarbons. The total particle concentration is usually less than 0.5 mg/m^3.

In the manufacture of the tread shown in the top processing line in Figure 28.28, rubber is fed from the feed mill to an extruder. In the extruder the material enters a heated screw feed barrel where it is compressed and forced out a die with a profile reflecting the shape and size of the tread. As the hot continuous extrudate comes out of the extruder there is some fume and vapor. The tread is picked up on a conveyor, cut to length, and a roll spreads rubber cement on the side that will bond to the tire carcass. The cement is a mix of rubber and a solvent, usually hexane or heptane. The total particulate exposure at the extruder operator's workstation is usually about 0.2 to 0.4 mg/m^3. Since the cement spreading tank is provided with local exhaust ventilation, the solvent concentrations are low.

9.2.3 Assembly

The assembly of the tire is now completed by the tire builder as noted in Figure 28.28. The three components, bead, plystock, and tread are positioned at the tire builders station in a file called a book. The tire building machine is basically a rotating cylinder sized for a certain tire that acts as a form for the construction of the tire. After the tire is built, the cylinder collapses so the tire can be removed. The two beads are first positioned on the drum and the inner plies are placed on the drum as it is rotated by an operator. The operator applies a small amount of tackifier (solvent) to the plies with a pad as he lays them up so that they will bond during vulcanization. This tackifier is usually a white gasoline that may have some aromatic content. Before the assembly is cured (vulcanized) it is called a green tire, a term is used throughout the rubber industry to identify an uncured rubber product. The particulate concentration in tire building is low, reflecting the general background in the tire plant of about 0.2 mg/m^3. Solvent exposures at the tire builders' workstation may be significant depending on the benzene content of the white gasoline used as a tackifier. If the benzene content is less than 0.1% the air concentrations of benzene will likely be less than 1 ppm.

9.2.4 Curing

In the final work sequence shown in Figure 28.29, the green tire is sprayed with a parting material prior to curing. Until the mid-1970s this spray was petroleum-based but, due to environmental concerns, it has been replaced with water-based materials. After this step the tire is taken to the curing area where the operation is conducted either manually or automatically. In manual curing, the operator places the tire form over a flexible bladder and the clamshell press is closed. The bladder is then inflated with live steam or hot water which drives the tire form against the inside of an aluminum mold with the tread profile. The curing press is heated to 100 to 200°C for 20 to 60 min. and curing, or the crosslinking of the rubber, takes place as a result of the action of the sulfur-based compound added at the mill operation. The process of vulcanization gives the tire its great strength and elasticity. In a high production tire plants the curing line is automatic. The operator spots the green tire at a loading station adjacent to the curing press. The tire is picked up automatically and positioned in the cavity of the press and the press closes and curing commences.

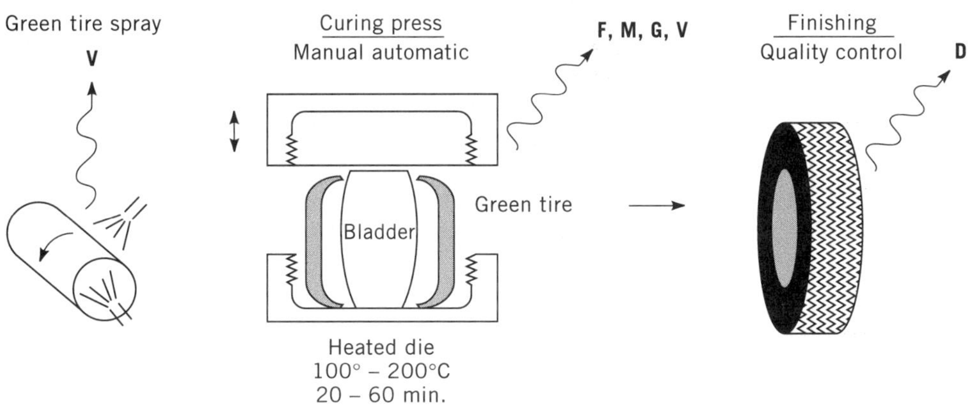

Figure 28.29. Schematic of the spraying of tire with a parting material followed by curing of the tire in a steam heated press.

The exposure in the press room is to the cloud released when the press opens after each cycle. The exact composition of the curing fume is not known but early studies by Rappaport (101) indicate a wide range of organic species. In an assay of the mutagenicity of particulates from air samples from this curing fume, seven out of eight samples were positive (98). The same bacterial test system demonstrated an increase in mutagenicity in urine samples of tire workers. In addition, a high density curing press line may result in heat stress to the operators.

All assembled rubber products require curing with heat. These procedures vary and may present specific hazards. Large conveyor belts are cured in flat, steam-heated presses. Water hose is cured by extruding a lead or nylon sheath over the hose and directing steam inside to cure the part. Continuous bath and drum curing are also conducted. Certain problems are common to all curing operations and include release of curing fume whose composition and concentration depends on the rubber stock and the curing temperature, heat stress due to the steam release and convective heat load, and noise exposure from air and steam release.

9.3 Control Techniques

9.3.1 Dust

There are some general dust control approaches that reside with the purchase of the various raw materials. In the last two decades the rubber, pharmaceutical, pigments, and dyestuff industries have given attention to the dustiness of the raw materials they produce and use. Dustiness testing has been extended and significant product design changes have been adopted to minimize dustiness and worker exposure while maintaining product acceptance. In tire manufacture at least a dozen chemicals in granular form are added in small quantities to the batch mix. In an effort to minimize worker exposure to dust from these chemicals, the British Rubber Manufacturing Association (102) has sponsored design of low dusting forms of these common chemicals. The properties of seven dust controlled forms of rubber chemicals have been reviewed by Hammond (103). The author notes that the disadvantages of the most effective approach, coating the chemical with a polymer, include the variability in active chemical content based on bulk weight, the reduced chemical content and the unsuitability of the polymer in the formulation.

In general, the more extensive the grinding or comminution of a granular material, the greater the dust hazard the material will present during transport, handling and processing. When possible, one should purchase the most coarse form of the chemical that is suitable for the process.

The use of water to suppress dust has been a basic tenet of occupational health for decades. Its value is seen in the application of antitack compounds. Until the 1960s, dry talc or limestone was frequently dusted on the slabs of masterbatch material to keep it from sticking. This operation resulted in poor housekeeping and a significant exposure to dust. The present technique, adopted in the 1960s, involves dipping the stock in a water slurry of the antitack compound before racking for storage. This simple change resulted in a significant reduction in worker exposure to dust.

9.3.2 *Solvent Vapors*

As noted earlier, to ensure a good bond between the components of a built-up assembly the rubber is made tacky by applying a small amount of solvent to the surface with a pad. For several decades this solvent was benzene. The worker exposure, probably in the range of 1 to 10 ppm, is believed responsible for the excess leukemia seen in older tire builders. Starting in the 1950s, the industry started to replace benzene with white gasoline. In studies completed in the 1970s by the Harvard School of Public Health Joint Rubber Studies Group, the residual benzene content in white gasoline was 4 to 7% and air sampling indicated that one-third of the air samples on tire builders exceeded 1 ppm (104). The OSHA PEL for benzene at that time was 10 ppm although it was anticipated that it would be dropped to 1 ppm (this change did occur in 1987). Technical grade chemicals commonly have significant impurities. The level of contamination should be identified and if sufficiently high, worker exposure should be evaluated. At that time the necessity for reduction in the impurity level can be determined.

9.3.3 *Curing Fume*

One of the most difficult air contaminant control problems in the rubber industry is the fume released from curing or compression molding presses. Control of this fume is necessary since, as mentioned in the Introduction, studies have confirmed excess lung cancer in curing room workers and tests have demonstrated that the fume is mutagenic. In addition, excess respiratory morbidity has been noted in this population. Two possible ventilation controls are available for this operation. The first, local exhaust, is the preferred approach but it is extremely difficult to apply on these presses. One can envisage a local hood which is activated only when the clamshell press opens. This approach would require rather sophisticated controls and represents a major innovation in control technology in this industry. Dilution ventilation is the conventional approach taken in most rubber plants. This would appear to be a reasonable application for this technique since there are a number of individual sources of contaminant, each press generates a relatively uniform mass of contaminant with minor temporal variations, and the contaminant is not highly toxic. In addition, the heat load from the presses require exhaust and this can be done effectively by dilution ventilation. There are the obvious limitations in dilution ventilation including high exhaust rates and the fact that the worker must be removed from the contaminant path. The initial installations in tire plants were straight dilution systems with roof exhausters without attempt at enclosure or designed replacement air supply. Later designs included enclosure by drop panels or curtains to form a large canopy over the curing press line. In some installations, the curtains dropped to within 10 feet of the floor with replacement air provided by general systems. Later designs introduced the replacement air directly at the edge of the curtained enclosure.

The efficiency of these systems varies considerably. The sampling of respirable mass concentrations does show differences by enclosure design, but the differences are overshadowed by the differences in contaminant level noted between manual and automatic presses. The first recommendation for control of fume in the curing area is therefore to convert to automatic presses.

If dilution ventilation is chosen as the approach the following general rules should apply:

- Fixed or flexible curtains should be dropped to the lowest possible level to reduce the face area.
- The worker should be positioned outside the hood.
- A minimum face velocity of 80 fpm should be achieved at the canopy face.
- Replacement air should be introduced adjacent to the curing line, but not directly at the canopy.

9.3.4 Specific Ventilation Designs

The ACGIH Ventilation Manual includes designs of local exhaust hoods for a range of rubber processing equipment (7). In addition, an excellent set of designs is available in a handbook published by the Rubber and Plastic Research Association of Great Britain (105).

9.3.5 Handling Practices

The British Rubber Manufacturers Association has devised a risk classification that permits the user to assign rubber accelerators, retarders and anti degradants to one of four toxicity categories (102). This approach permits the user to devise appropriate handling procedures for the chemicals in use.

ACKNOWLEDGMENTS

This material is based on previous Patty chapters on this topic by the author and on *Recognition of Health Hazards in Industry*, 2nd ed., Wiley, 1995, also by the author.

BIBLIOGRAPHY

1. *Industrial Health and Safety Criteria for Abrasive Blasting Cleaning Operations*, National Institute for Occupational Safety and Health, DHHS (NIOSH) Publication No. 75–122, Cincinnati, OH, 1975.
2. Annual Conference on Lead Removal From Industrial Structures, SSPC 412-268-2980. Structural Steel Painting Council, 1992.
3. B. Samini, H. Weill and M. Ziskind, *Arch. Environ. Health* **29** 61–66 (1974).
4. *NIOSH Alert, Request for Assistance In Preventing Silicosis and Deaths From Sandblasting*, National Institute for Occupational Safety and Health, DHHS (NIOSH) Publication No. 92-102, Cincinnati, OH, 1992.
5. G. R. Mackay, L. E. Stettler, C. Kommineni, and H. M. Donaldson, *Am. Ind. Hyg. Assoc. J.* **41**, 836–842 (1980).
6. DuPont Chemicals: Material Safety Data Sheet, Stauralite Products, Wilmington, DE July, 1990.

7. Committee on Industrial Ventilation, *Industrial Ventilation, A Manual of Recommended Practice*, 23rd ed., American Conference of Governmental Industrial Hygienists, Cincinnati, OH, 1998.
8. *Abrasive Blasting Respiratory Practices*. National Institute for Occupational Safety and Health, DHHF (NIOSH) Publication No. 74-104, Cincinnati, OH, 1974.
9. *Solvent Cleaning (Degreasing)*, Center for Emissions Control, Washington, DC, 1992, Pamphlet.
10. J. Van Dusan and J. W. Smith, (1989) *Am. Ind. Hyg. Assoc. J.* **50**, 252–256 (1989).
11. D. R. Blair, *Principles of Metal Surface Treatment and Protection*, Pergamon Press, 1972.
12. Technical Report, *Control Technology Assessment: Metal Plating and Cleaning Operations*, National Institute for Occupational Safety and Health, DHHS (NIOSH) Publication No 185-102, Cincinnati, OH, 1985.
13. W. A. Burgess, M. E. Ellenbecker, and R. D. Treitman, (1989), *Ventilation for Control of The Work Environment*, John Wiley, & Sons, Inc., New York, 1989.
14. *Users Guide to Powder Coating*, 3rd. ed. Association for Finishing Processes of the Society of Manufacturing Engineers, Dearborn, MI, 1993.
15. E. Olsen and L. Seedorff, *Ann. Occup. Hyg.* **34**, 371–378 (1990).
16. H. E. Myer, S. T. O'Block, and V. Dharmarajan, *Am. Ind. Hyg. Assoc. J.*, **54**, 663–670 (1993).
17. C. Winder and P. J. Turner, *Ann. Occup. Hyg.* **52**, 226–234 (1992).
18. M. K. Hansen, M. Larsen, and M. H. Cohr, *Scan. J. Work Environ. Health*, **13**, 473–485 (1987).
19. R. F. Herrick and T. J. Smith, *Scan. J. Work Environ. Health* **14**, Suppl. 1, 31–32 (1988).
20. J. B. D'Arcy and T. L. Chan, *Am. Ind. Hyg. Assoc. J.* **51**, 132–138 (1990).
21. T. L. Chan, J. B. D'Arcy, and R. M. Schreck, *Am. Ind. Hyg. Assoc. J.* **47**, 411–417 (1986).
22. M. W. Ackley *Am. Ind. Hyg. Assoc. J.* **41**, 309–316 (1980).
23. *Health Hazard Evaluation Report HETA 90-368-2137*, NIOSH Rockwell International, Newark, OH, 1990.
24. D. M. O'Brien and D. E. Hurley, *Am. Ind. Hyg. Assoc. J* **43**, 695–703 (1982).
25. D. M. O'Brien, *Am. Ind. Hyg. Assoc.* **45**, B-76, B-8 (1984).
26. L. W. Whitehead, G. L. Ball and L. J. Fine, *Am. Ind. Hyg. Assoc., J.* **45**, 767–772 (1984).
27. J. B. Hudnall, A. Suruda, and D. A. Campbell, *Am. Ind. Hyg. Assoc. J.* **54**, 32–35 (1993).
28. S. R. Woskie, S. K. Hammond, F. Rosenthal, M. F. Hallock and T. J. Smith, "Use of Stacked Filter Samplers to Estimate Coolant Aerosol Exposures," presented at the American Industrial Hygiene Conference, Montreal, 1987.
29. P. J. Ruane "Control of Potential Health Hazards of Formulation Practice," *Proceedings of a Conference on the Health Effects of Cutting Oils and Their Controls*, April 28 and 29, 1989, University of Birmingham, England, 1989, pp. 57–69.
30. M. Hallock, T. J. Smith, S. R. Woskie, and S. K. Hammond, *Am. J. Ind. Med.* **26**, 621–634 (1994).
31. O. Einarsson, E. Eriksson, G. Lindstedt, and J. E. Wahlberg, *Contact Dermatitis* **5**, 129–132 (1979).
32. National Institute for Occupational Safety and Health (NIOSH), *Health Hazard Evaluation Report, HETA 88-268-L1980*, 1989.
33. National Institute for Occupational Safety and Health (NIOSH), *Health Hazard Evaluation Report, HETA 87-092-1967*, 1989.

34. M. J. Evans, W. A. Hooper, A. J. Ingram, D. C. Pullen, and R. H. R. Aston, *Ann. Occup. Hyg.* **4**, 537–553 (1989).
35. T. L. Chan, J. B. D'Arcy, and J. Siak, *Appl. Occup. Environ. Hyg.* **5**, 162–170 (1990).
36. *Proceedings of a Conference on the Health Effects of Cutting Oils and Their Controls*, April 28 and 29, 1989, University of Birmingham, England, pp. 13–28.
37. *Proceedings of Metalworking Fluids Symposium*, Nov. 13–15, 1995, American Automotive Manufacturers Association, Dearborn, MI.
38. *Criteria For A Recommended Standard For Occupational Exposure to Metal Working Fluids* NIOSH Pub. No. 98-102, National Institute for Occupational Safety and Health, Cincinnati, OH, 1998.
39. R. J. G. Rycroft, (1989) "Dermatitis and Cutting Fluids" *Proceedings of a Conference on the Health Effects of Cutting Oils and Their Controls*, April 28 and 29, 1989, University of Birmingham, England, 1989, pp. 31–35.
40. P. J. Ruane, "Composition of Metal Working Fluids" *Proceedings of a Conference on the Health Effects of Cutting Oils and Their Controls*, April 28 and 29, 1989. University of Birmingham, England, 1989, pp 5–12.
41. E. Eisen, P. E. Tolbert, M. F. Hallock, R. R. Monson, T. J. Smith, and S. R. Woskie, *Am. J. Ind. Med.* **25**, 185–202 (1994).
42. P. S. Burge, "The Health Effects of Cutting Oils on The Lung", *Proceedings of a Conference on the Health Effects of Cutting Oils and Their Controls*, April 28 and 29, 1989, University of Birmingham, England, 1989, pp. 13–28.
43. P. Decoulfe, *J. Nat Cancer Inst.*, **61**, 1025–1030 (1978).
44. B. Jarvholm, B. Bake, B. Lavenius, G. Thiringer, and R. Vokmann, (1982), *J. Occup. Med.* **24**, 473–479 (1982).
45. J. E. Vena, H. A. Sultz, R. C. Fiedler, and R. E. Barnes, *Brit. J. Ind. Med.* **42**, 85–93 (1985).
46. S. M. Kennedy, I. A. Greaves, D. Kriebel, E. A. Eisen, T. J. Smith, and S. R. Woskie, *Am. J. Ind. Med.* **15**, 627–641 (1989).
47. R. B. Weaver, "Controlling microbiological contamination of cutting fluids", *Proceedings of a Conference on the Health Effects of Cutting Oils and Their Controls*, April 28 and 29, 1989, University of Birmingham, England, 1989, pp 57–69.
48. E. C. Hill, "Microbial Contamination of Metal Working Fluids", *Proceedings of a Conference on the Health Effects of Cutting Oils and Their Controls*, April 28 and 29, 1989, University of Birmingham, England, 1989, pp 29–30.
49. G. Lee, "Oil and Coolant Mists-Assessment Criteria and Control" *Proceedings of a Conference on the Health Effects of Cutting Oils and Their Controls*, April 28 and 29, 1989, University of Birmingham, England, 1989, pp 51–56.
50. U. Ulfvarson, *Scand. J. Work Environ. Health* 7 (Suppl. 2) 1–28 (1981).
51. R. M. Stern, (1983) Occupational Exposure in the Welding Industry: Health Risk Assessment, in *Health Effects of Combined Exposures to Chemicals in Work and Community Environments. Proceedings of a Course by the European Cooperation on Environmental Health Aspects of the Control of Chemicals, Lodz, Poland*, Oct. 18–22, 1982, World Health Organization, Copenhagen
52. IARC Monographs on the Evaluation of Carcinogenic Risks to Humans, Chromium, Nickel and Welding, Vol. 49, World Health Organization, International Agency For Research on Cancer, 1990.

53. *The Facts About Fume, A Welding Engineers Handbook*, The Welding Institute, Cambridge, UK, 1986.
54. M. Pantucek, *Amer. Ind. Hyg. Assoc. J.* **32**, 687 (1971).
55. J. Moreton and N. Jenkins, *Ann. Occup. Hyg.* **28**, 443–444 (1984).
56. H. Buck and Dessler, International Institute of Welding, Document III-311-67, Paris, 1967.
57. H. E. Stokinger, *Am. Ind. Hyg. Assoc. J.* **45**, 127–133 (1984).
58. *The Welding Environment*, American Welding Society, Miami, FL, 1973.
59. U. Ulfvarson, "Air Contaminants Involved in Welding in Swedish Industry—Sources of Variation in Concentrations," in R. M. Stern, A. Berlin, A. C. Fletcher, and J. Jarvisalo, eds, *Health Hazards and Biological Effects of Welding Fumes and Gases*, Excerpta Medica, Amsterdam, 1986, pp. 133–136.
60. W. H. Phoon, and K. Tan, *Occup. Health and Safety*, 19–25 (Jan. 1983).
61. L. Smith, *Ann. Occup. Hyg.* **10**, 113 (1967)
62. K. Dawada, and K. Twano, International Institute of Welding Document VIII-342-68 Paris, 1968.
63. B. Sjogren, L. Hedstrom, and G. Linstedt, *Brit. J. Ind. Med.* **41**, 192–196 (1984).
64. W. Kierst, J. Vselis, M. Graczyk, and A. Krynick, *Bull. Inst. Mar. Med. Gdansk* **15**, 149 (1964).
65. J. Steel, *Ann. Occup. Hyg.* **11**, 115 (1978)
66. C. E. Moss and W. E. Murray, *Welding J.* **9**, 37–46 (1979).
67. M. A. Stuchly and D. W. Lecuyer, *Health Physics.* **56**, 297–302 (1989).
68. R. M. Stern, A. Berlin, A. C. Fletcher, and J. Jarvisalo, eds, *Health Hazards and Biological Effects of Welding Fumes and Gases*, Excerpta Medica, Amsterdam, 1986.
69. R. Frant, *Ann. Occup. Hyg.* **6**, 113 (1963).
70. R. Lanau, *Ann. Occup. Hyg.* **10**, 175 (1967).
71. L. G. Wiseman and E. T. Chapman, "Test Program to Determine Whether Nickel Carbonyl Forms during the Welding of Nickel-containing Alloys: A Progress Report." in R. M. Stern, A. Berlin, A. C. Fletcher, and J. Jarvisalo, eds, *Health Hazards and Biological Effects of Welding Fumes and Gases*, Excerpta Medica, Amsterdam, 1986, pp. 133–1.
72. NIOSH, Health Hazard Evaluation Report, HETA 90-232-2138, Schulte Corp., 1990.
73. P. J. Hewitt and A. A. Hirst, *Ann. Occup. Hyg.* **35**, 223–231 (1991).
74. L. S. Waterman, C. J. Purnell, and P. W. P. Anderson, *Ann. Occup. Hyg.* **34**, 97–99 (1990).
75. B. Engstrom, M. L. Henriks-Eckerman, and E. Anas, *Am. Ind. Hyg. Assoc. J.* **51**, 561–565 (1990).
76. *Criteria for a Recommended Standard, Welding, Brazing and Thermal Cutting* DHHS (NIOSH) Public, No. 88-110a: PB 88-231774, NIOSH, Cincinnati, OH, 1988.
77. P. W. Glinsmann and F. S. Rosenthal, *Am. Ind. Hyg. Assoc. J.* **46**, 391–395 (1985).
78. *Safety in Welding, Cutting and Allied Processes*, ANSI Z49.1-1999, American Welding Society, Miami, FL, 1999.
79. *Optical Radiation Transmission Levels Through Transparent Welding Curtains*, National Institute for Occupational Safety and Health, NIOSH No. 88-110a PB, Cincinnati, OH, 1978.
80. J. W. Goller and N. W. Paik, *Am. Ind. Hyg. Assoc. J.* **46**, 89–63 (1985).
81. NIOSH Health Hazard Evaluation Report, HETA 89-040-2010, MDT/Castle, Rochester, NY, 1990b.
82. Lens Shade Selector, AWS Cat. 2.2-89.

83. W. Seabreak 1989, Personal communication.
84. R. W. Bohl and R. R. Langer, 1979, Personal communication.
85. C. Chelton, 1994, Personal communication.
86. R. D. Soule, in G. D. Clayton and F. E. Clayton eds., *Patty's Industrial and Toxicology*, 4th ed. Vol. 1, Part A, John Wiley & Sons, Inc., New York, 1991
87. P. F. Woolrich, in L. J. Cralley and L. V. Cralley eds. *Industrial Hygiene Aspects of Plant Operations* Vol. 2, Macmillan, New York, 1984.
88. J. Lipton and J. Linch, *Health Hazards Control in the Chemical Industry*, John Wiley & Sons, Inc., New York, 1987.
89. W. A. Burgess, "Philosophy and Management of Engineering Controls," in L. J. Cralley and L. V. Cralley, eds., *Patty's Industrial Hygiene and Toxicology*, John Wiley & Sons, Inc., New York, 1993.
90. Environmental Protection Agency, *Appl. Ind. Hyg.* **1**, R11–13 (1986).
91. C. L. Holzner, R. B. Hirsh, and J. B. Perper, *Am. Ind. Assoc. J.* **54**, 15–21 (1993).
92. A. D. Schwope, P. P. Costas, J. O. Jackson, and D. J. Weitzman, *Guidelines for the Selection of Chemical Protective Clothing*, 3rd ed., American Conference of Governmental Industrial Hygienists, Cincinnati, OH, 1987
93. F. F. Monson and K. K. Nakano, *Am. J. Epid.* **103**, 284–296 (1976).
94. F. F. Monson and K. K. Nakano, *Am. J. Epid.* **103**, 297–303 (1976).
95. E. Delzell and R. R. Monson, (1981), *J. Occup. Med.* **23**, 677–684 (1981).
96. L. Fine and J. Peters, *Arch. Environ. Health* **31**, 136–140 (1976).
97. H. G. Parkes, C. A. Veys, J. A. H. Waterhouse, and A. Peters, *Brit. J. Ind. Med.* **39**, 209–220 (1982).
98. "Genotoxic Health Hazards in the Rubber Industry," *Scan. J. Work Environ. Health* 9, Suppl. 2, 1–93 (1983).
99. T. M. Williams, R. L. Harris, E. W. Arp, M. J. Symons, and M. D. Van Ert, *Am. Ind. Hyg. Assoc. J.* **41**, 204–211 (1980).
100. P. Swuste, H. Kromhout, and D. Drown, *Ann. Occup. Hyg.* **37**, 117–134 (1993).
101. S. Rappaport, *Am. Ind. Hyg. Assoc. J.* **38**, 205–210 (1977).
102. *Toxicity and Safe Handling of Rubber Chemicals*, British Rubber Manufacturer's Association, England, 1978.
103. C. M. Hammond, *Ann. Occup. Hyg.* **23**, 95–109 (1980).
104. R. Treitman, 1976, personal communication.
105. *Ventilation Handbook for the Rubber and Plastics Industries*, Rubber and Plastics Research Association of Great Britain, England, 1979.

CHAPTER TWENTY-NINE

Philosophy and Management of Engineering Control

William A. Burgess, CIH (Ret.)

1 INTRODUCTION

This chapter will cover the engineering methods widely used to minimize adverse effects on health, well-being, comfort and performance at the workplace and in the community. Major attention will be given to controls for chemical airborne contaminants, although many of the control principles are also applicable to biological and physical hazards. Brief case studies will illustrate the applications of the techniques. The target for control may be a hand tool, a piece of equipment or a device, an integrated manufacturing process such as an electroplating line, or a complete manufacturing facility in a dedicated building such as a foundry.

The critical zones of contaminant generation, dispersion, and exposure are shown in Figure 29.1. Ideally the goal is to design each element of the process to eliminate contaminant generation. If it is impossible to achieve this goal and a contaminant is generated, the second defense is to prevent its dispersal in the workplace. Finally, if these defenses fail and the material released from the operation results in worker exposure, the back-up control is collection of the air containing the contaminant by exhaust ventilation. Frequently the contaminant is removed from the exhaust air stream by air cleaning before returning the air to the workplace or the general environment. Unfortunately, in the United States little emphasis is often given to the prevention of contaminant generation and release and we rely on ventilation as a principal remedy. In this chapter the author confirms the importance of ventilation but emphasis will be placed on the primary controls associated with process and material design to minimize generation and release of the contaminant.

Patty's Industrial Hygiene, Fifth Edition, Volume 2. Edited by Robert L. Harris.
ISBN 0-471-29754-2

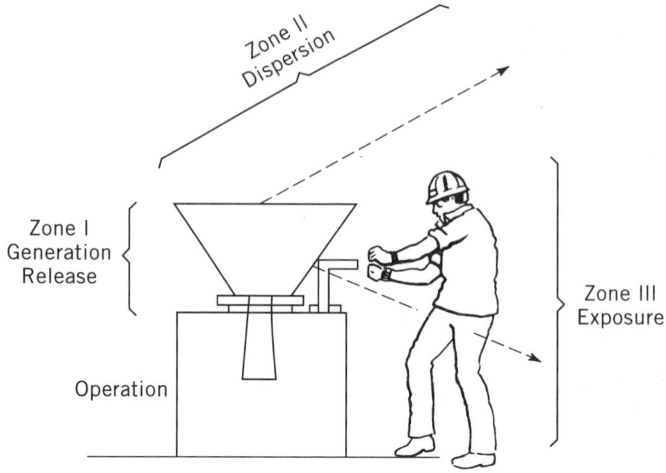

Figure 29.1. Contaminant generation, release, and exposure zones in the workplace.

The Occupational Safety and Health Administration (OSHA) considers engineering controls as any modification of plant, equipment, processes or materials to reduce employees' exposure to toxic materials and harmful physical agents. As noted in Table 29.1, authors have used a variety of approaches to classify the various engineering control techniques (1–4). These approaches provide a useful outline for discussion and act as checklists to ensure a comprehensive review. Sherwood and Alesbury (4) have also designed a comprehensive algorithm for control of air contaminants. Brandt (5) states: "Measures for preventing the inhalation of excessively contaminated air have been discussed by many authors and there are as many classifications of these methods as there are papers on the subject. The principles expounded, however, are always essentially the same." Brandt also reminds the reader that "The control of an atmospheric health hazard is rarely accomplished by a single measure. It usually involves a combination of measures."

In this chapter engineering controls will include all techniques except personal protective devices and administrative controls, those changes made in the work schedule to reduce the time weighted average exposure of the worker.

2 GENERAL GUIDELINES

The strategy to be used in solving a control technology problem varies depending on the setting. As described above, the control problem may involve a specific piece of equipment, a process, or an integrated manufacturing plant. In a common situation faced by many in the 1990s, a shop repairing electric motors must consider a change in solvents in a vapor-phase degreaser. The number of persons participating in the review is small, the decision will be made in a short time and it will probably rely heavily on vendor information. The

Table 29.1. Classifications of Methods for Control of Air Contaminants

First (1983)	*Burgess (1995)*
Material substitution	Toxic materials
Processing change	Eliminate
Equipment change	Replace
Local exhaust ventilation	Dust control
General (dilution) ventilation	Reduce impurities
Equipment enclosure	Equipment and process
Employee enclosure	Task modification
	Facility layout
Gideon et al. (1979)	Ventilation
	General ventilation
Control at source	Local exhaust ventilation
Material substitution	
Process or equipment change	*Sherwood and Alsbury (1983)*
Isolation	
Local exhaust ventilation	Automation, alternative methods
Work practice	Substitute materials, methods
Control at workplace	Plan layout, enclosures, remote
General exhaust ventilation	Local exhaust ventilation
Housekeeping	Dilution ventilation
Control at worker	Personal protective equipment
Isolation/personal protective equipment	

range of options presented in this chapter probably will not be thoroughly reviewed. An industrial hygienist will not be involved unless the service shop is a support facility operated by a major company or consultation is available from an insurance carrier, consulting industrial hygienist, or state OSHA consulting program.

A far more complex situation is the scale-up of a chemical process to manufacture a photographic dye. The research to develop the new molecule and synthesize a small quantity for initial tests is completed by a research chemist at a laboratory bench. At this stage little is known about the chemical or its toxicity, however, since gram quantities are handled in a well-controlled laboratory hood the risk is negligible. The next step involves a scale-up of the operation in a pilot plant facility designed to manufacture kilogram quantities of the material for testing purposes. This testing will provide limited data on the physical properties of the new chemical, its product potential, and toxicity data. If the results are encouraging, production scale-up is considered. A premanufacturing notification must be submitted to the Environmental Protection Agency (EPA) prior to production. This submission, which requires a description of air contaminant control technology and an estimation of worker exposure, involves a comprehensive review of the operation as noted in Table 29.2. It is decided to manufacture the chemical at an existing plant using conventional chemical processing equipment. If the industrial hygienist has been involved in the pilot plant operation he is well-placed to participate in a review of the control technology necessary for the full scale plant. As noted in Table 29.2, the range of talent available to

Table 29.2. Specialists Participating in Risk Assessment of New Chemical Manufacture[a]

	Adminis-trative	Legal	Sales	Marketing	Chemists	Engineers	Toxicol-ogists	Occupa-tional Hygienists	Environ-mental Scientists	Statis-ticians	Risk Analyst	Transpor-tation Systems Analysis	Econo-mists	Infor-mation Specialists
Part I: General Information														
A. Manufacturer identification	X	X												
B. Chemical identity					X									X
C. Marketing data			X	X						X			X	
D. *Federal Register* notice	X	X		X	X	X	X	X	X					
E. Schematic flow diagram					X	X		X	X					
Part II: Risk Assessment Data														
A. Test data		X			X		X		X	X	X			
B. Exposure from manufacture														
1. Worker exposure					X	X	X	X		X	X			
2. Environmental release					X	X			X	X	X			
3. Disposal					X	X	X	X	X		X			
4. By-products, etc.					X	X					X			X
5. Transportation		X		X		X					X	X		

C. Exposure from process operations												
1. Worker exposure					X	X	X	X		X	X	
2. Environmental release					X	X			X	X	X	
3. Disposal					X	X	X	X	X		X	
D. Exposure from consumer use	X	X	X	X		X			X		X	
Part III: Risk Analysis and Optional Data												
A. Risk analysis and optional data	X	X			X		X		X		X	
B. Structure activity relationships					X		X				X	
C. Industrial hygiene	X	X			X	X	X	X			X	
D. Engineered safeguards					X	X	X	X	X		X	
E. Industrial process and use restriction data					X	X						
F. Process chemistry					X	X						
G. Nonrisk factors: Economic and non-economic benefits	X	X	X	X	X	X					X	X

[a]From Arthur D. Little, Inc.

define the engineering controls is extensive and the time frame is extended. Hopefully step by step review to identify the appropriate engineering controls will minimize worker exposure. If exposure problems are not anticipated controls must be retrofitted at significant cost whereas early integration of controls are most easily accepted and economically implemented.

The most complex engineering control package is encountered in the design and construction of a major production facility. The project is complex due to the range of issues that are encountered; however, the design process is started with a clean design sheet and state of the art controls can be included in the design. In the example in Table 29.2 a large number of specialists are included in the review team. If the design is done in-house the industrial hygienist should be a key participant. If the design work is contracted to a design and construction firm, plant personnel are one-step removed from the design process. The company will have a project engineer to interface between the outside team and company engineering group. In this case clear communication must be established by the industrial hygienist to ensure involvement in the design review.

For the industrial hygienist facing his or her first process hazard review the job may seem awesome. An initial step is to identify your responsibility. Does it include merely health issues at the workplace or are environmental and fire protection problems to be included in your review? Once your responsibility has been defined, the issues of concern for this particular plant should be identified. Mature industries such as metallurgical, glass, and ceramics have been studied and extensive information is available on standard operations and sources of contamination. As noted in the Chapter 31, the EPA has published emission rates for many processes that are useful in process hazard reviews. Another source of information on control technology for mature industries is the series of over fifty technical reports published by the National Institute for Occupational Safety and Health (NIOSH) (see Appendix A).

If the company has similar operations elsewhere, obtain material and process data, industrial hygiene survey data and air sampling information from that plant to assist in the review. A step-by-step review of the operations is then undertaken. This health and safety review should be as rigorous as the review of percent yield on the process and produce a detailed process flow diagram showing the locations of contaminant loss in addition to the major physical stresses such as noise, heat, and ergonomics. It is important to also consider the ancillary systems including compressed air, refrigeration, cooling towers, and water treatment. Bulk handling of chemicals and granular minerals frequently presents a major industrial hygiene problem and should be given special attention. Finally a detailed description of the workers' tasks including required emergency response actions are required since work location and movement have a major effect on contaminant control. A comprehensive checklist for review of environmental and occupational health and safety issues in planning major plant construction is given by Whitehead (6).

An element that is frequently overlooked in the design of engineering controls is the worker. A review of a specific operation conducted by several workers frequently will identify subtle differences in the way the task is performed. A worker inspired by a wish to reduce lifting, provide "cool down" in the heat, or reduce dust exposure may modify the tasks to accomplish this end. Workers frequently have insight into details of the operation that the design engineer will overlook. The video–real time monitoring approach discussed in Section 3.2.1 is useful in defining these modifications in tasks.

It is difficult to recommend engineering controls for emerging technologies. Hopefully each industrial group will develop and publish its own engineering controls. Such is the case with advanced composite manufacturing, a technology that has seen expanding applications in recent years in the manufacture of aircraft. An association of companies involved in this work has published a volume describing the major hazards and the controls that should be applied (7).

Frequently the best control cannot be achieved directly on the design board. In the 1970s a number of cases of asbestosis in shipyard workers fabricating asbestos insulating pads for steam propulsion plants prompted the plant to install an integrated workstation incorporating a number of control features. Initial designs of the downdraft work table were fabricated and evaluated by workers resulting in a number of recommendations that were incorporated in the final design improving its acceptance and performance (Fig. 29.2).

3 CONTROL STRATEGY

The general strategy as noted earlier is to eliminate the generation source, if that is not possible then to prevent escape of the contaminant, and if it does escape to collect and

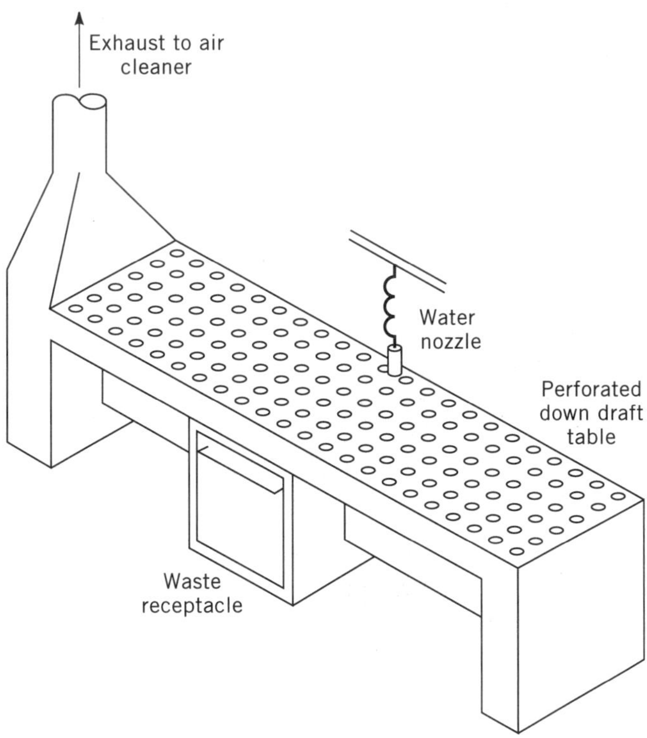

Figure 29.2. Integrated controls on an asbestos insulation fabrication workbench.

remove the contaminant before the worker is exposed. This strategy will be reviewed by considering action on toxic materials, equipment, processes, job tasks, plant layout, and exhaust ventilation.

3.1 Toxic Materials

The use of toxic materials range in complexity from simple wet degreasing to the synthesis of new chemicals as noted above. In degreasing the exposure is associated with the use of the cleaning agent or handling its waste. In synthesis of a new chemical, exposure may occur to raw materials, intermediates, by-products, the new chemical and waste streams. In minimizing potential worker exposure to these materials it is necessary to consider the following options.

3.1.1 Eliminate The Toxic Material

This option refers to elimination without replacement. Frequently toxic chemicals are used in a process to improve yield or in a product to improve function or appearance. In review, it may be possible to eliminate the chemical. The pressure to do so may come from potential health effects, the cost of workplace and environmental controls to introduce the toxic chemical into the plant, or it may arise from pressure from the marketplace. An example of the latter resulted from legislation in Switzerland requiring precautionary labelling of batteries containing mercury. The conventional Leclanche dry cell battery utilizes electrodes of manganese dioxide and zinc with an electrolyte of ammonium chloride. A characteristic of the cell is zinc corrosion with subsequent power loss. A mercury compound is added to inhibit this corrosion and resulting performance degradation. At least one manufacturer has determined the performance enhancement from mercury is not necessary and the mercury compound has been removed from the battery formulation (8). This change has eliminated exposure to mercury for several hundred workers in this company.

3.1.2 Replace with Alternative Material

This control approach holds the greatest potential for significant reduction in worker exposure to toxic chemicals. Historically, the approach has had success in the United States. One of the most effective steps was the cooperative action taken by the United States Public Health Service, the State of Connecticut Dept. of Public Health, industry, and labor unions in 1941 to prevent mercurialism by replacing mercuric nitrate with non-mercury compounds in the carroting process in felt hat manufacture.

The exposure of children to lead from lead-based paint and potable water supplies containing lead–tin–copper sweat joints has been a public health concern for decades. In 1986, the Amendments to the Safe Drinking Water Act effectively banned the use of lead-based solder. The acceptable alternative solders are alloys of tin in the range of 90–95% and the balance is copper, silver, zinc, or antimony (9). The introduction of the alternative solder will minimize exposure of manufacturing, construction and maintenance personnel to lead. The banning of mercury in hat manufacture demonstrates an action taken specifically for occupational health reasons; the action on lead-based solder was prompted by environmental concerns but will have a positive impact on the health of all workers.

A total of 10 materials have been banned for specific operations in the United Kingdom (10). An important element of the United Kingdom regulation is the prohibition of silica sand for abrasive blasting. Although there have been attempts to restrict the use of sand for this purpose in the United States, as the most frequently used blasting material it continues to present a major risk to workers.

Greater emphasis will be given to replacement control technology in the next decade. As discussed below, the impetus will continue to result from, not the workplace health issue, but the environmental health and ecological impact. Health and safety characteristics including toxicity, smog contribution, ozone depletion, and fire potential must be evaluated. In addition to the review of the performance of the replacement material, the changes in process and product design to satisfactorily change over to the new material must be considered in the final decision.

3.1.2.1 National Toxic Use Reduction Programs. Until the 1980s the EPA emphasis was pollution control and dealt principally with "end of pipe" abatement technology. In 1984 the Hazardous and Solid Waste Amendments to the Resource Conservation and Recovery Act of 1976 redirected environmental quality efforts from the conventional emphasis on waste treatment and disposal to waste minimization. This approach to waste management is characterized by source reduction, recycling and reuse. In 1986 the Office of Technology Assessment published a report that described the concept of prevention versus control (11). Since these environmental initiatives have great impact on the replacement of toxic chemicals in industry and, therefore worker exposure, it is important that the industrial hygienist understand the environmental management nomenclature (Table 29.3). The Pollution Prevention Act of 1990 consolidated this approach by specifying the following waste management hierarchy as national policy: first source reduction as the most desirable approach,

Table 29.3. Environmental Control Approaches[a]

Form of Control	Emphasis	End Rule
Pollution control	"End-of-pipe" control	Does not eliminate pollution, but merely transfers contaminant from one media to another. Incurs an environmental risk from transporting toxic chemicals.
Pollution prevention	Reduction in the production of contaminants	Direct waste reduction, covers all pollutants
Toxic use reduction	Reduction in pollutants generated in manufacturing by changes in plant operation procedures, production processes, materials, or end products Focus on a target list of chemicals	Priority setting permits control of high potential risk chemicals

[a]Adapted from Rossi et al. (15).

then recycling and reuse, then treatment, and finally disposal as the least desirable approach (12).

To initiate this strategy, in 1989 the EPA established the Waste Reduction Innovative Technology Evaluation (WRITE) Program. This program funded programs in six states and one county over a 3-year period to (*1*) provide engineering and economically feasible solutions to industry on specific pollution prevention problems (*2*) provide performance and cost information on these techniques and (*3*) promote early introduction of these pollution prevention programs into commerce and industry. A review of completed and on-going WRITE PROGRAMS that demonstrate the importance of replacement technology is shown in Table 29.4 (13). Although initiated for environmental quality reasons, each of these changes will have a major impact on worker health.

Denmark, one of the many European countries embarked on a national toxic use reduction program, base their work on the simple step model shown in Table 29.5 (14). The simplicity of this approach suggests that replacement technology can be done on an ad

Table 29.4. Examples of Material Replacement from EPA WRITE Program[a]

Operation	Original Material	Replacement
Anodizing aluminum parts	Chromic acid anodizing, hexavalent chromium exposure	Sulfuric acid anodizing, eliminates exposure to chromium
Removal of paint on defective parts	Methylene chloride-based paint stripper	Abrasive blasting with plastic beads, no vapor exposure
Flexograph printing	Application of alcohol-based ink labels	Water-based inks eliminate vapor exposure
Cleaning and deburring small metal parts	Vapor degreasing and alkaline tumbling	Automatic aqueous rotary washer. Eliminates vapor and caustic mist exposure
Cleaning flexographic plates	Solvent cleaning with Stoddard solvent, acetone, toluene, and alcohol	Cleaning with terpenes and aqueous cleaners

[a]Adapted from Harten and Licis (13).

Table 29.5. Step Model for the Substitution Process[a]

Step	
1	Problem identification
2	Identification and development of a range of alternatives
3	Identification of consequences of the alternatives
4	Comparison of the alternatives
5	Decision
6	Implementation
7	Evaluation of the result

[a]From Goldschmidt (14).

hoc basis, however, Goldschmidt warns that chemical replacement for occupational health reasons is a complex task and requires a systematic review of both the existing conditions and those prevailing after the replacement action. A review of 162 individual replacement actions by the Danish Occupational Health Services confirms that replacement is successful if done in a rational manner following the step model shown in Table 29.5. Among the impressive examples of this program is the replacement with water-based systems of the solvent-based paints in all indoor paints and most outdoor paints.

The principal advances in worker protection in the next decade will probably result from replacement technology accomplished as a result of such worldwide environmental health initiatives.

3.1.2.2 State Toxic Use Reduction Programs. The recent emphasis on pollution prevention and specifically on toxic use reduction in manufacturing is reflected in recent legislative action in 17 states and by the federal government (15). The Massachusetts Toxic Use Reduction Act of 1989 is described as a seminal program in that it does focus on toxic use reduction and excludes off-site recycling, off-site nonproduction unit recycling, transfer, or treatment. The five toxic use reduction techniques acceptable under the Massachusetts act are

> *In-process Recycling or Reuse.* Recycling, reuse, or extended use of toxics by using equipment or methods which become an integral part of the production unit of concern, including, but not limited to, filtration and other closed loop methods.
>
> *Improved Operations and Maintenance.* Modification or addition to existing equipment or methods including, but not limited to, such techniques as improved housekeeping practices, system adjustments, product and process inspections, or production unit control equipment or methods.
>
> *Changes in the Production Process.* Includes both the modernization of production equipment by replacement or upgrade based on the same technology or the introduction of a new production equipment of an entirely new design.
>
> *Input Design.* Replacement of a chemical used in production with a chemical of a lower toxicity.
>
> *Product Reformulation.* Redesign of the product to produce a product which is nontoxic or less toxic on use, release, or disposal.

Rossi (15) proposes two divergent routes the manufacturing group may take. The route for ease of implementation includes first, recycling and improved operations and maintenance, then production process changes, and finally replacement of toxic chemicals and product reformulation. However, for effective toxic use reduction the first step is to eliminate or replace the toxic chemical in the process or reformulate the product, then initiate production process changes and finally recycle and apply operations and maintenance control techniques.

An institute funded by industry assessment has been established at the University of Massachusetts at Lowell with the function to provide a teaching and research facility to encourage aggressive action in toxic use reduction.

3.1.2.3 Industry Programs. A number of company initiatives have been described which respond to specific legislation or broad environmental concerns. One company program has multiple goals to reduce the use of toxic chemicals, minimize toxic emissions to all environmental media, encourage recycling of chemicals, and reduce all waste (8). In the design of this program the company followed many of the recommendations of the Office of Technology Assessment report on the reduction of hazardous waste (11). In this company program specific goals for toxic use and waste reduction are set for all production managers and their annual performance reviews include this issue.

Each chemical used in the company is assigned to one of five Environmental Risk Categories (ERC). The materials in ERC I pose the greatest risk to the environment and ERC II, ERC III, and ERC IV represent progressively less risk. A fifth category includes plastic, steel, paper and other waste identified as rubbish, rubble, and trash. A critical issue in the design of the program was the decision logic for assigning categories. A number of options were considered including ranking chemicals based on workplace exposure guidelines such as threshold limit values (TLVs) or permissible exposure limits (PELs), label signal words (16), regulatory lists published by OSHA and EPA, and the mathematical weighting of a range of risk factors including toxicity and chemical and physical properties (17).

The company chose to assign the ERCs based on a "wise person" approach based on a battery of risk factors reflecting the materials' total occupational and environmental impact from release both as an isolated incident such as a spill or the chronic release of small amounts of the chemical to air, water, and soil. The majority of risk factors are toxicity-based and chosen to identify significant adverse health and environmental impact. Included in the group of risk factors are acute and chronic toxicity based on animal studies or structural analysis, human case history and epidemiological studies, carcinogenicity, mutagenicity, teratogenicity, and reproductivity effects. A limited number of physical and chemical properties associated with the "releasability" of the chemicals were also included in a second group of risk factors. The third group of risk factors reflected broad environmental impacts. The specific criteria for assignment to the five categories are shown in Table 29.6.

This company has a major chemical production facility involved in the manufacture of photographic dyes which require frequent changes in processes. Ahearn et al. (8) state that in the design of new processes and products the evaluation of toxic use and waste minimization technology is given equal importance to considerations of cost and performance of the new process or product. The engineering group developing the new processes rigorously review all materials and process alternatives to ensure that the target reductions are met. In this company the "end of pipe" abatement strategy is given secondary attention relative to optimization of process and product design for toxic use reduction.

An important part of the program is the goal of a 10% reduction in the use of toxic materials per unit of production each year for the first five years following the general strategy of elimination of use, replacement with a material of lower risk, reduction in

Table 29.6. Categories for a Corporate Toxic Use Reduction Program[a]

Categories I and II: Use to be reduced
Human or animal carcinogens, teratogens, or reproductive agents
Highly toxic chemicals
Chemicals with human chronic toxicity
Chemicals for which there are adverse environmental impacts
Category III: Transportation off-site to be reduced via source reduction or recovery and reuse on site
Suspected animal carcinogens
Moderately toxic materials
Chemicals that cause severe irritation of eyes and respiratory tract at low concentrations
Chemicals with limited chronic toxicity
Corrosive chemicals
Chemicals for which there are environmental considerations
Category IV: Disposal volume to be reduced via waste reduction or reuse following on-site recycling
All other chemicals not included in Category V
Category V: Any material such as paper, metal parts, and so on that is identified as rubbish, rubble, or trash.

[a]From Karger (17).

quantity used, and a hierarchal program of waste handling. The goal is to eliminate, replace or reduce the use of category I and II materials and assure the maximum reuse of categories III and IV materials.

In describing the success of the program the authors (8) present an example of a process in which a change resulted in elimination of a carcinogenic material. The original process required an aqueous oxidation step using a hexavalent chromium compound; the new process is based on a catalyzed air oxidation step which does not require the chromium compound (Figure 29.3). As a result of this change worker exposure to hexavalent chromium was eliminated and this process change saved the company over one million dollars. It is the conclusion of this company that the process that uses the least quantity of toxic chemicals is usually the most economical.

To gain entrance to the global market, manufacturers in the U.S. are implementing international management consensus standards. The most widely sought approval is that of the ISO 14,000 Environmental Management Systems Standard (18). To gain registration under this standard the firm is audited to ensure that the company has an effective management system in place so that appropriate decisions can be made regarding their impact on the environment. The ANSI International Advisory Council/Occupational Health and Safety Advisory Group is considering a comparable health and safety standard that will include recommendations on appropriate control technology.

3.1.2.4 Outcome of Replacement Technology. Although the replacement of toxic chemicals may provide impressive gains the replacement of a target material must be done with great caution. If the replacement technique is to be used successfully it must be carefully

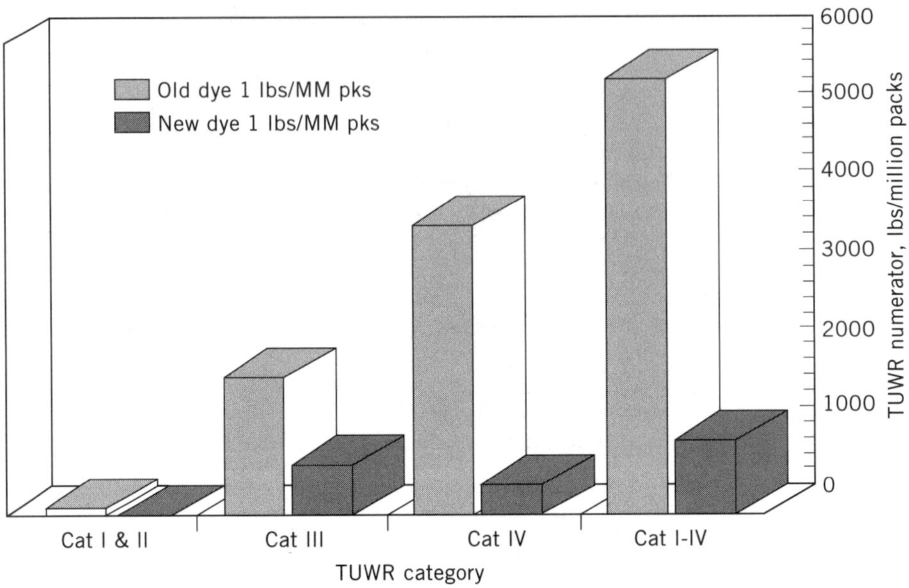

Figure 29.3. Impact of replacement technology on a chemical process resulting in the elimination of hexavalent chromium from the workplace and its impact on Toxic Use And Waste Reduction Categories (TUWR). Courtesy of Polaroid Corp.

reviewed to define its impact on worker health and environmental impact, product yield, and product quality. It is important that the impact be evaluated in both pilot plant operation and field trials before making the change.

Industrial hygienists practicing in the 1950s to the 1990s devoted more attention to solvent replacement for metal cleaning than any other control problem. It is useful to track the history of such cleaning agents during those years since it shows the complexity of the replacement technology approach and indicates pitfalls that may arise in using this technique. Wolf et al. review the major decisions on general degeasing solvents during this period as follows (19):

> In the 1960s and 1970s in particular, chlorinated solvents were widely substituted for flammable solvents because they better protected workers from flammability. As smog regulations were promulgated in the late 1960s and early 1970s users moved from the photochemically reactive flammable solvents and TCE to the other "exempt" chlorinated solvents. In the 1980s there was increased scrutiny of the chlorinated solvents: TCE, PERC, and METH were considered undesirable because of their suspect carcinogenicity, and TCE and CFC-113 were being examined for their contribution to stratospheric ozone depletion.

Each change starting in the 1950s, progressing from flammable solvents to chlorinated solvents to non-photochemical reactive materials to chlorofluorocarbon (CFC) solvents, introduced a new set of problems. The six major classes of solvents used in 1999 and the critical characteristics that determined their acceptability are shown in Table 29.7 (19). The alternatives to the common chlorinated solvents are not encouraging. The flammable solvents require rigorous in-plant fire protection controls. Both the flammable and combustible solvents including the terpenes are smog-producing, and therefore, are subject to tight federal and local air district regulations. The chlorinated solvents will continue to see

Table 29.7. Characteristics of Generic Solvent Categories[a]

Generic Solvent Category	ODP[b]	Photochemical Reactivity	GWP[c]	Flash Point[d]	Tested for Chronic Toxicity
Flammable solvents	—	Yes	—	F	
Isopropyl alcohol					Rule issued
Mineral Spirits					No
Combustible solvent[e]	—	Yes	—	C	
Terpenes					Limited[f]
DBE					No
NMP					Rule issued
Alkyl Acetates					No
Chlorinated solvents					
TCE	—	Yes	—	—	Yes
PERC	—	Yes[g]	—	—	Yes
METH	—	No	—	—	Yes
TCA	0.1	No	0.02	—	Yes
Chlorofluorocarbons					
(CFCs)		No		—	
CFC-11	1.0		1.0	—	Yes
CFC-113	0.8		1.4	—	Yes
Hydrochlorofluorocarbons					
(HCFCs)		No			
HCFC-123	0.02		0.02	—	In testing
HCFC-141b	0.1–0.18		0.09	—	In testing
HCFC-225	NA		NA	—	In testing
Hydrofluorocarbons					
(HFCs) and Fluorocarbons (FCs)					
Pentafluoropropanol	—	NA[h]	NA[h]	NA[h]	No

[a]From Wolf et al. (19). Reprinted by permission of *Journal of the Air & Waste Management Association.*
[b]ODP: The oxygen depletion potential (ODP) is the potential for ozone depletion of one kilogram of a chemical relative to the potential of one kilogram of CFC-11, which has a defined ozone depletion potential of 1.0.
[c]GWP: The global warming potential (GWP) of a chemical is the potential of one kilogram of the chemical to cause global warming relative to the potential of one kilogram of CFC-11 to cause global warming. CFC-11 has a GWP of 1.0.
[d]F refers to flammable; C refers to combustible.
[e]DBE is dibasic esters; NMP is *N*-methyl-2-pyrrolidone.
[f]One of the terpenes, *d*-limonene, has been tested.
[g]Although PERC is not photochemically reactive, it is not exempt under the Clean Air Act.
[h]NA means not available.

pressure for replacement due to the array of human toxicity and environmental effects. The increased use of chlorofluorocarbons (CFCs) in the 1970s and 1980s, rising to a level of 180×10^3 metric tons per year, was based on performance and reduced chronic toxicity. However, the impact of these chemicals on ozone depletion resulted in the ban on manufacturing of CFC-113 and 1,1,1-trichloroethane (TCA) worldwide in 1996. The hydrochlorofluorocarbons were considered interim candidates for industrial degreasing, but are now considered for banning in 2020 and 2040. The hydrofluorocarbons and the other candidate solvents not containing chlorine do not present an ozone depletion risk, but may present unacceptable toxicity concerns.

A vigorous research effort is pursued by dozens of companies to develop alternatives to the cleaning agents listed in Table 29.7. It is doubtful that a single agent will be found that will have the broad application of the chlorinated solvents, however progress is being made in nonozone-depleting, hydrocarbon-based systems.

3.1.3 Dust Control

A number of techniques have been proposed to minimize worker exposure to pneumonoconiosis-producing and toxic dusts based on changes in physical form or state of the material. The common techniques to reduce dustiness in this manner are discussed below.

3.1.3.1 Moisture Content of Material. The relationship between moisture content of granular materials and worker exposure to dust is well known. The use of water as a dust suppressant has been of great interest to the mining, mineral processing, and foundry industries. In the 1960s research by the British Cast Iron Research Association demonstrated that silica in air concentrations in foundries were maintained below existing exposure standards if the moisture content of foundry sand was kept above 30% (20). A sand handling technique that involves direct blending of new moist sand directly from the mixer with shakeout sand to add moisture to cool the shakeout sand and reduce dustiness is described by Schumacher (21). The author also demonstrated a correlation between results of a simple laboratory dustiness test and in-plant dust exposure. Goodfellow and Smith (22) have identified the importance of a dustiness index:

> Dustiness testing can be a useful testing tool in establishing the type and efficiency of dust control required for different materials and materials handling systems. Further field data and verification are required to develop this procedure into a useful tool for practitioners.

The Bureau of Mines and the National Industrial Sand Association have investigated the use of foams, steam, and water sprays for dust control in mineral processing plants (23). Foam was found to be approximately 20% more effective than water alone or water with a surfactant. Steam also produced impressive reductions in dust concentrations of 65%, however, as is the case with foam, it is more expensive to apply than water.

The British Occupational Hygiene Society (24) have also explored various dustiness estimation methods which may be useful in design of control systems. In chemical processing a solid product is frequently isolated in cake form that is wet with solvent or water.

Frequently this material must be dried to ensure product quality and shelf life and to permit packing and shipping. However, if the product is to be used "in-house" one should determine if the drying operation can be omitted permitting direct handling of the wet cake thereby saving money and reducing dustiness.

3.1.3.2 Particle Size. In general, the more extensive the grinding or comminution of a granular material, the greater the dust hazard the material will present during transport, handling, and processing. When possible, purchase the most coarse form of the chemical that is suitable for the process. There are production implications that may override considerations of worker exposure. As an example, if the material must be placed in solution the large particle size will slow this process.

3.1.3.3 Dust Controlled Forms. In the last two decades the rubber, pharmaceutical, pigments, and dyestuff industries have given attention to the dustiness of the raw materials they produce and use. Dustiness testing has been extended and significant product design changes have been adopted to minimize dustiness, worker exposure, and product acceptance. In tire manufacture at least a dozen chemicals in granular form are added in small quantities to the batch mix. In an effort to minimize worker exposure to dust from these chemicals, The British Rubber Manufacturing Association has sponsored design of low dusting forms of these common chemicals. The properties of seven dust controlled forms of rubber chemicals is reviewed by Hammond (25) in Table 29.8. The author notes that

Table 29.8. Dust-controlled Forms of Rubber Chemicals—Comparative Performance[a,b]

		Wax or Otherwise Bound			Polymer Bound		
Property	Untreated Powder	Soft Paste	Putty	Prills	Pellets/ Granules	Slab	Pellets/ Granules
Active content	5	3	4	5	4	4	4
Convenience of handling	3	1	2	4	5	2	5
Freedom from dust	1	5	5	4	4	5	5
General cleanliness and safety	1	1	2	3	3	5	5
Suitability for automatic weighing	3	1	1	4	4	1	5
Wastage	3	2	4	4	4	5	5
Ease of disposal of containers	3	1	3	3	4	4	4
Identification	—	—	—	—		—	—
Mill mixing behavior	3	1	3	3	3	5	4
Internal mixing behavior	5	2	3	5	5	5	5
Dispersion in rubber	5	5	4	3	4	4	4
Total	32	22	31	38	40	40	46

[a]From Hammond (25). Reprinted by permission of the British Occupational Hygiene Society.
[b]5 = excellent; 4 = good; 3 = average, 2 = below average; 1 = poor.

the disadvantages of the most effective approach, coating the chemical with a polymer, include the variability in active chemical content based on bulk weight, the reduced chemical content and the unsuitability of the polymer in the formulation.

3.1.3.4 Slurry Form. This application has limited application but in those cases where it can be used it does have great impact on dust concentrations. In the tire industry "master batch" rubber, rubber processed with all chemicals except the vulcanizing agent, is processed from the Banbury to the drop mill where it is "sheeted off" for storage. Until 1960 dry talc or limestone was dusted on the slabs of "master batch" material to keep it from sticking. This operation resulted in poor housekeeping and a significant exposure to talc. The present technique, adopted in the 1960s, involves dipping the stock in a slurry of talc in water before racking for storage. This simple change resulted in a significant reduction in worker exposure to talc.

3.1.4 Impurities in Production Chemicals

In low concentrations impurities or unreacted chemicals in raw or final product may represent a potential exposure that warrants attention.

3.1.4.1 Residual Monomer in Polymer. In polymer manufacture there is frequently unreacted monomer in the final product. Residual monomer had not been given much attention until the early 1970s when angiosarcoma, a rare liver cancer, noted in workers manufacturing the vinyl chloride polymer was attributed to the monomer exposure. Investigation showed that the polymer used in subsequent fabricating operations had unreacted monomer present in concentrations as high as 0.4% (26). This level of contamination prompted concern about the monomer exposure of workers handling the bulk polymer in plastic fabrication operations such as injection molding. As a result of this concern the vinyl chloride manufacturers modified the manufacturing process to reduce the monomer concentration to less than 1 ppm (27) thereby eliminating significant worker exposure. Residual monomer is frequently present in concentrations up to 1% in many of the common polymers and may warrant attention. If significant air concentrations are noted when handling the polymer, engineering control is first based on the removal or reduction of the monomer content in the polymer with other controls considered later if this is not adequate.

3.1.4.2 Solvent Impurities. Impurities may pose an unrecognized risk, especially in solvents of high volatility. In the manufacturing of automobile tires the various rubber components are "layed-up" on a tire building machine. To effect good bonding between the components the rubber is made tacky by applying a small amount of solvent to the surface with a pad. For several decades this solvent was benzene. The worker exposure, probably in the range of 1 to 10 ppm, may be responsible for the excess leukemia seen in older tire builders. Starting in the 1950s the industry started to replace benzene with white gasoline. In studies completed in the 1970s by the Harvard School of Public Health Joint Rubber Studies Group, the residual benzene content in white gasoline was 4–7% and air sampling indicated that 1/3 of the air samples on tire builders exceeded 1 ppm (28). The PEL for benzene at that time was 10 ppm although it was anticipated that it would be dropped to

1 ppm (this change did occur in 1987). Technical grade chemicals commonly have significant impurities. The level of contamination should be identified and if sufficiently high, worker exposure should be evaluated. At that time the necessity for reduction in the impurity level can be determined.

3.2 Equipment and Processes

In Section 3.1 a variety of engineering control options were considered focused on the choice of materials to minimize the generation and release of airborne contaminants. An equally important step is a review of the various alternatives in the choice of equipment and processes.

In the discussion on dusty materials in Section 3.1.3 techniques to determine the relative index of dustiness were mentioned to assist in the choice of material form and the dust suppression treatment. There is no such index for equipment and processes, however there are a number of operational insights that should be considered in choice of facility. Wolfson (29) has emphasized the importance of this step in stating the removal of the dispersal device should be a first step in the engineering control of air contamination.

3.2.1 Diagnostic Air Sampling

Contaminant control cannot be achieved until the significant operational elements of the process that generate and release the contaminant are identified. Occasionally this can be done simply by a critical review of the operations but usually diagnostic air sampling is necessary. The value of short interval, task-oriented air sampling using conventional integrated sampling with subsequent analysis has been clearly stated by Caplan (30). In describing this approach, illustrated in Table 29.9, Caplan states:

> For the job analyzed in Table 29.9 presumably a single sample would have shown a concentration of 0.21 mg/m^3. The task-oriented sampling, however, reveals several interesting things.

Table 29.9. Task-oriented Air Sampling[a]

Task	GA or BZ[b]	Minutes/Day	Concentration (mg/m^3)	Minutes/Day × Concentration
A. Charge pot	BZ	40	0.12	4.8
B. Unload pot	BZ	80	0.50	40.0
C. General survey	GA	250	0.16	40.0
D. Lab—sample trips	GA	20	0.05	1.0
E. Change room	GA	30	0.08	2.4
F. Pump room—repack	BZ	30	0.32	9.6
G. Lunch room	GA	30	0.07	2.1
		480		99.9

TLV = 0.2 mg/m^3 Wt. avg. $= \frac{99.9}{480} = 0.21$ mg/m^3

[a]*Source*: Caplan (30).
[b]GA = general air; BZ = breathing zone.

Column 5 shows that tasks B and C are the major contributors to the day's exposures and that a significant reduction in the concentration at either of those tasks would be adequate to bring the 8-hr exposure well below the TLV. This is true even though task C in itself is below the TLV. In addition, it shows that task F, well above the TLV concentration, is of such short duration that significant improvement in that part of the exposure would not have a large effect on the 8-hr exposure.

The difficulty in this approach is that frequently it is not possible to measure the air concentrations during brief individual tasks and activities due to the low air sampling rate and the limited sensitivity of the analytical procedures. To reveal important generation points in the job it is necessary to resolve the air concentration profile in a time frame of seconds. The advent of real-time, direct-reading air sampling instruments for particles, gases and vapors with response times of less one second now permit the investigator to identify these critical contaminant elements in process events and work practices. In the 1980s this air sampling technique saw expanded application with coincidental video taping of the worker during a work cycle. In its most sophisticated form, the time-coupled, real-time contaminant concentration at the workers breathing zone is superimposed on the video display permitting the viewer to analyze the data display to identify the specific time and location of release. Control technology can then be applied to those tasks or incidents.

The video display may also be used to identify work practices that may either positively or adversely affect worker exposure. In a talc bagging operation reviewed in Vermont in the 1970s, one individual consistently had the lowest dust exposure although visual inspection did not reveal any differences in equipment or work practice. If the real-time technique were available, the worker's "secret" could have been identified and applied to the other workers. If a "correct way of doing the job" can be identified the video-concentration format is a excellent educational tool for workers.

A series of studies by NIOSH investigators describe the application of this technique to a range of in-plant tasks (31, 32). The air sampling instrument is a real-time monitor with a response time much shorter than the period of the shortest worker activity or movement. The instrument is equipped with a data logger with a clock "locked-in" or synchronized with the video. The data logger is down-loaded to an IBM compatible computer and the data file is analyzed permitting a graphical overlay of air contamination data on the video tape. In Figure 29.4 the video display from a study of a bag dumping operation is re-created in a line drawing for clarity to show the time coupled concentration on the video screen display (33). Graphical representation of the air sampling data for three jobs with the concentration from the video overlay is shown in Figure 29.5 (34).

The application of this technique to control technology is shown (35) in a study of a chemical weighing and transfer station. The information obtained permitted the investigators to redesign the workstation controls based on a perimeter exhaust hood and an air shower to eliminate eddies induced by the worker's body. Effective control of the dust exposure was obtained with one-third the air flow of the original system. In addition to improved worker protection, the cost savings of the new system resulted in a payback period of 4.5 years.

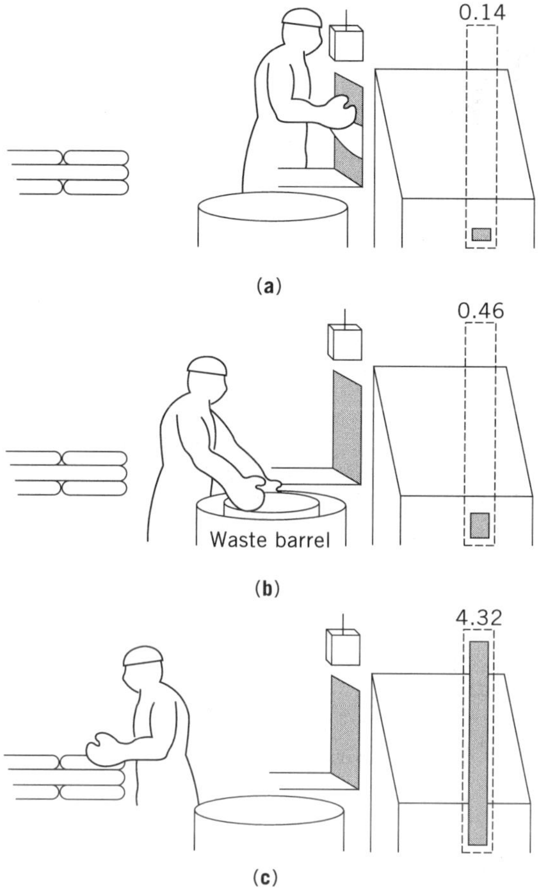

Figure 29.4. Relative dust exposure during bag dumping as determined by video-air sampling technique: (**a**) Operator slits bag and dumps the granular material into a hopper equipped with an exhaust hood with minor dust exposure, (**b**) Operator drops the bag into the waste barrel; there is a significant increase in dust concentration, (**c**) As the operator pushes the bag into the waste barrel a cloud of dust is released and the relative dust concentration increases to 4.32. (Reprinted with permission from *Applied Occupational and Environmental Hygiene*, **7**(4), 1992).

The Swedish National Institute of Occupational Health has adapted this technique to allow real-time mixing of the concentration and video portions by radio telemetry (36). The intermediate datalogging of exposure information is eliminated; the direct reading instrument data signal is converted to a radio frequency and broadcast with a transmitter to a receiver for processing by a microcomputer. The investigator receives immediate feedback on the effects of engineering controls and work practices. In the U.S., investigators at the National Institute for Occupational Safety and Health have modified this system for a series of control technology studies.

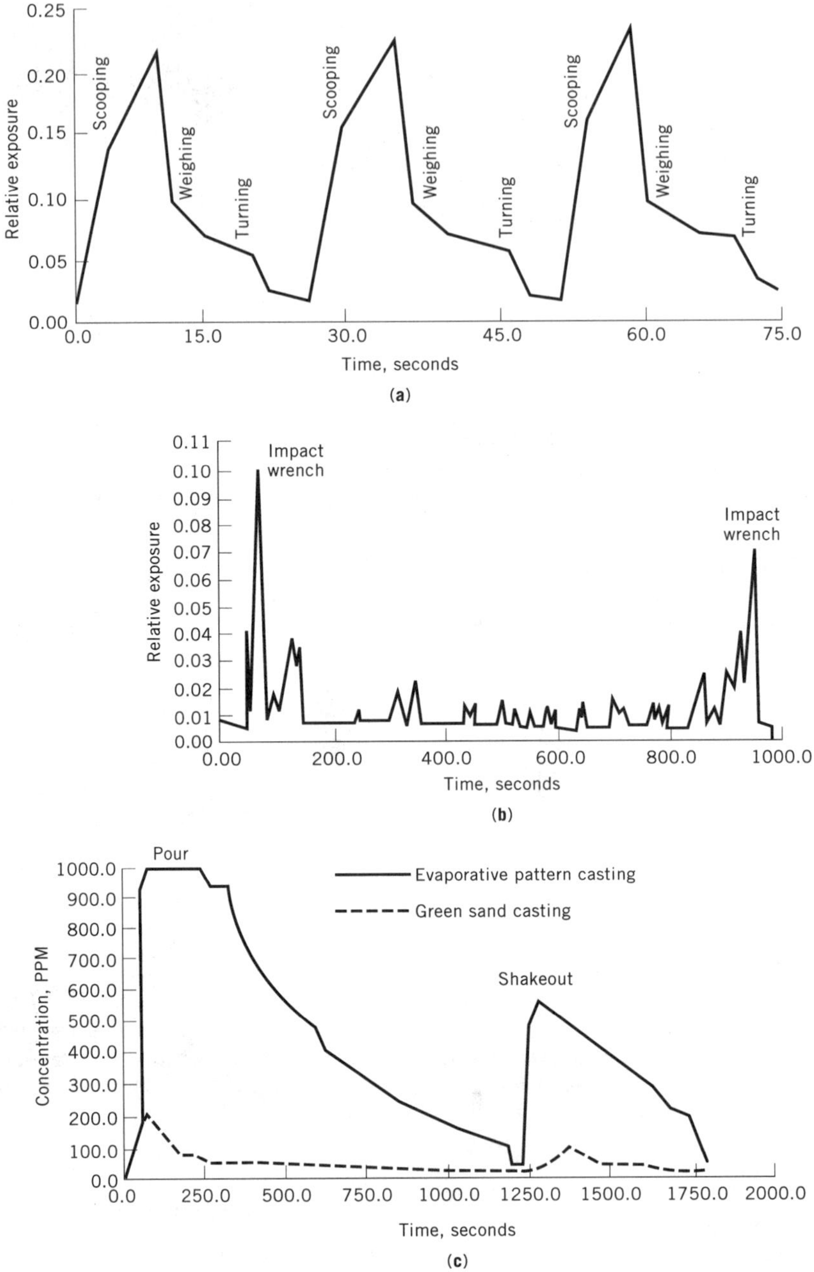

Figure 29.5. Peak dust exposures identified by video-real time monitoring, (**a**) Relative exposure versus time for three bags during manual weighout of powders. (**b**) Relative dust exposure during automotive brake servicing, (**c**) Carbon monoxide emissions from evaporative pattern casting and green sand processes. (From Ref. 34, reproduced by permission of *Applied Industrial Hygiene*.)

This diagnostic tool will see expanded application to workstation design in the next decade since it does permit the engineer to identify the specific tasks, equipment function, or worker movement that contribute to worker exposure.

3.2.2 Equipment

3.2.2.1 Ancillary Equipment. The stepwise review of the dispersal potential noted above should include all equipment, not only the major machinery. The importance of this approach is highlighted when one looks at the history of the simple air nozzle in industry. For most of the 1900s the widespread application of the air nozzle operating at line pressure (100 psi) was used to remove chips and cutting oil from machined parts and to dry parts after cleaning with solvent. In the 1970s OSHA required that the operating pressure for these nozzles for cleaning purposes be dropped from 100 to 30 psi This change had many positive effects including a reduction in noise, eye injuries, solvent mist, and vapor exposures.

Another common piece of equipment, albeit much more complex, is the centrifugal pump. The difficulty of equipment selection is typified by the experience of a chemical engineering team designing a modern chemical plant or refinery having hundreds of centrifugal pumps. A major issue in such plants is the impact of fugitive losses from rotating machinery such as pumps on the workplace and general environment (37). The total loss percentage from rotating machinery varies from 8 to 24% with overall uncontrolled emissions ranging from 31 to 3231 tons/year for a model chemical plant or a large refinery (38). The authors describe two possible control programs for pump fugitive losses—either a monitoring and maintenance program or engineering control by installation of dual seal pumps, seal-less pumps, and a closed exhaust hood with air cleaning.

The phase out of 1,1,1-trichloroethane and trichlorotrifluoroethane solvents in vapor phase degreasers accelerated the search for alternative solvents as described in Section 3.1.2.4. Initial steps to reduce emissions from this common equipment in the 1980s included the use of covers, increased freeboard height, refrigerated condenser fluids, and lower hoist speeds. Although these changes do reduce losses and therefore worker exposure, the survival of this type of equipment rests with the recent availability of vapor tight degreasers (39). In the past, closed system technology has been associated with major processing facilities. In the future individual job shop equipment will utilize this approach to meet critical workplace and environmental constraints.

3.2.2.2 Major Processing Equipment. The complexity of the control technology problem becomes apparent when one moves from choices of ancillary equipment to decisions on major pieces of processing equipment. As an example, a small chemical processing plant for organic synthesis is a multi-floor plant with a series of major operations staffed by six chemical technicians. The condensed version reviewed in this chapter (Fig. 29.6) permits discussion of equipment alternatives to minimize the generation and release of air contaminants. The three step process includes first, a reaction step to form a solid in suspension, conducted in a pressure vessel. In this case there is no alternative equipment. The reaction must be conducted in a closed pressure vessel equipped with the safety controls shown in Figure 29.6 to minimize the possibility of a chemical accident. The second step, the iso-

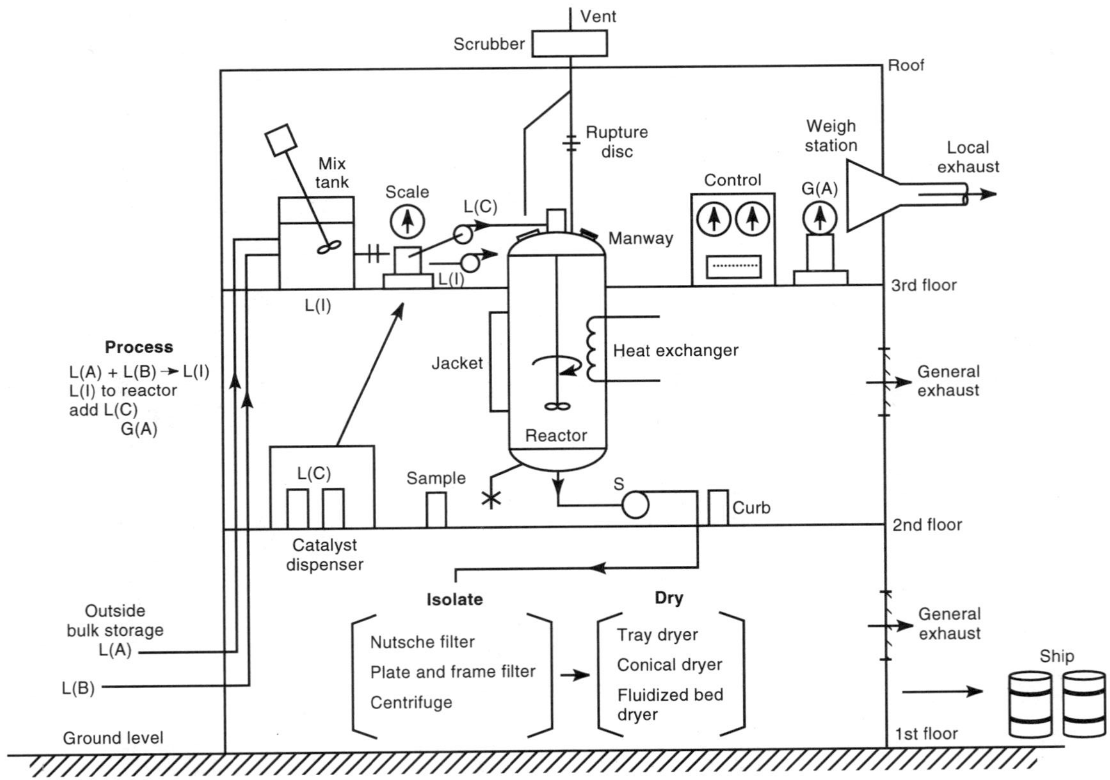

Figure 29.6. Chemical processing facility. Liquid reagents L(A) and L(B) are transferred to a mixing vessel. After mixing, the resulting liquid L(C) is transferred to the reactor by vacuum. Granular chemical G(A) is weighed under local exhaust ventilation (LEV) and dumped into the reactor through a charge port also equipped with LEV. A catalyst L(C) is transferred to the reactor by a positive displacement pump. After the reaction is complete the new chemical is isolated by one of three techniques—Neutsche filter; plate and frame filter, or centrifuge. The resulting wet cake is then dried by one of three methods—tray drier, conical drier or fluidized-bed drier. The isolation and drying techniques are ranked by environmental impact with the worst process noted first.

lation of the solid reaction product, can be done by at least three different techniques. The environmental contaminant potential of these processes varies considerably. However the choice cannot be made on this basis alone, since these techniques do have specific production capabilities that may be the major determinant in the choice. Finally the reaction product is dried and drummed for sale or use in a subsequent operation. Again there are several ways to accomplish the drying. The ranking of these isolation and drying operations based on their environmental impact is shown in Figure 29.6.

The responsibility for the choice of specific equipment in the processing plant will rest with the project design group and will be based on production capacity, ancillary services required, maintenance history, cost and delivery in addition to health and safety considerations. It is not expected that the industrial hygienist will have complete operational and application data on all chemical unit operations but their environmental impacts must be understood and made known to the engineering group.

It would be extremely helpful to the industrial hygienist if the individual equipment were assigned an index of contaminant generation/dispersal similar to the index of dustiness or fugitive losses from a pump. This approach is not currently realistic, however, a better characterization of the performance of such equipment will emerge in the next decade.

3.2.3 Processes

A review of the production engineering literature shows a range of processes which can be considered for a manufacturing facility. The choice is usually made based on production output and cost data. Industrial processes designed to accomplish a given task have one other parameter, the ability to contaminate the workplace; this characteristic of the process should be given equal weight to production criteria (40).

The development of technology in a given area such as welding is pushed by production needs not by health concerns. Since 1900 the technology has become more complex and frequently more environmentally challenging. Infrequently, studies are conducted to influence the process design to minimize air contamination. This was the case with Gray and Hewitt (41) who defined the welding operating parameters that influence fume generation and proposed operational configurations that minimize fume formation rate and the chemical composition of the fume thereby permitting improved efficiency of controls.

An exercise that demonstrates the value of a ranking approach for process environmental impacts is shown in the manufacture of an electronic cabinet (Fig. 29.7). As in the case of the chemical process shown above, the fabrication steps are simplified for ease of discussion. The four enclosure components are fabricated from light weight aluminum stock using sheet metal shear, brake, and punch. These operations may present a noise hazard hazard but do not generate air contaminants and will not be included in this review. The enclosure can be assembled and painted by a number of techniques; the processes are ranked from worst to best in terms of adverse effect from air contamination.

Another general approach that applies to many processes is the use of containment. This control approach was considered by an engineering control panel sponsored by the Environmental Protection Agency and the following containment ranking was proposed (42):

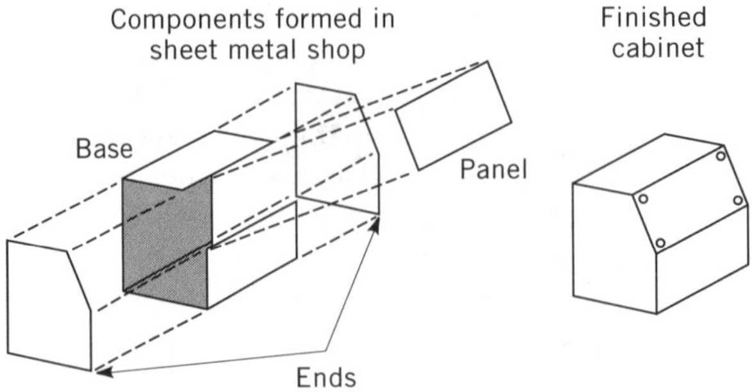

Finishing options

Metal cleaning	Welding	Painting
Wet degreasing with chlorinated HC	Torch welding	Air atomization
Vapor with chlorinated HC	Inert gas shielded arc welding	Airless spraying
Wet degreasing with terpenes	Submerged arc	Hot paint
Water with detergent	Resistance	Electrostatic
		Powder
		Dipping

Figure 29.7. Manufacture of an electronic cabinet. The operational options for cleaning, abrasive blasting and painting are ranked in terms of environmental acceptance with the worst process noted first.

Standard control measures may often be categorized for each process: a sealed and isolated system where airborne contaminant levels are very low (ppb range for vapors); a substantially closed system where airborne contaminant levels are still relatively low (fraction of ppm or low ppm range for vapors); a semi-closed system which is typical of nondedicated equipment used in job shop chemical processing facilities; and an open system.

An example of a sealed and isolated system is the modern refinery where such an approach is required both for production reasons and health and safety. A substantially closed system is typified by a semiconductor facility, and a semiclosed system by the chemical processing plant in Figure 29.6. In such fine chemical processing plants it is common for the product to be made by an elaborate chain of separate reaction steps with a number of intermediates formed. In many cases the intermediate chemical must be isolated and removed from the vessel. Each transfer of the intermediate poses a potential for release of contaminants. Since the reaction train may include dozens of steps and take weeks to complete a method that would streamline the process and reduce the number of transfers would reduce the exposure of the worker.

With the introduction of the pressure neutche to the United States it is now possible for chemical plants with relatively simple reaction trains to convert to a closed system. This reaction vessel permits the chemical engineer to design a reaction sequence including formation of the new molecule, cleaning of the product, drying and packaging all without removing the chemical from the vessel at each step with its resulting exposure to the chemical processing worker.

Even more impressive as a control technique is the introduction of a process called cascade chemistry (43). In this approach the organic synthesis reactions are conducted at room temperature and atmospheric pressure with carefully selected initial charging chemicals. Properly designed, one reaction cascades into another without the addition of new chemicals. In this system hazard is minimized by operating at low temperature and pressure while eliminating the isolation and transfer steps.

Most common facilities, ie foundries, electroplating, painting, welding, and machining are examples of open systems. The contaminant released as a mist from electrolytic plating operations can be trapped by a layer of plastic chips or a persistent foam blanket on the surface of the bath. Also surfactants may be added to the electrolyte to reduce mist escape. The plate-and-frame filter in the chemical processing industry is difficult to handle with local exhaust ventilation. Frequently a solvent wash of the cake with a highly toxic material results in high exposure when the filter is broken and the solvent-wet cake is removed. In some cases after the initial wash the cake can be washed with isopropanol to strip out the toxic solvent. It is then washed with water so that the exposure when the filter is broken and the product is removed is nil.

A series of process changes developed in the WRITE program is shown in Table 29.10.

3.3 Work Task Modification, Automation, Robotics

It is well known that simple changes in work tasks may have significant impact on job outcome. In the early 1900s, workplace time and motion studies were used to improve productivity. Later a job placement technique (44) based on a detailed analysis of the time and effort of the job tasks was effective in reducing on the job injuries. More recently, the

Table 29.10. Examples of Process Change Technology from EPA WRITE Program[a]

Operation	Process Change
Hand mixing of paint	Proportional mixer blends paint at the gun thereby eliminating handling paint and solvent at a mix operation.
Conventional manual air spray painting	Computer-controlled robotic painting with an electrostatic spray to reduce worker exposure to paint mist and solvent vapors
Performance testing of electronic parts with CFC-based cooling system	Installed compressed air cooling system

[a]Adapted from Harten and Licis (13).

analytical tools of the ergonomic specialist permits identifying difficult tasks contributing to occupational injuries and illness.

As indicated in Section 3.2.1, the industrial hygienist has techniques to investigate the source of contaminants, the generation mechanism, and to make a semiquantitative assessment of the generation rate. This approach has tremendous value in analyzing not only the critical generation points on the machine but in viewing the impact of specific worker actions and movement on air concentrations. Preliminary studies reveal that minor changes in work position and movement may offer significant reductions in exposure.

Occasionally the specific modification in work practice is dictated by knowledge of the mechanisms of generation and release of the contaminant. Such is the case with the flow of granular material at material transfer points. The generation mechanism is the air flow induced by the falling granular material (45). The induced airflow can be minimized by restricting the open area of the upstream face, reducing the free fall distance, and reducing the material flow rate as defined by Anderson's equation.

Automation is a general technique that separates the worker from the individual process. In the 1960s and the 1970s this was usually done by simple electromechanical equipment design. A good example of this procedure is the manufacture of asphalt roof shingles. In the 1970s, competition resulted in the automation of all parts of the process from the dipping of the stock to the bundling of the package.

Another example of a simple automation process that reduces worker exposure is tire curing. In the early plants the worker lifted the tire out of a curing press at the end of the curing cycle and placed another tire in the mold for curing. During this period he was directly exposed to the emissions released from the press. By 1970 most plants had automated this process. The worker now places the uncured tire on a holding rack in front of the curing press line. When the curing cycle is completed, the tire is ejected from the press to a belt conveyor and the next tire to be cured is transferred from the rack to the press without exposure of the worker.

The advent of robotic techniques has permitted almost all industrial procedures to be candidates for automation. The movement of the worker who buffs rubber boots can now be captured by a robotic system as shown in Figure 29.8. The ultimate application of robotic techniques is to spray painting. In this case the robotic system can reproduce the movement of a skilled painter. Although ventilation is still required on this job the worker overseeing the operation is separated from the point of release of the paint mist and solvent.

Robotics may have been a solution to a difficult problem in the 1950s. In a large generator shop the generator coils were preformed of copper bar stock. The coils were then wrapped with insulation tape and painted with an asphalt compound. Protective clothing notwithstanding, the workers had skin exposure to the asphalt which required agressive cleaning at the end of the shift. Dermatitis and photosensitivity were frequent occurences. At that time, coil winding experts devoted time and money to developing machine wrapping concepts without success. This type of problem can be solved by robotics in the 1990s.

3.4 Facility Layout

As stated earlier, the most efficient way to do a job is probably the one that impacts the least on the workplace environment. Certainly this is true insofar as overall plant layout

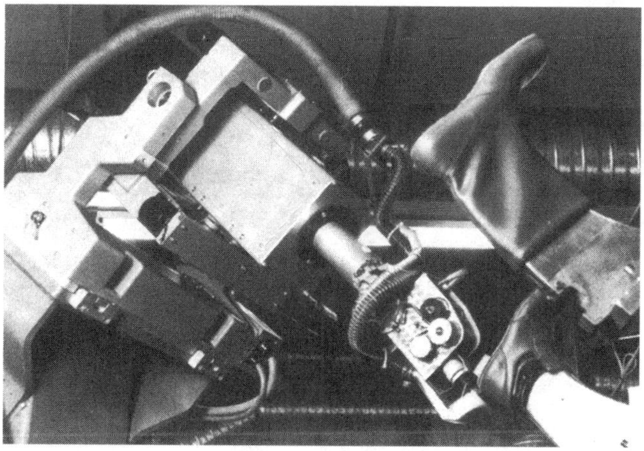

Figure 29.8. Robotic buffing of rubber boots. Photograph reproduced by permission of Matti Koivumaki.

is concerned (46). In the shop fabricating the enclosure in Figure 29.7 the desired flow of materials is from the incoming truck dock or railroad siding to sheet metal fabrication and then to assembly, finishing, inspection and finally shipping. In a facility manufacturing pharmaceuticals, the input chemicals are transported to bulk storage and then to chemical processing, packaging, inspection and shipping. In all manufacturing processes from handling metal to fine chemicals, it is important to minimize the distance the material is moved and the number of times it is picked up and transferred. If this rule is followed worker exposure to air contaminants will be minimized.

3.4.1 Material Transport

This issue is given major attention in industry where large quantities of raw material, intermediates and final products are handled. Examples are injection molding of children's toys, manufacture of automobile tires, and paint manufacture. In some cases material handling alone defines the plant layout. Frequently bulk storage is in large silos located outside the plant with delivery to the work stations by mechanical or pneumatic conveyors. This is true in the manufacture of plastic tape where tons of PVC granules are used each day (Fig. 29.9). This plant also requires large quantities of solvent delivered from bulk storage by piping to mix tanks and then directly to the coating heads. The intent of this system is to have all material handling to the individual coaters done in closed systems.

In organic synthesis it is common practice to have outside bulk storage of at least a dozen solvents. The solvents are transferred to an inside reaction vessel by piping to a reactor manifold with necessary valving and meters. Granular materials used in small quantities are stored in adjoining storage areas and delivered to the reactor and charged by manual dumping, with a dumping fixture, or occasionally with a transfer lock. When a large number of drums must be dumped a bulk handling system is used with direct delivery to the reactor vessel.

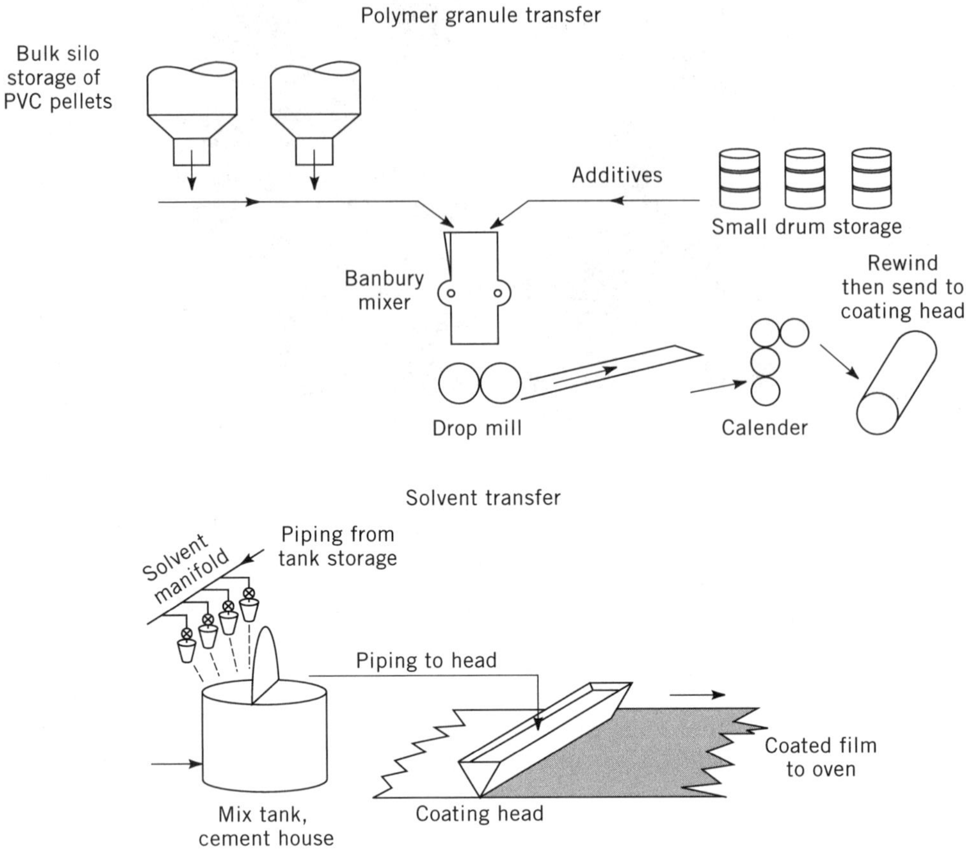

Figure 29.9. Manufacture of electrical tape showing the closed bulk transport of granular plastic and the piping of solvents to the mix tank.

3.4.2 General Considerations

The location of the process within the facility may influence the worker's exposure. This is true of operations such as foundries where jobs may be easily classed as clean or dirty and exposures vary greatly. The conventional iron foundry provides a good example of the importance of plant layout. It is common practice to define the optimal plant layout as one with air flow from the cleanest to the dirtiest operations with the exhaust focal point establishing this gradient [Fig. 29.10(**a**)]. The layout of a similar foundry which does not follow this guideline is shown in Figure 29.10(**b**). In the latter case the relatively clean molding line is positioned adjacent to the shakeout, an area where the control of airborne foundry dust and thermal degradation products is difficult. This poor layout results in silica and other contaminants released by shakeout moving into the molding area and exposing this work population. If the desired concentration gradient cannot be achieved by venti-

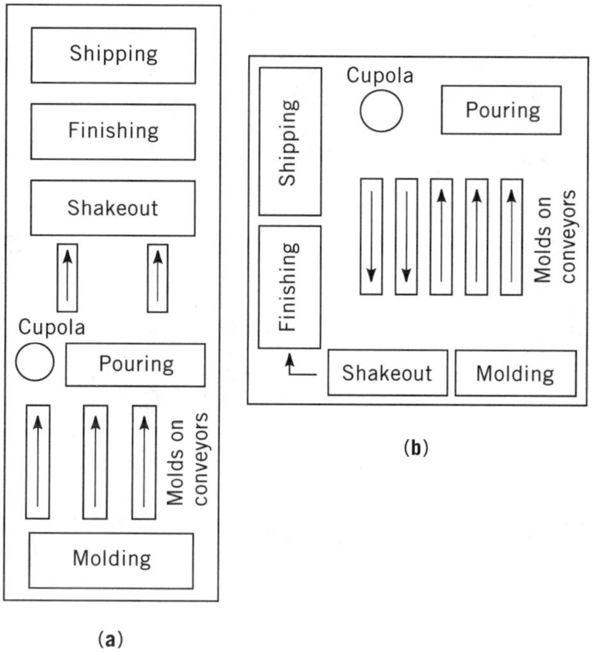

Figure 29.10. Layout of two gray iron foundries. Foundry in (**a**) is designed for straight process flow and gradual transition from activities that represent limited contamination to those of greater contamination. In foundry (**b**) the process flow reverses and the dirty shakeout ends up adjacent to the molding line.

lation or distance then structural walls or plastic barriers are a possibility. Although such barriers define the space, compartmentalizing complicates the design and application of local exhaust ventilation, replacement air, HVAC and other important services.

3.4.3 Workstations

A foundry in Finland faced problems not only of airborne contaminants but heat, noise, and housekeeping in casting cleaning. The problems were resolved by redesigning the open workspace to provide individual work stations designed with integrated services positioned for ease of work (Fig. 29.11). This layout resulted in improved material flow, better housekeeping, reduced air contamination, better ergonomics and improved productivity. The success of such an installation was due, in part, to close employee–employer involvement in the design.

3.4.4 Service and Maintenance

Frequently, close attention is given production area layout as noted in the above example, but rarely is attention given the working environment of the maintenance group. An ex-

(a)

(b)

Figure 29.11. Well-integrated work station in a foundry. (**a**) Overall layout of foundry finishing area; (**b**) individual work station with exhaust hood. Photograph reproduced by permission of Matti Koivumaki.

ample of an area where maintenance was not considered was tire-curing presses in the plants of the 1970s. The presses were arranged in double rows back to back with limited space between rows for steam, water and electrical services. While this layout and density represented optimal use of space, repairs of steam and water piping were done in an extremely tight space and controls for the worker were difficult to set up.

The best example of a planned layout that provides proper work space for the trades is noted in the semiconductor industry (Fig. 29.12). In this layout all transfer pumps, distri-

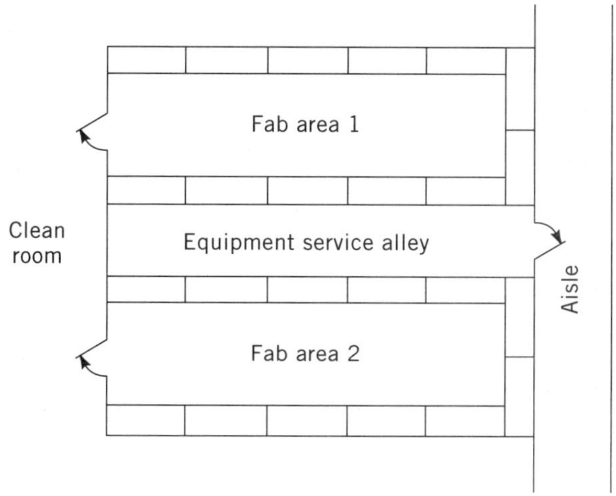

Figure 29.12. Layout of semiconductor facility showing service aisle isolated from, but adjoining the fabrication bays.

bution lines, and vacuum pumps are located in a service alley separated from the fabrication bay. This design reflects principal concern, not for the comfort of the maintenance worker, but rather to protect the fabrication area against contamination. Notwithstanding, it does result in adequate space for the trades to carry out their work in a safe manner (47).

An extension of this concern is the installation of a small field service bench with necessary tools and equipment in plant production areas where maintenance is frequently done. This arrangement permits the maintenance person to do many repairs at the site and not transport equipment back to a main facility with the potential for chemical spills.

3.4.5 Segregation of Operation

It is frequently necessary to segregate or remove an operation from the main production area as an engineering control measure. In the semiconductor and fiber optics industry each time a new facility is designed the engineers must choose whether to store small amounts of highly toxic gases close to the production tool or store large quantities at a segregated position some distance from the plant to be distributed by double-walled piping to the production tools. This latter segregation technique permits the plant to reduce the number of persons at risk while providing extensive controls at the work site. This option must be weighed against storing small quantities of gas directly at the tool thereby minimizing failures in the transfer systems but requiring frequent change of gas cylinders with the entailed risks.

3.4.6 Isolation of The Worker

The practice of isolating the worker as a control measure is placed under this section since it should be considered in conjuction with layout of the major equipment. This widely used technique is mandated by OSHA coke oven regulations which requires enclosures with clean, conditioned air. The larry car and pusher cars have controlled environment cab enclosures; the lidsmen have enclosures designed to remove them from the hostile coke oven environment during available rest periods.

In other industries this approach has been chosen to eliminate worker exposure to air contaminants while providing a comfortable working environment. These applications include enclosed booths on a variety of construction equipment, front end loaders in smelters, crane cabs in metallurgical industries, operator cabs in steel rolling mills, and pouring stations in foundries. Unfortunately off-the-shelf control booths are not available for the range of applications seen in industry, nor have engineering guidelines been published.

3.5 Ventilation

In the introduction to this chapter it was stated that ventilation is the third step in the control hierarchy of (*1*) do not generate the contaminant (*2*) if contaminant is generated, do not allow it to be released and (*3*) if it is released collect the contaminant before it reaches the worker. The design goal of industrial ventilation is to protect the worker from airborne contamination in the workplace. To the newcomer this may suggest installing a system that will reduce exposure below the Permissible Exposure Guidelines or an appropriate action level. This is not the case. The professional will design the system to meet the goal of "as low as reasonably practical" (48).

Within this control approach the effectiveness of the major ventilation techniques is shown in Table 29.11 (48). There is no agreement on the position of low-volume–high-velocity systems since its effectiveness varies greatly depending on the degree of integration with the tool.

Soule, in Chapter 30, "Industrial Hygiene Engineering Control" reviews the application of the two major ventilation control approaches, dilution and local exhaust ventilation (Table 29.12). In most industries dilution is not the primary ventilation control approach for toxic materials. It is accepted that local exhaust ventilation will not provide total capture of contaminant and dilution ventilation is frequently applied to collect losses from such systems. In addition, it is used for multiple, dispersed, low toxicity releases.

Table 29.11. Ventilation Control Hierarchy

Type of Ventilation[a]	Hood Type	Example
LEV	Total enclosure	Glove box
LEV	Partial enclosure	Laboratory hood
LEV	Low-volume-high-velocity, tool integrated	Portable grinder
LEV	Exterior hood	Welding hood
GEV	Mechanical exhaust	Roof ventilators
GEV	Natural	Wind induced

Source: Adapted from Ref. 48.
[a]LEV = local exhaust ventilation; GEV = general exhaust (dilution) ventilation.

Table 29.12. Application of Local Exhaust and General Exhaust Ventilation[a]

Local Exhaust Ventilation	General Exhaust Ventilation
Contaminant is toxic.	Contaminant has low order of toxicity.
Workstation is close to contaminant release point.	Contaminants are gases and vapors not particles.
Contaminant generation varies over shift.	Uniform contaminant release rate.
Contaminant generation rate is high with few sources.	Multiple generation sources, widely spaced.
Contaminant source is fixed.	Generation sites not close to breathing zone.
	Plant located in moderate climate.

[a]See also chapt. 30.

3.5.1 Design Phase

The specific design methods for both general exhaust ventilation and local exhaust ventilation are presented by Soule and in other volumes dedicated to ventilation control (49–51). An important pre-design phase identified by Burton (57) as problem characterization is frequently given little attention (Table 29.13). This is a topic that can be best addressed by an industrial hygienist who can provide data on emissions, air patterns and worker movement and actions.

If the process is new, a video tape of a similar operation with the same unit operations may be available. The best of all worlds would be the availability of a video-concentration tape as discussed in Section 3.2.1. If the facility is a duplicate of one in the company the industrial hygienist should visit the operation with the ventilation designer. Frequently the designer is an outside contractor. In this case it is important that the problem characterization approach be followed and an information package be provided the designer.

The precautions that should be reflected in the design have been discussed in detail elsewhere, but should include worker interface, access for maintenance, and routine testing. CAM/CAD technology now permits precise placement of equipment and ductwork so that *ad hoc* placement by the installer should be a thing of the past. As discussed earlier, it may be worthwhile to "mock-up" a specific design solution prior to final design and construction. This is especially true when a large number of identical workstations are to be installed as was the case in a shipyard asbestos insulation workroom as noted in Figure 29.2.

General cautions are appropriate on the use of available design data for control of industrial operations. The ACGIH Industrial Ventilation Manual provides the most comprehensive selection of design plates for general industry (51). Each of these plates provides four specific design elements: hood geometry, airflow rate, minimum duct velocity, and entry loss. The missing element is the performance of the hood in terms of percent containment or capture efficiency.

As noted by Burgess (52) there has been little consolidation and publication of successful designs by industrial groups. With the exception of the Steel Mill Ventilation volume published by the American Iron and Steel Institute in the 1960s (53) and the Foundry

Table 29.13. Information Needs for Problem Characterization

Emission Source Behavior

Location of all emission sources or potential emission sources.
Which emission sources actually contribute to exposure?
What is the relative contribution of each source to exposure?
Characterization of each contributor:
Chemical composition, temperature, rate of emission, direction of emission, initial emission velocity, continuous or intermittent, time intervals of emission.

Air Behavior

Air temperature
Air movement
Mixing potential
Supply and return flow conditions
Air changes per hour
Effects of wind speed and direction
Effects of weather and season

Worker Behavior

Worker interaction with emission source
Worker location
Work practice
Worker education, training, cooperation

[a]From Ref. 50. Reprinted with permission of D. Jeff Burton.

Environment Control Manual in 1972 (54), and more recently by the Semiconductor Safety Association (55) there is little evidence that companies in the United States wish to share ventilation control technology Major sources of information on the performance of ventilation systems are the technical reports on engineering control technology published by NIOSH (Appendix A). Many of these reports couple ventilation assessment with measurement of worker exposure.

This discussion indicates that ventilation control designs on standard operations in the mature industries have been published although performance has usually not been reported. It is important to evaluate performance by diagnostic air sampling both to ensure the worker is protected and to prevent overdesign. The latter was shown to be the case in the design for control of a push pull system for open surface tanks (56). A mock-up of one tank demonstrated that satisfactory control could be achieved at minimal airflow. Operating at the reduced airflow rate saved $100,000 in installation costs and $263,000 in annual operating costs.

3.5.2 Acceptance and Start-up Testing

It is important that new ventilation systems be inspected and tested on completion of the installation to ensure the system conforms to the design specifications and that worker exposure is kept as low as reasonably possible. Unfortunately it is the author's experience that only 10% of the new systems in general industry undergo such scrutiny. The situation is baffling given the importance of contaminant control, the general practice of industry to test other services before acceptance and the cost of ventilation systems (10 to 30 U.S. dollars per cfm to install and 1 to 4 U.S. dollars per cfm per year to operate).

3.5.2.1 Construction Details. The ventilation system design, completed by the plant facilities and engineering group or a consulting firm, is based on certain assumptions and specifications on hardware details such as elbows, entries, expansions, contractions, etc. The hood construction has been detailed by the draftsperson. If the hardware elements provided for installation are supplied by a standard manufacturer the losses will be approximately those used by the designer. If the system is large it may be worthwhile obtaining samples of the components to test for losses. The hoods warrant special attention. The entry losses on standard hood designs as noted in the ACGIH Ventilation Manual are quite reliable, however, if the hood design is not typical it may be difficult to estimate entry loss. In such a case the best solution is to have a single hood fabricated and tested by a laboratory to define the entry loss. This is especially important when a large number of hoods are to be installed. With a sample hood the coefficient of entry can be defined for use in routine hood static suction measurements to calculate airflow. This approach will be discussed in the next section.

An initial physical inspection of the completed system should be conducted to determine if the system is built according to the design. The size and type of fittings should be checked and the hood construction reviewed. The fan type and size should be as specified and the direction of rotation, speed and current drain should be noted. A detailed inspection sheet should be completed. This type of inspection may seem redundant but in the author's experience it is worthwhile. In one case a mitre elbow was added in a large submain to save space; the single elbow presented high losses which caused the system to malfunction.

The air cleaning component should be inspected to determine if it is installed according to specifications. If ancillary equipment such as pressure sensing or velocity-measuring equipment is included in the systems, it should also be inspected and calibrated.

All conventional ductwork and air cleaning equipment should be mounted on the suction side of the fan. If this is not done toxic contaminants may leak to the occupied space. In one such case an installation handling a volatile and odorous organic chemical with the duct under positive pressure contaminated the workplace. Due to the complexity of the system the cost of rectifying this problem was $50,000. This problem should have been picked up in the design review. If a duct run must be under positive pressure special design features must be utilized by the engineer.

3.5.2.2 Airflow. The general format for testing a new local exhaust ventilation system balanced without blast gates is shown in Figure 29.13. The airflow rate through each branch servicing a hood should be evaluated by a pitot-static traverse. The pitot-static tube is a primary standard and, if used with an inclined manometer, it does not require calibration. Care is required in the choice of traverse location and the device is limited at low velocities.

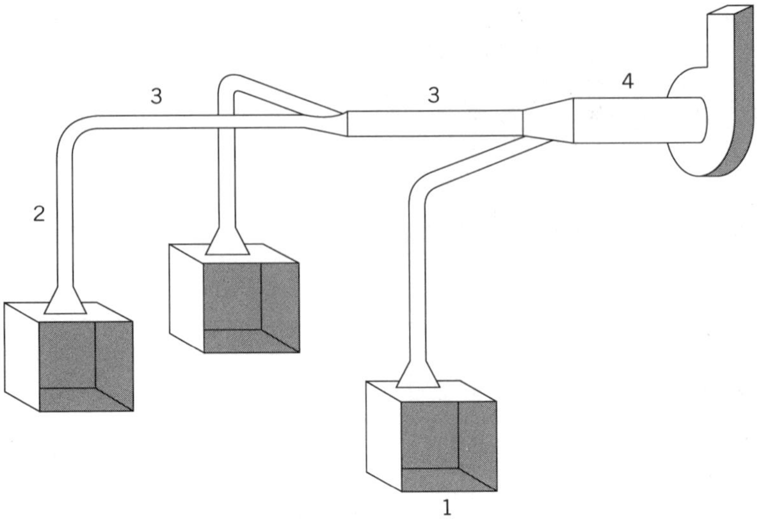

Figure 29.13. Locations for ventilation measurements: (1) at the face of booth-type hoods, (2) just downstream of a hood at a branch location, (3) in the main to define hood exhaust rate by difference, and (4) in the main to define total system airflow.

In the last decade electronic manometers with microprocessor-based instruments are available for direct reading of pressure and velocity. When used in conjunction with a small portable recorder the pitot traverse can be conducted by one person with ease.

One of two critical design velocities will be specified depending on the hood type. A partial enclosure such as a paint spray hood has a design velocity or control velocity at the plane of the hood face. Exterior hoods such as a simple welding hood will have a capture velocity specified at a certain working distance. The control and capture velocities may be calculated from the air flow rate and the hood dimensions. Frequently the actual velocities must be measured to satisfy plant or regulatory requirements. Face velocities at partial enclosures are usually evaluated by one of three direct reading anemometers: rotating vane, swinging vane, or heated element. These instruments have a wide range of applications but care should be taken to observe their limitations (49). These simple direct reading instruments can be used for the required inspections described in Section 3.5.2.

The measurement of hood static suction is also of value in periodic inspections and should be measured at the acceptance tests. The measurement, made 2–4 diameters downstream of the hood, provides a simple method of calculating airflow (49). The data collected in the acceptance tests are important baseline information for subsequent periodic inspections described in Section 3.6.3.

3.5.2.3 Performance. The term performance, as used in this discussion, describes the ability of the local exhaust ventilation to minimize the release of air contaminants from equipment serviced by the system. This measurement can be done by qualitative, semi-quantitative, and quantitativc methods.

3.5.2.3.1 Qualitative Methods. The use of smoke tracers released at the generation point, if done properly, provides an excellent qualitative method of exploring the performance of the hood. The control boundary can be established with this technique. In addition, it can be an excellent teaching tool for the worker who can view the impact of his own body and actions and that of external disturbances such as drafts from windows, doors, and traffic on the ventilation. Corrosive smoke cannot be used in semiconductor facilities; both dry ice and ultrasonic streams have been evaluated for this purpose. In Europe another qualitative technique, light scatter with back lighting, is widely used to identify the source and generation mode of particle contamination. Exquisite photographs of dust release based on this technique have been used as a design input for ceramic industry dust control.

3.5.2.3.2 Semi-quantitative. The local exhaust ventilation system is first assessed to determine if it meets good practice in terms of general design and airflow rate. If that is acceptable the system is checked qualitatively by smoke and then semi-quantitatively using a tracer (Fig. 29.14) (57). A series of release grids are designed which model the actual release area. An oil mist generator operating with corn oil is used to generate a submicron-sized aerosol. The generator is positioned to release the mist deep in the hood so that all mist is collected. The probe of a forward light scatter photometer samples the duct stream laden with the mist and the reading is defined as 100% containment. The grid is then positioned at the actual work release point and a new reading is taken in the duct. If all the mist is collected the reading will be 100%; if only ½ the oil mist is collected the reading of containment is 50%.

In one series of studies using this technique Ellenbecker (58) evaluated the capture efficiency of a series of local exhaust ventilation hoods in a chemical processing plant. The test aerosol was mineral oil with a count median diameter of 0.3 to 0.4 μm. In each operation the area source of contamination, such as a reactor charge port or the wet surface of the cake on a plate and frame filter, was simulated with a grid network of nozzles which

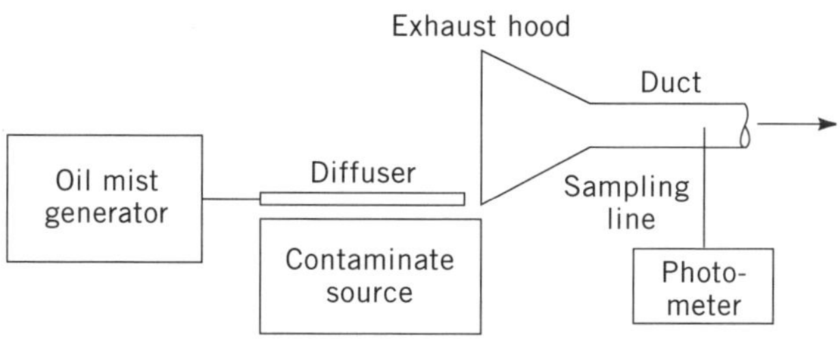

Figure 29.14. Containment fraction for a hood can be determined in the field using a tracer introduced first inside the hood to establish the 100% bench mark and then at the normal contaminant release point. From Ref. 57, reproduced with permission of the *Journal of the American Industrial Hygiene Association.*

could be varied from a point source to an area source several meters square. In this system the contaminant release for the grid was relatively passive and therefore one cannot duplicate the dynamics of the source such as the "throw". Nor can one duplicate the generation rate or other physical parameters such as temperature. For these reasons it must be identified as a semiquantitative procedure and extensive laboratory validation would be needed to upgrade the test to a quantitative procedure.

The performance of ten local exhaust hoods chosen for evaluation in this well maintained chemical plant were highly variable. Four hoods (two reactor, and two weigh scale hoods) with a capture efficiency of 100% were obviously operating satisfactorily. However in one case the exhaust rate was too high. The exhaust rate was reduced by a factor of four and a capture efficiency of 100% was still maintained. Such excessive flow rates will result in high operatings costs to exhaust the air and bring in tempered replacement air. Another four hoods achieving capture efficiencies of 90 to 100% (reactor, drum vent, centrifuge, and a weigh scale hood) were adequate for many operations but the investigator felt they might not be acceptable for highly toxic air contaminants. A single Neutsche filter hood had a capture efficiencies of 85%. Ellenbecker felt the design was unsatisfactory and the maintenance program must be reviewed and upgraded. A single bung hole exhaust hood with a capture efficiency below 70% was felt due to an inferior design.

This evaluation system and others now available using tracers such as sulfur hexafluoride are semiquantitative. They do not attempt to closely model the specific chemical or its generation rate. Although the tests are semiquantitative, the procedure is simple, the equipment suitable for in-plant tests, and the procedure can be done in 20 to 30 minutes. It can be effectively used by a technician to explore the effects in changes in hood geometry and placement, air flow rate, and baffle design. It is also an excellent training aid for chemical processors who must learn to use the exhaust system as a valuable tool.

3.5.2.3.3 Quantitative. The capture efficiency of a local exhaust ventilation system is usually impossible to establish in a plant using the process generated contaminant. To conduct such a test the system must be enclosed temporarily so that all the contaminant enters the hood to establish the 100% capture efficiency concentration. If that cannot be done both the fraction of the contaminant collected by the hood and that fraction lost from the hood must be independently determined. The fraction of the contaminant lost to the room is especially difficult to measure. Conroy (59) did however utilized this approach in the evaluation of capture efficiency on a degreaser hood. In her work both the local and general exhaust air flow rates were measured as was the volume of the work space. Solvent concentration at several locations in the general workplace and in the duct servicing the exhaust hood were measured periodically as was the velocity of the cross drafts in the workspace. Using mass balance models, the rate of solvent emission to the workspace was calculated based on the workspace concentrations and the ventilation characteristics of the workspace. The hood exhaust rate and the concentration in the duct determined the local exhaust emission rate. The ratio of the local exhaust ventilation emission rate to the total rate of solvent emission, that is, the sum of the loss to the room and that removed by the hood is the hood capture efficiency.

In most cases quantitative measurement of hood performance is made by air sampling on the worker. This procedure obviously does not measure capture efficiency but rather the ability of the local exhaust ventilation system to achieve a certain air concentration at the breathing zone of the worker. In such a case the most realistic test involves the operator using the actual chemical. Occasionally, as in the case of a Premanufacturing Notification (PMN) chemical, it may be useful to demonstrate performance of the controls using a substitute material. In one such test a simulant chemical was used for the actual PMN chemical. The simulant was a non-toxic blue dye that was inexpensive, reasonably dusty, and easy to sample and analyze. The test showed excellent control. A later test with the actual chemical confirmed the information obtained with the tracer.

3.5.3 Periodic Testing and Maintenance

The local exhaust ventilation system must be inspected periodically to ensure that the system meets performance and design standards and complies with various regulations. The inspection frequency varies and may range from a monthly to a semi-annual interval. The system is inspected visually. The motor-fan components are inspected to determine the condition of the fan blades, direction of rotation, belt tension, guarding and lubrication. The hoods and ductwork are inspected for corrosion damage, plugging, leaks, and any local modification by the plant. The airflow rate at each station is determined, preferably by the hood static suction method since it is accurate and fast. On semi-enclosing hoods such as paint spray booths the face velocity must be evaluated. Certain regulations may require measurement of capture velocity, that is the velocity at the contaminant release point. This measurement is difficult to conduct accurately. The measurement of airflow rate to achieve that velocity is the more accurate measurement.

The records management system for the inspection program is of great importance especially for facilities with hundreds of hoods. The need for such a system is described by Stott and Platts (60):

> The recording and manipulation of the necessary data is time-consuming, tedious and prone to inaccuracies. Most is recorded in note books which must be transcribed onto official forms which means that comparisons with previous measurements is often difficult. So at this stage, although a wealth of information is available regarding the extraction systems, the opportunity to analyze the data, initiate maintenance work, or diagnose faults is wasted.

These authors developed a computer-based system to satisfy these needs. The person conducting the field test uses a hand held data terminal which has been loaded with a program on the ventilation system under test. Each ventilation station is named, the position to be tested is identified, and a space is provided for entry of the test results. In the plant the tester calls up the station and conducts the velocity or pressure test measurement. The data are inserted into the portable data logger. On return to the office the data logger is down loaded to a personal computer. The data are scanned, and those stations where a preset minimum value is not achieved are identified by the computer for follow-up. Additional measurements are requested and based on these data a diagnosis of the ventilation system problem is identified by the computer for referral to plant engineering.

The importance of well-qualified personnel to design, install, operate and inspect local exhaust ventilation systems has been discussed in detail (50). It is especially important to develop an in-house training program for personnel who conduct the periodic inspections (52, 60).

The maintenance problems encountered in system operation and hopefully identified by the acceptance and periodic inspection schedule have been discussed (50). The cost of conducting an inspection program has not been published but in evaluating the cost-benefits of such a program one should recall the installation cost of the system and its operating cost in addition to the important role it plays in worker health. In many locations the required inspection frequency has prompted the installation of direct reading monitoring equipment such as pressure sensors at a hood static suction location, a pitot-static tube in the hood branch, a heated element sensor or a swinging vane anemometer at the hood face. In a nuclear energy facility an in-line orifice meter is permanently installed to measure flow through a glove box. This mass flow measuring technique has limited application in general industry due to energy loss, erosion of the orifice, and contamination.

In discussing the performance of local exhaust ventilation, Dalla Valle (61) stated:

> Whereas hoods are installed to eliminate a health hazard, tests should be conducted to establish their effectiveness. The ultimate criterion is not the provision of a "strong" suction but the handling of an air volume which reduces the concentration of the contaminants in question below the MAC level.

4 CONTROL TECHNOLOGY INFORMATION TRANSFER

The importance of control technology utilization by small and medium-sized enterprises (SMEs) has been clearly identified by both European and U.S. authorities. At the Third IOHA International Scientific Conference on Chemical Risks and Their Control, Haigh (62) stated that in Europe the SMEs (those facilities with less than 50 employees) comprised 98% of the 33,000 chemical companies and that nearly one-quarter of all SME employees are exposed to airborne chemical contaminants. A new Chemical Directive drafted by the European Union will require both assessment of risk and issuance of general principles for the prevention and control of such risks. To meet the needs of the SME, Haigh warns that advisory materials on contamination control technology must be simple yet comprehensive, useful and practical. In a recent initiative in the U.S., NIOSH has also identified SMEs as a critical target for effective communications on risk management. It is refreshing to see this recent concern for the need for control technology information for SMEs but how can it be achieved?

The conventional approach used by occupational health practitioners in developing a control strategy for handling toxic chemicals in industry is to examine the process to identify the contaminant, its physical form, contaminant release point, the dynamics of the release and a variety of other characteristics of the work environment (50). A range of contaminant control measures including process design, work methods, and engineering control can then be chosen to minimize the workers exposure so far as is reasonably practical. In practice, both in the U.S. and Europe, the appropriate occupational exposure limits (OELs) "drive" the selection of the control technology. In a mature occupational health program air sampling is conducted after the controls are installed to verify the efficacy of the controls. Unfortunately, in most SMEs this is not done.

For more than a decade a number of occupational health authorities have stated that although the OELs are invaluable to the occupational health professional they are not utilized to establish appropriate controls in the small plant setting. This scenario was clearly demonstrated in a UK survey by the Health and Safety Executive (63). The HSE conducted telephone interviews with representatives of 1,000 randomly selected users of chemicals and a smaller number of union health and safety specialists. The industries represented were predominantly SMEs, 40% had some use of chemicals and 60% used chemicals daily. The survey showed that these businesses had little understanding of the OEL concept, how the limits were developed, and most importantly how the OELs were to be used to choose optimal control technology. It was also apparent from the survey that the bulk of knowledge the firms did possess on controls for chemical handling was obtained from the manufacturer of the chemical or representatives who sold the product and not from health and safety professionals. The hierarchy of controls the HSE hoped to see in use namely first substitution then engineering and finally respiratory protection devices did not occur. In fact the priorities were reversed with respiratory protection the most widely used control followed by a variety of engineering methods. In general, substitution, the most effective control, was rarely used. The authors concluded that the knowledge of OELs by small and medium sized business was limited and played no direct part in workplace risk management (63).

This revelation, although forecast by anecdotal information, was disquieting since the program to establish the OELs in England is of high quality with OELs for only ten chemicals completed each year. It was obvious to HSE that although well executed, the OEL program is not assisting small and medium firms in the choice of control strategies. Faced with this situation, which is assuredly not restricted to the UK, the HSE is now implementing quite a different approach (64–66).

A basic element of the approach is a derivative of programs used by industry in both Europe and the U.S. to assign chemicals to risk categories. This proven technique has been used in both environmental toxic use reduction programs to define priorities and in the choice of control regimens for occupational settings. The system chosen by HSE is a simple ranking or "banding" of chemicals into one of five groups based on risk phrases (R phrases) to represent human health hazard (65). The R phrases are similar to the hazard signal words used in the U.S. to meet labelling regulations. The information is widely used in Europe and the U.S. and is well understood by all industry sectors including SMEs. The R phrase permits one to assign the chemical to one of the five hazard bands (A to E). A sixth category (S) includes chemicals that present a skin or eye hazard.

Exposure potential is established by knowledge of physical characteristics including dustiness of solids and volatility of liquids and the quantity of the chemical in use. The hazard banding and exposure potential are then used to establish the control approach as shown in Figure 29.15. The four general control approaches are shown in Figure 29.16 and the application of all factors to choose a control strategy is illustrated in Figure 29.17 for a dust with a Hazard Band of C. The detailed presentation of the control strategy by Maidment (66) is a textbook description of applicable control technology.

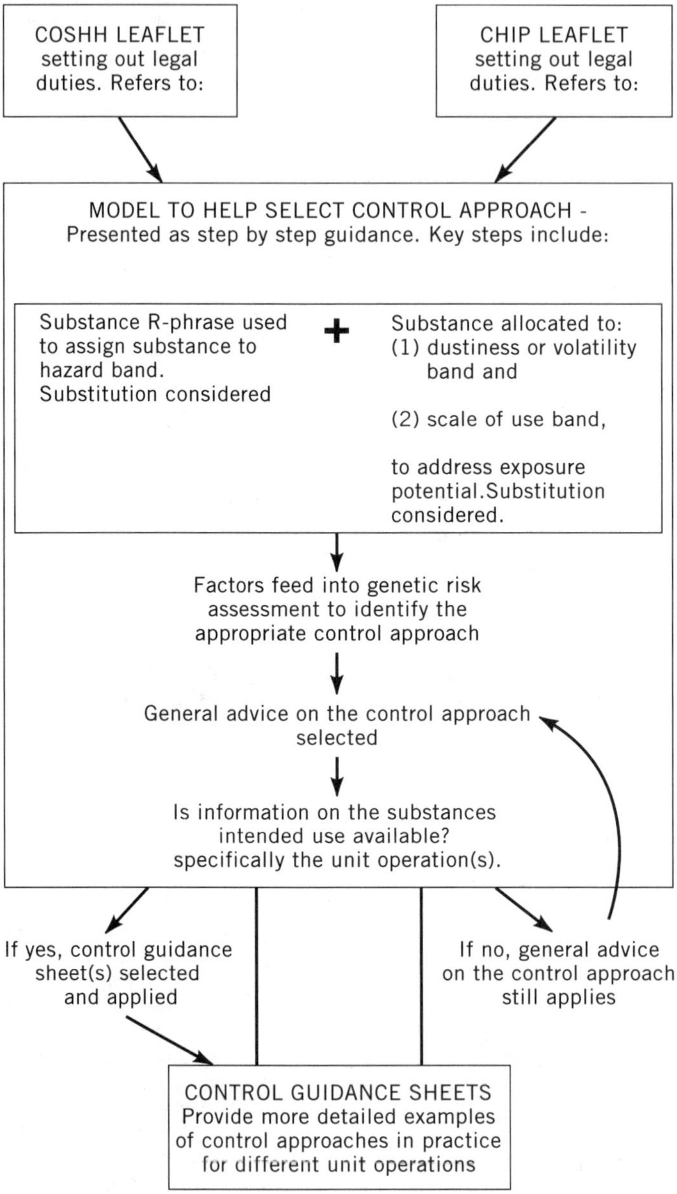

Figure 29.15. An overview of the UK scheme. From Russell et al. (64). Reprinted by permission of the British Occupational Hygiene Society.

Control approach 1 - General ventilation. This means a good standard of general ventilation and good working practices are needed. For many very small uses of some hazardous substances, ventilation from doors, windows and extractors are likely to be adequate.

Control approach 2 - Engineering control. This includes local exhaust ventilation ranging from a single point extract close to the source of hazards, to a ventilated partial enclosure. It also covers other engineering methods of control, e.g., cooling coils for vapors, but not complete containment.

Control approach 3 - Containment. This requires the hazard to be contained, or enclosed, but accepts that there may be limited, small scale breaches of containment in some instances.

Control approach 4 - Special. This means expert advice is needed in selecting the right control measure.

Figure 29.16. Control approaches used in the UK scheme. From Russell et al. (64). Reprinted by permission of the British Occupational Hygiene Society.

The HSE validated the use of the R banding scheme to effectively identify the appropriate control strategy for over 100 chemicals and found that scheme provides a level of exposure equivalent to or lower than that provided by health-based occupational exposure standard (65).

To support the control strategy HSE has designed control guidance sheets for a series of common chemical handling operations such as weighing and filling. For each unit operation, individual guidance sheets are tailored to the needs of small, medium and large chemical users. A total of 62 sheets are now available for distribution to the SMEs. The Health and Safety Executive plan to evaluate the field acceptance of the control guidance sheets in 1999. The control guidance sheets are an impressive communication device and could have major impact on the quality of control technology used by SMEs in the UK.

Will the goal of the UK program, improved implementation of effective control strategies by SMEs, be achieved with the result that workers exposure to toxic chemicals is minimized? An affirmative answer requires first, that the control strategies must have demonstrated effectiveness for a wide range of industries and applications. Secondly, the information must be widely disseminated to the SMEs in understandable form and be accepted by the industry as an integral part of the manufacturing process design.

The first requirement, that the efficacy of the recommended controls be demonstrated by in-plant studies, is a major hurdle. Unfortunately, few of the common controls widely used in chemical handling have undergone rigorous field validation. As noted earlier in this chapter, one of the primary engineering controls, local exhaust ventilation, displays variable collection efficiencies depending on process and local plant variables. The valuable "Ventilation Manual" published by the ACGIH includes a large number of hood designs suitable for chemically related processes. However, only a small fraction of these

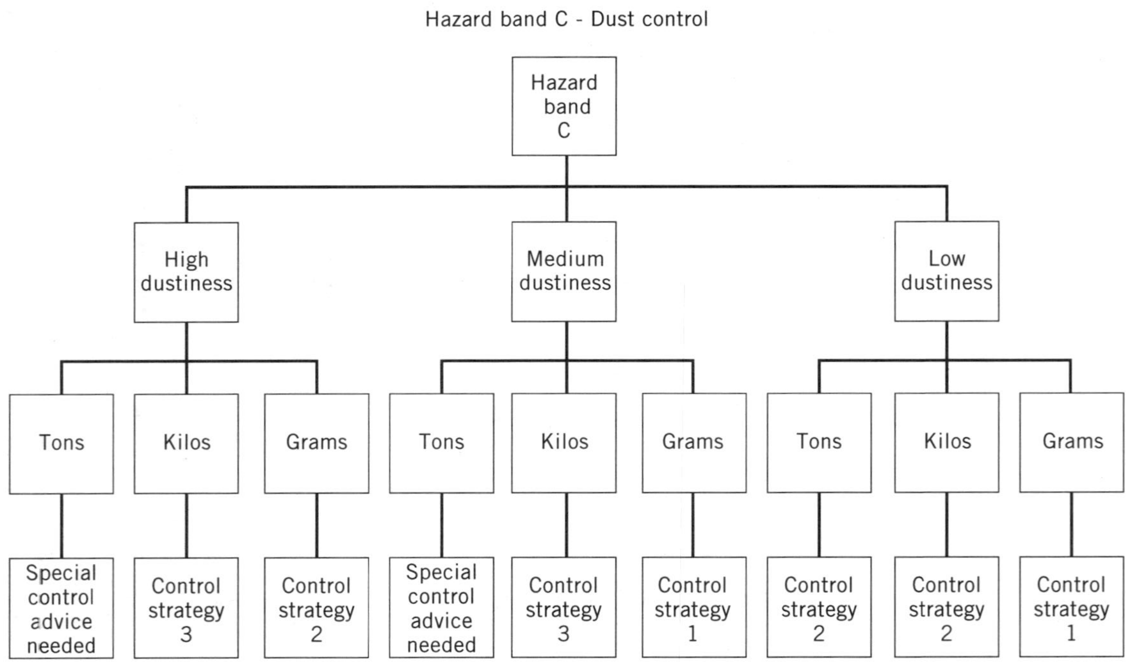

Figure 29.17. Combinations of factors affecting selection of the appropriate approach for solid chemicals assigned to hazard band C. From Russell et al. (64). Reprinted by permission of the British Occupational Hygiene Society.

designs have been validated under plant conditions. Additional funding for field investigations of the effectiveness of control technology is obviously needed.

The primary source of data on control technology rests with the chemical manufacturers. Release of contamination control design and performance information by manufacturers is an absolute imperative if this approach is to work. The HSE acknowledges this in their exposition of the program. Criticisms has been levelled at U.S. industry who, for a number of reasons have been hesitant to release control design data (52). This issue seems to less of a problem in the UK where excellent design data have been published by industrial associations.

Will the information on control technologies get to the intended user in SMEs? The HSE feels the anticipated involvement by chemical manufacturers, industry and trade associations and union groups will act as vehicles for effective dissemination of this information. The goal of the Product Steward in the chemical industry is to service each customer effectively by providing information not only the details on thc product but additionally provide comprehensive health and safety information. An extension of this program would be the acceptance by the manufacturer of the need to provide comprehensive control technology information.

In the US, dissemination of user-friendly information on control technology to SMEs has been of concern to NIOSH. Sheehy et al. (67) have displayed the effectiveness of such communications if properly prepared and distributed. The authors state the translation of detailed NOISH technical reports on control technology studies to provide useful information to SMEs can be done by "(*1*) developing clearly written materials that accurately describe engineering control methods in understandable language; and (*2*) identifying and utilizing existing private and government organizations that can help reach the small business owners and workers for distributing these materials". The authors cite the success of this approach using examples from embalming, radiator repair, brake repair, and furniture stripping.

5 SUMMARY

In this chapter recent advances in control technology are reviewed. The ability to identify critical exposure conditions has been enhanced by the coupling of sophisticated, direct-reading air sampling instruments with video taping of the job tasks. This technique permits the investigator to characterize the nature and origin of emissions in great detail with the subsequent application of effective controls. A second major shift during the decade is the emphasis on the reduction in the use of toxic materials with replacement by materials "kinder" to both the worker and the environment. Equipment advances during this period include robotic techniques which permit the worker to be separated from the point of greatest exposure. Finally the application of ventilation control is enhanced by improved knowledge of air flow patterns into hoods, new techniques for evaluating hood containment, and management tools for ventilation systems.

APPENDIX A NIOSH CONTROL TECHNOLOGY REPORTS[a]

RN: 00192317

TI: Engineering Health Hazard Control Technology for Coal Gasification and Liquefaction Processes. Final Report

AU: Anonymous

SO: Division of Physical Sciences and Engineering, NIOSH, U.S. Department of Health and Human Services, Cincinnati, Ohio, Contract No. 210-78-0084, 105 pages, 78 references

PY: 1983

RN: 00092812

TI: Control Technology Assessment: The Secondary Nonferrous Smelting Industry

AU: Burton-DJ; Coleman-RT; Coltharp-WM; Hoover-JR; Vandervort-R

SO: Division of Physical Science and Engineering, NIOSH, Cincinnati, Ohio, NIOSH Contract No. 210-77-0008, 393 pages

PY: 1979

RN: 00092805

TI: An Evaluation of Occupational Health Hazard Control Technology for the Foundry Industry

AU: Scholz-RC

SO: Division of Physical Sciences and Engineering, NIOSH, Cincinnati, Ohio, DHEW (NIOSH) Publication No. 79-114, Contract No. 210-77-0009, 436 pages, 56 references

PY: 1978

RN: 00145199

TI: Engineering and Other Health Hazard Controls in Oral Contraceptive Tablet Making Operations

AU: Anastas-MY

SO: Division of Physical Sciences and Engineering, NIOSH, U.S. Department of Health and Human Services, Cincinnati, Ohio, 94 pages

PY: 1984

RN: 00080675

TI: Development of an Engineering Control Research and Development Plan for Carcinogenic Materials

AU: Hickey-JLS; JJ-Kearney

SO: Applied Ecology Department, Research Triangle Institute, Research Triangle Park, North Carolina NIOSH Contract No. 210-76-0147, 168 pages, 19 references

PY: 1977

[a]*Source*: Courtesy of National Institute for Occupational Safety and Health.

RN: 00074602

TI: Engineering Control of Welding Fumes

AU: Astleford-W

SO: Division of Laboratories and Criteria Development, NIOSH, Cincinnati, Ohio, DHEW (NIOSH) Publication No. 75-115, Contract No. 099-72-0076, 122 pages, 13 references

PY: 1974

RN: 00074249

TI: Engineering Control Research and Development Plan for Carcinogenic Materials

AU: Hickey-J; Kearney-JJ

SO: Division of Physical Sciences and Engineering, NIOSH, Cincinnati, Ohio, Contract No. 210-76-0147, 167 pages, 19 references

PY: 1977

RN: 00052419

TI: Engineering Control Research Recommendations

AU: Hagoapian-JH; Bastress-EK

SO: Division of Physical Sciences and Engineering, NIOSH, Cincinnati, Ohio, Contract No. 099-74-0033, 210 pages, 181 references

PY: 1976

RN: 00182292

TI: Control Technology Assessment of Enzyme Fermentation Processes

AU: Martinez-KF; Sheehy-JW; Jones-JH

SO: Division of Physical Sciences and Engineering, NIOSH, U.S. Department of Health and Human Services, Cincinnati, Ohio, DHHS (NIOSH) Publication No. 88-114, 81 pages, 34 references

PY: 1988

RN: 00177551

TI: Minimizing Worker Exposure during Solid Sampling: A Strategy for Effective Control Technology

AU: Wang-CCK

SO: Division of Physical Sciences and Engineering, NIOSH, U.S. Department of Health and Human Services, 28 pages

PY: 1983

RN: 00177548

TI: A 3-E Quantitative Decision Model of Toxic Substance Control through Control Technology Use in the Industrial Environment

AU: Wang-CCK

SO: NIOSH, U.S. Department of Health and Human Services, Cincinnati, Ohio, 53 pages, 4 references

PY: 1982

RN: 00168700

TI: The Illuminating Engineering Research Institute and Illumination Levels Currently Being Recommended in the United States

AU: Crouch-CL

SO: The Occupational Safety and Health Effects Associated with Reduced Levels of Illumination. Proceedings of a Symposium, July 11–12, 1974, Cincinnati, Ohio, NIOSH, Division of Laboratory and Criteria Development, HEW Publication No. (NIOSH). 75-142, pages 17–27

PY: 1975

RN: 00133474

TI: Control Technology Assessment of the Pesticides Manufacturing and Formulating Industry

AU: Fowler-DP

SO: Division of Physical Sciences and Engineering, NIOSH, Cincinnati, Ohio, Contract No. 210-77-0093, 667 pages

PY: 1980

RN: 00148166

TI: NIOSH Technical Report: Control Technology Assessment: Metal Plating and Cleaning Operations

AU: Sheehy-JW; Mortimer-VD; Jones-JH; Spottswood-SE

SO: NIOSH, U.S. Department of Health and Human Services, Cincinnati, Ohio, Publication No. 85-102, 115 pages, 71 references

PY: 1984

RN: 00144935

TI: A Study of Coal Liquefaction and Gasification Plants: An Industrial Hygiene Assessment, a Control Technology Assessment, and the Development of Sampling and Analytical Techniques. Volume II

AU: Cubit-DA; Tanita-RK

SO: Division of Respiratory Disease Studies, NIOSH, U.S. Department of Health and Human Services, Morgantown, West Virginia, Contract No. 210-78-0101, 179 pages, 15 references

PY: 1983

RN: 00144934

TI: A Study of Coal Liquefaction and Gasification Plants: An Industrial Hygiene Assessment, a Control Technology Assessment, and the Development of Sampling and Analytical Techniques: Volume I

AU: Cubit-DA; Tanita-RK

SO: Division of Respiratory Disease Studies, NIOSH, U.S. Department of Health and Human Services, Morgantown, West Virginia, Contract No. 210-78-0101, 192 pages, 22 references

PY: 1983

RN: 00136362

TI: Control Technology Assessment of Selected Petroleum Refinery Operations

AU: Emmel-TE; Lee-BB; Simonson-AV

SO: Division of Physical Sciences and Engineering, NIOSH, U.S. Department of Health and Human Services, Cincinnati, Ohio, NTIS PB83-257-436, Contract No. 210-81-7102, 122 pages

PY: 1983

RN: 00136196

TI: Proceedings of the Second Engineering Control Technology Workshop, June 1981

AU: Konzen-RB

SO: Division of Training and Manpower Development, NIOSH, U.S. Department of Health and Human Services, Cincinnati, Ohio, NTIS PB-83-112-755, NIOSH Report No. 80-3794, 138 pages

PY: 1982

RN: 00133178

TI: Principles of Occupational Safety and Health Engineering, Instructor's Guide

AU: Zimmerman-NJ

SO: Division of Training and Manpower Development, NIOSH, U.S. Department of Health and Human Services, Cincinnati, Ohio, P.O. No. 81-3030, 262 pages, 17 references

PY: 1983

RN: 00135180

TI: Health Hazard Control Technology Assessment of the Silica Flour Milling Industry

AU: Caplan-PE; Reed-LD; Amendola-AA; Cooper-TC

SO: Division of Physical Sciences and Engineering, NIOSH, U.S. Department of Health and Human Services, Cincinnati, Ohio, 60 pages, 18 references

PY: 1982

RN: 00135171
TI: Engineering Noise Control Technology Demonstration for the Furniture Manufacturing Industry
AU: Hart-FD; Stewart-JS
SO: NIOSH, U.S. Department of Health, Education, and Welfare, Grant No. 1-RO1-OH-00953, 123 pages, 8 references
PY: 1982

RN: 00132081
TI: Control Technology Assessment for Chemical Processes Unit Operations
AU: Van-Wagenen-H
SO: NIOSH, Cincinnati, Ohio, Contract No. 210-80-0071, NTIS PB83-187-492, 19 pages
PY: 1983

RN: 00134239
TI: Engineering Control of Occupational Health Hazards in the Foundry Industry. Instructor's Guide
AU: Scholz-RC
SO: NIOSH, Cincinnati, Ohio, NTIS PB82-231-234, 156 pages, 49 references
PY: 1981

RN: 00133884
TI: Demonstrations of Control Technology for Secondary Lead Reprocessing
AU: Burton-DJ; Simonson-AV; Emmel-BB; Hunt-DB
SO: NIOSH, U.S. Department of Health and Human Services, Rockville, Maryland, Contract No. 210-81-7106, 291 pages
PY: 1983

RN: 00132005
TI: Pilot Control Technology Assessment of Chemical Reprocessing and Reclaiming Facilities
AU: Crandell-MS
SO: Engineering Control Technology Branch, NIOSH, Cincinnati, Ohio, NTIS PB83-197-806, 11 pages
PY: 1982

RN: 00130228
TI: Occupational Health Control Technology for the Primary Aluminum Industry
AU: Sheehy-JW
SO: Public Health Service, NIOSH, U.S. Department of Health and Human Services, Cincinnati, Ohio, DHHS Publication No. 83-115, 59 pages, 6 references
PY: 1983

RN: 00130213
TI: Control Technology Assessment in the Pulp and Paper Industry
AU: Schoultz-K; Matthews-R; Yee-J; Haner-H; Overbaugh-J; Turner-S; Kearney-J
SO: NIOSH, Public Health Service, U.S. Department of Health and Human Services, Cincinnati, Ohio, Contract No. 210-79-0008, 974 pages
PY: 1983

RN: 00106373
TI: Mechanical Power Press Safety Engineering Guide, Wilco, Inc., Stillwater, Minnesota
AU: Anonymous
SO: NIOSH, Division of Laboratories and Criteria Development, U.S. Department of H.E.W., Cincinnati, Ohio, 207 pages
PY: 1976

RN: 00123450
TI: Assessment of Engineering Control Monitoring Equipment. Volume II
AU: Anonymous
SO: Enviro Control, Inc., Rockville, Md., NIOSH, Cincinnati, Ohio, 393 pages, 23 references
PY: 1981

RN: 00123377
TI: Control Technology Assessment of Selected Process in the Textile Finishing Industry
AU: Collins-LH
SO: Bendix Launch Support Division, Cocoa Beach, Florida, NIOSH, Cincinnati, Ohio, 227 pages, 44 references
PY: 1978

RN: 00119991
TI: Phase I Report on Control Technology Assessment of Ore Beneficiation
AU: Todd-WF
SO: NIOSH, U.S. Department of Health and Human Services, Cincinnati, Ohio, 82 pages, 80 references
PY: 1980

RN: 00117613
TI: Symposium Proceedings, Control Technology in the Plastics and Resins Industry
AU: Anonymous
SO: NIOSH, U.S. Department of Health and Human Services, Cincinnati, Ohio, 333 pages, 23 references
PY: 1981

RN: 00117359

TI: Proceedings of the Symposium on Occupational Health Hazard Control Technology in the Foundry and Secondary Non-Ferrous Smelting Industries

AU: Anonymous

SO: NIOSH, U.S. Department of Health and Human Services, 401 pages, 45 references

PY: 1981

RN: 00116023

TI: Control Technology Summary Report on the Primary Nonferrous Metals Industry, Vol. IV, Appendix C

AU: Hoover-JR

SO: NIOSH, Center for Disease Control, Public Health Service, U.S. Department of Health, Education, and Welfare, 120 pages

PY: 1978

RN: 00116022

TI: Control Technology Summary Report on the Primary Nonferrous Metals Industry, Vol. 5, Appendix D

AU: Hoover-JR

SO: NIOSH, Center for Disease Control, Public Health Service, U.S. Department of Health, Education, and Welfare, 180 pages

PY: 1978

RN: 00115792

TI: Control Technology for Primary Aluminum Processing

AU: Sheehy-JW

SO: Department of Health and Human Services, Public Health Service, Center for Disease Control, NIOSH, Division of Physical Sciences and Engineering, Cincinnati, Ohio, pages 1–22

PY: 1980

RN: 00115530

TI: An Evaluation of Engineering Control Technology for Spray Painting

AU: O'Brien-DM; Hurley-DE

SO: NIOSH, Center for Disease Control, Public Health Service, U.S. Department of Health and Human Services, 117 pages, 52 references

PY: 1981

RN: 00114224
TI: Control Technology Summary Report on the Primary Nonferrous Metals Industry, Vol. II, Appendix A
AU: Coleman-RT; Hoover-JR
SO: Division of Physical Science and Engineering, NIOSH, 291 pages, 10 references
PY: 1978

RN: 00114223
TI: Control Technology Summary Report on the Primary Nonferrous Metals Industry, Vol. III, Appendix B
AU: Coleman-RT
SO: Division of Physical Science and Engineering, NIOSH, Cincinnati, Ohio, 102 pages
PY: 1978

RN: 00113373
TI: Control Technology Summary Report on the Primary Nonferrous Smelting Industry, Volume 1: Executive Summary
AU: Coleman-RT; Hoover-JR
SO: Division of Physical Science and Engineering, NIOSH, Cincinnati, Ohio, Contract No. 210-77-0008, Radian Corporation, Austin, Texas, 36 pages, 8 references
PY: 1978

RN: 00112367
TI: Control Technology Assessment of Raw Cotton Processing Operations
AU: Anonymous
SO: Envirocontrol Inc., Rockville, Md., NIOSH, Cincinnati, Ohio, Contract No. 210-78-0001, 351 pages
PY: 1980

RN: 00112366
TI: Engineering Control Technology Assessment for the Plastics and Resins Industry
AU: Anonymous
SO: Division of Physical Sciences, NIOSH, U.S. Department of Health, Education and Welfare, Contract No. 210-76-0122, Cincinnati, Ohio, 234 pages, 60 references
PY: 1977

RN: 00102819
TI: Proceedings of the Symposium on Occupational Health Hazard Control Technology In the Foundry and Secondary Non-Ferrous Smelting Industries
AU: Scholz-RC; Leazer-LD
SO: Department of Health, Education, and Welfare, Public Health Service Center for Disease Control, NIOSH, Cincinnati, Ohio, 447 pages, 10 references
PY: 1980

RN: 00094260

TI: Assessment of Selected Control Technology Techniques for Welding Fumes

AU: Van-Wagenen-HD

SO: Division of Physical Sciences and Engineering, NIOSH, Cincinnati, Ohio, NIOSH Publication No. 79-125, 29 pages, 17 references

PY: 1979

RN: 00094228

TI: Control Technology for Worker Exposure to Coke Oven Emissions

AU: Sheehy-JW

SO: Division of Physical Sciences and Engineering, NIOSH, Cincinnati, Ohio, NIOSH Publication No. 80-114, 29 pages, 26 references

PY: 1980

RN: 00092888

TI: Proceedings of NIOSH/University Occupational Health Engineering Control Technology Workshop

AU: Talty-JT

SO: Proceedings of the Workshop on Occupational Health Engineering Control Technology, May 16–17, 1979, Division of Physical Sciences and Engineering, NIOSH, Cincinnati, Ohio, 149 pages

PY: 1979

RN: 00092810

TI: Control Technology Summary Report on the Primary Nonferrous Metals Industry. Volume V. Appendix D: Review of the Testimony Presented at the 1977 OSHA Public Hearing on Sulfur Dioxide

AU: Hoover-JR

SO: Division of Physical Science and Engineering, NIOSH, Cincinnati, Ohio, NIOSH Contract No. 210-77-0008, 183 pages

PY: 1978

RN: 00092809

TI: Control Technology Summary Report on the Primary Nonferrous Metals Industry. Volume IV. Appendix C: Review of the Testimony Presented at the 1977 OSHA Public Hearings on Inorganic Lead

AU: Hoover-JR

SO: Division of Physical Science and Engineering, NIOSH, Cincinnati, Ohio, NIOSH Contract No. 210-77-0008, 123 pages

PY: 1978

RN: 00092808

TI: Control Technology Summary Report on the Primary Nonferrous Metals Industry. Volume III. Appendix B: Review of the Testimony Presented at the 1975/6 OSHA Public Hearing on Inorganic Arsenic

AU: Coleman-RT

SO: Division of Physical Science and Engineering, NIOSH, Cincinnati, Ohio, NIOSH Contract No. 210-77-0008, 105 pages

PY: 1978

RN: 00092807

TI: Control Technology Summary Report on the Primary Nonferrous Metals Industry. Volume II: Appendix A

AU: Coleman-RT; Hoover-JR

SO: Division of Physical Science and Engineering, NIOSH, Cincinnati, Ohio, NIOSH Contract No. 210-77-0008, 298 pages, 9 references

PY: 1978

The following reports are not yet listed in NIOSHTIC:

Control Technology for Ethylene Oxide Sterilization in Hospitals, V. D. Mortimer and S. L. Kercher, Division of Physical Science and Engineering, NIOSH, Cincinnati, Ohio, Pub No. 89-120, 167 pages, 82 references, 1989

Control of Asbestos Exposure During Brake Drum Service, J. W. Sheey, T. C. Cooper, D. M. O'Brien, J. D. McGlothlin, and P. A. Froelich, Division of Physical Science and Engineering, NIOSH, Cincinnati, Ohio, Pub No. 89-121, 69 pages, 48 references, 1989.

BIBLIOGRAPHY

1. M. W. First, *Am. Ind. Hyg. Assoc. J.* **44**, 621–626 (1983).
2. J. Gideon, E. Kennedy, D. O'Brien, and J. Talty, "Controlling Occupational Exposures—Principles and Practices", National Institute for Occupational Safety and Health, Cincinnati, OH, 1979.
3. W. A. Burgess, *Recognition of Health Hazard in Industry*, John Wiley & Sons, Inc., New York, 1995.
4. R. J. Sherwood and R. J. Alsbury, "Occupational Hygiene, Systematic Approach and Strategy of" in L. Parmeggiani, ed., *Encyclopaedia of Occupational Health and Safety*, International Labor Organization, Geneva, 1983.
5. A. D. Brandt, *Industrial Health Engineering*, Wiley, New York, 1947, pp. 50
6. W. Whitehead, *Appl. Ind. Hyg.* **2**, 79–86 (1987).
7. Suppliers of Advanced Composite Materials Association, "Safe Handling of Advanced Composite Materials," SACMA, Arlington, VA, 1991.

8. J. H. Ahearn, H. Fatkin, and W. Schwalm, *Pollution Prevention Review* **1** 13, (1991).
9. A. G. Kireta, *Heat./Pip./Air Cond.* **12**, 119–125 (1988).
10. Health and Safety Executive, "Control of Substances Hazardous to health Regulations 1988, Approved Codes of Practice," HMSO, London, 1988.
11. Office of Technology Assessment, "Serious Reduction of Hazardous Waste for Pollution and Industrial Efficiency", U.S. Government Printing Office, Washington, DC, 1986.
12. U.S. Congress, Pollution Prevention Act of 1990. *Congressional Record*, Section 6606 of the Budget Reconciliation Act, Oct. 26, 1990, p. 12517.
13. T. Harten and I. Licis, *J. Air Waste Manage. Assoc.* **41**, 1122–1129 (1991).
14. G. Goldschmidt, "Improvement of the Chemical Working Environment by Substitution of Harmful Substances. An Iterative Model and Tools." personal communication, 1991.
15. M. Rossi, K. Geiser, and M. Ellenbecker, *New Solutions* **2**, 25–32 (1991).
16. American National Standards Institute, "Precautionary Labeling of Hazardous Industrial Chemicals," ANSI Z129.1-1994", ANSI, New York, 1994.
17. E. Karger, personal communication, 1992.
18. International Organization for Standardization, ISO Central Secretariat., "International Standards Organization 14,000" Geneva, Switzerland 1996.
19. K. Wolf, A. Yazdani and P. Yates, *J. Air Waste Manage. Assoc.* **41**, 1055–1061 (1991).
20. British Cast Iron Research Association, Proceedings of the Working Environment in Iron Foundries, BCIRA, Birmingham, UK. 1981
21. J. S. Schumacher, *Am. Ind. Hyg. Assoc. J.* **39**, 73–78 (1978).
22. H. D. Goodfellow, and J. W. Smith, "Dustiness Testing—A New Design Approach For Dust Control" in J. H. Vincent ed., *Ventilation '88, Proceedings of the Second International Symposium on Ventilation for Contaminant Control*, Pergamon Press, Oxford, U.K., 1989.
23. J. C. Volkwein, A. B. Cecala, and E. D. Thimons. *Appl. Ind. Hyg.* **4**, 8 (1989).
24. British Occupational Hygiene Society, *Dustiness Estimation Methods for Dry Materials, Their Uses and Standardization*, Technical Guide No. 4, Science Reviews Ltd. Northwood, Middlesex, U.K., 1985.
25. C. M. Hammond, *Ann. Occup. Hyg.* **23**, 95–109 (1980).
26. P. Braun and E. Druckman, *N.E. J. Med.* **294**, 653–657 (1976).
27. A. R. Berens, *Pure Appl. Chem.* 365–375 (1981).
28. R. L. Treitman, personal communication, 1976.
29. H. Wolfson, "Is The Process Fit to Have Ventilation Applied?", in R. T. Hughes, H. D. Goodfellow, and G. S. Rajhans eds., *Proceedings of the 3rd International Symposium on Ventilation for Contaminant Control*, Sept. 16–20, 1991, Cincinnati, OH, 1993.
30. K. C. Caplan, "Philosophy and Management of Engineering Controls," in L. J. Cralley and L. V. Cralley, ed, *Patty's Industrial Hygiene & Toxicology*, John Wiley & Sons, Inc, New York, 1985, p. 625.
31. M. Gressel, W. A. Heitbrink, J. McGlothlin, and T. Fischbach, *Appl. Ind. Hyg.* **2**, 108–113 (1987).
32. D. O'Brien, T. Fischbach, T. Cooper, W. Todd, M. Gressel, and K. Martinez, *Appl. Ind. Hyg.* **4**, 238–243 (1989).
33. T. Cooper and M. Gressel, *Appl. Occup. Ind. Hyg.* **7**, 227–230, 1992.
34. M. Gressel, W. Heitbrink, J. McGlothlin, and T. Fischbach, *Appl. Ind. Hyg.* **3**, 316–320 (1988).

35. M. Gressel and T. Fischbach, *Appl. Ind. Hyg.* **4**, 227–233 (1989).
36. R. J. Kovein and P. A. Hents, *Appl. Occup. and Environ Health* **7**, 3 (1992).
37. British Occupational Hygiene Society, *Fugitive Emissions of Vapors from Process Equipment*, Technical Guide No. 3, Science Reviews Ltd, Northwood, Middlesex, U.K., 1984.
38. S. Lipton and J. Lynch, *Health Hazard Control in The Chemical Process Industry*, John Wiley & Sons, Inc., New York, 1987.
39. J. A. Mertens, *Envir. Prot.*, 25–29 (1991).
40. J. Peterson, "Selection and Arrangement of Process Equipment," in L. V. Cralley, L. J. Cralley and K. C. Caplan, eds., *Industrial Hygiene Aspects of Plant Operations*, Vol. 3, MacMillan, New York, 1985
41. C. N. Gray, and P. J. Hewitt, *Ann. Occup. Hyg.* **25**, 431–438 (1982).
42. EPA, *Appl. Ind. Hyg.*, **1**, R-11 to R-13 (1986).
43. N. Hall, *Science*, **266**, 32–33 (1994).
44. B. Hanman, *Physical Abilities to Fit The Job*, American Mutual Liability Insurance Company, Wakefield MA, 1968.
45. D. M. Anderson, *Ind. Med. Surg.* **34**, 168 (1964).
46. K. C. Caplan, "Building Types" in L. V. Cralley, L. J. Cralley, and K. C. Caplan, eds., *Industrial Hygiene Aspects of Plant Operations*, Vol. 3, MacMillan, New York, 1985.
47. M. Horowitz, personal communication, 1999
48. British Occupational Hygiene Society, *Controlling Airborne Contaminants in the Workplace*, BOHS Technical Guide No. 7, Science Reviews Ltd., Northwood, Middlesex U.K., 1987.
49. W. A. Burgess, M. J. Ellenbecker, and R. D. Treitman, *Ventilation for Control of the Work Environment*, John Wiley and Sons, Inc., New York, 1989.
50. D. J. Burton, *Industrial Ventilation Work Book*, DJBA, Inc., Salt Lake City, UT, 1997.
51. American Conference of Governmental Industrial Hygienists, Committee on Industrial Ventilation, *Industrial Ventilation, A Manual of Recommended Practice*, 23rd ed. ACGIH, Cincinnati OH, 1998.
52. W. A. Burgess, "The International Ventilation Symposia and the Practitioner in Industry" in R. T. Hughes, H. D. Goodfellow, and G. S. Rajhans, eds., *Ventilation '91, Proceedings of the Third International Symposium on Ventilation for Contaminant Control*, Sept. 16–20, 1991, ACGIH, Cincinnati, OH, 1993.
53. American Iron and Steel Institute, *Steel Mill Ventilation*, AISI, New York, 1965.
54. American Foundrymen's Society, *Foundry Environmental Control*, AFS, Des Plaines, IL, 1972.
55. M. Williams, D. Baldwin, and P. Manz, *Semiconductor Industrial Hygiene Handbook*, Noyes Publications, Park Ridge, NJ, 1995.
56. V. Sciola, "The Practical Application of Reduced Flow Push Pull Plating Tank Exhaust Systems", in R. T. Hughes, H. D. Goodfellow, and G. S. Rajhans eds., *Ventilation '91, Proceedings of the Third International Symposium for Contaminant Control*, Sept. 16–20, 1991, Cincinnati OH, 1993.
57. M. J. Ellenbecker, R. J. Gempel, and W. A. Burgess, *Am. Ind. Hyg. J.* **44**, 752–755 (1983).
58. M. Ellenbecker, personal communication, 1999.
59. L. M. Conroy, "Field Studies of Local Exhaust Hood Performance" in: A. Jansson and L. Olander, eds., *Ventilation '94—Proceedings of the 4th International Symposium on Ventilation for Contaminant Control*, Stockholm, Sweden, National Institute of Occupational Health, Sweden, 1994, pp. 333–338.

60. M. D. Stott and P. J. Platts, "The Ventdata Ventilation Plant Monitoring and Maintenance System," in H. D. Goodfellow, ed., *Ventilation '85 Proceedings of the First International Symposium for Contaminant Control*, Elsevier, Amsterdam 1986.
61. J. M. DallaValle, *Exhaust Hoods*, Industrial Press, New York, 1952.
62. R. Haigh, "The Problem in Europe: SMEs Need for Guidance on Control Selection" presented at the 3rd IOHA International Scientific Conference, Crans Montana, Sept. 1997.
63. M. D. Topping, C. R. Williams, and J. M. Devine, *Ann. Occup. Hyg.* **42**, 357–366 (1998).
64. R. M. Russell, S. C. Maidment, I. Brooks, and M. D. Topping, *Ann. Occup. Hyg.* **42**, 367–376 (1998).
65. I. M. Brook, *Ann. Occup. Hyg.* **42**, 391–400 (1998).
66. S. C. Maidment, *Ann. Occup. Hyg.* **42**, 391–400 (1998).
67. J. W. Sheehy, T. F. Schoenborn, and C. C. Gjesslug, "Dissemination of Information On Ventilation Controls To Small Businesses", in A. Jansson and L. Olander eds., *Ventilation '94—Proceedings of the 4th International Symposium on Ventilation for Contaminant Control*, Stockholm, Sweden, 1994, pp. 376–381, National Institute of Occupational Health, Sweden, 1994.

CHAPTER THIRTY

Industrial Hygiene Engineering Controls

Robert D. Soule, CIH, CSP

1 INTRODUCTION

One of the primary responsibilities of the industrial hygienist is to control health hazards in the work environment. Although diseases suspected of being related to the work environment were recognized more than 2500 years ago (e.g., lead poisoning of workers engaged in mining operations), the systematic application of industrial-hygiene-engineering controls is a recent technical development, that is, within the last few decades.

The importance of engineering control of the work environment among the alternative solutions to industrial hygiene problems is evident in every current industrial hygiene textbook. All list the hierarchy of control strategies in priority fashion as engineering controls, good work practices, administrative controls and, as a last resort, use of personal protective equipment. Many decisions of the Occupational Safety and Health Review Commission make it clear that engineering control of specific substances must be attempted even when complete control is not feasible just by engineering means. In other words, the use of personal protective equipment alone cannot be justified simply by analyzing a problem and determining that complete engineering control cannot be provided. Instead, the language of federal regulations has been interpreted as mandating that engineering controls be provided to the extent feasible and, in the event these are not sufficient to achieve acceptable limits of exposure, use of personal protection or other corrective measures may be considered.

It is clear that application of engineering controls, as a component of the total responsibility of the industrial hygienist, will remain a primary objective in the foreseeable future.

Patty's Industrial Hygiene, Fifth Edition, Volume 2. Edited by Robert L. Harris.
ISBN 0-471-29754-2

All professionals in the field of environment/safety/health (ESH) must be cognisant of the principles of engineering controls, and the engineering professional must be familiar with the details of implementing the control concepts. This chapter presents material useful to the industrial-hygiene engineer and nonengineer alike in applying their resources to the industrial-hygiene problems occurring in their respective settings. The focus here is on control of airborne contaminants; control of other types of stress, for example, noise, heat stress, and radiation, is discussed in the respective chapters elsewhere in this book.

2 PRINCIPLES OF CONTROL

Control principles available to the industrial-hygiene engineer are deceptively few: substitution, isolation, and ventilation, both general and localized. In a technological sense, an appropriate combination of these principles can be brought to bear on any industrial-hygiene-control problem to achieve satisfactory quality of the work environment. It may not be, and usually is not, necessary or appropriate to apply all of these principles to any specific potential hazard. A thorough analysis of the control problem must be made in order to ensure that a proper choice from among these methods is made. It must produce the proper control in a manner that is most compatible with the technical process, is acceptable to the workers in terms of day-to-day operation, and can be accomplished with optimal balance of installation and operating expenses.

With the tendency for acceptable exposure limits to become increasingly more stringent, the duties of the industrial hygiene engineer must include an ongoing analysis of installed control measures. Aside from problems associated with operation and maintenance of these provisions, the concept of the control provisions itself might become outdated by changes in process or regulations. The engineer then must be able to develop effective control methods and must have the capability to continue to evaluate the effectiveness of these methods regularly.

2.1 Substitution

It can be debated that substitution is not an engineering option. However, it frequently offers the most effective solution to an industrial-hygiene problem. There is a tendency to analyze any problem from the standpoint of correction rather than elimination. For example, the first inclination in considering a vapor-exposure problem in a degreasing operation is to provide or increase ventilation rather than substitute a solvent that has a much lower degree of hazard associated with its use. In its broadest sense, substitution includes replacing hazardous substances, changing from one type of process equipment to another, or in some cases even changing the process itself. In other words, material, equipment, or an entire process can be substituted to provide effective control of a hazard, often at minimal expense.

Some examples of substituting materials of lower toxicity are classics in the history of industrial hygiene. Most industrial hygienists are familiar with the substitution of red phosphorus for white in the manufacture of matches. Although this was done primarily in reaction to a tax law, the result was a markedly reduced potential hazard. The sequence of substitutions of degreasing solvents is an interesting one: from petroleum naphtha to carbon tetrachloride to chlorinated hydrocarbons to fluorinated hydrocarbons. Each of these substitutions alleviated one problem but resulted in a new one. This underscores a basic problem associated with using substitution as a control method, in that one hazard can be replaced by another inadvertently.

Many processes provide only limited opportunity to substitute materials. However, it might be possible to substitute, or at least modify, process equipment. This approach almost always is taken as a result of an obvious potential, usually physical, hazard. Applications to counter potential safety hazards are common: substituting safety glass for regular glass in enclosures, replacing unguarded equipment with properly guarded machines, replacing safety gloves or aprons with garments made of materials more impervious to chemicals being handled. Because substitution of equipment frequently is done as an immediate response to an obvious problem, it may not be recognized as an engineering control, even though the end result is every bit as effective.

Substituting one process for another may not be considered except in major modifications of a process. In general, a change in any process from a batch to a continuous type of operation carries with it an inherent reduction in potential hazard. This is true primarily because the frequency and duration of potential contact of workers with the process materials are reduced when the overall process approach becomes one of continuous operation. The substitution of processes can be applied on a fundamental basis. For example, substitution of airless spray for conventional spray equipment can reduce the exposure of a painter to solvent vapors. Substitution of a paint-dipping operation for the paint-spray operation can reduce the potential hazard even further. In any of these cases the automation of the process can further reduce the potential hazard.

2.2 Isolation

The principle of isolation frequently is envisioned as being limited to installation of a physical barrier between a hazardous operation and the workers. Fundamentally, however, this isolation can be provided without a physical barrier by appropriate use of distance and time. Perhaps the most common example of isolation as a control measure is associated with storage and use of flammable solvents. Large tank farms with dikes around tanks, underground storage of some solvents, detached solvent sheds, and fireproof solvent-storage rooms within buildings all are commonplace in American industry. Although the primary reason for the isolation of solvents is the risk of fire and explosion, the principle is no less valid as an industrial-hygiene measure.

Typically, application of the principle of isolation maximizes the benefits of additional engineering concepts such as local-exhaust ventilation. For example, the charging of mixers is the most significant operation in many processes that use formulated ingredients. When one of the ingredients in the formulation is relatively high in toxicity, it is worthwhile to isolate the mixing operation, that is, construct a separate mixing room, thereby confining the airborne contaminants potentially generated by the operation to a small area rather than having them influence the larger portion of the plant. Thus in ensuring isolation, the application of ventilation principles to control this contaminant at the source is much more effective.

2.3 Ventilation

To the industrial hygienist, ventilation is a method for providing control of an environment by strategic use of airflow. The flow of air may be used to provide either heating or cooling of a work space, to remove a contaminant near its source of release into the environment, to dilute the concentration of a contaminant to acceptable levels, or to replace air exhausted from an enclosure. Ventilation is by far the most important engineering-control principle available to the industrial hygienist. Applied either as general or local control, this principle has industrial significance in at least three applications: control of heat and humidity primarily for comfort reasons, prevention of fire and explosions, and most important to the industrial hygienist, maintenance of concentrations of airborne contaminants at acceptable levels in the workplace.

Principles and application of general and local-exhaust ventilation are presented in subsequent sections of this chapter. Use of these principles in solving industrial problems, whether general or localized, requires a basic understanding of the fundamentals of airflow. The scientific laws that define completely the motion of any fluid, including air, are complex, and except in the relatively simple case of laminar flow, we know relatively little about them. Nevertheless there are fundamental relationships that must be understood and conscientiously applied by the industrial-hygiene engineer. These are described briefly in the following sections; more elaborate discussions of these principles are given in several of the general references identified at the end of this chapter.

2.3.1 *Conservation of Mass*

The most basic principle of airflow is referred to as the "continuity equation" or principle of "conservation of mass." This relationship states that the mass rate of flow remains constant along the path taken by the fluid. For any two points in the fluid stream, therefore, we can express the relationship as

$$A_1 v_1 \delta_1 = A_2 v_2 \delta_2$$

where A = cross-sectional area (ft^2 or m^2)
v = linear velocity (ft/min or m/sec)
δ = specific weight (lb/ft^3 or g/m^3)

For most industrial applications, δ is relatively constant because the absolute pressure within a ventilation system varies over a narrow range and air remains essentially incompressible. Therefore,

$$A_1 v_1 = A_2 v_2 \text{ and } Q_1 = Q_2$$

where $Q = Av$, the volumetric rate of airflow (ft^3/min or m^3/sec).

2.3.2 *Conservation of Energy*

The energy content for steady flow of a frictionless, incompressible fluid along a single streamline can be described by Bernoulli's theorem, which states algebraically:

$$H + \frac{P}{\delta} + \frac{v^2}{2g_c} = C$$

where H = elevation above any arbitrary datum plane (ft or m)
P = absolute pressure (lb/ft^2 or g/m^2)
δ = specific weight (lb/ft^3 or g/m^3)
v = velocity (ft/sec or m/sec)
g_c = gravitational acceleration (ft/sec^2 or m/sec^2)
C = a constant, different for each streamline

The unit for each of the three terms in this expression is "foot-pounds per pound" (gram-meters per gram) of fluid, or "feet" of fluid, frequently referred to as elevation head, pressure head, and velocity head, respectively.

The elevation term H is usually omitted when Bernoulli's equation is applied to industrial-exhaust systems, because only relatively small changes in elevation are involved. Because all streamlines originate from the atmosphere, a reservoir of nearly constant energy, the constant C, is the same for all streamlines; the restriction of the equation to a single streamline thus can be removed. Furthermore, because the pressure changes in nearly all exhaust systems are at most only a few percent of the absolute pressure, the assumption of incompressibility may be made with negligible error.

2.3.3 Velocity Pressure

Velocity pressure is that pressure exerted by air in motion; it is analogous to kinetic energy and maintains air velocity. It exists only when air is in motion, acts in the direction of air flow, and is always positive in sign. In Bernoulli's equation, the term $v^2/2g_c$ represents the velocity head. The relationship between the velocity of air and velocity pressure is

$$v = \sqrt{2g_c h}$$

where v = velocity (ft/sec or m/sec)
g_c = gravitational acceleration (ft/sec^2 or m/sec^2)
h = head of air (ft or m)

2.3.4 Static Pressure

Static pressure overcomes the resistance in a system caused by friction of the air against the duct walls, produces initial air velocity, and overcomes turbulence and shock caused by a change in direction or velocity of air movement. Static pressure is analogous to potential energy, and it exists even where there is no air motion. It acts equally in all directions and tends either to collapse the walls of the duct upstream from the fan or to expand the walls of the duct on the downstream side. Static pressure usually is negative in sign upstream from a fan and positive downstream. It is measured as the difference between duct pressure and atmospheric pressure. The most common unit of static pressure is "inches of water" (in. H_2O).

2.3.5 *Total Pressure*

The driving force for the airflow is a pressure difference that is required to start and maintain flow. This pressure is called total pressure (*TP*) and has two components, velocity pressure (*VP*) and static pressure (*SP*); the relationship among the three is simply:

$$SP + VP = TP$$

Figure 30.1 shows the relationship among static pressure, velocity pressure, and total pressure at different points in a duct system.

If gas flowing through a duct system undergoes an increase in velocity, a part of the available static pressure is used to create the additional velocity pressure necessary to accelerate the flowing gas. Conversely, if the velocity is reduced at some point in a duct system, a portion of the kinetic energy or velocity pressure at that point is converted into potential energy or static pressure. Static pressure and velocity pressure are therefore mutually convertible. However, this conversion always is accompanied by a net loss of total pressure due to turbulence and shock; the conversion is never 100 percent efficient.

2.3.6 *Energy Losses*

Air in motion encounters resistance by all surfaces encountered by the flowing volume of air. Consequently, some of the energy of the air is given up in overcoming this friction and is transformed into heat. The rougher the surface confining the flow or the higher the flow rate, the higher the frictional losses.

Frictional loss in a duct varies directly with the length, inversely with the diameter, and directly with the square of the velocity of air flowing through the duct. This loss can be calculated from charts using the Fanning friction factor, which is an empirical function of Reynolds number, duct material, and type of construction (1).

Energy also is lost due to turbulence caused by a change in direction or velocity within a duct. The pressure drop in a duct system due to dynamic losses increases with the number of elbows or angles and the number of velocity changes within the system. The resulting pressure drop from these energy losses is expressed in units of "equivalent length." For example, an elbow of 10-in. diameter and 20-in. centerline radius is said to have an equiv-

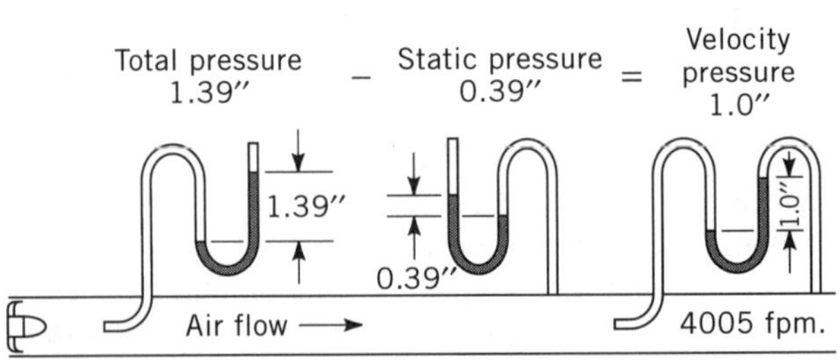

Figure 30.1. Relationship among velocity pressure, static pressure, and total pressure.

alent length of 14, meaning that the loss through the elbow will be the same as through 14 ft of straight pipe with the same diameter operating under the same conditions (2, 3).

Another method of defining the losses due to turbulence and friction is to express the losses in terms of velocity pressure. For example, a loss of $0.28VP$ in a transition or elbow means the incremental pressure drop is equal to 0.28 of the velocity pressure of the airstream at that point (2–4).

Acceleration and hood-entrance loss is a drop in pressure caused by turbulence when air is accelerated from rest to enter a duct or opening. Turbulence losses of this type vary with the type of opening.

This entry loss plus the acceleration energy required to move the air at a given velocity (1 VP) make up the hood-static pressure SP_h, which is expressed algebraically as

$$SP_h = h_e + VP$$

where h_e is hood-entry loss. The SP_h can be measured directly at a short distance downstream from the hood entrance. The calculation of SP_h is the first step in the design or evaluation of a local-exhaust system, which will be discussed later.

The coefficient of entry, C_e, is a measure of how efficiently a hood entry is able to convert static pressure to velocity pressure. The coefficient of entry is the ratio of rate of flow by the hood-static pressure to the theoretical flow if the hood-static pressure were completely converted to velocity pressure.

The result of the friction and dynamic losses to air flowing through ductwork is a pressure drop in the system. Bernoulli's theorem can be restated in a simplified expression of conservation of energy as follows:

$$SP_1 + VP_1 = SP_2 + VP_2 + \text{energy losses}$$

Static pressure plus velocity pressure at a point upstream in the direction of airflow equals the static pressure plus velocity pressure at a point downstream in the direction of airflow plus friction and dynamic losses.

3 GENERAL VENTILATION

As used in this chapter, the term general ventilation refers to the practice of supplying and exhausting large volumes of air throughout a work space. It is used typically in industry to achieve comfortable work conditions (temperature and humidity control) or to dilute the concentrations of airborne contaminants to acceptable limits throughout the work space. When properly used, general ventilation can be effective in removing large volumes of heated air or relatively low concentrations of low toxicity contaminants from several decentralized sources. General ventilation can be provided by either natural or mechanical means; often the best overall result is obtained with a combination of mechanical- and natural-air supply and exhaust.

3.1 Natural Ventilation

Natural ventilation can be provided either by gravitational forces, using thermal forces of convection primarily, or by anemotive forces, created by differences in "wind pressure." These two natural forces operate together in most cases, resulting in the natural displacement and infiltration of air through windows, doors, walls, floors, and other openings in an industrial building. Unfortunately, the wind currents and thermal-convection profiles on which natural ventilation depends are erratic and frequently unpredictable. Thus it is perhaps a misnomer to refer to natural ventilation as a "control" method, because to employ this technique requires dependence on, rather than control of, natural forces.

Indeed, there are applications for general natural ventilation. The pressure exerted on the upwind side and concurrent suction exerted on the downwind side of a building as a result of wind movement can be predicted for flat terrain fairly reliably. Thus the wind forces exerted on an isolated building in a relatively flat area permit the prediction of natural-ventilation forces. In the more common complex industrial building, however, the effects of the presence of one building on the others normally cannot be calculated; thus the use of wind-pressure models in the development of general ventilation systems is not feasible.

The impact and effectiveness of natural ventilation provided to a building depend both on the wind pressure profiles and on the thermal effects occurring within the structure. The warmer inside air tends to rise and leave the building through any available openings in the upper structure; cooler air tends to infiltrate the building by the reverse process in the lower structure. These thermal effects, which are much more predictable than the external wind forces, can be useful in design of a general ventilation scheme (5).

This chapter is devoted to engineering control and, because of the lack of control inherent in natural-ventilation systems, the remainder of this chapter presents concepts of ventilation and quality control of the work environment in terms of mechanically controlled ventilation systems. Obviously, many of the principles associated with use of mechanical-ventilation systems apply equally well to natural ventilation. However, the remainder of this chapter focuses on concepts and design principles for which it is assumed there is indeed control.

3.2 Mechanically Controlled Ventilation

A modern industrial complex characterized by a low profile building structure (i.e., large floor space, low height) as well as multistory buildings of masonry and glass construction present ventilation problems that were not found among older industrial plants. For the most part the industrial facilities constructed 30 or more years ago incorporated by design many features that permitted—in fact expected—natural ventilation of the work space. By contrast, modern buildings generally defy the exertion of natural-ventilation forces, and mechanical ventilation must be relied upon almost completely. Mechanical-ventilation systems exhaust contaminated air by mechanical means (exhaust fans) with the concomitant use of appropriate air supply to replace the exhausted air. The best method of achieving this in modern closed buildings is to supply air through a system of ductwork, distributing the air into the work areas in a manner that will provide optimal benefit to the workers for both comfort and control of contaminants.

3.3 Applications of General Ventilation

Normally, either or both of two end results are desired when applying general ventilation: an environment that is comfortable to the workers or one that is free of harmful concentrations of airborne contaminants.

3.3.1 Comfort Ventilation

Comfort ventilation, principally for heat relief, includes certain aspects of what is commonly considered so called air-conditioning engineering. Here, air is treated to simultaneously control temperature, humidity, and cleanliness and is distributed to maximize effects in the conditioned space. In the typical residential or office building, these requirements are primarily associated with comfort for the occupants. In many industrial situations, however, comfort conditions are impractical if not impossible to maintain, and the chief function of ventilation for comfort control is to prevent acute discomfort and the accompanying adverse physiological effects.

General ventilation also may be used to remove heat and humidity if a source of cooler air is available. If it is possible to enclose the heat source, as in the case of ovens or furnaces, a gravity or forced-air stack may be all that is necessary to prevent excessive heat from entering the workroom.

Air Conditioning. Complete, or integrated, air conditioning improves the control of temperature, humidity, radiation, air movements, or drafts, as well as the cleanliness and purity of the air. Under ordinary circumstances of comfort and health, the body temperature is kept approximately constant at its normal level of 98.6°F, through a balance between the internal production of heat (metabolism) and thermal equilibrium with the environment. The exchange of heat between humans and the environment is discussed in detail elsewhere in this book. According to the various equations that have been developed to express the thermal exchange between the human body and the environment, four external physical factors are of primary importance: air temperature, radiant temperature, air velocity, and the moisture content of the air. These four environmental factors are independently variable and may be combined in many different ways to produce the same thermal effect on humans. Thus an increasing velocity can offset a rise in air temperature, extreme radiant heat can be compensated somewhat by lowering the air temperature, and a considerable elevation of air temperature can be tolerated if the relative humidity is low. Combinations of thermal conditions that induce the same sensations are called thermoequivalent conditions. The effective temperature charts of the American Society of Heating, Refrigerating and Air-Conditioning Engineers (ASHRAE) give the combinations of equivalent conditions for clothed, sedentary individuals and for workers stripped to the waist. These charts are presented for illustration in Figure 30.2, Figure 30.3, and Figure 30.4. Although many indexes of heat stress and subjective response to this stress have been developed over the years, the effective temperature charts have been the primary tool of the air conditioning engineer. These scales were derived by experiments with test subjects exposed to a variety of thermal conditions by repeating trials with combinations of temperature, humidity, and air motion. Despite its limited subjective basis, subsequent laboratory and field experience with the effective-temperature scale has indicated that it is useful in defining thermal-

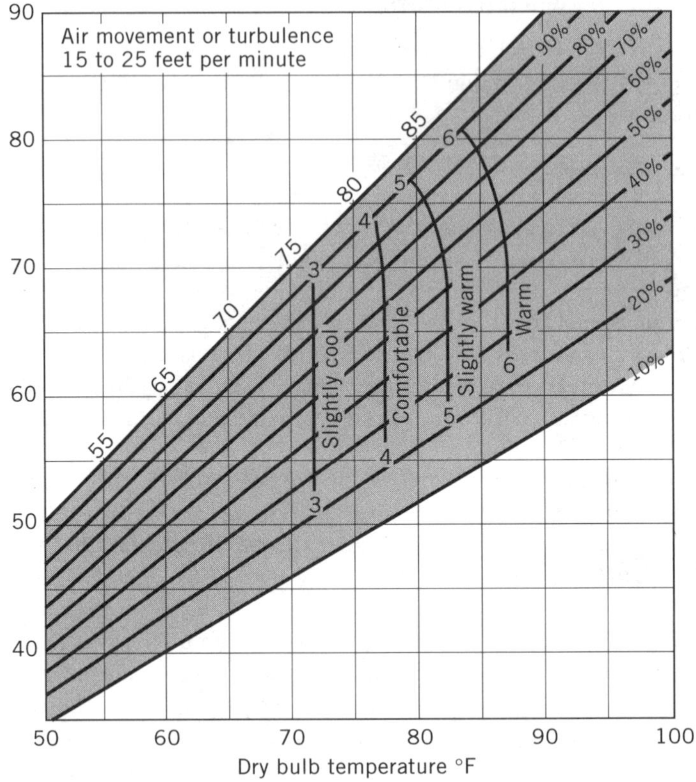

Figure 30.2. ASHRAE comfort chart for normally clothed people at rest in still air. Effective temperature (broken) lines indicate sensation of warmth immediately after entering conditioned space. Solid lines (3 to 6) indicate sensations experienced after 3-hr occupancy.

comfort conditions and, to a great extent, it correlates with objective measurements of physiological response to heat. Any other features of ventilation for purposes of comfort are similar to those associated with control of airborne contaminants and are discussed subsequently.

3.3.2 Dilution Ventilation

Dilution ventilation has as its primary function the maintenance of concentrations of airborne contaminants at or below acceptable-exposure limits, in terms of either potential fire and explosion or occupational-health considerations. If enough clean air is mixed with contaminated air, the concentration of the contaminant can be diluted to any reasonable level. Of course, dilution ventilation for purposes of fire and explosion control should not be employed in areas occupied by workers because the concentrations of concern from an occupational-health standpoint invariably are orders of magnitude below those of concern

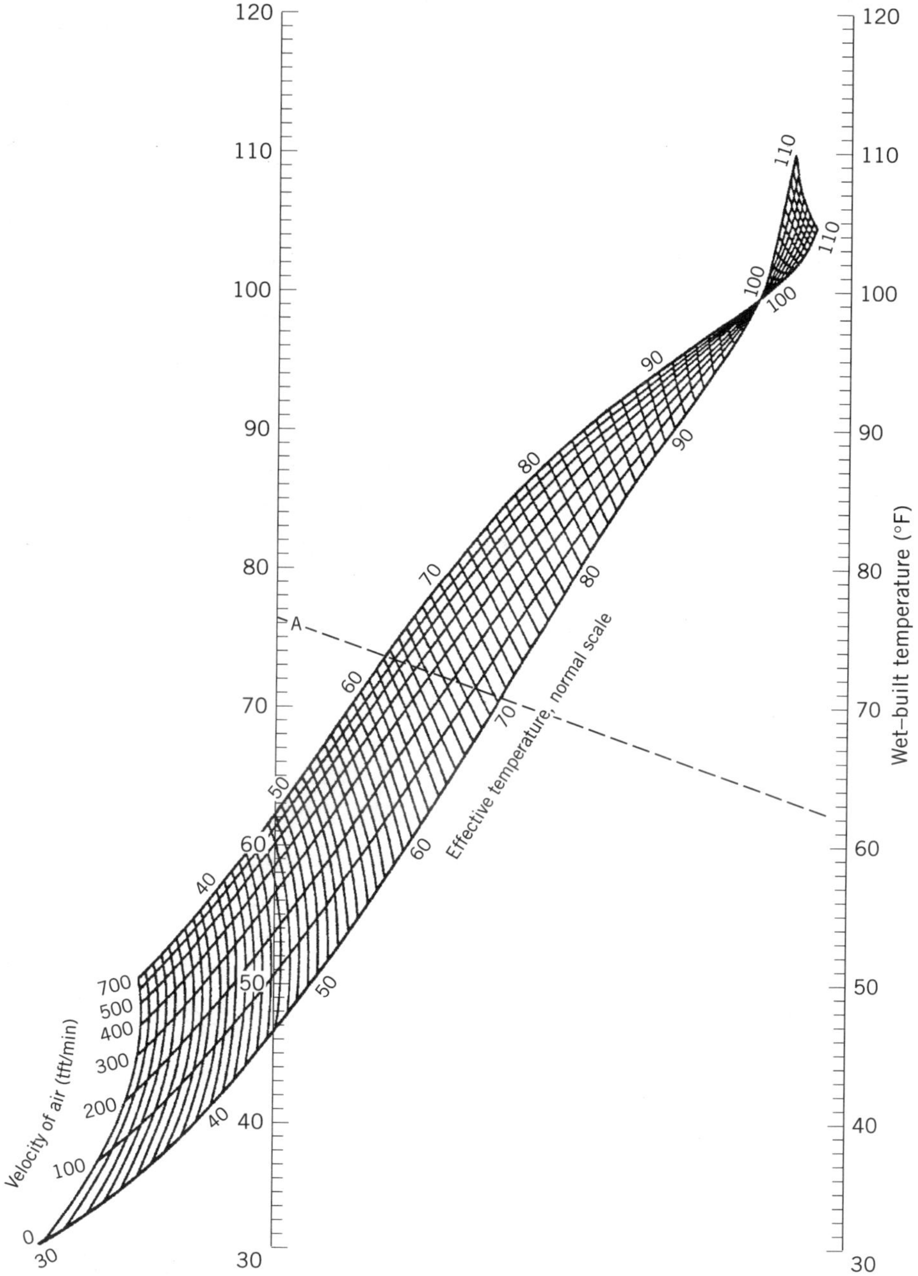

Figure 30.3. ASHRAE effective temperature scale for normally clothed persons at rest.

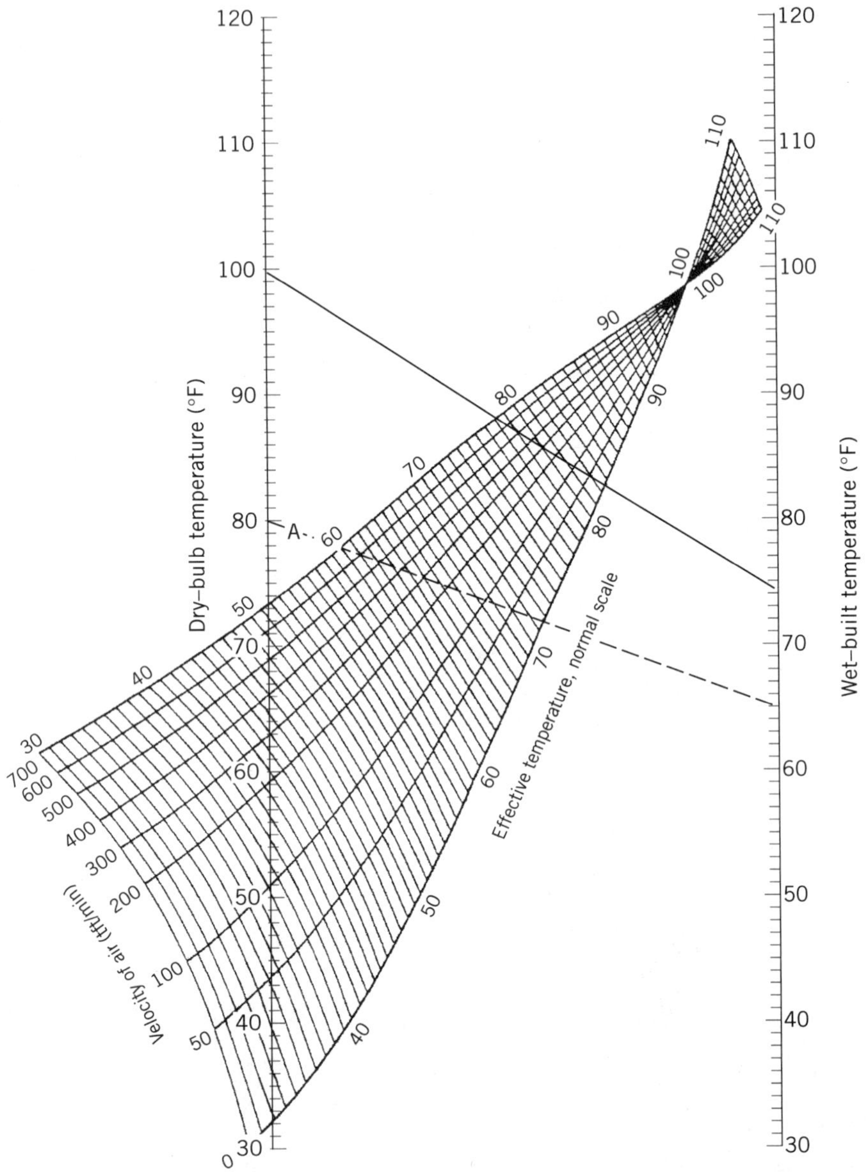

Figure 30.4. ASHRAE effective temperature scale for men stripped to the waist at rest or doing light physical work in rooms heated by convection methods. (From current edition of ASHRAE Handbook.)

for explosive limits. Exposures to atmospheres controlled to concentrations below the lower explosive limit or even a fraction thereof could cause narcosis, severe illness, or even death. Therefore it is extremely important not to confuse dilution-ventilation requirements for health-hazard control with those for fire and explosion prevention.

When considering whether dilution ventilation is appropriate for health-hazard control, it should be remembered that dilution ventilation has four limiting factors:

1. The quantity of contaminant generated must not be excessive, or the air volume necessary for dilution would be prohibitively large.
2. Workers must be far enough away from evolution of the contaminant or the contaminant must be of sufficiently low concentration, so the workers will not be exposed to excessive levels.
3. The toxicity of the contaminant must be relatively low.
4. The evolution or generation of the contaminant must be reasonable, uniform, and consistent.

A review of these factors indicates clearly that dilution ventilation is not normally appropriate for control of fumes and dust because the high toxicity often encountered requires excessively large quantities of dilution air. Moreover, the velocity and rate of evolution usually are very high, resulting in locally high concentrations, thus rendering the dilution ventilation concept inappropriate.

Rarely is dilution ventilation as satisfactory as local-exhaust ventilation for primary control of health hazards. Occasionally, however, dilution ventilation must be used because the operation of the process prohibits local exhaust. Dilution ventilation sometimes provides an adequate amount of control more economically than a local-exhaust system. This condition is an exception rather than the rule, and it should be kept in mind that the economic considerations of the long-term use of dilution-ventilation systems often overshadow the initial cost of the system because such a system invariably exhausts large volumes of heated air from the building. This work load can easily result in huge operating costs because of the need for conditioned-makeup air, and the general ventilation scheme would be much more expensive over an extended period of time.

In practice, dilution ventilation for control of health hazards is used to best advantage in controlling the concentration of vapors from organic solvents of relatively low toxicity. To apply the principle of dilution successfully to such a problem, data must be available on the rate of vapor generation or on the rate of liquid evaporation. Usually such data can be obtained from the plant records on material consumption.

Calculating Dilution Ventilation. The volume of a room to be ventilated and the ventilation rate are frequently related by taking the ratio of the ventilation rate to the room volume to yield a number of air changes per minute or number of air changes per hour. These terms, used quite frequently in discussion of ventilation requirements, are frequently employed incorrectly in industrial hygiene.

Ventilation requirements based on room volume alone have no validity. Calculations of the required rate of air changes can be made only on the basis of a material balance for the contaminant of interest. Similar calculations can be made for the rate of concentration increase or decrease; however, they require not only the air-change rate but also the rate of generation of contaminant. In the design of industrial ventilation, "x air changes" has valid application only very rarely. The term is useful when applied to meeting rooms, offices, schools, and similar spaces, in which the purpose of ventilation is simply the control of odor, temperature, or humidity, and the only contamination of air is the result of activity of people. Dilution ventilation requirements should always be expressed in cubic feet per minute or some other absolute unit of airflow, not in air changes per hour.

The concentration of a gas or vapor at any time can be expressed by a differential-material balance that, when integrated, provides a rational basis for relating ventilation to the generation and removal rates of a contaminant. Let C be concentration of gas or vapor at time, t, and take the following additional quantities:

G = rate of generation of contaminant
Q = rate of ventilation
K = design distribution constant, allowing for incomplete mixing
$Q' = Q/K$ = effective rate of ventilation, corrected for incomplete mixing
V = volume of room or enclosure

Starting with a fundamental material balance, and assuming no contaminant in the air supply, the following relationship applies:

rate of accumulation = rate of generation − rate of removal

This basic relationship can be applied in various ways.

Concentration Buildup. Rearranging the differential material balance, the buildup of contaminant can be expressed as follows:

$$\int_{C_1}^{C_2} \frac{dC}{G - Q'C} = \frac{1}{V} \int_{t_1}^{t_2} dt$$

$$\ln\left(\frac{G - Q'C_2}{G - Q'C_1}\right) = -\frac{Q'}{V}(t_2 - t_1)$$

If $C_1 = 0$ at $t_1 = 0$,

$$\ln\left(\frac{G - Q'C}{G}\right) = -\frac{Q'}{V}t$$

or

$$\frac{G - Q'C}{G} = e^{-(Q't/V)}$$

Rate of Purging. When a volume of air is contaminated but further contamination or generation has ceased, the rate of decrease of concentration over a period of time is as follows:

$$V\,dC = -Q'C\,dt$$

and

$$\int_{C_1}^{C_2} \frac{dC}{C} = -\frac{Q'}{V}\int_{t_1}^{t_2} dt$$

$$\ln \frac{C_2}{C_1} = -\frac{Q'}{V}(t_2 - t_1)$$

Maintaining Acceptable Concentration at Steady State. At steady state, $dC = 0$, and

$$Gdt = Q'Cdt$$

and

$$\int_{t_1}^{t_2} Gdt = \int_{t_1}^{t_2} Q'Cdt$$

At a constant concentration C, and uniform generation rate G,

$$G(t_2 - t_1) = Q'C(t_2 - t_1)$$

$$Q' = \frac{G}{C}$$

$$Q = \frac{KG}{C}$$

Therefore the rate of flow of uncontaminated dilution air required to reduce the atmospheric concentration of a hazardous material to an acceptable level can be easily calculated if the generation rate can be determined. Usually the acceptable concentration is considered to be the acceptable 8-hr time-weighted average (TWA) concentration. For liquid solvents the steady-state dilution ventilation requirement can be conveniently expressed as

$$Q = \frac{(6.71)(10^6)(\mathrm{SG})(\mathrm{ER})(\mathrm{K})}{(\mathrm{MW})(\mathrm{PEL})}$$

where Q = actual ventilation rate (ft^3/min)
SG = specific gravity of volatile liquid
ER = evaporation rate of liquid (pints/hour)
MW = molecular weight of liquid
K = design safety factor for incomplete mixing
PEL = permissible exposure limit (ppm)

Specifying Dilution Ventilation. The foregoing discussion introduced the concept of a design safety factor K for calculating dilution-ventilation requirements. The K factor is based on the following several considerations:

1. The efficiency of mixing and distribution of makeup air introduced into the room or space being ventilated
2. The toxicity of the solvent. Although PEL and toxicity are not synonymous, the following guidelines have been suggested for choosing the appropriate K value:

Slightly toxic	PEL $>$ 500 ppm
Moderately toxic	PEL $=$ 100–500 ppm
Highly toxic	PEL $<$ 100 ppm

3. A judgment of any other circumstances that the industrial hygienist determines to be of importance, based on experience and the individual situation. Included in these criteria are considerations such as seasonal changes in the amount of natural ventilation; reduction in operation efficiency of mechanical air-moving devices; duration of the process, operational cycle, and normal location of workers relative to sources of contamination; location and number of points of generation of the contaminant in the workroom or area; and other circumstances that may affect the concentration of hazardous material in the breathing zone of the workers. The K value selected usually varies from 3 to 10, depending on the foregoing considerations.

Table 30.1 lists the air volumes required to dilute the vapors of common organic solvents to the PEL level based on the liquid volume of solvent evaporated per unit time. These values must be multiplied by a K factor to allow for variations in uniformity of air distribution and for other considerations.

It must be emphasized that PELs are subject to revision, and the dilution values estimated from such tables become obsolete if the PELs are changed; therefore, such a table should be used with caution, and with reference to the latest PEL (or threshold limit) values (6, 7).

3.3.3 General Ventilation Airflow Patterns

General ventilation presents some disadvantages, the most significant one being that it permits the occupied space to become, in effect, a large settling chamber for the contaminants, even though the concentration may be within acceptable limits as far as potential health hazard is concerned. In some cases, the settling or separation of contaminants from the air may represent condensation of materials that were vaporized by high temperature processes and accumulated on surfaces after condensation. The undesirability of the settling of contamination onto surfaces within the plant has been dramatically demonstrated in plants handling highly combustible dusts in which, although the dust concentrations at any one time were not sufficient to constitute a hazard, accumulation over many months resulted in disastrous explosions. It is therefore necessary to maintain an effective ongoing

Table 30.1 Dilution Air Volumes for Vapors[a]

Liquid (TLV in ppm)[b]	Ft^3 of air (STP) Required for Dilution to TLV[c] (per pint evaporation)
Acetone (750)	7,350
n-Amyl acetate (100)	27,200
Benzene (10)	Not recommended
n-Butanol (butyl alcohol) (50)	88,000
n-Butyl acetate (150)	20,400
Butyl Cellosolve (2-butoxyethanol) (25)	Not recommended
Carbon disulfide (10)	Not recommended
Carbon tetrachloride (5)	Not recommended
Cellosolve (2-ethoxyethanol) (5)	Not recommended
Cellosolve acetate (2-ethoxyethyl acetate) (5)	Not recommended
Chloroform (10)	Not recommended
1,2 Dichloroethane (10) (ethylene dichloride)	Not recommended
1,2 Dichloroethylene (200)	26,900
Dioxane (25)	Not recommended
Ethyl acetate (400)	10,300
Ethyl alcohol (1000)	6,900
Ethyl ether (400)	9,360
Gasoline (300)	Requires special consideration
Isoamyl alcohol (10)	37,200
Isopropyl alcohol (400)	13,200
Isopropyl ether (250)	11,400
Methyl acetate (200)	25,000
Methyl alcohol (200)	49,100
Methyl *n*-butyl ketone (5)	Not recommended
Methyl Cellosolve (2-methoxyethanol) (5)	Not recommended
Methyl Cellosolve acetate (2-methoxyethyl acetate) (5)	Not recommended
Methyl chloroform (350)	11,390
Methyl ethyl ketone (200)	22,500
Methyl isobutyl ketone (50)	64,600
Methyl propyl ketone (200)	19,900
Naphtha (coal tar)	Requires special consideration
Naphtha VM & P (300)	Requires special consideration
Nitrobenzene (1)	Not recommended
n-Propyl acetate (200)	17,500
Stoddard solvent (100)	30,000–35,000
1,1,2,2-Tetrachloroethane (1)	Not recommended
Tetrachloroethylene (50)	79,200
Toluene (100)	38,000
Trichloroethylene (50)	90,000
Xylene (100)	33,000

Source: American Conference of Governmental Industrial Hygienists, Committee on Ventilation, *Industrial Ventilation—A Manual of Recommended Practice*, 20th ed., ACGIH, Lansing, MI, 1988, pp. 2–3.

[a]The following values are tabulated using the TLV values shown in parentheses, parts per million. TLV values are subject to revision if further research or experience indicates the need. If the TLV value has changed, the dilution air requirements must be recalculated. The values on the table must be multiplied by the evaporation rate (pts/min) to yield the effective ventilation rate (Q').

[b]See threshold limit values 1988–1989 in Appendix of manual.

[c]The tabulated dilution air quantities must be multiplied by the selected K value.

good-housekeeping program in conjunction with any broad-based engineering control such as general ventilation.

The routes by which the air will enter and leave the work space are important considerations in the design of general ventilation systems for a plant. The routes are of critical concern not only to the occupants of the building but potentially to the neighbors and the community at large. In general, the relative locations of air inlets and outlets should be considered in the implementation of a general-ventilation system.

Air Inlets. Sufficient combined or total-inlet area should be planned to accommodate the required volume of makeup air during the heating or cooling season, whichever needs the larger volume. This will prevent excessive inlet velocities that consume power, create drafts on workers, stir up dust, and interfere with performance of local-exhaust systems. The chosen inlet velocity is likely to represent an engineering compromise between low rates, which require large inlet areas, and higher rates, which facilitate rapid dilution of contaminants.

When widespread hazardous operations are controlled by dilution or general ventilation, inlets should be well distributed around the building to provide uniform circulation. If contaminants are controlled by localized dilution or by forceful diffusion into the general room-air reservoir, inlets may be purposely located to give nonuniform air-supply distribution.

Any thermal or convection effects within the building should be taken advantage of in location of air inlets. For this purpose the designer must avoid the location of inlets (and outlets) near the neutral zone of inside-outside pressure differentials, which is approximately midway between the floor and roof. Inlets should be located remote from stacks or ventilators discharging contaminants from the same or neighboring structures.

Air Outlets. Outlets should be located as far as possible from air inlets to prevent short-circuiting. This advice holds for both natural- and mechanical-ventilation systems.

If widely scattered operations are controlled by general dilution, outlets should be uniformly distributed around the building. Similarly, if contaminants are controlled by localized dilution, air outlets may be purposely located to short-circuit a corner of a large room, with the intent of creating a high rate of air change there without involving the atmosphere of the entire space. This method of localized-space ventilation requires a higher rate of exhaust air than supply air within the area to be controlled to prevent the spread of contaminants throughout the room.

Outlets, like inlets, should be placed to take full advantage of thermal effects. They should be protected against the direct force of prevailing winds, which reduce the capacity of any exhaust fans or destroy the anticipated airflow route planned on the basis of thermal effects inside the building.

Preferably, dilution air should enter the work space at approximate breathing zone height, pass through the workers' breathing zone, then through the zone of contamination; afterward, if it is not exhausted, it should enter a space of higher relative contamination. A useful concept in the interest of air-handling economy is progressive ventilation. The air removed from an industrial plant frequently may be directed in a way that will provide both local and general ventilation. In fact, the air may be routed through several areas in succession, always in the direction of increasing air contamination, as long as the exhaust from one area is acceptable as the supply for the next. Progressive ventilation saves energy in the form of airborne heat and horsepower to move the air.

3.3.4 *General Ventilation versus Local-Exhaust Ventilation*

Application of industrial-process ventilation has been somewhat evolutionary from the natural general ventilation achieved by design for hot, humid, and dusty operations to the highly effective local-exhaust hoods in current use in a variety of situations. Local ventilation prevents the spread of air contaminants throughout the building atmosphere with surprisingly small quantities of airflow in comparison with the volumes of air required by general ventilating systems. In spite of this characteristic efficient use of air in local-ventilation applications, there are highly persuasive reasons for the use of general ventilation in many industrial settings.

1. Simplicity and economy of natural-general ventilation
2. Relatively low first cost of mechanical-general ventilation
3. Absence of interference with manufacturing operations
4. Flexibility in plants with constantly changing layouts
5. Existence of contaminating processes throughout the entire plant
6. Desire for large volumes of air circulation in hot weather
7. Discovery that local-exhaust systems do not eliminate the necessity of supplying large volumes of heated air in the wintertime to replace that escaping by exfiltration from loosely constructed buildings. This volume of air change may be more than ample to control process air contaminants by the method of dilution, if means are provided to disperse contaminants from the immediate vicinity of the workers.

4 LOCAL EXHAUST VENTILATION

Local exhaust ventilation (LEV) is the concept of controlling a contaminant by capturing it at or near the place where it is generated and removing it from the work space. Local ventilation relies more heavily on mechanical methods of controlling airflow than does general ventilation. A local-exhaust system usually includes all of the following components: a hood or enclosure, ductwork, an air-cleaning device (where necessary for air pollution abatement purposes), and an air-moving device (usually an exhaust fan) to draw the contaminated air through the exhaust system and discharge it to the outside air. In general, local-exhaust systems consist of more individual components than do general-exhaust systems, and because they also offer more operational parameters that must be controlled within acceptable ranges, they require more maintenance and involve higher operating expenses as well.

When the primary purpose of the ventilation is to provide control of airborne contaminants, local-exhaust systems generally are superior to general ventilation. The advantages of the LEV system over general exhaust for any particular application will include many of the following:

1. Capture and control of a contaminant can be virtually complete with proper design of the system. Consequently, the exposure of workers to contaminants at the sources exhausted can be prevented. With general ventilation the contaminant is diluted when the exposure occurs, and at any given workplace this dilution may be highly variable and therefore inadequate at certain times.
2. Volumetric rates of required exhaust air are less with local ventilation; as a result, the volume of makeup air required is less. Local ventilation offers savings in both capital investment and heating costs.
3. The contaminant is contained in a smaller exhausted volume of air. Therefore if air pollution control is needed, it is less expensive because the cost of air pollution control is approximately proportional to the volume of air handled.
4. With particulate contaminants, local-exhaust systems can be designed to capture large, settleable particles or at least confine them within the hood, and thus greatly reduce the effort needed for good housekeeping.
5. Equipment and exposed surfaces in the workroom are better protected from such deleterious effects of the contaminant as corrosion and abrasion.
6. Fans used with local-exhaust systems usually have fairly high pressure characteristics to overcome pressure losses in the system. Therefore the performance of the fan system is not likely to be affected adversely by such influences as wind direction or velocity or inadequate-makeup air. This is in contrast to general ventilation, which can be affected greatly by seasonal factors.

4.1 Components of a LEV System

The four components of a simple local-exhaust system are illustrated in Figure 30.5: (1) a hood, (2) ductwork, (3) an air-cleaning device (cleaner), and (4) an air-moving device (fan). Typically, the system is a network of branch ducts connected to several hoods or

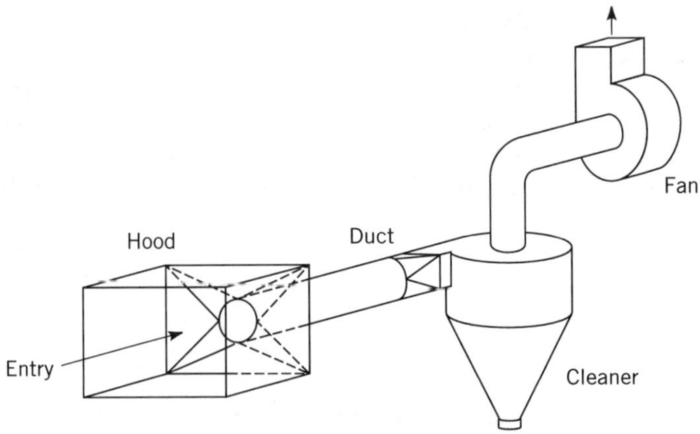

Figure 30.5. Interrelated components of a local exhaust system.

enclosures, main ducts, an air cleaner for separating the contaminants from the airstream, an exhaust fan, and a discharge stack to the outside atmosphere.

4.1.1 Hoods

A hood is a structure designed to partially enclose a contaminant-producing operation and to guide airflow in an efficient manner to capture the contaminant. The hood is connected to the ventilation system with ductwork that removes the contaminant from the hood. The design and location of the hood is one of the most critical aspects in the successful operation of a local exhaust system.

4.1.2 Ductwork

The ductwork in an exhaust system provides a path for flow of the contaminated air exhausted from the hood to the point of discharge. The following points are important in design of the ductwork:

1. Duct velocities must be high enough to prevent any dust in the air stream from settling out and plugging the ductwork.
2. When the air stream is relatively dust-free, the duct velocity should strike an economic balance between ductwork cost and fan, motor, and power costs.
3. Location and construction of the ductwork must furnish sufficient protection against external damage, corrosion, and erosion to maximize the useful life of the local-exhaust system.

4.1.3 Air Cleaner

Most exhaust systems installed for contaminant control need an air cleaner. Occasionally the collected material has some economic reuse value, but this is seldom the case. Collecting and disposing of the contaminant is usually inconvenient and certainly represents an added expense. Yet the growing concern with air-pollution control and the need to comply with legal restrictions on discharges from sources of atmospheric emissions place new importance on the air-cleaning device within a local-exhaust system.

4.1.4 Air-Moving Device

The fan, usually a centrifugal type, is the heart of the local-exhaust system. Wherever practicable, a fan should be placed downstream from the air cleaner so that it will handle uncontaminated air. In such an arrangement, the fan wheel can be the backward curved blade type, which has a relatively high efficiency and lower power cost. For equivalent air handling, the forward-curved blade impellers run at somewhat lower speeds, and this may be important when noise is a factor. If chips and other particulate matter have to pass through the impeller, the straight blade or paddlewheel-type fan is best because it is least likely to clog.

Fans and motors should be mounted on substantial platforms or bases and isolated by antivibration mounts. At the fan inlet and outlet, the main duct should attach through a vibration isolator, that is, a sleeve or band of very flexible material, such as rubber or fabric.

When the system has several branch connections, consideration should be given to using a belt drive instead of a direct-connected motor. Then if increased airflow is required at a future date, the need can be accommodated, to some degree, by adjusting the fan speed.

The importance and interaction of the various components of the exhaust system will become evident in the discussion on design of a ventilation system later in this chapter. Because of the importance of the role played by the exhaust hood in the overall success of the ventilation system, and because it is the component of the system that typically permits most innovation in concept, specific discussion of the application of the various types of hood follows.

4.2 Applications of Local-Exhaust Hoods

The hood is the point at which air first enters the LEV system. As such, the term can be used to apply to any opening in the exhaust system regardless of its shape or physical disposition. An open-ended section of ductwork, a canopy-type hood situated above a hot process, and a conventional laboratory booth-type hood all could be called hoods in the context of this discussion.

The hood captures the contaminant generated by a particular process or operation and causes it to be carried through the ductwork to a convenient discharge point. The quantity of air required to capture and convey the air contaminants depends on the size and shape of the hood, its position relative to the point of generation of the contaminant, and the nature and quantity of the air contaminant itself. It should be emphasized that there is not necessarily a standard hood that is correct for all applications of a particular operation because the methods of processing are unique to each operating plant. On the other hand, standard concepts for exhaust hooding have been developed and recommended for specific types of operations (2).

Hoods can be classified conveniently into four categories, based on the concept of contaminant capture/control: enclosures, booth-type hoods, receiving hoods, and exterior hoods.

4.2.1 Enclosure Hoods

Enclosure hoods normally surround the points of emission or generation of contaminant as completely as practicable. In essence, they surround the contaminant source to such a degree that all contaminant-dispersal action takes place within the confines of the hood itself. Because of this, enclosure hoods generally require the lowest rate of exhaust ventilation, and therefore are economical and quite efficient. They should be used whenever possible, and they deserve particular consideration when a moderately or highly toxic contaminant is involved.

4.2.2 Booth-Type Hoods

The common spray-painting enclosure and other booths are special cases of enclosure-type hoods. The common laboratory hood, in which one face of an otherwise complete enclosure is open, is typical of this application. Air contamination takes place within the enclosure, and air is exhausted from it in such a way, and at such a rate, that an average velocity is induced across the face of the opening sufficient to overcome the tendency of the contaminant to escape from within the hood. The three walls of the booth greatly reduce the exhaust-air requirements, although not to the extent of a complete enclosure.

4.2.3 Receiving Hoods

The term "receiving" refers to a hood in which a stream of contaminated air from a process is exhausted by a hood located near the source of generation of the contaminants specifically for purposes of control. Two examples of this type of hood are canopies situated above hot processes and hoods attached to grinders, positioned to take advantage of centrifugal and gravitational forces to maximize control of the dust generated by the process. Canopy hoods, frequently located above hot processes, are similar to exterior hoods in that the contaminated air originates beyond the physical boundaries of the hood. The fundamental difference between receiving and exterior hoods is that in the former, the hood takes advantage of the natural movement of the released contaminant, whereas in the latter, air is induced to move toward the exterior hood.

Receiving hoods are designed to be in the pathway of the contaminant as it is normally released by the operation. If space is limited, baffles or shields may be placed across the line of throw of the particles to remove their kinetic energy. Then the particles may be captured and carried into the hood by lower air velocities. Additional examples of receiving hoods include those associated with many hand-tool operations (surface grinders, metal polishers, stone cutters, and sanding machines).

4.2.4 Exterior Hoods

Exterior hoods must capture air contaminants generated from a point outside the hood itself, sometimes at quite a distance. Exterior hoods therefore differ from enclosure or receiving hoods in that their sphere of influence must extend beyond their own dimensions in capturing contaminants without the aid of natural forces such as natural drafts, buoyancy, and inertia. In other words, directional air currents must be established adjacent to the suction opening of the hood to provide adequate capture. Thus they are quite sensitive to external sources of air disturbance and may be rendered completely ineffective by even slight lateral movement of air. They also require the most air to control a given process and are the most difficult of the various hoods to design. Examples of exterior hoods include the exhaust slots on the edges of the tanks or surrounding a workbench such as a welding station, exhaust grilles in the floor or workbench below a contaminated process, and the common propeller-type exhaust fans frequently mounted in walls adjacent to a source of contamination.

4.3 Fundamental Concepts of LEV

4.3.1 Capture/Control Velocities

Local-exhaust hoods perform their function in one of two ways: capture or control. With the capture approach, air movement is created to draw the contaminant into the hood.

When the air velocity that accomplishes this objective is created at a point outside a nonenclosing hood, it is called capture velocity. Some exhaust hoods essentially enclose the contaminant source and create an air movement that prevents the contaminant from escaping from the enclosure. The air velocity created at the openings of such hoods is called the control velocity.

Successful design of any exhaust hood is based on correct determination of the two quantities, control velocity and capture velocity. The air velocity, which must be developed by the exhaust hood at the point or in the area of desired control, is based on the magnitude and direction of the air motion to be overcome and is not subject to direct and exact evaluation (Table 30.2). Many empirical ventilation standards, especially concerning dusty equipment such as screens and conveyor belt crossovers, are based on parameters such as "cubic feet per minute per foot of belt width." These so-called exhaust-rate standards are usually based on successful experience, are easily applied, and usually give satisfactory results if not extrapolated too far. In addition, they minimize the effort and uncertainty involved in calculating the fan action of falling material, thermal forces within hoods, and external-air currents. However, such standards have three major pitfalls:

1. They are not of a fundamental nature; that is, they do not follow directly from basic "laws."

Table 30.2. Range of Recommended Capture Velocities

Condition of Dispersion of Contaminant	Examples	Capture Velocity (fpm)
Released with practically no velocity into quiet air	Evaporation from tanks; degreasing, etc.	50–100
Released at low velocity into moderately still air	Spray booths; intermittent container filling; low speed conveyor transfers; welding; plating; pickling	100–200
Active generation into zone of rapid air motion	Spray painting in shallow booths; barrel filling; conveyor loading; crushers	200–500
Released at high initial velocity into zone of very rapid air motion	Grinding; abrasive blasting, tumbling	500–2000

Lower End of Range[a]	*Upper End of Range*[a]
Room air currents minimal or favorable to capture	Disturbing room air currents
Contaminants of low toxicity or of nuisance value only	Contaminants of high toxicity
Intermittent, low production	High production, heavy use
Large hood—large air mass in motion	Small blood—local control only

American Conference of Governmental Industrial Hygienists, Committee on Ventilation, *Industrial Ventilation—A Manual of Recommended Practice,* 20th ed., ACGIH, Lansing, MI, 1988, pp. 3–6.

[a]In each category, a range of capture velocity is shown. The proper choice of values depends on these several factors.

2. They presuppose a certain minimum quality of hood or enclosure design, although it may not be possible or practical to achieve the same quality of hood design in a new installation.
3. They are valid only for circumstances similar to those that led to their development and use. It should be clear that the nature of the process generating the contaminant will be an important determinant of the required capture velocity.

4.3.2 *Airflow Characteristics*

The flow characteristics at a suction opening are much different from the flow pattern at a supply or discharge opening. Air blown from an opening maintains its directional effect in a fashion similar to water squirting from a hose and, in fact, is so pronounced that it is often called "throw." However, if the flow of air through the same opening is changed so that the opening operates as an exhaust or intake with the same volumetric rate of airflow, the flow becomes almost completely nondirectional, and its range of influence is greatly reduced. As a first approximation, when air is blown from a small opening, the velocity 30 diameters in front of the plane of the opening is about 10 percent of the velocity at the discharge. The same reduction in velocity is achieved at a much smaller distance in the case of exhausted openings, such that the velocity equals 10 percent of the face velocity at a distance of only one diameter from the exhaust opening. (Figure 30.6 illustrates this point.) Therefore local-exhaust hoods must not be applied for any operation that cannot be conducted in the immediate vicinity of the hood.

4.3.3 *Airflow into Openings*

Airflow into round openings was studied extensively by Dalla Valle (8). His theory of airflow into openings is based on a point source of suction that draws air from all directions. The velocity at any point in front, distance *X*, of such a source is equivalent to the quantity

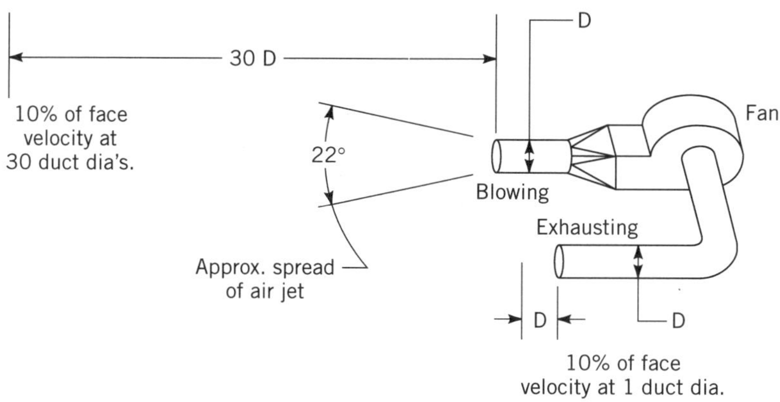

Figure 30.6. Airflow characteristics of blowing and exhausting.

of air Q flowing to the source divided by the effective area of the sphere of the same radius.

Conversely,

$$Q = VA$$
$$A = 4\pi X^2$$

Thus

$$Q = V(12.57X^2)$$

where Q = airflow (ft^3/min or m^3/sec)
V = velocity at point X (ft/min or m/sec)
X = centerline distance (ft or m)
A = duct area (ft^2 or m^2)
π = 3.14159 (dimensionless constant)

Taking the end of an open pipe to be approximated by a point source, Brandt (9) and Dalla Valle determined the actual velocity contours for a circular opening (Figure 30.7). These contours, or line of constant velocity, are best described by the following equation:

$$Q = V(10X^2 + A)$$

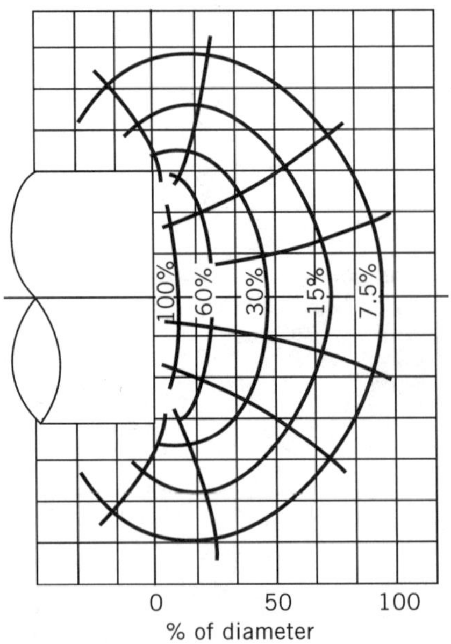

Figure 30.7. Velocity contours for unflanged circular opening.

4.3.4 Effects of Flanging

Flanges surrounding a hood opening force air to flow mostly from the zone directly in front of the hood. Thus the addition of a flange to an open duct or pipe improves the efficiency of the duct as a hood for a distance of about one diameter, as shown by the following equation:

$$Q = 0.75V(10X^2 + A)$$

For a flanged opening on a table or bench, the equation becomes

$$Q = 0.5V(10X^2 + A)$$

4.3.5 Slots

When the width-to-length ratio (aspect ratio) of an exhaust opening approaches 0.1, caution must be used in applying the generalized continuity equation because the opening behaves more like a slot. Using a line of reasoning similar to that of Dalla Valle, Silverman (10) considered the slot to be a line source of suction. Disregarding the end, the area of influence then approaches a cylinder and the velocity is given by

$$V = \frac{Q}{2\pi XL}$$

where L = length of slot (ft or m)
X = centerline distance (ft or m)
π = 3.14159 (dimensionless constant)

Correcting for empirical versus theoretical considerations, the design equation that best applies for freely suspended slots is

$$V = \frac{Q}{3.7XL}$$

Because flanging the slots will give the same benefits as flanging an open pipe, only 75 percent of the air is required to produce the same velocity at a given point. Therefore, for a flanged slot

$$V = \frac{Q}{2.8XL}$$

4.3.6 Air Distribution in Hoods

To provide efficient capture with a minimum expenditure of energy, the airflow across the face of a hood should be uniform throughout its cross section. For slots and lateral-exhaust applications, this can be accomplished by incorporating external baffles. Another method of design is to provide a velocity of 2000 to 2500 ft/min into the slot with a low velocity plenum or large-area chamber behind it. For large, shallow hoods, such as paint-spray booths, laboratory hoods, and draft-shakeout hoods, the same principle may be used. In these cases, unequal flow may occur, with resulting higher velocities near the takeoffs. Baffles provided for the hood improve the air distribution and reduce pressure drop in the hood, giving the plenum effect. When the face velocity over the whole hood is relatively high or the hood or booth is quite deep, baffles may not be required.

4.3.7 Entrance Losses in Hoods

Hood static pressure, SP_{h}, is the term used to refer to the negative static pressure exhibited in ductwork a short distance downstream from the hood. It represents the energy needed to accelerate the air from ambient velocity (often near zero) to the duct velocity and also to overcome the frictional losses resulting from turbulence of the air upon entering the hood and ductwork. Therefore,

$$SP_{\mathrm{h}} = VP + h_{\mathrm{e}}$$

where VP = velocity pressure in the duct
h_{e} = hood-entry loss

The hood entry loss, h_{e}, is expressed as a function of the velocity pressure VP. For hoods of most types, $h_{\mathrm{e}} = F_{\mathrm{h}}VP$, where F_{h} is the hood-entry-loss factor. For plain hoods, where the hood-entry loss is a single expression $F_{\mathrm{h}}VP$, the new VP referred to is the duct velocity pressure. The hood static pressure can be expressed as

$$SP_{\mathrm{h}} = VP_{\mathrm{duct}} + h_{\mathrm{e}}$$

or

$$SP_{\mathrm{h}} = VP_{\mathrm{duct}} + F_{\mathrm{h}}VP_{\mathrm{duct}} = (1 + F_{\mathrm{h}})VP_{\mathrm{duct}}$$

However, for slot, plenum, or compound hoods there are two entry losses, one through the slot and the other into the duct. Thus

$$\begin{aligned} SP_{\mathrm{h}} &= h_{\mathrm{e(slot)}} + VP_{\mathrm{duct}} + h_{\mathrm{e(duct)}} \\ &= F_{\mathrm{slot}}VP_{\mathrm{slot}} + VP_{\mathrm{duct}} + F_{\mathrm{duct}}VP_{\mathrm{duct}} \end{aligned}$$

The velocity pressure resulting from acceleration through the slot is not lost as long as the slot velocity is less than the duct velocity, and usually this is the case.

Another constant used to define the performance of a hood is the coefficient of entry, C_{e}. This is defined as the ratio of the actual airflow to the flow that would exist if all the static pressure were present as velocity pressure. Thus

$$C_e = \frac{Q_{actual}}{Q_{VP} = SP_h} = \frac{KA(VP)^{0.5}}{KA(SP_h)^{0.5}} = \left(\frac{VP}{SP_h}\right)^{0.5}$$

This quantity is constant for a given shape of hood and is very useful for determining the flow into a hood by a single-hood static-pressure reading. The coefficient of entry, C_e, is related to the hood entry loss factor, F_h, by the following equation only where the hood-entry loss is a single expression:

$$C_e = \left(\frac{1}{1 + F_h}\right)^{0.5}$$

Listings of the entry-loss coefficients, C_e and the entry losses, h_e, in terms of velocity pressure, *VP*, are available (11–14). Most of the more complicated hoods have coefficients that can be obtained by combining some of the more simple shapes illustrated.

4.3.8 Static Suction

The air volume for a hood can be specified by giving the hood-static pressure, SP_h, and duct size. For example, the hood-static pressure at a typical grinding wheel hood is 2 in. of water. This reflects a conveying velocity of 4500 ft/min and entrance coefficient, C_e, of 0.78. For other types of machinery where the type of exhaust hood is relatively standard, a specification of the static suction and the duct size can be found in various reference sources (2–4). Specification of the static suction without duct size is of course meaningless, because decreased size increases velocity pressure and static suction while actually decreasing the total flow and the degree of control. Therefore static-suction measurements for standard hoods or for systems in which the airflow has been measured previously are quite useful to estimate, in a comparative way, the quantity of air flowing through the hood.

4.3.9 Transport Velocities

In order to prevent particles of dust and fumes from settling and plugging ductwork, air velocities for transporting these materials must be relatively high. The minimum velocity, called transport velocity, typically is 3500 to 4000 linear feet per minute. At these velocities, frictional loss from air moving along the surface of the ducts becomes significant; therefore, all fittings, such as elbows and branches, must be wide-swept and gradual, having smooth interior surfaces. The cross-sectional area of the main duct generally equals the sum of the areas of cross sections for all branches upstream, plus a safety factor of approximately 20 percent. When the main duct is enlarged to accommodate an additional branch, the connection should be tapered, not abrupt.

Local-exhaust systems for gases and vapors may have lower duct velocities (1500 to 2500 ft/min) because there is little to settle and plug the ducts. Lower velocities markedly reduce the frictional and pressure losses against which the fan must operate, thereby realizing a saving in power cost for the same airflow.

4.4 Principles of Local-Exhaust Ventilation

Control of contaminants by application of local exhaust ventilation to specific problems is more effective if the following basic principles are followed.

4.4.1 Enclosure of the Source

Any operation to be exhausted by local ventilation should be enclosed as much as possible. This generally provides better control per unit volume of air exhausted. Nevertheless, the requirement of adequate access to the process must always be considered. An enclosed process may be costly in terms of operating efficiency or capital expenditure, but the savings gained by exhausting smaller air volumes may make the enclosure worthwhile.

4.4.2 Capture of Contaminants

The velocity of air through all openings must be high enough not only to contain the contaminant but also to remove the contaminant from the hood. The importance of optimal capture and control velocity was discussed in the preceding sections.

4.4.3 Keeping Contaminants Out of Breathing Zone

Hoods that do not completely enclose the process should be located as near to the point of contaminant generation as possible. Airflow should be provided in a direction away from the worker toward the contaminant source. This item, drawing upon the characteristics of blowing and exhausting from openings in ductwork, was considered in more detail in the preceding sections.

4.4.4 Providing Makeup Air

In order to keep the building from operating under negative pressure, all air exhausted from a building or enclosure must be replaced. This applies to local-exhaust systems as well as general-exhaust systems. Additionally, the incoming air must be tempered by a makeup-air system before being distributed inside the process area. Without sufficient makeup air, exhaust-ventilation systems cannot work as efficiently as intended. Makeup-air requirements are discussed subsequently.

4.4.5 Discharge of Exhaust Air

Air exhausted from a system must be discharged in such a way as to prevent its reentry into the work space. It is disconcerting to observe the frequency with which this principle is violated. The beneficial effect of a well-designed local-exhaust system can be offset by undesired recirculation of contaminated air back into the work area. Such recirculation can occur if the exhausted air is not discharged away from supply air inlets. The location of the exhaust stack, its height, and the type of stack-weather cap can all have a significant effect on the likelihood of contaminated air reentering through nearby windows and supply-air intakes.

4.5 Low Volume High Velocity Exhaust Systems

A unique application of the fundamentals of the LEV fundamentals discussed in this chapter have been applied in a unique way in what is referred to as low volume high velocity (LVHV) exhaust ventilation. With typical application of this approach, small volumes of air at relatively high velocities are used to control dust generated by portable hand tools and machining operations. Control is achieved by exhausting the air essentially at the point of generation by means of close-fitting, custom-designed hoods. To maximize the flexibility required by operators of the equipment, flexible, small-diameter, lightweight plastic hoses are used, being built into the overall design of the tool. Figure 30.8 illustrates a typical application for control of dust generated by a disk sander. The capture velocities in these situations are high; slot velocities typically are in excess of 20,000 linear ft/min. However, because of the short distances required for capture of the dust and the small-diameter tubing (ductwork) associated with the system, the total-airflow rate is low.

Although this technique does not have widespread applicability, there are situations for which it has optimized control of dust. Successful use of LVHV exhaust ventilation has been made in such diverse activities as rock drilling, conventional machining, machining of beryllium, wood sanding, metal-surface grinding, pneumatic chipping, internal-surface grinding, and chiseling. Development of these exhaust systems has been somewhat empirical. However, it is emphasized that proper design of such systems requires the same careful, deliberate consideration of ventilation principles as more conventional-exhaust systems, which is the topic of the following section.

5 DESIGN OF VENTILATION SYSTEMS

Over the span of a few decades, the design of ventilation systems has advanced from an art to a true science of engineering design. Many of the empirical guidelines or rules of thumb are still appropriately applied through continuing experimentation and satisfactory experience. These rules and guidelines have developed into special forms and charts published for the direct purpose of facilitating the design of ventilation systems.

In the industrial workplace there are two basic types of ventilation system: general and local exhaust. These concepts have been described earlier in this chapter. It is noted here that the primary concern of the industrial hygienist in the design of ventilation systems is to provide a safe and healthful environment for the worker. Although hygienists usually deal mainly with local-exhaust-ventilation considerations, general ventilation is also an integral part of their concern.

5.1 General Ventilation

As described earlier, general ventilation refers to the general flushing of a work environment with a supply of fresh air to maintain desirable temperature and humidity conditions or to dilute a contaminant or contaminants to acceptable levels.

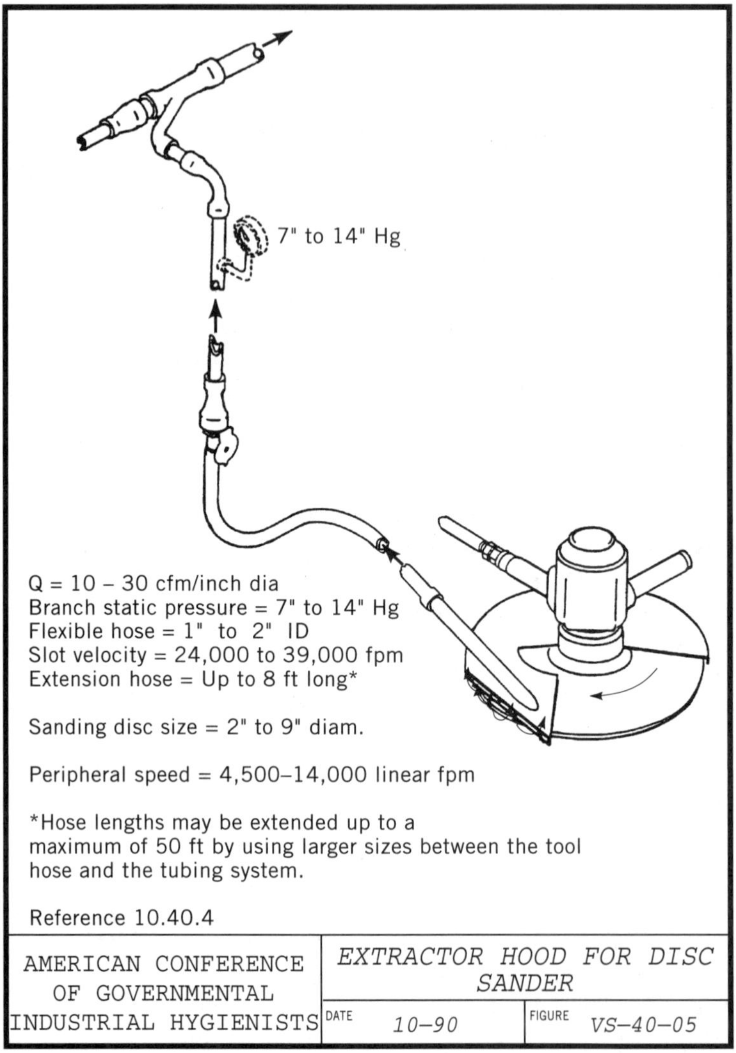

Figure 30.8. ACGIH-recommended arrangement for low volume high velocity hood.

5.1.1 Office Buildings

In nonproduction areas such as office buildings, the principal concern is worker comfort, as contrasted with general ventilation considerations for production areas where the focus is on maintaining contaminant levels at acceptable limits. Accordingly, it is necessary to consider the conditions that may affect worker comfort: air temperature, humidity, radiant-heat sources, concentrations of tobacco smoke or other potentially offensive odors, the possibility of body odors, and general airflow patterns desirable in the office space (13).

Various attempts have been made to develop indexes that describe a comfort zone, that is, a definition of a combination of parameters (usually temperature, humidity, and air velocity) that results in a comfortable environment. The comfort zone defined in the effective temperature charts (Figure 30.2) is one of the most frequently used indexes for this purpose. These comfort conditions can be met by proper selection of air-conditioning equipment for office spaces. A more pressing modern-day problem is that referred to commonly as tight building syndrome, resulting usually from insufficient supplies of fresh air to the interior of modern airtight buildings of masonry and glass. This condition of stale air as a unique industrial-hygiene problem is treated elsewhere in this series under general treatment of indoor-air quality. Generally the problem of odors is affected by the air supply, the space provided per occupant, and the odor-absorbing capacity of the air-conditioning process, as well as the temperature and humidity of the air.

In summary, the design of general office space ventilation requires that an analysis be made of the worker-comfort parameters (primarily temperature and humidity), the tolerable level of odor, the equivalent space allowed per person in the room, and the operating characteristics of the air conditioning unit to be installed. Table 30.3 summarizes the requirements for ventilation under a variety of conditions typical of office-type space.

5.1.2 Industrial Process Buildings

General-ventilation systems typically employed in process operations can be natural or mechanically controlled. As discussed earlier in this chapter, the two forces available for use as natural ventilation are wind pressure and thermal or convective currents produced within the building. Although many of the older production facilities were designed with features such as sawtooth or monitor-type roofs to provide maximum ventilation, these applications are becoming less common, particularly in light of the relatively straightforward application of mechanically controlled ventilation systems. In some industrial buildings such as warehouses and pump rooms, natural ventilation is sufficient to provide adequate movement of air, particularly because the number of people normally employed in

Table 30.3. Recommended Quantities of Outdoor Air Supplied to Occupied Space

Outdoor Air (ft^3/min per person)	Type of Space or Occupancy[a]
5–10	High-ceiling space such as bank, auditorium, church, department store, theater; room with no smoking
10–15	Apartment, barber shop, beauty parlor, hotel room; room with light smoking
15–20	Cafeteria, drug store with lunch counter, general office, hospital room, public dining room, restaurant; room with moderate smoking
20–30	Broker's boardroom, private office, tavern; room with heavy smoking
30–60	Conference room, night club; crowded room with heavy smoking

[a]Air-cleaning or odor-removing devices not used. (Space not less than 150 ft^3 per person or floor area not less than 15 ft^2 per person.) State and city codes should be consulted to make certain that minimum outdoor air requirements are followed.

such situations is limited. Despite the economic incentive associated with natural ventilation, however, it is somewhat restricted in its application, whereas ventilation by mechanically controlled devices is virtually without limit, particularly when used in conjunction with ductwork to distribute the air strategically throughout the space. A primary reason for applying general ventilation within a process area is to secure the dilution of the contaminant level to acceptable limits. Although this is accompanied by benefits, usually in terms of worker comfort, the chief concern is generally the elimination of a potential health hazard. In that respect, dilution ventilation has application chiefly for vapors or gases, and only under unusual circumstances is it appropriate to consider general ventilation as a control means when the contaminant of concern is a particulate material. In these instances, application of local-exhaust ventilation would be more appropriate. Table 30.1 summarizes the air volumes necessary to dilute the concentrations of specifically identified vapors to acceptable limits. These values, used in conjunction with the considerations presented earlier in this chapter concerning the airflow patterns created by the general-ventilation system, form the basis for the design of dilution-ventilation systems.

5.2 Design of Local-Exhaust Ventilation

When enclosure or other form of isolation or substitution is not appropriate or has been used to the practicable extent, the industrial-hygiene engineer looks to LEV as the primary-control measure. With proper application of LEV, the potential workroom contaminant is controlled at or near the source of generation or release of the substance. It should be kept in mind that local-exhaust systems are not necessarily designed to be 100 percent efficient in capturing and controlling contaminants released by the process. It is because of this residual or background level of contamination throughout the workroom that the general-ventilation provisions are included as a complementary aspect of the total ventilation within a plant.

Many approaches to the overall design of the LEV system are available to the engineer. However, they all consist of a series of similar steps. The following section presents the sequential procedure to be used in the design of local-exhaust systems. For illustrative purposes, an example of the complete procedure of the design of a local-exhaust system follows.

1. Prepare a sketch or layout of the workroom and operations within it. This will facilitate preparation of a sketch and/or layout of the contemplated exhaust system, using single-line drawings for ductwork. Changes in elevation need not be shown at this point. Although it is helpful if the layout is approximately to scale, subsequent steps in the procedure, such as calculating proper duct sizes, can be accomplished from even a very simple drawing. For purposes of illustration, consider the design of a local-exhaust system for a manual grinding station and a welding operation within the same general work area. The concept for the ventilation system to provide control of contaminants at these operations is illustrated in Figure 30.9.

2. Make a rough design or sketch of the desired hood for each operation, indicating direction and elevation of the outlet for the duct connections. The type of hood should be selected with the idea of performing the job with the least amount of air. This requires

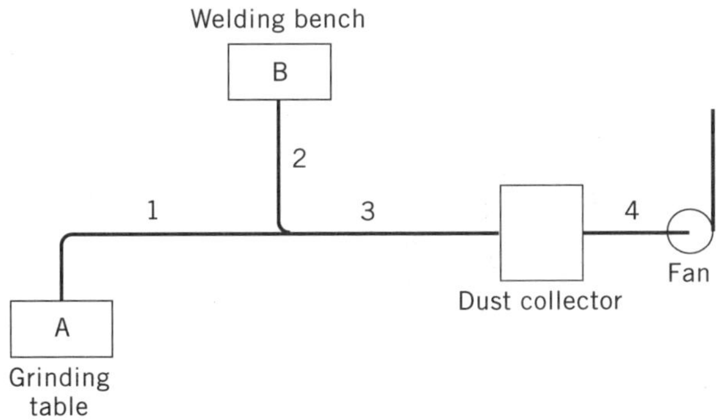

Figure 30.9. Simple sketch of local exhaust system.

some ingenuity in designing enclosures that will not interfere with production; they may be equipped with movable sections. There may not be a standard hood available for a particular application. For convenience, the more-or-less standard designs of local-exhaust hood for portable hand grinding and bench welding operations are used in this illustration. Figure 30.10 and Figure 30.11, respectively, depict the configuration and design parameters for these hood arrangements (2).

3. Estimate the required control or capture velocity for each operation and type of hood or enclosure considered. This part of the process involves some judgment, particularly with hoods of nonstandard design. For this illustration, the information presented in Figure 30.10 and Figure 30.11 is used in making preliminary estimates of appropriate velocities and volumetric-flow rates. Accordingly, an airflow rate of 200 ft^3/min per square foot (cfm/ft^2) of bench area for the grinding table and a slot velocity of 1000 (linear) ft/min for the welding bench are selected.

4. Estimate or compute the sphere of influence over which the selected capture velocities must function. It will be advantageous to consider the hood arrangements in three dimensions. In the illustrated problem, assume that work at both the grinding table and the welding bench will be conducted on a flat surface 2 $\times$ 4 ft and that most of the work will be conducted within 2 in. of the table surface.

5. Compute the total volumetric airflow needed for each hood and indicate this on the working sketch of the system or in an appropriate work sheet. From the information supplied in Figure 30.10 and Figure 30.11, the required volumetric airflows for the two hoods can be calculated. For this example, a volumetric-flow rate of 200 ft^3/min per square foot of work surface and a bench area of 8 ft^2 have been assumed. Thus a volumetric-flow rate of 1600 ft^3/min will be required for hood A, the grinding table. For the welding bench, the required volumetric rate will be 350 ft^3/min per linear foot of hood times 4 ft (the length of the hood), or a total of 1400 ft^3/min.

6. Summarize the volumetric airflow requirements for each section of the branch and main ductwork on the work sheet, or enter the air flows directly on the working sketch of the system. For this simple illustration, the summation of volumetric-rate requirements is straightforward:

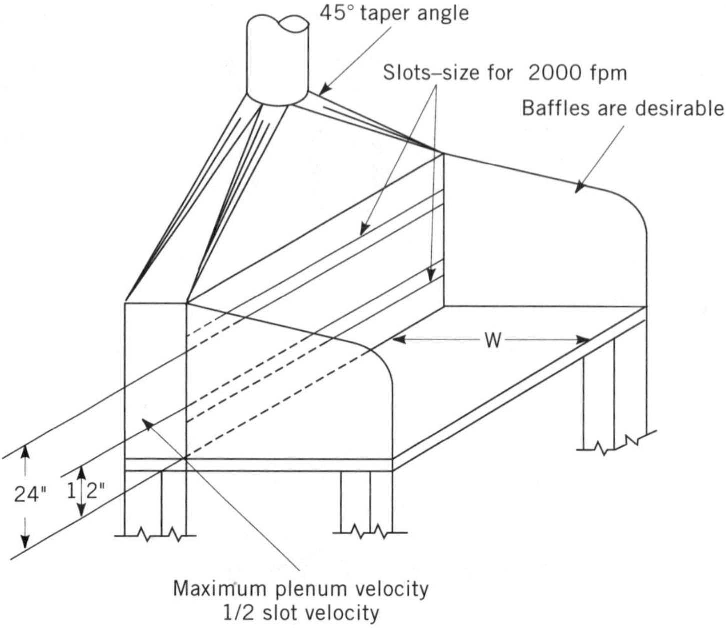

Q = 350 cfm/ft of hood length
Hood length = required working space
W = 24" maximum, if W> 24" see chapter 3
Minimum duct velocity = 2000 fpm
$h_e = 1.78\ VP_s + 0.25\ VP_d$

General ventilation, where local exhaust cannot be used:

Rod, diam	Cfm/welder
5/32	1000
3/16	1500
1/4	3500
3/8	4500

or

A. Far open areas. where welding fume can rise away from the breathing zone: cfm required = 800 × lb/hour rod used

B. For enclosed areas or positions where fume does not readily escape breathing zone: cfm required = 1600 × lb/hour rod used

For toxic materials higher airflows are necessary and operator may require respiratory protection equipment equipment.

Other types of hoods
Local exhaust: See VS–90–02
Booth: For design see VS–90–30
Q = 100 cfm/ft^2 of face opening
MIG welding may require precise air flow control

AMERICAN CONFERENCE OF GOVERNMENTAL INDUSTRIAL HYGIENISTS	*WELDING VENTILATION BENCH HOOD*		
	DATE *1–91*		FIGURE *VS–90–01*

Figure 30.10. ACGIH-recommended hood for portable hand grinding.

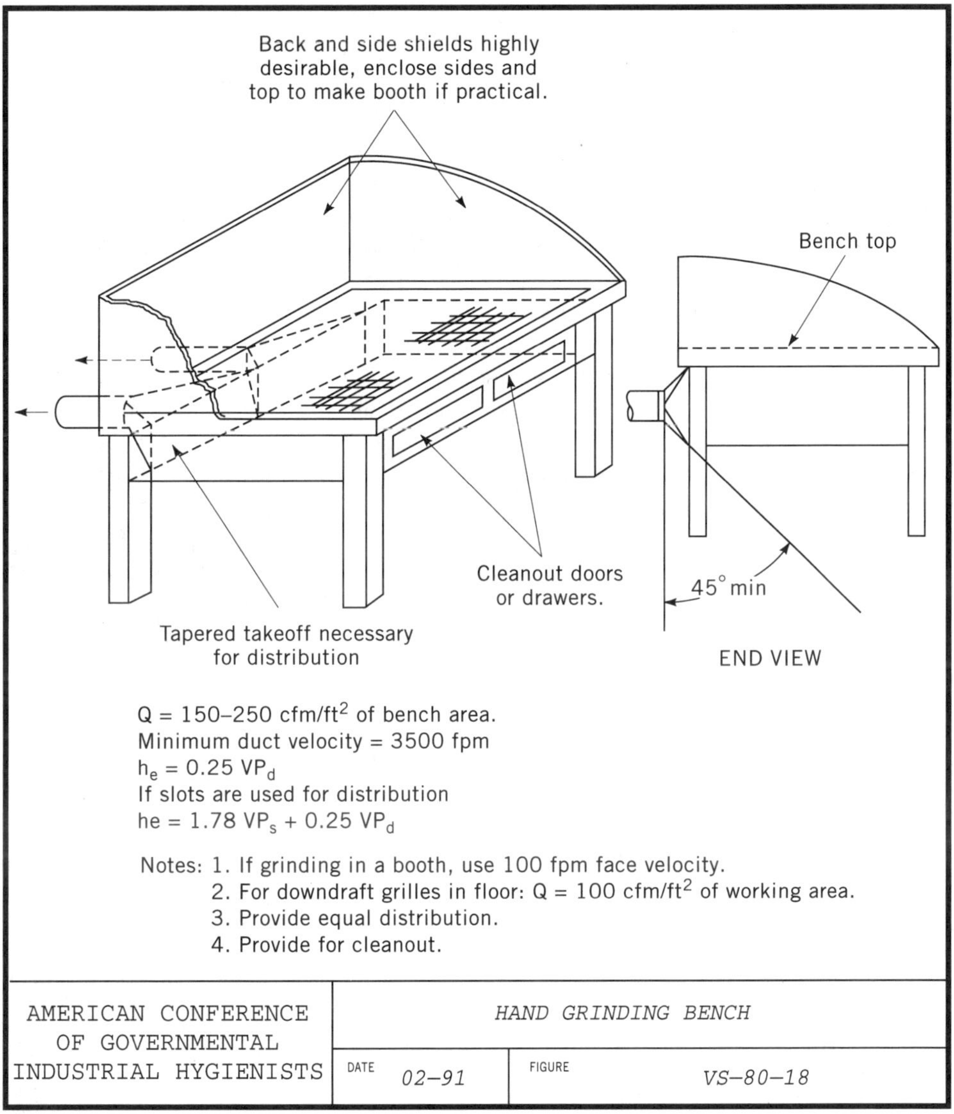

Figure 30.11. ACGIH-recommended hood for welding bench.

Duct section 1	1600 ft^3/min
Duct section 2	1400 ft^3/min
Duct sections 3, 4, and 5	3000 ft^3/min

7. Select the minimum effective-transport velocities for the contaminant-laden air being moved through the ductwork. For transporting particulate matter, data such as those in

Table 30.4 can be used. Velocities in the range of 1000 to 2000 linear ft/min commonly are used for gases and vapors, although in these cases duct velocities can be as low as is consistent with space and weight limitations. The recommended transport velocities presented in Table 30.4 indicate values of 4000 and 3500 linear ft/min for ductwork servicing the grinding table and welding bench hoods, respectively.

8. Determine the nearest practical duct diameter necessary to carry the required volumetric rate of air at the proper velocity. It is often necessary at this juncture to consider the standard sizes of ducts available from the fabricator to avoid the unnecessary expense of special fabrication. If the available duct sizes differ greatly from the theoretically required diameters, the actual transport velocity should be computed for use in subsequent design steps. In the example problem, dividing the volumetric-flow rates by the desired transport velocities, and assuming that standard ductwork is available in full-inch diameter increments only, the duct diameter nearest to the calculated necessary diameters is 8 in. for both the grinding table and the welding bench. Using 8-in.-diameter ductwork for these hoods would produce transport velocities of 4580 and 4010 linear ft/min for the grinding hood and welding hood, respectively.

9. Compute the velocity pressures corresponding to air velocities in each branch duct connected to an exhaust opening. Velocity pressures are calculated using the formula

10. Estimate the entry loss for each hood in percentage of velocity pressure in the branch duct. In the example problem, these relationships are given in Figure 30.10 and Figure 30.11.

11. Compute the entrance losses for all hoods, using information developed in items 9 and 10. Using the relationships presented in Figure 30.10 and Figure 30.11, the entrance losses for the grinding table hood and the welding bench hood can be calculated to be 0.327 and 0.357 in H_2O, respectively.

12. Compute the losses in elbows, the branch-to-main connections, or other transition points in the duct system. This can be done in terms of either equivalent straight-duct lengths or velocity pressures, whichever form corresponds to the data being used. The equivalent straight-duct method has the advantage that such transition losses can be added directly to the straight-duct sections for use with friction charts. The velocity-head method may be preferred by those who are especially familiar with the fundamentals of fluid mechanics. For this sample problem, refer to Figure 30.12 for the relative arrangement of equipment in the exhaust system. In computing friction and transition losses, assume that branches 1 and 2 contain two 90° elbows, that the centerline radius of the elbow turns is twice the duct diameter, that branch 2 merges with duct-section 3 on a 45° angle of entry, and that duct section 4 contains one 90° elbow also with a centerline radius of two diameters. On the basis of these assumptions, and using information such as that presented in Table 30.5 it can be shown that each of the elbows in duct sections 1 and 2 is equivalent to 10 ft of straight duct, that the elbow in duct section 4 is equivalent to 17 ft of straight duct, and that the transition point from duct section 2 to 3 is equivalent to 11 ft of straight duct.

13. Obtain the duct-friction losses from an appropriate chart, table, or formula. A friction chart like the one illustrated in Figure 30.13 can be used at this point. For the various

Table 30.4. Typical Recommended Transport Velocities

Material, Operation, or Industry	Minimum Transport Velocity (ft/min)
Abrasive blasting	3500–4000
Aluminum dust, coarse	4000
Asbestos carding	3000
Bakelite molding powder dust	2500
Barrel filling or dumping	3500–4000
Belt conveyors	3500
Bins and hoppers	3500
Brass turnings	4000
Bucket elevators	3500
Buffing and polishing	
Dry	3000–3500
Sticky	3500–4500
Cast iron boring dust	4000
Ceramics, general	
Glaze spraying	2500
Brushing	3500
Fettling	3500
Dry pan mixing	3500
Dry press	3500
Sagger filling	3500
Clay dust	3500
Coal (powdered) dust	4000
Cocoa dust	3000
Cork (ground) dust	2500
Cotton dust	3000
Crushers	3000 or higher
Flour dust	2500
Foundry, general	3500
Sand mixer	3500–4000
Shakeout	3500–4000
Swing grinding booth exhaust	3000
Tumbling mills	4000–5000
Grain dust	2500–3000
Grinding, general	3500–4500
Portable hand grinding	3500
Jute	
Dust	2500–3000
Lint	3000
Dust shaker waste	3200
Pickerstock	3000
Lead dust	4000
With small chips	5000
Leather dust	3500
Limestone dust	3500
Lint	2000

Table 30.4. (continued)

Material, Operation, or Industry	Minimum Transport Velocity (ft/min)
Magnesium dust, coarse	4000
Metal turnings	4000–5000
Packaging, weighing, etc.	3000
Downdraft grille	3500
Pharmaceutical coating pans	3000
Plastics dust (buffing)	3800
Plating	2000
Rubber dust	
Fine	2500
Coarse	4000
Screens	
Cylindrical	3500
Flat deck	3500
Silica dust	3500–4500
Soap dust	3000
Soapston dust	3500
Soldering and tinning	2500
Spray painting	2000
Starch dust	3000
Stone cutting and finishing	3500
Tobacco dust	3500
Woodworking	
Wood flour, light dry sawdust and shavings	2500
Heavy shavings, damp sawdust	3500
Heavy wood chips, waste, gren shavings	4000
Hog waste	3000
Wool	3000
Zinc oxide fume	2000

sections of ductwork associated with the example problem, the following friction losses can be derived by referring to Figure 30.13 or a similar chart:

Duct section 1	4.0 in. H_2O/100 ft of duct
Duct section 2	2.9 in. H_2O/100 ft of duct
Duct sections 3, 4, 5	1.0 in. H_2O/100 ft of duct

14. Summarize the losses for each branch to determine where to increase or decrease duct sizes to ensure that a balance of pressure drop will be maintained. The total-pressure drop from any points in the system to each upstream opening or exhaust point should be the same if balance is to be achieved. In actual operation, the airflow quantities through all open branches automatically adjust themselves to fulfill this requirement of balance. If the flow through any branch is too high, further closing of the damper increases the friction or energy loss through that branch, thus making it possible for a smaller volume of airflow to create the necessary pressure drop for a balanced condition. In summing the entry losses

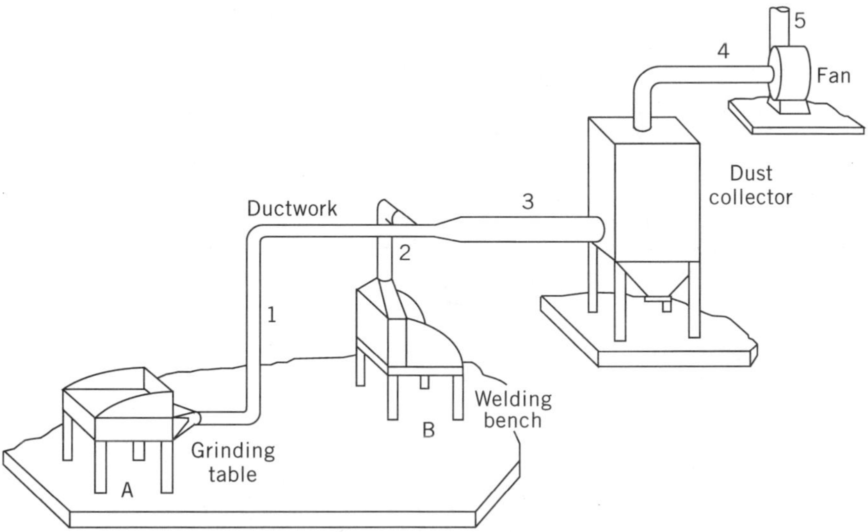

Figure 30.12. Representation of ventilation system.

for the hoods and the friction losses in the ductwork for duct sections 1 and 2, it can be shown that these two sections are balanced by virtue of the total-pressure drop at the common point (where duct section 2 merges with the main duct) being the same, namely, 1.60 in. H_2O.

15. After the system has been balanced on paper to a reasonable degree by adjusting round-duct sizes, any necessary conversion to rectangular-duct sizes can be made using appropriate conversion charts. In the sample problem, this step is not applicable.

16. Determine the adjusted total-pressure drop of the branch and main duct system by adding the pressure losses, beginning with any one exhaust opening and proceeding toward the fan or dust collector. If the branch-duct system has been reasonably balanced (in accordance with item "n"), it will be a simple matter to select the starting point for totaling the pressure losses. However, if no balancing has been attempted, the losses must be estimated for the single run of duct that is believed to have the greatest total resistance. Experienced designers, after a few moments of inspection, usually can determine the longest equivalent run in terms of friction loss—that is, the course that offers the greatest resistance to airflow, not necessarily the longest one in terms of physical length. In summing the total-pressure drop of the branch- and main-duct system, it can be calculated that the total-pressure drop up to the point of entry into the dust collector is 1.9 in. H_2O.

17. Add the resistance of accessory equipment such as air cleaners and weather caps. For the most part, this information should be obtained directly from the manufacturers, because equipment varies greatly in resistance to airflow. For this example, assume that the pressure drop across the dust collector is 2.0 in. H_2O.

18. Select a fan of proper type, size, and speed to produce the highest operating efficiency, or lowest horsepower, consistent with other considerations. Use the calculated total

Table 30.5. Equivalent Resistance of Elbows and Duct Connections

EQUIVALENT RESISTANCE IN FEET OF STRAIGHT PIPE

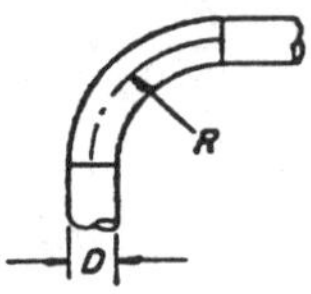

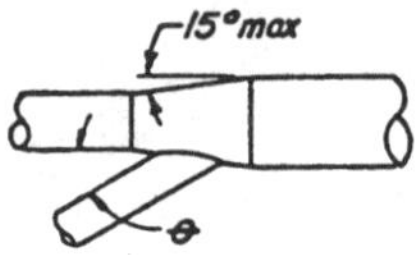

See Fig. 6-24

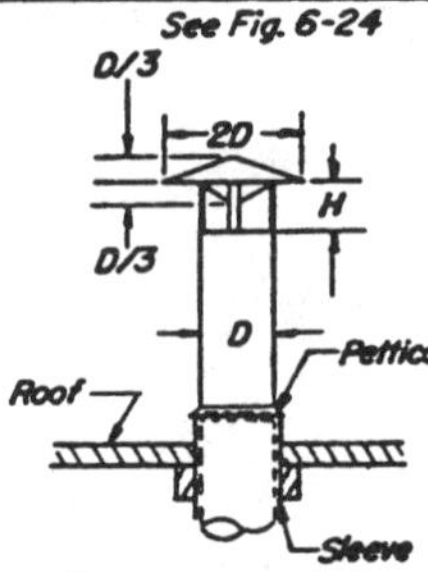

Not recommended

Pipe D	90° Elbow * Centerline Radius			Angle of Entry		H, No of Diameters		
	1.5 D	2.0 D	2.5 D	30°	45°	1. D	.75 D	.5 D
3"	5	3	3	2	3	2	2	9
4"	6	4	4	3	5	2	3	12
5"	9	6	5	4	6	2	4	16
6"	12	7	6	5	7	3	5	20
7"	13	9	7	6	9	3	6	23
8"	15	10	8	7	11	4	7	26
10"	20	14	11	9	14	5	9	36
12"	25	17	14	11	17	6	11	44
14"	30	21	17	13	21	7	13	53
16"	36	24	20	16	25	9	15	62
18"	41	28	23	18	28	10	18	71
20"	46	32	26	20	32	11	20	80
24"	57	40	32			13	24	92
30"	74	51	41			17	31	126
36"	93	64	52			22	39	159
40"	105	72	59					
48"	130	89	73					

*For 60° elbows — 0.67 x loss for 90°
45° elbows — 0.5 x loss for 90°
30° elbows — 0.33 x loss for 90°

AMERICAN CONFERENCE OF GOVERNMENTAL INDUSTRIAL HYGIENISTS

DUCT DESIGN DATA

DATE 1-76 | Fig. 6-11

[a]For 60° elbows: 0.67 × loss for 90°
45° elbows: 0.5 × loss for 90°
30° elbows: 0.33 × loss for 90°

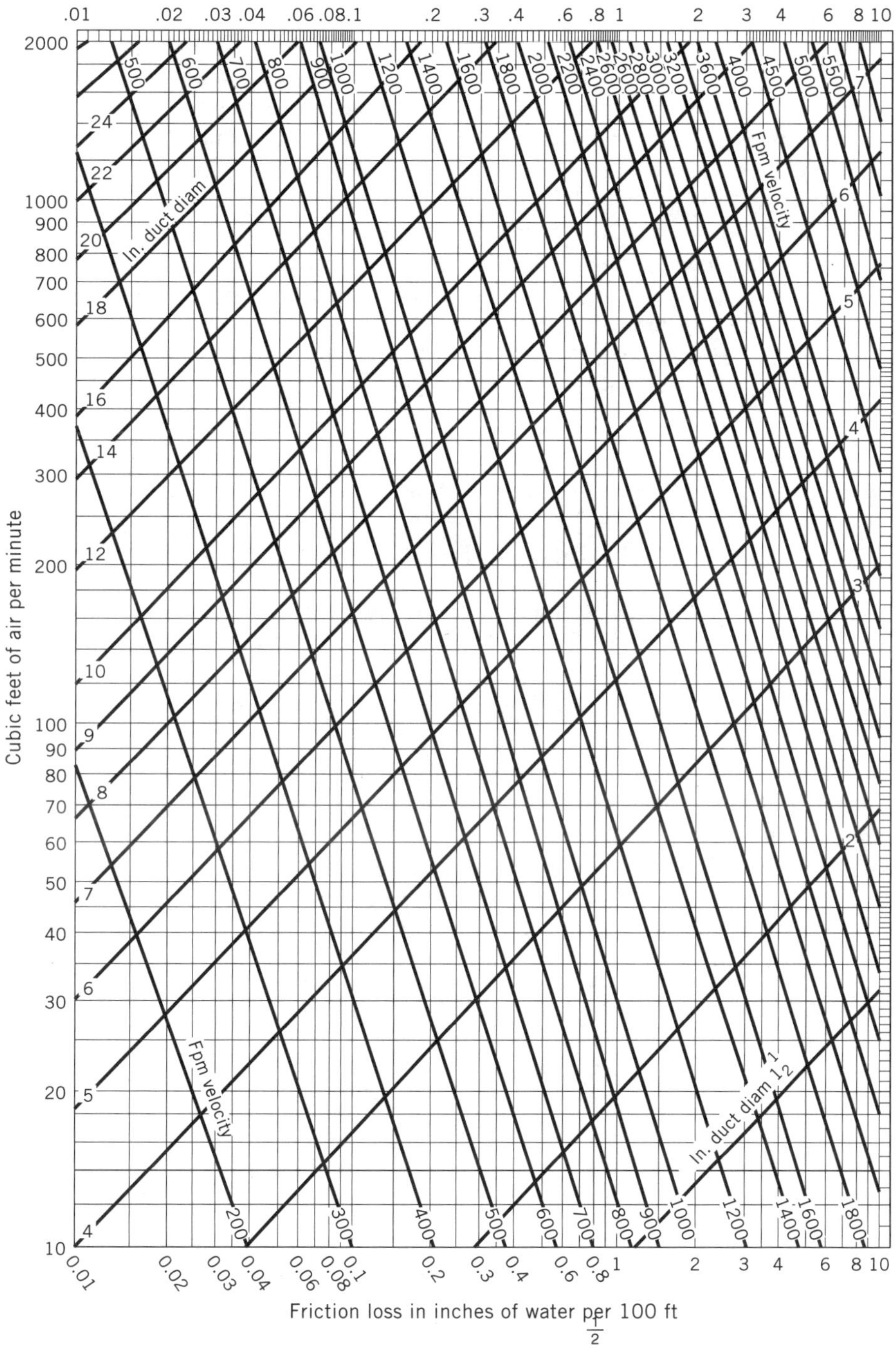

Figure 30.13. Friction of air in straight ducts.

airflow in cubic feet per minute and the total-system resistances or static pressure for fan-selection purposes. In selecting a fan, it is necessary to be familiar with the fan laws and system curves, that is, how the static pressure and volumetric airflow are related. A fan curve or system curve shows graphically the possible combinations of volumetric flow and static pressure for a given application. Because the fan and system each can operate only at a single point on its own curve, the combination can operate only where these curves intersect (Figure 30.14). Thus a fan must be chosen on the basis of its characteristics and the requirements of the system in which it will be applied. Each fan is characterized by the volume of gas flow, the pressure at which the flow is produced, the speed of rotation, the power required, and the efficiency. These quantities are measured by manufacturers with standard testing methods, and the results are plotted to furnish the characteristic "fan curves" of most manufacturers. Tables such as Table 30.6 are available to assist the design engineer in selecting a fan. The fan static pressure is calculated using the following equation:

$$
\begin{aligned}
\text{fan } SP &= SP_{(\text{fan inlet})} + SP_{(\text{fan outlet})} - VP_{(\text{fan inlet})} \\
&= 4.20 + 0.10 - 0.91 \\
&= 3.39 \text{ in. } H_2O
\end{aligned}
$$

Thus the fan and motor selected for this system should be capable of handling a static pressure of 3.5 to 4.0 in. H_2O.

5.3 Ventilation Regulations

Some applications of ventilation, both general and local, are covered specifically by occupational safety and health regulations. The federal standards as established by the Occupational Safety and Health Administration (OSHA) now include specific ventilation requirements for the applications identified in Table 30.7.

In addition to the specific references to ventilation requirements in the aforementioned standards, most occupational-health standards for specific substances, for example, ben-

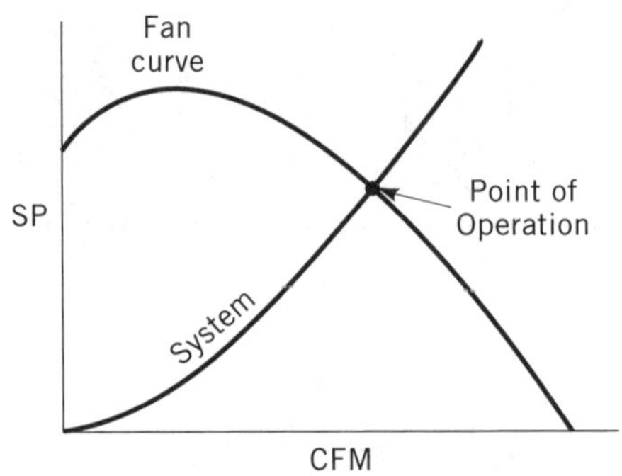

Figure 30.14. Fan and system curves.

Table 30.6. Typical Multirating Table for Fans

Volume	Outlet Velocity	Velocity Pressure	Static Pressure, SP (in. H_2O)																	
			1 in.		2 in.		3 in.		4 in.		5 in.		6 in.		7 in.		8 in.		9 in.	
(ft^3/min)	(ft/min)	(in. water)	rpm	bhp	rpm	bhp	rpm	bhp	rpm	bhp	rpm	bhp	rpm	bhp	rpm	bhp	rpm	bhp	rpm	bhp
2,520	1000	0.063	437	0.63	595	1.27	728	2.00	837	2.66										
3,120	1200	0.090	459	0.85	610	1.55	735	2.30	842	3.10										
3,530	1400	0.122	483	1.05	626	1.87	746	2.72	847	3.57	943	4.60								
4,030	1600	0.160	513	1.33	642	2.18	759	3.17	858	4.12	950	5.21	1030	6.29						
4,530	1800	0.202	532	1.61	666	2.56	774	3.63	876	4.63	964	5.82	1040	6.92	1125	8.18				
5,040	2000	0.250	572	2.00	688	2.97	797	4.12	890	5.30	976	6.50	1052	7.75	1134	8.96	1208	10.15	1270	11.67
5,540	2200	0.302	603	2.36	712	3.43	816	4.66	910	5.93	999	7.38	1068	8.60	1145	9.93	1210	11.18	1279	12.82
6,040	2400	0.360	637	2.79	746	3.99	840	5.33	926	6.73	1017	8.17	1088	9.50	1160	10.88	1230	12.25	1288	13.92
6,550	2600	0.422	670	3.27	762	4.62	866	6.05	954	7.83	1032	9.08	1095	10.50	1171	11.98	1245	13.50	1298	15.10
7,060	2800	0.489	708	3.81	795	5.32	892	6.72	963	8.78	1050	9.97	1125	11.60	1188	13.06	1257	14.70	1310	16.48
7,560	3000	0.560	746	4.42	833	6.05	920	7.70	993	9.32	1068	11.00	1142	12.75	1210	14.28	1277	15.98	1328	17.80
8,060	3200	0.638			866	6.96	943	8.71	1020	10.40	1097	12.10	1168	14.02	1228	15.50	1292	17.36	1340	19.15
8,560	3400	0.721			900	7.93	964	9.80	1053	11.48	1120	13.30	1188	15.35	1248	16.93	1310	19.00	1360	20.90
9,070	3600	0.808					1010	11.00	1078	12.70	1148	14.65	1213	16.70	1270	18.42	1335	20.75	1380	22.60
9,570	3800	0.900					1038	12.25	1108	14.15	1170	14.90	1240	18.80	1292	19.46	1355	22.35	1405	24.40
10,080	4000	0.998					1162	13.60	1138	15.40	1200	17.35	1270	19.70	1320	21.70	1380	23.15	1430	26.40
10,580	4200	1.100							1168	16.90	1230	19.05	1283	21.50	1348	23.50	1405	26.10	1450	28.45
11,100	4400	1.210							1198	18.58	1258	20.55	1322	22.50	1373	25.40	1430	27.95	1478	30.60
11,600	4600	1.310							1232	20.30	1290	22.50	1355	23.80	1405	27.40	1450	30.15	1500	32.90
12,100	4800	1.450							1270	21.00	1321	24.40	1383	25.65	1432	29.60	1482	32.40	1528	35.20
12,600	5000	1.570							1301	24.20	1355	26.40	1410	28.80	1462	31.80	1513	34.60	1555	37.80
15,120	6000	2.230													1622	45.90	1670	49.00	1702	51.50

Table 30.7. Industrial Activities Covered by OSHA Standards Specifying Ventilation Requirements

Industrial Application	Standard Reference
Abrasive blasting	29 CFR 1910.94(a)
Asbestos	29 CFR 1910.1001(c)
Bulk gas plants	29 CFR 1910.106(f)
Confined spaces	29 CFR 1910.146
Dip tanks containing combustible or flammable liquids	29 CFR 1910.108(b)
Electrostatic spraying	29 CFR 1910.107(i)
Flammable or combustible liquids in storage rooms and enclosures	29 CFR 1910.106(d)
Grinding, polishing, and buffing	29 CFR 1910.94(b)
Hydrogen	29 CFR 1910.103(b)
Laundries	29 CFR 1910.262(c)
Open surface tanks	29 CFR 1910.94(d)
Oxygen	29 CFR 1910.104(b)
Powder coatings	29 CFR 1910.107(i)
Sawmills	29 CFR 1910.265(c)
Spray-finishing operations (general)	29 CFR 1910.107(d)
Surface coating	29 CFR 1910.94(c)
Textiles	29 CFR 1910.262(rr)
Welding, cutting, and brazing	29 CFR 1910.252(f)

zene, ethylene oxide, formaldehyde, and lead, incorporate language clearly meant to include ventilation considerations. In many recent OSHA health standards there is a section titled "Methods of Compliance," which contains a statement inferring that the employer must institute engineering controls and work practices to reduce and maintain employee exposure to or below the PEL-TWA, except to the extent that such controls are not feasible. It is inconceivable that such engineering controls would not rely heavily on ventilation.

For the most part, the ventilation requirements incorporated into OSHA standards are based on consensus-type standards developed by organizations such as the American National Standards Institute (ANSI) and the National Fire Protection Association (NFPA). Although the past decade has not seen much regulatory activity regarding ventilation requirements, the number and degree of specificity of ventilation standards likely will increase in the future.

Many of the occupational-health standards being developed by OSHA are couched in language that makes it clear that engineering control of the specific contaminants covered by the standard is the primary responsibility of the employer. In fact, engineering controls (chiefly ventilation in the case of airborne contaminants) must be applied to the extent feasible even if they are not sufficient by themselves to reduce the airborne concentration of the contaminant to the specified acceptable level. Thus in addition to the occupational health standards dealing with specific aspects of ventilation, there are indirect references to the requirement for installation of ventilation systems to maintain the work environment within stated acceptable limits. Accordingly, the need for application of sound engineering in the control of workplace contaminants undoubtedly will continue to increase.

5.4 Computer Assistance

The preceding sections have dealt with the process of designing ventilation systems, using calculations based on determination of a balanced system. Thus, to a large extent, the calculations use trial-and-error solutions to optimize the dimensions and other quantitative aspects of the system. For example, there generally is an interrelationship among the key variables affecting airflow through a section of ductwork: duct diameter, air velocity, volumetric-flow rate, and resistance to airflow. These relationships are depicted in graphs or friction charts, such as that in Figure 30.13. One would expect, therefore, that a means of number crunching, that is, repetitive calculations using similar sets of data, would facilitate the ventilation-design process.

Similarly, because once the parameters of the system have been selected the calculations are relatively straightforward, a mechanical calculator would be helpful. For these reasons, then, it is not surprising that use of computer programs has become routine in the design of ventilation systems (15–18). These programs typically follow the same procedure outlined in the preceding sections and therefore assume the user has a basic understanding of ventilation design. Perhaps the most useful of these programs incorporate the equivalent-foot method in which all elements of the system (elbows, fittings, and transitions) are converted to equivalent feet of straight duct, thus simplifying the analytical process. These programs, both in the classroom in ventilation-design courses and in actual use by engineers, have been found to be useful and fairly reliable with only minor modifications of the programs. Programming for laptop and hand-held computers is readily available, facilitating even more the design process for the ventilation engineer.

Many large companies and architect/engineering firms, of course, have incorporated ventilation design and specification preparation aspects into their computer-assisted engineering work. Access to computer-generated optimal characteristics of a ventilation system, together with use of auto-CAD (computer-assisted drawing) programs, makes proper ventilation design a relatively simple, rapid process. With this rapidly advancing computer technology, availability of such assistance, will enhance the industrial-hygiene practitioner's general awareness of control principles.

6 MAKEUP-AIR REQUIREMENTS

Adequate supplies of outside air are necessary to ensure proper operation of ventilation systems. Properly designed exhaust systems will remove toxic contaminants by themselves; however, they should not be relied upon to draw air into the building to replace that which is exhausted, because this will result in negative pressure within the building, leading to undesirable effects. Mechanically controlled systems for supplying the air to the work space are preferred and in most modern industrial facilities are necessary to achieve the overall desired ventilation. Most outside air-supply systems currently installed in industrial plants have been placed in operation only to replace the exhausted air, with little or no regard to the overall positive effects that can be achieved by the air-supply system in the total environmental-control system. There is a definite trend toward consideration of the environmental-control potential of air-supply systems, that is, ability to beneficially affect the air quality, temperature, and humidity of the workplace. Control of air volume, velocity, and temperature is important for a satisfactory work environment; a proper air-supply system can provide both replacement air and some degree of environmental control.

6.1 Need for Adequate Makeup Air

Air enters a work space in an amount equal to the volume actually exhausted, regardless of whether mechanical provisions are made for replacement of the air. Of course in some cases, particularly when the total-exhaust volume is small, the replacement air enters the building with no adverse effects, even in the absence of a mechanical makeup-air system. When the exhaust volumes are relatively large in comparison to the size of free inlet area of the building, however, undesirable effects will be experienced. Older industrial buildings with large void spaces in the enclosing walls may provide significant openings, and air leakage into the building can be quite pronounced. Modern windowless plants of masonry construction, however, can be essentially airtight, with the result that the building is starved for air if there is a significant volume of exhaust ventilation. In a building that is relatively open, an influx of outside air is particularly undesirable in northern climates, where the incoming air cools the perimeter of the building, workers exposed to the incoming air are subjected to drafts, temperature gradients are produced within the building, and the internal-heating system is usually overtaxed. Although the incoming air eventually may be tempered to acceptable conditions by mixing with interior air, it is an ineffective heat-transfer process, usually resulting in wasted energy.

Reasons for considering an adequate supply of makeup air in conjunction with a total ventilation system include the following:

1. Lack of adequate makeup air creates a negative pressure within the building; this increases the static pressure that must be overcome by the exhaust fans if they are to function properly. This effect can result in a reduction of the total volume of air exhausted from the enclosure, and it is particularly serious with low pressure fans such as the common wall fans and roof exhausters of the propeller type.

2. Cross drafts created by the infiltration of uncontrolled makeup air interfere with proper operation of exhaust hoods and also may disperse contaminated air from one area of the building to another; moreover, they can seriously interfere with proper operation of process equipment. The relatively high velocity cross drafts through windows and doors created by the infiltration of air can even dislodge settled material on beams and other horizontal surfaces, resulting in resuspension of solid contaminants in the workroom.

3. If the negative pressure created within the building is significant enough, there may be actual backdraft on flues and other natural draft stacks, posing a potential health hazard from release of combustion products into the workroom. Such backdrafting can occur in natural-draft stacks at very low negative pressures. Secondary problems associated with the backdraft phenomenon include difficulty in maintaining pilot lights in burners, erratic operation of temperature controls, and potential corrosion damage in stacks and heat exchangers due to condensation of water vapor from the flue gases.

4. The cold drafts resulting from infiltration of uncontrolled makeup air can produce worker discomfort, leading to reduced work efficiency attributable to lower ambient temperatures and the drafts impinging on the workers.

5. The differential pressures created across internal partitions may make doors difficult to open or close if the amount of infiltrated makeup air is high enough. In some extreme cases, this can pose a safety hazard when doors are moved by the force of the pressure gradient itself in an uncontrolled manner.

6. The conditions of cooler ambient temperatures near the building perimeter often lead to decisions to correct the problem by installing more heating equipment in those areas when installation of a proper makeup-air system would eliminate the problem. The heaters warm the air only after it has entered the building, usually too late to be of benefit to the workers in the perimeter. Then the overheated air moves toward the building interior, making those areas often uncomfortably warm. Often attempts are then made to alleviate this problem by installing more exhaust fans to remove the excess heat, but this further aggravates the problem of temperature gradients within the building. The net result is the unnecessary wasting of heat without solving the problem. The fuel consumption associated with attempts to achieve comfortable temperatures within the building by this approach is much higher than it would be with a properly installed system for providing makeup air.

6.2 Principles of Supplying Air

The volume of makeup air provided to a work space generally should equal the total volume of air removed from the building by exhaust-ventilation systems, process venting, and combustion processes. The determination of the actual amount of air removed from the building usually requires no more than a simple inventory of air-exhaust locations, with measurements of the amount of air being exhausted and compilation of the exhaust volumes. It frequently is desirable to incorporate any reasonable projections for additional exhaust requirements in the near future, certainly if process changes or additions or modifications to the plant are being contemplated. Often it is desirable to install an air-supply system of a capacity slightly larger than actually needed to allow for future needs. In most cases the drive mechanism on the air-supply system can be modified to supply only the desired quantity of air.

The locations of the makeup-air units, with the consequent distribution of air and creation of airflow patterns within the building, often do not receive sufficient thought before installation, and in many instances this aspect of the system is neglected altogether. However, successful operation of the total-ventilation system depends upon proper distribution of makeup air. The following principles for supplying makeup air should be incorporated into the design of an air-supply system:

1. Locate the inlet for the fresh air away from any sources of contamination, such as exhaust stacks or furnace exhausts. It is advisable to filter the fresh air, in order to protect equipment and to ensure maximum efficiency of heat exchange in the tempering systems.

2. Incorporate a mechanically operated air mover, that is, a fan or blower, in the air-supply system to prevent the development of negative pressure within the room or building.

3. Design the locations of the makeup-air entries so as to maximize utilization of the air within the work space. Properly located makeup-air-supply inlets can be used to provide general dilution-ventilation first, after which the air provides makeup for exhaust systems operated within the building. Obviously, this concept does not apply for specialized types of makeup air such as those installed for spot-cooling purposes, where the air is introduced at temperatures significantly below room temperature, and at specific work stations. In general, the air distribution pattern created by the combination of makeup-air supply and process exhaust must be engineered carefully to provide effective coverage within the building without creating excessive drafts, which could interfere with process operations or compromise the comfort of workers.

4. Introduce the makeup air into the plant at a height of approximately 8 to 10 ft wherever practicable. In this way, the air is used first by the workers, maximizing the results of general ventilation. Such an arrangement for distribution of the incoming air also permits closer control of the ambient workplace temperature.

5. Heat or cool the makeup air, as necessary, to approximately the desired inside temperature. Accordingly, the temperature range for incoming air typically varies between 65 and 80°F, depending on the geographic location of the building and the climatological and meteorological conditions characteristic of the location and time of year.

6.3 Recirculation of Exhaust Air

Typical application of LEV includes discharge of the contaminated air to the atmosphere, with or without benefit of an air-cleaning device. In recent years, industries have begun actively investigating exhaust systems that will clean the air of contaminants to a degree sufficient to permit discharge of the air directly to the workplace. This approach is desirable because air-cleaning or pollution control devices have been employed to substantial degrees to meet air quality or emission standards. Additionally, the heating and/or cooling of makeup air needed to replace exhausted air is an expensive item, particularly because costs of energy are increasing and in some cases energy to operate new installations is not even available.

The acceptability of recirculating air systems obviously depends to a great extent on the health hazard associated with the contaminant being exhausted. It has been a fairly consistent policy of official occupational-health agencies not to condone the recirculation of exhausted air if the contaminant may have an adverse effect on the health of the workers. Even though the air cleaner used is efficient enough to sufficiently clean the air for health protection, incorrect operation or poor maintenance of the system could result in return of the contaminated air.

6.3.1 Circumstances Permitting Recirculation of Air

Recirculation of exhausted air is sometimes a feasible method of supplementing ventilation within the workplace. In some situations, the economic and energy-conservation factors may be important enough to warrant the additional capital and operating expenditures of assuring the safe provision of recirculated air. In general, recirculation of exhausted air may be permitted under the following circumstances:

1. The ventilation system must be furnished with an air-cleaner system efficient enough to provide an exit concentration, that is, concentration of contaminant in the air recirculated to the workroom, not exceeding an allowable value, which may be calculated for equilibrium conditions using the following equation:

$$C_R = \frac{(\text{PEL} - C_O)Q_T}{2Q_R\text{K}}$$

where C_R = concentration of contaminant in exit air from the collector before mixing
Q_T = total ventilation flow through the affected work space (ft^3/min)
Q_R = recirculated air flow (ft/min)
K = a mixing factor, usually varying between 3 and 10 (3 = good mixing conditions)
PEL = permissible exposure limit for contaminant
C_O = concentration of contaminant in workers' breathing zone with local exhaust discharged outside

This air-cleaning system is referred to as the primary system.

2. A secondary air-cleaning system of efficiency equal to or greater than the primary system should be installed in series with the primary cleaner. As an alternative to this approach, a reliable monitoring device may be installed to furnish a representative sample of the recirculated air. This monitoring system must be fail-safe with respect to failure of the power supply, environmental contamination, or the obvious results of poor maintenance.

3. A warning signal should be provided to indicate the need for attention to the secondary air-cleaning system, or the air monitor should indicate when the concentrations have exceeded predetermined allowable limits.

4. There should be a mechanism for either immediate bypass of the recirculated air to the outdoors or complete shutdown of the contaminant-generating process to become operative under conditions that activate the system's warning device.

Although the application of recirculated air systems in industry has been quite limited, the potential for and anticipated future applications of this approach are quite extensive. NIOSH (19) indicated that 356 of the 514 compounds constituting the ACGIH list of threshold limit values at the time of the study were potentially recirculatable. The compounds excluded from consideration were primarily those identified as having carcinogenic properties and those for which a ceiling exposure limit had been established. This preliminary analysis of design and operating criteria for recirculation systems emphasized that air monitoring equipment should be on-line, automatic, and specific for the individual contaminant. Although a preliminary effort, this study did indicate that use of recirculated air in industrial workplaces will become much more commonplace than it is today. Good summaries of the considerations for recirculated air have been published (20, 21).

6.3.2 *Design Considerations for Recirculated Air*

With the full expectation that interest in and development of refined recirculated air systems will continue, the following factors and considerations should be incorporated into analysis of the appropriateness of recirculated air systems in any particular setting.

1. Usually it is necessary to provide general-ventilation air in addition to that recirculated to ensure continual dilution of the contaminants in the recirculated airstream. When it is proposed or possible that all the supplied ventilation air be that recirculated from exhaust systems, all possible contaminants in the airstream must be evaluated, not just the most significant. Of particular concern would be minor contaminants in the airstream that might pass through the primary and secondary air-cleaning devices. For example, a fabric filter-high efficiency particulate filter combination, although providing essentially complete collection of particulate contaminants, could allow the concentration of gases and vapors to build up during the course of operation from relatively insignificant to potentially hazardous levels.

2. Wherever practicable, the recirculating air system should be designed to permit bypass to the outdoors when weather conditions permit. For example, if the system is intended to conserve heat during winter months, it can discharge outdoors in warmer weather if windows and doors are designed to permit sufficient makeup air when opened. Of course continuous operation of the bypass mode would not be desirable where the work space is conditioned or where mechanically supplied makeup air is required at all times.

3. Air recirculated through wet collectors may pose a problem in that the humid air from such equipment usually causes uncomfortably high humidity within the workplace and possible condensation problems. Excessive humidity may be prevented through the use of auxiliary ventilation equipment.

4. Design data and testing programs preceding installation of recirculating air systems should consider all operational time periods of the system for it can be reasonably expected that the concentration of contaminants exiting typical collectors will vary over time.

5. As with design and layout of any air supply system, the ductwork from the recirculating air system should provide adequate mixing with other air supplies to avoid uncomfortable drafts on workers or air currents that could adversely affect the performance of local-exhaust hoods.

6. In general, even when a monitoring device is available, the installation of a secondary air-cleaner system is preferable because it usually lends a greater degree of reliability to the system and requires a less sophisticated maintenance program.

7. Although the primary concern for the quality of the recirculated air centers on acceptable exposure limits, odors or nuisance values for contaminants should be considered as well. In fact, in some locations, adequately cleaned recirculated air, provided by a system with appropriate safeguards, can be of better quality than the makeup-air supply entering the building from outside.

8. Routine testing, maintenance procedures, and records should be developed for the circulating air system. In addition, the workroom air should be tested periodically.

The foregoing considerations for recirculating air systems do not appear to impose any insurmountable obstacles to continue development of proper criteria for design and operation of such systems. With the probable continuing concern for energy conservation and air quality (both within and outside the industrial plant), it is likely that the next generation of industrial-hygiene engineers will routinely deal with recirculating air systems.

BIBLIOGRAPHY

1. R. H. Perry and D. Green, eds., *Perry's Chemical Engineers' Handbook*, 8th ed., McGraw-Hill, New York, 1993.
2. American Conference of Governmental Industrial Hygienists, Committee on Industrial Ventilation, *Industrial Ventilation: A Manual of Recommended Practice*, 23rd ed., ACGIH, Lansing, MI, 1998.
3. H. J. McDermott, *Handbook of Ventilation for Contaminant Control*, 2nd ed., Butterworth, Stoneham, MA, 1985.
4. J. L. Alden and J. M. Kane, *Design of Industrial Ventilation Systems*, 5th ed., Industrial Press, New York, 1982.
5. American Society of Heating, Refrigerating, and Air-Conditioning Engineers, *ASHRAE Guide and Data Book-Fundamentals and Equipment*, ASHRAE, New York, 1987.
6. OSHA General Industry Safety and Health Regulations, *U.S. Code of Federal Regulations*, Title 29, Chapter XVII, Part 1910.1000.
7. American Conference of Governmental Industrial Hygienists, *Threshold Limit Values for Chemical Substances and Physical Agents in the Workroom Environment*, ACGIH, Cincinnati, OH, 1989.
8. J. M. Dalla Valle, *Exhaust Hoods*, Industrial Press, New York, 1944.
9. A. D. Brandt, *Industrial Health Engineering*, Wiley, New York, 1947.
10. L. Silverman, "Velocity Characteristics of Narrow Exhaust Slots," *Ind. Hyg. Toxicol. J.* **24**, 267 (1942).
11. American Society of Heating, Refrigerating and Air-Conditioning Engineers (ASHRAE), Handbook of Fundamentals, ASHRAE, Atlanta, GA, 1993.
12. American National Standards Institute, Fundamentals Governing the Design and Operation of Local Exhaust Systems, ANSI Z9.2 Committee, New York, 1979.
13. American Society of Heating, Refrigerating and Air Conditioning Engineers (ASHRAE), Ventilation for Acceptable Indoor Air Quality, ASHRAE, Atlanta, GA, 1989.
14. D. J. Burton, *IAQ and HVAC Workbook*, IVE, Inc., Salt Lake City, UT, 1997.
15. D. E. Clapp, D. J. Groh, and C. M. Nenadic, "Ventilation Design by Microcomputer," *Am. Ind. Hyg. Assoc. J.* **43**, 212 (1982).
16. American Conference of Governmental Industrial Hygienists, Microcomputer Applications in Occupational Health and Safety, Lewis Publishers, Chelsea, MI, 1987.
17. J. J. Loeffler, "Using Programmable Calculators for Ventilation System Design, II;" *Plant Eng.* **Sept. 14**, (1978).
18. H. P. Shotwell, "A Ventilation Design Program for Hand-Held Programmable Computers," *Am. Ind. Hyg. Assoc. J.* **45**, 749 (1984).

19. National Institute for Occupational Safety and Health, Recirculation of Exhaust Air, NIOSH Report No. 76-186, Cincinnati, OH, 1976.
20. B. A. Plog, J. Niland, and P. J. Quinlan, eds., *Fundamentals of Industrial Hygiene*, 4th ed., National Safety Council, Itasca, IL, 1996.
21. L. V. Cralley and L. J. Cralley. *In-Plant Practices for Job Related Health Hazards Control*, Vol. 2, Engineering Aspects, Wiley, New York, 1989.

CHAPTER THIRTY-ONE

The Emission Inventory and Dilution Ventilation

Robert L. Harris, Ph.D., CIH and Earl W. Arp, Jr., Ph.D., CIH

1 INTRODUCTION

An emission inventory for an industrial or commercial enterprise is a compilation of information from which one can calculate or estimate the rates (quantity per unit time) at which pollutants are released to the environment. For purposes of this chapter, only the emissions that contaminate workroom or community air are considered; emissions to surface or ground water, to soil, or to other environmental receptors are treated only as they may, in turn, result directly in emissions to workplace or community air.

An emission inventory may be simple or complex. A rudimentary inventory may consist of source location, date, identification of process, a qualitative listing of materials used, and an index of size (e.g. annual production rate) for the subject enterprise. Such an inventory, along with emission factors generated by studies of similar processes, will permit the making of an estimate of annual emissions. A comprehensive emission inventory, on the other hand, may contain sufficient detail to permit quantification of emissions, including temporal variations, for a number of specific materials from each point of release in a complex industrial process.

An inventory of emissions, along with various other kinds of companion information, discussed later in this chapter, permits the making of estimates of the nature, and sometimes the intensity, of exposures to airborne agents in workplaces or in the community. The level of detail needed and achievable for the inventory depends both on the purposes for which it is to be used and the data sources, or data generating efforts, which can be utilized.

Patty's Industrial Hygiene, Fifth Edition, Volume 2. Edited by Robert L. Harris.
ISBN 0-471-29754-2

2 ELEMENTS OF AN EMISSION INVENTORY

The compilation of emission inventories is a well-established, specifically identifiable, activity in the field of air pollution control. Practices and procedures have been highly developed and descriptions of the technology are available (1–5). Emission inventory has been practiced in industrial hygiene for many years, but has not been identified as a categorical work area in this field to the extent that it has been in the field of air pollution control. Although the types of data used and the techniques for obtaining them vary somewhat, the same basic elements appear in emission inventories in both fields of work.

2.1 Identification of Agents

Anticipation, recognition, evaluation, and control of hazards are the four basic steps in the practice of industrial hygiene and community air pollution control. An emission inventory, regardless of whether it is specifically identified as such, is necessary in all four steps and is particularly important in the first three, anticipation, recognition, and evaluation. It is clear that if a hazard is to be dealt with, it must first be anticipated and/or recognized; this recognition, and the identification of an emission, is a rudimentary emission inventory. Evaluation requires more than identification. In situations involving release of chemical agents to the air, evaluation may include obtaining additional information such as quantity, character, and temporal variations of emissions, all of which are part of the emission inventory. The design and implementation of emission control requires detailed information about the emission source that goes beyond the level of detail usually required for an emission inventory.

The federal Toxic Substances Control Act of 1976 (6), among other things, provides for the collection of information regarding chemical substances manufactured, imported, or produced in the United States. Such information will permit preliminary assessment of potential exposures and possible effects on health and the environment. Compliance with the act requires identification, by process and location, of many chemical agents in industry and commerce. The notification, reporting, and record keeping provisions of the act and its implementing regulations facilitate the agent identification component of a comprehensive, plant-wide emission inventory.

Not all materials handled in industry and commerce are hazardous. According to the American Chemical Society's Chemical Abstract Service, its registry listed more than 19 million compounds as of the end of March, 1999 (7), up from about 13 million in 1992 (8). The number of registered compounds has been growing at a rate of about 20,000 per week in the past decade. Of the millions of compounds that exist, more than 70,000 are considered, for purposes of the Toxic Substances Control Act, to be in commercial use in 1998 (9). Some toxic dose information, based on experimental animal work or other observations, is available on about 139,700 compounds as of July 1997 (10), up from about 89,000 in 1987 (11). Some of these compounds are relatively nontoxic, others are not in common use, and for many the toxicity information is fragmentary. Probably fewer than 1000 compounds have been identified with occupational health or community air pollution problems sufficiently to permit development of workplace or air quality standards. The identification, by means of an emission inventory, of materials that are released from a process or operation, however, is a fundamental step in the recognition of those that represent a potential hazard in the workplace or the community.

For manufacturing processes, preliminary identification of potentially hazardous agents often can be based upon the identification of process raw materials, intermediate and by-product materials, and the process end products. In commercial enterprises the identification of materials that are handled permits the singling out of those that may represent potential hazards. For combustion sources, information on the composition of fuel and the type of combustion equipment used permits qualitative identification of pollutant components.

The evaluation phase of an industrial hygiene or community air pollution problem requires, in addition to identification of the agents of concern, a number of other kinds of information that can be obtained in an emission inventory. Among the most fundamental of these is the identification and description of the site or location at which the contaminant is released to the air.

2.2 Identification of Emission Sites

The most cursory emission inventory may identify the site or location of an emission source only as a particular plant or commercial establishment. Such location information is generally useful only for preliminary surveys of community air pollution or for indicating the need for more thorough workplace exposure evaluation. Any emission inventory use other than agent identification alone requires more specific emissions location information. For example, in air pollution control the identification of specific stacks, vents, and other points of emission is necessary for diffusion modeling and for most impact evaluations and emission regulatory activities. For industrial hygiene purposes the location of a specific workplace, process point, and perhaps even a particular process equipment opening (e.g., a mixer charging port) or work practice (e.g., the handling of shipping bags after use) may be needed. Such location information is vital to the hazard assessment process. It is necessary for identifying the workers subject to exposure from the particular source and for identifying alternatives from which to select the means for control of any hazard caused by the emission.

An emission inventory that identifies both the materials emitted from a process and the specific locations at which these materials are released can serve as the first step, and perhaps the only step necessary, in the evaluation of a hazard and initiation of a control effort.

2.3 Time Factors in Emissions

Time resolution in emission inventories may be yearly, seasonal, monthly, weekly, daily, hourly, or even less than hourly, depending on specific needs and the availability of data. For initial surveys of community air pollution, yearly average emissions by plant site may be satisfactory. At the other extreme, the assessment of emissions to workplaces that involve cyclic or intermittent operations may require use of time intervals of 15 minutes or less. For short time or intermittent operations—for example, the taking of materials samples at process sampling ports—the actual time interval of emission and the frequency with which the operation occurs should be recorded in the inventory.

In some cases the interval of record for the quantity of material released is relatively long—for example, a monthly record of solvent use—even though actual release may be cyclic or may occur over short intervals. In such cases the emission inventory record should contain a sufficiently detailed description of the process or operation to permit estimation of actual emission intervals and the quantities of materials released during these intervals.

It is important that the emission inventory record include both the date on which the inventory was done and the calendar interval for which it applies. When a change in process or operation occurs that materially affects the composition, quantity, or condition of an emission, the emission inventory record should be updated to reflect the change. In the absence of any substantial change, the inventory should be revalidated at convenient intervals, perhaps annually, or as may be required by governmental regulation. When an emission inventory record is updated or revalidated, the old record should be retained; a sequential inventory file over a long period may be invaluable in future retrospective environmental epidemiologic studies.

3 QUANTIFYING EMISSIONS

Emissions can be quantified either by direct measurement, such as source testing, or by indirect means. Indirect means include techniques such as process materials balance or the determination of an index parameter—for example, a production rate—that can be related empirically to emissions.

3.1 Source Sampling

Source sampling is ordinarily associated with measurement of air pollutant emissions. Under some circumstances the techniques can be applied to industrial hygiene investigations as well. The techniques of air pollutant source testing have been described in detail by Paulus and Thron (12) and updated by Thron (3); the first (1977) chapter, "Stack Sampling", lists 72 references, the updated chapter (1986) lists 90. The Environmental Protection Agency (EPA) has published stepwise procedures on source sampling for particulates (13); this publication contains a number of data recording forms that are useful in sampling not only for particulates but for other agents as well.

Source sampling ordinarily consists of withdrawing a representative sample from a contaminant-bearing gas stream in a duct or stack. Analysis of the sample yields data on concentration of the contaminant in the gas stream. Concentration data, combined with companion data on gas flow rates in the ducts or stacks, yield values for contaminant emission rates for gas streams released to the atmosphere.

The critical concern in source sampling is the representativeness of the sample. Both composition and flowrate of a contaminated gas stream may vary as the processes and operations that generate it vary. Thus, representativeness of a source sample depends very much on the representativeness of processes and operations at the time of sampling.

When the contaminant is particulate, the collection of a representative sample requires isokinetic sampling and use of an unbiased sampling traverse pattern. Isokinetic sampling is performed by taking the sample at such a flowrate that the sampled gas stream enters the inlet nozzle of the sampling probe with velocity equal to that which prevails at the specific point in the cross section of the stack or duct from which the sample is being withdrawn. When the velocity of gas at the sampling point in the duct or stack is greater than that in the sampling nozzle, part of the approaching gas stream is deflected around the nozzle. Smaller particles tend to follow the deflected gas stream while larger ones, by virtue of their momentum, tend to continue their trajectories and enter the nozzle; this results in a nonrepresentative overabundance of larger particles in the sample. When the velocity of the gas stream at the sampling point in the duct or stack is lower than the velocity entering the sampling nozzle, the gas stream converges into the nozzle inlet, carrying with it the smaller particles, but losing some of the larger ones which are carried past the nozzle by their momentum; the sample then is nonrepresentative because of a deficiency of larger particles.

The velocity of the gas stream in a duct or stack is not uniform throughout its cross section. For this reason, and because particles are not necessarily uniformly distributed within a duct or stack, a specific traverse pattern is ordinarily used in source sampling and in the measurement of velocity for determination of flowrates. For purposes of a sampling traverse, the cross-sectional area of the duct or stack is divided into equal sized subareas; sample increments and velocity readings are taken at the centers of these subareas. Sampling time should be the same for each traverse point in a duct or stack, and whenever possible the entire sampling traverse should be accomplished during some interval of time during which the stack or duct flowrate and the composition of emissions do not change. Isokinetic sampling and equal area traverses are discussed in detail, including descriptions of apparatus and calculation methods, in the source sampling references cited earlier (3, 12, 13).

The sampling of gases and vapors differs from sampling of particulates in that isokinetic sampling is not required unless concentrations differ from place to place in the duct cross section. The collection apparatus and reagents used in a gas sampling train also differ from those used for particulates. Filtration, inertial size classification, and impingement with capture in liquid media, are the collection mechanisms used for particulates source sampling; liquid absorption, adsorption on solids, and freeze-out are the methods usually employed for gas and vapor sampling. Information on sampling apparatus, collecting media, and analytic methods for a number of gases and vapors have been tabulated by Paulus and Thron (12) and updated by Thron (3).

As mentioned earlier, the techniques of source sampling are most often applied in air pollution emission measurements. They can, in some circumstances, be applied for inplant industrial hygiene purposes as well. When a workplace is served by general dilution ventilation in the exhaust mode, the techniques of source sampling can be applied to the exhausted airstreams to determine the rate at which contaminants are released to the workplace air. The calculation methods described in Section 5.3.1 of this chapter can be used to estimate emissions when the concentrations of contaminant in the supply and exhaust air, and the ventilation rate are known with the system at equilibrium. In applying the equations to exhaust air streams the factor K of equation 17 has the value 1, and the contaminant generation rate is the sum of the generation rates calculated for all exhaust discharges. The procedure is most applicable when general dilution ventilation in the exhaust mode is the sole, and controlling, ventilation regimen for the workplace, that is, when there is no mechanical local exhaust ventilation. If local exhaust ventilation is used in the workplace, its influence as general dilution ventilation must be taken into account when using this technique to estimate workplace emissions.

Source sampling techniques may also be used to quantify emissions from individual points of release in a workplace. The emissions may be captured using a temporary exhaust ventilation setup, and the amount of material released may be determined by sampling from that exhaust stream. Application of this technique to a single source or emission point in a space that contains several sources of the air contaminant requires either elimination of the influence of the other sources or correction for them. A mechanical arrangement can be provided to supply contaminant-free outside air to the test source. The exhaust stream from the test source will then contain only contaminant from that source and will be unbiased by contaminant from other sources.

Alternatively, monitoring of the concentration of contaminant in room air that supplies the source test exhaust system permits correction for other sources; emissions from the test source can be determined by the difference in contaminant concentration in supply and exhaust air of the test system.

A method for measuring emission rates of gases or vapors from point or surface sources using tracer gas has been reported (14). Tracer gas is released at a measured rate at the location of the emission source; the emission rate is calculated from the tracer gas emission rate and the correlation of the concentrations of the emitted gas or vapor and the tracer gas measured at the same locations close to the source. Experience in application of the tracer gas method to quantify emissions of vapors in an industrial setting, and precautions to be observed in using the method have been reported (15). Persons contemplating use of the method are urged to examine the cited reference for its listed precautions. When aerodynamic behavior of the pollutant gas or vapor emitted from the source and that of the tracer gas are the same, the calculation of emission is quite straight-forward:

$$q_P = q_T(C_P/C_T) \tag{1}$$

where q_P = pollutant gas emission rate, mg/sec
q_T = tracer gas flow rate, mg/sec
C_P = measured pollutant gas air concentration, mg/m^3
C_T = measured tracer gas concentration, mg/m^3, at pollutant gas sampling location.

As is discussed in the paragraph above, correction must be made in the source emission rate if the supply air is contaminated or if there are other sources of the emitted gas or vapor in the workplace.

3.2 Materials Balance

In some cases knowledge of processes and operations permits determination of the amount of material released to the air of a workplace without emission measurements. If it is known that a gas or vapor is generated by chemical reaction or otherwise, and is released to workplace air in proportion to the use of a raw material or a production rate, that index of generation can be used to determine the release rate. Examples include the generation of products of combustion by unvented open flames, as is the case with direct fired unit heaters. Here fuel composition and use rate are indices of contaminant emissions. Uses of volatile solvents in which the solvents do not become part of a product, but evaporate completely into the workplace air, are common in industry. Solvent use rate, in such cases, is also an emission rate. When exhaust ventilation is applied to some operations in a workplace and not to others, distinction must be made between that portion of the material which is captured by exhaust ventilation and that which is released in the occupied workplace; only the portion that is released directly to the workplace air is used to estimate workplace exposures. For estimating community air pollution emission rates, the total amount of volatile material that evaporates into the atmosphere is taken as the emission rate regardless of whether the material is released to workroom air or through exhaust ventilation systems.

The American Petroleum Institute (API) has reported mathematical relationships that describe evaporation losses of petroleum products from tanks during loading and unloading (16). The materials balance concepts of the API procedure can be applied to estimating vapor emissions from the loading of volatile liquids into vessels that are vented to workroom air. The mass of vapor expelled by displacement when a volatile liquid is transferred into a vessel is

$$M = 1.37VSP_v \frac{mw}{T} \tag{2}$$

where M = mass of vapor expelled (lb)
V = volume of liquid transferred to the vessel (ft^3)
S = fraction of vapor saturation of expelled air
P_v = true vapor pressure of the liquid (atm)
mw = molecular weight of vapor
T = temperature of the tank vapor space (°R)

Except for S, the fraction of vapor saturation, the various parameters of equation 2 are ordinarily known or measurable quantities. For splash filling of a vessel that was initially vapor free, or for the refilling of a vessel from which the same liquid has just been withdrawn, the value of S can ordinarily be taken as 1 (16).

Equation 2 may overestimate the mass of vapor emission if the vessel walls are substantially colder than the volatile liquid. The true vapor pressure depends on the temperature of the liquid; in the case of the cold vessel, however, some vapor may condense on the inner wall surfaces and fail to escape through the vent into the workroom air.

When complete evaporation of a volatile material does not take place, some index other than total use is needed for quantifying emissions. In the simple case, when the amount used and the amount remaining can be determined, the amount released as vapor can be obtained by difference. When this is not possible, more sophisticated means such as exhaust air sampling must be employed.

With complete evaporation of a mixture of volatile materials, the quantity of each component that vaporizes is merely the quantity of that material in the mixture. Partial evaporation of a mixture, however, does not necessarily yield vapor quantity of each component in proportion to the quantity of that component in the liquid mixture. According to Raoult's law, the equilibrium partial pressure of each component of a perfect solution is the product of the vapor pressure of the pure liquid and its mole fraction in the solution

$$p_n = P_n X_n \tag{3}$$

where p_n = partial pressure of component n
P_n = vapor pressure of pure liquid n
X_n = mole fraction of component n in the liquid mixture

Thus, vapor yielded by partial evaporation from a mixture of volatile materials is richer in the more volatile and leaner in the less volatile components than is the original liquid solution. The use of Raoult's law permits estimation of emission rates of components of a solution when partial evaporation takes place. The composition of the parent solution in each case must be known; values for vapor pressures of pure liquids can be found in chemical handbooks. When a substantial fraction of a liquid mixture evaporates, the change in its composition as the fractions of more volatile components decrease should be taken into account in applying Raoult's law.

Raoult's law should be used with caution in estimating emissions from partial evaporation of mixtures; not all mixtures behave as perfect solutions. Elkins, Comproni, and Pagnotto measured benzene vapor yielded by partial evaporation of mixtures of benzene with various aliphatic hydrocarbons, chlorinated hydrocarbons, and common esters, as well as partial evaporation of maphthas containing benzene (17). Most measurements for all four types of mixtures showed greater concentrations of benzene vapor in air than were predicted by Raoult's law. Of five tests with naphtha-based rubber cements, one yielded measured values of benzene concentration in air in agreement with calculated values, the other four showed measured benzene concentrations in air to be 3 to 10 times greater than those calculated using Raoult's law.

Substantial deviation from Raoult's law is not always the case, however, even with benzene. Runion compared measured and calculated concentrations in air of benzene in vapor mixtures yielded by evaporation from a number of motor gasolines and found excellent agreement (18).

In the absence of other more certain means, Raoult's law can be used to estimate emissions generated by partial evaporation of mixtures that approximate ideal solutions or mixtures in which the solution is nearly pure in one component. The applicability of Raoult's law to the mixtures being assessed should be validated, or quantitative measurements of emissions should be done, if accurate emission values are needed.

For dilute solutions the partial pressure of the component present in lower concentration is given by Henry's law, expressed as follows:

$$p_n = H_n X_n \tag{4}$$

where p_n = partial pressure of component n
H_n = Henry's law constant
X_n = mole fraction of component n in the liquid mixture

Henry's law is also applicable to the solubility of a gas in dilute liquid solution, and solubilities of gases in liquid may be expressed in terms of Henry's law constants. These constants and the applicable concentration ranges for valid use of Henry's law can be determined only empirically. Henry's law constants for solutions of some common gases in water, and equilibrium data for some gases in a few other solvents may be found in handbooks such as *Chemical Engineers' Handbook* (McGraw-Hill).

Application of materials balance concepts for determination of emissions other than those for combustion, chemical reaction, or evaporation of volatile materials, ordinarily requires engineering analysis on a case-by-case basis. A procedure for estimating emission rates from multiple indoor sources has been reported (19). The same modeling procedure has been used for estimating emissions from open-top vapor degreasers (20).

3.3 Emission Factors

The need exists for emissions estimates for large numbers of sources in community air pollution studies. The impracticability of source-by-source emissions tests has led to the development of emission factors. An emission factor is a pollutant emission rate for a particular type of emission source expressed as a quantity of pollutant released per unit of activity of that type source. The unit of activity chosen in each case is one that can be determined and can be related quantitatively to emissions; it may be ton of product, million Btu of heat produced, mile of vehicle travel, or other such index unit. Emission factors represent typical emissions from a class of sources and ordinarily cannot be applied with confidence to individual sources. In the absence of other information, however, emission factors for a particular source type can give useful insights into the character and general levels of emissions from individual sources of that type.

The most reliable emission factors are those based on a combination of emission measurements, process data, and engineering analysis for a large number of sources. Those that do not have a theoretical basis and are derived from only one type of data, or from data from only a few sources, should be used with caution. In some cases sufficient knowledge or information is available to permit development of empirical or analytic relationships between emission rate and some process parameter such as material composition or stream temperature. Such factors are generally the most reliable of all and can even be applied to individual sources with reasonable confidence.

Several thousands of individual air pollutant emission factors for a large number of source types have been tabulated and reported by the EPA's Office of Air and Waste Management (21). Process descriptions and emission control practices, along with typical collection performance for various types of control, are presented for most of the source types covered. Table 31.1 lists major source types for which emission factors appear in the current publication. In a separate document the EPA has published emission factors

Table 31.1. Major Source Types for Which Air Pollution Emissions Factors Have Been Adopted[a]

1. External Combustion Sources
   - Bituminous and subbituminous coal combustion
   - Anthracite coal combustion
   - Fuel oil combustion
   - Natural gas combustion
   - Liquefied petroleum gas combustion
   - Wood waste combustion in boilers
   - Lignite combustion
   - Bagasse combustion in sugar mills
   - Residential fireplaces
   - Residential wood stoves
   - Waste oil combustion
2. Solid Waste Disposal
   - Refuse combustion
   - Sewage sludge incineration
   - Medical waste incineration
   - Municipal solid waste landfills
   - Open burning
   - Automobile body incineration
   - Conical burners
3. Stationary Internal Combustion Sources
   - Stationary gas turbines for electricity generation
   - Heavy duty natural gas-fired pipeline compressor engines
   - Gasoline and diesel industrial engines
   - Large stationary diesel and all stationary dual-fuel engines
4. Evaporation Loss Sources
   - Dry cleaning
   - Surface coating
     - Nonindustrial surface coating
     - Industrial surface coating
       - General industrial surface coating
       - Can coating
       - Magnet wire coating
       - Other metal coating
       - Flat wood interior panel coating
       - Paper coating
       - Polymeric coating of supporting substrates
       - Automobile and light duty truck surface coating operations
       - Pressure sensitive tapes and labels
       - Metal coil surface coating
       - Large appliance surface coating
       - Metal furniture surface coating
       - Magnetic tape manufacturing
       - Surface coating of plastic parts for business machines
   - Wastewater collection, treatment and storage
   - Polyester resin plastic products fabrication
   - Asphalt paving operations
   - Solvent degreasing
   - Waste solvent reclamation
   - Tank and drum cleaning
   - Graphic arts
     - General graphic printing
     - Publication gravure printing
   - Commercial/consumer solvent use
   - Textile fabric printing
5. Petroleum Industry
   - Petroleum refining
   - Transportation and marketing of petroleum liquids
   - Natural gas processing
6. Organic Chemical Process Industry
   - Carbon black
   - Adipic acid
   - Explosives
   - Paint and varnish
   - Phthalic anhydride
   - Plastics
     - Poly(vinyl chloride)
     - Poly(ethylene terephthalate)
     - Polystyrene
     - Polypropylene
   - Printing ink
   - Soap and detergents
   - Synthetic fibers
   - Synthetic rubber
   - Terephthalic acid
   - Lead alkyl
   - Pharmaceuticals production
   - Maleic anhydride
   - Methanol
   - Acetone and phenol
   - Propylene
   - Benzene, toluene, and xylenes
   - Butadiene

Table 31.1. Continued

- Cumene
- Ethanol
- Ethyl benzene
- Ethylene
- Ethylene dichloride and vinyl chloride
- Ethylene glycol
- Ethylene oxide
- Formaldehyde
- Glycerine
- Isopropyl alcohol

7. Liquid Storage Tanks
   - Organic liquid storage tanks
8. Inorganic Chemical Industry
   - Synthetic ammonia
   - Urea
   - Ammonium nitrate
   - Ammonium sulfate
   - Phosphate fertilizers
     - Normal superphosphate
     - Triple superphosphate
     - Ammonium phosphate
   - Hydrochloric acid
   - Hydrofluoric acid
   - Nitric acid
   - Phosphoric acid
   - Paint and varnish
   - Phosphoric acid
   - Sulfuric acid
   - Chlor-alkali
   - Sodium carbonate
   - Hydrogen cyanide
9. Food and Agricultural Industries
   - Tilling operations
   - Growing operations
     - Fertilizer applications
     - Pesticide applications
     - Orchard heaters
   - Harvesting operations
     - Cotton harvesting
     - Grain harvesting
     - Rice harvesting
     - Cane sugar harvesting
   - Livestock and poultry feed operations
     - Cattle feedlots
     - Swine feedlots
     - Poultry house
     - Dairy farms
   - Animal and meat products preparation
     - Meat packing plants
     - Meat smokehouses
     - Meat rendering plants
     - Manure processing
     - Poultry slaughtering
   - Dairy products
     - Natural and processed cheese
   - Cotton ginning
   - Preserved fruits and vegetables
     - Canned fruits and vegetables
     - Dehydrated fruits and vegetables
     - Pickles, sauces, and salad dressings
   - Grain processing
     - Grain elevators and processes
     - Cereal breakfast food
     - Pet food
     - Alfalfa dehydrating
     - Pasta manufacturing
     - Bread baking
     - Corn wet milling
   - Confectionery products
     - Sugar processing
       - Sugarcane processing
       - Sugar beet processing
     - Salted and roasted nuts and seeds
       - Almond processing
       - Peanut processing
   - Fats and oils
     - Vegetable oil processing
   - Beverages
     - Malt beverages
     - Wines and brandy
     - Distilled spirits
   - Miscellaneous food and kindred products
     - Fish processing
     - Coffee roasting
     - Snack chip deep-fat frying
     - Yeast production
   - Tobacco products
   - Leather tanning
   - Agriculture wind erosion
10. Wood Products Industry
    - Lumber
    - Chemical wood pulping
    - Pulp bleaching
    - Papermaking

Table 31.1. Continued

Plywood manufacturing	Perlite manufacturing
Reconstituted wood products	Abrasives manufacturing
Waferboard and oriented strand manufacturing	12. Metallurgical Industry
Particleboard manufacturing	Primary aluminum production
Medium density fiberboard	Coke production
Charcoal	Primary copper smelting
Wood preserving	Ferroalloy production
11. Mineral Products Industry	Iron and steel production
Hot mix asphalt plants	Primary lead smelting
Asphalt roofing	Zinc smelting
Bricks and related clay products	Secondary aluminum operations
Calcium carbide manufacturing	Secondary copper smelting and alloying
Refractory manufacturing	Gray iron foundries
Portland cement manufacturing	Secondary lead processing
Ceramic products manufacturing	Secondary magnesium smelting
Clay and fly ash sintering	Steel foundries
Western surface coal mining	Secondary zinc processing
Coal cleaning	Storage battery production
Coal conversion	Lead oxide and pigment production
Concrete batching	Miscellaneous lead products
Glass fiber manufacturing	Leadbearing ore crushing and grinding
Fruit manufacturing	Electric arc welding
Glass manufacturing	Electroplating
Gypsum manufacturing	13. Miscellaneous Sources
Lime manufacturing	Wildfires and prescribed burning
Mineral wool manufacturing	Fugitive dust sources
Construction aggregate processing	Paved roads
Sand and gravel processing	Unpaved roads
Crushed stone processing	Heavy construction operations
Lightweight aggregate manufacturing	Aggregate handling and storage piles
Phosphate rock processing	Industrial wind erosion
Diatomite processing	Abrasive blasting
Taconite ore processing	Explosives detonation
Metallic minerals processing	Wet cooling towers
Clay processing	Industrial flares
Talc processing	14. Greenhouse Gas Biogenic Sources
Feldspar processing	Emission from soils—Greenhouse gases
Vermiculite processing	Termites—Greenhouse gases
Alumina manufacturing	Lightning emissions—Greenhouse gases
	Enteric fermentation—Greenhouse gases

[a]EPA 1995, with Supplements through 1996.

for arsenic, asbestos, beryllium, cadmium, manganese, mercury, nickel, and vanadium for processes involving these materials (22).

Emission factors in the EPA tabulation generally apply to identifiable point sources in processes or operations from which pollutants are released to the atmosphere through vents or stacks. As such, they have limited applicability to in-plant industrial hygiene assessments. They do, however, identify some of the air contaminants that are generated by these processes; the same contaminants may represent potential in-plant exposures.

3.4 Fugitive Sources

Contaminant emissions from point sources such as tank vents or transfer points, and from processes and operations that clearly involve release of a process material—for example, release of volatile components in cementing or painting operations—are ordinarily capable of identification and quantification. There are other types of sources not so easily accommodated and sometimes neglected in emission inventories. Such sources, often called fugitive sources, may be intermittent, temporary, or unpredictable; many are unrecognized or ignored in emission inventories. Fugitive sources and the generation of secondary pollutants deserve attention in emission inventories, however, even though all of them may not be capable of quantification or prediction.

Fugitive sources that are of consequence primarily in the field of air pollution, and for which emission factors have been developed, include unpaved roads, aggregate handling and storage piles, heavy construction operations, and paved roads (21). Other sources that are consequential from the standpoints of both air pollution and industrial hygiene include the following:

1. Urban fires.
2. Industrial process fires.
3. Materials spills (accidents, equipment failure or malfunction).
4. Sample collection and analysis.
5. Process leaks (flanges and piping, valves, packing glands, conveyors, pumps, compressors, tanks and bins, etc.).
6. Relief valves and control device bypasses.
7. Maintenance activities (tank and vessel cleaning, filter cleaning, replacing piping, pumps, etc.).
8. Emissions from waste streams and reemissions of collected materials.
9. Secondary reactions (nonproduct process reactions, extraprocess reactions).

Urban structural fires are intermittent phenomena predictable only in the aggregate and not as individual events; they are considered to be emission sources primarily in the air pollution sense. An industrial hygiene consequence of urban fires, however, is the exposure to toxic materials of firefighters and others who may be involved in rescue or control activities. A number of toxic atmospheric contaminants are generated by structural fires; these tend to vary from one fire to another. Of concern in all structural fires, however, is emission of carbon monoxide. Burgess et al. (23) have described exposures of emergency personnel to carbon monoxide emissions in real fire situations; the maximum sustained air concentrations of carbon monoxide to which these persons were exposed in the events in which it was measured was about 2%. Materials balance using pyrolysis and oxidation processes offer one means of estimating the quantities of pollutants that may be generated in any particular case.

Estimates of emissions from industrial fires require individual analysis. The quantities and characteristics of combustion products depend on the nature of the materials that burn or are present in the fire, and on the circumstances of combustion (e.g., whether open or confined). Again, consideration of pyrolysis and oxidation phenomena may permit estimation of the nature and quantities of air contaminants generated by any particular event. The likelihood and consequences of accidental fires is an appropriate consideration in industrial emission inventories.

Material spills in manufacturing, transporting, and uses of industrial materials are not infrequent occurrences. Studies of processes and operations with specific attention to spills can reveal the frequencies and magnitudes of spills if they are usual occurrences (24). Such spills can then become part of an emissions inventory. When spills are not a usual occurrence, an emission inventory can do little more than trigger consideration of the possibilities and consequences of such events. Guidance for protection of workers when a spill of a hazardous substance occurs, developed as a joint project of the National Institute for Occupational Safety and Health (NIOSH), the Occupational Safety and Health Administration (OSHA), the U.S. Coast Guard, and the U.S. Environmental Protection Agency (EPA), has been published (25).

Process leaks are sometimes of major consequence from the standpoint of industrial hygiene. Control of process leaks, for example, has been a major factor in achieving acceptable working conditions in vinyl chloride monomer and polymerization plants. Process leaks can be identified by inspection or instrumental methods; timely repair may obviate quantifying for an emission inventory. Should quantification be necessary, the techniques of source testing or materials balance may suffice.

Predicting and quantifying emissions of particulate material from industrial sources is generally more difficult and more technically demanding then is assessment of emissions of gases and vapors. Quantifying emissions of fugitive dusts, for example, has generally been a difficult exercise. A chamber technique for quantifying fugitive dust emissions from handling of granular materials has been reported (26). A series of research papers from one group of investigators report predictive equations and validation exercises for emissions of dusts from transfer points granular materials being dropped for various heights at transfer points (27–30). This research has yielded a statistically derived mathematical relationship combining factors which are fundamental in generation of dust from parent granular material:

$$G_i = F_i^a(S^b/B^c) \tag{5}$$

where G_i = amount of dust of size i generated
F_i = fraction of particles of size i in the test material
S = particle separation forces

B = interparticle binding forces
a, b, c = exponents to be determined by measurement of fundamental properties of the parent material

Interested persons should examine reference 30 for values of coefficients and exponents derived from almost 600 gravitational transfer exercises with various parent granular materials. The investigators suggest that with additional work procedures to measure and characterize separation forces in other dust emission processes may allow application of the fundamental conceptual relationships presented in these reports to other processes such as mixing, grinding, rotary drying, or crushing (30).

As acceptable worker exposure levels become lower, attention to detail in emission evaluation takes on added importance. Routine tasks such as process sample collection may present a potential for excessive exposure when such sampling involves open manhole covers, open sample containers, or unconfined sampling streams. Bell et al. addressed this problem for vinyl chloride sampling and suggested techniques that may find application in other industries, particularly in petrochemical operations (31).

Release of materials to community air or to a workplace through process relief valves occurs from time to time. The frequency of operation of these devices and the magnitude of releases, obtained from plant records or other sources, can be used for emission inventory purposes. Recognition of the existence of relief valves or control device bypasses in a process is important in an emission inventory. A common source of emissions in processing plants is leakage from seals of pumps used to transport volatile liquids. It has been reported that emission rates can increase as much as five-fold when a pump stops; the reason for this is not known (32). Pumps designed to prevent leakage through shaft seals are now available.

Kletz has stated that many of the foregoing sources of fugitive emissions from plant processes can be controlled with emergency isolation valves (33). In addition to suggesting a method of control, this author also presents a leakage profile for both an olefin processing plant and an aromatic processing plant that could serve as a basis for developing a fugitive emissions checklist for use at similar operations.

Volatile air contaminants may be released to the atmosphere from process sewers, drainage ditches, and/or collecting ponds. Such releases are of concern from the standpoint of both air pollution and industrial hygiene. Two examples are offered. Consider a case in which a gravity separator is used in an enclosed benzene recovery system to separate the organic and aqueous streams. Water from this separation may be reused for other purposes elsewhere in the plant. Even though benzene is only slightly soluble in water, such water reuse can result in measurable concentrations of benzene in the air at workplaces in the plant where benzene is not used. Consider another case in which alkaline scrubbing water is used to remove fluorides from a process waste gas stream. Discharge of the scrubbing water to a waste pond in which the pH is low permits release to the atmosphere of the collected fluorides. The use of the air pollution control scrubber in such a case only relocates the site of fluoride emission from the process stack to the waste pond. Air sampling and materials balance techniques are means for assessing emissions from waste streams and reemissions of collected materials.

Descriptions of processes and products do not always reveal whether by-products are formed or whether any by-products and their parent chemicals are stable throughout the manufacturing process and in the environment. This issue has been discussed, and an approach to assessing the significance of by-products and secondary reactions have been presented (34). The likelihood and consequences of hydrolysis, pyrolysis, oxidation, and other reactions are factors in the assessment. The identification of processes and principal materials is the first step in exploring for secondary pollutant emissions in an emission inventory. Literature review and perhaps laboratory exercises may yield indicators of the presence of secondary contaminants; air sampling may be required for validation.

Although it is not always possible to make quantitative, or perhaps even qualitative, assessments of fugitive sources and secondary pollutants, the possibilities of their occurrence and the opportunities for intervention to protect workers and the community merit scrutiny in emission inventories.

4 IN-PLANT EMISSION ESTIMATES FROM RECORDS AND REPORTS

In the absence of in-plant measurements of emissions, estimates derived from secondary sources of information can be used for inventory purposes. In some cases data are available from similar processes, and estimates by analogy may prove to be sufficient for a given purpose. In other cases unique features of a process or operation may make estimates by analogy inappropriate. An alternative to measurement data or analogy makes use of records and reports whose basic purposes were other than for industrial hygiene, but whose content may be extracted, combined, or otherwise manipulated into usable estimate of emissions and potential exposure. Vital information may be recorded in a variety of business documents including engineering, accounting, production, quality control, personnel, and governmental records. Although these records represent a rich source of information for emission inventory purposes, they can serve other purposes in the practice of industrial hygiene and occupational health research as well. Some of these purposes are mentioned from time to time in the following discussions of specific types of records.

4.1 Research and Development Records

Industrial research and development units often issue a variety of reports to producing units as aids to bringing a new product on stream, to ensure a required level of uniformity in a given item produced by different plants, or to make changes or improvements in operating equipment, materials, or techniques. Valuable insight with respect to materials and work practices employed in the past may be gleaned from several of these records. Among such records are the following:

Product Specification. Final product and intermediate component data included in a product specification often given dimensions, weight, materials of construction, processing aids, processing conditions, tools and fabricating equipment, and special notes.

Standard Operating Procedure. This record may exist under a variety of names; SOP, Standard Practice, Uniform Methods, and (Company Name) Manual of Operations appear among the myriad of titles. Many, however, share fairly specific operational instructions, usually in a step-by-step approach by processing sequence. Vital data from this record can include:

1. Fabrication equipment.
2. Authorized materials of construction.
3. Acceptable processing aids.
4. Processing conditions.
5. Alternate materials for cyclic operations (e.g. summer vs winter stocks).
6. Change notices affecting materials or techniques.
7. Instructions regarding protective equipment.

Formulations and Raw materials. Detailed data concerning the qualitative aspects of potential exposures can be derived from study of information describing the raw materials employed in a process.

1. Raw material specifications may contain composition data, restrictions on contaminants, and acceptance testing schedules.
2. A listing of authorized vendors can offer clues to potential contaminants based on knowledge of a vendor's source of supply or processing methods.
3. Formulation records frequently contain information on the composition of mixtures as well as the methods, equipment, quantity, and conditions of processing.

4.2 Analytical Laboratory Records

Quality control considerations often dictate that incoming raw materials be monitored for selected chemical and physical properties, that supplies from prospective vendors be subjected to acceptance testing, and that production quality be assayed by sampling at intermediate stages of the process. A single journal entry of analytical results may include a wealth of intelligence applicable to an emissions inventory. Analytical laboratory records may include some of the following:

Raw Material Test Results. Recorded in either the analyst's journal or on prepared forms will be entries such as:

1. Analytical results vis-a-vis specifications.
2. Date of analysis, and perhaps date of receipt.
3. Quantity of the shipment.
4. Vendor.
5. Method of analysis.
6. Analyst.

Authorized Vendors. Quality control laboratories often maintain vendor-supplied information regarding purchased materials. Material Safety Data Sheets, product specifications, production methods, and quality control statements all offer bits of useful data.

4.3 Process and Production Records

Records generated by production units relate to such matters as scheduling, quality control, cost control, and to a lesser extent, training exercises. Depending on the product and process, production-type records can be of considerable value in generating an emissions inventory.

Flow Charts. A detailed diagram of a process is particularly useful in identifying points and potential for release of air contaminants.

Operating Conditions. Time, temperature, and pressure data are useful in assessing possible release of reaction products, by-products, degradation products, and unreacted raw materials.

Scheduling. Shift, cyclic, and seasonal variations may influence the emissions of contaminants. Such variations can often be ascertained through production scheduling records. Included under this heading are records of batch versus continuous production modes. Gantt chart records offer a particularly attractive systematic source of scheduled activities which can be both precise and detailed.

4.4 Plant Engineering Records

Virtually all industrial organizations include an engineering component which is responsible for the physical plant and supporting facilities. This responsibility has been known to result in a plant archive that is particularly useful in emissions inventory activity and in research that requires reconstruction of past environmental conditions.

Plant Layout. The plant floor plan, coupled with elevation drawings, affords a visualization of the process, materials flow, occupied areas, and possibly insights relating to contaminant generation and control. Furthermore, many engineering drawings are cross-referenced to other drawings of equipment, emissions control features, and adjacent work areas. Useful characteristics of plant engineering files include:

1. Processing equipment type, extent, and location on layouts.
2. Department floor space showing geographic boundaries and physical barriers.
3. Engineering controls, particularly ventilation systems, are depicted in mechanical equipment layouts, which normally show the location, site, type, and rating of air-moving equipment, as well as a reference to the detailed plate covering that system.
4. Dates of change or modification are often included on drawings, either by direct entry or by reference to a new set of plates.

Equipment Specifications. Files on processing equipment installed by or under the supervision of plant engineers often include specifications that supply clues to potential problems such as process emissions, sound power rating and directivity data, or emission control features incorporated in the equipment.

Project Records. Each major project of construction, modification, or installation normally proceeds as an integral project with an individual(s) assigned as coordinator. Usually this project officer accumulates records that include data on:

1. Equipment specifications.
2. Performance checks and acceptance test results.
3. Prime contractor and subcontractors.

Material Balance. A detailed accounting of materials, products, and side streams, either measured or theoretical, can serve as a basis for a quantitative estimate of emissions.

4.5 Industrial Engineering Records

The practice of industrial engineering includes the analysis of work procedures to improve either work conditions or productivity or both. Especially for the period following World War II, industrial engineering records contain information on work methods and standards that are useful for compling emission inventories and for estimating past emissions and exposures that may have resulted from them. Such records include:

Job Descriptions.

1. Description numbers, department listings, and simple descriptive job titles can provide the link between the tasks performed and the work history recorded in an individual's personnel file.
2. Job location within a plant, thus the potential for various exposures, may be ascertained from the geographic area, processing equipment, or department listing contained in a job description.
3. The listing in a job description of tools, equipment, and processing aids can help establish the nature of emissions and characteristics of potential exposures (e.g. inhalation, skin absorption).
4. Task evaluations often include judgments concerning such health-related items as work conditions, safety requirements, hazards, skills, effort, and responsibility.

Time and Motion studies. Task breakdowns are useful to establish contact with materials of interest, and the fraction of time spent on certain tasks coupled with cycle time can afford an estimate of the extent of exposure.

Process Analysis. Evaluations of an entire process focus on tasks performed by all the members of a crew rather than those of individual workers. Particularly useful in process analysis reports are listings of fractions of a workday spent at different workstations by members of the operating crew, including break time and sequencing of tasks.

4.6 Accounting Records

Accounting records cover virtually all phases of an operation for both outside reporting and internal control purposes. Major areas of interest for emission inventory purposes include records on incoming materials, internal use of these raw materials, and plant production factors (35).

Purchasing. Raw materials for both production and support operations, as well as processing equipment, are ordinarily obtained from outside the company, resulting in some acquisition record. Even when raw materials are drawn from captive sources, records of resource depletion may be available.

1. Listings of materials, quantities, and supplier may be available through the purchasing ledger, often found in the accounts payable portion.
2. Capital equipment is normally amortized by a recognized accounting procedure, the evaluation of which can be employed to determine the date of acquisition. Alternatively, many organizations maintain a property book that contains similar information.

In-plant Issues of Materials

1. Data on raw materials consumption often can be developed through accumulation of issues to departments, cost centers, or plants, or through collected expense or work-in-process accounts.
2. Material variance accounts may serve as surrogates for material balance data if the latter are unavailable. The absolute variance value between actual and standard material usage may be of marginal value, but changes in this quantity can be important in establishing trends.
3. Nonproduction supplies are often issued to maintenance and other support departments through a "general store" account, with the receipt listing the item and quantities issued.

Inventory. Inventory control may be maintained through stock records that list material, vendor, date, and quantity received, date and quantity of issue, department to which issued, and location and quantity on hand.

Production. Finished goods and work-in-process production accounts may include an identification of the work schedule, days operated, and average production level.

4.7 Personnel and Other Personnel-type Records

Personnel records, *per se*, provide some insights into the nature and locations of contaminant releases applicable to emission inventories. In addition they are vital to other industrial hygiene and occupational health research activities. In any study of causal association between conditions of work and health experience, the work experience of each individual is an integral component. Thus, accurate and reliable personnel-type records are vital. Organizational units which normally can develop, or assist in developing, useful records include the personnel, payroll, industrial relations, safety, and medical departments.

Personnel Records. The file on a particular employee typically lists department and job assignments chronologically and provides a brief description of tasks, limitations, and periods of absence. Earnings records sometimes provide a clue to potential exposures through a listing of special pay or rates for hazardous tasks, and exposure duration or regimens by shift differentials or overtime pay.

Industrial Relations. Seniority listings and work force distribution charts are useful in establishing the locations of jobs and personnel currently assigned or employed at various times in the past.

Safety and Medical. Insurance carrier survey reports, fire and explosion inspection reports, sickness–accident–compensation records, administrative control documents that limit time at a particular task or require worker rotation, along with written SOPs for activities such as entering tanks and vessels and other hazardous tasks, all represent the prospect for a record of emission sites and past working conditions. These records, however subjective, may be suitable to rank-order jobs by exposure potential to certain agents.

Union Records. Seniority listings and union membership rolls can help to identify the working populations in the past, Records of negotiated agreements between labor and management often contain provisions for work schedules, protective clothing and equipment, provisions for personal hygiene, clothing changes, and other health-related clauses; all of which help to describe conditions of work.

4.8 Government Mandated Records and Reports

The 1980s may be described as the decade of disclosure regarding potential exposure to toxic and hazardous chemical agents. As part of the general "right-to-know" movement, OSHA, EPA, and state agencies promulgated numerous regulations intended to ensure that employees, customers, the general public and government agencies would have the necessary information with which to evaluate potential exposures resulting from chemical releases to the work place and ambient environment. The resultant records and reports provide valuable information to compiling the emission inventory, and will sometimes afford insights to potential exposure conditions. Copies of past reports of this type are likely to be filed in the plant or corporate office(s) which deal(s) with environmental affairs and/or which deals with government regulatory agencies.

Some of the sentinel government requirements are briefly described below along with insights to potential exposure conditions.

4.8.1 OSHA Hazard Communication

OSHA Hazard Communication Standard (**29** *CFR* 1910.1200) (36) requires that employers inform their employees of potential hazards associated with chemical agents in their places of work. It also requires that manufacturers and distributors of regulated chemicals make hazard information available to their downstream customers. Chemical manufacturers, distributors, and processors have achieved compliance through written programs, employee training, container labeling and Material Safety Data Sheets (MSDS). Firms will generally maintain documentation of all sectors of their program for compliance purposes. As a result of hazard communication requirements being effective in all OSHA regulated sectors, MSDSs are now ubiquitous in work places across America.

4.8.2 Access to Medical (and Exposure) Records

In 1980, OSHA promulgated its access to medical records rule (**29** *CFR* 1910.20) which requires employers to maintain exposure records for 30 years, and to inform employees of their right to review their individual exposure records (36). Many firms routinely advise workers of monitoring results through work place postings or personal letter reports. This general rule is applicable to all exposure records not the subject of a specific OSHA requirement.

4.8.3 OSHA Standards Governing Specific Chemicals

Several OSHA standards governing exposure to specific chemical agents include requirements for work place monitoring and maintaining exposure records beyond the provisions of the general rule identified in the preceding section. Examples of these requirements are found in the standards for benzene (**29** *CFR* 1910.1028), lead (**29** *CFR* 1910.1025), and formaldehyde (**29** *CFR* 1910.1048).

4.8.4 Toxic Substances Control Act

The Toxic Substances Control Act (6) imposed on chemical manufacturers and distributors a number of new testing, recordkeeping and reporting obligations. Under Section 8(e) of TSCA, anyone observing a new substantial risk to health or the environment resulting from chemical exposure must report this finding to the USEPA within 15 days of discovery. EPA in turn, summarizes the results of TSCA 8(e) reports and makes these summaries available to the general public.

Under the requirements of TSCA Section 8(c), chemical manufacturers and distributors are required to maintain records of significant adverse reactions alleged to be caused by a (specific) chemical substance or mixture whether or not this allegation can be substantiated by the firm. These internal company files are available to the EPA for inspection purposes.

Of particular usefulness in in-plant emission inventories are agent listings prepared for compliance with regulatory requirements under the Toxic Substances Control Act (6). Regulations under TSCA require the listing of any of some 70,000 agents (current number), by CAS number or trade name, which are used in a plant or which may be present in plant processes (37). Industrial hygienists and engineers with knowledge of plant processes can make judgments regarding the potential for release of each agent in these plant-specific lists in each workplace in a plant. The determination may be either qualitative, i.e., a yes/no determination that a particular agent is present at a particular workplace, or semiquantitative with ordinal estimates of frequency of release and of magnitude of emissions for each agent and workplace. Lynch (38) has described the application of such a system for estimation of workers' exposures in petrochemical plants.

4.8.5 SARA Title III

The Emergency Planning and Community Right to Know Act (EPCRA), also known as Title III of the Superfund Amendments and Reauthorization Act of 1986 (39) established comprehensive requirements for public reporting of chemical inventories and releases to the environment. Reports are required of facilities operating in the manufacturing sector (SIC Codes 10, 12, 20–39 and some parts of 49, 51, and 73), which employ 10 or more full-time employees and which manufacture, process or use any listed chemical agent in quantities equal to or greater than threshold quantities established by the federal government. The list of regulated substances in July 1999 stands at 579 chemicals plus 28 chemical categories. Threshold quantities necessary to trigger reports are 25,000 pounds per year of manufactured or processed chemical agents, and 10,000 pounds of any listed material otherwise used on an annual basis. Some agents characterized as especially hazardous have lower threshold quantities. The EPA maintains a SARA Title III hotline, 800 424-9346 for questions about SARA, and reporting forms, lists of regulated chemicals, and reporting requirements can be found at the internet address ⟨*www.epa.gov/opptintr/tri*⟩.

SARA reports fall into two basic categories: on-site inventory (Sections 311 and 312) and releases to the environment (Section 313 and Form R). Under the provisions of Section 311, regulated facilities are required to submit either a list of chemical substances stored or an individual MSDS for each regulated chemical. Section 312 requires owners or operators to file inventory reports listing the amounts and locations of regulated chemicals either by hazard class (Tier I report) or by specific chemical agents (Tier II reports). Inventory reports are filed with the Local Emergency Planning Commission (LEPC), the local fire department and the State Emergency Response Committee (SERC) (**40** *CFR* Part 370, 52 FR38344).

Section 313 of SARA requires an annual report estimating quantities of any of the some 600 regulated toxic chemicals released to the environment (**40** *CFR* Part 372, 53 FR4500). Regulated toxic chemical release inventory reports must be filed with the SERC in the state wherein the reporting facility is located and the EPA. Individual facility, toxic release inventory (TRI) reports are compiled by EPA and made available to the public for a nominal fee. Under SARA, the list of regulated toxic chemicals and threshold reporting quantities are subject to change from time to time. Beginning in 1992, additional reporting requirements were added to Form R as a result of the Pollution Prevention Act of 1990 (EPA Form 9350-1; revised 5/14/92).

Industrial hygienists may be able to use reports required under SARA Title III for estimating in-plant emissions and potential for workplace exposures in the same ways they may use TSCA reports as described above.

4.8.6 Resource Conservation and Recovery Act

Owners and operators of Transportation, Storage and Disposal (TSD) facilities managing Resource Conservation and Recovery Act (40) regulated hazardous wastes incur numerous obligations to maintain records and submit reports detailing materials handled and conditions in effect during periods of operation. Some of the more instructive reports for purposes of the emission inventory include manifest and shipping papers (**40** *CFR* 264.71), inspection records (**40** *CFR* 264.15) and the biennial report of operations (**40** *CFR* 262.4).

4.9 Governmental Records and Reports

Data so broad as to characterize an entire industry, or so specific as to deal with a single chemical compound, are routinely and systematically collected, assembled, and published by various government agencies. Most of these records are available to the enterprising investigator, usually at little or no cost. A few of these record types are mentioned here as illustrations.

Nationwide Statistics. The U.S. Tariff Commission publishes an annual listing of Production and Sales of Synthetic Organic Chemicals. It includes listings of raw materials, the production chemicals by use category (surface-active agents, etc.), and a directory of manufacturers of each material. Time series analyses of the entries can indicate the dates of introduction, extent of use, and dates of decline in use of a given material.

Industry-wide Data. Government directed or controlled operations such as government owned-contractor operated plants compile data for required reports, many of which find their way into the public domain. For example, information on virtually all facets of the synthetic rubber industry during the 1940–1945 period are available through the various reports of the Rubber Reserve Company (41, 42). Included are flow charts, raw materials, formulations, plant capacities, and locations.

Specific Products. On June 15, 1844, Charles Goodyear received U.S. Patent 3633 for a compound consisting of gum elastic, sulfur, and white lead (43). Thus, the approximate period of time at which rubber compounders could be considered as potentially exposed to white lead as a rubber accelerator is established. More than five million U.S. patent numbers have accrued since the mid-nineteenth century, doubtless including many materials, processes, and devices of interest to health investigators.

4.10 Record Location

Within many corporate industrial organizations, records of major additions or changes, as well as events affecting the entire operating units, are likely to be available at a central location, such as the division or corporate offices. Examples of these types of records are overall production levels, descriptions of uniform practices, and product specifications. More specific records, or those dealing with minor plant alterations, are ordinarily available only at the local plant level. Current records are ordinarily found in the departments identified in Sections 4.1 through 4.8 of this chapter; historical records may be found in those departments or in a central plant storage facility. At the plant level project officers often retain personal copies of papers related to their projects; if a plant record has been discarded, such personal records may provide valuable historical data.

4.11 Output from Records

Measures derivable from secondary sources of data are limited by the nature and detail of the record content. In general, the more distant the time of interest, the less complete will be the record, and the lower will be the confidence in its accuracy. When data are sparse and only plant-wide data are available, estimated measures such as gallons per cubic foot of plant building volume per month, or pounds used per employee per day, may be the limit of detail for emission estimates, but this may serve for plant-to-plant contrasts. More detailed data such as quantities of materials used in specific departments and descriptions of controls for specific work areas permit more detailed estimates. Material generation rates coupled with general area ventilation rates may enable one to estimate an area average concentration. Such data may be particularly useful for within-plant contrasts or for plant-to-plant studies of particular areas or processes.

Records pertaining to the detailed nature of tasks, such as standard operating procedures or job descriptions, make possible a ranking of jobs—thus a ranking of increments of time in individuals' work histories—by potential for exposure to any particular agent or combination of agents of interest. In this fashion personnel with increments of time in their work histories having similar exposures, e.g. homogeneous exposure groups, (HEGs) (44), can be grouped for comparison of health experience with other workers who share different common exposure profiles, i.e., who have histories of work in other HEGs.

As an example of the use of secondary information, an illustration of the type of emissions output that has been accomplished for retrospective research purposes through the use of such sources of information follows.

In Figure 31.1 the standard operating procedure is the pivotal document that identifies both tasks and materials for a step in manufacturing. A formulation record, or series of records, for each material used in that particular manufacturing step identifies each raw material constituent of each material. Examination of the raw material specification for each constituent then permits identification of agents (e.g. coal tar naphtha) that may be emitted in the workplace, often with additional compositional detail (e.g., aromatic content or benzene content). Because formulations, constituents, and specifications change from time to time, the listing of potential emissions thus identified must show inclusive dates on an agent by agent basis.

The tasks portion of the SOP includes a job description, which in turn has a specific job code. The job code is listed in the personnel record of each person who has held that particular job. Examination of the personnel records of the plant's current and past workers

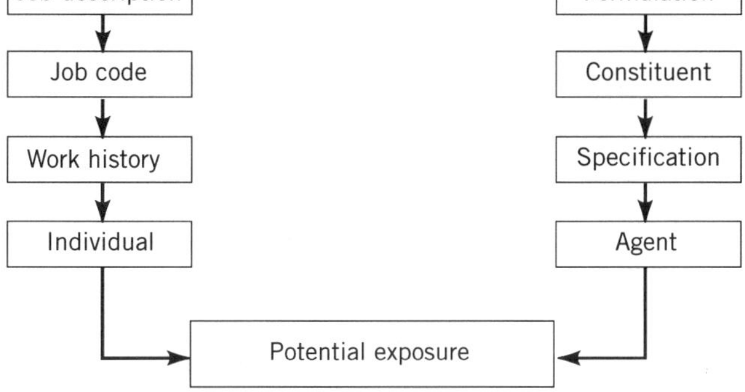

Figure 31.1. Synthesis of potential exposures associated with a manufacturing step for which a standard operating procedure is available.

identifies complete work history, including dates and duration of each job code assignment for each individual who has held the specific job of interest.

In this way the potential for contact with a particular chemical agent by an identified individual, including dates and duration of the contact, can be developed, even to events that occurred in the distant past.

5 USES OF EMISSION INVENTORIES

Emission inventories are fundamental to the identification, evaluation, and control of industrial hygiene and community air pollution hazards associated with chemical agents. Not all uses for inventory information can be examined here, but a few are discussed.

5.1 Regulatory Requirements

Various governmental regulatory programs in occupational safety and health, air pollution, and toxic substances control have reporting requirements that involve emission inventory information. Regulations regarding the nature and content of these reports, the agents covered, and the types of establishment with reporting obligations change periodically. Persons responsible for industrial or commercial enterprises that involve emissions to community air, or the handling or production of agents that may be covered in occupational health or toxic substances regulatory programs, should keep abreast of the specific requirements imposed by state and federal regulations. Emission inventory programs can then be so designed and operated that the emissions data necessary to satisfy reporting requirements are obtained.

5.2 Community Air Pollution Dispersion Estimates

Data on emissions of a pollutant to community air and companion meteorological data permit estimation and prediction of ground level concentrations of that pollutant at any desired location in the community. Methods and procedures for manual calculations to estimate the impact of individual sources (45), and a basic computer program for calculating concentration resulting from emissions from large numbers of point and area sources (46) have been reported. The mathematical relationships and computational techniques are too extensive to be repeated here. Application is best undertaken with the workbook or guide in hand. Suffice it to say that both methods can be used by engineers and industrial hygienists when the necessary data on emissions, source locations, receptor locations, and meteorology are available.

5.3 Workplace Exposure Estimates

Qualitative estimates of the potential for emissions and consequent exposures of workers can be made when agents have been identified and conditions of their use are known. Even the most rudimentary emission inventory—the identification of agent, process, and location of use—permits this kind of assessment. An operating program in which the industrial hygiene department of a large research facility systematically obtains descriptions of all uses of the many hundreds of chemical agents available at the facility has been described (47). The system includes a hazard rating for each agent based on its toxicity, flammability, reactivity, and special properties such as may be found in the agent's Material Safety Data Sheet. Use information does not identify emissions *per se*, but retrieval of use information for a particular agent permits rapid subjective assessment of its emissions potential throughout the facility. Quantitative information on emissions, either measured, calculated, or estimated, can be combined with data on workplace ventilation to estimate workroom concentrations of an agent. Such estimates may occasionally be of use in validating sampling data, but are more likely to have value in such applications as assessing growth or decay of concentrations in nonequilibrium situations, in predicting the consequences of new processes or changes in operations, or in estimating retrospectively the levels of a contaminant to which particular groups of workers may have been exposed. The latter use is of particular interest in reconstructing emergency or accident episodes, or in retrospective epidemiological studies of worker populations.

5.3.1 Gases and Vapors

Emission rates of gases and vapors are used in the mathematical relationships that yield values for concentrations of contaminants in workroom air. In addition to emission rates, one must know physical dimensions or volume of the workplace and the ventilation rate. The mathematical relationships (48) were derived using the concept of materials balance. The rate of change in the quantity of contaminant in the air of a workplace is described by equation 6. All emissions of a contaminant to workroom air comprise a contaminant generation rate, G.

$$V(dX/dt) = G + QX_s - QX \tag{6}$$

Integration of equation 6 yields:

$$X_t = X_i \exp(-Qt/V) + (X_s + G/Q)[1 - \exp(-Qt/V)] \tag{7}$$

where V = volume of the ventilated workplace (m^3).

G = rate of contaminant generation as volume per unit time of contaminant gas or vapor (m^3/min).

Q = effective exhaust ventilation rate, volume/unit time of air plus contaminant (ordinarily the volume rate of contaminant, G, can be ignored in Q; furthermore, only with perfect mixing can the actual ventilation rate be used in dilution ventilation calculations, see the following discussion of the factor K) (m^3/min).

t = interval of time elapsed since conditions represented by X_i and t_0, i.e., $t = 0$ (min).

X = fraction of contaminant in workplace air (10^{-6} ppm)

X_i = fraction of contaminant present in workplace air at time $t = 0$.

X_t = fraction of contaminant present in workplace air a time t.

X_s = fraction of contaminant in incoming dilution air.

Equation 7 applies only for intervals of time during which G and Q are constant. If either G or Q changes, then a new value of X_i and a new t_0 must be established, i.e., X_t at the end of one time interval of constant G and Q becomes X_i for the beginning of the time interval which immediately follows with a new t_0.

At equilibrium, when t is great, equation 7 simplifies to:

$$X = X_t = X_s + G/Q \tag{8}$$

or with equilibrium and clean air supply:

$$X = X_t = G/Q \tag{9}$$

When the concentration of the contaminant in the workplace air increases with time, the maximum concentration for the time interval t is X_t; for decay situations the maximum concentration for the interval t is X_i. Exposure guides and standards may list both maximum allowable short-term concentrations and allowable average concentrations. The average concentration over time t, also derivable from materials balance, is

$$X_{\text{av}} = X_s + G/Q + (V/Qt)(X_i - X_s - G/Q)[1 - \exp(-Qt/V)] \tag{10}$$

As is the case with equation 7, at equilibrium, when t is great, equation 10 simplifies to equation 8.

Equation 7 may be used to calculate the concentration of a gas or vapor remaining in a vessel or tank at any time after purging ventilation (decay) or contaminant introduction (growth) has begun. It may also be used in the following form to estimate the interval of time required for a particular ventilation rate to cause a particular concentration X_t to be reached:

$$t = V/Q \ln[(G + QX_s - QX_i)/(G + QX_s - QX_t)] \tag{11}$$

In some repetitive operations the generation rate for a contaminant may vary in a cyclical manner, being at one rate for one interval of time and at a different rate for another interval. If, in such a case, the greater generation rate is identified as G, the lower rate as G', and the time intervals t and t' are companion to G and G', respectively, when the cycle has equilibrated so that the workroom concentration pattern has become repetitive, the maximum concentration that occurs during the cycle is represented by:

$$X_{\text{max}} = \frac{(X_s + G/Q)[1 - \exp(Qt/V)] - (X_s + G'/Q)[1 - \exp(-Qt'/V)]}{\exp(-Qt'/V) - \exp(Qt/V)} \tag{12}$$

For the same cyclic operation after equilibrium has been reached, the average concentration over a large number of cycles is

$$X_{\text{av}} = X_s + (Gt + G't')/(Qt + Qt') \tag{13}$$

In the simple case when there is no generation during the time interval t' the term G' becomes zero in equations 12 and 13.

In 1988, for use in a research proposal, Dr. Michael R. Flynn, Department of Environmental Sciences and Engineering, University of North Carolina at Chapel Hill, derived equations which describe change in concentration with time, and expected inhalation dose, when a volatile solute aerates from a nonvolatile solvent during a batch process when there is no replacement of solute. In such a batch-type process the concentration of solute in solution decays with a half-time characteristic of the solute and the process. The instantaneous generation rate, G, is proportional to the concentration of solute in the solution, so the value of G will decay with time. The value of G at any time, t, after the initiation of such a batch process is represented by:

$$G_t = G_o \exp(-K't) \tag{14}$$

where G_t = the generation rate at time t
G_o = the initial generation rate at t = 0
K' = 0.693/half-time of solute concentration in the solution

The substitution of $G_o \exp(k't)$ for G in equation 6 and integration yields for air concentration at time t after beginning of a batch process:

$$X_t = X_s + [G_o/(Q - K'V)][\exp(-K't)] + \{X_i - X_s - [G_o/(Q - K'V)]\}[\exp(-Qt/V)] \tag{15}$$

Equation 15 is applicable over any time interval following the beginning of a batch process with decaying generation rate described by Equation 14, which has an initial value G_o, and for which Q is constant.

The integration of Equation 15 over time, t, during which the workspace is occupied after beginning of the batch process yields the concentration–time product (i.e., the inhalation dose for a worker who is present) for decaying generation rate (batch process):

$$Xt = X_s t + [G_o/K'(Q - K'V)][1 - \exp(-K't)] + (V/Q)[X_i - X_s - G_o/(Q - K'V)][1 - \exp(-Qt/V)] \tag{16}$$

In real workroom situations instantaneous perfect mixing does not take place. Usually when dilution ventilation is applied, the contaminant is emitted at locations that are in the occupied portion of a workroom; not all the ventilation air mixes with the contaminant while it is in this occupied space. With known or estimated true values of ventilation rate an empirical factor K may be used in calculation of estimated exposures to account for departures from perfect mixing. The definition of K is

$$K = Q_{\text{actual}}/Q \tag{17}$$

where Q_{actual} = ventilation rate, volume/unit time, as measured or obtained from engineering records (m^3/min).

Q = ventilation rate, volume/unit time, that is effective in determining the concentration at an exposure site, i.e. the value to be used in calculations (m^3/min).

The effective ventilation rate, Q, is to be used in equations 7 through 16 for calculation of exposure estimates. The choice of a value for K is a matter of judgment for an investigator; it is ordinarily in the range of 1 to 10 (49). Calculations using engineering records for ventilation rates, actual emission rates, and a number of concentration measurements of three different contaminants in each of two actual dilution ventilation situations, have yielded K values ranging from 2.2 to 5.0 with a mean value of 3.25 (50). In these two dilution ventilation situations, solvent vapors were being emitted in a number of locations in the workrooms at elevations that were approximately the same as breathing zone and sampling elevations. They may be considered to represent generally good mixing of emissions and ventilation air in actual application of dilution ventilation in industrial situations. The observed values are consistent with the lower values of the range of typical K values; the greater the departure from efficient mixing of emissions and ventilation air, the greater would be the value of K. It should be noted that K is not a safety factor; it is an index of departure from perfect mixing. For example, for a dilution ventilation situation in which all dilution air and all emissions pass through a single exhaust fan, the K factor for the discharge air at the fan location has the value 1. Jayjock describes a box model for estimating workplace exposures to volatile materials and discusses dealing with departures from perfect mixing (51), he also discusses the relatively ineffectiveness of dilution ventilation in controlling exposures from sources with large vaporizing surface area to room volume ratios (52).

Investigations in the choice of, and use of, the K factor continue; the results of a large number of computational simulations were reported at the 1999 AIHCE (53). Although knowledgeable use of a K factor is a convenient, and often adequate, means for dealing with departures from perfect mixing of contaminants and dilution ventilation air, more quantitative means have been described. A two-zone model in which quantitative concentration relationships air in the lower zone ordinarily occupied by workers, and in the upper ordinarily unoccupied zone where ventilation air may circulate, has been described (54). This paper contains appendices in which mathematical relationships for build-up and decay of contaminant concentrations in the two zones of a two-zone room are derived.

In an excellent analysis of dilution ventilation relationships Roach (55) has shown that the variance of emission, G, is inversely proportional to the length of the time interval t over which emissions take place, and that the variance in concentration, X, is then inversely proportional to the product of ventilation rate Q and V^2, the square of room volume. Thus, for a given average emission rate, the smaller the workspace and the smaller the ventilation rate, the greater will be the fluctuations in concentration, while the greater the volume of the space and the greater the ventilation rate, the smaller will be the fluctuations.

Roach has also described how the effectiveness of exhaust ventilation for reducing concentrations in the vicinity of employees is influenced by turbulent diffusion as well as by the bulk air movement (56). This reference is very worthwhile reading for anyone engaged in design or assessment of exhaust or dilution ventilation facilities.

With perfect mixing in a dilution ventilation situation, measurement of workplace concentration and application of materials balance equations (e.g., Eqs. 7–9) permits calculation of a contaminant generation rate. In many practical workroom situations, however, in which concentration may vary from one location to another, even with constant generation rate, these calculations probably will not yield a true generation rate. Application of turbulent diffusion relationships in the context of near-field far-field concepts (near-field may be defined as that space within, for example, one meter radius of the point of interest [i.e. an exposed worker or a generating source], and far-field may be the entire workplace volume) can address this matter (57). This reference presents a hemispherical diffusion model and mathematical relationships for dealing with turbulent diffusion from a point source.

5.3.2 Particulates

Caution must be exercised in the application of dilution ventilation theory and rationale, as described for gases and vapors, to particulate contaminants that are released to workroom air. Gravitational settling may be an appreciable factor in the decay of air concentrations of particulates. The smaller the particles, the greater is the likelihood that they will remain airborne and the more applicable are dilution ventilation relationships. When emissions of submicrometer sized particles—for example, most components of smoke and metal oxide fumes—are known, concentrations can be approximated by use of the dilution ventilation equations for gases and vapors.

Dusts with a wide range of particle sizes do not necessarily behave in air in the same way as fumes. Estimates of the relative concentrations of respirable dusts may be made with dilution ventilation equations when particulate emissions data are sufficient for quantification of the respirable fraction.

In assessing exposures to particulates, particles larger than those considered respirable should not be ignored. Of these larger particles, those that are inhaled and deposited in the nasopharynx and tracheobronchial system before they can reach the pulmonary spaces of the lungs represent ingestion exposures. Airborne concentrations of such particles generally decay at appreciable rates because of gravitational settling; dilution ventilation relationships do not apply to them.

6 RECORDS RETENTION PROGRAM

Emissions records may be generated for a variety of purposes. Some records are for purposes that can be served in a relatively short time; then the record may be considered a candidate for discarding. Other emissions records may be generated specifically for purposes such as health research, which require long retention. Emissions records made for various short- or long-term purposes may have secondary use in long-term community air pollution or occupational health research. Whether they represent typical operations or unusual events, such records may be invaluable in the future for reconstructing conditions of work to compare with long-term health experience of persons who engaged in that work. Since not all specific agents of future interest are identifiable in advance, it is prudent to retain emissions records for as many agents as possible.

Emissions records, or records from which emissions estimates can be made, such as those identified in Section 4 of this chapter, should be retained for a long time. A single specific age at which a record no longer has value cannot be stated. For research purposes exposure records have value equal to, or greater than, that of medical examination records. Because long latent periods are associated with health effects of some agents, early exposure records may be invaluable in the search for causal associations. Exposure records, or emission records, that identify contact with agents or from which exposure estimates can be made, should be retained long enough to be used in studies of the mortality experiences of the worker populations to which they apply.

The need for long-term retention of environmental records is recognized in the record keeping provisions of the Occupational Safety and Health Administration regulation of toxic and hazardous substances (36). Time periods for which employee exposure records must be retained vary with the substance being regulated. For example, records of exposures to asbestos must be retained for at least 20 years, those for vinyl chloride must be retained for not less than 30 years, and those for coke oven emissions must be retained for the duration of employment plus 20 years, of for 40 years, whichever is longer. Emission data are pertinent to these records. Such retention is appropriate for all records of emissions of air pollutants or of chemical agents released in places of work.

The responsibility for retention of emission records for any establishment should be clearly defined. This responsibility may well be placed with the same organizational unit that has responsibility for retention of air monitoring records and/or health experience records. Records that serve as secondary sources of health related information, such as those identified in Section 4 of this Chapter, may be kept in different locations. A checklist of all such records that are needed for emission records purposes should be assembled and kept up to date for each establishment and retained by the unit having responsibility for emissions records. Responsibility for retention of each of these secondary records should be clearly defined and this responsibility, along with location of each record set, should appear on the records checklist.

BIBLIOGRAPHY

1. USEPA, *U.S. Environmental Protection Agency, A Guide for Compiling a Comprehensive Emission Inventory*, APTD 1135, EPA, Research Triangle Park, NC, 1973.
2. J. R. Hammerle, A. C. Stern, ed., in *Air Pollution*, Vol. 3, 3rd ed., Academic Press, New York, 1977, pp. 717–784.
3. R. W. Thron, A. C. Stern, ed., in *Air Pollution*, Vol. 7, 3rd ed., Academic Press, New York, 1986, Chapt. 4, pp. 163–217.
4. R. B. Engdahl, R. E. Barrett, and D. A. Trayser, in A. C. Stern, ed., *Air Pollution*, Vol. 7, 3rd. ed., Academic Press New York, 1986, Chapt. 8, pp. 339–393.

5. J. C. Berry et al. in A. C. Stern, ed., *Air Pollution*, Vol. 7, 3rd. ed., Academic Press, New York, 1986, Chapt. 9, pp. 395–508.
6. TSCA, Toxic Substances Control Act, Public Law 94-469, 15 USC 2601 et seq. (1976).
7. Chemical Abstract Service, American Chemical Society, P.O. Box 3012, Columbus, OH 43210, 1999.
8. Chemical Abstract Service, American Chemical Society, P.O. Box 3012, Columbus, OH 43210, 1992.
9. TOSCA Hot Line—202 554–1404, 1999.
10. NIOSH, *Registry of Toxic Effects of Chemical Substances*, on-line reference service, Cincinnati, OH (800) 638-8480, FAX (301) 480-3537, (1999). Internet: httpt://www.nlm.nih.gov/publications/factsheets/online_databases.html.
11. NIOSH, *Registry of Toxic Effects of Chemical Substances*, on-line reference service, Cincinnati, OH (most recent hard-copy publication, 1987), 1992.
12. H. J. Paulus and R. W. Thron, in A. C. Stern, ed., *Air Pollution*, Vol. 3, 3rd ed., Academic Press, New York, 1977, pp. 525–587.
13. USEPA, U.S. Environmental Protection Agency, Administrative and Technical Aspects of Source Testing for Particulates, EPA 450/3-74-047, EPA, Research Triangle Park, NC, 1974.
14. A. B. Antonsson, *AIHAJ* **51**, 352–355 (1990).
15. D. Bemer, J. M. Dessagne, and G. Aubertin, *AIHAJ* **60**, 354–362 (1999).
16. American Petroleum Institute, *Evaporative Loss from Tank Cars, Tank Trucks, and Marine Vessels*, API Bulletin 2514, API Washington, DC, 1959 (out-of-print).
17. H. B. Elkins, E. M. Comproni, and L. D. Pagnotto, *AIHAJ* **24**, 99–101 (1963).
18. H. E. Runion, *AIHAJ* **36**, 338–350 (1975).
19. J. E. Franke and R. A. Wadden, *Envr. Sci. Technol.* **21**, 45–51 (1987).
20. P. A. Scheff et al., *Appl. Occup. Environ. Hyg.* **7**, 127–134 (1992).
21. U.S. Environmental Protection Agency, *Compilation of Air Pollutant Emission Factors, A.P. 42*, 5th. ed., Vol. 1, *Stationary Point and Area Sources*, (to Supplement E, 1999), USEPA, Research Triangle Park, NC 27711 (1995).
22. D. Anderson, *Emission Factors for Trace Substances*, EPA 450-2-73-001, EPA, Research Triangle Park, NC, 1973.
23. W. A. Burgess et al., *AIHAJ* **38**, 18–23 (1977).
24. M. Smith, *Investigations of Passenger Car Refueling Losses*, Scott Research Laboratories, Inc., San Bernardino, CA, National Technical Information Service, PB-121 592, Washington, DC, 1972.
25. D. R. Streng et al. *Hazardous Waste Sites and Hazardous Substances Emergencies*, DHHS (NIOSH) Pub. No. 83-100, PHS CDC, NIOSH, Cincinnati, OH, 1983.
26. D. A. Lundgren, *J. Aerosol Sci.* **17**, 632–634 (1986).
27. M. Plinke, D. Leith, D. Holstein, and M. G. Boundy, *AIHAJ* **52**, 521–528 (1991).
28. M. Plinke, D. Leith, R. G. Goodman, and F. Loffler, *Part. Sci. Technol.* **12**, 71–87 (1994).
29. M. Plinke, D. Leith, R. Hathaway, and F. Loffler, *Bulk Solids Handling* **14**, 101–106 (1994).
30. M. Plinke, D. Leith, M. G. Boundy, and F. Loffler, *AIHAJ* **56**, 251–257 (1995).
31. Z. Bell et al. *Chem. Eng. Prog.* **71**(9), 54 (1975).
32. R. K. Flitney and B. S. Nau, *Ann. Occup. Hyg.* **30**, 241–247 (1986).
33. T. Kletz, *Chem. Eng. Prog.* **71**(9) (1975).

34. E. Sowinski and I.H. Suffett, *AIHAJ*, **38**, 353–357 (1977).
35. R. N. Anthony, *Management Accounting, Text and Cases*, 4th ed, Irwin, Homewood, IL, 1970.
36. OSHA, "*Occupational Safety and Health Standards, Subpart Z-Toxic and Hazardous Substances*" **29** *CFR* 1910 (with amendments through June 8, 1992, 57 FR24310), Department of Labor, Washington, DC, 1992.
37. P. L. Lubs, E. J. Kerfoot, and J. M. McClellan, *AIHAJ* **43**, 418–422 (1982).
38. J. R. Lynch, "*Implementation and Validation of an Exposure Estimating System*" paper 96 presented at the 1991 AIHCE, Salt Lake City, UT, 1991.
39. SARA, Superfund Amendments and Reauthorization Act, Title III, 42 USC 11001 et seq., 1986.
40. RCRA, Resource Recovery and Conservation Act, 42 USC 6901, and **40** *CFR* Parts 260–271 (1976).
41. Rubber Reserve Company, Report on the Rubber Program 1940–1945, Feb. 24, 1975.
42. G. S. Whitby, ed., *Synthetic Rubber*, Wiley, New York, 1954.
43. G. D. Babcock, History of the United States Rubber Company, A Case Study in Corporate Management, Indiana Business Report 39, Graduate School of Business, Indiana University, 1966.
44. N. C. Hawkins, S. K. Norwood, and J. C. Rock, eds., *A Strategy for Occupational Exposure Assessment*, American Industrial Hygiene Association, Akron, OH, 1991.
45. D. B. Turner, *Workbook of Atmospheric Dispersion Estimates*, U.S. Public Health Service Publication 999-AP-26, National Air Pollution Control Administration, Cincinnati, OH, 1969.
46. A. D. Busse and J. B. Zimmerman, *User's Guide for the Climatological Dispersion Model*, EPA-R4-024, EPA, Research Triangle Park, NC, 1973.
47. W. E. Porter, E. L. Hunt, and N. E. Bolton, *AIHAJ* **38**, 51–56, (1977).
48. R. L. Harris, *Industrial Hygiene Engineering Training Course*, U.S. Public Health Service, Occupational Health Field Headquarters, Cincinnati, OH, 1960.
49. *Industrial Ventilation: A Manual of Recommended Practice*, 23rd. ed, ACGIH Inc., Cincinnati, OH, 1998.
50. J. C. Baker Jr., *Testing a Model for Predicting Solvent Vapor Concentrations in an Industrial Environment*, M.S.E.E. Technical Report, Department of Environmental Sciences and Engineering, University of North Carolina, Chapel Hill, NC, 1977.
51. M. A. Jayjock, *AIHAJ* **49**, 380–385 (1988).
52. M. A. Jayjock, *AIHAJ* **55**, 230–235 (1994).
53. C. Feigley, J. Bennett, and J. Kahn, *AIHCE Abstract No. 338*, AIHCE Annual Meeting, Toronto, 1999.
54. M. Nicas, *AIHAJ* **57**, 542–550 (1996).
55. S. A. Roach, *Ann. Occup. Hyg.* **20**, 65–84 (1977).
56. S. A. Roach, *Ann. Occup. Hyg.* **24**, 105–132 (1981).
57. M. Nicas, *AIHAJ* **59**, 680–688, (1998).

CHAPTER THIRTY-TWO

Respiratory Protective Equipment

Warren R. Myers, Ph.D., CIH

1 HISTORICAL REVIEW

The respirator is historically one of the oldest types of equipment used for personal protection in the occupational setting. It is uncertain when the use of primitive forms of respiratory protection began. However, Pliny the Elder recorded the first known use of such a device early in the 1st century AD. Although respirators have been used for almost 2000 years, the genesis of their technical development is actually the mid-19th century. During that century, respiratory technology advanced rapidly as evidenced by the:

1. Appearance of the first early ancestors of modern atmosphere-supplying respirators.
2. Recognition that a different type of respiratory equipment was needed for protection against particulate hazards than was needed for protection against gaseous or vaporous hazards.
3. Rapid advancement of particulate filter technology after the discovery of Brownian motion (first observed 1827).
4. Discovery of the adsorptive properties of activated charcoal for many types of organic vapors.

Despite these major advances in knowledge and technology, respirator development continued slowly. Chemical warfare agents, introduced in World War I, focused attention on the need for adequate respirators. In the United States, the Bureau of Mines successfully carried out this work for the army. The forerunners of present-day respirators began to

Patty's Industrial Hygiene, Fifth Edition, Volume 2. Edited by Robert L. Harris.
ISBN 0-471-29754-2

appear after World War I. In those early years, misuse of surplus army gas masks by civilians and employers, who had sole discretion about deciding when and how respirators were used, highlighted the need for respiratory protection standards.

The Bureau of Mines in 1919 initiated the first respirator certification program in the United States and certified their first respirator, a self-contained breathing apparatus in 1920 (1). However, it had no statutory authority to regulate the use of respirators in the United States. The use of respirators continued unregulated until the Federal Coal Mine Health and Safety Act was enacted in 1969. As a result of this legislation, regulations were promulgated in Title 30 of the *Code of Federal Regulations Part 11* (30 *CFR* Part 11) governing the certification and use of respirators in the mining industry (2). Responsibility for the respirator certification program was housed within the Bureau of Mines.

Closely following passage of the 1969 Federal Coal Mine Health and Safety Act was the 1970 enactment of the Occupational Safety and Health Act. This legislation established the Occupational Safety and Health Administration (OSHA) and the National Institute for Occupational Safety and Health (NIOSH). The Occupational Safety and Health Administration is given the regulatory authority to require and monitor the use of respirators in general industry (Title 29 *Code of Federal Regulations* Part 1910.134 (29 *CRF* Part 1910.134)) and construction (29 *CFR* 1926.103) (3, 4). The OSHA promulgated a revision for 29 CFR Part 1910 and 1926 in January of 1998 (5, 6).

Responsibility for the respirator certification program was transferred from the Bureau of Mines to NIOSH in 1972 by a memorandum of understanding between the Departments of Labor and Health and Human Services. As a result, NIOSH shared a joint certification function with the Mine Safety and Health Administration (MSHA) and certification labels carried the designation of both agencies prior to 1995 (e.g., NIOSH/MSHA approved for . . .).

In 1995, NIOSH promulgated new certification regulations, Title 42 *Code of Federal Regulations* Part 84 (Part 84) to replace 30 *CFR* Part 11 (7, 8). At the present time, Part 84 addresses only revised performance tests for particulate filters used on or as a negative pressure facepiece (originally those devices tested under Subpart K, 30 *CFR* Part 11). Designation as "NIOSH-certified" indicates that a respirator has met the minimum performance standards, in 42 *CFR* Part 84, applicable to its particular class of RPE. Knowing the minimum performance standards and recognizing the limitations of the NIOSH certification tests are important when selecting and using a "NIOSH-certified" repirator. Achieving proper respiratory protection with NIOSH-certified RPE requires that (*1*) the RPE be properly selected for the environment in which it will be used; (*2*) its face fitting characteristics be adequately evaluated and considered in the selection process; and (*3*) workers receive periodic training that adequately prepares them to conscientiously and properly wear and use the RPE.

Three recognized standard-producing groups have produced guidance or standards on establishing and maintaining the elements of a good respiratory protection program. They are:

A. The American National Standards Institute (ANSI), "Practices for Respiratory Protection," 1992 (9).

B. The National Institute for Occupational Safety and Health (NIOSH) respirator approval regulations (42 *CFR* Part 84) and respirator use recommendations.
C. The Occupational Safety and Health Administration (OSHA) respirator use regulations for general industry (29 *CFR* 1910.134) and substance- or standard-specific respirator use regulations (e.g., lead 29 *CFR* 1910.1025, inorganic arsenic 29 *CFR* 1910.1018, benzene 29 *CFR* 1910.1028, etc.).

The primary objective of industrial hygiene is to prevent contaminants from escaping into the atmosphere. Hazard control should start at the outset of process, equipment, and plant design stages so those potential exposure sources can be engineered out. But very often exposure control becomes an issue after the process(es) are operational and therefore control becomes a more difficult and costly proposition. However, even then consideration should be given to the use of effective engineering controls to eliminate and/or reduce exposure to respiratory hazards. The preferred priority of controls steps are (*1*) substitution of a less toxic substance, (*2*) encapsulation or isolation of the process, (*3*) use of local exhaust ventilation or general ventilation in conjunction with filters and scrubbers to control the effluents, (*4*) use of personal protective equipment (PPE).

Respiratory protective equipment should be used only when all higher priority control steps are not technically or financially feasible. Real world circumstances however demonstrate that engineering controls often do not reduce exposures sufficiently to eliminate a respiratory hazard. As a result respiratory protective equipment (RPE) becomes the only technically and economically feasible adjunct to engineering controls to minimize and control respiratory exposures. According to recent estimates reported by the OSHA Office of Regulatory Analysis, roughly 5 million workers, 19–20% of the mining/manufacturing/construction workforce, in 1.3 million establishments use or have access to some type of respiratory protective equipment (5). Thus, RPE continues to be an important component in many respiratory exposure control plans. As a result, selection, use and evaluation of RPE have become more important as well.

2 LIST OF DEFINITIONS

Aerodynamic diameter. The diameter of a unit density sphere having the same settling velocity as a particle of arbitrary shape and density.

Airborne bacteria and virus. Airborne suspensions of living organisms usually found in the size range from 0.001 to 15 micrometers.

Air purifying respirator (APR). A respirator employing an air-purifying element to remove specific air contaminants by passing ambient air through the air-purifying element.

Air-purifying element. A component used in respirators to remove solid or liquid aerosols or gases or vapors from air.

Assigned Protection Factor (APF). The minimum expected workplace level of respiratory protection that would be provided by a properly functioning respirator or class of respirators, to a stated percentage of properly fitted and trained users.

Atmosphere-supplying respirator. A respirator that supplies the respirator user with breathing air from a source independent of the ambient atmosphere, and includes supplied-air respirators (SARs) and self-contained breathing apparatus (SCBA) units.

Canister or cartridge. A container with a filter, sorbent, or catalyst, or combination of these items, which removes specific contaminants from the air passed through the container.

Compressed-breathing gas. Oxygen or air stored in a compressed state and supplied to the wearer in gaseous form.

Demand respirator. An atmosphere-supplying respirator that admits breathing air to the facepiece only when a negative pressure is created inside the facepiece by inhalation.

Disposable respirator. A respirator that is discarded after the end of its recommended period of use, after excessive resistance or physical damage, or when odor breakthrough or other warning indicators render the respirator unsuitable for further use.

Dust. A solid mechanically produced particle usually found in the harmful size range of from 0.5 to 10 micrometers particle diameter.

Emergency situation. Any occurrence such as, but not limited to, equipment failure, rupture of containers, or failure of control equipment that may or does result in an uncontrolled significant release of an airborne contaminant.

End-of-service-life indicator (ESLI). A system that warns the respirator user of the approach of the end of adequate respiratory protection, for example, that the sorbent is approaching saturation or is no longer effective.

Escape-only respirator. A respirator intended to be used only for emergency exit.

Elastomeric facepiece or mouthpiece. A respirator component designed to provide a gas-tight or dust-tight fit with the face and may include headbands, valves, and connections for canisters, cartridges, filters, or respirable gas source.

Filter. An air-purifying element designed to remove solid or liquid aerosol from air.

Filtering facepiece. A particulate respirator with a filter as an integral part of the facepiece or with the entire facepiece composed of the filtering medium (see disposable respirator).

Fit test. The use of a protocol to qualitatively or quantitatively evaluate the fit of a respirator on an individual.

Fume. A solid condensation particle, generally of size range from 0.1 to 1-micrometer aerodynamic diameter.

Gas. An aeriform fluid that is in a gaseous state at 0° Centigrade and 760 millimeter of mercury.

Hazardous atmosphere. (*1*) Any atmosphere containing a toxic or disease producing gas, vapor, dust, fume, mist, or pesticide, either immediately or not immediately dangerous to life or health; or (*2*) any oxygen deficient atmosphere.

Helmet or hood. A respiratory inlet covering. A hood completely covers the wearer's head and neck, and may also cover portions of the shoulders and torso. Helmet refers to a rigid respiratory inlet covering that also provides head protection against impact and penetration. Both are supplied with incoming respirable air for the wearer to breathe. They may include a headharness and connection for a breathing tube.

High efficiency particulate air (HEPA) filter. A filter that is at least 99.97% efficient in removing monodispersed particles of 0.3 micrometers in diameter. The equivalent NIOSH 42 *CFR* Part 84 particulate filters are the N100, R100, and P100 filters.

Immediately dangerous to life or health (IDLH). An acute respiratory exposure that poses an immediate threat to loss of life, immediate or delayed irreversible, adverse effects on health or acute eye exposure that would prevent escape from a hazardous atmosphere.

Loose-fitting facepiece. A respiratory inlet covering that is designed to form a partial seal with the face.

Mist. A liquid condensation particle with a size ranging from submicroscopic to macroscopic.

Negative pressure respirator (tight fitting). A respirator in which the air pressure inside the facepiece is negative during inhalation with respect to the ambient air pressure outside the respirator.

Not Immediately dangerous to life or health (non-IDLH). Any hazardous atmosphere which may produce physical discomfort immediately, chronic poisoning after repeated exposure, or acute adverse physiological symptoms after prolonged exposure.

Oxygen-deficient atmosphere. An atmosphere with oxygen content below 19.5% by volume at sea level or a partial pressure of less than 148 millimeters of mercury.

Positive pressure respirator. A respirator in which the pressure inside the respiratory inlet covering exceeds the ambient air pressure outside the respirator.

Powered air-purifying respirator (PAPR). An air-purifying respirator that uses a blower to force the ambient air through air-purifying elements to the respiratory inlet covering.

Pressure demand respirator. A positive pressure atmosphere-supplying respirator that admits breathing air to the facepiece when the positive pressure is reduced inside the facepiece by inhalation.

Qualitative Fit Factor (QLFF). A qualitative measure of the minimum assured fit of a specific respirator to a particular individual, when a validated qualitative fit test is used.

Qualitative fit test (QLFT). A pass/fail fit test to assess the adequacy of respirator fit that relies on the individual's response to the test agent.

Quantitative fit factor (QNFF). A quantitative estimate of the fit of a particular respirator facepiece to a particular individual defined under the conditions of a quantitative fit test.

Quantitative fit test (QNFT). An assessment of the adequacy of respirator fit by numerically measuring the amount of leakage into the respirator.

Respiratory inlet covering. A portion of a respirator that forms the protective barrier between the user's respiratory tract and an air-purifying device or breathing air source or both. It may be a facepiece, helmet, hood, suit, or a mouthpiece respirator with nose clamp.

Self-contained breathing apparatus (SCBA). An atmosphere-supplying respirator for which the breathing air source is designed to be carried by the user.

Service life. The period of time that a respirator, filter or sorbent, or other respiratory equipment provides adequate protection to the wearer.

Supplied-air respirator (SAR) or airline respirator. An atmosphere-supplying respirator for which the source of breathing air is not designed to be carried by the user.

Tight-fitting facepiece. A respiratory inlet covering that forms a complete seal with the face.

Respiratory tract. Includes the anatomical features of the nose and mouth, pharynx, larynx, trachea, bronchi and lungs.

Simulated workplace protection factor (SPF). Surrogate measure of the workplace protection provided by a respirator done in a laboratory simulation that has been shown to have a stated correlation to workplace protection factors.

Smoke. A chemically generated particulate matter of organic origin usually found in the size range from 0.01 to 0.3 micrometer.

Threshold limit values (TLVs). Occupational exposure guidelines published annually by the American Conference of Governmental Industrial Hygienists.

Vapor. The gaseous state of a substance that is solid or liquid at 0° Centigrade and 760 millimeters of mercury.

Workplace protection factor (WPF). A measure of protection provided in the workplace, under the conditions of that workplace, by a properly functioning respirator that is correctly selected, fit tested, worn and used.

3 REGULATORY UPDATE

In the last several years both NIOSH and OSHA have promulgated significant changes to their standards governing respiratory protection. In 1998 OSHA revised 29 *CFR* Part 1910.134 and 1926.103 and in 1995 NIOSH promulgated a new respirator standard 42 *CFR* Part 84 revising parts of 30 *CFR* Part 11 that became effective in 1998.

3.1 OSHA Revision to 1910.134

On January 8, 1998, OSHA promulgated a new standard for respiratory protection covering 29 *CFR* 1910.134 and 29 *CFR* Part 1926.103. The revisions apply to general industry as well as construction, shipyard, longshoring and marine terminal workplaces and biohazards, except *M. tuberculosis*. Respiratory protection against *M. tuberculosis* is covered under 29 *CFR* 1910.139. Some of the major changes and additions to the respiratory protection standard are briefly discussed.

The new standard replaces all references to MSHA/NIOSH certification or 30 CFR Part 11 with NIOSH certification under Part 84. It also includes a number of important definitions used in the standard, e.g., air-purifying respirator, canister or cartridge, end-of-service-life indicator, fit factor, immediately dangerous to life or health, etc.

The requirements for an acceptable respiratory protection program were modified and expanded. They were modified to reflect advances in respiratory protection technology, such as the widespread use of modern methods of fit testing. OSHA also noted that several of the provisions of the previous standard were vague and had caused compliance difficulties for employers.

The standard requires employers to designate a person as program administrator and to ensure that this person is qualified to perform the responsibilities of the position. Appropriate training or experience or both can qualify this individual. Qualifications need to be commensurate with the complexity of the program and the responsibility to oversee and evaluate it.

The standard allows an employer to provide respirators on a voluntary basis on request by employees or permit employees to use their own respirators, if it is determined that respirator use itself will not create a hazard. The standard in these situations still requires the use of a NIOSH certified respirator for the contaminant of concern. It also stipulates, except for the use of filtering facepiece respirators that the employer must establish and implement those elements of a written respirator protection necessary to ensure that:

1. any employee using a respirator voluntarily is medically able to use that respirator, and
2. it is cleaned, stored, and maintained so that its use does not present a health hazard to the user.

OSHA states in the preamble of the standard that it allows this exception for filtering facepiece because it concludes "There are no medical limitations on the use of these respirators, so employers who allow their use, need only ensure that masks are not dirty or contaminated, that their use does not interfere with employees ability to work safely and that they provide the employees with the information contained in Appendix D of the final rule" (5).

The standard requires the employer to provide effective training to employees who are required to use respirators. The training must be comprehensive, understandable, and recur annually or more often if necessary. It should provide the employee with sufficient knowledge to demonstrate an understanding of:

a. Why the respirator is necessary and how improper fit, usage, or maintenance can compromise the protective effect to the respirator.
b. What the limitations and capabilities of the respirator are.
c. How to use the respirator effectively in emergency situations including situations in which the respirator malfunctions.
d. How to inspect, put on and remove, use, and check the seals of the respirator.
e. What the procedures are for maintenance and storage of the respirator.
f. How to recognize medical signs and symptoms that may limit or prevent the effective use of respirators.

The standard requires the employer to conduct evaluations of the workplace to ensure that the written respiratory protection program is being properly implemented, and to consult employees to ensure that they are using the RPE properly. Any problems that are identified during this assessment must be corrected. Factors to be assessed in the evaluation are respirator fit, appropriate respirator selection of the hazard, proper respirator use and proper respirator maintenance.

The standard recognizes that using RPE may place a physiological burden on employees. This burden is a function of the type of RPE worn, the job and workplace conditions in which the RPE is used and the medical status of the employee. Of interest to the industrial hygienist is the requirement of paragraph (e)(5). Supplemental information for the Physician or Other Licensed Health Care Professional (PLHCP). This subparagraph requires that the following information must be provided to the PLHCP before the PLHCP can make a recommendation concerning an employee's ability to use RPE.

a. The type and weight of the RPE to be used by the employee.
b. The duration and frequency of RPE use, including use for rescue and escape.
c. The expected physical work effort.
d. Additional protective clothing and equipment to be worn.
e. Temperature and humidity extremes that may be encountered.

The PLHCP must also be provided with a copy of the written respiratory protection program and a copy of paragraph (e)(5)(iii).

The employer is required to obtain a written recommendation regarding the employee's ability to use the RPE from the PLHCP. The recommendation can provide only the following information:

a. Any limitations on RPE use related to the medical condition of the employee, or relating to the workplace conditions in which the RPE will be used, including whether or not the employee is medically able to use the RPE.
b. The need, if any, for follow-up medical evaluations.
c. A statement that the PLHCP has provided the employee with a copy of the PLHCP's written recommendation.

Annual medical evaluations are not required but at a minimum the employee shall provide additional medical evaluations if:

a. An employee reports medical signs or symptoms that are related to ability to use RPE.
b. A PLHCP, supervisor or the respirator program administrator informs the employer that an employee needs to be reevaluated.
c. Information from the respiratory protection program including observations made during fit testing and program evaluation, indicates a need for employee reevaluation.
d. A change occurs in workplace conditions (e.g., physical work effort, protective clothing and temperature) that may result in a substantial increase in the physiological burden placed on an employee.

Other major changes in the standard include:

a. Establishment of a cartridge change schedule when gas or vapor respirators are used that do not have end of service life indicators.
b. Requirements for fit testing all tight-fitting respirators with a repeat frequency of at least once a year.
c. Specific protocols for conducting qualitative and quantitative fit testing.

3.2 NIOSH Adoption of 42 *CFR* Part 84

In July 1995, NIOSH replaced 30 *CFR* Part 11 with a new respirator certification regulation, 42 *CFR* Part 84. NIOSH recognized that some of the particulate certification tests incorporated into 30 *CFR* Part 11 in 1972, particularly in Subpart K, were Bureau of Mines particulate test procedures, some dating from the 1930s. NIOSH felt that particulate filter certification procedures in Subpart K of 30 *CFR* Part 11 should be revised based on laboratory and field research and testing conducted in the 1970s and 1980s as well as new and improved manufacturing technology. The new Part 84 reflects changes made to the performance requirements for particulate filters or particulate filtering facepieces used as certified, negative pressure, nonpowered, air purifying, particulate respirators. Certification test requirements for PAPRs using particulate filters were not addressed in the 1995 revision. NIOSH has indicated it plans future revisions to Part 84 to address PAPRs using particulate filters. The remaining certification test requirements were incorporated into Part 84 without change for exhalation valve leakage, gas and vapor cartridges and canister atmosphere supplying devices, etc. A new sequence of approval numbers (TC-84A-xxxx) is used for the particulate, air-purifying, nonpowered, negative pressure respirators certified under Part 84. Since the certification requirements for PAPRs and all other respirator types did not change they continue to use the sequence of approval numbers previously used with 30 *CFR* Part 11. Under Part 84, PAPR certification has been restricted to using only HEPA filters.

This rule also eliminates the combination categories of paint spray and pesticide respirator approvals; however, other combination respirators (e.g., particulate and acid gases or organic vapors) can be certified under Part 84. All particulate respirators have a certification label bearing the NIOSH and the Department of Health and Human Services (DHHS) emblems when certified under Part 84.

The new Part 84 certification regulations eliminate the silica dust, silica mist, lead fume, and thermally generated, dioctyl phthalate (DOP) filter efficiency tests. These tests challenged the filter with different aerosols having different particle sizes and had a pass/fail criterion based on mass efficiency, except for the hot DOP test. These tests have been replaced by an "oil aerosol", cold DOP or a "particulate aerosol", sodium chloride (NaCl) filter penetration test. NIOSH established the new test performance criteria to "simulate worst-case respirator use and very severe test conditions" (8). NIOSH states "that filters certified with the new criteria can be used without particle size analysis or filter penetration testing in the workplace" (8).

The Part 84 certification regulation eliminates the dust, mist, dust and mist and dust, fume and mist class respirators and replaces them with nine classes of filters. The nine classes are based on three levels of filter efficiency, 95%, 99%, and 100% (actually 99.97%), and three categories of resistance to efficiency degradation, as evaluated with cold DOP oil aerosol.

3.3 Information Access Explosion on the WW

In the mid 1990s access to and management of information was forever changed with the development of the world wide web (www) and software tools to conveniently access and use it. The explosion of industrial hygiene and safety information from both free and subscription venues on the www is nothing short of phenomenal. Table 32.1 contains a listing of a small sampling of web sites that contain information on RPE. The information sources are divided into three groups: government (NIOSH and OSHA), manufacturers, and others. Web addresses given may change.

4 RESPIRATOR PERFORMANCE

The performance of RPE is typically evaluated in two ways. The first consists of NIOSH certification examinations that evaluate individual components (e.g., filter efficiency, exhalation valve leakage, etc.) or performance parameters (e.g., minimum flow rate for an air supply respirator) of the RPE.

The second consists of evaluations on completely assembled and worn RPE. These evaluations measure and compare total inward leakage (TIL) or the in-facepiece concentration, C_i, of the RPE with the contaminant concentration outside the RPE or the outside concentration, C_o. They consist of laboratory tests, (e.g., fit tests or simulated workplace studies) or workplace studies (e.g., workplace protection factor studies). Typically *TIL* or C_i is recognized to be the sum of all sources of leakage into the facepiece:

$$C_i = TIL = L_{\mathrm{fs}} + L_{\mathrm{ape}} + L_{\mathrm{ex}} + L_{\mathrm{n}} \tag{1}$$

where L_{fs} = leakage through the face seal
L_{ape} = leakage through the air-purifying element
L_{ex} = leakage through the exhalation valve
L_{n} = leakage through other pathways

One of the first reported evaluations using TIL to assess RPE performance was by the Bureau of Mines in its 1937 regulations governing the testing and approval of supplied air respirators (10). These regulations specified that the RPE was to be tested in a test atmosphere of flint (50 ± 10 mg/m^3; $D_{\mathrm{p}} = 0.6$ μm and GSA ≤ 1.9) for thirty minutes while the test subject carried out each of the following activities for five minutes apiece:

1. Walking, turning head, and dipping chin.
2. Pumping air with a tire pump into a 1-cubic foot cylinder to pressure of 25 psi or equivalent work.

Table 32.1. Sampling of Websites for Information on RPE

Group	Item	Web Address (http://www.—)
NIOSH	Home page	cdc.gov/niosh
	Databases	cdc.gov/niosh/database.html
	Certified Equipment List	
	IDLH Values Documentation	
	Pocket Guide to Chemical Hazards	
	NIOSH Documents	cdc.gov/niosh/topreg.html
	NIOSH Guide to Industrial Respiratory Protection	
	NIOSH Respirator Decision Logic	
OSHA	Home page	osha.gov
	Standards and Interpretation—(search on respirators)	osha-slc.gov/oshdoc/oshdoc/toc_interps.html
	Directives (search on respirators)	osha-slc.gov/oshdoc/toc_directive.html
Manufacturers	Mine Safety Appliances	msanet.com
	Scott Aviation	corptech.com.
	Draeger Safety, Inc.	drager-usa.com
	E.D. Bullard, Co.	bullard.com
	3M	mmm.com
	UVEX Safety Inc.	uvex.com
	North Safety Products	SafetyOnline.net/north/
	Hornell Speedglas, Inc.	speedglas.com
	Glendale	glendale-laser.com
	Racal Health and Safety Inc.	recalhealth.com
	Neoterick Health Technologies, Inc.	nell.com/data/o1oo/htm/nroo546.htm
	Sellstrom Mfg. Co.	selstrom.com
Others	International Society for Respiratory Protection	llnl.gov/isrp

3. Resting.
4. Repeat of the 1st activity.
5. Repeat of the 2nd activity.
6. Repeat of the 3rd activity.

During the 30-minute period of testing, air was withdrawn continuously at the rate of 32 Lpm from the respiratory inlet covering at a point as near as convenient to the wearer's nostrils. The in-facepiece sample was analyzed gravimetrically to determine the total mass of flint collected. That amount was then compared to the total mass collected from sampling the supply of air that was feeding the respirator. Performance of the RPE was considered acceptable if the difference in mass estimates did not exceed 0.05 mg.

4.1 Protection Factors

Total inward leakage evaluations generally express performance as a ratio of the outside concentration and the in-facepiece concentration. The U.S. Bureau of Mines, in Approval Schedule 21B, issued in 1965, made reference to the term "Decontamination Factor" (11). It was defined as "the ratio of the concentration of dust, fume, or mist present in the ambient atmosphere to the concentration of dust, fume or mist within the facepiece while the RPE is being worn." This term is no longer used in the respirator literature, but an analogous term, "Protection Factor" (PF) has taken its place.

The protection factor, like the decontamination factor, represents an expression of the performance of RPE based upon the ratio of the two generalized concentration variables, C_o and C_i. These concentration variables, in addition to being used to define a protection factor, can be used to express the penetration (P) or efficiency (E) of the RPE. Equality exists between protection factor, penetration, and efficiency as follows (12):

$$P = C_i/C_o$$

$$E = (C_o - C_i)/C_o$$

$$PF = C_o/C_i = 1/P = 1/(1 - E)$$

As such, the protection factor is a generic term that lacks specificity of definition associated with the conditions around which C_o and C_i are collected. The respirator committee of the American Industrial Hygiene Association adopted a set of protection factor definitions to provide more specificity with the use of the term (13). They are used where applicable in this chapter and some are listed in the Definitions and Abbreviations section of the chapter.

4.1.1 *Assigned Protection Factors*

The first major step in the selection of RPE is to determine what type(s) of respiratory protection is suitable for control of the contaminant of concern. The second major step in the selection process is to determine how much of the contaminant is present and how much protection is required to reduce contaminant exposures to acceptable levels. Given that other considerations of the selection process are met, then the level of protection needed is matched with the assigned protection factor (APF) given to different types of respirators.

Since the protection factor is an expression of the penetration or efficiency of the RPE, they serve to link the level of protection the RPE is thought capable of providing with the level of protection required by the worker. The ability to evaluate respirators and provide some assessment of their capacity to provide protection is thus a very key element for successful respirator selection.

Assigned protection factors also are used by regulatory agencies and occupational safety and health professionals to determine, for different classes of RPE, the limiting concentration of contaminants in the ambient environment against which the RPE would afford adequate protection to the user (12). For example, if a respirator has an assigned protection factor of 10, it would be considered suitable (given it was properly selected, fitted and used) to prevent workers from being overexposed at concentrations of up to 10 times the OSHA permissible exposure limit (PEL).

Assigned protection factor levels for different classes of RPE are recommend by NIOSH, the American National Standards Institute (ANSI) Z88.2 committee and OSHA health standards, e.g., cadmium (9, 14, 15). Historically the basis for these recommendations was Hyatt's 1976 APF recommendations for different classes of RPE. His recommendations were based on quantitative fit factors (QNFF) determined by a quantitative fit test (QNFT) and use of a safety factor (16). NIOSH and OSHA during the 1970s incorporated those APFs into their own recommendations and regulations (17). For example, in the NIOSH–OSHA standard completion project an APF of 1000 was adopted for PAPRs.

To use quantitative fit factors to establish levels of protection requires the fundamental assumption that they are related to or are indicative of the protection level provided by the RPE in the workplace. This assumption has been seriously questioned by a series of workplace studies that began in the early 1980s. Those studies will be discussed under the section on workplace testing. As a result, as more meaningful data become available NIOSH, OSHA, and ANSI have revised their APF recommendations, most notably, those for PAPRs and continuous flow SAR (9, 14, 15).

NIOSH states that "APFs based solely on laboratory fit testing should be viewed and applied with particular caution, even when the laboratory testing involves a simulated work regimen" (14). It also considered field performance data, when available, to make its APF recommendations. American National Standards Institute (ANSI) Z88.2-1992, *Practices for Respiratory Protection* states that when WPF data were available they were used to establish its APFs recommendations.

For the practicing occupational health and safety professional it is very important when considering selection of RPE to understand the potential limitations of the APF level given to a particular class of RPE. The most recent APF recommendations by NIOSH, ANSI, and OSHA are noted in Table 32.2. In its new respirator standard OSHA announced that it would be revising it's respirator APF requirements at a later date.

4.2 Fit Testing

The amount of protection provided by a respirator can be no better than its fit to the face. Face seal leakage, particularly for APRs, is often the largest component of the TIL. Given

Table 32.2. Assigned Protection Factor Levels Recommended by NIOSH and ANSI Z88.2-1992 or Required by OSHA in the Cadmium Standard for Different Classes of Respiratory Protective Equipment

Class of Respirator	NIOSH	ANSI-1992	OSHA[a]
Half facepiece			
Any dust filter	10	10	
HEPA or 100% filter	10	10	10
Chemical cartridge	10	10	
Full facepiece			
Any dust filter	10	100	
HEPA or 100% filter	50	100	50
Chemical cartridge	50	100	
Powered air-purifying respirator			
Half facepiece			
Any particulate filter	50	50	
HEPA or 100% filter			50
Chemical cartridge	50	50	
Full facepiece			
Any dust filter	50	100	
HEPA or 100% filter Chemical cartridge	50	1,000	250
Helmet/Hood			
Any particulate filter	25	100	
HEPA 100% or Filter	25	1,000	25
Chemical cartridge	25	1,000	
Loose-fitting facepiece			
Any particulate filter	25	25	
HEPA 100% or Filter	25	25	25
Chemical cartridge	25	25	
Supplied air respirator—continuous flow			
Half facepiece		50	50
Full facepiece		1,000	250
Hood or helmet	25	1,000	25
Loose fitting facepiece	25	25	
Supplied air respirator—demand flow			
Half facepiece	10	10	
Full facepiece	50	100	
Hood or helmet	25		
Loose fitting facepiece	25		
Supplied air respirator—pressure demand or positive pressure			
Half facepiece	1,000	50	1,000
Full facepiece	2,000	1,000	1,000
Supplied air respirator—pressure demand or positive pressure full facepiece with auxiliary SCBA operated in pressure demand or positive pressure	1,000		>1,000
SCBA—pressure demand or positive pressure with full facepiece	10,000	[b]	>1,000

[a]APF values taken from the cadmium standard 1910.1027 but they do vary from standard to standard.

[b]Although positive-pressure respirators are currently regarded as providing the highest level of respiratory protection, a limited number of recent stimulated workplace studies concluded that all users may not achieve protection factors of 10,000. Based on this limited data, a definitive assigned protection factor could not be listed for positive-pressure SCBAs. For emergency planning purposes where hazardous concentrations can be estimated, an assigned protection factor of no higher than 10,000 should be used.

that a respirator is properly selected and used and conscientiously worn, fit is the crucial determinant of whether adequate respiratory protection is possible. If the facepiece does not fit, it does not protect regardless of how effectively other components of the respirator work. For example, any benefits to be achieved by using a HEPA filter that has penetration of only 0.03% would be quickly lost if the facepiece was not capable of providing a fit of comparable quality. Interest in respirator fit led to a study of facial measurements from which three measurements were chosen for use in facial sizing (18). The principal dimensions chosen in relation to respirator face fit are face length (Menton-nasal root depression), face width (bizygomatic breadth), and lip length (Fig. 32.1). The dimensions of face length and face width are used for full-face masks. The dimensions of face length and lip length are used for half and quarter masks.

Utilizing U.S. Air Force anthropometric studies, facial sizes for two 25-person panels were established to represent the working population (19). The panels in Figure 32.2(**a**) and (**b**) for full and half facepieces respectively present the population distribution for facial sizes related to face length and face width and lip length and face width, respectively. These panels have been widely utilized to determine how respirators fit the general population. Today most manufacturers supply half and full facepiece respirators in multiple sizes that cover these facial groupings.

Over the years researchers have identified a number of factors that influence the degree and variability of the fit of RPE particularly, half and full facepiece APRs. They are (1) the design of the respirator's facepiece, (2) facial characteristics of the user, and (3) the conditions of use. As early as 1962, observations were made that the shape of the nose influenced the degree of fit achieved with a half facepiece (20). Hyatt also concluded that design of the facepiece influenced the degree of fit achieved with the respirator (21).

In his *Guide to Industrial Respiratory Protection*, Pritchard notes that facial characteristics including lack of teeth or dentures, scars, hollow temples, prominent cheek bones,

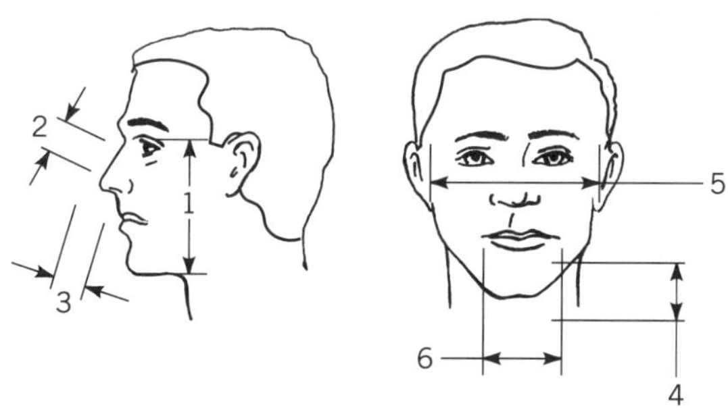

1. Nasion - Menton (face length)
2. Length of Nasal Bone
3. Pronasale - Subnasale (Nose protrusion)
4. Mandible Height (chin length)
5. Bizygomatic Diameter (face width)
6. Bichelion Diameter (mouth width)

Figure 32.1. Facial anthropometric dimensions for sizing half and full facepieces.

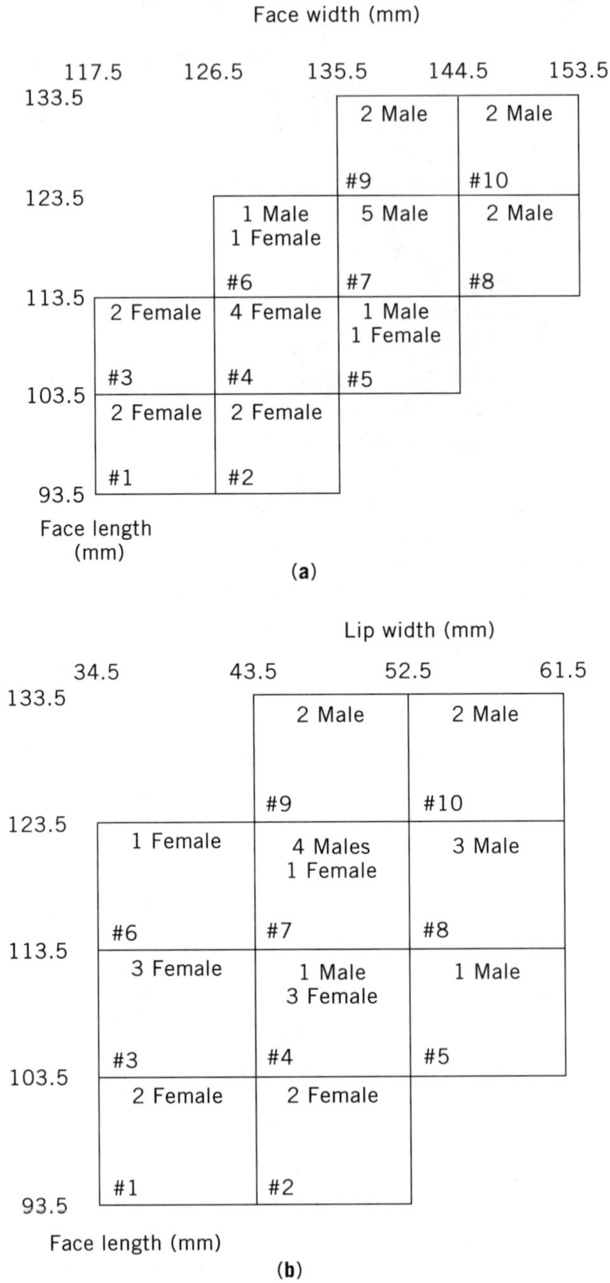

Figure 32.2. Twenty-five person respirator fit test panels developed by Los Alamos National Laboratory. (**a**) Full facepiece panel. (**b**) Half facepiece panel.

deep skin creases and changes caused by weight loss or gain can influence the level of fit achieved with RPE (22). Studies by Oestenstad et al. found that approximately 70% of the face seal leaks occurring with negative-pressure, half facepiece, respirators occurred at the nose (23–26). They found a significant correlation between face size, facial characteristics and the site of face seal leakage (nose, cheek, or chin) that varied by gender. These authors conclude that respirator fit test panels (e.g., the LANL 16 or 25 member half facepiece fit test panel) and respirator selection guidance should consider nasal dimensions.

Conditions of use also effect fit. These include presence of facial hair in the form of beard stubble, mustaches, goatees or sideburns that interfere with the face seal surface of the respirator during wearing. Other factors of use such as temperature, humidity, breathing pattern and rate, head and body movements, perspiration, etc., have been noted in the literature to influence face seal leakage.

Two generic procedures of assessing respirator fit have been developed that are widely used by general industry. They are the quantitative fit test (QNFT) and the qualitative fit test (QLFT). The quantitative fit test provides a numerical measure of fit, the QNFF that is assessed independently of interpretation by the wearer. The qualitative fit test on the other hand measures fit based on the wearer's interpretation as to whether the presence of a fit test agent is detected.

Both fit test procedures utilize a positive or negative (+/−) "fit check" that must be successfully conducted according to the respirator manufacturer's instructions before the actual fit test can be conducted. The fit check is not a fit test and should not be used in place of a recognized QNFT or QNLT. Myers et al. evaluated whether +/− fit checks were an effective aid in helping users of RPE achieve a good fit when donning the RPE (27). They evaluated three filtering facepiece respirators and one elastomeric facepiece respirator. Two randomly selected groups of 32 people each were used. One group was trained to don the RPE using the manufacturer's +/− check procedure as an aid, while the second group was trained to don the RPE without conducting a +/− check. The data obtained from the experiment suggested that, in general, RPE users obtained fewer unsuccessful donnings and more consistent donnings when fit checks were used as an aid in donning both types of RPE. Fit checks were found to be fairly useful, easy-to-learn tools for respirator wearers to discriminate between good and poor donnings.

4.2.1 Qualitative Fit Tests

Qualitative fit tests, use simpler equipment than that required for QNFT and depend on the subjective opinion of the user as to whether the respirator leaks.

The QLFT methods expose the respirator user to an atmosphere with a contaminant, and the respirator is acceptable if the wearer can not detect the contaminant. These methods depend on the subjective opinion of the user to determine whether the respirator fits.

OSHA has accepted and published in Appendix A of the 1998 Respiratory Protection standard 1910.134, QLFT fit test protocols for isoamyl acetate, saccharin, Bitrex™ (denatonium benzoate), and irritant smoke (stannic chloride). When selecting a QNFT protocol, consideration must be given to properly matching the QNFT test agent to the type of RPE to be fit tested. For example, a P95 particulate, filtering facepiece respirator could be tested with saccharin or Bitrex but not isoamyl acetate or irritant smoke. If the respirator program manager adopts QNFT, OSHA requires that the applicable protocol, as published in Appendix A, be followed.

4.2.2 Quantitative Facepiece Fit Testing

During the last 30 years, a substantial amount of respirator research has been focused on developing quantitative methods to evaluate how well a respirator fits. One of the first reported attempts to measure face seal leakage quantitatively was by Guyton and Lense (28). They developed a fit test involving *B. globigii* bacteria. The bacteria were aerosolized from suspension in saline into a test chamber. The amount of leakage into the mask was determined by collecting the bacteria on a cotton filter that was held in the mouth of the wearer of the RPE. Subjects were instructed to inhale through their mouth so that all the breath was filtered as it passed through the mouth-held filter. Exhalation was done through the nose. Thus, the collection of bacteria was accomplished only during the inhalation. By comparing estimates of the concentrations of bacteria in the chamber and inside the mask, a quantitative estimate of penetration (fit) could be made.

Burgess, Silverman, and Stein proposed another procedure to test the fit of RPE quantitatively that utilized uranine dye aerosol with a geometric mean particle size of 0.2 micrometers and a standard deviation of approximately 2 (29). Due to its fluorescent properties, the uranine was detectable in quantities as small as 0.1 nanogram per mL of water by photometric techniques. The uranine aerosol that leaked into the mask was sampled only during inhalation of each respiratory cycle. This was accomplished with a sampling detector, which used the output of a pressure sensor to initiate and terminate sampling in sequence with inhalation and exhalation. The open-face filter holder used to sample uranine leakage was placed inside the respirator so as to occupy a sampling position in the cavity between the facepiece and subject's face. This was one of the earliest reports of air sampling from inside a respirator using extractive sampling techniques.

Hounan et al. described a quantitative fit test method based on using an atmosphere containing submicrometer sodium chloride (NaCl) aerosol (30). The authors note that even though sodium chloride is hygroscopic (a potential limitation with inhalation studies), it is nontoxic and readily detectable with flame photometry. The quantitative assessment of the respirator's performance was based upon estimates of the concentration of NaCl aerosol in the exhaled air. The exhaled air was routed to a 2-L reservoir where it was mixed with dry air to prevent condensation and to dry the NaCl particles. An air sample was then collected from the reservoir volume and fed to a flame photometer. To deal with the problems of lung retention, the lung deposition of NaCl aerosol was measured on each test subject before testing began.

Flyger reported on a quantitative fit test method that utilized helium as the challenge agent (31). Helium was selected as the tracer gas because it has a low solubility in the lungs. Therefore, he reasoned, the helium concentration in the facepiece cavity would be independent of the breathing cycle so sampling could be done over the complete cycle without corrections for lung retention. In discussing the use of a gas challenge agent, the author made the following observation. "The penetration of particles through a leak is limited by interception, impaction and diffusion of the particles during passage of the leak, while the penetration of helium varies only with the air flow through the leak. The penetration of helium must consequently be considered the upper limit to the penetration of particles."

White and Beal extended the NaCl quantitative fit method described by Hounan et al. to include various head and face movements and normal and deep breathing patterns into the test procedure (32). Burgess and Shapiro reported on a test method developed to evaluate the protection provided by a PAPR against daughter products of radon (33). The test subject was exercised at a 622 kp-m/min work rate. The radon gas was allowed to attach to cigarette smoke that was reported to have a mean particle size of 0.25 μm. This is possibly the first article in which fit test results are used to draw inference about what level of protection the respirator would provide during actual use conditions.

In 1962, Adley and Wisehart reported using dichlorodifluoromethane gas in a QNFT developed to evaluate SARs and APRs (34). In 1969, Adley and Uhle reported on a quantitative fit method utilizing Freon (35). Testing was done with the subject standing motionless, talking, breathing deeply to simulate hard work conditions, and lastly moving his/her head. This series of "exercises" was a forerunner to the "exercises" now commonly used in quantitative fit testing. Lowry et al. and Bentley et al. have used sulfur hexafluoride (SF_6) to compare gaseous and aerosol fit testing procedures (36, 37). Bentley et al. compared SF_6, argon, and sodium chloride as a fit test agent with SARs and SCBAs and found good correlation between them.

In 1972 Hyatt et al. at Los Alamos National Laboratory described quantitative fit methods using an oil aerosol of dioctyl phthalate (DOP) and an improved quantitative fit methods using polydispersed NaCl (38). The unique aspect of both test methods was that they used in-facepiece sampling in conjunction with fast response detectors and recorders to provide a real-time measurement of leakage into a respirator. A forward light scattering photometer was used to measure the aerosol penetration in units of percent of the concentration of the DOP aerosol in the test chamber. While wearing the respirator in the test chamber, the test subject carried out the following exercises: (*1*) normal breathing; (*2*) deep breathing; (*3*) nodding head up and down; (*4*) turning head side to side; (*5*) bending forward and touching toes; (*6*) talking; (*7*) smiling if the respirator was equipped with a half-mask facepiece, or frowning if the respirator was equipped with a full facepiece; and (*8*) normal breathing.

The average of the peak DOP penetrations leaking into the interior of the respirator was determined for each exercise. The mean of the average penetration for all exercises was used in determining a respirator protection factor for each test. The authors felt that to assign a half-facepiece respirator to an individual test subject, the protection measured by the fit test had to equal or exceed 10. Similarly, to assign a full facepiece to an individual test subject, the protection factor measured by the fit test had to equal or exceed 100. These researchers used the data from in-facepiece sampling not only to assess the fit of individual respirators but also to compare the fit provided to a specific individual by different brands of respirators. As a result, they felt that the "best fitting" respirator could be provided to an individual. The methods described by Hyatt were rapidly accepted and viewed as models for quantitative fit testing in the U.S.

Hack et al. and Douglas et al. also assessed, within the constraints of this QNFT procedure, the protection provided by different classes of RPE (39–43). They found tight-fitting SARs and SCBAs operated in the demand mode provided the user with protection similar to that provided by the equivalent tight fitting, air-purifying respirator; continuous flow SAR provided protection lower than pressure demand SAR; and pressure demand SCBA provided the best protection to the wearer. Because of questions about the toxicity of DOP different oils, most notably corn oil and poly(ethylene glycol), are presently being used for the oil mist QNFT.

Condensation nucleus counter (CNC) technology was adapted for use in fit testing through a cooperative effort between the U.S. Army and private industry. The commercial model of the CNC is the Portacount Plus™. It is small, lightweight, portable and since it measures the condensation nuclei in room air it does not require a separate test chamber. Air drawn into the device by a small sampling pump passes through a heated chamber saturated with alcohol, and enters a cooled chamber that causes the alcohol vapor to condense on to particles in ambient air. The particles grow to about 10 μm in size during the alcohol condensation and enter a forward light scattering chamber for counting. The Portacount Plus™, Model 8020 is available from TSI, Inc. St. Paul, MN.

Until the 1990s all QNFT methods used in-facepiece sampling methods to detect the leakage of either an aerosol or gas/vapor challenge agent into the facepiece. Crutchfield et al. developed a QNFT method that does not depend on measuring aerosol or gas/vapor leakage (44). His method measures the amount of air leaking through the face seal while maintaining a predetermined level of pressure in the facepiece. The method is called controlled negative pressure (CNP). The CNP fit test instrument unit is small, lightweight, and easily portable and does not require a test chamber. It does require special airtight cartridge replacement manifolds. These are available for most brands of respirators. The method is based on exhausting air from a temporarily sealed respirator facepiece at a rate sufficient to generate and then sustain a constant negative pressure inside the facepiece while the wearer holds her/his breath. The magnitude of the negative pressure is preselected to replicate the mean inspiratory pressure inside the facepiece during normal wear. With the inhalation valve pathways temporarily blocked, measurement of the exhaust flow rate yields a synonymous measure of the leakage flow rate into the mask during inspiration under normal use conditions. The authors state that a major advantage of the method is that a worker can be fit tested with her/his assigned respirator because the method does not require a destructive sampling probe. The controlled negative pressure fit test instrumentation is available, as the Dynatech Nevada FitTester 3000 from Dynatech Nevada Inc. Carson City, NV.

OSHA has accepted and published in Appendix A of the 1998 Respiratory Protection standard 1910.134, QNFT fit test protocols for (5):

a. use of nonhazardous test aerosols (such as corn oil, polyethylene glycol 400 [PEG 400], di-2-ethylhexyl sebacate [DEHS], or sodium chloride) generated in a test chamber and employing instrumentation to quantify the fit of the respirator;
b. use of ambient aerosol as the fit test agent and appropriate instrumentation (condensation nuclei counter) to quantify the fit of the respirator; and
c. use of controlled negative pressure and appropriate instrumentation to measure the volumetric leak rate of a facepiece to quantify the fit of the respirator.

4.2.3 *Problem and Limitations of QNFT*

A major problem with QNFT techniques that rely on extracting a sample from inside the facepiece is that research has shown face seal leakage does not mix instantaneously and uniformly within the facepiece cavity. This leads to in-facepiece sampling biases of significant magnitude for both half and full facepiece respirators. Early researchers that developed QNFT methods commonly assumed that face seal leakage mixed uniformly and rapidly within the facepiece cavity (45). As a result, consideration given to issues such as where to locate the sampling probe, how deeply it should be extended into the facepiece cavity, etc. were based on professional judgement.

Studies by Myers et al. (45–47), Holton et al. (48, 49), Campbell and Myers (50), and Oestenstad et al. (24) have found that in-facepiece sampling will not provide representative samples of TIL. Myers et al. and Oestenstad et al. demonstrated that face seal leakage could streamline within the facepiece cavity clearly illustrating one reason for face seal leakage not mixing rapidly or uniformly (24, 45). These conditions lead to large, variable sampling biases in concentration or penetration measurements made by in-facepiece sampling. Variations in several parameters of the person–respirator system have been identified to cause significant changes in the sampling bias (45–48). They are

a. Location of the probe on the midline of the respirator.
b. Depth at which the probe is inserted into the facepiece cavity.
c. Breathing being done all through the mouth or all through the nose.
d. Area where the face seal leakage occurs.
e. An interaction between the area where face seal leakage occurs and whether breathing is all through the nose or mouth.
f. Size of the challenge aerosol.

A key observation from these works is that design features of the facepiece, for either half or full facepiece, negative pressure APR, can have a significant effect on the magnitude of the bias associated with using in-facepiece sampling with aerosol based QNFT. The important point to be made from this research is that it is difficult to determine if a higher QNFF obtained with one respirator brand and size over another brand and size is due to a "truly better fit" or sampling error. These observations may explain in part why a correlation between QNFF and WPF has not been demonstrated.

Myers and Hornung studied different in-facepiece sampling techniques for half and full facepieces (51). They compared a common in-facepiece sampling method using a probe mounted virtually flush on the wall of the facepiece in the general area between the nose and mouth situated on or near the midline. The flow rate typically used was from 1 to 2 Lpm. The method was compared to four other in-facepiece sampling methods. The mean bias was found to be -21.3% and -25.9% respectively for full and half facepieces. This average bias was the worst of all the methods evaluated. By simply having the wearer breathe through the mouth, increasing the flow rate and locating the sampling probe approximately 1/2 inch in front of the mouth the sampling bias decreased to -3.4% and -12.5% respectively for full and half facepieces. As a result of this study, they recommend for those using an aerosol based QNFT system that the in-facepiece sampling probe be located, if possible, on the midline of the respirator, opposite the mouth and extended into the half facepiece cavity at least 1/4 inch. The wearer should be instructed to breathe through the mouth. If the sampling rate can be adjusted use something between 3 to 5 Lpm. When using full facepieces they recommend that the sampling probe be located in the nose cup if one is used and a sample rate of 3 to 5 Lpm. If the full facepiece has no nose cup the sampling probe should be on the midline opposite the mouth and extended to within approximately 1/2 inch to 1 inch of the mouth. Again a flow rate from 3 to 5 Lpm should be used.

Studies by Hinton (52), Myers et al. (53–56) and Lenhart and Campbell (57) have found no correlation between the QNFF obtained by a respirator–person combination and the WPF achieved in the workplace. Based on these studies, quantitative fit test data do not appear to be good predictors of workplace protection even when the respirator is used with a good respirator program. The relationship between quantitative fit testing and workplace protection is yet uncertain. Figure 32.3(**b**) is a plot of paired QNFF and WPF values measured on two brands of PAPRs. The QNFFs were determined immediately before the workers went into their workplace for the WPF measurements. The data plot clearly illustrates that no correlation exists between the data.

Two important questions were raised by these studies. (1) Will a respirator that gives the "best fit" in a quantitative fit test be the respirator that will provide the "best protection" in the workplace? and (2) Should QNFF data be the basis upon which assigned protection factors are established?

The answer to the first question is not clear. Considering all available data, it appears that there is little justification for making a respirator selection decision based solely on comparing QNFF values that a worker may achieve with different brands of respirators. The approach better supported by the data is to allow a worker to first select what is to her/him the most comfortable brand, size, etc. and then compare the QNFF to an accept/reject screening level. Historically, a commonly used screening level for QNFFs is 10 times the APF of the RPE. A recent laboratory study affirms the appropriateness of incorporating a safety factor of 10 in setting pass/fail criteria levels for QNFT (58). NIOSH advises that "When QNFT is used for fit screening, the fit factor screening level should be chosen with caution and with recognition of the uncertainty of its effectiveness. As appropriate, periodic evaluation of the effectiveness of each respirator during use in the workplace should be conducted to ensure that each wearer is being provided with adequate respiratory protection" (14). The OSHA respirator standard indicates that an acceptable pass/fail screening level for half and full facepiece is respectively 100 and 500 (5).

In regard to the second question both NIOSH and ANSI have used WPF data, when available, to develop their APF recommendations. NIOSH states in its respirator decision logic that "APFs based solely on laboratory fit testing should be viewed and applied with particular caution, even when the laboratory testing involves a simulated work regimen. To date no relation has been demonstrated between laboratory QNFF and measured workplace performance" (14). As NIOSH points out the problem is that QNFF results do not seem to correlate with the protection that is actually achieved with the RPE. However in late 1998, Coffey et al. reported observing a strong relationship between QNFF measures by some QNFT methods and dose in laboratory studies (58). They compared QNFF mea-

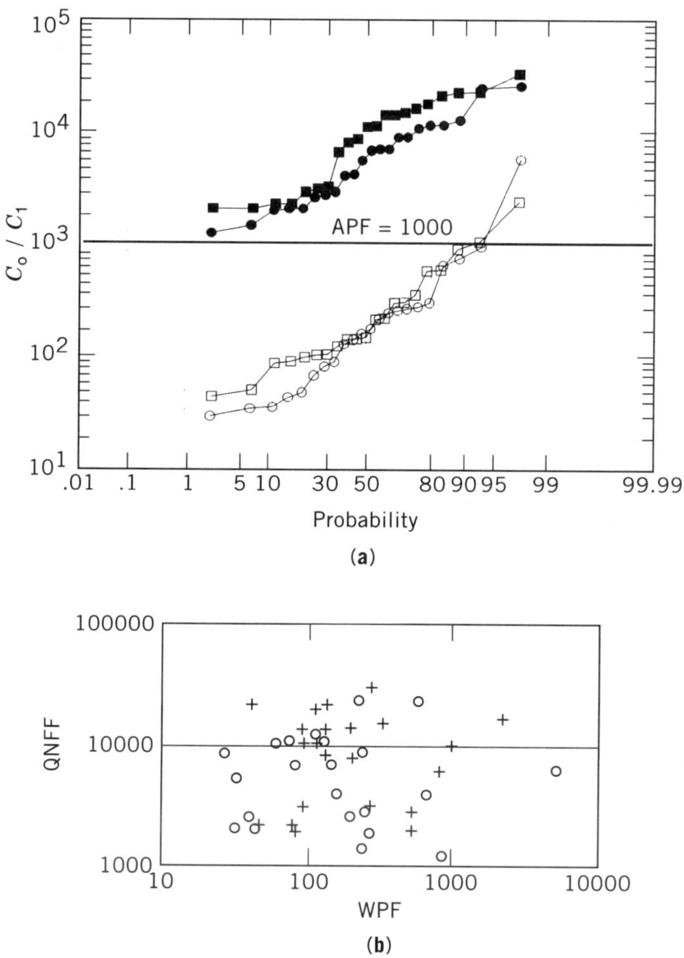

Figure 32.3. (**a**) Log probability plot of QNFF and WPF values for two brands of PAPR equipped with helmets and high efficiency filters; (**b**) Plot of paired QNFF and WPF values for the same two brands of PAPR + PAPR 1, ○ PAPR 2.

surements with a paired measurement of a respirator wearer's actual exposure assessed by end-exhaled air analysis for 1,1,2-trichloro-1,2,2,-trifluoroethane (Freon 113) under the same conditions. The QNFF measurements were then correlated with the paired Freon-113 exposures. They found that a corn oil–based QNFT had the strongest relationship (R Squared = 0.81) between QNFF measurements and total Freon 113 exposure dose. The corn oil QNFT utilized a chamber and light scattering photometry coupled with an in-facepiece sampling method that employed a deep probe position opposite the upper lip and a sample flow rate of 5 Lpm. The QNFT method representing that used with the Portacount Plus™, available from TSI, Inc. (St. Paul, MN) had the second strongest relationship (R Squared = 0.78) between the QNFF measurements and total Freon 113 exposure dose. This study represents the first research evidence that under controlled laboratory study some QNFT methods produced fit factors which strongly correlated with actual exposure.

It seems that some of the reasons postulated for observed differences between QNFFs and protection are becoming clear while others are yet undefined. The majority of the published work dealing with issues in this area are reviewed in the following.

Gaboury and Burd conducted a WPF study on negative-pressure, half-mask respirators (59). The QNFT was done with the TSI Portacount. They found that the QNFF values did not correlate with the WPFs they measured. Sampling bias associated with in-facepiece sampling on half and full facepiece respirators was evaluated in a series of studies (45–47). On one half facepiece the average sampling bias was found to be −16.6% and ranged from −99% to +98% which caused the measured QNFF values to vary from 44 to 4,728 while the actual QNFF was 87. On full facepieces average biases as large as −53% were found. These large biases on the full facepieces resulted from the sampling probe being in a position that clean inhalation air flowed over it. Circular or slit leak geometry was found to influence the TIL of half facepiece respirators.

Chen and Willeke, and Xu et al., using aerosol challenge found that a leak with a slit geometry or several circular leaks produced lower face seal leakage than a single circular leak of the same cross-sectional area (60, 61). They hypothesized that this was due to the complexity of the leak geometry. Of particular note was the observation a high-efficiency filtering facepiece respirator having the same number of face seal leaks as a dust and, may actually provide less protection at low breathing rates than the dust-mist filtering facepiece. This is due to the greater pressure drop (breathing resistance) of the high-efficiency respirator allowing the aerosol to take the path of least resistance through the face seal rather than through the filter. Campbell in a theoretical study of filter pressure drop and face seal leakage predicted that such could be the case (62). It has been suggested that the underlying assumptions of the quantitative fit tests may not be valid (63, 64). These assumptions are, (*1*) that aerosol fit test results are applicable to other types of contaminants, (*2*) aerosol samples collected with conventional in-facepiece sampling techniques are representative of the actual TIL, and (*3*) QNFT exercise protocols make the QNFT more predictive. Crutchfield et al. concluded that there may be aerosol loss, and thus, overestimation of protection factors using aerosol quantitative fit tests (65). They measured QNFFs by both an aerosol method and the CNP method. Consistently the QNFF values obtained by the aerosol method were a magnitude higher than those obtained by the CNP method. They concluded such factors as leak penetration losses, incomplete mixing due to streamlining, and lung retention may be causing the overestimation of fit factors by the aerosol QNFT methods.

A number of studies have evaluated facial anthropometric parameters and the type and location of face seal leaks on QNFT results. Oestenstad et al. using a florescent dye studied the shape and location of face seal leak sites on groups of respirator wearers with different types of half facepiece respirators (24, 25). They measured 12 different facial anthropometric dimensions on 73 subjects and then, by conducting QNFT, determined if QNFFs could be correlated with anthropometric dimensions. No significant correlation was found for the dimensions that were used to define the LANL half facepiece fit test panel [lip width and Menton-Nasion length (face length)]. These authors noted that leaks at the nose, singularly or in combination with leaks at the chin or cheek accounted for approximately 79% of all occurrences of observed face seal leakage. About 71% of the significant differences in facial dimensions for leak site subsets were attributed to gender. The authors concluded that nasal dimensions should be considered when defining a respirator fit test panel and selecting a respirator for an individual.

Liau et al. also evaluated the relationship of facial anthropometric dimensions on QNFT results (66). They evaluated nasal root breadth and face-length, nose length, nose protrusion, chin length, mouth width, and face width. They studied 243 workers that wore four brands of RPE that came in multiple sizes. Regression analysis indicated that mouth width and face width were the two most important of respirator fit. However the correlation coefficient for these facial anthropometric determinants were only 0.22 and 0.30. These low correlation coefficients suggest that a particularly strong association does not exist. The authors do suggest though that these measurements should be made to aid in the proper selection of RPE.

A similar experiment was conducted with three filtering facepiece dust and mist respirators and three elastomeric facepiece equipped with dust and mist, dust, fume, and mist, and high-efficiency filters (67). The respirators, sealed to a mannequin, had face seal leaks introduced through the use of tubes and wires. The tubes and wire created seal leaks of different geometry. This study found that face seal leakage was dependent upon the aerodynamic diameter of the aerosol and the pressure drop characteristics of the respirator at different flow rates. Myers et al. (68) in a laboratory study confirmed that the measurement of face seal leakage was affected by the tidal volume of the respirator wearer as well as the aerosol size as originally reported by Tuomi (69), Hines (70), and Holton (48, 49). In addition to using different size polystyrene latex spheres they also used acetone vapor as a challenge agent and concluded that using a vapor as the challenge agent might be a more critical test of face seal leakage than a test using an aerosol. The effect of lung retention on WPF measurements has been evaluated mathematically and correction factors recommended for lung deposition and respirator dead space (70, 71).

Another source of error could be that the APF of a particulate respirator depends upon the particle size of the ambient aerosol. Leaks having a slit geometry, as compared to other types of leaks, allow fewer particles of all sizes into the facepiece and especially those particles having a size of 2 micrometer or larger (48, 49). It was noted that as the probe moved further away from the leak site a smaller number of particles were counted inside the mask. The maximum leakage through holes in nine of the respirator facepieces tested occurred with an aerosol having a size of 0.2 to 1.0 micrometers. Particles that were larger or smaller do not enter the leak site as easily. They suggest that aerosol QNFT procedures should take into account the size distribution of the test aerosol and of the aerosol in the workplace to which the respirator will be exposed. This helps to ensure that the measured leakage during the quantitative fit test reflects the actual leakage in the workplace.

Respirator field studies have demonstrated this to be true for PAPRs where the APF based on QNFF data was 20 to 40 times higher than the protection actually achieved (36, 53–57). Consequently NIOSH and ANSI have lowered their recommended APF levels for this type of RPE. On the other hand, numerous WPF studies involving negative pressure, half facepiece respirators (both filtering facepiece and elastomeric) appear to support the recommended APF of 10 (57, 72–79). A review of half facepiece WPF studies by Nelson supports an APF of 10 for half facepiece APRs (80). He found from analyzing 390 WPF observations that the best estimate of the fifth percentile WPF was 13, with a 95% confidence interval of 10 to 18. Myers and Zhuang, using a probabilistic analysis of 205 half facepiece WPF data points, estimated the probability of a worker being over exposed (81). Their analysis found the probability of being overexposed, with a 95% confidence interval was less than 2.7%. If they considered an in-facepiece sampling error and lung loss error totaling 50% of the probability of being overexposed, with a 95% confidence interval was less than approximately 6%. They conclude, based on the studies reviewed, that an APF of 10 for half facepieces seemed to be appropriate. However, Nicas and Spears, using other assumptions and statistical models, analyzed a variety WPF data and concluded that an APF of 5 for half facepiece respirators would be more prudent (82, 83).

4.3 Workplace Testing

A limited number of workplace studies of respirator performance were carried out in the 1970s. Reist et al. evaluated the level of respiratory protection obtained by coal miners using half facepiece dust respirators (84). Revoir evaluated the protection provided by single-use (filtering facepiece) respirators against cotton dust (85). Samimi et al. measured the respiratory protection provided to sandblasters by continuous flow SAR equipped with hoods or abrasive blast helmets (86, 87). Moore and Smith evaluated three brands of half facepiece respirators for protection against SO_2 in a copper smelter (88). Toney and Barnhart evaluated the protection provide by SAR equipped with half and full facepieces in paint spraying operations (89). In the early 1980s, NIOSH began to conduct workplace studies. Their studies measured the amount of protection provided by the respirator by measuring the concentrations of workplace contaminants inside and outside of the respirator when the respirator was properly fit, correctly used and conscientiously worn.

The first of these measured the workplace protection factors of tight-fitting, full facepiece, PAPRs used during silica bagging operations. The workplace protection factors obtained ranged from 16 to 215 (53). Follow up studies measured WPFs on two brands PAPRs equipped with a helmet and high efficiency filters at a secondary lead smelter and a battery manufacturing plant and a tight fitting, high efficiency PAPR and a half facepiece respirator at a primary lead smelter (54–57). In the secondary lead smelter study, workers were quantitatively fit tested in their PAPR before the shift in which the WPF was determined (55, 56). The WPF for one brand of PAPR ranged from 28 to 5500 with a geometric mean of 165 and a geometric standard deviation of 3.57 while, the quantitative fit factors ranged from 1,200 to 24,600 with a geometric mean of 5,100 and geometric standard deviation of 2.4 (Fig. 32.3(**a**)). The other brand of PAPR had WPFs ranging from 42 to 2,323 with a geometric mean of 205 and geometric standard deviation of 2.83 and quantitative fit factors ranging from to 2,000 to 31,500 (Fig. 32.3**a**). When the two mean WPFs were compared to each other, no significant difference was found. No significant difference was found when the two mean quantitative fit factors were compared to each other. However, when the mean WPF values were compared to the mean quantitative fit values, a significant difference was found. This study found no correlation between the QNFF measured on the workers at the start of the shift and subsequent WPFs measurements. At the time of these studies the recommended APF for PAPRs was 1000 or higher (17, 90).

The tight fitting, high-efficiency, PAPR and half facepiece study was conducted in the sinter plant and the blast furnace area of a large primary lead smelter (57). The study used 25 male workers who used respirators in their normal job activities. The workers were quantitatively tested before and after each shift. The workplace protection factors for both the negative-pressure and the powered air-purifying respirators were found to be log normally distributed. For the negative-pressure respirator, about 98% of the WPFs were above 10 while 75% were above 100. It should be noted that these numbers were for a population that had attained protection factors of at least 250 with the quantitative fit tests before the workplace testing. For the powered air-purifying respirators 98% of the WPFs were above 50, 90% above 110, 75% above 200, and only 25% above 1000. No correlation was found between QNFFs and WPFs. The APF for each class of respirator was calculated using a one-sided lower 90–95% tolerance limit. At a confidence level of 90%, approximately 95% of the negative pressure half-masks WPFs exceeded a value of 10 and 95% of the WPFs for the PAPR exceeded a value of 50. A study involving PAPRs used at a brass foundry measured average WPF values of approximately 100 (91). This series of studies strongly suggested that the APF level of 1000 used by NIOSH and OSHA for PAPRs was in gross error and served as the basis for NIOSH, OSHA and ANSI to lower their APF recommendations for PAPRs.

Dixon and Nelson conducted a field study involving half facepiece APRs used in a lead atmosphere. Each worker passed the isoamyl acetate QNLT fit test (72). The 37 workplace protection factor measurements made during this study had a geometric mean of 3,400 and a geometric standard deviation of 3.8. The WPF values ranged from 94 to 27,000, The authors concluded that the APF of 10 for a half-mask negative pressure respirator was appropriate since all the WPFs exceeded 10. Galvin et al. measured the performance of a chemical cartridge respirator used for styrene exposures in a fiberglass reinforced bathtub and shower stall manufacturing facility (75). Workers in two different jobs were studied. The workers were fit tested using the irritant smoke QNLT and trained in respirator use. The inside facepiece sampling data was corrected for pulmonary retention of the styrene. The study determined that the workplace protection factors did not vary between jobs. However, they did vary between workers and on single workers during different wearing periods. The geometric standard deviation between workers was 1.92, within worker 2.93, and the total deviation was 3.51. Gaboury et al. studied half facepiece negative pressure APRs equipped with combination organic vapor/acid gas/dust/fume/mist filters and PAPRs equipped with a helmet and combination organic vapor/HEPA filter cartridges at a primary aluminum smelting plant (59). Workers wearing half facepiece respirators were fit tested those wearing the PAPRs were not. Based on exposure to benzo-alpha-pyrene, the WPFs for the PAPRs were found to range from 371 to 8,658 with a geometric mean of 1,414 while for the half facepiece APR the WPFs ranged from 13 to 410 with a mean of 47. Ninety five percent of the WPFs for the PAPR exceeded 275 while 95% of the WPFs for the half facepiece respirators exceed 9. Wallis et al. evaluated the workplace performance of a filtering facepiece APR against manganese dust (76). Other studies of exposures to mineral sand and cadmium have been conducted on half-mask respirators (92, 93).

A number of workplace studies have also been conducted in Europe. Hery et al. measured the WPFs of six European respirators, two filtering facepiece and four elastomeric APRs (94, 95). All of the APRs provided geometric mean WPFs in the range of 8 to 13. Tannahill et al. conducted a WPF study on three brands of negative-pressure, full facepiece respirators used for asbestos stripping operations (96). The data indicated that the APF of 900 times set by the British Health and Safety Executive was too high. The three brands of respirators had geometric mean WPFs of 200, 577, and 120.

Popendorf et al. evaluated several classes of APR in a WPF study of workers engaged in indoor swine production, poultry production, and grain handling (97). The results showed that the filtering facepiece APR had a mean WPF of 13, the half facepieces 19, and the PAPRs 30. These values are similar to the values obtained in other studies. They also investigated the effect of relatively small random errors (imprecision) within multiple, independent variables and larger variability due to bias and in accuracy in the result. They found that small random errors put a practical limit on the precision of measuring high WPF values but did not present the same limit when measuring low WPF values. This is an area that still requires further investigation. Myers et al. measured the performance of three elastomeric and one filtering facepiece, dust/fume/mist class APRs in three brass foundries (77). To use one or more of the APRs the workers had to pass a QNFT using the TSI Inc., Portacount™ or the saccharin QNLT. The overall geometric mean WPF values for all four APRs were respectively 166, 80, and 75 for foundries 1, 2, and 3. No statistical difference was found between the WPF values for the elastomeric and filtering facepiece APRs. The 5th percentile estimate for the pooled WPF data was 9. Zhuang et al. reported on a WPF study conducted during aircraft paint-spraying operations (78). Three elastomeric APRs, equipped with combination organic vapor/HEPA cartridges, were evaluated during two different painting applications. They found a geometric mean of 3982 with a 5th percentile of 388. Elastomeric and filtering facepiece dust/mist class APRs have been evaluated in various operations of steel making (79). The geometric mean WPFs for all brands of RPE (three elastomeric and two filtering facepieces) were found to exceed 100 and all 5th percentile WPF values exceeded 24.

Based on a review of eight WPF studies of half facepiece APRs, Nelson found differences between mean WPF and the type of filter (dust, HEPA, etc.) used on the half facepiece APR (80). Elastomeric and filtering facepiece APR, of the same filter class, had WPF values that were not significantly different.

5 RESPIRATOR CLASSIFICATION AND CERTIFICATION

The term "certified respirator" has become a part of the language of industrial hygiene. In 1920 the Bureau of Mines (BOM) began to publish "Approval Schedules" for specific types of respirator, and approved their first respirator the same year (98). Manufacturers voluntarily submitted their devices for testing, which, if satisfactory, could be sold as BOM certified devices. In 1972 the testing and certification program was assumed by NIOSH, in its laboratories at Morgantown, West Virginia. The approval categories and some of the approval requirements are given below.

Respiratory protective devices are broadly classified as air purifying and air supplying devices. Each class of device can be equipped with a variety of respiratory inlet coverings. Respiratory inlet coverings can be configured as a half or full facepiece or a loose fitting hood or helmet. The loose fitting hood or helmet can be configured with a shroud that may include just the face, the entire upper body or in other cases an air supplied suit. The tight-fitting facepiece is intended to adhere snugly to the skin of the wearer. It is available in three varieties: quarter mask, half mask, and full-face mask. The quarter mask covers the nose and mouth; the half mask covers the nose, mouth, and chin; and the full face mask covers the entire face from chin to hairline and from ear to ear. A device that does not fit easily into the foregoing categories is the mouthpiece respirator. A mouthpiece is held in the wearer's mouth, and a clamp is placed over the nostrils. The lips are pursed tightly around the mouthpiece, and all air comes through the filtering device or air supply attached to the mouthpiece. It has been used for mine rescue and is now used as an emergency escape device.

The facepiece is usually held in place by elastic or rubber straps. Half facepiece RPE are secured to the face with two straps attached at two points on each side of the facepiece, that is, four-point suspension. The four straps, instead of being attached to tabs on the edge of the facepiece, may be part of a yoke that is fastened to the facepiece by one or two points in the front of the facepiece. Full-face masks have a head harness attached to the facepiece at four, five, or six points. The large sealing surface of the full-face mask and the distribution of the headband attachment assists in maintaining a stable facepiece with less slippage than is experienced with quarter or half masks.

One of the best-known loose-fitting, hood respirators is the SAR equipped with a hood such as used in abrasive blasting. Air is provided through a hose leading into the hood. Because the hood is not tight fitting, it is important that sufficient air be provided to maintain an outward flow of air and prevent contaminants from entering the hood. The hood concept can be extended to cover the entire body and provide whole body protection. Disposable impervious clothing is available for this purpose. There are no NIOSH approval regulations for supplied-air suits. An example of a helmet device is a PAPR equipped with a helmet incorporating a face shield with or without a neck cuff. A fan draws or pushes air through an air-purifying device that is then supplied to the rear of the helmet where it travels inside the top and down inside the face shield.

5.1 Positive Pressure Classification

Respirator regulations and literature frequently contain reference to "positive pressure". The term is used to describe a mode of operation in which the pressure within a respiratory inlet covering is higher, i.e., positive, relative to ambient pressure outside the respiratory inlet covering. It is the consensus of the respiratory community that such operation affords the highest possible level of respiratory protection. The rational is that if a face seal opening(s) occurs the positive pressure would force air out of it rather than allow inward leakage of contaminant. Over the years pressure demand SCBA and SAR, continuous flow SAR and PAPRs have all been considered positive pressure RPE. However it must be noted that NIOSH has no approval criteria for positive pressure devices *per se* in Part 84. It does, however, have approval criteria for pressure demand devices which requires that pressure inside the respiratory inlet covering stay positive relative to atmospheric pressure during the certification test. That test is conducted with a breathing machine using a 622 kp.-m./min. work rate cam, operated at 24 respiration/min. The peak instantaneous flow rate under these test conditions is approximately 115 Lpm. So a device certified by NIOSH, as pressure demand is in fact a positive pressure device. As work rate increases both minute volume and peak instantaneous flow rate increase. The question of whether a device is positive pressure becomes one of defining a work rate and therefore an air delivery rate to which the RPE should be engineered. For continuous flow SAR and PAPRs the minimum flow rate required by NIOSH is 115 Lpm for tight fitting facepieces and 170 Lpm for loose fitting facepieces.

It is apparent from reported field studies (54–57, 91, 97) and laboratory studies (99, 100) that PAPRs equipped with helmets do not operate as positive pressure devices. These laboratory studies found that PAPRs equipped with a helmet may be susceptible to room air currents and point up the usefulness of a neck cuff of cape. Based in part on these field and laboratory studies, manufacturers of PAPRs equipped with helmets have modified their designs to incorporate a neck cuff with the shield of the helmet. Field studies on PAPRs equipped with tight-fitting facepiece are less numerous but the data do suggest that they are not operating as positive pressure devices (57). However, laboratory testing of PAPRs equipped with tight-fitting facepieces and HEPA filters under simulated strenuous work level continue to report QNFFs of at least 1000 (99, 100).

Under laboratory conditions, Hack et al. studied the performance of supplied air hoods (40). Using mild work exercises and an air supply of 170 Lpm, they found that of all 16 supplied-air hoods achieved a QNFF of 500, 15 achieved a QNFF of 1000, and 14 achieved a QNFF of 2000. All hoods achieved a QNFF of 2000 at 400 Lpm. The laboratory evidence does indicate negative pressure potential is created during the time of inhalation if the instantaneous inspiratory flow rate exceeds 170 Lpm, the minimum flow rate required for NIOSH certification. However Skaggs et al. testing supplied air hoods under simulated work conditions with an air supply of 170 Lpm measured a QNFF of 20,000 (100). They did note that reduced protection occurred when a fan blew contaminant under an unsecured cape. Raven et al. has measured negative pressure within the facepiece of pressure demand SARs at work rates of 35% maximum work rate (MWR), 65% MWR, 70% MWR, and 80% MWR (101). At higher work rates the time duration that the facepiece stays negative increases. Although the pressure demand SCBA is still considered to provide the best respiratory protection, it also has been found that negative facepiece pressure may occur at high work rates. Stengel and Rodrigues bench-tested fourteen, pressure demand, SCBA on a breathing machine set at a high work rate (minute volume of 100 Lpm) (102). Ten of the 14 pressure demand SCBA maintained a positive pressure in the facepiece when the cylinder pressure was full. At the point when cylinder pressure dropped low enough to activate the low-pressure alarm, positive pressure was maintained in only one of the six. Myhre et al. tested subjects wearing pressure demand SCBA at 65 and 80% of MWR (103). They found the pressure in the facepiece to be negative 8% and 36% of the time, respectively. Dahlback and Balldin, testing pressure demand SCBA, found negative pressure in the facepiece at 80% of MWR (104). Wilson et al. recommend that SARs and SCBAs be designed to accommodate peak flows of 400 Lpm because there are respirator users in the work force with lung function capabilities sufficient to produce instantaneous inspiratory flow rates of this magnitude (105). For pressure demand SCBA used in fire fighting, the National Fire Protection Association has recommended that positive pressure should be maintained in the facepiece when tested on a breathing machine operating at 30 respiration/min. producing a minute volume of 103 Lpm (106). These conditions produce a peak instantaneous demand of approximately 330 Lpm. They feel these test parameters are sufficient to meet the respiratory demands of approximately 98% of the fire fighter population.

5.2 Respirator Facepiece Valves

5.2.1 Inhalation Valves

Inhalation valves prevent moisture-laden exhaled air from coming in contact with sorbents and filters. Exhaled air tends to increase breathing resistance of the filters and cause them to be changed more frequently. NIOSH requires that inhalation valves must be provided where necessary to prevent excessive exhaled air from adversely affecting filters unless the filters are specifically designed to resist moisture. With full facepieces, malfunctioning inhalation valves that allow significant amounts of leakage during exhalation result in maintaining higher than expected levels of contaminant in the facepiece cavity (107). Brosseau evaluated inhalation and exhalation valves from half facepiece, negative pressure RPE for leakage during an 8-hour cyclic breathing test using two work rates, 415 and 622 kg-m/min, and two particle sizes, 0.3 and 0.8 micrometer (108). Inhalation valves averaged 20% leakage for all experiments. The penetration of 0.3 micrometer particles was less than the penetration of 0.8 micrometer particles (13% vs. 27%). No inhalation valve failures occurred. Current NIOSH certification criteria include no performance tests for the inhalation valve(s).

5.2.2 Exhalation Valves

With the exception of some disposable respirators, NIOSH regulations require that tight fitting respiratory inlet coverings be designed to include an exhalation valve(s), which provide a one-way exit for exhaled air. This important component of the RPE, while out of sight and often neglected, needs frequent checking to ensure that it is working. A malfunctioning exhalation valve can render useless both positive and negative pressure respirators. If an obstruction such as a hair or a dirt particle comes between the valve and the valve seat, a serious leak can occur. NIOSH requires that exhalation valves be protected against damage and external influence. This is commonly done with a valve cover. In addition to protecting the valve the valve cover retains a small reservoir of exhaled air. A finite time elapses between the beginning of inhalation and the closure of the exhalation valve. During this period the clean, recently exhaled air prevents toxic material from entering the valve. Exhalation valves are evaluated by NIOSH as part of the certification evaluation. Test criteria are contained in several Subparts of Part 84, eg Subpart K paragraph 84.182. The valve(s) can not have a leakage rate of greater than 30 mL/min. when subjected to a suction of 25 mm of water column height. In a study of inhalation and exhalation valves, Brosseau reported that exhalation valve leakage ranged from 0.0 to 0.03% for all test conditions (108). No failure of exhalation valves occurred during the eight-hour test. At a low work rate (415 kg-m/min), particle size had no effect on exhalation valve penetration but at a high work rate (622 kg-m/min), the 0.3 micrometer particles were found to be less penetrating than the 0.8 micrometer particles (0.035 vs. 0.06%).

5.3 Air-Purifying Respirators

Air-purifying devices remove contaminants from the atmosphere and can be used only in atmospheres containing sufficient oxygen to sustain life (at least 19.5% by volume at sea level) and within specified concentration limitations of the specific device. Air-purifying respirators incorporate an air-purifying element that is capable of removing aerosol, gas, or vapor contaminants from the air before the air can reach the respiratory tract. With the exception of filtering facepiece respirators, the air-purifying elements used with respirators are designed for replacing. The useful life of an air-purifying device is dependent upon the concentration of the contaminant(s), the breathing volume of the wearer, and the capacity of the air-purifying medium. Air-purifying units can be divided into two functions: aerosol removal, and gas and vapor removal. For aerosol removal the aerodynamic size and charge of the aerosol particles and the face velocity of the air entering the filter influence the efficiency of filtration. Gas and vapor removal requires a suitable sorbent material. The air-purifying element is composed of a filter, a gas or vapor sorbent or a combination of both. NIOSH certified air-purifying RPE is only designed for use during entry into or escape from atmospheres not IDLH. Air-purifying RPE can not be used for entry into IDLH atmospheres or atmospheres that are oxygen deficient.

Nonpowered APR are negative pressure devices and must have a tight-fitting facepiece. They rely on the negative pressure produced within the facepiece cavity during inhalation to draw ambient air through the air-purifying elements. NIOSH certifies half, full face, or mouthpiece respirators as a negative pressure, APR. The mouthpiece respirator is a special APR that is held securely in the mouth of a user and used for escape purposes only. Powered air purifying respirators can be equipped with tight fitting as well as loose fitting hoods or helmets.

5.3.1 Particulate Respirators

Air-purifying, particulate respirators utilize filters to remove aerosols from the air. Filters capture aerosols using a combination of filtration mechanisms. These mechanisms are impaction, interception and diffusion. Filters relying solely on these mechanisms are commonly referred to as "mechanical filters." Electrostatic attraction, resulting from charge on the aerosol or fibers of the filter or both, is often used to enhance these filtration mechanisms. Filters that use an engineered method to add an electrostatic charge to fibers to enhance electrostatic attraction are commonly referred to as "electrostatic filters."

Regardless of the chemical composition, it is the aerodynamic particle size of the aerosol that determines which of the mechanisms is predominating. The efficiency of aerosol removal is related to its aerodynamic size, with approximately 0.3-micrometer aerodynamic diameter aerosols being the most penetrating. This most penetrating particle size can shift slightly by changes in the face velocity challenging the filter. The efficiency of all mechanisms is improved as the filter fiber diameter is decreased and the fiber packing density is increased. The resistance to airflow per unit of surface area of the filter usually increases as the filter fiber diameter decreases and fiber-packing density increases. A filter with high collection efficiency for small aerosols usually has a higher resistance to airflow than a filter having lower collection efficiency for small aerosols. The face area of the high-efficiency filter must be larger to keep the inhalation flow resistance equal to that of the lower-efficiency filter. During use the efficiency of filters improve, and breathing resistance increase, as the aerosol loads on the filter surface and begins to act like a sieve. Electrostatic charge can be added onto fibers by using resins or be engineered into synthetic fibers as permanent dipoles. Electrostatic filters, when compared to mechanical filters of similar efficiency without an electrostatic charge, have lower resistance to airflow without decreasing filter efficiency. In general, oil aerosols are more penetrating of electrostatic filters than particulate aerosols. It is hypothesized that the oil degrades the electrostatic charge of the resin or coats the fiber and in effect shields its internal dipole.

5.3.1.1 Filter Classification. NIOSH no longer classifies filters for use in nonpowered APRs as dust, dust and mist, dust, mist and fume, or high efficiency. In Part 84, Subpart K NIOSH now classifies nonpowered air-purifying particulate filters into three series, N (Not resistant to oil), R (Resistant to oil), and P (oil Proof). The N-series filters are restricted to use in those workplaces free of oil aerosols. The R- and P-series filters are intended for removal of any particulate that includes oil-based liquid aerosols. The filters of each series are further classified by efficiency against a most penetrating particle size as: (*1*) 100 class filters which have a minimum efficiency ≥99.97%; (*2*) 99 class filters which have a minimum efficiency ≥99%; (*3*) 95 class filters which have a minimum efficiency ≥95%. These tests incorporate an approximately 0.3 micrometer aerodynamic diameter particle that filtration theory indicates should be a most penetrating particle size. Other significant changes in certification test performance parameter include using:

a. An air flow rate of 85 ± 4 liters that simulates a high work rate (42.5 ± 2 liters per minute through each filter for paired filters).
b. DOP and NaCl particles at Botzmann charge equilibration.
c. Measurement of instantaneous penetration rather than a time weighted average.
d. Using filter number efficiency rather than mass efficiency.
e. High total filter loading (up to 200 mg for N- and R-series filters, and continued loading until there is no further decrease in efficiency for P filters).

It should be noted that 42 *CFR* Part 84 still requires an evaluation of initial breathing resistance at 85 ± 2 Lpm however the requirement for an evaluation of final breathing resistance after filter loading was dropped. The test conditions and aerosols for the nonpowered respirator filters are summarized in Table 32.3. The degradation category (N, R, or P) is determined by using either DOP or NaCl as the test aerosol. Sodium chloride is only slightly degrading to filter efficiency, whereas DOP is very degrading. No particle size limits apply to respirators with Part 84 filters. Protection for the user is based on the efficiency of the filter and the PEL of the contaminant, usually determined by an industrial hygienist.

Table 32.3. Description of Filter Classes and Key Test Parameter for Filters Certified under Title 42 Code of Federal Registrations Part 84.

Class of Filter	Efficiency (%)	Test Agent	Test Maximum Loading (mg)	Type of contaminant	Service Time[a]
			N-series		
N 100	99.97	Sodium chloride	200	Solid or water-based particulate but *not* aerosol containing oil mist	Nonspecific[b,c]
N 99	99				
N 95	95				
			R-series		
R 100	99.97	Dioctyl phthalate (DOP) oil	200	Any	One work shift[b,d]
R 99	99				
R 95	95				
			P-series		
P 100[e]	99.97	Dioctyl phthalate (DOP) oil	Stabilized efficiency	Any	Nonspecific[b]
P 99	99				
P 95	95				

[a]If research indicates additional service time limitations may be recommended by NIOSH for specific workplace conditions.

[b]Limited by considerations of hygiene, filter damage, and breathing resistance.

[c]High (200 mg) filter loading in the certification test is intended to address the potential for filter efficiency degradation by solid or water-based (i.e., non-oil) aerosols in ther workplace. Accordingly, there is no recommended service time limit in most workplace settings. In workplaces with high aerosol concentrations, service life should only be extended beyond 8 hours of use (continuous or intermittent) by performing an evaluation in specific workplace settings that demonstrates: that extended use will not degrade the filter efficiency below the certified efficiency level, or that the total mass loading of the filter is less than 200 mg (100 mg per filter for dual-filter respirators).

[d]No specific service time limit when oil aerosols are not present. In the presence of oil aerosols, service time may be extended beyond eight hours of use (continuous or intermittent) by demonstrating that extended use will not degrade the filter efficiency below the certified efficiency level, or that the total mass loading of the filter is less than 200 mg (100 mg per filter for dual-filter respirators).

[e]The P-100 filter must be color-coded magenta. The Part 84 Subpart KK HFPA filter on a PAPR will also be magenta, but the label will be different from the P I00 filter, and the two filters cannot be interchanged.

Filters designated as N series are tested with NaCl and are not resistant to efficiency degradation by oils. Therefore, N-series filters may be used for any solid or non-oil containing particulate contaminant. In workplace conditions that result in high filter loading (i.e., 200 mg), service life for N-series filters should be limited to 8 hours. Extending service life beyond 8 hours of use (continuous or intermittent) should only be done after an evaluation of the specific workplace. This evaluation should demonstrate that:

a. service life exceeding 8 hours will not degrade the filter's efficiency below the level specified in Part 84, or
b. the total mass loading of the filter(s) is less than 200 mg.

Reevaluation is necessary when conditions change or modifications are made to processes that could change the type of particulate generated.

R-series filters may be used for any particulate contaminant. If used for an oil-containing particulate, a one-shift use limit applies. Service life for the R-series filters can be extended using the same two methods and qualification described for N-series filters.

P-series filters may be used for any particulate contaminant. Recent NIOSH laboratory studies indicate the efficiency of P-series filters may be significantly reduced with long term use in the presence of oil aerosols (109). Heavy oil loading reduces the efficiency of P100 filter levels much less than that required of P95 filters. The reduction in filter efficiency was not always accompanied by an increase in breathing resistance that would signal the user to replace the filter, or filter element. They found that the reduction in filter efficiency varied significantly from model to model and therefore could not make a single filter change recommendation that was appropriate for all models. NIOSH has requested that each manufacturer of P-series filters establish service time recommendations as part of their user instructions and their customer support programs.

The service life of all filters is limited by considerations of hygiene, damage, and breathing resistance. All filters should be replaced whenever they are damaged, soiled, or causing noticeably increased breathing resistance.

5.3.2 Gas and Vapor Respirators

Gas and vapor respirators are commonly referred to a "chemical cartridge respirators" or "gas masks". The difference is that gas masks utilize only full facepieces and the air purifying element, called a canister, is larger than the cartridge element used with chemical cartridge respirators.

Gas and vapor removal is accomplished with sorbents that have collection efficiencies that are dependent on the chemical composition and physical properties of the gas or vapor. Sorbents are porous and have a large surface area per unit weight-up to 1500 m^2/g. As the gas or vapor passes through the sorbent bed, the molecules are sorbed on the surface. Adsorption, chemisorption, absorption, or catalysis captures gases and vapors when passing through a particulate (sorbent) bed of chemicals. Adsorption is the physical attraction of a vapor to an adsorbent surface. Chemisorption occurs when the sorbent surface and the vapor bond chemically. Absorption occurs when the vapor penetrates below the surface of the absorbent and chemically reacts with it. If the solid material causes the vapor to change chemically without affecting the solid then it is a catalytic reaction.

The sorbent most widely used is activated carbon or charcoal. Alone it is an excellent adsorbent of many organic vapors. Impregnating it with specific materials increases its retention efficiency for additional gases and vapors. Where there is reason to suspect that a sorbent has a high heat of reaction with a substance, use of that sorbent is not recommended. For such a substance, only nonoxidizable sorbents should be allowed.

NIOSH certified chemical cartridges and canisters are not for use against gases or vapors with poor warning properties, unless the respirator is approved with an effective ESLI or those that generate high heats of reaction with sorbent materials in the cartridge or canister. Specific MSHA or OSHA standards may permit use of chemical cartridges or canister for gases or vapors that do not have adequate warning properties. Generally, warning properties are defined according to odor, taste, eye irritation, or respiratory irritation. Adequate warning properties imply that the gas or vapor of interest has a persistent odor or irritant effect at concentrations at or below the OSHA PEL or other occupational exposure limit if a PEL does not exist. Gas and vapor APR cannot be used in IDLH atmospheres or in atmospheres containing less then 19.5% oxygen by volume. Gas masks (canister respirators) may be used for escape if the atmosphere is not oxygen-deficient. Gas and vapor APRs should not be allowed for either entry into or escape from hazardous environments when supporting evidence exists to demonstrate that unreasonably short service life would occur at the maximum use concentration. Although limited in number, there are special purpose gas and vapor APRs respirators, e.g., vinyl chloride or formaldehyde, that are approved by MSHA/NIOSH for protection against gases and vapors when respirators approved for a given class of contaminants (e.g., organic vapors) cannot be used due to sorbent deficiencies.

OSHA's respirator standard allows users to establish respirator change out schedules. At this time OSHA has not defined acceptable protocols for establishing change out schedules (5). However it is generally recognized that such schedules could be based on data obtained from breakthrough studies conducted in the workplace or in the laboratory under "worst" conditions. Definition and guidance for establishing change out schedules waits further development from OSHA, NIOSH, and ANSI.

5.3.2.1 Chemical Cartridge Respirators. NIOSH-certified chemical cartridges or canisters are permissible for use against gas or vapor contaminants up to a concentration that is the lesser of either the concentration from multiplying the respirator's APF classification by the contaminant's OEL, or the IDLH concentration of the contaminant, given an O_2 deficiency does not exist. Specific types of chemical cartridge respirators are approved under Part 84, Subpart L. Canisters can be equipped with particulate filters to form combination gas/vapor and particulate RPE.

Gas masks are similar to chemical cartridge respirators but the quantity of available sorbent material is greater. They are available with half or full facepieces and hoods or helmets. Although some of the approval test gases and vapors for chemical cartridge respirators are the same as those used to test gas masks, the test concentrations and the minimum sorbent life span differ.

5.3.2.2 Gas Masks. Gas masks are defined as full facepieces attached by a breathing tube to a canister that can be carried on the front or back of the wearer or mounted on the chin of the facepiece. An exception is the escape gas mask, which may have a half mask or a mouthpiece and nose clip. Gas masks provide respiratory protection against certain specific gases and vapors in concentrations up to 2% by volume (20,000 ppm) or as specified on the canister label and against particulate matter. Specific gas masks are approved under Part 84, Subpart I. NIOSH certifies an escape gas mask that is designed for use during escape only from non-IDLH atmospheres.

5.3.2.3 Gas and Vapor Cartridges and Canisters. The effectiveness of any sorption cartridge or canister should be established for the particular gases or vapors to be captured. The certification tests used by NIOSH do not test the sorption efficiency for all gases and vapors found in the workplace. The certification bench tests for gas and vapor canisters and cartridges are in Subpart I and L respectively of Part 84. The challenge gas or vapor and its concentration, test flow rate, maximum allowable penetration, and minimum service life for these devices are summarized in Table 32.4. The bench test requirements for escape canisters are similar, but not identical to those for chin-style canister listed in Table 32.4. The specific test criteria for escape gas masks can be found in Subpart I of Part 84. The NIOSH recommended maximum use concentrations for chemical cartridges and canisters are listed in Table 32.5.

Canisters and cartridges are evaluated by NIOSH against a relatively few gases and/or vapors so breakthrough information for many chemicals is not readily available and must be obtained from other sources such as the technical literature and respirator manufacturers Nelson and Harder did extensive work in determining the service life for organic vapor cartridges exposed to various organic vapors (110). They exposed commercial organic vapor cartridges to high concentrations of organic vapors and recorded their breakthrough times. Using the NIOSH procedure for organic vapor cartridge testing, they tested the cartridges against 107 organic vapors. Nineteen of the 107 materials tested failed in less than 50 min, which is the minimum time the cartridge must last under NIOSH test conditions. Eleven of the 19 failures occurred in the chlorinated hydrocarbon group.

The ability of a person to detect a given odor is strongly influenced by their innate olfactory powers, past experience with the odor and their level of attention (111). This is why NIOSH does not approve the use of cartridges for materials with poor warning properties NIOSH will and has approved sorption cartridges for materials with poor warning properties if they have an end of service life indicator on the cartridge. Such ends of service life devices are available for carbon monoxide, mercury vapor, and vinyl chloride. Without an end of service life indicator, information concerning the effectiveness of the sorbent with the contaminant should be obtained, and a conservative time of service set. The nose can then be used as a secondary alarm.

5.3.2.4 Combination Cartridges and Canisters. Cartridges and canisters can be certified for more than one class of gas or vapor (e.g., organic vapor and acid gas) or they also may be used in combination with an aerosol filter (e.g., organic vapor and particulate R 95 filter). A chemical cartridge or canister certified for more than one class of gas or vapor may contain more than one type of sorbent media. Combination canisters for gas masks must meet the minimum breakthrough times at the specific challenge concentrations and conditions noted in Table 32.4. Combination cartridges must meet one-half the minimum

Table 32.4. Bench Test Requirements for NIOSH Certified Gas and Vapor Cartridges, and Gas Mask Canisters

Device Type	Test Condition	Gas or Vapor	Concentration (ppm)	Flow Rate (Ipm)	Maximum Allowable Penetration (ppm)	Minimum Service Life[a] (minutes)
			Cartridge(s)[b]			
Ammonia	As received	NH_3	1,000	64	50	50
	Equilibrated	NH_3	1,000	32	50	50
Chlorine	As received	Cl_2	500	64	5	35
	Equilibrated	Cl_2	500	32	5	35
Hydrogen chloride	As received	HCl	500	64	5	50
	Equilibrated	HCl	500	32	5	50
Methylamine	As received	CH_3NH_2	1,000	64	10	25
	Equilibrated	CH_3NH_2	1,000	32	1	25
Organic vapors	As received	CCl_4	1,000	64	5	50
	Equilibrated	CCl_4	1,000	32	5	50
Sulfur dioxide	As received	SO_2	500	64	5	30
	Equilibrated	SO_2	500	32	5	30
			Chin-style Gas Mask Canister			
Acid gas	As received	SO_2	5,000	64	5	12
	Equilibrated	SO_2	5,000	32	5	12
	As received	Cl_2	5,000	64	5	12
	Equilibrated	Cl_2	5,000	32	5	12
Organic	As received	CCl_4	5,000	64	5	12
Vapor	Equilibrated	CCl_4	5,000	32	5	12
Ammonia	As received	NH_3	5,000	64	50	12
	Equilibrated	NH_3	5,000	32	50	12

Carbon monoxide	As received	CO	20,000	64		60
	Equilibrated[c]	CO	5,000	32		60
	Equilibrated[c]	CO	3,000	32		60
Combination of 2 or 3 chin-style types[d]						
Combination of all chin-style types[e]						
		Front or Back Mounted Gas Mask Canisters				
Acid gas	As received	SO_2	20,000	64	5	12
	Equilibrated	SO_2	20,000	32	5	12
	As received	Cl_2	20,000	64	5	12
	Equilibrated	Cl_2	20,000	32	5	12
Organic	As received	CCl_4	20,000	64	5	12
Vapor	Equilibrated	CCl_4	20,000	32	5	12
Ammonia	As received	NH_3	20,000	64	50	12
	Equilibrated	NH_3	20,000	32	50	12
Carbon monoxide	As received	CO	20,000	64		60
	Equilibrated[c]	CO	5,000	32		60
	Equilibrated[c]	CO	3,000	32		60
Combination of 2 or 3 front or back mounted types[d]						
Combination of all front or back mounted types[e]						

[a]Minimum life will be determined at the indicated maximum allowable penetration.

[b]Where a respirator is designed for respiratory protection against more than one class of gas or vapor, as for use in ammonia and in chlorine, the minimum service life shall be one-half that shown for each type of gas or vapor. Where a respirator is designed for respiratory protection against more than one gas of a type, as for use in chlorine and sulfur dioxide, the stated minimal life shall apply.

[c]Relative humidity of test atmosphere will be $95 \pm 3\%$; temperature of test atmosphere will be $25 \pm 2.5°C$.

[d]Test conditions and requirements are as shown in the table.

[e]Test conditions and requirements are as shown in the table except that the minimum service lives for acid gas, organic vapor and ammonia will be six minutes.

Table 32.5. NIOSH Recommended Maximum Use Concentrations[a] for Gas and Air Purifying Elements

	Classification of Gas and Vapor Air-Purifying Elements		
Type of Gas or Vapor	Cartridge(s)	Chin-Style Canister	Front- or Back-Mounted Canister
Organic vapors	1,000[b]	5,000†	20,000[c]
Acid gases			
Sulfur dioxide (SO_2)	50	100	100
Chlorine (Cl_2)	10	25	25
Hydrogen chloride (HCl)	50	100	100
Ammonia (NH_3)	300	500	500
Methyl amine (CH_3NH_2)	100	—	—
Carbon monoxide (CO)	NA[d]	NA[d]	1,500

[a]Expressed in ppm.
[b]Maximum use concentration will be 1,000 ppm or equal to or lower than the immediately dangerous to life or health (IDLH) value for the specific organic vapor, whichever is lower.
[c]Maximum use concentration for "entry into" will be limited to the value listed or to the IDLH value for the specific organic vapor, whichever is lower.
[d]NA = not available.

breakthrough times at the specific challenge concentrations and conditions noted in Table 32.4.

A chemical cartridge or canister combined with a particulate filter provides respiratory protection where exposure can be to mixed gas/vapor and particulate contaminants. In addition to meeting the requirements for gas and vapor certification these combinations must also pass the applicable particulate filter tests and meet a final breathing resistance limit, after filter loading, of 70 mm H_2O at a flow rate of 85 ± 2 Lpm.

5.3.2.5 Carbon Monoxide. Carbon monoxide is not sorbed, but by catalytic action is oxidized into carbon dioxide. The carbon monoxide canister is filled with a mixture of manganese and copper oxides, known as Hopcalite. This mixture catalytically converts carbon monoxide to carbon dioxide without the presence of moisture. Moisture stops the reaction. To prevent the entry of moisture carbon monoxide canisters have layers of drying agents on both sides of the catalytic agent. Because carbon monoxide has no warning odor, an end of service life indicator shows when the drying agent is saturated and the life of the canister is ended.

5.3.3 Powered Air Purifying Respirators

The PAPR is designed with a battery-operated blower to push or pull air through the air-purifying element(s) and supply it to the user. As of July 10 1998, NIOSH only certifies PAPRs equipped with high efficiency (HEPA) particulate filters that have been tested against thermally generated monodispersed DOP aerosol under the test parameters described in Part 84, Subpart KK, paragraph 84.1151. Powered air-purifying respirators equipped with a tight-fitting facepiece must provide a minimum of 115 Lpm of air to the facepiece. Those equipped with a hood or helmet must provide a minimum of 170 Lpm to the loose fitting facepiece. The PAPR is available in three configurations: tight-fitting facepieces, hoods, and helmets. Because the PAPR is an APR, they must not be used in an IDLH environment or in the presence of atmospheric contaminants for which the filters or sorbents are not designed. They should not be considered as positive pressure devices.

5.4 Atmosphere Supplying

5.4.1 Self-Contained Breathing Apparatus

Self-contained breathing apparatus provide complete breathing protection for various periods of time based on the amount of breathing air or oxygen supplied and the breathing demand of the wearer. The SCBA respirator carries an air supply of compressed air, compressed or liquid oxygen, or oxygen-generating chemicals. No external connection to an air supply is necessary. Self-contained breathing apparatus are classified as open circuit and closed circuit devices. In open circuit devices the exhalation flow is dumped to the atmosphere. In closed circuit devices, the exhalation flow is kept in the respirator, scrubbed of excess CO_2 and moisture, and reconditioned with fresh O_2. Self-contained breathing apparatus are certified under the test requirements of Part 84 Subpart H. The inhalation flow is engineered to operate in demand or pressure demand modes.

Pressure demand SCBA is often used in very high concentrations of contaminants. At such high concentrations it must be recognized that certain materials, particularly some gases and vapors, are capable of harming the body by means other than through the respiratory tract. The possibility of dermal exposure must also be considered. To avoid that possibility, consideration must be given to selecting the appropriate protective clothing that should be worn in addition to just selecting the proper respiratory protection. For example, ammonia, in concentrations of approximately 3% or higher, can cause skin burns, particularly on moist skin. Hydrocyanic acid, a gas at slightly above room temperature, is capable of penetrating the skin and causing systemic poisoning, although to do so, concentrations considerable higher than those required for poisoning through the respiratory tract must be present.

5.4.2 Open-Circuit Self-Contained Breathing Apparatus

The most widely used SCBA is an open-circuit device consisting of a compressed air cylinder, airflow regulator, airline and facepiece, all of which must weigh less than 35 lb. The open-circuit SCBA must, at an inhalation resistance of 51 mm H_2O and an air pressure of 500 psi, deliver a minimum air flow of 200 Lpm. The duration of air supply can be certified at different intervals from a minimum of 3 min to a maximum of 240 min. Combination SAR and auxiliary SCBA units can be approved for use in IDLH atmospheres. In the United States SCBAs are available in units with air cylinders pressurized to 2200 or 4500 psi. The air supply cylinders may be steel, aluminum, fiberglass wrapped aluminum, or a composite cylinder of an aluminum bladder inside a filament wound fiberglass cylinder. The fiberglass cylinder and the fiberglass wrapped aluminum cylinders reduce the weight of the air cylinders significantly. The airflow regulator functions in the

SCBA and the SAR in a similar manner. It opens to admit air into the facepiece when the pressure in the facepiece is decreased. The airflow can vary depending on the inhalation pressure. In contrast to the SAR regulator, it may be a two-stage pressure-reducing regulator because the air supply cylinder is under such high pressure. In addition, the regular must be designed to have a manual bypass valve for emergency use by the wearer if the regulator were to malfunction.

The SCBA customary used for entry into hazardous atmospheres is rated at 30 or 60 minutes; however, escape only SCBA units can have rated service life of 3, 5, 10, or 15 min. Service life ratings given with this equipment are nominal and should not be regarded as absolute. High work rates combined with the stress of emergency conditions will markedly reduce the actual use time of the air cylinder. The SCBA can be obtained in demand or pressure demand mode. Only the pressure demand units should be used because they provide the highest degree of protection of any respirator. Only pressure demand SCBA respirators are recommended for use in environments that may be immediately dangerous to life or health (IDLH).

5.4.3 Closed-Circuit Self-Contained Breathing Apparatus and Self-Contained Self-Rescuer

Closed-circuit devices, available with up to a 4-hour service life, reuse and condition the exhaled air by circulating it through a chemical to remove the carbon dioxide. The exhaled air then flows to the breathing bag, the breathing air reservoir for the facepiece. Because oxygen is consumed during cellular respiration and the carbon dioxide is removed, a small amount of oxygen is added back into the system. The oxygen can be added from a cylinder of liquid or compressed oxygen. Oxygen may also be stored as a solid chemical compound for use with a closed circuit SCBA. The chemical, usually a type of peroxide, reacts in the presence of moisture and carbon dioxide to produce pure oxygen. Because moisture and carbon dioxide are both components of exhaled breath, circulating exhaled breath through the oxygen-rich chemical forms a usable breathing mixture. The carbon dioxide is removed, as it becomes part of the reaction to release oxygen. The air then enters a breathing bag for mixing before being drawn into the facepiece. A disadvantage of this unit is that once the reaction has been started in the canister, it cannot be terminated until the chemical reaction is complete. Canisters are available for up to 2-hr duration. Certain types of these units referred to as self-contained self-rescuers are used as emergency escape devices for miners. Most closed circuit SCBAs are demand mode devices. The units have been used successfully for many years in mine rescue operations. The rescue teams are given extensive training in the use of the equipment to ensure proper use.

5.4.4 Supplied-Air Respirators

Supplied-air respirators deliver breathing air from a remote storage site, through a supply hose to the respiratory inlet covering. Supplied-air respirators do not have air-purifying filters and cartridges; instead, they depend on delivering acceptable quality breathing air to the user from an external source. With the exception of hose masks equipped with blowers, these devices should be used only in atmospheres not immediately dangerous to life or health. Supplied-air respirators are certified under the test requirements of Part 84

Subpart J. They are divided into three classes A, B, and C. Class A and B SAR are hose masks and not widely used. Class C SAR can be equipped with half masks, full face masks, and hoods or helmets supplied by a hose with air under pressure not to exceed 125 psi. Class C may be approved in demand, pressure demand, or continuous flow modes. A minimum airflow delivery of 115 Lpm is required for use with a tight-fitting facepiece and 170 Lpm with a hood or helmet. The maximum allowable airflow for SARs is 425 Lpm. NIOSH certification of type "CE" refers to a device for abrasive blasting that has outer coverings. Hoses for class C devices can be obtained in 15-, 25-, or 50-ft lengths up to a maximum of 300 ft. The maximum inhalation and exhalation resistances are both 50 mm H_2O.

5.4.4.1 Hose Masks, Powered and Nonpowered. The oldest SAR unit is the type "B" hose mask. It consists of a large-diameter hose, anchored in an area with acceptable quality breathing air and connected to the wearer's facepiece. The diameter of the hose is large, which reduces inhalation resistance. Another version is the type "A" hose mask, which has a hand- or electric-powered air mover connected to the fresh air source to assist in supplying air. The Bureau of Mines started to approve hose masks in 1929. Considering the wide variety and use of airline SAR equipment today, hose masks are a rarity and there are really no use situations where they are the RPE of choice.

5.4.4.2 Type C SAR. The type "C" supplied-air respirator (often referred to as an airline respirator) differs from hose masks in that the air is delivered to the facepiece under pressure up to 125 psi. The NIOSH-approved SAR includes the facepiece, breathing tube, belt-mounted air control valve, and air hose from 15 to 300 ft long. An airflow regulator is also required for demand mode and pressure demand equipment. The user is responsible for providing the air supply, any necessary airline filters, manifold with airflow control valve for the air hose connection, and the airline pressure gauge. The airflow, set by a valve at the supply manifold, can be changed only by manually adjusting the airflow control valve. This is in contrast with the open-circuit self-contained breathing apparatus, which has an airflow that varies automatically with the inhalation pressure of the user. Problems can be encountered in ensuring the minimum required flow is delivered. Changes such as an airline that is longer than design length, has a smaller diameter, or has been spliced can result in a drastic reduction in the air supply with the same setting of the pressure gauge. Airflow can be engineered to occur in a demand mode, pressure demand mode, or continuous flow mode.

The demand mode SAR, as the name implies, supplies air to the user on demand. When the user inhales, the respirator airflow regulation valve opens and air under pressure enters the respirator. The negative pressure in the facepiece may reach 50 mm H_2O before the air starts to flow. Airflow ceases when inhalation ceases. The facepiece pressure is comparable to the negative pressure present when using an air-purifying respirator. The SAR may have a half or full facepiece. In each case, because of the negative pressure in the facepiece during inhalation, the degree of facepiece leakage is comparable to that of an air-purifying respirator with the same facepiece. Because demand mode SARs do not offer protection to the user that is superior to negative pressure, air-purifying respirator, their selection as RPE of choice is very limited except for conditions where only an O_2 deficiency exists.

The problem of leakage into negative pressure facepieces led to the positive pressure facepiece, which greatly reduces contaminant leakage into the respirator facepiece. The pressure demand SAR is similar to the demand mode SAR except for the exhalation valve and the airflow regulator. The exhalation valve is spring loaded so that a pressure of up to 38 mm H_2O can be maintained inside the respirator facepiece. During inhalation the pressure in the facepiece still drops but at some point while still above atmospheric pressure the airflow regulator opens providing airflow. Normally when inhalation occurs, the positive pressure inside the facepiece is reduced but remains positive. It is possible under high work rates for the user to demand a higher instantaneous airflow than the SAR can deliver. Under this condition, negative pressure may occur in the facepiece for a short period of time. Regardless, the PDM provides the greatest degree of protection for the user of SAR. The only respirators carrying the positive pressure.

Continuous flow SARs can be equipped with tight fitting and loose fitting facepieces. The loose fitting SAR may consist of a supplied air hood, helmet, blouse, or full suit. It requires a minimum continuous airflow of 170 Lpm when in use. The tight-fitting SAR is simply a facepiece with a continual flow of air through the mask. It requires a minimum continuous airflow of 115 Lpm when in use. The hood of the SAR introduces the air supply through a manifold above the head of the user. The air flows downward and exits under the edges of the hood. Better designs have neck cuffs. NIOSH certification regulations specify a minimum airflow of 170 Lpm for SARs equipped with hoods. The literature contains many references to the positive pressure nature of continuous flow devices. However, users are advised that numerous field and laboratory studies, reviewed earlier in this chapter, indicated that at minimum certification air flow rates (115 Lpm for tight fitting facepieces or 170 Lpm for loose fitting facepieces) these devices do not maintain positive pressure.

5.5 Supplied Air Suits

There are no official regulations governing the design, sale, and use of supplied air suits, but the Department of Energy (DOE) has a supplied-air suit protocol for use in reviewing suits used by DOE contractors (109). Much is unknown about the design and use of airline suits. The traditional figure of 170 Lpm for supplied air hoods may not provide sufficient airflow to prevent inward leakage for these devices. Airline suits should be tested to determine their efficiency prior to use in atmospheres of high toxicity.

5.6 Escape Respirators

Escape devices have a single function: to allow a person working in a normally safe environment sufficient time to escape from a suddenly developing respiratory hazard(s). Escape devices can be separated into two categories: air-purifying respirators and self-contained breathing apparatus. Air-purifying respirators include the escape gas mask respirator, the gas mask (canister) respirator, and the filter self-rescuer. The escape gas mask consists of a half-mask or a mouthpiece respirator. The mouthpiece respirator can be used for short periods of time to escape from low concentrations of organic vapor or acid gas. The escape gas mask, which utilizes a half-mask, removes contaminants from the air. These respirators may also be used to escape from low concentrations of organic vapor or acid gas. Escape gas mask respirators equipped with full facepieces can also be used for escape from IDLH conditions but not from oxygen-deficient atmospheres. No air-purifying device is suitable for escape from a potentially oxygen-deficient atmosphere. A self-contained breathing apparatus (SCBA) provides air to the user for escape from oxygen-deficient environments. Escape SCBA devices are commonly used with full facepieces or hoods and, depending on the supply of air, are usually rated as 3 to 60 minute units. All SCBA devices can be used in oxygen-deficient atmospheres. Self-contained self-rescuer (SCSR) devices are certified by MSHA/NIOSH for escape from mines, but these devices may also have application in other similar environments. SCSRs are mouthpiece respirators that provide a source of oxygen-enriched air for up to 60 minutes.

When selecting escape apparatus, careful consideration must be given to potential eye irritation. This consideration is important for determining whether a gas mask or SCBA equipped with a full facepiece should be selected rather than a device equipped with a half-mask or mouthpiece. The majority of gas masks or escape gas masks can be used in situations involving gas, vapor, or particulate. For escape from particulate-contaminated environments, an air-purifying element must be selected that will provide protection against the given type of particulate. The information in Table 32.6 should be used to select the appropriate escape apparatus.

6 RESPIRATOR SELECTION AND USE

The primary objective of industrial hygiene it to prevent contaminants from escaping into the atmosphere. Obviously, hazard control should start at the process, equipment, and plant design levels where effluents can be effectively controlled at the outset. With operating processes, the problem becomes more difficult. However, in all cases, consideration should be given to the use of effective engineering controls to eliminate and/or reduce exposure to respiratory hazards. The preferred priority of controls steps is (*1*) substitution of a less toxic substance, (*2*) encapsulation or isolation of the process, (*3*) use of local exhaust ventilation or general ventilation in conjunction with filters and scrubbers to control the effluents, and (*4*) use of personal protective equipment (PPE). Respiratory protective equipment should be used only when all higher priority control steps are not technically or financially feasible. Since there are many types of RPE, it is important that they be selected with utmost care to ensure that workers receive proper protection. To select a respirator properly, the Industrial Hygienist must base her/his recommendation on an evaluation of all respiratory hazards in the workplace, and an identification of all relevant workplace and user factors for a particular workplace. This requires thorough knowledge of the process, related equipment, raw materials, end products, and by-products that can possibly create an exposure hazard. When assessing the exposure potential for a respiratory hazard(s), consideration should be given to potential emergency conditions that could produce unexpectedly high contaminant concentrations or even IDLH levels. This helps assure that proper emergency RPE and procedures are available and understood by affected personnel.

Table 32.6. Selection Options for Escape Respirators

Escape Conditions	Type of Respirator
Short distance to exit, no obstacles (no oxygen deficiency)	Any escape gas mask[a] or gas mask[b] Any escape self-contained breathing apparatus having a suitable service life[c] Any acceptable device for entry into emergency situations
Long distance to exit or obstacles along the way (no oxygen deficiency)	Any gas mask[b] Any escape self-contained breathing apparatus having a suitable service life[c] Any self-contained, self-rescuer having a suitable service life
Potential oxygen deficiency	Any escape self-contained breathing apparatus having a suitable service life[c] Any self-contained self-rescuer having a suitable service life

[a]An escape gas mask is a respirator designed for use during escape only from immediately dangerous to life or health (IDLH) or non-IDLH atmospheres. It may consist of a half mask facepiece or mouthpiece, appropriate air-purifying element for the contaminant, and associated connections, maximum use concentrations for these types of respirators are designated by the manufacturer.
[b]A gas mask consists of a full facepiece and either chin-style or front- or back-mounted canisters with associated connections. Maximum use concentrations for canister air-purifying elements are listed in Table 34.5
[c]Escape self-contained breathing apparatus can have rated service lives of 3 to 60 minutes. All acceptable devices for entry into emergency situations can also be used.

Examples of the types and sources of information that need to be gathered are:

Material safety data sheets.

Specific regulations or guidelines for the use of this material.

Determination of contaminants physical, chemical, and toxicological properties of the contaminant(s).

Airborne concentration.

IDLH concentration.

PEL.

Does the material have adequate warning properties below it's exposure limit concentration?

Does the material irritate the skin, nose, or eyes?

Does the material readily pass through the skin?

If the air contaminant is a gas or vapor, can the material be absorbed by a gas or vapor canister?

Any service life information available (for cartridges and canisters)?

Does the material have a carcinogen classification?

Does, or is there potential for, an oxygen deficient atmosphere to exist?

APF required (contaminant concentration/PEL)?

The information obtained on general use conditions for respirators should include a description of the actual job task, including the duration and frequency, location, physical demands, and industrial processes, as well as the comfort of the respirators. Some general use conditions may preclude the use of specific types of respirators in certain circumstances because the individual must be medically and psychologically suitable to wear a given respirator for a given task, particularly if the respirator is a self-contained breathing apparatus (SCBA).

Information obtained on the service life of the cartridge/canister under conditions of intended use should be evaluated regardless of the odor warning properties of the chemicals. These evaluations should be based on the gas and or vapor contaminants and the temperature and relative humidity extremes (high and low) existing in the workplace. NIOSH recommends when conducting a laboratory-based service life test that the challenge concentrations of the gases and vapors should be at least 10 times the maximum use concentration of the respirator. The service life obtained from these tests should be used to determine how long a cartridge/canister could provide protection under actual use conditions. This information can be used to set up cartridge replacement schedules and should be used in conjunction with sensory warning properties. Workers should be trained to exit the contaminated area whenever they detect the odor of the contaminants. Obtaining complete information on all criteria needed to select RPE may be difficult. When conflicting or inadequate data are found, experts should be consulted before RPE selection decisions are made.

The adequacy of the respirator selected is very dependent on the validity of the exposure limit used. As a result, the assessment of the atmosphere for oxygen content and concentration levels of particulate and/or gaseous contaminants is critical. Furthermore, NIOSH, OSHA, and ANSI Z88.2 recommend that conditions in the work area and worker's exposure to respiratory hazards should be monitored to assure that they have not significantly changed. This necessitates periodic monitoring of the air contaminant concentration to which the respirator wearer is exposed. Many things such as changes in the operation or process, air movement, temperature, or humidity, affect the concentration of a substance in the work area atmosphere. Therefore, the air contaminant should be sampled. Preferably, sampling should be in the respirator wearer's breathing zone. Both the time-weighted average and peak concentrations of the contaminant should be determined. However, the air sampling campaign must be correctly planned and executed and the data properly analyzed for useful results to be obtained. The information in this section presents basic, but not all-inclusive, information useful for properly selecting RPE. More complete and detailed information is available from the following sources:

OSHA General Respirator Regulations, 29 CFR1910.134, 29 CFR 1926 and respirator regulations specific to a particular contaminant;

American National Standard Practices for Respirator Protection, ANSI Z88.2-1992 Available from the American National Standards Institute, Inc., 1430 Broadway, New York, NY, 10018.

Respirator Decision Logic, National Institute for Occupational Safety and Health, DHHS (NIOSH) Publication No. 87-108

Respiratory Protection: A Manual and Guideline, 3rd ed. Available from the American Industrial Hygiene Association, 2700 Prosperity Avenue, Suite 250, Fairfax, VA 22031.

Breathing Apparatus for the Fire Service Available from National Fire Protection Association, 470 Atlantic Avenue, Boston, MA 02110.

Threshold Limit Values, Available from Secretary-Treasurer of the American Conference of Governmental Industrial Hygienists, 1330 Kemper Meadow Drive, Cincinnati, OH 45240-1634.

6.1 Restrictions and Requirements for All Respirator Use

To ensure that the RPE selected will provide adequate protection under the conditions of intended use the following requirements and restrictions should be considered:

1. An OSHA compliant respiratory protection program should be instituted. OSHA requires that a respiratory protection program must contain the following basic elements. Each of these program elements are reviewed and background information is given and discussed in later portions of the section. It is important to note that the new OSHA respirator standard specifies that the respirator program administrator be qualified by appropriate training or experience that is commensurate with the complexity of the program he/she has to administer or oversee and conduct the required evaluations of program effectiveness.
   a. Procedures for selecting respirators for use in the workplace.
   b. Medical evaluation of employees required to use respirators.
   c. Fit testing procedures for tight-fitting respirators.
   d. Procedures for proper use of respirators in routine and reasonably foreseeable emergency situations.
   e. Procedures and schedules for cleaning, disinfecting, storing, inspection, repairing,discarding and otherwise maintaining respirators.
   f. Procedures to ensure adequate air quality, quantity and flow of breathing air for atmosphere-supplying respirators.
   g. Training of employees in the respiratory hazards to which they are potentially exposed during routine and emergency situations.
   h. Training the employees in the proper use of respirators, including putting on and removing them, any limitation on their use and their maintenance.
   i. Procedures for regularly evaluating the effectiveness of the program.

2. NIOSH recommends that whenever possible, quantitative evaluation of the protection factor in the workplace should be performed to confirm the actual degree of protection provided by the respirator to each worker.
3. Qualitative or quantitative fit testing of tight fitting facepieces used with air-purifying and atmosphere supplying devices should be done as part of the selection process. When QNFT is used the pass/fail screening level should be chosen with

caution and with the recognition of the uncertainty of its effectiveness since no studies have demonstrated what QNFF values correlate with actual protection. OSHA requires as a minimum a QNFF of 100 for tight fitting half facepieces and 500 for tight fitting full facepieces.

4. Negative pressure RPE should not be used when facial scars or deformities interfere with the face seal.
5. Negative pressure RPE should not be used for entry into oxygen deficient atmospheres.
6. No RPE (including positive pressure) should be used when facial hair interferes with the face seal.
7. The RPE should be properly maintained, correctly used, and conscientiously worn.
8. The usage limitations of the air-purifying element particularly gas and vapor cartridges, should not be exceeded.
9. OSHA requires the use of RPE certified by NIOSH.
10. Workers should be instructed to leave the contaminated area immediately upon suspicion of respirator failure and then to determine the problem.
11. Workers are not exposed to a single unvarying concentration of a hazardous substance, rather individual exposures may vary throughout a work shift and between days. The highest anticipated concentration should therefore be used to compute the required protection factor for each respirator wearer.
12. Respirator wearers should be aware of the variability in human responses to the warning properties of hazardous substances. When warning properties must be relied on as part of a respiratory protection program, each prospective wearer should be screened for the ability to detect the warning properties of the hazardous substances at exposure concentrations that are less than the exposure limit for each given substance.

6.2 Respiratory Hazards

In occupational settings, toxic materials generally enter the body in three ways: (*1*) through the gastrointestinal tract, (*2*) through the skin, and (*3*) through the respiratory tract. Of these the respiratory tract presents the quickest and most direct entry of contaminants into the body and the circulatory system. The two basic classes of respiratory hazard are oxygen deficiency and chemical airborne contaminants that can be in the physical form of a solid, liquid, gas, vapor or a combination of these.

6.2.1 Oxygen Deficiency

The normal concentration percent of oxygen in air at sea level is 20.9% by volume, which is equivalent to a partial pressure of approximately 160 mm Hg. The NIOSH and OSHA define an oxygen-deficient atmosphere as any atmosphere containing oxygen at a concentration below 19.5% at sea level or approximately 148 mm Hg. At oxygen concentrations below 16% at sea level, decreased mental effectiveness, visual acuity, and muscular coordination occur. At oxygen concentrations below 10%, loss of consciousness may occur, and below 6% oxygen, death will result. Individuals exposed to low concentrations of oxygen note often only mild subjective changes and collapse can occur without warning (22). The minimum requirement of 19.5% oxygen at sea level provides an adequate amount of oxygen for most work assignments and includes a safety factor. The safety factor is needed because oxygen-deficient atmospheres offer little warning of the danger, and the continuous measurement of an oxygen-deficient atmosphere is difficult. Since oxygen-deficient atmospheres are IDLH only the most reliable respirators are recommended. The most reliable respirators are SCBA units or SAR units equipped with auxiliary SCBA units. Because a high protection factor is not necessary to ensure an adequate supply of oxygen even in an atmosphere containing no oxygen, any NIOSH certified SCBA is adequate. Air-purifying RPE or SAR without auxiliary SCBA is not for use in oxygen deficient atmospheres. In assessing exposure conditions, it is important to remember that oxygen deficiency can occur in confined spaces by displacement of air by other gases and vapors; or by means of oxidation processes such as fire, rusting, aerobic bacteria, etc., where oxygen is consumed.

6.2.2 Chemical Airborne Contaminants

Chemical airborne contaminants include particulate matter in the form of discrete particles of solids or liquids, gaseous material in the form of a gas or vapor, or a combination of both gaseous and particulate matter.

6.2.2.1 Particulate Hazards. Particulate contaminants may be classified according to their physical and chemical characteristics and their biological effect on the body. One of the key attributes of an aerosol's ability to enter the respiratory system is it size. More specifically its aerodynamic particle size. Particles below 10 micrometer aerodynamic diameter are generally considered to be respirable. That means that they are small enough to pass through the nose and the pharyngeal area of the respiratory tract and enter the trachea, bronchi and upper ciliated regions of the lung. In the healthy lung, particles from 5 to 10 micrometers aerodynamic diameter are generally removed from the respiratory system by the constant cleansing action of the ciliated epithelium in the upper respiratory tract. Particles below 5 micrometers aerodynamic diameter are more apt to reach the nonciliated regions of the deep lung and alveolar ducts and sacs. However, with excessive "dust" exposures, diseased systems, smoking etc. the efficiency of the cleansing action can be markedly reduced.

Occupationally related aerosols are generally classified as dust, mist, fume, smoke, airborne living organisms and allergens, or various combinations of these aerosols. The fate of particles that enter the respiratory tract depends on their solubility, aerodynamic particle size, chemical characteristics, and metabolism in the body. Hence, particulate contaminants are also classified according to their biological effects as:

1. Inert aerosol that produce minor irritation or discomfort but in sufficient quantity can overwhelm the protective mechanisms of the respiratory tract.
2. Allergy producers that cause severe reactions with some sensitized individuals.

3. Chemical irritants that can damage mucous membranes or lung tissue by chemical reaction.
4. Fibrosis producers that cause development or scar tissue in the lung such as silica or coal dust.
5. Carcinogens or suspect carcinogens such as asbestos, chrome VI or radioactive particulate matter.
6. Systemic poisons that have sites of action in additional organ systems of the body such as lead, cadmium, or arsenic.
7. Fever producers such as the fumes of zinc and copper.

6.2.2.2 Gaseous Contaminants. Gaseous contaminants occurring as true gases such as sulfur dioxide, carbon monoxide, etc., or vapors from organic liquids such as benzene or methyl ethyl ketone can likewise be classified according to their chemical characteristics and biological effect on the body. Chemically, gaseous contaminants may be classified as follows:

1. Inert gas which do not metabolize in the body but can produce an oxygen deficiency by displacement of air, e.g., helium, argon, neon, etc.
2. Acid gases which are acids or produce acids by reaction with water, e.g., sulfur dioxide, chlorine, hydrogen chloride etc.
3. Alkaline gases which are alkales or produce alkales by reaction with water, e.g., ammonia, phosphine etc.
4. Organic compounds which exist as true gases or vapors from organic liquids.
5. Organometallic compounds that are organic gases or vapors that have metals attached to organic groups within the molecule, e.g., tetraethyl lead and organic phosphates.

Gaseous contaminants may be classified according to biological effects as follows:

1. Asphyxiants that interfere with the uptake, transport or utilization of oxygen in the body. Asphyxiants can be simple asphyxiants that create an oxygen deficiency by air displacement, e.g., nitrogen, methane, hydrogen, etc. or chemical asphyxiants that interfere with the uptake and transport of oxygen in the blood system (e.g., carbon monoxide) or that interferes with "internal respiration" or oxidation of the tissue cells (e.g., hydrogen cyanide).
2. Chemical irritants that can irritate the respiratory system and cause the development of pulmonary edema or fluid in the lung, e.g., acid and alkaline gases.
3. Anesthetics that can cause loss of feeling, unconsciousness, and death, e.g., chloroform, ether, etc.
4. Systemic poisons that can damage critical organs and systems of the body, e.g., metallic mercury vapor, hydrogen sulfide, etc.

6.3 Selection

Respiratory protective devices vary in design, application, and protective capability. The user must, therefore, assess the inhalation hazard and understand the specific use and limitations of available equipment to assure proper selection. After collecting the information specified in earlier portions of this section it can be applied to a decision logic to help "walk through" the respirator selection process. Figure 32.4 contains an example of a decision logic adopted, with revisions and additions from NIOSH (14).

The purpose of establishing an IDLH exposure level in RPE selection is to ensure that the worker can escape from a given contaminated environment in the event of failure of the RPE. The IDLH is considered a maximum level above which only a highly reliable breathing apparatus providing maximum worker protection is permitted. Any appropriate approved respirator may be used to its maximum use concentration up to the IDLH concentration. Some specific IDLH guidelines are provided in the literature such as the American Industrial Hygiene Association (AIHA) Hygienic Guides and the NIOSH Pocket Guide for Hazardous Chemical Substances (113).

Eye protection in the form of respirators with full facepieces, helmets, or hoods is required for routine exposures to airborne contaminants that cause any irritation to the mucous membranes of the conjunctivae or the cornea or cause any reflex tearing. Eye protection is required for contaminants that cause minor subjective effects as well as for those that cause any damage, including disintegration and sloughing of conjunctival or corneal epithelium, edema, or ulceration. For escape, some eye irritation is permissible if the severity of irritation does not inhibit the escape and if no irreversible scarring or ulceration of the eyes or conjunctivae is likely.

When selecting escape apparatus, careful consideration must be given to potential eye irritation. This consideration is important for determining whether a gas mask or SCBA equipped with a full facepiece should be selected rather than a device equipped with a half-mask or mouthpiece. Escape gas masks can be used in situations involving gas, vapor, or particulate contaminants if the atmosphere is not oxygen-deficient. The information in Table 32.6 should be considered when selecting appropriate escape apparatus.

6.4 Training

The proper respirator is of no value without training for the user. The training experience should provide the user, in addition to the training elements listed subsequently, an opportunity to handle the RPE, to receive donning instructions and to undergo size selection and fit testing. Competent persons should instruct both supervisors and workers. Minimum training shall include the following components:

1. Instruction in the nature of the hazard, whether acute, chronic, or both, and an honest appraisal of what may happen if the proper device is not used.
2. Explanation of why respiratory protection instead of engineering controls is being used. If engineering controls are being used, explain why they are not adequate, and what efforts are being made to reduce or eliminate the need for respirators. Explanation of why more positive control is not immediately feasible.
3. Explanation of the respirator program and OSHA regulations.
4. A discussion of why this is the proper type of unit for the particular purpose.
5. A discussion of the device's capabilities and limitations.

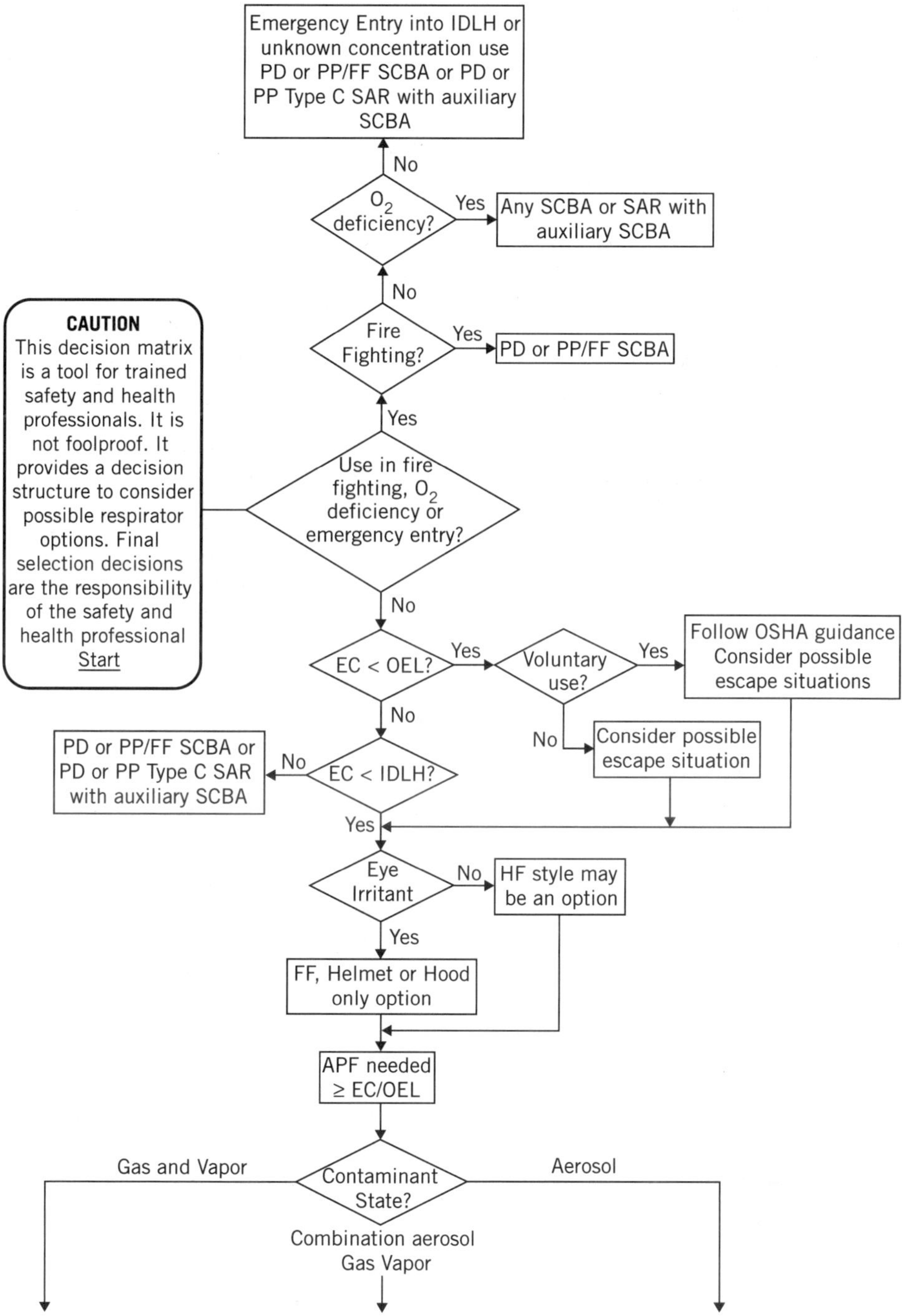

Figure 32.4. Respirator Decision Logic.

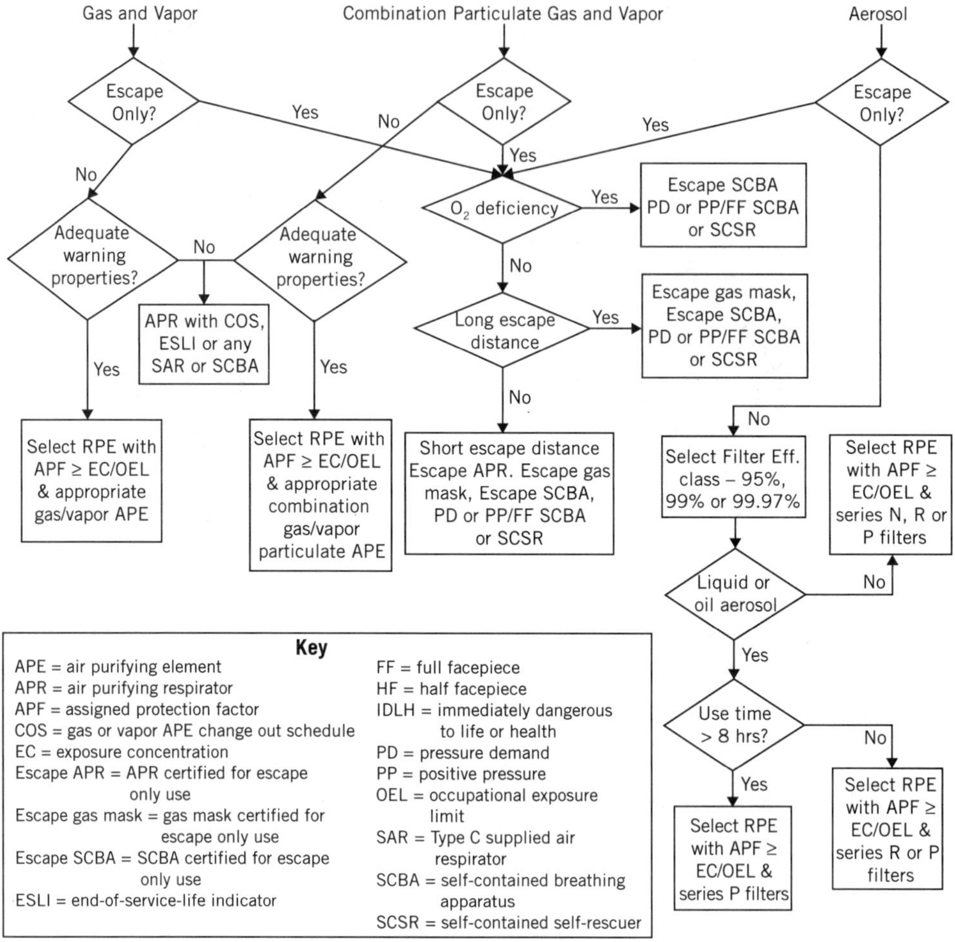

Figure 32.4. Continued.

6. Explanation of how to inspect, don, check the fit, and wear the respirator selected by the user.
7. Instruction and training in actual use (especially a RPE for emergency use) and close and frequent supervision to assure that it continues to be properly used.
8. Explanation and if needed field training of how to recognize and cope with emergency situations.

6.5 Medical Surveillance

Workers should never be assigned to any operations requiring respiratory protection until a physician has determined that they are capable physically and psychologically to perform the work using the respiratory protective equipment. The OSHA respirator standard requires that no one should be assigned to tasks requiring use of RPE unless he/she has been found physically able to do the work while wearing the respirator. Readers are referred to the OSHA standard 1910.134 (e) and the OSHA respirator medical evaluation questionnaire in Appendix C of the standard.

Some users experience breathing distress or claustrophobia while wearing a respirator (114). Recognition of such users should be made prior to entering life-threatening environments. Morgan and Raven, using an anxiety profile, are able to predict those who may experience problems wearing respirators in high work rate environments (115).

6.6 Evaluation of Respirator Program Effectiveness

OSHA 1910.134 states that respirator program effectiveness shall be inspected and evaluated regularly. Periodic monitoring is necessary to ensure that workers are adequately protected. The program should be evaluated at least annually, and the written operating procedures should be modified to reflect the evaluation results if necessary. Frequent inspection of respirator use determines whether the correct respirators are being used and worn properly. Examination of respirators in use and in storage indicates how well they are maintained. Wearers should be consulted periodically about their acceptance of respirators, including the discomfort, resistance to breathing, fatigue, interference with vision and communication, restriction of movement, and interference with job performance, and their confidence in the respirator's effectiveness. The results of periodic inspections of respirator use, consultations with wearers, measurements of hazard levels in work areas, and medical surveillance of wearers should be reviewed, studied, and analyzed to determine the effectiveness of the respirator program. Evidence of excessive exposure to hazards should be followed up to determine why inadequate protection was provided, and action should be taken to remedy the problem. The results of the program evaluation should be presented in a written report that should list plans to correct faults and the target dates for their implementation.

6.7 Maintenance and Care of RPE

Proper inspection, maintenance, and repair of RPE are necessary to ensure that it stays in good working order. The maintenance program differs on the basis on plant size, the complexity of the RPE used etc. However, the goal is to maintain the equipment in a condition that provides the same effectiveness it had when manufactured. To ensure that the respirator remains serviceable a maintenance program must be in place the first day of respirator use. Respirators must be kept clean, sanitary, and in good working order. The RPE should be cleaned and disinfected following the manufacturers recommended procedures if those procedures are of equivalent effectiveness as the cleaning and disinfecting procedures detailed in Appendix B-2 of 1910.134 (5). Respirators used by a single wearer should be cleaned and disinfected as often as necessary. Respirators used by different individuals should be cleaned and disinfected between each use. The RPE should be stored in accordance with any applicable manufacturer instructions. This should be to protect them from damage, contamination, dust, sunlight, extreme temperatures, excessive moisture and damaging chemicals. They should also be packaged or stored to prevent deformation of the facepiece and exhalation valve. The RPE used in routine situations should be inspected before each use and during cleaning. RPE maintained for use in emergency situation should be inspected monthly according to the manufacturer instructions and checked for proper function before and after each use. Escape only RPE should be inspected before being carried into the workplace for possible use. The inspection results should be dated and recorded and maintained as a historical file. Never should RPE that fails inspection or is found defective be used. Only personnel with adequate training to ensure the equipment is functionally sound after the work is accomplished should do repair or replacement of other than disposable parts. Only parts supplied by the manufacturer for the product being repaired should be used otherwise the NIOSH certification will be voided.

6.8 Breathing Air Quality

Air compressors or compressed air bottles can supply breathing air for SARs and SCBAs. OSHA and NIOSH require that compressed breathing air meet the Type 1 Grade D standards of the Compressed Gas Association (116). Grade D standards are met if the following contaminant levels are not exceeded: 10 PPM carbon monoxide, 1000 PPM carbon dioxide, and 5 mg/m^3 condensed hydrocarbons. Generally there are two sources of contamination for compressed breathing air, the air intake and the air compression system. The air in the compressor system will contain the contaminants present at the air intake. Care should be taken to see that the air intake is not located near any point source emissions such as diesel truck exhaust in shipping areas. In urban areas with serious air pollution problems it may be difficult to provide an air intake location free from contamination. The most common air compressor used for breathing air is the oil-lubricated compressor, which add the oil mist and hydrocarbon vapors to the compressed air. If for some reason the compressor overheats, it is possible for partial combustion of the oil to occur and generate carbon monoxide. There are filters and sorbents available to remove the oil mist and hydrocarbon vapors from the compressed air, but carbon monoxide removal from the compressor air is more complex. Carbon monoxide catalytic removal units contain Hopcalite to convert CO to CO_2. Because moisture ruins Hopcalite, drying units protect the Hopcalite. Dual drying towers are installed on some Hopcalite units so that while one is removing CO from the air supply, the other is being dried. When dry it will automatically cycle to replace the other unit. Carbon monoxide catalysis units are not mandatory on oil-lubricated systems, but periodic monitoring for CO is required. In addition, the compressor must be equipped with a high-temperature alarm to notify personnel that the compressor has overheated. Although not required, continuous-reading CO monitors should be installed to monitor the CO concentration in the breathing air. A solution to the potential problem of carbon monoxide is to use a non-oil-lubricated compressor. The diaphragm pump, oil-less compressors with graphite or Teflon piston rings, and water-lubricated compressors are available. Pressure demand or demand SAR does not require a continuous flow of air and can use stored compressed air instead of an on-site air compressor. Breathing air certified to meet Type 1, Grade D standards, can be purchased in cylinders holding more than 6000 liters. For users needing small quantities of breathing air, purchasing bottled air means no need for a breathing air compressor or to install breathing air monitors and filters.

6.9 Special Problems with Selecting and Using Respirators

6.9.1 *Respirators and Hair*

Respiratory protective devices should never be worn when a satisfactory face seal cannot be obtained or maintained. Facial hair is a very common cause of an unsatisfactory seal. It is imperative that clean, smooth skin be in contact with the sealing edge of the facepiece because even a mild growth of facial hair may interfere with this seal. Stobbe et al. in a review of facial hair and respirator fit studies, found that the face seal of negative pressure respirators was adversely affected by facial hair (117). Restrictions on facial hair is an emotional subject to some, and company policies prohibiting facial hair, not carefully planned and carried out, have been successfully challenged (118). A policy statement prohibiting facial hair that interferes with wearing a respirator should be concise and include only respirator users. It must be part of a respirator program, but not the only part enforced.

6.9.2 *Cold Temperatures*

In cold temperatures problems of facepiece flexibility, visibility, and frozen valves must be considered. Those who work in cold environments should ask the manufacturer about respirators for cold weather. The facepiece will fog in cold weather when exposed to warm exhaled breath. A nose cup, which must be used at temperatures below 32°F, is available for all full facepiece respirators to keep the warm air away from the visor (119). Valves and regulators should be carefully checked before the respirator is used to ensure that they are functioning satisfactorily.

6.9.3 *Corrective Lenses*

Temple bars on eyeglasses interfere with the seal of full facepiece respirators and are not acceptable. Eyeglass kits are available from all manufacturers for use with full facepieces. If half and quarter masks interfere with proper seating of eyeglasses, the respirator fit is not satisfactory, and either another respirator or another style of eyeglass must be selected.

6.9.4 *Communications*

Communication through a facepiece is difficult, and trying to speak loudly may affect the face seal. There are several devices available to assist the respirator wearer. Some respirators have built-in speaking diaphragms. Although they are protected, care must be taken that thin wire or hot sparks do not pierce the thin diaphragm. Wireless units are available that can be placed inside full facepiece respirators for radio communication. Throat and ear microphones are also available. None of these units require penetration of the facepiece that can affect the respirator approval. Use of any electrical device must also consider potential explosive problems if flammable vapors are present.

ACKNOWLEDGMENTS

This revision makes substantial use of the excellent information base provided by Darrel Douglas in the 4th Edition of *Patty's Industrial Hygiene and Toxicology*. The major areas of revision deal with updating information on fit testing and workplace performance studies, access to information on the World Wide Web, NIOSH promulgation of new respirator filter performance criteria in 42 *CFR* Part 42, and OSHA revision of its respirator standard 29 *CFR* Part 1910.134.

BIBLIOGRAPHY

1. Procedure for Establishing a Test of Permissible Gas Masks, Schedule 14 U.S. Department of the Interior, (Approved May 22, 1919), p. 7.
2. "Respiratory Protective Devices: Tests for Permissibility," *Code of Federal Regulations*, Title 30, Part 11, 1972.
3. "Occupational Safety and Health Standards, Toxic and Hazardous Substances, Air Contaminant" *Code of Federal Regulations*, Title 29 Part 1910, 1971.
4. "Safety and Health Standards for Construction," *Code of Federal Regulations*, Title 29 Part 1926, 1971.
5. "Respiratory Protection," *Code of Federal Regulations*, Title 29 Part 1910.134, 1998
6. "Respiratory Protection," *Code of Federal Regulations*, Title 29 Part 1926.103, 1998.
7. "Respiratory Protective Devices: Tests for Permissibility." *Code of Federal Regulations*, Title 30 Part 11, 1989.
8. "Approval of Respiratory Protective Devices" *Code of Federal Regulations*, Title 42 Part 84, 1996.
9. *Practices for Respiratory Protection (ANSI Z88.2-1992)* American National Standards Institute, New York, 1992.
10. *Procedure for Testing Supplied Air Respirators for Permissibility*, Schedule 19A U. S. Department of the Interior, (Approved August 9, 1937), p. 15.
11. Title 30, Subchapter B, Part 14, Filter—Type Dust, Fume and Mist Respirators—Requirements for Investigation, Testing and Certification, Schedule 21B U. S. Department of the Interior, (Approved January 19, 1965),1965). p. 7.
12. W. R. Myers, Evaluation of In-Facepiece Sampling Bias on Respirator, Ph.D. Dissertation, West Virginia University, 1986.
13. "American Industrial Hygiene Association Respiratory Protection Committee: Respirator Performance Terminology, Letter to the Editor," *Amer. Ind. Hyg. Assoc. J.* **46**, B22–B24 (1985).
14. W. R. Myers, N. J. Bollinger, T. K. Hodous, N. A. Leidel, S. H. Rabinovitz, and L. D. Reed, *Respirator Decision Logic*, National Institute for Occupational Safety and Health, DHHS(NIOSH) Publication No. 87-108, 1987.
15. "Cadmium," *Code of Federal Regulations*, Title 29 Part 1910.1027, 1998.
16. E. C. Hyatt, "Respirator Protection Factors," Los Alamos Scientific Laboratory Report No. LA-6084-MS, Jan. 1976.
17. Standards Completion Program, *Respirator Decision Logic*, National Institute for Occupational Safety and Health, Occupational Safety and Health Administration, 1975.

18. J. T. McConville, and E. Churchill, *Human Variability and Respirator Sizing*, National Institute for Occupational Safety and Health, March, 1976.
19. A. L. Hack and J. T. McConville, *Amr. Ind. Hyg. J.* **39**(12), 970–975 (1978).
20. F. E. Adley and D. E. Wisehart, *Amr. Ind. Hyg. Assoc. J.* **23**(4) 251–256 (1962).
21. E. C. Hyatt, *Am. Ind. Hyg. Assoc. J.* **24**, 295 (1963).
22. J. Pritchard, *Guide to Industrial Respiratory Protection*, National Institute for Occupational Safety and Health, DHHS (NIOSH) Publication NO, 76-189, 1976.
23. R. K. Oestenstad, J. L. Perkins and V. E. Rose, *Amr. Ind. Hyg. Assoc. J.* **51**(5), 280–284 (1990).
24. R. K. Oestenstad, H. K. Dillon and J. L. Perkins, *Amer. Ind. Hyg. Assoc. J.* **51**(5), 285–290 (1990).
25. R. K. Oestenstad and J. L. Perkins, *Amr. Ind. Hyg. Assoc. J.* **53**(10), 639–644 (1992).
26. R. K. Oestenstad and J. B. Graffeo, *J. Int. Soc. Resp. Protec.* **11**(3), 6–14 (1990).
27. W. R. Myers, M. Jaraiedi and L. Hendricks, *J. Appl. Ind. Hyg.* **10**, 898–905 (1995).
28. H. G. Guyton and F. T. Lense, *AMA ARC'H Ind.,Health.* **14**, 236 (1956).
29. W. A. Burgess, L. Silverman, and F. Stein, *Am. Ind. Hyg. Assoc. J.* **22** (6), 422 (1961).
30. R. F. Hounam, *A Method for Evaluating the Protection Afforded When Wearing a Respirator*, AERE-R4125, Atomic Energy Research Establishment, Harwell, U.K., 1962.
31. H. Flyger, *Health Physics.* **11**, 223 (1965).
32. J. M. White and R. J. Beal, *Am. Ind. Hyg. Assoc. J.* **27**(3), 239 (1966).
33. W. A. Burgess and J. Shapiro, *Health Physics.* **15**, 115 (1968).
34. F. E. Adley and D. E. Wisehart, *Am. Ind. Hyg. Assoc. J.* **23**, 251 (1962).
35. F. E. Adley and R. J. Uhle, *Am. Ind. Hyg. Assoc. J.* **30**(4), 355 (1969).
36. P. L. Lowry, L. D. Wheat and J. M. Bustos, *Am. Ind. Hyg. Assoc. J.* **40**, 291–299 (1979).
37. R. A. Bentley, G. J. Bostock, D. J. Longson and M. W. Roff, *J. Int. Soc. Resp. Prot.* **2**(4), 313–337 (1984).
38. E. C. Hyatt and C. P. Richards, *A Study of Facepiece Leakage of Self-Contained Breathing Apparatus by DOP Man Tests Progress Report, July 1, 1971 through February 29, 1972*, Los Alamos Scientific Laboratory Report No. LA-4927-PR, April 1972.
39. A. L. Hack, A. A. Trujillo, O. D. Bradley, and K. Carter, NUREG/CR-1586, LA-8432-MS, *Evaluation and Performance of Escape Type Self-Contained Breathing Apparatus*, Los Alamos Scientific Laboratory, July 1980.
40. A. L. Hack, A. A. Trujillo, O. D. Bradley, and K. Carter, NUREG/CR-2652, LA-9266-MS, *Evaluation and Performance of Closed-Circuit Breathing Apparatus*, Los Alamos Scientific Laboratory, April 1982.
41. A. L. Hack, O. D. Bradley and A. Trujillo, *Am. Ind. Hyg. Assoc. J.* **41**(5), 376–380 (1980).
42. A. L. Hack, A. Trujillo, O. D. Bradley and K. Carter, NUREG/CR-1235, LA-8188-MS, *Evaluation and Performance of Open-Circuit Breathing Apparatus*, Los Alamos Scientific Laboratory, January 1980.
43. D. D. Douglas, P. R. Hesch, and P. L. Lowry, LA-NUREG-6612-MS, *Supplied-Air Hood Report*, Los Alamos Scientific Laboratory, Dec. 1976.
44. C. D. Crutchfield, R. M. Murphy, and M. D. Van Ert, *Am. Ind. Hyg. Assoc. J.* **52**(4), 172–176 (1991).
45. W. R. Myers, J. Allender, R. Plummer, and T. Stobbe, *Am. Ind. Hyg. Assoc. J.* **47**(2), 106–114 (1986).

46. W. R. Myers, J. Allender, W. Iskander and C. Stanley, *Ann. Occup. Hyg.* **32**(3), 345–359 (1988).
47. W. R. Myers and J. Allender, *Annals of Occupational Hygiene* **32**(3), 361–372 (1988).
48. P. M. Holton, D. L. Tackeft and K. Wilieke, *Am. Ind. Hyg. Assoc. J.* **48**(10), 848–854 (1987).
49. P. M. Holton and K. Wilieke, *Am. Ind. Hyg. Assoc. J.* **148**(10), 855–860 (1987).
50. D. L. Campbell and W. R. Myers, *J. Intern Soc Respiratory Protection* **7**(2), 1–8 (1989).
51. W. R. Myers and R. W. Hornung, *Ann. Occup. Hyg.* **37**(2), 151–166 (1993).
52. J. J. Hinton, "Reliability of Quantitative Fit Protection Factors in Assessing Face-to-Facepiece Seals," University of Texas, Health Science Center, Houston, TX 1980, Unpublished thesis.
53. W. R. Myers and M. J. Peach III, *Ann. Occup. Hyg.* **27**(3), 251 (1983).
54. W. R. Myers, M. J. Peach III, and J. Allender, *Am. Ind. Hyg. Assoc. J.*. **45**(4), 236–241 (1984).
55. W. R. Myers, M. J. Peach III, K. Cutright, and W. Iskander, *Am. Ind. Hyg. Assoc. J.* **45**(10), 681–684 (1984).
56. W. R. Myers, M. J. Peach III, K. Cutright, and W. Iskander, *J. Intern. Soc. Respir. Prot.* **4**(1), 62–89 (1986).
57. S. W. Lenhart, D. L. Campbell, *Ann. Occup. Hyg.* **28**(2), 173–82 (1984).
58. C. C. Coffey, D. L. Campbell, W. R. Myers, and Z. Zhuang, *Am. Ind. Hyg. Assoc. J.* **59**(12), 862–870 (1998).
59. A. Gaboury, D. H. Burd, and R. S. Friar, *App. Occ. Environ. Hyg.* **8**(1), 19–25 (1993).
60. C. C. Chen and K. Willeke. *Am. Ind. Hyg. Assoc. J.* **53**(9), 533–539 (1992).
61. U. Xu, K. Willeke, A. Juozaitis, T. Myojo, G. Talaska, and R. Shuka, *Am. Ind. Hyg. Assoc. J.* **55**, 309–314 (1994).
62. D. L. Campbell, *J. Int. Soc. Respir. Prot.* **2**, 198 (1984).
63. C. D. Crutchfield and M. D. Van Ert, *J. of the Intern. Soc. Respir. Prot.* **11**(2), 5–18 (1993).
64. C. D. Crutchfield, R. M. Murphy, and M. D. Van Ert, *Am. Ind. Hyg. Assoc. J.* **54**(1), 10–14 (1993).
65. C. D. Crutchfield, R. W. Murphy, and M. D. Van Ert, *J. Int. Soc. Resp. Protect.* **11**(2), 5–18 (1993).
66. Y. Liau, A. Bhaftacharya, H. Ayer, and C. Miller, *Am. Ind. Hyg. Assoc. J.* **43**(12), 897–899 (1982).
67. W. C. Hinds and P. Bellin, *Appl. Ind. Hyg.* **13**(3), 158–164 (1988).
68. W. R. Myers, H. Kim and N. Kadrichu, *J. Inter. Soc. Resp. Protect.* **1**(9), 6–21 (1991).
69. T. Tuomi, *Am. Ind. Hyg. Assoc. J.* **46**, 308–312 (1985).
70. W. C. Hinds and P. Bellin, *Am. Ind. Hyg. Assoc. J.* **53**, 711–722 (1993).
71. P. Hewett, B. G. Pallay and J. F. Gamble, *Am. Ind. Assoc. J.* **54**(4), 142–149 (1993).
72. S. W. Dixon and T. J. Nelson, *J. Int. Soc. Respir. Prot.* **2**, 347–361 (1984).
73. G. S. Fergin, *Am. Ind. Hyg. Assoc. J.* **45**, 53–537 (1984).
74. L. D. Reed, S. W. Lenhart, R. L. Stephenson, and J. R. Allender, *Appl. Ind. Hyg.* **2**, 53–56 (1987).
75. K. S. Galvin, S. Selvin and R. C. Spear, *Am. Ind. Hyg. Assoc. J.* **1**(51), 625–631 (1990).
76. G. Wallis, R. Menke and C. Chelton, *Am. Ind. Hyg. Assoc. J.* **54**(10), 576–583 (1993).
77. W. R. Myers, Z. Zhuang and T. Nelson, *Am. Ind. Hyg. Assoc. J.* **57**, 166–174 (1996).
78. Z. Zhuang and W. R. Myers, *Am. Ind. Hyg. Assoc. J.* **157**, 50–57 (1996).
79. W. R. Myers and Z. Zhuang, *Am. Ind. Hyg. Assoc. J.* **59**, 789–795 (1998).

80. N. J. Nelson, *Am. Ind. Hyg. Assoc. J.* **56**, 717–724 (1995).
81. W. R. Myers and Z. Zhuang, *Am. Ind. Hyg. Assoc. J.* **59**, 796–801 (1998).
82. M. Nicas and R. C. Spear, *Am Ind. Hyg. Assoc. J.* **53**, 411–418 (1992).
83. M. Nicas and R. C. Spear, *Am. Ind. Hyg. Assoc. J.* **53**, 419–426 (1992).
84. P. C. Reist, W. C. DeSieghardt, H. E. Harris and W. A. Burgess, *Ann. of the N.Y. Acad. of Sci.* **200**, 698 (1972).
85. W. H Revoir, *Am. Ind. Hyg. Assoc. J.* **35**, 503–510 (1974).
86. B. Samimi, H. Weill and M. Ziskind, *Arch. Envirn. Health.* **29**, 61–66 (1974).
87. B. Samimi, A. Meilson, H. Weill and M. Ziskind, *Am. Ind. Hyg. Assoc. J.* **36**(2), 140 (1975).
88. D. E. Moore and T. J. Smith, *Am. Ind. Hyg. Assoc. J.* **37**(8), 453–458 (1976).
89. C. R. Toney and W. L. Barnhart, *Performance Evaluation of Respiratory Protective Equipment Used in Paint Spraying Operations*, U.S. Department of Health and Human Services, NIOSH. HEW Publication No. (NIOSH) 76-177.
90. *Practices for Respiratory Protection (ANSI Z88.2-1980).*, American National Standards Institute, New York, 1980.
91. S. S. Que Hee and P. Lawrence, *Am. Ind. Hyg. Assoc. J.* **44**(10), 746–751 (1983).
92. G. S. Hewson and Ralph Mi. *Am. Ind. Hyg. Assoc. J.* **53**, 713–720 (1992).
93. T. J. Smith, W. C. Ferrell, M. O. Varner, and R. D. Putnam, *Am. Ind. Hyg. Assoc. J.* **41**, 624–628 (1980).
94. M. Hery M, J. P. Meyer, M. Villa, G. Hubert, J. M. Gerber, G. Hecht, D. Franc, and J. Herrault, *J. Int. Soc. Resp. Protect.* **11**(4), 15–39 (1993).
95. M. Hery, M. Villa, G. Hubert and P. Martin, *Annals Occ. Hyg.* **35**, 181–187 (1991).
96. S. N. Tannahill, R. J. Willey and M. H. Jackson, *Annals Occ. Hyg.* **34**, 547–552 (1990).
97. W. Popendorf, J. A. Merchant, S. Leonard, L. F. Burmeister, and S. A. Olenchock, *App. Occup. Environ. Hyg.* **10**(7), 595–605 (1995).
98. B. J. Held, "History of Respiratory Protective Devices in the U. S., Pre World War I," Lawrence Livemore National Laboratory, Energy Research and Development Administration, contract W-7405-Eng-48.
99. R. A. da Roza, C. A. Cadena-Fix, and J. E. Kramer, "Powered Air-Purifying Respirator Study Final Report," Lawrence Livermore National Laboratory, Report Number UCRL-53757, July 1986.
100. B. J. Skaggs, J. M. Loibell, K. D. Carter and E. C. Hyatt, *Effects of Temperature and Humidity on Respirator Fit under Simulated Work Conditions*, Los Alamos National Laboratory Report Number NUREG/CR-5090 LA-1 1236, July 1988.
101. P. B. Raven, O. Bradley, D. Rohm-Young, F. L. McClure, and B. Skaggs, "Physiological Response To 'Pressure-Demand' Respirator Wear," *Am. Ind. Hyg. Assoc. J.* **43**(10), 773–781 (1982).
102. J. Stengel and R. Rodrigues, *J. Int. Soc. Respir. Protect.* **2**(4), 362–368 (1984).
103. L. G. Myhre, R. D. Holden, F. W. Baumgardner, and D. Tucker, "Physiological Limits of Firefighters," ESL-TR-97-06, Engineering and Services Laboratory, U.S. Air Force Engineering and Services Center, January 1979.
104. G. O. Dahlback and U. I. Balldin, *Am. Ind. Hyg. Assoc. J.* **45**(3), 177–181 (1984).
105. J. B. Wilson, P. B. Raven, W. P. Morgan, S. A. Zinkgraf, R. G. Garmon, and A. W. Jackson, *Am. Ind. Hyg. Assoc. J.* **50**(2), 85–94 (1989).

106. "Open-Circuit Self-Contained Breathing Apparatus for Firefighters" (NFPA 1981) National Fire Protection Association, Quincy, MA, 1987.

107. D. L. Campbell, J. R. Allender, and W. R. Myers, Journal of the International Society for Respirator Protection. 8(1), 26–32, (1990).

108. L. M. Brosseau, *Am. Ind. Hyg. Assoc. J.* **59**(3), 173–180 (1998).

109. "Letter to All Users of P-Series Particulate Respirators—NIOSH Service Time Recommendations for P-Series Particulate Respirators," NIOSH Respirator User's Notice (May 2, 1997).

110. G. O. Nelson and C. A. Harder, *Am. Ind. Hyg. Assoc. J.* **35**(7), 491–510 (1974).

111. J. E. Amoore and E. Hautala, "Odor as an Aid to Chemical Safety: Order Thresholds Compared with Threshold Limit Values and Volatilities for 214 Industrial Chemicals in Air and Water Dilutions," *J. Appl. Toxicol.* **3**(6), 272–290 (1983).

112. O. D. Bradley, A. G. Trujillo, and R. Henins, *Testing of Supplied Air Suits Accepted for use by DOE Contractors for O_2 Decrease and CO_2 Increase with Air Supply Shut Off.* Los Alamos National Laboratory, Los Alamos, NM, 1987.

113. *NIOSH Pocket Guide to Chemical Hazardous*, National Institute for Occupational Safety and Health, DHHS(NIOSH) Publication No. 97-140, June 1997.

114. W. P. Morgan, "Psychological Problems Associated with the Wearing of Industrial Respirators," *Am. Ind. Hyg. Assoc. J.* **44**(9), 671–675 (1983).

115. W. P. Morgan and P. B. Raven, "Prediction of Distress for Individuals Wearing Industrial Respirators," *Am. Ind. Hyg. Assoc. J.* **46**(7), 363–368 (1985).

116. "Commodity Specification for Air" CGA Specification G-7.1, Compressed Gas Association, Inc. Arlington, VA (1989).

117. T. J. Stobbe, R. A. da Roza, and M. A. Watkins, "Facial Hair and Respirator Fit: A Review of the Literature," *Am. Ind. Hyg. Assoc. J.* **49**(4), 199–203 (1988).

118. G. L. Holt, "Employee Facial Hair Versus Employer Respirator Policies," *J. Appl. Ind. Hyg.* **2**(5), 200–203 (1987).

119. NIOSH Emergency Bulletin on the Use of Self-Contained Breathing Apparatus in Low Temperatures, J. A. Oppold, Director Division of Safety Research NIOSH/CDC/DHHS, Jan 15, 1982.

CHAPTER THIRTY-THREE

Asbestos Management in Buildings

Michael A. Coffman, CIH, CSP and Jaswant Singh, Ph.D, CIH

1 INTRODUCTION

Since the late 1970s, asbestos, perhaps more so than any other environmental or occupational health issue, has had a profound impact on our society. The management of asbestos in buildings has been the subject of considerable litigation, legislation and regulation not only in the United States, but worldwide. In a broader sense, asbestos has provided a focal point for the national and international debate on how best to evaluate and respond to environmental health risks.

It is likely that this debate will continue for many years to come and it is not the intent of this chapter to provide further input to this debate, but rather to deal with the issue on a more practical basis. For the owner or manager of a building containing asbestos, the issue is not one of intellectual or philosophical curiosity or of regulatory agendas. For the building owner or manager, asbestos is an issue which must be dealt with on a day-to-day basis. The building owner or manager is faced not only with the realities of effectively managing a known human carcinogen, but also in dealing with a public perception (or misperception) which has been heavily influenced by two decades of media attention and complying with a myriad of federal, state and local regulations.

Under the Asbestos Hazard Emergency Response Act (AHERA), the United States Congress required that the Environmental Protection Agency (EPA) to conduct a study to determine the extent of danger to human health posed by asbestos in public and commercial buildings. In its February 1988 report to Congress (1), EPA concluded that there were approximately 750,000 public and commercial buildings in the United States that contained

Patty's Industrial Hygiene, Fifth Edition, Volume 2. Edited by Robert L. Harris.
ISBN 0-471-29754-2

asbestos. Of these, EPA estimated that 500,000 contained partially damaged asbestos and more than 300,000 buildings contained some asbestos that was significantly damaged. The EPA estimated the cost of abating asbestos at private and commercial properties to be $50 billion.

Management of asbestos in buildings requires the input of professionals in many disciplines, including industrial hygiene, toxicology, medicine, law, risk management, engineering, architecture, construction, and administration. The professional industrial hygienist, as a key member of this team, is uniquely qualified to coalesce all input into meaningful informed decisions.

In asbestos management, the industrial hygienist plays multiple roles:

- As a health professional, the industrial hygienist understands the health risks of asbestos and can relate that risk to human exposure.
- As an air monitoring specialist, the industrial hygienist has a firm grasp of the sampling and analytical techniques to ensure a clear and meaningful interpretation of sampling results.
- As an engineer, the industrial hygienist has an understanding of the practical issues involved in contamination control, work area isolation, work practices, and the use of personal protective equipment.
- As a teacher, the industrial hygienist is routinely called upon to communicate technical and health risk information to laypersons and to instruct workers in safe work practices.
- As a project manager, the industrial hygienist has experience in working with a team to achieve specific goals.

Industrial hygienists can also be effective facilitators in asbestos abatement. In its publication "Recommendations for Asbestos Abatement Projects," the American Industrial Hygiene Association (2) listed 12 critical points requiring industrial hygiene input (Table 33.1).

1.1 Mineralogical Classification of Asbestos

The term asbestos is a commercial name referring to a group of naturally occurring fibrous silicates. The asbestos minerals belong to two mineralogical groups, serpentines and amphiboles. The asbestiform varieties of serpentine and amphibole are shown in Table 33.2.

The most abundant type of asbestos is chrysotile, which is the only member of the serpentine group. Chrysotile is composed of curly fibers that can shear into smaller fibrils.

Unlike chrysotile's curled shape, amphiboles crystallize in straight double chains, resulting in needlelike structures. Amphiboles are typically more brittle than chrysotile and tend to cleave longitudinally. The amphiboles of greatest commercial importance are crocidolite and amosite, whose individual fibrils tend to be thicker than those of chrysotile. Other amphiboles, which are similar structurally but differ in chemical composition, include tremolite, anthophyllite, and actinolite.

Table 33.1. Asbestos Management Activities Requiring Industrial Hygiene Input[a]

Building surveys
Pre-abatement sampling
Abatement recommendations
Cost estimates for remedial alternatives
Warning, notification, and record keeping
Asbestos management programs
Asbestos abatement specifications
Asbestos abatement contractor selection
Pre-abatement project meeting
Pre-abatement inspection
Monitoring compliance with specifications
Post-abatement inspections

[a]From Ref. 2.

Table 33.2. Minerological Classification of Asbestos

Mineral Group	Mineral	Chemical Formula
Serpentine	Chrysotile	$(Mg,Fe)_6(OH)_8Si_4O_{10}$
Amphibole	Actinolite	$Ca_2Fe_5(OH)_2Si_8O_{22}$
Amphibole	Tremolite	$Ca_2Mg_5(OH)_2Si_8O_{22}$
Amphibole	Amosite	$Mg_7(OH)_2Si_8O_{22}$
Amphibole	Anthophyllite	$(Mg,Fe)_7(OH)_2Si_8O_{22}$

1.2 Historical Perspective

To understand fully the magnitude of the asbestos management problem, it is useful to review the history of asbestos use and its eventual decline. The unique fire-resistant properties of asbestos, a naturally occurring mineral, have been known since ancient times. Roman historians wrote about asbestos. Plutarch described the asbestos lamp wicks that were used in temples. Herodotus mentioned the use of asbestos as a cremation cloth that made it possible to salvage the ashes of the honored dead from their funeral pyres. Pliny described asbestos fabric as "a rare and costly cloth, the funeral dress of kings." He also described the use of a transparent bladder skin as a respirator to avoid the inhalation of asbestos dust by slaves, perhaps the earliest, if oblique, inference of asbestos-related disease (3).

Later, even Charlemagne found a use for asbestos. Anticipating war with an Arabic prince, Harun-al-Raschid, Charlemagne had a tablecloth woven of asbestos fiber. He invited the prince and his coterie to a feast, after which Charlemagne entertained his guests by throwing the soiled asbestos tablecloth into a fire to "clean" it. Apparently, this feat so amazed the prince's followers that they advised their leader not to wage war against such a powerful magician as Charlemagne.

In modern times, the practical uses of asbestos were first demonstrated during the 1850s in Quebec by a young shopkeeper named Henry Ward Johns. His neighbors, who were farmers, complained of a curious fibrous rock which made plowing nearly impossible. The merchant, however, recognized the rock as asbestos and began tinkering to find a practical application. Johns mixed the asbestos with burlap, pitch, and manila paper, ran the mixture through his wife's clothes wringer, and developed the first asbestos roofing product. In an era of wooden buildings heated by wood or coal fires, Johns's fire-resistant roofing became an instant success. Next he turned his attention to gasket materials and eventually to pipe and boiler insulation. In 1898, before introducing his pipe and boiler insulation products, Johns died. It is ironic that Henry Ward Johns, the man who pioneered the modern use of asbestos fiber, apparently died of an asbestos-related disease. According to his death certificate and obituary, Johns died of "dust phthisis pneumonitis" (4) [the term asbestosis was first used to describe a fibrotic lung disease in 1927 by Dr. W. E. Cooke (5)]. Following his death, Johns's company was purchased by C. B. Manville to form the Johns-Manville Company.

Heralded as the "magic mineral" (6) asbestos grew in use throughout the first half of the twentieth century. The unique properties of asbestos, its high tensile strength, temperature resistance, and chemical resistance, plus its abundance in nature led to its use in a growing list of products.

In World War II, when the United States mobilized an unprecedented shipbuilding effort, asbestos was considered to be a "strategic material" and its production was controlled by the U.S. government. Not only was asbestos used to insulate the boiler and piping systems on these ships, but later in the war, an asbestos insulation product was sprayed onto the inside framework of the vessels. Following World War II, a similar type of sprayed-on insulation product was developed for fire protection on structural steel in multistory buildings. In 1970, sprayed "inorganic fiber" was reportedly used as a fireproofing agent in well over half of all large multistory buildings (7).

In the early 1970s, the use of asbestos began to decline following the publication of various health studies, particularly the work of Dr. Irving Selikoff and his associates at the Mt. Sinai School of Medicine in New York. Responding to Selikoff's and other studies, as well as growing public concern, the City of Philadelphia in 1970 enacted the first ban on asbestos fireproofing products (8). Other cities across the United States followed suit, and in 1973, the EPA enacted a nationwide ban on spray-applied fireproofing.

Since then, the use of asbestos in the United States has continued to decline (see Table 33.3). From a peak in 1973 of almost 800,000 metric tons, U.S. consumption of asbestos plummeted to approximately 29,000 metric tons in 1994. While U.S. consumption of asbestos has continued to decline, its use is reportedly on the increase in some countries, particularly developing countries such as Brazil, China, India and South Korea (see Table 33.3).

The sharp decline in asbestos consumption in the United States parallels the rise in litigation against the manufacturers of asbestos products. In 1982, Manville (formerly Johns-Manville), the company started nearly a century earlier by Henry Ward Johns, filed for bankruptcy. Manville's financial problems were attributed to lawsuits brought by asbestos victims and property owners. By 1985, the claims against Manville totaled more than $80 billion (11).

Table 33.3. Consumption of Asbestos Fibers by Country[a]

Nation	Tons Consumed		
	1985	1988	1994
Former USSR	1,963,010	2,080,825	—
Japan	265,358	324,000	195,000
Brazil	151,934	191,025	190,000
China	150,543	157,100	220,000
India	110,038	101,500	123,000
Italy	122,408	96,000	—
South Korea	61,843	74,500	85,000
Former W. Germany	60,106	54,171	—
United States	154,232	83,000	29,000

[a]From Refs. 9 and 10.

1.3 Applications of Asbestos in Buildings

The EPA (12) estimated that at its peak use, asbestos was used in as many as 3000 products. Table 33.4 lists some of the most common applications in building products. In building environments, the greater concern is for "friable" materials. Friable materials are defined by the EPA (13) as those that may "be crumbled, pulverized or reduced to powder by hand pressure." The friability of a product is generally thought to be related to its fiber release potential, with the more friable materials posing the greatest risk of fiber release. Of the products described in Table 33.4, four are generally considered to be friable, including fireproofing, acoustical plasters, pipe and boiler insulation, and ceiling tiles. In a 1984 survey conducted for EPA, 231 buildings in 10 cities were inspected to determine the

Table 33.4. Asbestos-containing Materials Commonly Used in Building Construction

Type of Material	EPA Category
Fireproofing	Surfacing materials
Decorative acoustic plasters	
Pipe and pipe fitting insulation	Thermal system insulation
Boiler and tank insulation	
HVAC duct insulation[a]	
Gaskets	
Floor tiles	Miscellaneous materials
Roofing felts	
Roof coatings	
Asbestos/cement pipe	
Asbestos/cement sheet	
Fabrics	

[a]HVAC = heating, ventilation, and air conditioning.

prevalence of friable asbestos-containing materials in U.S. buildings. Based on the findings of that study, EPA and its contractors (14) estimated that friable, asbestos-containing products are present in approximately 20% of all buildings in the United States. Further findings of this study by type of product are summarized in Table 33.5.

The use of asbestos products is not restricted to commercial or public buildings; single-family homes may also contain one or more asbestos-containing building materials. In a study of 62 selected homes in three U.S. cities, Rock (15) reported damaged asbestos-containing materials in 45 of the homes. The type of asbestos-containing material was found to vary according to geographic location with soft acoustic plasters prevalent in homes in the San Francisco area and thermal system insulation found in homes in Cleveland and Philadelphia.

2 UNDERSTANDING THE HEALTH RISKS

2.1 Asbestos-related Diseases

The inhalation of asbestos fibers has been shown to increase the risks of lung disease, including asbestosis, lung cancer, and mesothelioma.

2.1.1 Asbestosis

It has been known since the early part of this century that workers who inhaled large amounts of asbestos dust sometimes developed a disabling or fatal fibrosis of the lungs (16, 17). This condition is commonly called asbestosis. Asbestosis, a diffuse interstitial pulmonary fibrosis or fibrotic pneumoconiosis, occurs almost exclusively among workers who have chronic exposure to high concentrations of airborne asbestos fibers. The disease, which on an average requires 15 or more years to develop, is characterized by the formation of scars or collagen tissue in the lungs, and is associated with progressive deterioration of pulmonary function and work capacity.

Table 33.5. Prevalence of Asbestos-containing Materials[a]

		Prevalence of Asbestos-containing Friable Material (%)			
Type of Building	Number of Buildings Surveyed	Sprayed- or Troweled-on	Ceiling Tile Insulation	Pipe/ Boiler	All Types
Federal govt.	66	16	2	25	39
Residential (10 or more rental units)	55	18	0.5	44	59
Private (nonresidential)	110	4	0	12	16
All buildings (combined)	231	5	0.1	16	20

[a]From Ref. 14.

2.1.2 Lung Cancer

Asbestos exposure was first linked to bronchogenic carcinoma (lung cancer) in 1935, but it was not until 1947 that epidemiologic evidence of a causal connection was established (18). Since then, numerous studies have confirmed an association between occupational exposure to asbestos and significantly elevated rates of bronchogenic cancer. This form of lung cancer generally occurs 20 to 30 years after exposure begins, and a strong relationship between bronchogenic cancer and asbestosis has been demonstrated. Smoking and asbestos exposure act synergistically to increase the risk of lung cancer.

2.1.3 Mesothelioma

The first link between asbestos exposure and pleural mesothelioma (a cancer of the lining of the lung cavity) was reported in 1943. A strong association was not established until 1960 when a study in a crocidolite mining area of South Africa found that 31 of 33 persons with pleural mesothelioma had experienced some asbestos exposure (19). Additional evidence of a causal relationship between asbestos exposure and pleural mesothelioma has since accumulated. This type of carcinoma is virtually unknown in the general population; the incidence rate among asbestos workers has been estimated at approximately 5 to 10%. An estimated 1600 cases of mesothelioma were reported in the United States in 1980 (20). Mesothelioma is invariably fatal, usually within two years of diagnosis. The risk of contracting mesothelioma appears to be independent of smoking.

Epidemiologic evidence has not clearly demonstrated mesothelioma to be dose-related, although in general, exposure levels associated with this disease tend to be less than those that produce asbestosis. Mesothelioma has been reported in family members of asbestos workers and in persons living near asbestos mining areas (20).

2.1.4 Other Cancers

A relationship between asbestos exposure and other forms of cancer, in particular gastrointestinal cancer, has been reported in some studies. However, this relationship has not been clearly established.

2.1.5 Pleural Plaques

A relatively benign disorder related to asbestos exposure is the occurrence of pleural plaques. Pleural plaques refer to calcification of the pleura on the inner surface of the rib cage. They do not interfere with pulmonary function to a significant extent, nor do they necessarily indicate the presence of pulmonary fibrosis. The occurrence of pleural plaques has been strongly associated with past asbestos exposure (occupational or nonoccupational) (21) and is often considered a marker of such exposure.

2.2 Mechanisms of Fiber Toxicity

Estimating the health risks from low-level asbestos exposures (such as those typical in building environments) is complicated by several factors including the long latency period associated with asbestos-related diseases, the relationship of fiber size and type of asbestos with health risk, and concurrent exposures to other agents, most notably cigarette smoke.

2.2.1 *Synergy with Tobacco Smoke*

The relationship of asbestos and tobacco smoke with the incidence of lung cancer has been well documented. In studies of asbestos insulation workers, Selikoff et al. (22) reported that workers who smoked were 10 times more likely to die of lung cancer than nonsmoking asbestos workers (Table 33.6). In studies of factory workers exposed to amosite asbestos, Selikoff et al. (23) estimated that the combination of cigarette smoking and asbestos exposure increased the risk of lung cancer death about 80 times.

Several mechanisms have been proposed to explain this synergistic relationship. Sanchis et al. (24) and Cohen et al. (25) have suggested that toxic and irritant properties of cigarette smoke cause dysfunction and loss of ciliated and secretory cells lining the respiratory tract, preventing these cells from effectively capturing and clearing deposited particles. Gerde and Scholander (26) suggest that the asbestos fiber acts as a carrier for carcinogens in the smoke. They proposed that the adsorption of phospholipids from the lung onto the surface of asbestos fibers allows lipophilic carcinogens in the tobacco smoke (e.g., polycyclic aromatic hydrocarbons) to diffuse more readily across the bronchial lining and through cell membranes of the bronchial epithelium.

Though the mechanism of this synergistic relationship is not completely understood, the management implications of this phenomenon are clear. Many contractors in the asbestos-abatement industry have chosen selectively to hire nonsmokers as a means of reducing their companies' long-term risk. Smoking abatement workshops have been used as an effective tool by many companies to assist employees occupationally exposed to asbestos to quit smoking, thereby reducing their overall lung-cancer risk.

2.2.2 *Type of Asbestos*

A controversial issue is the relative risk of different forms of asbestos, particularly chrysotile versus amphiboles. While some countries have established different exposure limits for the various forms of asbestos, reflecting a belief that amphiboles (particularly crocidolite) pose a greater health risk than chrysotile, the United States Occupational Safety and Health Administration (OSHA) has taken the view that there is insufficient evidence available to define clearly the relative risks of the different forms of asbestos and have therefore established a single limit (27).

Table 33.6. Asbestos and Smoking Age-Standardized Lung Cancer Death Rates[a]

Group	Asbestos Exposure	Smoker	Death Rate	Mortality Ratio
Control	no	no	11.3	1.00
Asbestos workers	yes	no	58.4	5.17
Control	no	yes	122.6	10.85
Asbestos workers	yes	yes	601.6	53.24

[a]From Ref. 23.

Generally, studies on the relative risks of different forms of asbestos have focused on mesothelioma. The difference in the carcinogenicity between chrysotile fibers and the amphiboles (which show greater incidence of microthelioma) (28) has been attributed to the greater pulmonary retention of amphibole fibres. For all fibers exceeding 10 mm in length, the longer the fiber, the more readily it dissolves.

Hughes and Weill (29) concluded that there is strong evidence that crocidolite and amosite pose a greater risk of mesothelioma than chrysotile. By contrast, the Committee on Nonoccupational Health Risks of Asbestiform Fibers of the National Research Council (30) concluded that the available epidemiologic literature "does not present a clear picture . . . on the relative ability of different fiber types to cause disease." The committee further concluded that experimental and animal studies have also failed to identify discernible differences in the carcinogenicity of different types of asbestos fibers.

2.2.3 *Size of Asbestos Fibers*

Another controversial issue is the size of asbestos fibers and the relative risk of large versus small fibers. This issue has become particularly important as transmission electron microscopy (TEM) has become the preferred method for analyzing air samples in nonoccupational environments. Because TEM is able to detect the smallest asbestos fibers (fibrils), researchers are now able to measure concentrations in buildings of fibers that were previously undetected. Some experts disagree, however, on the significance of TEM sampling results as an indication of health risk.

It is well accepted in the practice of industrial hygiene and occupational medicine that particle size determines the extent to which particles are able to enter the respiratory tract and penetrate into the lung. The size of asbestos fibers may also determine how readily, once they reach the lung, they are engulfed by macrophages. Dupre et al. (31) reported that fibers shorter than 5 mm are readily engulfed by macrophages, whereas longer fibers are not.

In studies in which various sizes of fibers were implanted onto the pleura of rats, Stanton et al. (32, 33) found that the biological activity of such fibers was directly related to their dimension. Other investigators have reported similar findings (34) suggesting that the long, thin asbestos fibers are most apt to be pathogenic. In a comprehensive review of this issue, Lippmann (35) concluded that asbestosis is most closely related to the surface area of retained fibers, mesothelioma is most closely associated with numbers of fibers longer than 5 μm and thinner than 0.1 μm, and lung cancer is most closely associated with fibers longer than 10 μm and thicker than 0.15 μm.

2.3 Levels of Exposure

Occupational and general public asbestos exposure levels are summarized in Table 33.7, Table 33.8, and Table 33.9. These exposure data clearly show the large differences that exist between historical occupational exposures in shipyards, mining and milling operations, and commercial insulating and the levels being reported in building environments. The levels found in building environments are generally several orders of magnitude lower than levels in occupational environments. It is also interesting to note that the results of

Table 33.7. Summary of Historical Asbestos Exposure Levels In Industry

Type of Exposure	Analysis[a] Method	Mean Concentration (f/cc) Fibers >5 μm	Fibers <5 μm[b]	Total	Ref.
Asbestos mining (worker exposures)	PCM	1.3	NR	—	36
Asbestos milling (worker exposures)	PCM	10.0	NR	—	36
Asbestos mining (worker exposures)	PCM	1.8	NR	—	37
Shipbuilding (worker exposures)	PCM	250	NR	—	38
Textile manufacturing (worker exposures)	OM	3.4	NR	—	39

[a]PCM = phase-contrast microscopy; OM = optical microscopy.
[b]NR = not reported.

samples collected in occupied buildings invariably showed predominantly small fibers (shorter than 5 μm in length).

2.4 Extrapolation of Historical Data/Risk Models

All of the epidemiologic studies described previously involved occupationally exposed populations. These individuals were exposed to fiber concentrations far in excess of those which occur in building environments. Most estimates of risk at low levels of exposure, such as nonoccupational exposures in buildings, are mathematical extrapolations from data representing much higher levels of exposure. Some of the health risk (dose–response) models are summarized in Table 33.10. These models vary considerably in their estimate of health risks from asbestos exposures, particularly at the relatively low concentrations that may occur in public and commercial buildings.

In a February, 1988 report to Congress, EPA (1) estimated the number of asbestos-related deaths of noncustodial occupants of commercial and public buildings. Between the years 1988 and 2118, a period of 130 years, EPA estimated that there would be 4280 deaths resulting from exposure to asbestos in buildings. This translates to an average of 33 deaths per year in the United States. Using its estimate of 35.6 million people exposed, EPA's estimated mortality rate may be expressed as 0.092 deaths for every 100,000 persons exposed per year.

2.5 Occupational Exposure Limits

Until 1986, asbestos exposure in the United States was limited to 2.0 fibers per cc. In 1986, OSHA published asbestos standards for general industry and construction, both of which included a permissible exposure limit (PEL) for all fiber types of 0.2 f/cc. On August 10, 1994, OSHA published its final standards to amend the 1986 regulations further reducing the PEL to 0.1 f/cc.

Table 33.8. Summary of Asbestos Exposure Levels In Construction-Related Occupations

Type of Exposure[a]	Number of Samples	Mean Fiber Conc.	Range of Fiber Concentrations <5 μm	Ref.
Commercial insulating (worker exposure)		9		40
Commercial insulating (worker exposure)		5–99		41
Commercial insulating (worker exposure)		12		42
Building renovation				43
Carpenter		0.13		
Electrician		0.13		
Sheet metal worker		0.19		
Painter		0.08		
Building Renovation				
Carpenter	41	0.16	<0.01–0.69	44
Electrician	39	0.11	<0.01–1.6	
Pipefitter	6	0.12	0.04–0.53	
HVAC worker	30	0.14	0.01–2.8	
Gen. laborer	21	0.14	0.02–0.87	
Installation of A/C sheet in cooling towers	102	0.10	<0.02–0.5	44
Removal of insulation from pipes using glove bags	15	0.01	<0.01–0.02	44
Wet removal of ACM				44
Fireproofing	255	1.1	<0.01–170	
Pipes and boilers	57	0.46	<0.01–8.0	
Industrial ovens	23	0.22	0.03–3.0	
Refinery pipes	11	0.09	<0.01–0.57	
ACM removal (misc.)	479			
Wet methods	—	0.9	NR[b]	45
Dry removal	—	17.	NR[b]	

[a]All sample results reported above are for phase-contrast microscopy. HVAC = heating, ventilation, and air conditioning; ACM = asbestos-containing material.
[b]NR = not reported.

3 CONDUCTING BUILDING SURVEYS

To determine whether asbestos-containing materials were used in the original construction or subsequent renovations of a building, a detailed building survey, including the collection of bulk samples of suspect materials, is required.

It should be noted that in many U.S. states, persons performing such inspections must first be licensed and the industrial hygienist planning to perform such inspections should always contact the local authorities for information before undertaking such an investigation.

Table 33.9. Summary of Asbestos Exposure Levels in Nonoccupational Environments[a]

Type of Exposure	Analysis Method[a]	Mean Concentration (f/cc) Fibers >5 μm[b]	Fibers <5 μm[b]	Total	Ref.
Prevalent levels in 17 commercial and public buildings	TEM	NR	NR	0.001[c]	46
	PCM	0.007	—	0.007	
Prevalent levels in 25 schooling buildings	TEM	NR	NR	0.002[a]	47
Prevalent levels in 19 Ontario buildings	TEM	<0.001	0.021	0.021	48
Prevalent levels in 43 U.S. government buildings	TEM	NR	NR	0.00059 to 0.00073	49
Prevalent levels in 10 commercial buildings	TEM	ND	0.0006	0.0006	50

[a]TEM = transmission electron microscopy; PCM = phase-contrast microscopy.
[b]NR = not reported; ND = none detected.
[c]Results were originally reported in units of nanograms per cubic meter. To allow comparison, these data were converted to fibers per cubic centimeter using a factor of 30 fibers/nanogram.

Table 33.10. Mortality Estimates at Low Levels of Asbestos Exposure

Ref. Source	Estimated Exposures Concn. (f/cc)	Duration (years)	Age at First Exposure	Mortality per Million Persons Exposed
51	0.001	10	22	16
	0.001	10	7	31
52	0.01	10	20	15–150
53	0.1	20	25	2260
54	0.006	35	25	140
55	0.0004	73	0	33
56	0.0005	20	20	10
57	0.002	30	NR[a]	120
58	0.002	50	20	46

[a]NR = not reported.

3.1 Building Inspection

As the initial step of a building survey, the investigator should review building plans, remodeling records, and related documents to determine if asbestos-containing materials were specified. It should be emphasized that such information is not a substitute for a thorough building inspection. Information on the type and location of building materials that might potentially contain asbestos will help focus the inspection effort on asbestos-containing materials in the building. A building inspection might start with those areas where asbestos-containing materials are reported to be present and then expand to all parts of the building. If suspect materials are found, they should be sampled and analyzed as discussed below.

Materials that contain asbestos may be sprayed or troweled onto surfaces in either a relatively loose or cementitious form. Among the most common materials are asbestos-containing insulation products used on pipes and boilers. Pipes and boilers are not typically directly accessible to building occupants other than maintenance or custodial workers. It is not unusual, however, for these types of materials to be disturbed when repairing leaks, rebuilding valves, or during other routine maintenance and repair of mechanical and plumbing systems. Pipe and boiler insulation may look like chalk, bricks, cement, plaster, or corrugated cardboard. Often these insulation materials are enclosed in a fabric, metal, or plastic jacket. The fabric jacket may also be made of asbestos fiber. The boiler wrapping and pipe jackets should be carefully inspected for damage that could result in the release of asbestos fibers.

Nonfriable asbestos-containing materials include asbestos cement products, floor tile, and gaskets. These products normally release fibers only when cut, drilled, sanded, or removed during remodeling or demolition.

In performing a building survey, the investigator must visit and inspect every room in a building including the custodian's storage closets, steam tunnels, boiler room, attics, and so on. A person familiar with the building such as the building engineer or a custodian is an ideal guide in performing this survey.

3.2 Bulk Sampling and Analysis

As the building inspection is performed, bulk samples of suspect materials should be collected. At a minimum, three replicate samples should be collected of each encountered material. The investigator, however, should always consult appropriate regulations for sampling requirements.

It is important to collect bulk samples to the full depth of the suspect material to ensure that a complete and representative sample is collected. In many situations, additional layers of nonasbestos insulation may have been applied over the top of an older asbestos-containing material.

A convenient tool for bulk sampling is the cork borer. This sharpened metal tube is forced into the suspect material, and a plug is removed. Samples can also be collected using a pocket knife. Prior to sampling, the area should be gently wetted with a spray bottle to minimize fiber release. The cork borer must be cleaned between samples to prevent cross-contamination. Each sample is transferred from the cork borer to a labeled plastic vial and submitted to the laboratory for analysis.

In collecting bulk samples, precautions should be taken to minimize exposure of the investigator and other building occupants to asbestos fibers. Bulk samples should be collected at a time when the building is vacated.

The accepted method of analysis of building materials for the presence of asbestos is polarized light microscopy (PLM). The EPA (59, 60) has specified this technique in regulations developed for schools. The PLM procedure is based on the spectrographic method used by minerologists for many years to identify and classify different types of minerals. Briefly, the technique involves:

- Examination of the bulk material under a stereomicroscope (100X magnification) to determine the fibrous nature of the material, the color and morphology of the fibers, and the types of fibers present.
- Mounting of each type of fiber on microscope slides in liquids of appropriate refractive indexes.
- Examination of the mounted slides under a polarized light microscope at 100X magnification and with dispersion-staining objectives.

The fibers are identified based on (*1*) the morphology (the shape, color, and general appearance) of the fibers, (*2*) the optical characteristics of the fiber when viewed using polarizing filters, and (*3*) characteristic colors produced by various fibers under polarized light microscopy using the dispersion-staining objective. The relative abundance (percent composition) of each fiber type is estimated while examining the sample through the stereomicroscope.

3.3 Evaluating Exposure Risks

In addition to identifying asbestos-containing materials, a comprehensive building survey should include an evaluation of the potential exposure risk, for example, the likelihood that a particular asbestos-containing material will release asbestos fibers into the building air during day-to-day activities or during foreseeable maintenance or renovation activities. In evaluating this risk, several factors must be considered, including the condition of the asbestos material, its friability, and its accessibility to building occupants.

3.3.1 Condition

The condition of the asbestos-containing material may determine the likelihood of fibers being released into the air. An assessment of the condition depends upon a combination of the quality of the installation, adhesion of the material to the underlying substrate, deterioration, vandalism, and other damages including water damage.

3.3.2 Friability

The products containing asbestos vary widely in their relative degree of friability. The more friable the material, the greater the potential for asbestos fiber release.

3.3.3 Accessibility

If the asbestos-containing material can be easily reached, it is subject to accidental or intentional contact and damage. Material that is readily accessible (within reach) is most likely to be disturbed. Accessibility therefore is one of the most important indicators of exposure potential. The close proximity of the friable material to heating, ventilation, lighting, and plumbing systems requiring maintenance or repair indicates high accessibility. Existing damage is an obvious indicator of accessibility.

Table 33.11 describes one strategy for rating each of these factors that has been used by the authors in assessing potential exposure risks in commercial and public buildings.

EPA and other groups have developed algorithms by assigning numerical scores to the factors discussed above and to other factors. Algorithms proposed by EPA, the U.S. Navy, the Massachusetts Public Schools, and the Colorado Public Schools and the criteria used in each algorithm are summarized in Table 33.12. When results are compared between buildings, and for similar materials, such algorithms may provide a relative measure of the overall fiber release potential. Such algorithms, however, are based on individual subjective judgments. These algorithms should not be considered a substitute for professional judgment and for routine air monitoring as the means of establishing asbestos exposure risks.

3.4 Record Keeping

Careful documentation of field observations made by the investigator is another important aspect of a building survey. Survey records should include:

- Identification of all areas/rooms which were inspected whether or not suspect materials were identified.
- The condition of the materials, their accessibility, and friability on an area-by-area basis.
- Descriptions of all sampling locations.
- Descriptions of each unique material observed and sampled.

In describing areas of the building or mechanical systems, it is best to rely on descriptions provided in mechanical drawings or by facility personnel. In some cases, names given to building areas or systems may be unique to that building. By using these names, it is easier to identify and relocate specific materials, sampling points, and so on. Wherever possible, building drawings should be annotated to document location(s) of asbestos-containing materials. In commercial buildings, the location description should not rely on room numbers or names of office occupants or tenants, for such buildings undergo frequent renovation, often involving the reconfiguration of entire floors.

4 AIR SAMPLING AND ANALYSIS

The preceding sections outlined procedures for determining whether asbestos-containing materials are present in a building and for assessing the likelihood that these materials will release fibers into the air. To evaluate the extent to which building occupants may be exposed to asbestos fibers, it is necessary to collect air samples.

Table 33.11. Exposure Risk Factors for Asbestos-Containing Materials in Buildings

Factor	Description
	Condition
Good	Material is intact and shows no signs of deterioration
Fair	Visual inspection and physical contact indicate that the material is breaking up into layers or beginning to come loose from the substrate. There may be small areas (<10% of the total area) where the material is deteriorating. There may be signs of accidental or intentional damage
Poor	The material is noncohesive. Pieces are dislodged and debris in the area is evident. Parts of the material may be hanging loosely or may have fallen to the floor. The possibility of severe accidential or intentional damage should be investigated
	Friability
Low	Material that is difficult yet possible to damage by hand. This would include most troweled-on materials and manufactured items such as ceiling tiles
Moderate	Fairly easy to dislodge and crush or pulverize by hand. Material may be removed in small or large pieces
High	The materials is fluffy, spongy, or flaking and may have pieces loose or hanging
	Accessibility
Staff only	The material is located above a suspended ceiling or is concealed by ducts or piping. The building occupants cannot touch the material. Only maintenance staff would normally have access to the asbestos-containing material
Moderate	The material is touched only during infrequent maintenance or repair activities. Building occupants rarely touch the material or throw objects against it
Easy	Material is touched frequently owing to routine maintenance, and/or the building occupants can touch the material during normal activity

4.1 Sampling Strategies

Air sampling, as it is applied within the context of the building environment, may be divided into several basic categories.

4.1.1 Prevalent Level Sampling

Prevalent level or background air monitoring may be used to establish concentrations of airborne fibers in a building under "normal" building conditions. Such sampling is performed to establish background levels prior to asbestos abatement or renovation or may be a part of a long-term management plan. Prevalent level monitoring may be repeated annually to evaluate any significant change in fiber concentrations that may be indicative of deterioration of the asbestos-containing product. To obtain results that are representative of the entire building, at least one sample should be collected for every 10,000 ft^2. At a minimum, one sample should be collected per floor. Sampling media should be placed at "breathing zone" height (3 to 5 ft above floor level). Sampling should be performed during

Table 33.12. Comparison of Evaluation Criteria for Several Algorithms[a]

Criterion Used	Algorithm			
	EPA[b]	Navy[c]	Massachusetts[d]	Colorado[e]
Condition	X	X	X	X
Accessibility	X	X	X	X
Friabity	X	X	X	X
% Asbestos content	X	X	X	X
Activity/movement	X	X	X	X
Air plenum/air movement	X		X	X
Water damage	X			
Exposed surface				
No. of occupants		X		
Occupancy duration		X		

[a]Adapted from Ref. 65
[b]From Ref. 61
[c]From Ref. 62
[d]From Ref. 63
[e]From Ref. 64

normal building occupancy; this is essential in obtaining representative results. Generally, it is necessary to use high-volume pumps (flow rate of 8 to 10 L/min). When samples are to be analyzed by electron microscopy, ambient (outdoor) air samples should also be collected for comparison.

4.1.2 Personal Sampling

Personal sampling is used during maintenance, renovation, or abatement work to determine employee exposure (without consideration to the use of a respirator) to airborne fibers. The results of personal sampling are used to select proper respiratory protection and for documentation required under many state and federal regulations. Air sampling results can also be used as a means of evaluating those work practices that result in the lowest employee exposure.

4.1.3 Perimeter Sampling

During an abatement or renovation project, samples are collected from locations outside the work area (but inside the building) to determine how well asbestos fibers are being contained. These samples are especially important in situations where unprotected people occupy other areas of a building in which such work is being conducted.

The number of perimeter air samples collected varies with the size and nature of the project. Priority should be placed on sampling at potential leakage points outside the containment barriers. A typical daily sampling scheme might include two or three samples inside the building, one sample in the clean room of the decontamination unit and one sample outside the building. As discussed previously, the latter is necessary for comparison purposes, particularly when using electron microscopy for analysis.

4.1.4 *Clearance Sampling*

Prior to dismantling containment barriers for either abatement or renovation projects, "clearance" air samples are collected. Clearance air monitoring is performed in conjunction with visual inspection to ensure that the work area is free of asbestos contamination. Final clearance samplers are placed inside the work area with approximately one sampler for every 10,000 ft^2. A minimum of three samples, or at least one per room, should be collected.

As in prevalent level sampling, a large air volume is necessary to obtain the desired detection limit. Such air samples may be collected using passive or aggressive sampling techniques. Passive sampling implies monitoring an area as is, without creating any additional disturbance in the air. For clearance monitoring performed in school buildings, EPA requires that clearance air sampling be performed using "aggressive" methods. The objective of aggressive sampling is to resuspend fibers that may have settled onto the floor and other surfaces prior to sampling. Briefly, this technique uses a 1-horsepower leaf blower to blow off all walls, ceilings, floors, ledges, and other surfaces in the room. A 20-in. fan is then used to provide constant air movement (one fan for every 10,000 ft^2). Samples are collected while the fans are operating.

4.2 Analysis

Analysis of air samples is performed using one of three techniques: phase-contrast microscopy (PCM), scanning electron microscopy (SEM), or transmission electron microscopy (TEM). The relative capabilities of each of these three procedures are described in Table 33.13.

4.2.1 *Phase-Contrast Microscopy*

A phase-contrast microscope is a light microscope equipped with phase annuli in the objective and condensor lenses to provide enhanced contrast between the fibers and the background. Samples for PCM analysis are collected on cellulose ester membrane filters, which are then rendered transparent (cleared) with a chemical solution or solvent so that trapped particulate material can be viewed through the microscope at a magnification of approximately 400X. This method cannot distinguish between asbestos and other fiber types and can detect only those fibers longer than 5 μm and wider than about 0.25 μm. Because of these limitations, results of PCM analysis provide only a measure of the total airborne fiber concentration. As the proportion of the airborne fibers less than 0.25 μm in diameter increases, PCM becomes a less reliable analytical tool.

A principal advantage of PCM is the ability to perform analysis on-site. During asbestos removal or renovation projects, it is critical that air sampling results be available immediately. This is especially important in perimeter sampling where the results are used to determine the effectiveness of containment barriers. On-site analysis allows the project manager to make this determination within a few hours, and if a problem is discovered, it can be quickly corrected.

PCM is the method of choice for evaluating occupational exposures to asbestos.

Table 33.13. Comparison of Analysis Procedures for Asbestos Fibers[a]

	Analysis Procedure		
Comparison Factor	PCM	SEM	TEM
Standard methods	NIOSH P&CAM 239 method	No standard method	EPA provisional method and update
Quality assurance	Proficiency analytical testing program	No lab testing program	Proposed NVLAP accreditation program
Cost	\$25–50	\$100–300	\$300–800
Availability	Readily available	Less available	Readily available
Time requirements	1 hr preparation and analysis, <6 hr turnaround	4 hr preparation and analysis, 6–24 hr turnaround	4–24 hr preparation and analysis, 2–7 days turnaround
Sensitivity (thinnest fiber visible)	0.15 μm at best; 0.25 μm typical	0.05 μm at best; 0.20 μm typical	0.0002 μm at best; 0.0025 μm typical
Specificity	Not specific for asbestos	Definitive for asbestos when used to its fullest capabilities	Definitive for asbestos when used to its fullest capabilities

[a]PCM = phase-contrast microscopy; SEM = scanning electron microscopy; TEM = transmission electron microscopy; NVLAP = National Voluntary Laboratory Accreditation Program; P&CAM = physical and chemical analysis methods.

4.2.2 Scanning Electron Microscopy

SEM is a technique that directs an electron beam onto the sample surface and collects the reflected beams. A magnified image is produced on a viewing screen. Under the best conditions, SEM can provide a sharp, three-dimensional image of the filter surface. Air samples for SEM analysis may be collected on polycarbonate or cellulose ester filters.

When connected to an energy-dispersive X-ray analyzer (EDS), SEM can distinguish between asbestos and nonasbestos fibers by performing an elemental analysis of individual fibers observed on the viewing screen. SEM can detect fibers with diameters as small as 0.05 mm.

4.2.3 Transmission Electron Microscopy

TEM is a technique that focuses an electron beam onto a thin sample. As the beam transmits through certain areas of the sample, an image, resulting from varying densities of the sample, is projected onto a fluorescent screen. Air samples are collected on polycarbonate or cellulose ester filters (depending on the specific protocol being followed).

TEM is the only method of these three techniques that can detect the smallest diameter asbestos fibers (fibrils). TEM does not, however, produce the same high-quality image as SEM. The images produced by TEM are essentially shadows of the particles being examined. TEM can be equipped with either or both EDS and Selected Area Electron Diffraction (SAED) for fiber identification. The latter procedure provides information on the crystalline structure of individual fibers viewed.

4.3 Direct-Reading Instruments

Fibrous aerosol monitors (FAMs) are direct-reading devices designed for analyzing airborne fiber concentrations. Similar to other types of particulate monitors that rely on light scattering, FAMs have been modified to discriminate between fibrous and nonfibrous particulates. Results are reported as a fiber count. These devices operate by drawing air through a tube where two pairs of electrodes create electric fields perpendicular to each other and to the axis of the tube. The two fields are alternately activated, causing the fibers to align and oscillate as they travel through the tube. A laser beam shines along the axis of the flow tube and the fibers scatter this light in directions perpendicular to their long axis. A photomultiplier tube, positioned in the wall of the flow tube, gives a pulsed response as the scattered light scans over it. Fiber size is determined from the shape of these pulses.

As discussed previously, the conventional methods for measuring concentrations of airborne asbestos fibers involve collecting fibers on filters and analyzing them with either a light microscope or an electron microscope. Results of comparative testing using the fibrous aerosol monitors and conventional techniques do not always show good agreement (66). FAMs have nevertheless found considerable application in the asbestos abatement industry where, equipped with alarms, they are used frequently to warn of "barrier" failures.

Accurate, defensible results depend on careful calibration of direct-reading aerosol monitors. Unfortunately, calibration of these monitors is difficult. Aerosols cannot be obtained in commercial calibration mixtures nor are there aerosol-generation systems available in which the concentrations can be determined a priori. As a result, in routine use, calibration is performed by comparing field monitoring results with "reference" techniques involving filter collection and microscopic analysis of the filters.

5 CONTROL OF ASBESTOS IN BUILDINGS

5.1 Control Options

5.1.1 Encapsulation

The term "encapsulation" describes the spray application of a sealant to the asbestos-containing materials. The application of encapsulants helps bind the asbestos fibers together to reduce the fiber release potential.

Usually encapsulation involves the application of multiple coats using two types of sealants. A penetrating encapsulant is applied first. The penetrant is intended to infiltrate into the friable material to help bind the fibers together from within. The second coat is called a "bridging encapsulant." A bridging encapsulant forms an impervious skin at the exposed surface of the asbestos-containing material. Encapsulation is typically less expensive than removal. It is not, however, a permanent solution to the asbestos problem. Encapsulation does not prevent the release of fibers from direct mechanical disturbances of the friable material.

The performance of an encapsulant with any particular asbestos application is unpredictable. The most frequently encountered problem is a cohesive or adhesive failure of the original asbestos product. Many friable asbestos-containing products are not capable of supporting the additional weight of the encapsulant. Encapsulation should not be considered without first testing the cohesion/adhesion strength of the asbestos-containing product. A procedure for cohesion/adhesion testing has been established by the American Society for Testing and Materials (ASTM) (67).

5.1.2 Enclosure

Enclosure involves the construction of a rigid barrier between asbestos-containing products and building occupants. The purpose of enclosure is to prevent damage to the asbestos-containing material, thereby reducing the likelihood of fiber release. Enclosure can be an effective means of control where frequent entry into the enclosure is not necessary. Suspended ceilings may seem like an enclosure; however, in most cases, they do not provide effective control because:

- Maintenance workers frequently work above a suspended ceiling to gain access to electrical, ventilation, or lighting equipment.
- The space above the suspended ceiling often functions as a return-air plenum for the building ventilation system.

Two types of enclosure systems are routinely employed: (a) fixed, mechanical barriers, and (b) spray-applied barriers. A fixed, mechanical barrier might consist of a plywood or drywall enclosure around the asbestos-containing material. Spray-applied barriers, often called encasements, differ from encapsulation in that the "cured" sealant forms a hard external surface. Also, commercially available encasement systems typically include some means of mechanically fastening the barrier to the substrate (e.g., pipe, structural steel) thus preventing delamination.

Enclosure should not be used when (a) there is a high probability of water damage, (b) where the asbestos-containing material is already damaged, or (c) in situations where it will frequently be necessary to work above or behind the enclosure.

5.1.3 Removal

When competently performed, and with adequate protection for workers and building occupants, removal should minimize the risks of asbestos exposure in a building. Serious questions have been raised, however, about the effect of asbestos removal on subsequent asbestos fiber concentrations. In studies conducted in nine buildings before and after asbestos removal, Pinchin (68) reported equal or higher asbestos fiber concentrations after removal than were present before removal (Table 33.14).

Another concern with asbestos removal is the availability and relative safety of replacement products. In most cases, the asbestos product must be replaced with a suitable substitute. At a minimum, the substitute must provide the same acoustical or thermal insulation

Table 33.14. Concentrations of Asbestos Fibers Before and After Removal[a]

	Before Asbestos Removal[a]			After Asbestos Removal[b]		
Project Number	PCM (f/cc)	TEM (f/cc)	TEM (fibers >5 μm/cc)	PCM (f/cc)	TEM (f/cc)	TEM (fibers >5 μm/cc)
1	0.022	0.027	<0.005	0.07	0.057	<0.004
2	<0.1	0.016	0.003	0.1	0.054	0.008
3	<0.1	0.015	0.001	<0.1	0.026	0.005
4	<0.1	0.023	0.003	<0.1	0.010	0.001
5	0.003	0.037	0.002	0.039	0.53	0.028
6	0.004	0.012	0.001	0.022	0.119	0.002
7	0.002	0.016	<0.001	0.003	0.182	0.012
8a	0.03	—	0.004	0.01	—	0.01
8b	<0.01	0.019	<0.004	0.03	0.256	0.012

[a]Adapted from Ref. 69
[b]PLM = phase-contrast microscopy; TEM = transmission electron microscopy.

properties as the original material. Indeed, in some cases building owners are discovering that, as a result of local building codes changes since the original construction, they must replace the asbestos product with a product exhibiting superior insulation properties. The most widely used asbestos substitutes are fibrous glass, rock wool, and slag wool, commonly referred to as man-made mineral fibers. Since their introduction about 90 years ago, it has been recognized that these man-made mineral fibers can irritate the skin and eyes and, under dusty conditions, may also irritate the nose and throat. Recent concerns, however, have focused on the possible carcinogenicity of asbestos substitutes. Man-made mineral fibers are discussed in greater detail in MAN-MADE MINERAL FIBERS (Chapter 5).

Despite these limitations, asbestos removal and replacement may be the best option when the asbestos-containing material is badly deteriorated. In the United States, removal of friable asbestos is required by EPA prior to the demolition of a building (70).

5.1.4 Management in Place

In 1990, the U.S. Environmental Protection Agency published the most recent in a series of Guidance Documents for asbestos in buildings (71). In this document, EPA placed a greater emphasis on the option of leaving asbestos-containing materials in place and use of work practice and engineering controls to minimize asbestos exposure during those activities most likely to disturb ACM.

The objective of an Asbestos Management program is to minimize asbestos exposures of building occupants and workers by controlling access to the asbestos-containing material, maintaining the material in good condition, and using work practices that minimize fiber release.

An Asbestos Management Program is a set of specific procedures applied to routine activities such as cleaning, maintenance, and renovation to maintain the building as free of asbestos contamination as possible. Such a program must be developed for each specific building and in the United States, should incorporate the requirements of the U.S. Occupational Safety and Health Administration (OSHA).

Elements of an effective Asbestos Management Program will vary with the circumstances of each building but generally will include the items listed below.

5.1.4.1 Periodic Surveillance. Because the condition of asbestos-containing materials in a building may change with time owing to water damage, accidental disturbance, or aging, routine surveillance of these materials is an essential element of an asbestos management program. The incidental reporting of damage by maintenance or custodial workers or building occupants should be documented. Routine surveillance should be performed by an individual trained in the assessment protocol, and the results should be carefully documented.

5.1.4.2 Routine Air Monitoring. Routine prevalent level monitoring is another type of periodic surveillance that should be included in a management program. Such monitoring is useful for assessing changes that may have occurred in the condition of the materials as well as for further documenting the airborne concentrations during normal building occupancy. During routine air monitoring, maintenance and renovation activities underway should be documented.

5.1.4.3 Notification. Occupants of buildings, including maintenance and custodial personnel, should be informed of the location(s) of asbestos-containing materials, the potential health risks associated with asbestos exposure, and the air monitoring results. Further, occupants should be instructed to avoid contact with asbestos-containing materials and to advise building management of any observed damage.

5.1.4.4 Employee Protection. In most buildings it will sometimes be necessary for maintenance personnel to work in close proximity to or disturb asbestos-containing materials in order to effect a repair or adjust heating, ventilation, and air conditioning or lighting. With the use of proper equipment and good work practices, maintenance employees may perform such work with minimal exposure to asbestos.

5.1.4.5 Medical Surveillance Programs. Those employees whose jobs may necessitate occasional contact with asbestos should be included in a medical surveillance program including routine medical examination, chest X-rays, and pulmonary function testing.

5.1.4.6 Training. All employees who may potentially come into direct contact with asbestos-containing materials should be trained in following the correct work procedures, using personal protective equipment, and understanding the health risks of asbestos. Under the OSHA requirements, such training; must include:

- Methods of recognizing asbestos.
- The health effects associated with asbestos exposure.
- The relationship between smoking and asbestos in producing lung cancer.

- The nature of operations that could result in exposure and importance of engineering controls, work practices, respirators, housekeeping procedures, hygiene facilities, protective clothing, decontamination procedures, emergency procedures, and waste disposal procedures, and any necessary instruction in the use of these controls and procedures.
- The contents of EPA 20T-2003, "Managing Asbestos In-Place" July 1990 or its equivalent in content (71).
- The purpose, proper use, fitting instructions, and limitations of respirators.
- The appropriate work practices for performing the asbestos job.
- Medical surveillance program requirements.
- The content of the OSHA standard including appendices.
- The names, addresses and phone numbers of public health organizations which provide information, materials and/or conduct programs concerning smoking cessation.
- The requirements for posting signs and affixing labels and the meaning of the required legends for such signs and labels.

5.1.4.7 Special Work Practices for Maintenance Activities. Even relatively routine maintenance procedures may involve disturbance of asbestos-containing materials. This applies especially to work above suspended ceilings in buildings where the steel structural members are sprayed with asbestos fireproofing. Special equipment and work practices, such as those shown in Table 33.15 are necessary to prevent exposures of the maintenance employees as well as other building occupants.

5.1.4.8 Special Work Practices for Renovation. Plans for routine renovation in a building should be carefully evaluated to determine if such activities will disturb the asbestos-

Table 33.15. Asbestos-Control Procedures For Routine Maintenance Activities

Minimal occupancy	Whenever possible, maintenance activities involving possible disturbance of asbestos-containing materials should be performed when the affected area is unoccupied
Cleaning	Any asbestos-containing materials dislodged during such work should be cleaned up with a high-efficiency particulate air filter-equipped vacuum cleaner or by wet mopping and never by sweeping. The entire work area should be cleaned thoroughly using a combination of wet mopping and vacuuming at the completion of the maintenance work, but before the area is reoccupied
Personal protection	For such maintenance activities, employees should be provided with approved respirators and disposable coveralls with hoods or hats. The use of respiratory protection necessitates the development of a respirator program. Respirators should be individually fitted, and employees should be trained in their correct use and limitations
Disposal	Asbestos-contaminated wastes generated during such activities must be disposed of in accordance with applicable laws and regulations
Wet methods	Wet methods should be employed if the work involves direct disturbance of the asbestos-containing material

containing materials. If so, precautions such as those described in Table 33.16 must be taken. The cost impact of such controls may be minimized through careful planning and scheduling of the renovation project.

5.1.4.9 Episodic Releases. In the event of an unplanned disturbance of an asbestos-containing material, a quick response is essential to prevent exposures to occupants. A contingency plan, including making arrangements with appropriate outside resources (e.g., abatement contractors, consultants), should be developed for each building in which asbestos-containing materials are present.

5.2 Selecting the Best Option

For the property owner and manager, as well as for the industrial hygienist, selecting the best asbestos control option can be a complex decision. The potential penalties for a poor

Table 33.16. Asbestos-Control Procedures For Routine Renovation Activities

Isolation	The renovation area should be isolated from the remainder of the facility. Typically, this involves the construction of a temporary plastic barrier with an airlock for entry of workers. Ventilation in the work area should also be temporarily isolated and a portable exhaust fan [equipped with a high-efficiency particulate air (HEPA) filter] should be used to place the renovation area under a slight negative pressure with respect to the rest of the building. Wherever possible, this fan should discharge out-of-doors
Warning signs	Signs should be posted at all entrances to the renovation area to warn unprotected persons from entering
Cleaning	Any asbestos-containing materials dislodged during such work should be cleaned up with a HEPA filter-equipped vacuum cleaner or by wet mopping and never by sweeping. The entire work area should be cleaned thoroughly using a combination of wet mopping and vacuuming at the completion of the renovation work, but before the area is reoccupied
Personal protection	For such activities, renovation workers should be provided with approved respirators and disposable coveralls with hoods or hats. The use of respiratory protection necessitates the development of a respirator program. Respirators should be individually fitted, and employees should be trained in their correct use and limitations
Disposal	Asbestos-contaminated wastes generated during such activities must be disposed of in accordance with applicable laws and regulations
Wet methods	Wet methods should be employed if the work involves direct disturbance of the asbestos-containing material
Air monitoring	At regular intervals during the renovation work, air monitoring should be performed along the perimeter of work area to demonstrate the efficacy of the isolation and ventilation controls at preventing fibers from escaping. After the work area has been thoroughly cleaned, but before the barrires are removed, clearance air monitoring should be performed

decision may include regulatory fines, civil court actions, and potential health risks. Because each building is unique, asbestos management decisions must be made on a building-by-building basis. The risks posed by the presence of asbestos in buildings can vary considerably depending on:

- The condition, location, and friability of the asbestos-containing material(s).
- The way(s) in which the building is used (e.g., school, office, retail store).
- The type of ventilation systems in the building.
- The amount and extent of maintenance and renovation work anticipated.

The industrial hygienist must recognize and consider not only the actual health risks, but also the risks perceived by the building occupants, parents of schoolchildren, and the general public.

From a practical point of view, there are only two alternatives: immediate removal or deferred removal. If the removal is deferred, the building owner must implement an on-going asbestos management program, which may include controls such as enclosure or encapsulation.

Removal is generally the only viable alternative when asbestos-containing materials are badly deteriorated. For buildings in which the asbestos-containing materials are in good condition and do not pose an immediate health risk, the decision may be based on economic and practical considerations.

Deferring removal may postpone a large capital expense. The ongoing operating costs, however, of an asbestos management program may also be considerable. In addition to the costs of monitoring and controlling asbestos, future renovation/remodeling costs in an asbestos-containing structure will also be higher. In commercial properties, the loss or reduction of rental income may be a significant factor.

As with any critical business or technical decision, the key is to evaluate carefully all of the pertinent factors discussed above before making an "informed decision."

5.3 Planning an Abatement Action

No matter which control technique is chosen, it is important to employ specially trained and qualified professionals to perform this work and to require strict adherence to established operating procedures. All persons performing asbestos abatement work should be properly protected, trained, and supervised so that building tenants and other persons outside of the containment areas are not exposed to unacceptable concentrations of asbestos fibers as a result of abatement activities.

Before an asbestos abatement project is undertaken, the following steps are necessary:

- Contract specifications tailored to the individual project.
- A thorough evaluation of contractors and their previous performance prior to awarding the contract.
- On-site surveillance of contractor performance and air monitoring during abatement activities by an independent third-party expert working directly for the building owner or manager.

Location: :

Audit Date: *Investigator:*

Has the facility been surveyed by an accredited inspector or by a certified industrial hygienist (CIH) for asbestos-containing materials?	yes	no	NA
Did the results identify friable Asbestor-Containing Materials (ACM) in the building?	yes	no	NA
If not, stop here.			
Have building occupants and employees been advised of the presence of ACM in the building?	yes	no	NA
Have procedures been developed to advise (in writing) contractors working in the building whose activities may potentially disturb ACM?	yes	no	NA
Have signs been posted at the entrance to mechanical rooms/areas containing ACM which employees or other building occupants reasonably can be expected to enter identifying the material, its location, and appropriate work practices?	yes	no	NA
Where feasible, have installed asbestos products been labeled?	yes	no	NA
Do maintenance of custodial employees perform work involving direct disturbance of ACM?	yes	no	NA
If so, what types of asbestos-related work activies do these employees perform:			
• *Class I* asbestos work" (activities involving the removal of Thermal System insulation and surfacing ACM)	yes	no	NA
• *Class II* asbestos work" (activities involving the removal of other types of ACM including wallboard, floor tile and sheeting, roofing and siding shingles, and construction mastics)	yes	no	NA
• *Class III* asbestos work" (repair and maintenance operations where ACM is likely to be disturbed)	yes	no	NA
• *Class IV* asbestos work" (maintenance and custodial activities during which employees contact but do not disturb ACM and activities to clean up dust, waste, and debris resulting from Class I, II, III activities)	yes	no	NA
Has the facility developed written procedures for asbestos-related work activities?	yes	no	NA
For Class I work, do these procedures include:	yes	no	NA
• HEPA-filter equipped vacuum cleaners?	yes	no	NA
• Wet methods?	yes	no	NA
• Prompt clean-up and disposal of asbestos wastes and debris?	yes	no	NA
• Ventilation?	yes	no	NA
• Enclosure or isolation?	yes	no	NA
• Respiratory protection?	yes	no	NA
• Supervision by "competent person"?	yes	no	NA

(a)

Figure 33.1. Asbestos Management Program audit example checklists, United States.

- Sealing of all surfaces? yes no NA
- Isolation of HVAC Systems? yes no NA
- Negative Pressure Enclosure (NPE) system or glove bags or glove boxes or a Water Spray Process System? yes no NA
- Air monitoring/personal exposure monitoring? yes no NA

For Class II work, do these procedures including:

- HEPA-filter equipped vacuum cleaners? yes no NA
- Wet methods? yes no NA
- Prompt clean-up and disposal of asbestos wastes and debris? yes no NA
- Ventilation? yes no NA
- Enclosure or isolation? yes no NA
- Respiratory protection? yes no NA
- Supervision by a "competent person" yes no NA
- Air monitoring/personal exposure monitoring? yes no NA
- Critical barriers or isolation? yes no NA
- Sealing of all surfaces? yes no NA

(Vinyl and asphalt flooring materials)

1. Sanding prohibited? yes no NA
2. HEPA-filter equipped vacuum? yes no NA
3. Wet methods? yes no NA
4. Dry sweeping is prohibited yes no NA
5. Mechanical chipping is prohibited (except in a negative pressure yes no NA

enclosure)? yes no NA

6. Intact removal of tiles if feasible?

(Roofing material) yes no NA

1. Intact removal if feasible? yes no NA
2. Wet methods? yes no NA
3. Continuous misting of cutting machines? yes no NA
4. HEPA dust collector w/power roof cutter? yes no NA
5. Lowering of materials via covered, dust-tight chute, crane or hoist yes no NA

ASAP and no later than end of shift?

6. Shutdown and isolation of roof level HVAC equipment?

yes no NA

(Asbestos-cement products) yes no NA

1. Prohibit cutting, abrading or breaking? yes no NA
2. Wet methods?
3. Immediate lowering of unwrapped material via covered dust-tight chute, crane or hoist? yes no NA yes no NA
4. Lowering by the end of shift sealed materials?
5. Nails cut with flat, sharp instruments?

yes no NA

(Gaskets) yes no NA

1. If deteriorated, use of a glovebag? yes no NA
2. Immediately place in dispose container?
3. Any scraping performed wet?

For Class III work, do these procedures include: yes no NA

yes no NA

- HEPA-filter equipped vacuum cleaners? yes no NA
- Wet methods? yes no NA
- Prompt clean-up and disposal of asbestos wastes and debris? yes no NA
- Ventilation? yes no NA

(b)

Figure 33.1. Continued.

• Wet methods?	yes	no	NA
• Drop clothes/mini-enclosures/glove bags as needed?	yes	no	NA
• Enclosure/isolation?	yes	no	NA
For class IV work, do these procedures include:			
• HEPA-filter equipped vacuum cleaners?	yes	no	NA
• Wet methods?	yes	no	NA
• Prompt clean-up and disposal of asbestos wastes and debris?	yes	no	NA
• Ventilation?	yes	no	NA
• Enclosure or isolation?	yes	no	NA
• Respiratory protection?	yes	no	NA
Has the facility developed an asbestos-mangement plan?	yes	no	NA
Have employees received training in asbestos appropriate to the type of work they are involved in?	yes	no	NA
Did training include:			
• Methods of recognizing asbestos	yes	no	NA
• The health effects associated with asbestos exposure	yes	no	NA
• The relationship between smoking and asbestos in producing lung cancer	yes	no	NA
• The nature of operations that could result in exposure and importance of engineering controls, work practices, respirators, housekeeping procedures, hygiene facilities, protective clothing, decontamination procedures, emergency procedures, and waste disposal procedures, and and necessary instruction in the use of these controls and procedures?	yes	no	NA
• The contents of EPA 20T-2003, "Managing Asbestos In-Place" July 1990 or its equivalent in content?	yes	no	NA
• The purpose, proper use, fitting instructions, and limitations of respirators?	yes	no	NA
• The appropriate work practice for performing the asbestos job?	yes	no	NA
• Medical surveillance program requirements?	yes	no	NA
• The content of the OSHA standard including appendices?	yes	no	NA
• The names, addresses and phone numbers of public health organizations which provide information, materials and/or conduct programs concerning smoking cessation.	yes	no	NA
• The requirements for posting signs and affixing labels and the meaning of the required legends for such signs and labels?	yes	no	NA
Has the plant instituted a medical surveillance program for employees who for a combined total of 30 or more days per year are engaged in Class I, II, and III work or are exposed at or above a permissible exposure limit?	yes	no	NA
Has the plant designated a "competent person" to supervise asbestos related work? If so, who is this person? ____________ ____________	yes	no	NA
Has the "competent person" received appropriate training?	yes	no	NA
Auditors Assessment and Recommendations:			

(c)

Figure 33.1. Continued.

- Clearance and other performance criteria to determine successful project completion.
- In addition to an industrial hygienist and the abatement contractor, planning for an asbestos abatement project requires input from several experts including:
  - Architects.
  - Mechanical and electrical engineers.
  - Contract specialists.
  - Building managers.
  - Building maintenance people.

Although often overlooked in the planning process, the building maintenance people may provide valuable insight owing to their intimate knowledge of the building and building systems.

Figure 33.1 is a checklist designed to assist a building owner in determining applicability of and compliance to the requirements of the U.S. Occupational Safety and Health Administration.

BIBLIOGRAPHY

1. L. M. Thomas, "EPA Study of Asbestos-Containing Materials in Public Buildings: A Report to Congress," U.S. Environmental Protection Agency, Washington, DC, Feb., 1988.
2. F. W. Boelter, and R. E. Sheriff, *American Industrial Hygiene Association Recommendations for Asbestos Abatement Projects*, AIHA, Akron, OH Sept. 1986.
3. D. Hunter, *The Diseases of Occupations*, 4th ed., Little, Brown, Boston, 1969, p. 1009.
4. P. Brodeur, "Annals of Law; The Asbestos Industry On Trial; 1. A Failure to Warn," *The New Yorker*, (June 10, 1985).
5. W. E. Cooke, "Pulmonary Asbestosis," *Br. Med. J.* **2**, 1024 (Dec. 1927).
6. M. S. Badollet, "Asbestos, a Mineral of Unparalleled Properties," *Can Mining Metall. Bull.* (April 1951).
7. W. B. Reitze et al., "Application of Sprayed Inorganic Fiber Containing Asbestos: Occupational Health Hazards," *Am. Ind. Hyg. Assoc. J.* **33**, 179–191 (1972).
8. "City Orders a Ban on Asbestos Sprays in Building Industry," *Philadelphia Inquirer*, (Dec. 3, 1970).
9. The Asbestos Institute, *Chrysotile Asbestos: an Overview*, Montreal, Quebec, 1999.
10. M. Prevost, The Asbestos Institute, Montreal, Quebec, personal communication.
11. "Manville Accord Would Settle Asbestos Claims," *Wall Street Journal* (Oct. 16, 1985).
12. "Commercial and Industrial Use of Asbestos Fibers; Advance Notice of Proposed Rulemaking." *Fed. Reg.* **44**(202) (Oct. 17, 1989).
13. D. Keyes and B. Price, *Guidance for Controlling Friable Asbestos-Containing Materials in Buildings*, EPA 560/5-83-002, U.S. Environmental Protection Agency, Office of Pesticide and Toxic Substances, Washington, DC, March 1983.
14. J. Strenio, P. C. Constant, Jr., M. Gabriel, D. R. Rose, and D. Lentzen, *Asbestos in Buildings: A National Survey of Asbestos-Containing Friable Materials*, EPA360/5-84-006, United States Environmental Protection Agency, Office of Toxic Substances, Washington, DC, Oct. 1984.

15. A. R. Rock, "Preliminary Results from the Study of Asbestos in 45 Homes," in *Asbestos-Its Health Risks, Analysis, Regulation and Control*, American Pollution Control Association, Pittsburgh, PA, 1987.
16. R. R. Sayers, and W. C. Dreesen, "Asbestosis," *Am. J. Public Health* **29**(3), 467–479 (March 1939).
17. G. W. Wright, "Asbestos and Health in 1969," *Am. Rev. Respir. Dis.* **100**(4), 467–479 (Oct. 1969).
18. I. J. Selikoff and D. H. K. Lee, *Asbestos and Disease*, Academic Press, New York, 1978.
19. J. C. Wagner, C. A. Sleggs, and P. Marchand, "Diffuse Pleural Mesothelioma and Asbestos Exposure in the North Western Cape Province," *Br. J. Ind. Med.* **17**, 260–271 (1960).
20. I. J. Selikoff, H. A. Anderson, and H. Seidman, "Asbestos Disease Among Household Contacts of Asbestos Workers" in I. J. Selikoff, ed., *Disability Compensation for Asbestos-associated Disease in the U.S.*, Environmental Sciences Laboratory, Mount Sinai School of Medicine of the City University of New York, 1982, pp. 73–76.
21. R. J. Levine, *Cancer Control Monograph: Asbestos*, SRI International, Palo Alto, CA, April 1978.
22. I. H. Selikoff, E. C. Hammond, and H. Seidman, "Asbestos Exposure, Cigarette Smoking and Death Rates," *Ann. N.Y. Acad. Sci.* **330**, 473–490 (1979).
23. I. H. Selikoff, H. Seidman, and E. C. Hammond, "Mortality Effects of Cigarette Smoking Among Amosite Asbestos Factory Workers," *J. Natl. Cancer Inst.* **65**(3), 507–1513 (Sept. 1980).
24. J. Sanchis, M. Dolovich, R. Chalmers, and M. T. Newhouse, "Regional Deposition and Lung Clearance Mechanisms in Smokers and Nonsmokers," in W. H. Walton, ed., *Inhaled Particles III*, Vol. I, Unwin Brothers, Surrey, England, 1971, pp. 183–191.
25. D. Cohen, S. F. Arai, and J. D. Brain, "Smoking Impairs Long-Term Dust Clearance from the Lung," *Science* **204**, 514–517 (May 1979).
26. P. Gerde and P. Scholander, "A Hypothesis Concerning Asbestos Carcinogenicity: The Migration of Lipophilic Carcinogens in Adsorbed Lipid Bilayers," *Ann. Occup. Hyg.* **31**(3), 395–400 (1987).
27. U.S. Department of Labor, Occupational Safety and Health Administration "Preamble to the OSHA Asbestos Standard," **29** *CFR* Parts 1910 and 1926, "Occupational Exposure to Asbestos, Tremolite, Anthophyllite, and Actinolite; Final Rules," *Fed. Reg. Bol.* **51**(119) (June 20, 1986).
28. J. C. McDonald, "Epidemiological Significance of Mineral Fiber Persistence in Human Lung Tissue." *Environ Health Perspect.* **102** (Suppl. 5): (1994).
29. J. M. Hughes and H. Weill, "Asbestos Exposure-Quantitative Assessment of Risk," *Am. Rev. Respir. Dis.* **135**, 5–13 (1986).
30. National Research Council, Committee on Nonoccupational Health Risks of *Asbestiform Fibers, Asbestiform Fibers: Nonoccupational Health Risks*, National Academy Press, Washington, DC, 1984, p. 132.
31. J. S. Dupre, J. F. Mustard, R. J. Uffen, D. N. Dewees, J. S. Dupre, J. J. Laskin, and L. B. Kahn, *Report of the Royal Commission on Matters of Health and Safety Arising from the Use of Asbestos in Ontario*, Vol. 1, 1984.
32. M. F. Stanton and C. Wrench, "Mechanisms of Mesothelioma Induction with Asbestos and Fibrous Glass," *J. Natl. Cancer Inst.* **48**(3) (March 1972).
33. M. F. Stanton and M. Layard, "The Carcinogenicity of Fibrous Minerals," in *Proceedings of the Workshop on Asbestos: Definitions and Measurement Methods*, Gaithersburg, MD, July 18–20, 1977, NBS Special Publication 506, Washington, DC, 1978.

34. F. Pott, K. H. Friedrichs, and F. Huth, "Results of Animal Experiments Concerning the Carcinogenic Effect of Fibrous Dusts and Their Penetration with Regard to the Carcinogenesis in Humans," *Zbl. Bakt. Hyg.* **162**, 467–505 (1976).
35. M. Lippmann, Review: "Asbestos Exposure Indices," *Environ. Res.* **46**(1) (June 1988).
36. L. A. Schultz, W. Bank, and G. Weems, *Airborne Asbestos Fiber Concentrations in Asbestos Mines and Mills in the United States*, Bureau of Mines Health and Safety Program, Technical Progress Report 72, Ps-222 611, U.S. Department of the Interior, Denver, CO, 1973.
37. D. R. Roberts, *Industrial Hygiene Report: Asbestos at Jaquays Mining Corp, PS81-241846*, National Institute of Occupational Safety and Health, Cincinnati, OH, 1980.
38. P. G. Harries, "Experience with Asbestos Disease and its Control in Britain's Naval Dockyards," *Environ. Res.* **11**, 261–267, 1977.
39. A. R. Daly, A. J. Zupko, J. L. Hebb, and R. F. Weston, *Technological Feasibility and Economic Impact of OSHA Proposed Revision to the Asbestos Standard*, Asbestos Information Association/North America, Washington, DC, 1976.
40. U.S. Environmental Protection Agency, Office of Toxic Substances, *Support Document for Final Rule on Friable Asbestos-containing Materials in School Buildings: Health Effects and Magnitude of Exposure*, Washington, DC, 1982.
41. W. B. Reitze, W. J. Nicholson, D. A. Holaday, and I. J. Selikoff, "Application of Sprayed Inorganic Fiber Containing Asbestos: Occupational Health Hazards," *Am. Ind. Hyg. J.* **31**, 178–191 (1972).
42. W. D. Nicholson, *Dose–Response Relationships for Asbestos and Inorganic Fibers*, Environmental Sciences Laboratory, Mount Sinai School of Medicine of the City University of New York, 1981.
43. N. W. Paik, R. J. Walcott, and P. A. Brogan, "Worker Exposure to Asbestos during Removal of Sprayed Material and Renovation Activity in Buildings Containing Sprayed Material," *Am. Ind. Hyg. Assoc. J.* **44**(6) (1983).
44. M. Coffman, and R. Walcott, "Asbestos Exposure in the Construction Industry," presented at the American Industrial Hygiene Conference, Las Vegas, NV, May, 1985.
45. W. M. Ewing and G. J. Simpson, "Air Sampling at 52 Asbestos Abatement Projects," presented at the American Industrial Hygiene Conference, Philadelphia, PA, May 24, 1983.
46. W. J. Nicholson, A. N. Rohl, and I. Weisman, *Asbestos Contamination of the Air in Public Buildings*, EPA Publication No. EPA-450/3-76-004, Oct. 1975.
47. P. C. Constant, Jr., F. J. Bergman, G. R. Atkinson, D. R. Rose, D. L. Watts, E. E. Logue, T. D. Hartwell, B. P. Price, and J. S. Ogden, *Airborne Asbestos Levels in Schools*, EPA Publication No. EPA 560/5-83-003, June, 1983.
48. D. J. Pinchin, *Asbestos in Buildings*, Royal Commission on Asbestos Study Series No. 8, Royal Commission on Asbestos, Toronto, Table 1982, p. 6.7.
49. J. Hatfield, J. Stockrahm, and J. Chesson, *Draft Report for Task 2-31: Preliminary Analysis of Asbestos Air Monitoring Data for the U.S. Environmental Protection Agency*, Office of Toxic Substances, Washington, DC, 1987; taken from L. M. Thomas, *EPA Study of Asbestos-Containing Materials in Public Buildings, a Report to Congress*, EPA, Washington, DC, Feb. 1988.
50. M. Coffman, "Asbestos Management in a Changing Environment," *Proceedings of Symposium on Asbestos in Buildings: Management, Measurement and Risks*, Clayton Environmental Consultants, Inc., Southfield, MI, July 28, 1988.
51. Ontario Royal Commission on Asbestos, *Report of the Royal Commission on Matters of Health and Safety Arising from the Use of Asbestos in Ontario*, Queen's Printer, Toronto, 1984.

52. Chronic Hazard Advisory Panel on Asbestos, *Report to the U.S. Consumer Product Safety Commission*, U.S. Consumer Product Safety Commission, Washington, DC, 1983.
53. U.S. Occupational Safety and Health Administration, *Quantitative Risk Assessment for Asbestos-Related Cancers*, Washington, DC, 1983.
54. Putnam, Hayes and Bartlett, Inc., *Cost and Effectiveness of Abatement of Asbestos in Schools*, draft report for U.S. Environmental Protection Agency, Office of Pesticides and Toxic Substances, 1984, Washington, DC.
55. National Academy of Sciences, Committee on Nonoccupational Health Risks of Asbestiform Fibers, National Research Council, *Asbestiform Fibers: Nonoccupational Health Risks*, National Academy Press, Washington, DC, 1984.
56. R. Doll and J. Peto, *Effects on Health of Exposure to Asbestos*, U.K. Health and Safety Commission, Her Majesty's Stationery Office, London, 1985.
57. L. M. Thomas, *EPA Study of Asbestos-Containing Materials in Public Buildings: A Report to Congress*, U.S. Environmental Protection Agency, Washington, DC, Feb. 1988.
58. P. E. Enterline, "Extrapolation from Occupational Studies; a Substitute for Environmental Epidemiology," *Environ. Health Perspect.* **42**, 39 (1981).
59. U.S. Environmental Protection Agency, *Fed. Reg.*, **40** *CFR* Part 763 (Oct. 30, 1987).
60. D. E. Lentzen, E. P. Brantly, Jr., K. W. Gold, and L. E. Myers, *Interim Method for the Determination of Asbestos in Bulk Insulation Samples*, EPA-600/M4-82-020, U.S. Environmental Protection Agency, Office of Pesticides and Toxic Substances, Washington, DC, December 1982; U.S. Environmental Protection Agency, *Fed. Reg.* **40** *CFR* Part 763 (Oct. 30, 1987).
61. *Asbestos Exposure Assessment in Buildings Inspection Manual*, U.S. Environmental Protection Agency, Oct. 1982.
62. "U.S. Navy Procedure for Risk Evaluation: Asbestos Hazard Index," in *Report R883, Management Procedure for Assessment of Friable Insulating Materials*, Civil Engineering Laboratory, Naval Construction Batallion Center, Port Hueneme, CA, Chapt. 5.
63. K. S. Irving, R. G. Alexander, and H. Bavley, "Asbestos Exposures in Massachusetts Public Schools," *AIHA J* (April, 1980).
64. Baldwin, Beaulieu, Buchan, and Johnson, "Asbestos in Colorado Schools," *Public Health Rep.* (July–Aug. 1982).
65. L. Simkins and W. Hill, "Algorithms: Their Usefulness in Hazard Assessment," presented at the Annual Meeting of the National Asbestos Counsel, Oakland, CA, Sept. 1987.
66. W. Phanprasit, V. E. Rose, and R. K. Oestenstad, "Comparison of the Fibrous Aerosol Monitor and the Optical Fiber Count Technique for Asbestos Measurement," *Appl. Ind. Hyg.* **3**(1) (Jan. 1988).
67. American Society for Testing and Materials, *ASTM Method E736-86*, Philadelphia, PA.
68. D. J. Pinchin,*Asbestos in Buildings*, Royal Commission on Asbestos Study Series, No. 8, Royal Commission on Asbestos, Toronto, 1982.
69. D. J. Pinchin, *"Asbestos in Buildings," Royal Commission on Asbestos Study Series, No. 8, Royal Commission on Asbestos, Toronto*, 1982, Ref. 68, pp. 7–12, Table III.
70. "National Emission Standard for Hazardous Air Pollutants (NESHAPS)," **40** *CFR* part 265, subparts A & M, (1973).
71. United States Environmental Protection Agency, *Managing Asbestos in Place—A Building Owner's Guide to Operations and Maintenance Programs for Asbestos-Containing Materials.* EPA 20T-2003, July 1990.

Index